Steven C. Anderson

The Lizards of Iran

Map of
PERSIA
divided into
ZOOLOGICAL PROVINCES
 Persian plateau
 Caspian provinces
 Wooded part of Zagros Mts &
 Persian Mesopotamia
 Baluchistan and shores of
Scale of English Miles
50 25 0

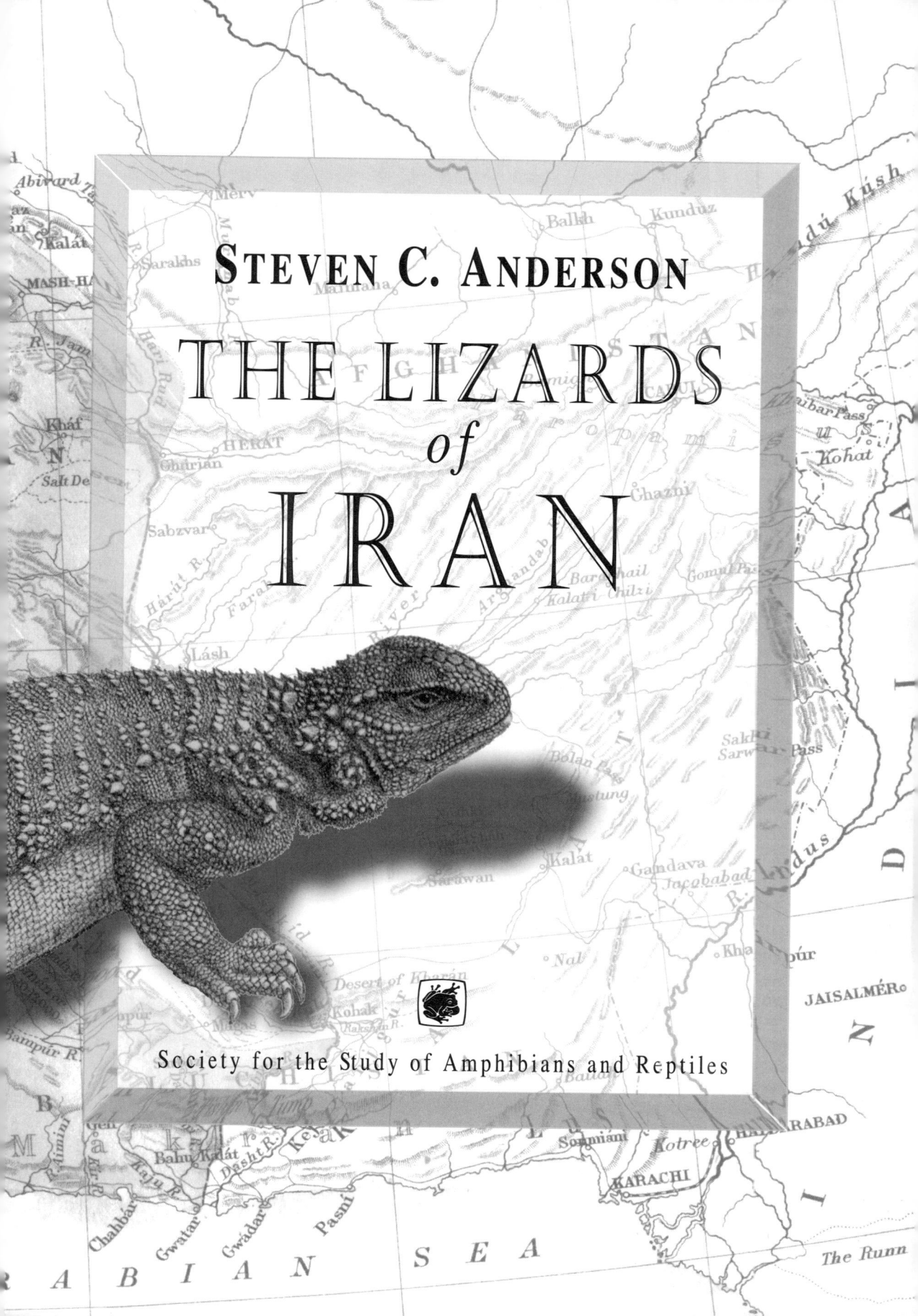

EDITOR'S NOTE

The Society is greatly indebted to Alan E. Leviton for his editorial assistance in every phase of the production of this book. Two new species of lizards were described after the text had been typeset. It was decided to incorporate these into the main text rather than put them in an appendix, which resulted in a somewhat unorthodox arrangement.

The drawing of an Iranian spiny-tailed lizard (*Uromastyx asmussi*) by George Henry Ford and the map depicting Iran's zoogeographical provinces on the title page were both taken from William T. Blanford's book, "Eastern Persia . . . volume II. The Zoology and Geology." Macmillan and Co., London, 1876.

CONTRIBUTIONS TO HERPETOLOGY, VOLUME 15

KRAIG ADLER, *Editor*   TIMOTHY D. PERRY, *Associate Editor*

Volumes in the *Contributions to Herpetology* series can be purchased from the Publications Secretary, Robert D. Aldridge, Department of Biology, Saint Louis University, 3507 Laclede, Saint Louis, Missouri 63103, USA (*telephone*: area code 314, 977–3916 or 977–1710; *fax*: area code 314, 977–3658; *e-mail*: ssar@slu.edu). A list of all Society publications, including those of The Ohio Herpetological Society and the *Catalogue of American Amphibians and Reptiles*, is printed at the end of this book; additional copies of this list are available from Dr. Aldridge. Volumes in the *Contributions* series are published irregularly and ordered by separate subscription, although Society members receive a substantial pre-publication discount. Authors who wish to have manuscripts considered for publication in the *Contributions* series should contact the Editor: Kraig Adler, Cornell University, Department of Neurobiology and Behavior, Seeley G. Mudd Hall, Ithaca, New York 14853–2702, USA.

Members of the Society receive a quarterly technical journal (*Journal of Herpetology*) and a quarterly news-journal (*Herpetological Review*). Currently, dues are US$30.00 for students, $50.00 for all others, world-wide; institutional subscriptions are $95.00. Additional $35.00 for airmail delivery outside the USA. Society members receive substantial discounts on *Herpetological Circulars* and *Facsimile Reprints* and on books in the *Contributions* and *Herpetological Conservation* series. The *Catalogue* is available by separate subscription. Apply to the Society's Treasurer, Robert D. Aldridge (address above). Overseas customers can make payments in USA funds or by International Money Order. All persons may charge to MasterCard or VISA (include account number and expiration date).

© 1999 Society for the Study of Amphibians and Reptiles
Library of Congress Catalog Number: 99-70849. ISBN: 0-916984-49-4.
Production specifications are given on the last page of this book.

DEDICATION

**This work is dedicated to the memory of three field zoologists
who began the scientific study of the Iranian herpetofauna**

Filippo De Filippi
1814–1879

William Thomas Blanford
1832–1905

Nikolai Aleksyevich Zarudny
1859–1919

FOREWORD

During the 1960s and 70s, Iran began a broad and extensive program of conservation embracing wildlife protection, the preservation of habitats, and the establishment of a large network of national parks and protected regions. In the course of implementation of this program, the Department of the Environment of Iran employed numerous experts in various fields from many countries and sponsored research by numerous scientists interested in Iran's fauna and flora. We were cognizant of the importance of assessing the biological diversity of all regions of the country and of preserving this heritage for future generations. Dr. Steven Anderson was one of the visiting scientists whom we were pleased to sponsor during that period. I believe he was the first American scientist to conduct extensive field research on an entire group of our fauna — in this case the suborder of lizards. He was, moreover, a scientist who combined expertise and mastery in his field with an understanding of and sympathy for Iran's long history and culture.

In the past two decades, world events have focused greater attention and fiscal resources on other social issues, but among the benefits reaped from the seeds sown in past decades, the current book on the lizards of Iran is an outstanding example. This book, while designed for a specialized scientific audience, is written in a style which should make it accessible to the next generation of students of Iranian natural history and to conservation workers of the Department of the Environment, particularly those in the parks and protected regions. It encompasses what is currently known about the lizard fauna and provides a rich base for future studies of biological diversity. Such studies must continue for all faunal and floral groups if our renewable biological resources are to be managed sustainably for the generations to come.

Many of the scientists from other countries who have worked in Iran in the past remain committed to the study of our flora and fauna. Their cooperation with a new generation of Iranian scientists now being educated at our universities will be crucial in an age when information and technology are expanding at a geometric rate. Let us hope that studies of biodiversity and conservation biology will be one focus for continued international cooperative efforts.

ESKANDAR FIROUZ
Former Deputy to the Prime Minister of Iran and
Director, Department of the Environment

TABLE OF CONTENTS

LIST OF PLATES	6
INTRODUCTION	9
Acknowledgments	13
A SHORT HISTORY OF HERPETOLOGY IN IRAN WITH AN INTRODUCTION TO THE LITERATURE	15
THE LIZARD FAUNA OF IRAN	42
The Geography of Iran Relative to Lizard Distribution	42
Ecological and Behavioral Influences on Lizard Distribution	52
Distributional Patterns of Iranian Lizards in Southwest Asia	59
Relationships of the Iranian Lizard Fauna to Neighboring Areas	60
Paleogeographical Influences on Lizard Distribution	61
Physiographic and Climatic Barriers to Distribution	65
SYSTEMATIC SECTION	66
Key to the Lizard Families of Iran	66
AGAMIDAE	67
Key to the Genera of Agamidae of Iran	67
Genus *Calotes* Cuvier	68
Calotes versicolor (Daudin)	68
Genus *Laudakia* Gray	69
Key to the Species of *Laudakia* in Iran	70
Laudakia caucasia (Eichwald)	70
Laudakia erythrogastra (Nikolsky)	73
Laudakia melanura Blyth	75
Laudakia microlepis (Blanford)	76
Laudakia nupta (De Filippi)	78
Genus *Phrynocephalus* Kaup	81
Key to the Species of *Phrynocephalus* in Iran	82
Phrynocephalus arabicus J. Anderson	84
Phrynocephalus clarkorum S. Anderson and Leviton	85
Phrynocephalus helioscopus (Pallas)	87
Phrynocephalus interscapularis Lichtenstein	88
Phrynocephalus luteoguttatus Boulenger	89
Phrynocephalus maculatus J. Anderson	90
Phrynocephalus mystaceus (Pallas)	92
Phrynocephalus ornatus Boulenger	93
Phrynocephalus persicus De Filippi	94
Phrynocephalus raddei Boettger	96
Phrynocephalus scutellatus (Olivier)	97
Genus *Trapelus* Cuvier	99
Key to the Species of *Trapelus* in Iran	99
Trapelus agilis (Olivier)	100
Trapelus persicus (Blanford)	104
Trapelus ruderatus (Olivier)	107
ANGUIDAE	111
Key to the Anguidae of Iran	111
Genus *Anguis* Linnaeus	111

    Anguis fragilis Linnaeus 111
    Genus *Ophisaurus* Daudin 113
        Ophisaurus apodus (Pallas) 113
EUBLEPHARIDAE 116
    Genus *Eublepharis* Gray 116
    Key to the Species of *Eublepharis* in Iran 117
        Eublepharis angramainyu S. Anderson and Leviton 118
        Eublepharis cf. *macularius* (Blyth) 120
        Eublepharis turcmenicus Darevsky 122
GEKKONIDAE 125
    Key to the Genera of Gekkonidae in Iran 127
    Genus *Agamura* Blanford 129
        Agamura persica (A. H. A. Duméril) 130
    Genus *Asaccus* Dixon and S. Anderson 132
    Key to the Species of *Asaccus* in Iran 134
        Asaccus elisae (F. Werner) 134
        Asaccus griseonotus Dixon and S. Anderson 136
        Asaccus kermanshahensis Rastegar-Pouyani 137
    Genus *Bunopus* Blanford 138
    Key to the Species of *Bunopus* in Iran 138
        Bunopus crassicaudus Nikolsky 139
        Bunopus tuberculatus Blanford 140
    Genus *Carinatogecko* Golubev and Szczerbak 143
    Key to the Species of *Carinatogecko* 144
        Carinatogecko aspratilis (S. Anderson) 144
        Carinatogecko heteropholis (Minton, S. Anderson, and J. A. Anderson) 145
    Genus *Crossobamon* Boettger 146
        Crossobamon eversmanni (Wiegmann) 147
    Genus *Cyrtopodion* Fitzinger 149
    Key to the Species of *Cyrtopodion* in Iran 151
        Cyrtopodion agamuroides (Nikolsky) 153
        Cyrtopodion brevipes (Blanford) 154
        Cyrtopodion caspium (Eichwald) 155
        Cyrtopodion gastrophole (F. Werner) 156
        Cyrtopodion heterocercum (Blanford) 158
        Cyrtopodion kachhense (Stoliczka) 159
        Cyrtopodion kirmanense (Nikolsky) 160
        Cyrtopodion kotschyi (Steindachner) 161
        Cyrtopodion longipes (Nikolsky) 163
        Cyrtopodion russowii (Strauch) 165
        Cyrtopodion sagittifer (Nikolsky) 166
        Cyrtopodion scabrum (Heyden) 167
        Cyrtopodion spinicauda (Strauch) 169
        Cyrtopodion turcmenicum (Szczerbak) 170
    Genus *Hemidactylus* Oken 171
    Key to the Species of *Hemidactylus* in Iran 172
        Hemidactylus flaviviridis Rüppell 172

INTRODUCTION

Hemidactylus persicus J. Anderson 173
Hemidactylus turcicus (Linnaeus) 174
Genus *Pristurus* Rüppell 175
  Pristurus rupestris Blanford 176
Genus *Ptyodactylus* Goldfuss 177
Key to the Southwest Asian Species of *Ptyodactylus* 178
  Ptyodactylus sp. indet. 178
Genus *Rhinogecko* de Witte 178
Key to the Species of *Rhinogecko* in Iran 179
  Rhinogecko femoralis (M. Smith) 179
  Rhinogecko misonnei de Witte 180
Genus *Stenodactylus* Fitzinger 181
Key to the Species of *Stenodactylus* in Iran 182
  Stenodactylus affinis (Murray) 182
  Stenodactylus doriae (Blanford) 183
Genus *Teratoscincus* Strauch 185
Key to the Species of *Teratoscincus* in Iran 185
  Teratoscincus bedriagai Nikolsky 186
  Teratoscincus microlepis Nikolsky 187
  Teratoscincus scincus (Schlegel) 188
Genus *Tropiocolotes* W. Peters 190
Key to the Species of *Tropiocolotes* in Iran 191
  Tropiocolotes helenae (Nikolsky) 191
  Tropiocolotes latifi Leviton and S. Anderson 193
  Tropiocolotes persicus (Nikolsky) 195
  Tropiocolotes cf. *steudneri* (W. Peters) 196
LACERTIDAE 199
Key to the Genera of Lacertidae in Iran 199
Genus *Acanthodactylus* Fitzinger 200
Key to the Species of *Acanthodactylus* in Iran .... 201 (see also 256)
  Acanthodactylus blanfordi Boulenger 201
  Acanthodactylus boskianus (Daudin) 257
  Acanthodactylus grandis Boulenger 203
  Acanthodactylus micropholis Blanford 205
  Acanthodactylus nilsoni Rastegar-Pouyani 258
  Acanthodactylus opheodurus Arnold 258
  Acanthodactylus schmidti Haas 206
Genus *Eremias* Fitzinger 207
Key to the Species of *Eremias* in Iran 208
  Eremias acutirostris (Boulenger) 211
  Eremias andersoni Darevsky and Szczerbak 211
  Eremias arguta (Pallas) 212
  Eremias fasciata Blanford 213
  Eremias grammica (Lichtenstein) 215
  Eremias intermedia (Strauch) 216
  Eremias lalezharica Moravec 217
  Eremias lineolata (Nikolsky) 218

- *Eremias nigrocellata* Nikolsky 219
- *Eremias nigrolateralis* Rastegar-Pouyani and Nilson 220
- *Eremias persica* Blanford 221
- *Eremias pleskei* Bedriaga 223
- *Eremias scripta* (Strauch) 224
- *Eremias strauchi* Kessler 225
- *Eremias velox* (Pallas) 227

Genus *Lacerta* Linnaeus 228
Key to the Species of *Lacerta* in Iran .... 228 (see also 259)
- *Lacerta brandtii* De Filippi 231
- *Lacerta cappadocica* (F. Werner) 232
- *Lacerta chlorogaster* Boulenger 234
- *Lacerta defilippii* (Camerano) 236
- *Lacerta media* Lantz and Cyrén 237
- *Lacerta mostoufi* Baloutch 238
- *Lacerta praticola* Eversmann 239
- *Lacerta princeps* Blanford 240
- *Lacerta raddei* Boettger 243
- *Lacerta steineri* Eiselt 245
- *Lacerta strigata* Eichwald 246
- *Lacerta valentini* Boettger 248
- *Lacerta zagrosica* Rastegar-Pouyani and Nilson 260

Genus *Mesalina* Gray 249
Key to the Species of *Mesalina* in Iran 249
- *Mesalina brevirostris* Blanford 249
- *Mesalina watsonana* (Stoliczka) 251

Genus *Ophisops* Ménétriés 254
- *Ophisops elegans* Ménétriés 254

Addendum to Lacertidae 256
Revised Key to the Species of *Acanthodactylus* in Iran 256
- *Acanthodactylus boskianus* (Daudin) 257
- *Acanthodactylus nilsoni* Rastegar-Pouyani 258
- *Acanthodactylus opheodurus* Arnold 258

Revised Key to the Species of *Lacerta* in Iran 259
- *Lacerta zagrosica* Rastegar-Pouyani and Nilson 260

SCINCIDAE 262
Key to the Genera of Scincidae in Iran 262
Genus *Ablepharus* Fitzinger 263
Key to the Species of *Ablepharus* in Iran 263
- *Ablepharus bivittatus* (Ménétriés) 263
- *Ablepharus pannonicus* (Lichtenstein) 264

Genus *Chalcides* Laurenti 267
- *Chalcides ocellatus* (Forsskål) 268

Genus *Eumeces* Wiegmann 269
Key to the Species of *Eumeces* in Iran 269
- *Eumeces schneideri* (Daudin) 269
- *Eumeces taeniolatus* (Blyth) 272

- Genus *Mabuya* Fitzinger 274
  - Key to the Species of *Mabuya* in Iran 274
    - *Mabuya aurata* (Linnaeus) 274
    - *Mabuya vittata* (Olivier) 277
- Genus *Ophiomorus* A. M. C. Duméril and Bibron 278
  - Key to the Species of *Ophiomorus* in Iran 279
    - *Ophiomorus blanfordi* Boulenger 279
    - *Ophiomorus brevipes* (Blanford) 280
    - *Ophiomorus nuchalis* Nilson and Andrén 282
    - *Ophiomorus persicus* (Steindachner) 283
    - *Ophiomorus streeti* S. Anderson and Leviton 283
    - *Ophiomorus tridactylus* (Blyth) 285
- Genus *Scincus* Laurenti 287
  - *Scincus scincus* (Linnaeus) 287
- UROMASTYCIDAE 289
  - Genus *Uromastyx* Merrem 289
  - Key to the Species of *Uromastyx* in Iran 289
    - *Uromastyx aegyptius* (Forsskål) 290
    - *Uromastyx asmussi* (Strauch) 291
    - *Uromastyx loricatus* (Blanford) 293
- VARANIDAE 295
  - Genus *Varanus* Merrem 295
  - Key to the Species of *Varanus* in Iran 295
    - *Varanus bengalensis* (Daudin) 295
    - *Varanus griseus* (Daudin) 297
- BIBLIOGRAPHY OF THE HERPETOLOGY OF IRAN 301
- APPENDIX I. ABBREVIATIONS USED IN TEXT 345
- II. LOCALITIES AND MATERIAL EXAMINED 349
- III. GAZETTEER 381
- INDEX 415

LIST OF PLATES

PLATE 1	SATELLITE PHOTOGRAPH OF IRAN
PLATE 2	AGAMIDAE I
PLATE 3	AGAMIDAE II
PLATE 4	AGAMIDAE III
PLATE 5	AGAMIDAE IV
PLATE 6	ANGUIDAE, EUBLEPHARIDAE, GEKKONIDAE I
PLATE 7	GEKKONIDAE II
PLATE 8	GEKKONIDAE III
PLATE 9	GEKKONIDAE IV
PLATE 10	GEKKONIDAE V
PLATE 11	LACERTIDAE I
PLATE 12	LACERTIDAE II
PLATE 13	LACERTIDAE III
PLATE 14	LACERTIDAE IV
PLATE 15	LACERTIDAE V
PLATE 16	LACERTIDAE VI, SCINCIDAE I
PLATE 17	SCINCIDAE II
PLATE 18	SCINCIDAE III
PLATE 19	UROMASTYCIDAE, VARANIDAE
PLATE 20	HABITATS I
PLATE 21	HABITATS II
PLATE 22	HABITATS III
PLATE 23	HABITATS IV
PLATE 24	HABITATS V
PLATE 25	HABITATS VI

NOTE: **The plates are grouped following page 300.**

THE LIZARDS OF IRAN

INTRODUCTION

My work on the herpetofauna of Iran began with field studies in southwestern Iran in 1958 and continued during subsequent years through studies of museum collections. In 1975, I was invited to carry out further field studies in Iran under the auspices of the Iran Department of the Environment. During a four-month period, I visited all provinces of the country and was able to gain first-hand experience with most of the major habitats and biotopes. I was fortunate in being able to encounter many of the lizard species in the field.

Owing to subsequent sociopolitical events in Iran, the future of biological field work in that country has become uncertain and few Western zoologists have visited Iran since the Islamic Revolution, so far as I am aware. It seems especially timely to summarize our present knowledge of the Iranian fauna, to list all documented localities (and to identify these as precisely as possible, as many geographic names and provincial boundaries have changed as a result of the Islamic Revolution), to call attention to all relevant literature, and to point out recognized systematic problems. I began this project with an earlier paper (S. Anderson, 1979) on the turtles, crocodiles, and amphisbaenians. Mahmoud Latifi of the Razi Institute published a handbook to the snakes of Iran (Latifi, 1985 [Farsi version], 1991 [English translation and emended checklist], 1992 [second Farsi edition]). I published a brief account of the amphibians of the country in the *Encyclopaedia Iranica* (S. Anderson, 1985).

The present volume is intended primarily to serve the needs of the professional herpetological community by presenting a synopsis of current knowledge of the lizard fauna of Iran from a biosystematic perspective. However, I have tried to serve a broader audience, including field workers, herpetoculturists, biologists concerned with various aspects of biodiversity, and, particularly, researchers in those areas where there is little access to the primary literature of herpetology.

In the species accounts I have proceeded as follows:

Order of presentation: Because no linear arrangement can represent phylogeny, I have arranged accounts alphabetically, first by family, genera within families, species within genera, etc., even though this results in the wide separation of uromastycids from agamids, varanids from anguids, and so forth.

Species name: Species preceded by an asterisk (*) have not yet been documented for Iran, but are considered likely to occur there. In addition to the Latin names, I have provided English vernacular names for most taxa. I have followed the vernacular usage of Ananjeva, *et al.* (1988), but have also gleaned common names from the popular literature and other sources readily available to me. I have not made an exhaustive search for every English name that has been used. In some cases, no suitable vernacular name could be found and I have provided names of my own invention, usually approximations of the Latin names or translations of Russian or German common names. I have not adhered to the vernacular usage proposed by Frank and Ramus (1995) as many of their names for Iranian species seem inappropriate.

Synonymy: In addition to the primary synonymy for the taxon, I have included every literature reference known to me that mentions Iranian material in any significant way. These citations include taxonomic works, reports on collections, catalogues and checklists, and papers dealing with the natural history of the species. I have not included references that deal only with non-Iranian specimens or observations; thus, some important synonyms based on extralimital material are not included in the synonymies, but these are often mentioned in the remarks.

Diagnosis: This statement only provides those characterizations which are sufficient to distinguish a species from all other currently recognized species in the genus and I have included autapomorphies, where recognized.

Color pattern: I have provided information on the color in life if known to me; otherwise, the statement is based on Iranian specimens fixed initially in formalin and preserved in ethanol, or occasionally, on statements in the literature.

Sexual dimorphism: Wherever possible, I have given information to make possible determination of gender without dissection.

Size: The snout-to-vent length and the tail length of the largest Iranian specimen are given, and wherever possible, these measurements are given for the largest specimen of each sex. If relatively few Iranian specimens have been available and they seem to be significantly smaller than specimens known from elsewhere in the range as a probable result of inadequate sampling, I have given the maximum measurements for extralimital specimens. In many instances, such measurements are from the literature, the gender not having been provided.

Habitat: I have tried to present a comprehensive statement of the known habitat for each species, based on my own observations or statements in the literature pertaining to Iran. In cases where nothing is known of the habitat in Iran, I have cited information for the species elsewhere in its range.

Distribution: I have given a broad general statement about the total range of each species, and a more detailed statement of its distribution within Iran. A distribution map showing the Iranian collecting localities is provided for each species.

Iranian localities and material examined: (See Appendix II). I have included all localities known to me from museum specimens and from the literature, in which case the locality is followed by the initials of the author and the last two digits of the date in parentheses (e.g., WB 76 = William Blanford, 1876 in the synonymy and bibliography — see Appendix I) and I have provided a museum citation for the specimens I have examined from each locality (museum abbreviations follow Leviton, et al., 1985, 1988 — see Appendix I). Wherever possible, I have associated the localities in the literature with a standardized spelling in the "United States Board on Geographic Names Gazetteer no. 19, *Iran*," (1956), and only this spelling has been listed under each species in the interest of saving space. To avoid repeating the latitude and longitude coordinates for place names, these have been incorporated into the cross-referenced gazetteer (Appendix III), which attempts to include all Iranian place names and variant spellings in the herpetological literature. In cases where the reader questions my interpretation of the species identification of a previous author, it may be necessary to check the locality cited in the original paper. Distances and elevations appear as given in publications and on museum labels. Conversion of miles and feet to kilometers and meters appears in brackets. I have not converted the English system of measurements to the metric system exclusively, because to do so gives a false impression of precision. For convenience in locating place names, the localities have been listed under the ostan (administrative province) in which they occurred as of 1968 (Fig. 1 [Fisher, 1968: map 2; S. Anderson, 1974: 28, fig. 1]). I am aware that the boundaries, names, and numbers of administrative units in Iran have changed since the Islamic Revolution (Fig. 2). Although it has been expedient to stabilize this organization of localities during the long span of years over which I have been working on this project, hindsight indicates that organization by physiographic province would have been more realistic.

Principal collectors of previously unreported material are as follows: CAS 102479–102572 and FMNH 141263–141710 (William S. and Janice K. Street Expedition, 1962–

FIGURE 1. Map of Iran showing provinces (ostans) as of 1968 (based on Fisher, 1968, redrawn). 1 Tehran; 2 Gilan; 3 East Azarbaijan; 4 West Azarbaijan; 5 Kordestan-Kermanshah; 6 Khuzestan-Lorestan; 7 Esfahan; 8 Fars; 9 Kerman; 10 Baluchistan-Sistan; 11 Khorasan; 12 Mazanderan.

1963); FMNH 170854–171213 (Street Expedition, 1968); UMMZ 129779–129823 (Douglas Lay, 1969); UMMZ 130555–130586 (Richard W. Redding, Jr., 1971); UMMZ 133277–133293 (Redding, 1973); CAS 141018–141302 and MMTT 721–1473 (Steven C. Anderson Expedition, 1975); MMTT 1–720 (various collectors associated with Iran Department of the Environment, various dates); CAS 111930–111941, 140544–140824, and 142092–143300 (Razi Institute, Iran, via Mahmoud Latifi, various dates); BMNH (various collectors and dates, prominent among them, W. K. Loftus [1853, 1855, SW Iran], Gen. F. J. Goldsmidt [1873, Sistan], E. Lort Phillips [1887, Gesham Id., Bushire], S. Butcher [1894, Persian Gulf Coast], R. B. Woosnam [1905, western Iran], H. E. J. Biggs [1936, Central Plateau], G. Popov [1951, Kerman and Baluchistan], L. Cornwallis [1966, Shiraz area]).

Distribution maps: All Iranian localities for which I have sufficiently precise information have been plotted. Open symbols (O) are literature records; solid symbols (●) represent material examined; rectangles (■) indicate type localities.

Text-figures: I have tried to provide figures that will assist in the identification of specimens. Many of these show characters not mentioned in the keys or diagnoses. In addition, I have reproduced figures from historic sources that are not readily available, especially where this book is likely to be used in the field. These include illustrations of type specimens, illustrations showing variation, problematic specimens, and others. Owing to limited funds and to my limited artistic abilities, few of the illustrations are new, and thus, the keys and taxonomic accounts lack consistency of illustration.

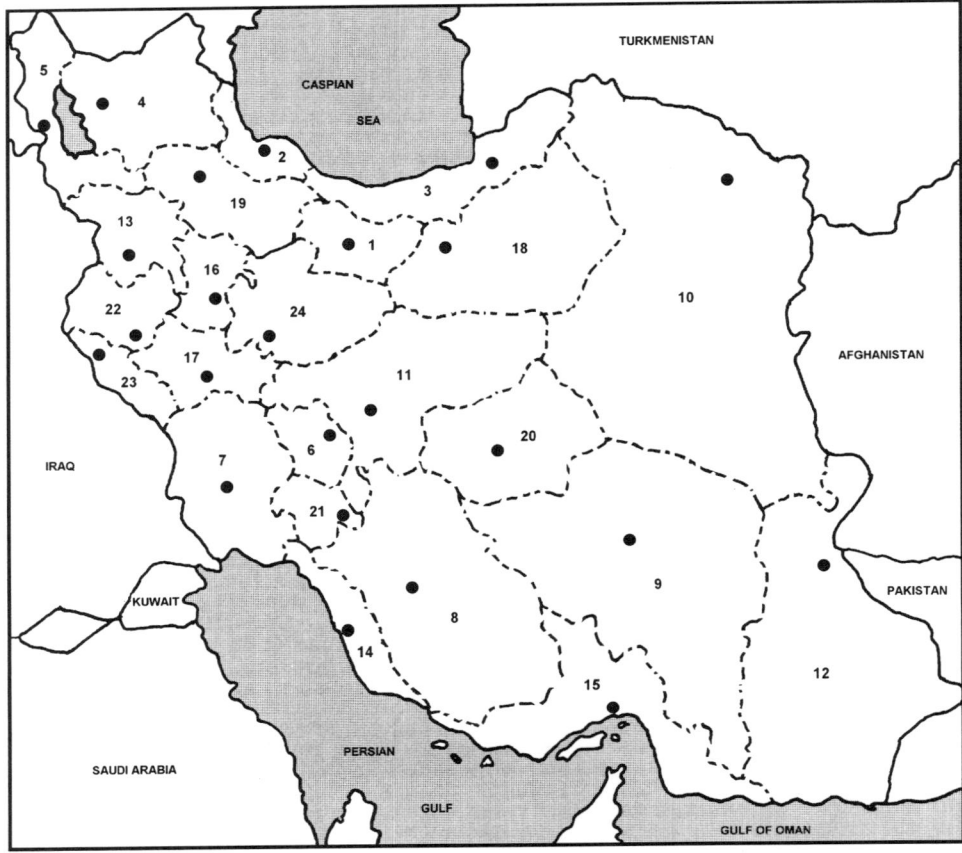

FIGURE 2. Map of Iran showing provinces (ostans) as of 1991 (based primarily on Shenassi, 1985, redrawn, and Latifi, 1991). Format: Ostan/Province administrative center (provincial capital). 1 Tehran/Tehran; 2 Gilan/Rasht; 3 Mazanderan/Gorgan; 4 Azarbayjan-e Sharqi/Tabriz; 5 Azarbayjan-e Gharbi/Orumiyeh; 6 Chahrmahal-va-Baktiyari/Shar-e Kord; 7 Khuzestan/Ahvaz; 8 Fars/Shiraz; 9 Kerman/Kerman; 10 Khorasan/Mashhad; 11 Esfahan/Esfahan; 12 Sistan-va-Baluchistan/Zahedan; 13 Kordestan/Sanandaj; 14 Bushehr/Bandar-e Bushehr; 15 Hormozgan/Bandar-e 'Abbas; 16 Hamadan/Hamadan; 17 Lorestan/Khorramabad; 18 Semnan/Semnan; 19 Zanjan/Zanjan; 20 Yazd/Yazd; 21 Boyerahmad-va-Kogiluye/Boyerahmad; 22 Bakhtaran/Bakhtaran; 23 Ilam/Ilam; 24 Markazi/Arak.

Color plates: In selecting photographs, I have given priority to Iranian examples of the represented species and to those for which I have the most data, including gender and museum register number, sometimes in preference to photographs of better quality with less information. For several species I have included more than one photograph in order to show gender or age-class differences, dorsal and ventral views, color pattern variation, or other variations. In instances where I lack Iranian examples or have photographs of insufficient quality, I have used extralimital examples of Iranian species. I have selected photographs of biotopes representative of most of the physiographic provinces of Iran. I have cited these as habitats of only those species which were actually collected or observed at the locality represented, or in similar habitats in the near vicinity. Where I have cited latitude and longitude, these coordinates have been extrapolated from maps and must be considered approximations.

Descriptions of the physiographic provinces of Iran relevant to understanding the ecology and biogeography of the herpetofauna are to be found in papers by Lay (1967), S. Anderson (1966a, 1968), and Fisher (1968). Vegetation is discussed by Zohary (1963) and Bobek (1968). Climate is reviewed by Adle (1960a, 1960b) and Ganji (1968). The several papers by Bobek (1952a, 1952b, 1963) and Butzer (1956, 1957, 1958a, 1958b, 1961, 1963) should be consulted by those concerned with recent paleogeography and paleoclimatology in relation to the development of contemporary floras and faunas of Southwest Asia.

The continuing publication (in fascicles) of *Encyclopaedia Iranica* is providing coverage of the fauna, flora, and geography of Iran in addition to articles of cultural interest.

ACKNOWLEDGMENTS

Once again, I wish to acknowledge my debt to those persons listed in two previous papers (S. Anderson, 1974, 1979). These include all of the collectors of museum materials, all of the curators who loaned me specimens and/or made me welcome at their institutions, my associates and colleagues at the California Academy of Sciences, many personnel of the Iranian Oil Exploration and Producing Company, who facilitated my field work in 1958, officials and staff of the Iran Department of the Environment, all those who accompanied me in the field at various times, and, above all, my family.

Portions of the work were supported by a grant from the American Philosophical Society (Grant no. 4959 – Penrose Fund, 1968), by five Faculty Research Grants (1971, 1972, 1975, 1989, 1995) from the University of the Pacific, Eberhardt Summer Scholar Award (1989), Eberhardt Scholar Award (1994), Francis Hunter Faculty Research Grant, Biology Department (all University of the Pacific), and by travel and field assistance from the Iran Department of the Environment (1975). The University of the Pacific made possible the 1975 field work through a faculty development leave, and such leaves in 1980 and 1984 made possible work on the manuscript. A faculty development leave in 1989 enabled me to do museum research at the Natural History Museum in London and at the Royal Museum of Scotland in Edinburgh. The bulk of the systematic research was carried out at the California Academy of Sciences while I was Associate Curator of Herpetology there, and through use of the collections and facilities of the Department of Herpetology and the library of that institution, where I have been a Research Associate after 1970.

Virginia Hudson typed portions of the manuscript at various stages of its preparation. Over the past quarter century Marina Tarala and Monica Schutzman of the interlibrary loan staff of the University of the Pacific have tracked down obscure references in at least ten languages and obtained copies and loans for me.

The following people generously contributed photographs: Kraig Adler, Herman A. J. in den Bosch, J. C. Briggs, Ilya Darevsky, Josef Eiselt, Eskandar Firouz, Michael Golubev, Haji Gholi Kami, Alan E. Leviton, Sherman A. Minton, Jiří Moravec, Göran Nilson, William Ross, and Jean Vasserot.

I thank Jeromie Anderson, who read the eublepharid and gekkonid sections, E. N. Arnold, who read the agamid and lacertid sections, Aaron Bauer, who read the eublepharid and gekkonid sections, Michael Golubev, who read the agamid, eublepharid, and gekkonid sections, J. Robert Macey, who read parts of the agamid section, Nasrullah Rastegar-Pouyani, who read the agamid and lacertid sections, and Darrel Frost, who corrected my Latin misusages with respect to the genus *Cyrtopodion*. Kraig Adler and Alan Leviton read the entire manuscript. All made useful and thought-provoking comments and criticisms and saved me from several instances of embarrassment. I did not always heed their advice and, of course, they are absolved from all responsibility for my errors. Penultimately, Alan Leviton also set the manuscript in camera-ready page proof.

The following curators and curatorial staff have generously provided museum services over the nearly 40 years that I have been working on this project, loaned me specimens, made me welcome at their institutions, called my attention to Southwest Asian material, checked register numbers and data, carried out correspondence regarding material in their collections, and shared their ideas about systematics and biogeography: N. B. Ananjeva, E. N. Arnold, the late J. C. Battersby, the late C. M. Bogert, F. W. Braestrup, C. Cole, I. Darevsky, A. Dubois, R. Drewes, M. Golubev, A. Grandison, J. Guibé, the late G. Haas, J. Eiselt, R. Inger, A. Kluge, M. Latifi, A. Leviton, H. Marx, C. McCarthy, R. Murphy, the late J. Peters, J. Simmons, A. Stimson, G. Swinney, the late N. Szczerbak, W. Tanner, F. Tiedemann, R. Tuck, J. Vindum, Y. Werner, the late E. Williams, G. Zug, and R. Zweifel. Jens Vindum, California Academy of Sciences, checked register numbers and locality data for many of the Academy's specimens.

Geologists of the Iranian Oil Exploration and Producing Company allowed me to accompany their field explorations in 1958: Howard Anderson, Rufus Cook, Gerald A. James, Dory Little, and William O. Williams. They actively assisted my collecting activities on several occasions.

A special note of thanks is owing to Eskandar Firouz, then Director of the Department of Environment, and to Robert G. Tuck, Jr., then advisor to the Department of Environment, who invited me to Iran in 1975 as a Visiting Scientist and made my field studies possible. Field companions in 1975 were Reza Khaza'i, Ruben McCullers, Abbas Mirili, and Robert Tuck.

My greatest debt is to my family: to my late father, Howard T. Anderson, who always encouraged my interest and education in natural history and who frequently discussed ideas about geology, paleogeography, and evolutionary theory with me; to my mother, Lois B. Anderson, who was not only supportive of my interests, but who shared her home with captive members of several local faunas; to my wife, Kay, with whom I've shared the hard times and the good times for 39 years; and to my son, Malcolm, who sacrificed the usual father-son sports, scouting, etc., for natural history field trips.

A SHORT HISTORY OF HERPETOLOGY IN IRAN WITH AN INTRODUCTION TO THE LITERATURE

Reports on Travels, Collections, and Field Work in Iran

Natural history exploration has often been tied to geopolitical events. Naturalists, frequently trained as physicians in the 18th and 19th centuries, often accompanied military expeditions, boundary commissions, railway and telegraph surveys, and diplomatic missions. Throughout western Asia, "native" collectors (meaning Indian, in the case of British expeditions) were usually hired to obtain and prepare zoological and botanical specimens. The early period of natural history collecting by Europeans in Iran coincided with the expansionist activities of three great powers, the Ottoman, Russian, and British empires. The latter two, particularly, saw the independent states and khanates of Southwest Asia merely as buffers between their spheres of territorial hegemony and sought to establish well-delimited borders that would prevent direct military conflict. Each feared the efforts of the other to establish increased trade and economic influence in its realm. Fears that the Russians would seek a warm-water port in the Persian Gulf or Indian Ocean have determined Western foreign policy in Southwest Asia to the present day, while the Russians early sought to maintain control over the trade routes and trade centers of Central Asia. The resulting intrigues of the "Great Game" have been well-chronicled (for a recent example, see Hopkirk, 1994). A history of Iran during this period is provided in volume 7 of *The Cambridge History of Iran* (Avery, *et al.*, 1991). Herpetological publications have been tied to geopolitical events even into the present decade, two handbooks having been hastily prepared to coincide with the Gulf War operations (Leviton, *et al.*, 1992; U.S. Department of Defense, 1992), in part to aid in identification of venomous snakes (always a potential hazard to military field operations), in part to encourage herpetological collecting on the part of military personnel, and, finally, to help in the ecological survey and clean-up after the war.

The approximate present boundaries of Iran have been more-or-less well established since the 16th century, although the precise delimitation of those borders has been the source of much conflict and arbitration since that time. An outline history of the establishment of Iran's boundaries is presented in *Encyclopaedia Iranica* (McLachlan, 1989; Planhol, 1989; Balland, 1989; Kechichian, 1989; Schofield, 1989).

Iran had been in conflict with the Ottoman Empire since the early 16th century and had lost much of Kurdistan and Mesopotamia as a result. However, the present boundaries had been established as a poorly-defined frontier zone in 1639. A border commission composed of representatives of Britain, Russia, and the Ottoman government was established in 1843, resulting in the Treaty of Erzurum, signed in 1847. Topographic mapping of the border was completed in 1869. British interest in Persian oil and the formation of the Anglo-Persian Oil Company as well as a German proposal for a Basra rail link increased foreign power interests in delimiting a firm border, including control of shipping and navigation of the Shatt al'Arab (the confluence of the Tigris and Euphrates rivers, flowing into the Persian Gulf).

Following the fall of the Ottoman Empire in 1918, Iraq became an independent state under British mandate in 1921. Disputes over control of the Shatt al'Arab, Kurdish issues in northeastern Iraq, and separatist Arabs in Khuzestan have strained relations between Iran and Iraq since this time and the absence of an effective border accord served as one of many excuses for the invasion of Iran by Iraq in 1980, initiating an eight-year war and continuing unsettled disputes.

The current border between Turkey and Iran was firmly established in 1939 as a result

FIGURE 3. George Albert Boulenger (1858–1937). (Courtesy George Sprague Myers/Alan E. Leviton Portrait File, California Academy of Sciences.)

of arbitration after incursions by both countries across the Perso-Turkish border, which had been established in 1914. The principal herpetological contributions to result from the establishment of the western boundaries of Iran were those of the Mesopotamian Expeditionary Force 1915–1919 reported by Boulenger (1920e, 1920f) (Fig. 3) and Procter (1921).

Boundary problems between Russia and Iran began with the arrival of the Russians in the Caspian area after the conquest of Astrakhan in 1556. The first incursion of the Russians into the Caucasus and northern Iran occurred in 1772, at the end of the Safavid Dynasty and during the Afghan invasion of Iran, the troops of Peter the Great having been invited in by the governor of Rasht, who was resisting the Afghans. The Russians occupied Darband, Baku, Gilan, Mazanderan, and Astarabad, all within Iranian territory. These areas were subsequently relinquished by Russia following the death of Peter the Great and the reestablishment of Persian power by Nader Shah. The Russians gained access to Transcaucasia in 1783 when Georgia signed a treaty with them during a weakening of central power in Iran under the Zands, and Georgia subsequently was annexed by the Russian Empire in 1801.

Following two Russo-Iranian wars (1801–1813 and 1826–1828), the present boundary was established west of the Caspian Sea, with Russia occupying all conquered territory, the border being defined by the course of the Arax River and the crest of the Talysh Mountains, topographic features which had served to halt the advance of Russian troops. There were border incidents for decades thereafter, because nomadic tribes habitually crossed the borders in the Moghan Steppes, which had been divided arbitrarily. It was not until 1957 that a Russo-Iranian protocol was signed that precisely defined the border.

During the period of territorial dispute in the Caspian and Transcaucasian areas during the late 18th and early 19th centuries, a number of natural history collections were made in northern and western Iran.

The first European naturalist to visit Iran was Samuel Gottlieb Gmelin, employed by the Russian government to explore parts of Gilan and Mazanderan in 1770–1772 (see Gmelin 1774, 1784; Pallas, 1831 [Fig. 4]). Gmelin lived at Enzeli for several months and collected natural history specimens in the vicinity. He eventually died in captivity in the Caucasus. Subsequent explorers who made zoological collections in Iran included G.-A. Olivier to Esfahan and Tehran in 1796 as a representative of the French government (Olivier, 1804, 1807); Edouard Ménétriés (Fig. 5), employed by the Russian government (1832) in the Talysh Mountains; Carl Eduard von Eichwald (Fig. 6), also representing the Russians (1834–1837, 1841), who touched at two or three places on the Iranian coast during his exploration of the Caspian shores from 1825 to 1826; Aucher-Eloy, a French botanist who visited Shiraz, Bushire, Bandar 'Abbas, the Bakhtiari Mountains, Hamadan, Tehran, and Tabriz, made zoological collections, unfortunately without precise locality data; W. K. Loftus, attached as geologist to the commission which surveyed the borders between Turkey and Iran in 1849–1852, collected some zoological specimens along the Iranian-Turkish

FIGURE 4. Peter Simon Pallas (1741–1811). (Courtesy of Kraig Adler.) FIGURE 5. Edouard Ménétriés (1802–1861). (Courtesy of Kraig Adler.) FIGURE 6. Carl Eduard Ivanovich von Eichwald (1795–1876). (Courtesy of Kraig Adler.)

border region; and some reptiles, without adequate locality data (including *Hemipodion persicum* Steindachner, 1867), were brought back to the museum at Vienna from Iran by Theodor Kotschy, an Austrian botanist.

East of the Caspian, Russian expansion into lower Central Asia did not come into direct conflict with Iranian power. The Treaty of Golestan (1813) reserved to Russia a monopoly of naval power on the Caspian Sea and the Russians maintained a naval station at Ashurada at the entrance to the Bay of Astarabad until the end of the 19th century. A first Russian border post was established at Chikishlyan, north of the mouth of the Atrak River.

During the 17th and early 18th centuries, Safavid regimes had resettled Kurdish and Azeri populations, known for their fierce fighting abilities, in Khorasan to serve as guardians of the natural frontier in the eastern Alborz and the Kopet Dag against incursions by Turkoman tribes. Exploration of eastern Iran was undertaken by the Russians during the mid-19th century, and a few reptiles were collected by Count Eugen von Keyserling during the mission led by N. de Khanikoff in 1858–1859 and subsequently described by Strauch (1863, and subsequent memoirs on various reptile groups). "Pacification" of Turkomen tribes by the Russians in the 1880s established the frontier at the foot of the mountains of Khorasan. The Treaty of Tehran in 1881 defined the border precisely, incorporating almost all of the Turkoman populations into the Russian Empire. In 1884, the Russians occupied the old city of Sarakhs on the eastern bank of the Tajan River and a contact point was established here facing the Iranian post on the western bank (see Curzon, 1892:I:183–209). Substantial economic exchange through Khorasan followed the pacification of the Turkoman tribes of Ashkhabad and Merv and the subsequent construction of the Caspian railway and the automobile route from Ashkhabad to Quchan and, subsequently, other routes through the Kopet Dag.

Except for the few reptiles and fishes collected by Count Keyserling in Khorasan, extensive collections in this region seem not to have been made until Zarudny's travels at the end of the century (see below).

Apparently, the first attempt to compile a list of all of the vertebrates known from Iran was that of Filippo De Filippi (1865) (frontispiece) of Turin, listing 30 mammals, 167 birds, 39 reptiles, 3 amphibians, and 22 fishes. De Filippi accompanied an Italian embassy to Iran

FIGURE 7. John Anderson (1833–1900). (Courtesy of the Library of the Natural History Museum, London.)

in 1862 along the route from Poti on the Black Sea via Tiflis and Tabriz to Kazvin and Tehran, returning via Rasht, the Caspian Sea, and Russia. His zoological and geological collections were supplemented by those of his companion, Marquis Giacomo Doria, later director of the Museo Civico di Storia Naturale at Genoa, who journeyed to southern Iran and collected reptiles as well as other zoological specimens. Many of Doria's specimens were acquired by the Natural History Museum, London, in 1869, and others came in with the Lataste collection in 1920.

In a report on amphibians and reptiles added to the Indian Museum, Calcutta, collections, John Anderson (1872) (Fig. 7) recorded 17 reptiles (of which, six were described as new) and three amphibians. Most were from the Shiraz, Resht, and Tehran areas, and although the collectors were not recorded in Anderson's paper, Blanford (1876) stated that this was a collection made by Major Oliver B. St. John during the years 1869–1871.

An updated and much more thorough systematic synopsis of the vertebrates, exclusive of fishes, was that of Blanford (1876). William Thomas Blanford (frontispiece), a geologist and naturalist, was appointed to the Geological Survey of India in 1855. He travelled widely, made many published contributions, descriptive and theoretical, to the geology and zoology of southern Asia, and deposited his many collections in the Indian Museum, Calcutta, and in the Natural History Museum, London. He was the geologist and naturalist appointed by the Government of India to travel in the company of St. John from Gwadar (in present-day Pakistan) through Baluchistan and southern Iran as far west as Shiraz and thence to Tehran via Esfahan in 1872. He did some collecting in the Alborz Mountains north of Tehran as well. His collecting was assisted and supplemented by St. John. In addition to preliminary papers describing new taxa (Blanford, 1874a, 1874b, 1875a), his researches resulted in his comprehensive *Zoology and Geology* of Persia, published as Vol. II of *Eastern Persia, an Account of the Journeys of the Persian Boundary Commission, 1870–71–72* (Blanford, 1876). Narratives of the travels of this commission, along with an account of the physical geography of Persia, maps, and itineraries, were published in Vol. I, edited by Major General Sir Frederic J. Goldsmid. Blanford listed 89 mammals, 383 birds, 92 reptiles, and 9 amphibians. This author also provided an account of scientific natural history in Iran prior to 1876. Blanford included a brief descriptive analysis of the fauna and considered the relationships of the faunas of various regions of Iran to those of neighboring geographic regions. No other author, before or since, has attempted such a comprehensive consideration of the Iranian fauna.

Blanford's treatise was the most extensive zoological work to result from the period of boundary arbitration in Iran. Goldsmid's expedition was connected with the establishment of Afghanistan as a buffer state by the European powers, and in this case, resolving political and geographic disputes between Iran and Afghanistan over the control of Sistan, the most densely populated region on the Perso-Afghan border and thus of great strategic importance. The Sistan endorheic basin encompasses the outlet of the Helmand (or Hirmand) River and several less important (mainly seasonal) streams. The Sistan Arbitration Commission was under the command of (then) Col. Goldsmid, who had recently (1871) demarcated the Makran boundary between Iran and the Khanate of Kalat in Baluchistan.

Continued struggles over water relations necessitated a second Sistan Arbitration Commission, and the British served this function once again, this time under the command of Col. A. H. McMahon, who had previously been involved in the establishment of the southern boundary of Afghanistan with the Afghan Delimitation Commission (see below). This Sistan expedition remained in the field for more than two years (1903–1905), but the Goldsmid boundaries were only slightly amended. Disputes over water have continued in this region to the present day.

The present political division of the Iranian Plateau along the Perso-Afghan border goes back to the 18th century, when two rival powers emerged from the ruins of the Safavid empire, the Dorrani empire on the east and the Qajar empire on the west. The boundaries of their respective spheres of influence were established when Dost-Mohammad Khan, Amir of Kabul, took Herat in 1863. The formal process of border delimitation was long (1872–1935). South of the bend in the Harirud, the contested district of Dasht-e Hashtedan was divided between the two countries in 1889 through the arbitration of Major General C. S. McLean, then British consul general for Khorasan and Sistan. The central part of the frontier was not definitively demarcated until a Turkish commission under General Fahreddin Altay drew a boundary more than 400 km long between Hashtadan and Sistan in 1934–1935.

In 1881, Blanford reported on a collection of reptiles, collector unknown, purchased by the British Museum from the natural history dealers Watkins and Doncaster. Most were from the vicinity of Bushire and on the route from Bushire to Esfahan, via Shiraz. This collection included six species of lizards and 12 species of snakes. Blanford described *Agama persica, Scincus conirostris*, and *Hydrophis temporalis* in this report.

Following Blanford's work, the pace of studies of the Iranian fauna increased. The zoology of the Afghan Delimitation Commission, 1884–1885, was reported by J. E. T. Aitchison (1887), a medical doctor whose natural history specialty was botany, assigned as naturalist attached to the mission. The expedition travelled in areas of what are now Pakistan, Afghanistan, Iran, and Turkmenistan; natural history specimens were collected along the entire route. The areas of Iran visited by the Commission were Sistan and eastern Khorasan. A total of 35 species of reptiles and two species of amphibians were collected; these were worked up by George Albert Boulenger of the the "British Museum (Natural History)," who described six species (*Stenodactylus lumsdeni, Agama isolepis, Phrynocephalus ornatus, P. luteoguttatus, Scapteira acutirostris,* and *Lytorhynchus ridgewayi*). In addition, a seventh new species was present (*Phrynocephalus clarkorum*, confused among *P. ornatus*) but not recognized until some 80 years later (S. Anderson and Leviton, 1967).

James A. Murray (1884b) reported on the zoological collections from Iran in the Karachi Museum. The amphibians and reptiles were collected by W. D. Cumming of the Persian Telegraph Commission in the early 1880s and in 1876–1877 by some unknown exploring commission. The collection included eight species of lizards, one snake, and one anuran.

Gustav Radde (1886), founder and curator of the Caucasian Museum and Library in Tiflis, published on the natural history of the southwest Caspian coast, where he had travelled along the Iranian border.

In 1889, Robert Günther published the natural history of the Lake Urmia area in northwestern Iran, and in 1899, Boulenger published on a collection of amphibians and reptiles from the region and islands within the lake (5 species of frogs, 5 lizards, 2 snakes) resulting from Günther's expedition. In 1916, the Lake Urmia region was visited by an

FIGURE 8. Aleksandr Mikhailovich Nikolsky (1858–1942). (Courtesy of Kraig Adler.)

expedition from the Caucasian Museum; 16 species of amphibians and reptiles were reported by V. Rostombekov (1928, 1938, in Georgian, with an English summary).

The most important collections of vertebrates made in Iran around the turn of the century were those of the ornithologist Nikolai Aleksyevich Zarudny (see frontispiece). He made a series of expeditions to eastern Iran, investigating the mountains and desert lowlands of Khorasan, Sistan, and Baluchistan in 1884, 1892, 1896, 1898, 1900–1901, and in 1903–1904 he travelled to western Iran, entering in Mazanderan, crossing west-central Iran and the Zagros Mountains, proceeding south to Khuzestan, returning through the same regions, but visiting Gilan before departing. Zarudny published itineraries for most of these expeditions (Zarudny, 1899, 1901, 1902a, 1902b, 1904). He published the herpetological results of his 1896 and 1898 expeditions (Zarudny, 1897, 1903), and in 1903 and 1911, he also published his important ornithological work, *The Birds of Eastern Persia*. New herpetological taxa and detailed accounts of Zarudny's herpetological collections, which were deposited in the museum of the Academy of Sciences in St. Petersburg, were reported by Aleksandr M. Nikolsky (1886, 1896, 1897, 1899, 1900, 1903, 1907a) (Fig. 8). In 1906, Zarudny became curator of the museum at Tashkent.

Franz Werner (Fig. 9) made a number of contributions to the herpetology of Iran, the first in 1895, reporting on materials collected by the botanist Josef Bornmüller during his travels in 1893–1894. The reptiles from Iran included one species of turtle, 11 lizards, and five snakes. In this paper he described *Phyllodactylus elisae* (from Iraq), *Hemidactylus bornmuelleri* (from Iraq), and *Agama microtympanum*. In 1902, Bornmüller visited the Alborz Mountains and the southern Caspian coast, and again, Werner (1903) reported his collection, which included one turtle, nine lizards, and one snake, which he described as a new species, *Zamenis bornmüllerorum* (at publication, he realized that he had redescribed *Contia* [=*Eirenis*] *collaris*). In 1917, Werner recorded reptiles collected by Prof. Andreas in Fars Province in 1877, 1878, and 1905 in the Kazerun-Shiraz-Persepolis regions and deposited in the Zoological Museum of the University of Göttingen (later transferred to Zoologisches Forschungsinstitut und Museum Alexander Koenig, Bonn).

FIGURE 9. Franz Werner (1867–1939). (Courtesy of Kraig Adler.)

This collection included 16 species of lizards (including *Gymnodactylus gastropholis*, described in this paper), 16 snakes (one species described as new, *Zamenis andreana*), and one turtle. Werner (1929) reported on the collection made in western Asia by Alfons Gabriel, including one species of turtle, four lizards, three snakes, and one frog collected in Iran. In 1936, another collection made by Gabriel was recorded by Werner. This collection included two species of turtles, 14 lizards, and five snakes. This paper included the description of *Gymnodactylus gabrielis* and a checklist of the fishes, amphibians, and reptiles known from

Iran at that time. Finally, in 1938, some 43 years after his first contribution to Iranian herpetology, Werner published yet another collection of reptiles by Gabriel, including one species of turtle, eight lizards, and four snakes, and a small collection made by K. H. Rechinger (five species of lizards, one snake). In this paper, he described *Bunopus biporus* as new.

The Afghan-Baluch Boundary Commission, in 1896, in the course of demarcating the border between Baluchistan (in present-day Pakistan) and Afghanistan, made a collection of reptiles under the guidance of Dr. F. P. Maynard and Captain A. H. McMahon. Most of the species of this region are shared with Iranian Baluchistan, although a number (*Phrynocephalus euptilopus, P. luteoguttatus, Eremias aporosceles*) have not yet been found in Iran, while others have been discovered in Iran comparatively recently (*Lytorhynchus maynardi, Eristicophis macmahonii*). This collection was published by A. Alcock and F. Finn (1896), with notes on the geography of the route and on the animals by Maynard.

Louis A. Lantz (1918) published a report on a small collection of reptiles in the Zoological Museum of Moscow University from the Tajan River, just inside present-day Turkmenistan at the Iran-Afghan-Turkmenistan border. These specimens, representing 16 species, were collected by N. V. Meriakri between April and September, 1914. All are species occurring in Iran, as well, and the paper represents one of the first characterizations of the fauna of this region.

Joan B. Procter (1921) published a brief account of amphibians and reptiles from Iran and Iraq collected by the British Expeditionary Forces in Mesopotamia. Only two species (snakes) were actually collected within Iran, although five species of lizards were from Kuretu, Iraq, on the Iranian border.

In 1921, L. D. Morich (Moritz or Moritch) (1922) visited the southern Central Asian deserts and neighboring Iran, reporting six species of reptiles from northern Iran. Again, in 1928, he travelled in Turkmenistan and northeastern Iran, collecting 27 species of reptiles in Iran (Morich, 1929).

The archaeologist/anthropologist, Henry Field, carried out extensive field research in Southwest Asia, including Iran, and continued to make contributions to the natural history of the region as well as to his own professional fields. Among his many published works were a bibliographic series covering the botany and zoology of Southwest Asia, and accounts of his expeditions' travels and discoveries (Field, 1953, 1955, 1956a, 1956b, 1957, 1958, 1959a, 1959b, 1960, 1962, 1963, 1966; Field and Clifton, 1970; Field and Laird, 1968, 1969, 1970, 1972). Herpetological specimens collected by Field and his associates were deposited in the Field Museum of Natural History, Chicago, and at the Museum of Comparative Zoology, Harvard University (MCZ). His early collections (1928–1934) were reported by Karl P. Schmidt (1939) (Fig. 10) and these included 32 species from western Iran, collected by Henry Field, Richard A. Martin, and Ernst Herzfeld. A list of the animals collected by the 1934 expedition was published by Sanborn (1939). During the Peabody Museum–Harvard University Expeditions of 1950 and 1955, Henry Field assembled a collection of amphibians and reptiles in Iran, Iraq, Syria, Saudi Arabia, Bahrain, Trucial Oman, and Pakistan. The squamates were reported by Georg

FIGURE 10. Karl Schmidt (1890–1957). (Courtesy of the Field Museum, negative #Z 86242, Chicago.)

FIGURE 11. Georg Haas (1905-1981). (Courtesy George S. Myers/Alan E. Leviton Portrait File, California Academy of Sciences.)

FIGURE 12. Yehudah Werner. (Courtesy George S. Myers/Alan E. Leviton Portrait File, California Academy of Sciences.)

Haas (Fig. 11) and Yehudah Werner (1969) (Fig. 12). This report included eight species of lizards and nine species of snakes from Iran. The specimens from this expedition are deposited in the Museum of Comparative Zoology, Harvard University.

The Danish Scientific Investigations in Iran, 1935–1938, resulted in a collection of amphibians and reptiles deposited in the Universitetets Zoologiske Museum at Copenhagen. The principal collector of herpetological specimens was E. Kaiser, with additional specimens collected by H. Blegvad, M. Køje, B. Løppenthin, K. Paludan, and G. Thorson. Nine species of sea snakes collected in the Persian Gulf and Gulf of Oman were reported by Helge Volsøe (1939). This was the first paper to record precise localities for sea snakes taken in these waters. In addition to scale counts, color pattern, and remarks on morphology, Volsøe provided a brief zoogeographical summary and some general remarks on the biology of sea snakes. The amphibians and terrestrial reptiles were reported by K. P. Schmidt (1952, 1955) and consisted of 29 species, including three new taxa.

Malcolm Smith (Fig. 13) published a brief characterization of the Iranian herpetofauna in the British *Admiralty Handbook for Persia* in 1945. His three earlier volumes (Smith, 1931, 1935, 1943) on reptiles in the *Fauna of British India* series have served as a primary reference for eastern and southern Iranian reptiles since their publication.

In 1948, the botanists Paul Aellen and Karl Rechinger undertook a botanical/zoological expedition to Iran. From early April through early December, they collected in the Reza'iyeh Basin (= Urmia Basin) and in many localities on the Central Plateau. To the south, they reached Bandar 'Abbas, on the Persian Gulf, and Iranshahr in Baluchistan to the east. A brief account of this expedition, along with an itinerary, two maps, and a list of collecting stations was published by Aellen (1950) and the herpetological results separately by Lothar Forcart (1950). In his account of the amphibians and reptiles, Forcart included material collected in Mazanderan in 1931 by the geologists A. Erni and R. Buxtorf, as well as specimens collected in April 1949 by G.

FIGURE 13. Malcolm Arthur Smith (1875–1958). (Courtesy George S. Myers/Alan E. Leviton Portrait File, California Academy of Sciences.)

Charif between Zahedan and Chah Bahar, Baluchistan (although these specimens lacked precise locality data). This material is deposited in the Naturhistorischen Museum Basel.

Otto Wettstein (1951) (Fig. 14) published the herpetological results of the Austrian Expedition to Iran in 1949–1950. The excursions of this expedition, which began in April 1949 and ended in July 1950, covered some 24,000 km, during which the provinces of the

FIGURE 14. Otto Wettstein-Westerscheimb (1892–1967). (From Eiselt, 1967.)

FIGURE 15. Jean Guibé. (Courtesy George Sprague Myers/Alan E. Leviton Portrait File, California Academy of Sciences.)

Central Plateau, the Zagros Mountains, and the Caspian Sea were visited. Specimens were collected by Jens Hemsen, Alfred Kaltenbach, Heinz Löffler, and Ferdinand Starmülner and are deposited in the natural history museum in Vienna. This collection included three species of frogs, four turtles, 27 lizards, and 10 snakes from Iran and some additional species from Lebanon and Iraq. Apart from a few individual specimens, I have not examined this collection. In this paper, Wettstein provided a brief descriptive zoogeography of Iran, based on reptilian distribution. In 1953, Wettstein added *Vipera renardi* to the Iranian fauna.

In 1957, Johann Popp assembled a small collection of amphibians and reptiles at Shush and in the vicinity of the Rud-e Dez, Khuzestan. These were deposited in the Zoologischen Staatssammlung in Munich and reported by Walter Hellmich (1959).

Jean Guibé (Fig. 15) published three papers concerning the herpetofauna of Iran, the first (Guibé, 1957) resulting from collections made by J. Francis Petter during two expeditions of the Pasteur Institute of Iran, the second (Guibé, 1966a) a list of specimens collected by André Villiers during a Franco-Iranian expedition, mainly in southeastern Iran, and a third (1966b), revising the genera *Tropiocolotes* and *Microgecko*. These collections are in the Museum National d'Histoire Naturelle, Paris.

In 1959, I spent nine months (mid-February to mid-December) in Khuzestan, based at Masjed Soleyman in the western Zagros foothills. During this time, I made brief forays to the lowlands of the eastern Mesopotamian Plain and to the Bander 'Abbas region and Persepolis (S. Anderson, 1963).

The first record of crocodiles in Iran, apparently, was that of Jahanbani and Saram in 1960, in a popular article on crocodile hunting in Baluchistan, which appeared in the Farsi magazine *Game and Nature*.

In 1961, Howard Stutz undertook a botanical trip in western Iran (Tehran-Qom-Esfahan-Abadeh-Persian Gulf coast), during the course of which he made a herpetological collection, subsequently deposited in the museum of Brigham Young University. Richard J. Clark and Erica D. Clark collected herpetological specimens while passing through northern Iran in June and July, 1964, on their way to Afghanistan, where they did substantial herpetological collecting and made useful observations (Clark, *et al.*, 1969), and again on their return from Afghanistan in September of that year. These two collections were reported by Clark, Clark, and S. Anderson (1966). Again, in 1968, the Clarks travelled from Greece to carry out investigations in Afghanistan (Clark, 1990) via northern Iran, collecting 16 lizard species and seven snake species in the latter country (Clark, 1991). The Clarks' materials from Turkey (Clark and Clark, 1973), Iran, and Afghanistan are in the California Academy of Sciences. Maps showing their collecting sites are to be found in their papers.

In 1962–1963, G. L. Ranck and L. H. Hermann collected small vertebrates for the United States Army Medical Research and Development Command under the direction of H. W. Seltzer. These specimens were deposited in the United States National Museum (now the National Museum of Natural History), Washington, D.C. In 1963–1965, this work was continued by J. W. Neal and Robert G. Tuck (Fig. 16). These field workers travelled extensively in Iran and Pakistan, collecting over 900 reptiles and amphibians from 113 localities in all provinces of Iran and most physiographic regions. Among their important contributions (Tuck, 1971b) was the rediscovery of *Tropiocolotes helenae*. Tuck (1971a) published an annotated list of the Iranian herpetological specimens in the National Museum of Natural History, with maps of the collecting sites and a gazetteer.

William S. Street and Janice K. Street (Fig. 17) led two expeditions to Iran, the first in 1962–1963, the second in 1968 (Fig. 18). The primary purpose of these expeditions was to collect mammal specimens and to contribute to knowledge of the biology, systematics, and distribution of these animals in Iran. However, as on their Afghan expedition in 1965 (Hassinger, 1968; S. Anderson and Leviton, 1969), amphibians and reptiles were also collected. Douglas M. Lay (Fig. 17), a mammalogist, accompanied the first expedition, published an account of the mammals collected, and a summary of the mammals known from Iran (Lay, 1967). In this paper, he gave an account of the expedition, including a narrative of the expedition routes, descriptions of the collecting sites, a map showing collecting sites, a gazetteer of localities from which mammals have been collected (historically, as well as those of the expedition), and a summary physical geography of Iran. Lay's account is particularly important in recording the biotopes in which vertebrates were collected. The Streets wrote a popular account of this expedition (Street and Street, 1986). An account of the turtles and lizards collected by the first expedition was included in my Ph.D. dissertation (S. Anderson, 1966).

The second Street expedition collected mammals and their ectoparasites in western Iran from mid-July to mid-December, 1968. Mammalogists accompanying the expedition were Daniel Womochel (Fig. 19) and Anthony F. DeBlase (Fig. 20), along with parasitologist Richard Rust. Again, amphibians and reptiles were also collected. DeBlase (1980), in his

FIGURE 16. Robert G. Tuck. (Courtesy of Brevard Community College, Cocoa, Florida.)

FIGURE 17. William S. and Janice K. Street and Douglas M. Lay (right) in the field in Iran. (Courtesy © The Field Museum of Natural History, Chicago [negative #GN 79794].)

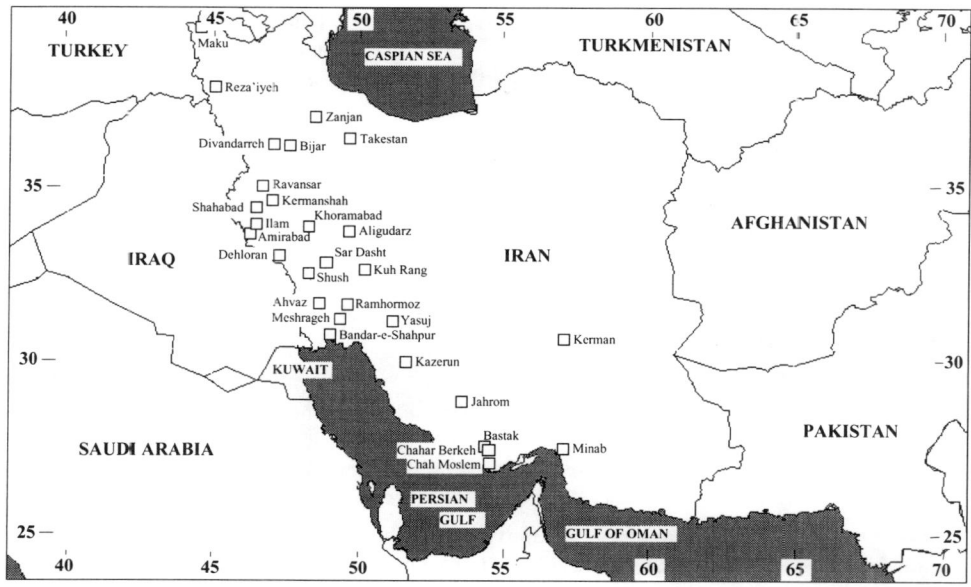

FIGURE 18. Reference place names for the Second Street Expedition to Iran, 1968. Generalized localities shown on map; details provided below.

3 mi SE Aligudarz.
25 mi SW Aligudarz.
3 mi N Amirabad (20 mi N Mehran), 1000 ft.
30°25′N 49°05′E, Bandar-e Shahpur, sea level.
3 mi NW Bastak, 1300 ft.
6 mi NW Bastak, 1300 ft.
2 mi N Bijar.
26°44′N 54°35′E, Chah Moslem, 50–100 ft.
7 mi E Chah Moslem, 100 ft
27°05′N 54°35′E, Chahar Berkeh, 700 ft.
2 mi W Dehloran, 900 ft.
35°55′N 47°02′E, Divandarreh, 5800 ft.
1 mi SE Divandarreh.
1 mi S Divandarreh.
2 mi S Divandarreh.
3 mi S Divandarreh.
3 mi SE Divandarreh.
9 mi SE Divandarreh.
20 mi N Divandarreh
30 mi S Divandarreh.
20 mi S Divandarreh.

Ghala Raihand Mts, W Divandarreh.
3 mi E Harivan, Kordestan.
33°38′N 46°26′E, Ilam, 5000 ft.
29.9 mi NW Ilam.
29.9 km SW Ilam.
70 km SW Ilam.
28°31′N 53°33′E, Jahrom, 3200 ft.
18 mi NE Kazerun.
70 mi N Kazerun.
34°19′N 47°04′E, Kermanshah.
4 mi N Kermanshah.
5 mi N Kermanshah.
12 mi E Kermanshah, 4290 ft.
33°30′N 48°20′E, Khoramabad, 4100 ft.
32°18′N 50°13′E, Kuh Rang.
4.2 mi NW Kuh Rang.
5 mi NW Kuh Rang.
6 mi NW Kuh Rang, 8500 ft.
2.2 mi W Maku, 3925 ft.
Vicinity of Maku.
21 mi E Maku.
25 mi E Maku, 3925 ft.
30°54′N 49°24′E, Meshrageh, 53 mi SE Ahvaz., 200 ft.

Jarrahi River at Meshrageh.
27°09′N 57°05′E, Minab., 100 ft.
2 mi SW Minab, 100 ft.
4 mi SW Minab.
40 mi SE Minab.
63 mi SE Minab, 2000 ft.
5 mi W Puldasht, West Azarbaijan.
10 mi W Puldasht.
31°16′N 49°36′E, Ramhormoz, Khuzestan.
12.9 mi N Ravansar (50 mi NE Kermanshah), 5000 ft.
15.9 mi N Ravansar.
37°33′N 45°04′E, Reza'iyeh.
5.8 mi SW Reza'iyeh.
32°32′N 48°52′E, Sar Dasht, near Lordegan, 7000 ft.
38.5 mi from Shahabad, Kermanshah, direction not stated.
4.2 mi N Takestan.
4.7 mi N Takestan.
8.8 mi N Takestan.
30°50′N 51°10′E, Yasuj.
5 mi NE Yasuj.
24 mi NW Zanjan.
87 mi N Zanjan.

FIGURE 19. Daniel Womochel. (Courtesy © The Field Museum, Chicago [Negative #Z 89846].)

FIGURE 20. Anthony DeBlase. (Courtesy © The Field Museum, Chicago [Negative #Z 89845A].)

study of the bats of Iran, provides a brief account of the second expedition and a gazetteer of all of the localities from which bats have been collected in Iran. His study is unquestionably the most thorough systematic account of a single vertebrate group occurring in Iran. The lizards collected by the 1968 expedition are recorded in the present book for the first time; the turtles of both expeditions were reported in my paper on the turtles, crocodiles, and amphisbaenians of Iran (S. Anderson, 1979). No account has yet been written of the amphibians and snakes of the two expeditions. The collections were deposited in the Field Museum of Natural History, cosponsor of the expeditions. A representative sample of the amphibians and reptiles collected by the first expedition is in the California Academy of Sciences.

Peter C. H. Pritchard (1966) published some notes on turtles in Iran, resulting from his observations during the Oxford University Expedition to northern Persia, July-September, 1963.

In 1968, I published a lengthy descriptive zoogeographic analysis of the lizard fauna (S. Anderson, 1968).

Josef Johann Schmidtler (Fig. 21) and his son, Josef Friedrich Schmidtler (Fig. 22), visited the Zagros regions of Kermanshahan, Lorestan, Khuzestan, and Fars in 1968 and 1970. Their investigations resulted in several important papers on Iranian amphibians and reptiles, including one on *Bufo surdus* and a key to Iranian *Bufo* (J. J. and J. F. Schmidtler, 1969), *Eublepharis angramainyu* (J. J. and J. F. Schmidtler, 1970), *Tropiocolotes helenae* (J. J. and J. F. Schmidtler, 1972), observations on *Batrachuperus persicus* in captivity (J. J. and J. F. Schmidtler, 1971), and a revision of the salamandrid genus *Neurergus* (J. J. and J. F. Schmidtler, 1975). Their collected material is in the Zoologischen Staatssammlung München and in the private collections of the junior Schmidtler.

FIGURE 21. Josef Johann Schmidtler (1910–1983). (Courtesy of J. F. Schmidtler.)

Michael J. Casimir (1970, 1971) wrote general articles on the herpetofauna of Iran and Afghanistan for readers of

FIGURE 22. Josef Friedrich Schmidtler. (Courtesy of J. F. Schmidtler.)

Die Aquarien und Terrarien-Zeitschrift, following his trip in 1968 through those two countries during which he collected herpetological material for the Senckenberg Museum.

Hans Steiner also travelled to Iran in 1968, discovering the first hynobiid salamander from Iran (*Batrachuperus persicus* Eiselt and Steiner, 1970), and again in 1971, collecting additional specimens and recording observations on distribution, ecology, and variation (Steiner, 1973).

During 1970 and 1971, Marc Bosch, Steven Bullock, Wayne Kinunen, and Peter S. Walczak carried out studies on crocodiles and marine turtles for the Division of Research and Development of the Iran Game and Fish Department. This work resulted in a series of unpublished Job Project Reports submitted to the Iran Game and Fish Department (Bosch, et al., 1970; Bullock and Kinunen, 1971; Kinunen and Bullock, 1970; Kinunen and Walczak, 1971; Walczak, 1971; Walczak and Kinunen, 1970; Walczak and Kinunen, 1971), which were cited and summarized by S. Anderson (1979).

FIGURE 23. Josef Eiselt. (Courtesy of Kraig Adler.)

Eiselt (1971b) (Fig. 23) published a brief history of the contributions of the museum in Vienna to the natural history of Iran.

In the 1970s, until the Islamic Revolution, Robert G. Tuck was an advisor to the Iran Department of Environment charged with establishing collections and training Iranian personnel for the Musé-ye Melli'i Tarikh-e Tabi'i (Iran National Museum of Natural History — MMTT) in Tehran. He was *de facto* Acting Director of the museum during this period, and in addition to initiating and building collections, he facilitated the travels and researches of visiting naturalists, including herpetologists. He also found time to publish a number of papers in herpetology (Tuck, 1974 [8 titles], 1977, 1979). Among these were articles in Farsi to introduce native speakers to the fauna.

The decade of the 1970s was a particularly important period for Iranian natural history and conservation; the Iran Game and Fish Department became the Iran Department of Environment and, under the directorship of Eskandar Firouz, established National Parks, Protected Regions, and International Biosphere Reserves throughout Iran, protecting representative and critical habitats. With the help of foreign advisors, a national botanical garden and herbarium and a national natural history museum were established and plans were developed for a zoological park and aquarium (see Firouz, 1974). Following the change in government after the revolution, these projects assumed a lower national priority and their future is still uncertain, although, in the late 1980s and 1990s a renewed awareness of environmental issues has resulted in the implementation of conservation initiatives begun in the 1970s. Nevertheless, a population growth rate of nearly four percent annually has doubled the human numbers since the revolution, with all that implies for environmental deterioration (see Firouz, 1998 for a review of enviromental conservation in Iran).

The most detailed amphibian study thus far is that of Josef Eiselt and Josef Friedrich Schmidtler (1973) on the frog fauna of Iran.

Hans-Hermann Schleich travelled in Iran in 1973, 1974, and 1975 and published three papers on the reptile fauna: the first on his field observations of *Phrynocephalus helioscopus* (Schleich, 1976), the second his publication of two composite distribution maps attempting to summarize the distributions of all reptiles known from Iran (Schleich, 1977), and the third his field observations of *Trapelus agilis, Laudakia caucasia,* and *L. erythrogastra* (Schleich, 1979).

Josef Eiselt also traveled to Iran in 1973, and among other herpetological specimens added to the collections in Vienna, was his new species, *Phyllodactylus ingae*, published in December 1973 (a synonym of *Asaccus griseonotus* Dixon and S. Anderson, published in November 1973). The Vienna collection also contains specimens from Iran collected by Eiselt in 1961, 1967, 1968, 1971, 1974, 1976, and 1977.

The University of Michigan Museum of Zoology has a small collection of amphibians and reptiles collected in 1971 and 1973 by Richard W. Redding, Jr., who accompanied an archeological expedition to western Iran. The lizards of this collection are included in the present report.

FIGURE 24. Ilya S. Darevsky. (Courtesy of Kraig Adler.)

Ilya Darevsky (Fig. 24) was the first herpetologist to visit Iran under the Iran Department of the Environment Visiting Scientist program, in autumn 1974. His collections, deposited in the Muze-ye Melli-ye Tarikh-e Tabii, Tehran (MMTT) and the Zoological Institute, Saint Petersburg, resulted in the discovery of a new species of *Eremias* (Darevsky and Szczerbak, 1978).

In 1975, I was invited by the Government of Iran to be a Visiting Scientist with the Iran Department of the Environment. During a four-month period, I was based at the natural history museum in Tehran and was provided research space and a vehicle and drivers. A number of field investigations enabled me to visit many of the National Parks and Protected Regions and to observe and collect in most of the biotopes and geographic regions throughout the country (Fig. 28). In addition, I examined all of the specimens in the herpetological collections of the museum in Tehran. Observations resulting from this trip are reported in the present work. The materials collected during my 1958 trip are in the California Academy of Sciences, and those from the 1975 trip in the California Academy of Sciences and in the MMTT.

In 1976, Göran Nilson and Claes Andrén (Fig. 25) of the Göteborg Natural History Museum collected amphibians and reptiles in northern Iran. Results of their work included a new species of toad, *Bufo kavirensis* Andrén and Nilson (1979), and a new skink, *Ophiomorus nuchalis* Nilson and Andrén (1978), from the Kavir Protected Region. They also made observations at Jonstan, a locality on the southern slopes of the Talysh Mountains, and in (former) Mohammad Reza Shah National Park, east of the Alborz Mountains. They reported on the herpetofauna of the Kavir Protected Region (Nilson and Andrén, 1981) and their material is deposited in Göteborg. Together with European colleagues, these

FIGURE 25. Göran Nilson (left) and Claes Andrén (right). (Photo by S. Anderson, 1995.)

FIGURE 26. Jiří Moravec. (Photo by S. Anderson.)

authors have been engaged in studies of the systematic relationships of snakes in the genus *Vipera*, particularly those of Southwest Asia and the Caucasus (e.g., Nilson and Andrén, 1984, 1985a, 1985b, 1986, 1992). In 1984, they published a study of the Iranian ratsnakes of the *Elaphe longissima* complex.

The Third Czechoslovak-Iranian Entomological Expedition to Iran took place in 1977 from 26 March to 12 August under the leadership of L. Hoberlandt. The expedition made entomological collections in 154 localities that are situated in six geographically and floristically well-characterized areas. A small herpetological collection was made, which yielded at least one new species of lizard, *Eremias lalezharica* Moravec, 1994 (Fig. 26), and the apparent rediscovery of the small gecko, *Tropiocolotes latifi* (Moravec and Černy, 1994). Hoberlandt (1983) provided a route map and brief descriptions of the localities visited.

S. Anderson (1985) presented a brief synopsis of the Iranian amphibian fauna, listing six species of salamanders (a seventh, *Salamandra salamandra semenovi*, has since been recorded), and 17 frogs from Iran.

Günter Schultschik and Sebastian Steinfartz traveled to Iran in March and April 1995 on a herpetological excursion. They reported on the amphibian results (Schultschik and Steinfartz, 1996), including visits to the type localities of *Neurergus kaiseri* and *Batrachuperus persicus* and the habitats of *N. microspilotes* and *Salamandra salamandra semenovi* for purposes of characterizing these localities. Excursions into western Iran in 1996 and 1997 have been undertaken by the Dutch investigators Herman A. J. in den Bosch (Bosch, 1996) and John and Jan Mulder (Mulder, 1998).

In 1996, yet another Czech expedition collected amphibians and reptiles in Iran (5 amphibian species, 3 turtles, 34 lizards, and 26 snakes). A recent account of this expedition (Frynta, et al., 1997) characterizes the localities using 30 environmental variables. Černy (1996a-d) published a popularized account of herpetological collecting on this expedition.

Several Iranian workers have made important contributions to herpetology in the modern period. Mahmoud Latifi (1985; 1992; English translation, 1991) (Fig. 27) of the Razi Institute compiled a handbook to the snakes of Iran, including keys for identification, descriptions, illustrations, distributions, and information on snake bite. In the English edition, the editors, A. E. Leviton and G. Zug, provided an emended list of Iranian snakes, a total of 75 taxa.

Latifi, recently retired from the Razi Antivenom Institute, in addition to his handbook of the snakes of Iran (see above), has published on venoms and Iranian venomous snakes (Latifi and Manhouri, 1966; Latifi and Farzanpay, 1973; Latifi, 1975; Latifi, 1984). Mehdi Rai' (1969) of the University of Tehran wrote his dissertation on Iranian colubrid snakes. Mohammad Baloutch (Fig. 29), in addition to training the next generation of Iranian herpetologists, has published short papers on the

FIGURE 27. Mahmoud Latifi. (Courtesy George Sprague Myers/Alan E. Leviton Portrait file, California Academy of Sciences.)

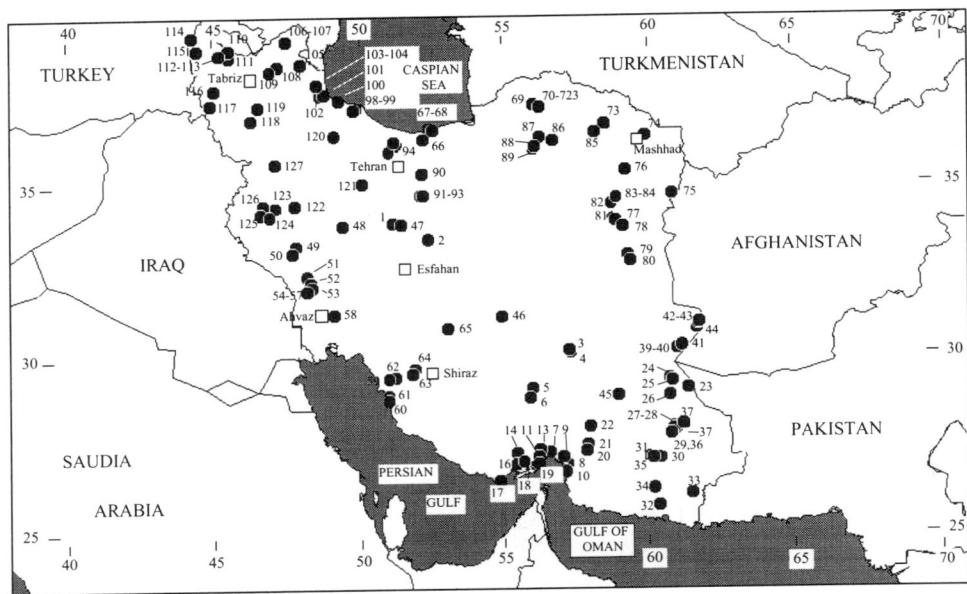

FIGURE 28. Collecting stations in Iran for Steven C. Anderson, 1975.

1: 33°59′N 51°25′E, Kashan, just W of the city, W of enclosed public gardens (Bagh-Shah). – **2**: 33°26′N 52°20′E, 6 km by road NW Ardestan. – **3**: 30°15′N 57°14′E, 2 km on dirt track branching left from Kerman-Bam road, 13 km E eastern edge of city of Kerman, 2000 m. – **4**: 30°15′N 57°12′E, Kerman-Bam road 10 km E eastern edge of city of Kerman, 1700–2000 m. – **5**: 29°06′N 56°00′E, 50 km S Sirjan on Kerman-Bandar 'Abbas road, 1680 m. – **6**: 28°54′N 55°52′E, 80 km S Sirjan on Kerman-Bandar 'Abbas road, 1860 m. – **7**: 27°19′N 56°33′E, 19 km NE Shaqu on inland road to Jask, 50–100 m. – **8**: 27°00′N 57°08′E, 17 km SSE Minab on inland road to Jask, 50–100 m. – **9**: 27°11′N 57°03′E, Date grove 5 km NNW Minab. – **10**: 26°49′N 57°02′E, 43 km S Minab, 2 km N Kuhestak, slightly above sea level. – **11**: 27°23′N 56°16′E, 20 km N Shaqu to turnoff to Kuh-Genu, westerly 8 km on track to base of foothills of Kuh-e Genu, Kuh-e Genu Protected Region, 300 m. – **12**: 27°13′N 56°08′E, 32 km on road up Kuh-e Genu, Kuh-e Genu Protected Region, 1000 m. – **13**: 27°14′N 56°10′E, Kuh-e Genu, 12 km by road below peak, 1700 m. – **14**: 27°17′N 56°29′E, 12 km NE Shaqu on road from Bandar 'Abbas to Bandar-e Lengeh, 0-50 m. – **15**: 27°01′N 55°43′E, 109 km W Bandar 'Abbas-Kerman road on road to Bandar-e Lengeh, 0-50 m. – **16**: 26°57′N 55°29′E, 136 km W Bandar 'Abbas-Kerman road on road to Bandar-e Lengeh, 0-50 m. – **17**: 26°31′N 54°47′E, 4 km on road 1 km W Bandar-e Lengeh airport, ruined buildings on beach, slightly above sea level. – **18**: 26°32′N 54°43′E, 9 km W Bandar-e Lengeh airport, 0-50 m. – **19**: 26°57′N 56°14′E, Qeshm Island, sandstone ridge and sand dunes above town of Qeshm, 0–50 m. – **20**: 21 km northerly from Rudan on road to Jiroft, 575 m. – **21**: Dahaneh Nais, 37 km northerly from Rudan on road to Jiroft, 450 m. – **22**: 29°06′N 57°56′E, 75 km northerly of Jiroft on road to Bam, 1850 m. – **23**: 29°08′N 61°20′E, 52 km southerly on Zahedan-Mirjaveh road from point where railroad crosses the road south of Zahedan. – **24**: 29°28′N 60°41′E, 32 km W Zahedan on road to Cheshme Ziarat, 1900 m.

25: 29°23′N 60°49′E, 13 km S Zahedan on road to Khash, 1500 m. – **26**: 28°59′N 60°42′E, 66 km S Zahedan on road to Khash, 1760 m. – **27**: 28°10′N 61°11′E, 10 km S Khash on road to Iranshahr, 1425–1450 m. – **28**: 28°01′N 60°51′E, 49 km S Khash on road to Iranshahr, 1400 m. – **29**: 27°52′N 60°46′E, 81 km S Khash on road to Iranshahr, 1200 m. – **30**: 27°12′N 60°25′E, 3 km W Bampur, Jaz Murian depression, 600 m. – **31**: 27°10′N 60°10′E, 30 km W Bampur, Jaz Murian depression, 540 m. – **32**: 25°45′N 60°24′E, 83 km S Nikhshahr on road to Chah Bahar, 200 m. – **33**: 26°09′N 61°27′E, on Rud-e Sarbaz, across from Hudar, 320-380 m. – **34**: 26°13′N 60°13′E, Nikhshahr, 500m. – **35**: 27°10′N 60°09′E, 31 km W Bampur, Jaz Murian depression, 470 m. – **36**: 27°54′N 60°48′E, 93 km N Iranshahr on road to Khash, 1350 m. – **37**: 28°03′N 60°53′E, 231 km N Iranshahr (47 km S Khash), 1480 m. – **38**: 28°02′N 60°55′E, 39 km S Khash on road from Iranshahr, 1460 m. – **39**: 30°15′N, 60°57′E, 11 km NE junction with Zahedan-Mashhad road, on road to Zabol, 500m. – **40**: 30°16′N 61°01′E, 12-18 km NE junction with Zahedan-Mashhad road, on road to Zabol, 500 m. – **41**: 30°24′N 61°08′E, 33-42 km NE junction with Zahedan-Mashhad road, on road to Zabol, 500 m. – **42**: 31°05′N 61°42′E, dunes along river where road from Zabol to Dust-e Mohammad Khan crosses Rud-e Hirmand, 500 m. – **43**: 31°03′N 61°38′E, 10 km SW Rud-e Hirmand, abandoned village SE of road from Zabol to Dust-e Mohammad Khan, 450 m. – **44**: 30°51′N, 61°39′E, 7 km W Zehak, between Zehak and Khamak, 500 m. – **45**: 28°59′N 58°57′E, 26 km W Mil-e Naderi, 6 km E Fahraj on road between Bam and Zahedan, 650 m. – **46**: 31°13′N, 54°55′E, 98 km S Yazd on road to Rafsanjan, 1600 m. – **47**: 33°54′N 51°30′E, 8 km S Kashan on road to Yazd, and 6 km W of road, 1175 m. – **48**: 33°53′N 49°29′E, Pol Doab, 40 km W Arak on road to Khorramabad, 1850 m. – **49**: 33°19′N 47°53′E, Afrineh, 77 km S Khorramabad on road to Andimeshk, 1100 m.

50: 33°04′N 47°44′E, 10 km S Pol-e Dokhtar on road to Andimeshk, 1150 m. – **51**: 32°25′N 48°16′E, 14 km S Andimeshk on road to Ahvaz. – **52**: 32°12′N 48°23′E, along Rud-e Dez, 10 km W junction of Andimeshk-Ahvaz road with Shush road. – **53**:

32°05′N 48°22′E, 3 km E Haft Tappeh. – **54**: Near ferry on Rud-e Dez, 6 km E of road to Choga Zanbil, Dez Protected Region. – **55**: 32°01′N 48°16′E, HQ Karkheh National Park, 2 km W Andimeshk-Ahvaz road. – **56**: 32°01′N 48°16′E, *Tamarix* thicket, Karkheh National Park, 6 km W Andimeshk-Ahvaz road. – **57**: 32°01′N 48°16′E, 2 km W Andimeshk road on road to Karkheh National Park. – **58**: 31°16′N 49°10′E, 1 km E Kupal, dunes on Ahvaz-Behbehan road, 25–50 m. – **59**: 29°24′N 51°02′E, 41 km SE Bandar-e Rig on road to Borazjan, 40 m. – **60**: 28°53′N 51°01′E, Alchangi, 32 km ESE Bushehr, 0-50 m. – **61**: 28°53′N 51°02′E, near Alchangi, 33 km SE Bushehr, 0-50 m. – **62**: 29°28′N 51°21′E, 5 km N Delaki on road from Bushehr to Shiraz at point where coastal plain meets foothills of Zagros Mountains, 100-110 m. – **63**: 29°34′N 51°53′E, crest of Mian Kotal pass, ca 20 km W Dasht-e Arzhan on road from Bushehr to Shiraz, 2000+ m. – **64**: 29°42′N 52°03′E, 12 km E Dasht-e Arzhan on road to Shiraz, 1880 m. – **65**: 30°51′N 53°06′E, 28 km N Deh Bid, on road between Shiraz and Esfahan, 2160 m. – **66**: 36°17′N 52°21′E, 26 km S Amol on road to Tehran, 450 m. – **67**: 36°42′N 52°36′E, 4 km W Babolsar on road along coast of Caspian Sea. – **68**: 36°41′N 52°29′E, 14 km W Babolsar, public beach on Caspian Sea. – **69**: 37°21′N 56°01′E, 10 km W Mohammad Reza Shah National Park HQ, 7 km E of park entrance, near stream S of road, 450 m. – **70**: 37°21′N 56°12′E, Mohammad Reza Shah National Park, 9 km E park HQ, 800 m. – **71**: 37°19′N 56°16′E, 3 km S on road 22 km W Robat-e Qareh Bil, Mohammad Reza Shah National Park, 1000–1300 m. – **72**: 37°20′N 56°15′E, 22 km W Robat-e Qare Bil at road junction, Mohammad Reza Shah National Park, 1000–1300 m. – **73**: 36°48′N 58°30′E, 22 km S junction of Quchan-Mashhad road with road to Soltanabad, on road to Soltanabad, 1570 m. – **74**: 36°26′N 59°53′E, 29 km N Gold Mosque in Mashhad on road to Kalat-e Naderi,1130 m.

75: 34°47′N 60°47′E, 4 km N communications station, Tayyebat, 900 m. – **76**: 35°29′N 59°12′E, 8 km S Robat-e Sang on road to Torbat-e Heydariyeh, 1760 m. – **77**: 34°05′N 58°50′E, 36 km S Bidokht on road to Birjand from Torbat-e Heydariyeh, 1540 m. – **78**: 33°49′N 59°07′E, 50 km S Shah Abad (12 km N Qaen) on road to Birjand from Torbat-e Heydariyeh, 1420 m. – **79**: 33°03′N 59°19′E, 23 km N Birjand on road from Torbat-e Heydariyeh, 1850 m. – **80**: 32°49′N 59°26′E, 20 km S Birjand on road to Zahedan and NE about 1 km on dirt track toward mountains, 1610 m. – **81**: 34°05′N 58°50′E, 36 km S Bidokht on road from Birjand, 1550 m. – **82**: 34°35′N 58°43′E, semi-abandoned village 26 km N Gonabad on road to Torbat-e Heydariyeh, 900 m. – **83**: 34°39′N 58°47′E, 35 km N Gonabad on road to Torbat-e Heydariyeh, 850 m. – **84**: 34°41′N 58°48′E, 39 km N Gonabad on road to Torbat-e Heydariyeh, 850 m. – **85**: 36°35′N 58°11′E, 23 km N Soltanabad on road to Quchan, 1250 m. – **86**: 36°21′N 56°43′E, salt flats 5 km W Kahak, 110 km W Sabzevar on road to Shahrud, Turan Protected Region, 800 m. – **87**: 36°24′N 56°17′E, 16 km S 'Abbasabad in Turan Protected Region on dirt track, 880 m. – **88**: 36°09′N 56°08′E, 34 km S 'Abbasabad on dirt road into Turan Protected Region, 870 m. – **89**: 36°05′N 56°03′E, Turan Protected Region, 42 km SE 'Abbasabad, 920 m. – **90**: 35°16′N 52°12 E, 14 km W Garmsar on road to Tehran, 900 m. – **91**: 34°44′N 52°10′E, vicinity of Shah 'Abbas Caravanseray near N foot Siah Kuh, Kavir Protected Region, 1000 m. – **92**: 2-4 km SE Shah 'Abbas Caravanseray, foot of Siah kuh, Kavir Protected Region, 1000+ m. – **93**: 1–2 km WSW Shah 'Abbas Caravanseray, Kavir Protected Region, 1000+ m. – **94**: 35°59′N 51°05′E, 37 km N Karaj on road to Chalus, above Karaj Dam Lake, Central Alborz Protected Region, 1800 m. – **95**: 27 km N Asara on road to Chalus at junction with road to Gajareh, Central Alborz Protected Region, 2135 m. – **96**: 36°12′N 51°20′E, 64 km S Chalus on road to Karaj, Central Alborz Protected Region, 1670 m. – **97**: 36°19′N 51°16′E, 39 km S Chalus on road to Karaj, 600 m. – **98**: 37°15′N 49°58′E, 2 km E Astaneh Ashrafiyeh on road from Chalus to Rasht, –25 m. – **99**: 37°16′N 49°54′E, 2 km S road between Chalus and Rasht on dirt road 4 km W Astaneh Ashrafiyeh, –50 m.

100: 37°29′N 49°23′E, Mordab, 2 km S road from Bandar Pahlavi to Astara on road to Safari Lodge, 9 km W Bandar Pahlavi, below sea level. – **101**: 37°42′N 48°52′E, 8 km W Bandar Pahlavi-Astara road on road to Khalkhal 2 km N Asalom (69 km W and N Bandar Pahlavi), 110 m. – **102**: 37°38′N 48°42′E, 15 km E Khalkhal on road from Asalom, 1990 m. – **103**: 78 km N Khalkhal on road to Ardabil, 1720 m. – **104**: 38°00′N 48°34′E, Neur Lake, 35 air km ESE Ardabil (16 km easterly by dirt road from Budalalu, which is 34 km S Ardabil on road to Khalkhal), 2400 m. – **105**: 38°30′N 48°02′E, 18 km W Razi on road to Meshkinshahr, 1150 m. – **106**: 10 km S Arax River (16 km S Alireza-abad) on road to Meshkinshahr, 220 m. – **107**: 39°14′N 47°29′E, 1 km S Qaraqachid on road to Meshkinshahr, 900 m. – **108**: 38°26′N 47°15′E, 22 km E Ahar on road from Meshkinshahr, 1800 m. – **109**: 38°17′N 46°57′E, 31 km S Ahar on road to Tabriz, 2150 m. – **110**: 38°56′N 45°35′E, 6 km W Jolfa, Marakan Protected Region, 1350 m. – **111**: 38°44′N 45°36′E, 26 km S Jolfa on road to Marand, 1725 m. – **112**: 38°51′N 45°13′E, 92 km W Marand on road to Maku, Marakand Protected Region, 1520 m. – **113**: 38°52′N 45°10′E, 21 km E Maku on road to Marand, 1090 m. – **114**: 39°20′N 44°17′E, 11 km NE 'Arab-e Dizehsi on road from Maku, 1900 m. – **115**: 38°56′N 44°28′E, 55 km SE 'Arab-e Dizehsi on road to Qaeh Zia'ed Din, 2080 m. – **116**: 37°50′N 45°03′E, 1 km E Goltappeh at margin of Lake Reza'iyeh (Urumiyeh), dirt road 39 km N Reza'iyeh, 1425 m. – **117**: 37°25′N 44°56′E, 5 km W Band (15 km W Reza'iyeh) on dirt track to waterfalls, 1500 m. – **118**: 36°56′N 46°17′E, 19 km ESE Miandoab on road to Shahindezh, 1350 m. – **119**: 37°22′N 46°30′E, Maragheh Paleo Site, 4 km SW Chalilvan (29 km E Maragheh), 1800 m. – **120**: 36°29′N 49°08′E, 5 km N toward mountains on road 77 km E Zanjan and 25 km W Abhar, 1760 m. – **121**: 35°07′N 50°08′E, 24 km W Saveh on road to Hamadan, 1160 m. – **122**: 34°28′N 47°46′E, 22 km E Kangavar on old abandoned road parallel to road to Kangavar, 1640 m. – **123**: 34°24′N 47°09′E, 2 km E and above village of Taq-e Bostan, 1450 m. – **124**: 34°17′N 46°57′E, 12 km W Kermanshah on road to Shahabad, 1600 m. – **125**: 34°13′N 46°41′E, creek near turnoff to microwave station 42 km W Kermanshah on road to Shahabad, 1640 m. – **126**: 34°29′N 46°41′E, 31 km NW Kermanshah on road to Nowsud, 1440 m. – **127**: 41 km SW Bijar, 1940 m.

FIGURE 29. Mohammad Baloutch and students. (Courtesy of Mohammad Baloutch.)

FIGURE 30. Nasrulla Rastegar-Pouyani. (Photo by S. Anderson.)

herpetofauna (Baloutch, 1972; Baloutch and Thireau 1986) and continues his research. B. Bastani (1979) published a clinical review of snakebite in Iran with an emphasis on Fars Province.

FIGURE 31. Haji Gholi Kami. (Courtesy of H. G. Kami.)

Currently, Nasrullah Rastegar-Pouyani (Fig. 30) is studying with Göran Nilson in Göteborg, working on reptiles that he has collected in western Iran as well as preparing papers on agamids. He has published a new species of *Asaccus* (Rastegar-Pouyani, 1996) and is working on additional newly discovered Iranian taxa (Rastegar-Pouyani, 1995). Haji Gholi Kami (Fig. 31), a former student of Mohammad Baloutch, is now at Gorgan University where he has established a zoological museum, and is working on Iranian salamanders, particularly of the genus *Batrachuperus*. Baloutch and Kami (1995) published a book on the amphibians of Iran in Farsi (Persian). These younger workers represent the next generation of herpetologists in Iran, and it is to be hoped that, living in the area of study, they will bring a higher level of detail and sophistication to our understanding of this fauna.

Brief Papers Describing New Taxa from Iran

A number of relatively brief papers devoted principally to descriptions of new taxa from Iran have been published. These include: Brandt, 1838 (new species of snakes collected by M. Karelin from northern Iran); Strauch, 1863 (*Centrotrachelus asmussi, Teratoscincus keyserlingii*); Steindachner, 1867 (*Hemipodion persicum*); Blanford, 1874a (8 new species of lizards); Blanford, 1874b (8 new lizards, 2 new snakes, and one new anuran); Blanford, 1875 (*Uromastyx microlepis, Centrotrachelus loricatus*); Nikolsky, 1896 (*Testudo zarudnyi, Gymnodactylus longipes, Teratoscincus zarudnyi, Phrynocephlus spiniventris, Stellio erythrogaster, Eremias nigrocellata, Scapteira lineolata, Bufo oblongus*); Nikolsky, 1899 (*Teratoscincus microlepis, T. bedriagai*); Nikolsky, 1903 (*Alsophylax persicus, Contia bicolor, Bufo persicus*); Annandale, 1906 (*Testudo baluchiorum*); Boulenger, 1908 (*Lacerta chlorogaster*); Nesterov, 1916 (*Rhithrotriton derjugini derjugini, R. d. microspilotus, Salamandra semenovi*); Boulenger, 1917 (*Lacerta viridis woosnami*); Boulenger, 1920c

(*Zamenis hotsoni*); Boulenger, 1920d (*Testudo buxtoni*); Boulenger, 1920b (*Contia condoni*); Lantz and Suchow, 1934 (*Apathya cappadocica urmiana*); Suchow, 1936 (*Lacerta princeps kurdistanica*); S. Anderson and Leviton, 1966 (*Eublepharis angramainyu*); Minton, et al., 1970 (*Tropiocolotes persicus bakhtiari*); Eiselt and Steiner, 1970 (*Batrachuperus persicus*); Darevsky, 1970 (*Rhynchocalamus melanocephalus satunini*); Eiselt and J. F. Schmidtler, 1971 (*Bufo viridis kermanensis, Rana macrocnemis pseudodalmatina*); Eiselt, 1971 (*Eirenis rechingeri*); Leviton and S. Anderson, 1972 (*Tropiocolotes latifi*); Szczerbak, 1972 (*Eremias strauchi kopetdaghica*), Dixon and S. Anderson, 1973 (*Asaccus griseonotus*); Eiselt, 1973 (*Phyllodactylus ingae* [=*Asaccus griseonotus* Dixon and S. Anderson, 1973]); S. Anderson, 1973 (*Bunopus aspratilis*); Darevsky and Szczerbak, 1978 (*Eremias andersoni*); Clergue-Gazeau and Thorn, 1979 (*Batrachuperus gorganensis*); de Witte, 1980 (*Rhinogecko misonnei*); Szczerbak, 1990 (*Eumeces taeniolatus parthianicus*), Moravec, 1994 (*Eremias lalezharica*); Eiselt, 1995 (*Lacerta steineri*); Rastegar-Pouyani, 1996 (*Asaccus kermanshahensis*); Rastegar-Pouyani, 1998 (*Acanthodactylus nilsoni*); Wischuf and Fritz, 1996 (*Mauremys caspica ventrimaculata*); Rastegar-Pouyani and Nilson, 1998 (*Lacerta zagrosica*).

The "Classical" Literature of Herpetology

Many species which occur in Iran were originally described by European herpetologists in the "classical" literature of scientific natural history. Kraig Adler (1989) has provided brief biographies of most of these early workers, and of more recent European and American naturalists, now deceased, from before the time of Linnaeus to the late 1980s. These biographies include the following naturalists who made taxonomic contributions related to the Iranian herpetofauna: Carl Linnaeus (1707–1778), Professor at the University of Uppsala, who developed the system of binomial nomenclature; Josephus Nicolaus Laurenti (1735–1805), Viennese physician, who published the first major herpetological review after Linnaeus, his doctoral thesis, *Specimen Medicum, Exhibens Synopsin Reptilium Emendatam cum Experimentis circa Venena*; Johann Gottlob Theaenus Schneider (1750–1822), German philologist and naturalist, Professor of Philology at the University of Frankfurt (Oder) and, later, University Librarian at Breslau, best known for his *Historiae Amphibiorum* (two volumes, 1799 and 1801); Bernard-Germain-Étienne de la Ville-sur-Illon, Comte de Lacepède (1756–1825), held the chair covering fishes and reptiles at the Muséum National d'Histoire Naturelle in Paris, published *Histoire Naturelle des Quadrupèdes Ovipares et des Serpens* (1788–1789); Blasius Merrem (1761–1824), German naturalist at the University of Marburg, best known for *Tentamen Systematis Amphibiorum*, covering all known species at that time; Georges Cuvier (1769–1832), comparative anatomist and paleontologist, Professor at the Muséum National d'Histoire Naturelle, responsible for building the great collections at that museum and author of *Le Règne Animal*, a summary of all animals; François-Marie Daudin (1774–1804), author of *Histoire Naturelle, Générale et Particulière des Reptiles*, in eight volumes (1801–1803), the standard herpetological reference of its day; Johann Georg Wagler (1800–1832), associated with the Bavarian Academy of Science in Munich, author of *Natürliches System der Amphibien* (1830); Leopold Joseph Franz Johann Fitzinger (1802–1884), prolific Viennese naturalist, published *Neue Classification der Reptilien* (1826) and *Systema Reptilium* (1843), described many genera and fixed the type species of others; Arend Friedrich August Wiegmann (1802–1841), Professor at the University of Berlin, where he was in charge of the herpetological collection, described many genera and founded the important natural history journal, *Archiv für Naturgeschichte*;

Hermann Schlegel (1804–1884), prominent naturalist associated with the Rijksmuseum van Natuurlijke Historie in Leiden, author of the first truly scientific treatise on snakes, *Essai sur la Physionomie des Serpens* (1837) and the first to use trinomial nomenclature; André-Marie-Constant Duméril (1774–1860), Professor of Ichthyology and Herpetology at the Muséum National d'Histoire Naturelle, regarded as the greatest taxonomic herpetologist of his era, a time when the natural history spoils of empire were flowing into the museum at Paris, and who, together with Gabriel Bibron, initiated the nine-volume (plus atlas) *Erpétologie Générale ou Histoire Naturelle Complète des Reptiles* (1834–1854); Gabriel Bibron (1806–1848), collaborator with Duméril, responsible for determination of specimens, synonymy, and description of species in the *Erpétologie Générale*; John Edward Gray (1800–1875), initial developer of the zoological collections of the British Museum, adding one million specimens during his tenure, and the author of some 1200 titles in zoology and the initiator of the series of catalogues covering the collections; Wilhelm Carl Hartwig Peters (1815–1883), collected extensively in East Africa over a five-year period, became Director of the Zoological Museum and Professor of Zoology at the University of Berlin, developing the holdings there into one of the world's three major zoological collections of the time, and described numerous species from throughout the world in nearly 150 herpetological papers; Edward Blyth (1810–1873), London-born vertebrate zoologist, became first curator of the Asiatic Society of Bengal and described many Asian species during his 22 years of service in India; Auguste-Henri-André Duméril (1812–1870), collaborator with and eventually successor to his father, André-Marie-Constant Duméril, with whom he completed volumes 7 and 9 of the *Erpétologie Générale* (1854) and the *Catalogue Méthodique de la Collection des Reptiles* (1851); Albert Carl Ludwig Gotthilf Günther (1830–1914), German-born ichthyologist and herpetologist, joined the staff of the British Museum in 1857 and succeeded Gray as Keeper in 1875, published catalogues of the museum's herpetological and fish collections, and, in 1864, *The Reptiles of British India*, added nearly a million specimens to the zoological collections and moved the collections to the new natural history museum in South Kensington in 1882; Edward Drinker Cope (1840–1897), vertebrate paleontologist, anatomist, and foremost American herpetologist, Curator of Herpetology at the Philadelphia Academy of Natural Sciences, revolutionized herpetological classification through his anatomical and paleontological studies, described the Anatolian-Zagrosian salamander genus *Neurergus*, although most of his new taxa were American; Alexander Alexandrovich Strauch (1832–1893), the first eminent Russian herpetologist, Curator, and later, Director of the Zoological Museum of the Imperial Academy of Sciences, St. Petersburg, developed the museum as a major world center of herpetology, published revisionary studies of world-wide scope, described a number of Transcaucasian and Transcaspian taxa which also occur in Iran; Giorgio Jan (1791–1866), naturalist of Hungarian descent, born in Vienna, became the founding head of the Museo Civico di Storia Naturale in 1838 and began his *Iconograpie Général des Ophidiens* in 1853 in collaboration with his illustrator, Ferdinando Sordelli (1837–1916), a great folio of 300 plates in 51 parts (1860–1882) of the snakes of the world; he published the snakes of De Filippi's Iranian travels (Jan *in* De Filippi, 1865); Franz Steindachner (1834–1919), Austrian ichthyologist and herpetologist associated with the Naturhistorisches Museum in Vienna and eventually its director, built the collection into one of the world's largest, and did field work in the Americas, Africa, and the Middle East; John Anderson (1833–1900), Scottish physician and field naturalist, became Curator of the newly-created Indian Museum in Calcutta in 1865, and collected in southern and eastern Asia, authored a collection of papers assembled to form *Contribution to the Herpetology of*

Arabia (1896), amassed large collections in Egypt following his retirement, and published the herpetological volume of his *Zoology of Egypt* series in 1898.

George Albert Boulenger (1858–1937), the most productive taxonomic herpetologist of his era, Belgian by birth, in charge of lower vertebrates at the British Museum, published nearly 900 papers and a nine-volume series of catalogues of the amphibians and reptiles, the most comprehensive review ever published of the world's herpetofauna; Oskar Boettger (1844–1910), German herpetologist and malacologist who established the Senckenberg Museum, Frankfurt am Main, as a world center for herpetology, described many new genera and species from all areas of the world; Jacques Vladimir von Bedriaga (1854–1906), born of Russian nobility, studied in Germany, becoming a specialist in Eurasian herpetology and informally associated with Nikolsky and the museum at St. Petersburg, where he worked on Central Asian collections and authored a catalogue of the reptiles of western highland Asia, a revision of the Lacertidae, and other important papers on the herpetology of the Russian Empire; Lajos Méhelÿ (1862–1952?), Hungarian zoologist, in charge of herpetology at the Hungarian National Museum, wrote some 90 papers on herpetology, most on Hungarian species, but many on distant places, including Armenia and Persia; Aleksandr Mikhailovich Nikolsky (1858–1942), Strauch's successor at the Zoological Museum in St. Petersburg, participated in numerous expeditions, including to Persia and Turkestan, and in addition to his several treatises on the herpetofaunas of the Russian Empire and adjacent countries (1905, 1915, 1916, 1918), published on reptiles and amphibians of the several expeditions of the ornithologist N. A. Zarudny in Iran; Franz Werner (1867–1939), Professor at the University of Vienna, who described 24 genera of reptiles and amphibians and over 400 species and subspecies, including many which occur in Iran; Frank Wall (1868–1950), British medical officer and leading student of snakes of the Indian Empire, served with the Mesopotamian Expeditionary Force in Iraq during the First World War, made many contributions to the taxonomy of Indian snakes and reported on a collection of snakes from Maidan-e Naftun in southwestern Iran (Wall, 1908), deposited his types and skull collection in the British Museum (Natural History); Thomas Nelson Annandale (1876–1924), zoologist and anthropologist born in Edinburgh, founded the Zoological Survey of India in 1916, conducted faunal surveys throughout the Indian Empire and elsewhere in Asia, including the Sistan Basin; Géza Gyula Imre Fejérváry von Komlós-Keresztes (1894–1932), Hungarian herpetologist, succeeded Méhelÿ as curator at the National Museum, published on frog osteology and fossil lizards as well as Hungarian species and a few papers on exotic species; Malcolm Arthur Smith (1875–1958), British-born physician to the Royal Court of Siam and medical officer to the British Legation in Bangkok, spent most of his life studying the amphibians and reptiles of Southeast Asia and India, conducting research at the British Museum (Natural History) after retirement in 1925, publishing his *Monograph of the Sea-Snakes (Hydrophiidae)* in 1926 and a three-volume series on the reptiles of the Indian and Indochinese regions, extending westward to the Iranian border (1931, 1935, 1943), still the most comprehensive herpetological work for South Asia; Karl Patterson Schmidt (1890–1957), founded the herpetology department of the Field Museum of Natural History in Chicago and developed it into a leading research center, with a collection particularly rich in specimens from Southwest Asia, including Iran, and among his publications were several dealing with the Southwest Asian fauna, including Iran (e.g., Schmidt, 1939, 1952, 1955); Robert Friedrich Wilhelm Mertens (1894–1975), Curator of Herpetology and, later, Director of the Senckenberg Museum in Frankfurt (Main), probably the most prolific modern European herpetologist with cosmopolitan interests in herpetology, published nearly 800 scientific titles, including many on Southwest Asia, and a number on Iran in particular, including remarks on snakes from northern Iran (Mertens, 1940), in which he called attention to a concentration of melanistic specimens of several species in eastern Mazanderan, reports on two collections of amphibians and reptiles from Iran, based on the entomological expeditions of Willi Richter in 1954 and 1956 and E. Schüz in 1956 (Mertens, 1956, 1957), and the description of a new viper, *Vipera latifi* (Mertens, Darevsky, and Klemmer, 1967); Georg Haas (1905–1981), morphologist, paleontologist, and herpetologist born in Vienna, emigrated to Palestine and the Hebrew University of Jerusalem, wrote a number of papers on the herpetology of Southwest Asia, one, written with Yehudah Werner, containing Iranian localities (Haas and Y. Werner, 1969); Sergius Alexandrovich Chernov (1903–1964), Curator of the Department of Herpetology at the Academy of Sciences in Leningrad (now St. Petersburg) and Paul Victorovich Terentjev (1903–1970), Chair of Zoology, Leningrad State University, two prominent Russian herpetologists during the Soviet period, in addition to their individual papers and books on the herpetofauna of the arid and semiarid regions of the southern Soviet Union, coauthored *Synopsis of the Reptiles and Amphibians of the USSR* in three editions (1936, 1940, 1949), including many taxa which also occur in Iran; Andrei Grigoryevich Bannikov (1915–1985), vertebrate zoologist and conservationist, Chair of Zoology at the Moscow Veterinary Academy, was senior author of *Field Guide to Amphibians and Reptiles of the USSR* (1977, with I. S. Darevsky, V. G. Ischenko, A. K. Rustamov, and N. N. Szczerbak), which gives accounts of species which also occur in Iran; Alphonse Richard Hoge (1912–1982), Director of the Division of Biology of the Instituto Butantan in São Paulo, of Belgian descent, a specialist on the snakes and lizards of South America, coauthored a paper with M. Latifi and M. Eliazan (1968) on the venomous snakes of Iran; Ion Eduard Fühn (1916–1987), author of major reviews of Rumanian herpetology, in charge of the sections of arachnology and herpetology at the Rumanian Academy of Sciences, Bucharest, revised the skink genus *Ablepharus*; Muhtar Başoğlu (1913–1981), the first scientific Turkish herpetologist, Chair of the Section of Systematic Zoology, published three guides to the herpetofauna of Turkey, *Amphibians of Turkey* (1973, with N. Özeti), *Reptiles of Turkey*, parts I and II (1977, 1980, with İ. Baran).

Checklists, Catalogues, Keys, and Handbooks
Useful to the Study of Herpetology In Iran

Jacques Vladimir von Bedriaga (Fig. 32) published a catalogue of reptiles of western highland Asia in 1879 and a work on the amphibians and reptiles of the Middle East, also in 1879, both of which include species found in Iran.

The earliest and assuredly most comprehensive of modern herpetological catalogues, checklists, and keys is the multivolume series of British Museum catalogues compiled by George Albert Boulenger (anurans, 1882; salamanders, 1882; turtles and crocodilians, 1889; lizards, 1885a, 1885b, 1887a; snakes, 1893, 1894, 1896). These were revisions and up-dates of the earlier catalogues by John Edward Gray (1838–1873) and Albert Günther (1858–1881). Although long out of date, these catalogues contain vast amounts of useful data, and are still essential to anyone doing serious systematic work in herpetology.

In 1892, Oskar Boettger (Fig. 33) produced a catalogue of the amphibians in the collection of the Senckenberg Museum; this was followed by a volume on the turtles, crocodiles, and lizards in 1893 and the snakes in 1898.

Willi Wolterstorff (1898) published a work on the urodeles of southern Asia, including taxa inhabiting Iran, and in 1925 he published a catalogue of the amphibians in the collection of the museum in Magdeburg.

FIGURE 32. Jacques Vladimir von Bedriaga (1854–1906). (Courtesy of Kraig Adler.)

FIGURE 33. Oskar Boettger (1844–1910). (Courtesy of Kraig Adler.)

Kenneth R. G. Welch's (1983a) book, *Herpetology of Europe and Southwest Asia, a Checklist and Bibliography of the Orders Gymnophiona and Urodela* and (Welch, 1984b) *Herpetology of Europe and Southwest Asia: a Checklist and Bibliography of the Orders Amphisbaena, Sauria and Serpentes*, and Keith A. Harding and K. R. G. Welch (1980) *Venomous Snakes of the World: a Checklist* provide provisional lists of species of these groups for the areas covered, but cannot be relied upon for details of distribution, current nomenclature, or the older literature (see S. Anderson, 1983 for review of Welch 1984b).

Ulrich Joger's (1984) synopsis of the venomous snakes of the Near and Middle East contains keys useful in the identification of Iranian species; Philippe Golay (1985) has published checklists and keys to the terrestrial proteroglyphs of the world; and the United States Department of Defense (1991) published a handbook to the venomous snakes of the Middle East with color photographs and maps.

Snakes of the World. Volume 1. Synopsis of Snake Generic Names, by Kenneth L. Williams and Van Wallach (1989) is a useful compendium, although promised additional volumes covering the species have not appeared.

Two useful world checklists have been published by the Association of Systematics Collections, *Amphibian Species of the World*, edited by Darrel R. Frost (1985) and *Crocodilian, Tuatara, and Turtle Species of the World* (King and Burke, 1989).

Two books by J. W. Steward, one on the salamanders (Steward, 1969) and one on the snakes of Europe (Steward, 1971) cover species that extend into northern and western Iran, as does the more comprehensive *A Field Guide to the Reptiles and Amphibians of Britain and Europe*, by E. N. Arnold and John A. Burton (1978).

World coverage of turtles is provided by Peter C. H. Pritchard (1979), whose encyclopedia provides keys, photographs, species accounts, and natural history information; by Carl H. Ernst and R. W. Barbour (1989), who include species accounts and keys; and by John Iverson (1992), whose checklist includes spot locality distribution maps for each species as well as an extensive bibliography.

Synoptic Studies of the Amphibians and Reptiles of Countries Adjacent to Iran

Pakistan: Three major synoptic works exist for the herpetofauna of this region, the three-volume work on the reptiles of British India by Malcolm A. Smith (turtles and crocodilians, 1931; lizards, 1935; snakes, 1943), Minton's (1966) herpetology of then-West Pakistan, closely followed by that of Robert Mertens (1969). These fundamental compilations have been added to by several workers, notably M. S. Khan and coworkers (1968–1997), and by Auffenberg and coauthors (1979–1993).

FIGURE 34. Alan E. Leviton (1983). (Photo by Allen Greer.) FIGURE 35. Steven C. Anderson (1996). (Photo by Kay Anderson.)

Afghanistan: The only synoptic treatment of the fauna is the checklist and key by Alan E. Leviton (Fig. 34) and Steven C. Anderson (1970) (Fig. 35), following on their several prior papers on the amphibians and reptiles of this country.

Turkmenistan: The first synoptic treatment of the reptiles specifically of this republic was that of Oleg P. Bogdanov (1962). Sakhat Shammakov (1981) and Chary A. Ataev (1985) updated and extended this work. The handbook of the amphibians and reptiles of the former U.S.S.R. by A. G. Bannikov, I. S. Darevsky, V. G. Ischchenko, A. K. Rustamov, and N. N. Szczerbak (1977), with keys, descriptions, and distribution maps, also covers this fauna. Because these works are in Russian, they have not been widely accessible to Western readers. Particularly valuable to English-speakers are the chapters by Szczerbak on the zoogeography of the reptiles and by Ataev, Rustamov, and Shammakov on the reptiles of the Kopetdagh in *Biogeography and Ecology of Turkmenistan* edited by Fet and Atamuradov (1994).

*Azerbaidzhan** (Russian transliteration): In 1929, Sobolevsky published an account of the amphibians and reptiles of the Talysh Mountains and the Lenkoran lowlands; the herpetofaunas are similar to those of topographically contiguous regions in Iran. The most recent synoptic treatment of the amphibians and reptiles is that of Abdullah M. Alekperov (1978) as well as the handbook for the Soviet Union by Bannikov, *et al.* (1977), both in Russian.

Armenia: Three synoptic works (in Russian) cover the herpetofauna of Armenia, a field guide to snakes, lizards, and turtles by Sergius A. Chernov (1937), *Animal Kingdom of the Armenian SSR, Vol. 1. Vertebrates* by S. K. Dalj (1954), and a work on the venomous snakes by Ilya S. Darevsky (1960), in addition to the handbook by Bannikov, *et al.* (1977).

Turkey: The only recent synopses of the Turkish herpetofauna are the three-volume work by Muhtar Başoğlu and colleagues (Başoğlu and Baran, 1977, turtles and lizards; 1980, snakes; Başoğlu and Özeti, 1973, amphibians). Although these books are in Turkish, they have English summaries and keys. These have been supplemented by İbrahim Baran's bibliography of the amphibians and reptiles of Turkey in *Zoological Bibliography of Turkey*, edited by Kasparek (1986) and, more recently, Baran and Atatur's *Turkish Herpetofauna*, (1998), which covers all amphibians and reptiles in Turkey. Much of the original work on the herpetofauna of Turkey has been published in English or German.

Iraq and Kuwait: The most recent synoptic treatment of the Mesopotamian lowlands and the bordering foothills is the handbook by A. E. Leviton, S. C. Anderson, Kraig Adler, and S. A. Minton (1992). This work contains keys, brief species accounts, and color photographs; it covers the areas engaged in the Gulf War, including western Iran and the

---

* The traditional transliteration for the Iranian province has been Azarbaijan, while that for the republic to the north has been variously Azerbaijan, Azerbaidzhan, or Azerbaidjan. The map by Shenassi (1991) (see Fig. 2) uses Azerbayjan for the Iranian province. In the text of this work, I use Azarbaijan for the Iranian province and Azerbaidzhan for the republic.

Gulf Coast of the Arabian Peninsula. Much of this fauna extends into Jordan (Disi, 1996) and Syria (Disi and Böhme, 1996).

Arabian Peninsula and the Gulf Emirates: The first synoptic treatment of this large area was that of John Anderson (1896), still an important work for systematists. More modern works include the handbook by Leviton, *et al.* (1992) mentioned above; the amphibians are treated by Emilio M. Balletto, Maria A. Cherchi, and John Gasperetti (1985); E. N. Arnold (1986c) summarized the lizards and amphisbaenians with a key and checklist; Norman L. Corkill and J. A. Cochrane (1966) briefly covered the snakes of Arabia and Socotra, this work superseded by the more thorough and comprehensive treatment by Gasperetti (1988), containing descriptions, keys, and illustrations, including color photographs. Michael Gallagher (1971) published a brief account of the amphibians and reptiles of Bahrain, and in 1990 a synopsis of the snakes of the Arabian Gulf and Oman with color photographs. A. E. Leviton and S. C. Anderson (1967) published on the reptiles of Abu Dhabi.

Revisionary Studies of Family and Generic Taxa

George Albert Boulenger's (1920a, 1921a) two-volume monograph of the lacertids served as the foundation for future systematic work on this family, and contains much information that is still essential to the study of these lizards.

Malcolm A. Smith (1926) monographed the sea snakes, many species of which occur in the Persian Gulf and the Gulf of Aden. This work has been the basis for all subsequent systematic work on these snakes (e.g., Volsøe, 1939, on the sea snakes of the Persian Gulf; Voris, 1977, on the phylogeny of sea snakes).

The lacertid genus *Eremias* has been revised three times in this century, first by G. A. Boulenger (1918d), then by L. A. Lantz (1928), and most recently by Nikolai N. Szczerbak (1974). All authors recognized subgeneric groupings and the latter author separated the Palearctic species from African species previously assigned to this genus, and recognized the genus *Mesalina* for the Saharo-Sindian species.

The boid snake genus *Eryx* has been reviewed three times this century, first by Olive Stull (1935), then by Andrew F. Stimson (1969) in his *Das Tierreich* contribution on the Boidae (in which he followed Stull's synopsis of *Eryx*), and by Anatoly Tokar (1989) in a more thorough treatment based on osteology.

The genera of the family Salamandridae were revised by Wolterstorff in 1935.

The only comprehensive study of the skink genus *Eumeces* was published in 1936 by Edward H. Taylor.

In 1942, Robert Mertens published his three-volume monograph of the Varanidae, still the most comprehensive and complete work on these lizards.

In 1959, Hymen Marx wrote a review of the Southwest Asian colubrid snake genus *Spalerosophis*; previously (Marx, 1953), he had published the only paper devoted to the monotypic elapid genus *Walterinnesia*, an Arabian genus and species which extends into Khuzestan. In 1965, Marx and George Rabb published their detailed analysis of relationships and zoogeography of viperine snakes, which covered the genera of vipers that occur in Iran.

In 1966, S. Anderson and Leviton reviewed the genus *Ophiomorus* for the first time since Boulenger (1887b) and described three new species, one of which (*O. streeti*) came from Iran.

In 1967 (English translation in 1978), Ilya Darevsky published *Rock Lizards of the Caucasus*, a major systematic treatise on the lizards of the genus *Lacerta*, subgenus *Archaeolacerta*, including information on ecology and natural history. This work has served as the basis for numerous subsequent studies of this group by Darevsky and his coworkers. It includes the Iranian species of this subgenus.

Since 1967, Arnold Kluge (1967; 1983; 1985; 1987; 1991; 1993) has been involved in systematic studies of higher categories of geckos, and these studies have contributed much to an understanding of Southwest Asian gekkonid genera and clarified the nomenclature.

In 1968, Robert Thorn published a synopsis of Palearctic salamanders, which included the Iranian species known at that time.

Ion Fühn (1969a) revised the skinks of the genus *Ablepharus*; he removed the Australian species and he restricted the name to the Southwest Asian–southeastern European species. V. K. Eremchenko and N. N. Szczerbak (1986) undertook a more thorough revision of the ablepharine skinks of the former Soviet Union and adjacent countries.

Alan E. Leviton and Steven C. Anderson (1970) reviewed the colubrid snake genus *Lytorhynchus*, a Saharo-Sindian group, three species of which occur in Iran.

The cosmopolitan anuran genus *Bufo* has been the subject of many revisionary studies. Boulenger (1880) analyzed the Palearctic and African species; the Eurasian species were studied morphologically by Inger (1972) and immunologically by Linda Maxson (1981); Josef J. and Josef F. Schmidtler (1969) and Josef Eiselt and Josef F. Schmidtler (1973) used morphological criteria in studying the Iranian species.

Edwin Nicholas Arnold has made many contributions to the herpetology of Southwest Asia, to lizard systematics in general, and to the Lacertidae in particular. In 1973, he published an important study of the relationships of Palearctic species of the genus *Lacerta* and related genera. In 1983, his study of the Saharo-Sindian lacertid genus *Acanthodactylus*, based on osteology and genitalia, was published, and in 1989 his study of the phylogeny and biogeography of the family Lacertidae appeared.

Another revision of *Acanthodactylus*, this one by Alfredo Salvador, was published in 1982. Significantly, this study, using some of the same material, but different characters, arrived at nearly the same systematic conclusions as that of Arnold. This genus had previously been revised by Boulenger in 1918.

E. N. Arnold and A. E. Leviton (1977) revised the North African–Arabian genus *Scincus*. The most widely distributed species of this genus has one subspecies (*Scincus scincus conirostris*) which extends into the sand dune areas of lowland Khuzestan.

Hahn (1978a) published a review of the Scolecophidia (Leptotyphlopidae and Typhlopidae) of the world in *Das Tierreich*, and, in the same year (Hahn, 1978b), a review of the species of *Leptotyphlops* of Asia.

Ronald S. Whiteman (1978) wrote his Masters thesis on the osteology of the agamid genus *Phrynocephalus* and included a biogeography and evolutionary history based on this study. Because this study was never published, it has received little attention in the literature.

Scott M. Moody's PhD dissertation (1980) on the relationships of the Agamidae, although unpublished, has been a major influence on subsequent studies of this family, and his conclusions have served as the basis for nomenclatural changes (e.g., Moody, 1987; Leviton, *et al.*, 1992).

In 1981, Michael Golubev and Nikolai Szczerbak created the genus *Carinatogecko* for two little-known species of geckos, *Bunopus aspratilis* and *Tropiocolotes heteropholis*.

The Afro-Asian viperid genus *Echis* has been the subject of revisionary studies (Cherlin, 1981; 1983a; 1983b; 1990; Cherlin and Borkin, 1990). Some of the nomenclatural changes recommended by these authors have not been widely accepted by other workers, in part because *Echis* is a genus of wide medical importance with a considerable literature and there is a concern that changes in nomenclature would introduce confusion into the medical and pharmacological literature.

In 1984, Darevsky, Eiselt, and Lukina reviewed the rock lizards of the *Lacerta saxicola* group that occur in northern Iran, distinguishing among the species and providing a key and distribution map.

In 1985, I. B. Dotsenko resurrected *Pseudocyclophis* Boettger, 1888, as a monotypic genus for the widely distributed snake, previously known as *Eirenis persicus* (J. Anderson, 1872). In 1989, he reviewed the Southwest Asian genus *Eirenis*, recognizing two subgenera.

Lee Grismer (1988) reviewed the eublepharids and restricted the generic name *Eublepharis* to the Southwest Asian species.

Stephen D. Busack, *et al.* (1988) studied relationships in the salamandrid genus *Triturus*, using immunological data. These salamanders occur in the northern and northwestern provinces of Iran having a positive water balance. A subsequent revision incorporating cytogenetics and reproductive interactions with comparative biochemistry was published by MacGregor, *et al.* (1990).

Ulrich Joger (1986) reviewed *Uromastyx* based on immunological distance, and Scott M. Moody (1987) published a cladistic study of the genus, based on morphology. The latter author raised *Uromastyx*, together with *Leiolepis*, to family level (Uromastycidae).

Göran Nilson and Claes Andrén (1986a; 1986b) reviewed the mountain vipers of the Middle East and the *Vipera xanthina* complex. These are the most detailed reviews to date that include the Iranian vipers.

Nikolai N. Szczerback and Michael L. Golubev (1986; 1996 English translation) published their revisionary studies of Palearctic geckos in a book which will doubtless remain the standard work on the geckos of western Asia (see Bauer, 1987; S. Anderson, 1997 for reviews). Although there will be differences in interpretation of relationships and altered nomenclature in the future, the information contained in this book will continue to contribute to subsequent gekkonid studies.

Yet another study of the intrafamilial relationships of the family Lacertidae is that of Mayer and Benyr (1994) based on albumin evolution. Their conclusions are at odds with those of previous revisions (e.g., Boulenger, 1920–1921; Arnold 1973, 1989; Szczerbak, 1974) in a number of important points which have biogeographic significance. Some of these points are mentioned in the systematic section.

THE LIZARD FAUNA OF IRAN

The Geography of Iran Relative to Lizard Distribution

The biogeography of any region can be viewed from three interrelated perspectives.

(1) *Descriptive*: the distribution of organisms in relation to physiographic features of the region under study. This involves study of the morphology of the animals and taxonomic evaluation of available material and literature records. It is subjective in that it involves judgments based on the worker's knowledge of the groups involved, and includes all of the bias inherent in the taxonomic system employed.

(2) *Ecological*: the distribution of organisms relative to environmental factors. This, too, is largely descriptive, containing the same pitfalls mentioned above, and usually it must be highly inferential, as available data are often diffuse and scanty, and knowledge of physiological responses to the physical factors of the environment is far from complete for any one species. Nevertheless, rough correlations are possible. These can at least prove stimulating to our speculations regarding the causes of present distribution patterns, and allow us to develop testable hypotheses.

(3) *Historical*: this is often highly speculative, particularly in the absence of a fossil record. It involves a consideration of all discernible factors, in the dimension of time, which have produced present patterns of distribution. It is, therefore, an attempt to analyze observed distributional data in the light of available information regarding geomorphology, paleogeography, paleoclimatology, paleoecology, and the evolutionary development of the organisms involved. The development of cladistics in systematics has provided a useful tool in biogeography in that it provides phylogenetic diagrams of relationships that can be compared with historical geographical data. The cautious use of biochemical systematic data has also provided hypotheses regarding the times (or at least relative times) of divergence of taxa, which, again, can be compared with the historical record. Cladistic and biochemical systematic studies are only just beginning for Southwest Asian taxa. The most speciose and widespread genera, such as *Laudakia, Phrynocephalus, Trapelus, Cyrtopodion, Tropiocolotes, Acanthodactylus, Eremias,* and *Ophiomorus* should prove most instructive in future biogeographic analyses.

Iran consists of a complex of mountain chains enclosing a series of interior basins that lie at altitudes of 300 to 1,500 meters above sealevel (Plate 1). These mountain ranges rise sharply from sealevel on the north and south, and from the flat, low-lying plain of Mesopotamia in the west. Eastward, and in the northwest, the highlands extend beyond Iran in the form of largely continuous and uninterrupted features: in the east they are prolonged as the massifs of Afghanistan and Baluchistan (Pakistan), and in the northwest as the plateau uplands of Azarbaijan and eastern Asia Minor. This entire upland is spoken of by some writers as the Iranian plateau, despite the fact that politically it includes most of Afghanistan and a large part of the territory of Pakistan. This is the usage of the term in this book.

Iran has been characterized as a bowl, with a high outer rim surrounding an irregular and lower, but not low-lying, interior. The rim is formed by various groups of mountain chains, some of which, especially in the west and north, are not only high and bold but also extensive in ground area; those of the south and east are narrower, lower in general height, more interrupted by lowland basins, and therefore less of a barrier, climatically and biogeographically. In the east, however, climatic effects — chiefly aridity, with the accumulation of sand and rock debris — reinforce the diminished importance of relief. Hence, the concept of a physiographical rim can be maintained. (Fisher, 1968:5–6.)

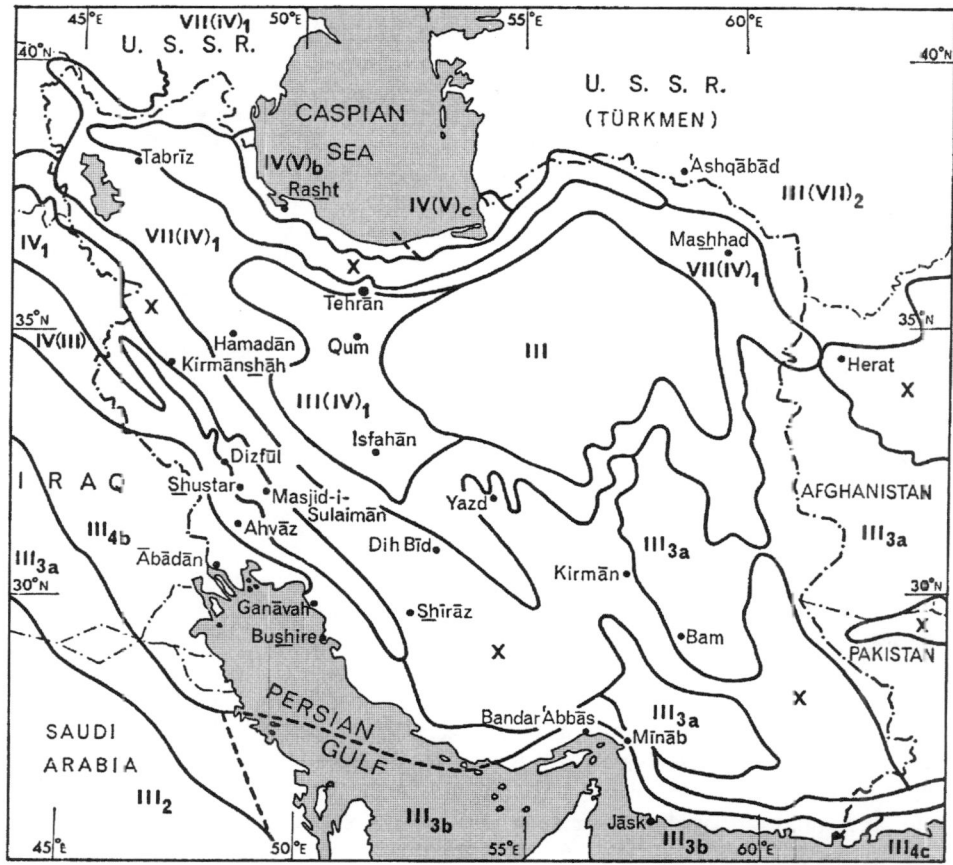

FIGURE 36. Climate map of Iran, using the climatic type designations of Walter and Lieth (1960): III subtropical, hot and arid; IV Mediterranean, winter rains; V warm-temperate, humid; VI humid, with cold season, VII arid, with cold season; X mountain areas in other regions. (After S. Anderson, 1968.) Reprinted with the permission of Cambridge University Press.

The climate of Iran has been described by Adle (1960), Ganji (1968), and Ehlers (1992). There is a general increase in mean annual temperature from the northwest to the southeast, reflecting the respective geographical positions and elevations of the different regions. Precipitation zones correspond with temperature zones, the west, northwest, and north receiving both higher annual averages and seasonal distribution of precipitation suitable to growth of vegetation, whereas the part of the Iranian Plateau situated in the rain shadow of neighboring mountains receives much less (Ehlers, 1992). For summary purposes in considering the lizard habitats, I have used the climatic designations of Walter and Lieth (1960) (see Fig. 36) which are based on a number of parameters of potential ecological importance: mean annual temperature, mean annual sum-total of precipitation, mean daily minimum of coldest month, absolute minimum, monthly means of temperature, monthly means of precipitation, arid period prevailing when precipitation falls below temperature curve, humid period prevailing when precipitation is above temperature curve, period of frost, where mean minimum temperature of month falls below 0°C, and period where absolute minimum falls below 0°C (see S. Anderson, 1968, Table 6 for details).

Overviews of the vegetation and flora of Iran have been provided by Zohary (1963) and Bobek (1968).

Nearly 30 years have passed since my previous attempt to analyze the lizard fauna of Iran (S. Anderson, 1966, 1968). Since that time, new material has accumulated in museums, the taxonomy of many species has been revised, and the general acceptance of plate tectonics, together with a cladistic approach to systematics, has changed our views of biogeography. On the other hand, little more has been added to our understanding of the natural history and ecology of Iranian lizard communities. Much of what follows here is a summary revision of my 1968 paper, while additional remarks on natural history, ecology, and faunistics are based on field observations incidental to collecting in 1975 and the study of museum specimens. What follows should be viewed as an attempt to point the way to the many studies that need to be undertaken rather than as a definitive analysis.

For purposes of discussing the geography of Iranian lizards I consider 13 physiographic regions of Iran.

The Central Plateau   (Plates 20A–B, 21C, 23E, 24A)

The term Central Plateau is used here to designate the internal drainage basin of the Iranian Plateau lying entirely within the confines of the Iranian borders, and rimmed by mountains. Certain species range broadly over this entire region: *Trapelus agilis, Laudakia microlepis, Phrynocephalus maculatus, P. scutellatus, Agamura persica, Teratoscincus scincus, Eremias persica, Mesalina watsonana, Eumeces schneideri*, and *Varanus griseus caspius*.

On the north and west of the plateau occur certain taxa which are found also in contiguous regions: *Laudakia caucasia, Trapelus ruderatus, Phrynocephalus persicus, Bunopus crassicauda, Ophisops elegans, Ablepharus bivittatus*, and *Mabuya aurata transcaucasica*. The plateau distribution of these species is largely confined to the inner slopes of the Zagros and Alborz Mountains bordering the plateau, and coincides with the Kordish-Khorasan rainfall pattern of spring and winter precipitation in excess of 200 mm annually.

Largely confined on the plateau to the southeastern portion are several lizards, most of which occur also in the highlands of Baluchistan or in Sistan: *Laudakia nupta, Uromastyx asmussi, Cyrtopodion agamuroides, C. kirmanense, Acanthodactylus blanfordi, A. micropholis, Eremias fasciata*, and *Ophiomorus brevipes*.

Species confined to the northeast of the plateau (principally Khorasan) include: *Phrynocephalus mystaceus, Crossobamon eversmanni, Cyrtopodion caspium, Eremias lineolata, E. velox velox, E. nigrocellata*, and *Eumeces taeniolatus parthianicus*.

One of the least known areas of the Iranian Plateau, and yet one of the most interesting and most important faunistically, is the north-south chain of upland masses separating the interior plateau basin from Afghanistan, including the drainages of both the Hari Rud and the Sistan basin. This region has been termed the Qa'in and Birjand highlands. Many of the species recorded from this region are known from single records, or from a few localities all within this border district. Other species are shared with the Kopet Dagh. Species of lizards known from this region are: *Trapelus agilis, Laudakia caucasia, L. microlepis, L. erythrogastra erythrogastra, L. nupta, Phrynocephalus maculatus, P. mystaceus, P. ornatus, P. scutellatus, Eublepharis* cf. *macularius, Agamura persica, Cyrtopodion agamuroides, C. caspium, C. kirmanense, C. longipes longipes, Teratoscincus bedriagai, T. scincus, Eremias fasciata, E. grammica, E. lineolata, E. nigrocellata, E. persica, Mesalina watsonana, Ablepharus pannonicus*, and *Varanus griseus caspius*.

Various factors account for the presence of this relatively large and varied fauna within a rather narrow strip of land. In this region several faunistically distinct areas are in contact. With southern Turkmenistan it shares such species as *Laudakia caucasia, L. erythrogastra, Phrynocephalus mystaceus, Cyrtopodion caspium, Eremias grammica*, and *E. lineolata*. On the eastern slopes of the central section of this region certain species are shared with the Helmand Basin. Most of these are species confined to the eastern half of the Iranian Plateau, that occupied by Afghanistan and upland Pakistan. Included in this group are *Phrynocephalus ornatus, Teratoscincus bedriagai, T. microlepis, T. scincus keyserlingii,* and *Eremias fasciata*. Most of the species on the western flanks of this upland are wide-ranging plateau species which cross the passes into Afghanistan and highland Pakistan. Included are *Trapelus agilis, Phrynocephalus maculatus, P. scutellatus, Agamura persica, Eremias persica, Mesalina watsonana,* and *Varanus griseus caspius*. Characteristic mountain species of the northern and central ranges include *Laudakia caucasia, L. microlepis, L. erythrogastra,* and *Cyrtopodion caspium. Laudakia nupta* is a species characteristic of the southern ranges.

The road from Mashhad to Zahedan runs the length of this border region. The terrain crossed by this road is one of plains and mountain masses, the road running through the mountains in some places, along the eastern front in others, thus traversing a considerable variety of habitats. A good general account of this road has been provided by Lay (1967).

A number of species or subspecies of lizards occur only within the basins or in the mountain ranges of the Central Plateau: *Laudakia microlepis, Cyrtopodion agamuroides, C. kirmanense, C. longipes longipes, Rhinogecko misonnei, Tropiocolotes latifi, Eremias andersoni, E. lalezharica, Lacerta mostoufi,* and *Ophiomorus nuchalis*.

The interior basins cover more than half the total land area of Iran. Much of the present surface was, during pluvial periods, occupied by large lakes which formed a fairly continuous system that extended into Afghanistan and Central Asia. Strandlines indicate the former levels of these lakes of interior drainage basins. The largest of these basins, the Dasht-e Kavir or Kavir-e Bozorg and the Dasht-e Lut or Southern Lut are virtually devoid of vegetation and have not been adequately explored for fauna. The upper elevations of the interior ranges support dry juniper forest and remnants of the dry pistachio-almond-maple forest, which once covered the elevated parts of the interior plateau next to the Zagrosian oak forest. The lower elevations of the plateau are covered with steppe associations where rainfall permits, and desert vegetation in the less watered areas. Natural vegetation throughout the plateau has been severely degraded through cultivation, pastoralism, and woodcutting. (Bobek, 1968).

The Reza'iyeh (Urumiyeh) Basin (Plate 25C–D, G)

This is the largest of the downthrow basins (structural basins of tectonic origin, generally formed by downward movement along bounding faults or overthrusts that produce boundary highlands) of the northern Zagros region. It has no drainage outlet and the shallow lake itself lies at 1,290 m above sealevel, and is highly saline. Much of the territory around the lake is limestone with extensive igneous intrusions.

The following lizards are known to occur in the area that drains into Lake Reza'iyeh: *Laudakia caucasia, Trapelus ruderatus, Phrynocephalus persicus, Eremias pleskei, E. strauchi strauchi, Lacerta brandti, L. cappadocica urmiana, L. media, L. raddei raddei, L. strigata, Ophisops elegans, Ablepharus bivittatus, Eumeces schneideri princeps,* and *Mabuya aurata transcaucasica*. The faunal affinities of this region are with highland eastern Anatolia and the Transcaucasian republics.

The Sistan Basin (Plate 21B)

The oval-shaped Sistan lowland is a complex downthrow zone with a steep boundary on its eastern side, formed by a narrow ridge between Neh and Nosratabad. The northeastern rim lies in Afghanistan and is made up by the Hindu Kush ranges. The largest expanse of freshwater on the Iranian Plateau, the Hamun-e Helmand, lies largely within Persian territory in the lowest part of the basin. The Helmand River, flowing out of Afghanistan, is the main feeder stream of the basin, supplemented by a few torrential seasonal flows for a few days each year plunging down the steep western flanks of the basin. Although lying within an internal drainage basin, the lake itself does not become brackish, because for much of the year there is a natural drainage via the Shalaq Rud into the still lower swamp, the Gaud-e Zirreh, in Afghanistan. (Fisher, 1968.)

Species and subspecies known from the Iranian part of the Sistan basin: *Trapelus agilis, Laudakia nupta, Phrynocephalus maculatus, P. ornatus, P. scutellatus, Agamura persica, Crossobamon eversmanni, Cyrtopodion agamuroides, C. caspium, C. longipes, C. russowii, C. scabrum, Teratoscincus bedriagai, T. microlepis, T. scincus keyserlingii, Eremias acutirostris, E. fasciata, E. persica, Mesalina watsonana, Ophiomorus tridactylus,* and *Varanus griseus caspius.*

The majority of these species belong to two faunal categories: widely distributed Iranian Plateau forms, such as *Trapelis agilis, Laudakia nupta, Phrynocephalus maculatus, P. scutellatus, Agamura persica, Cyrtopodion caspium, C. russowii, Teratoscincus scincus keyserlingii, Eremias persica, Mesalina watsonana, Eumeces schneideri,* and *Varanus griseus caspius,* primarily occupying the uplands, and a sand-adapted Helmand fauna, *Phrynocephalus ornatus, Teratoscincus bedriagai, T. microlepis, Eremias acutirostris, E. fasciata,* and *Ophiomorus tridactylus. Crossobamon eversmanni* is a sand-adapted gecko of extreme eastern Iran and southern Turkmenistan, *Cyrtopodion agamuroides* is an apparently narrowly distributed upland form, and *C. scabrum* is a broadly distributed form of the low southern deserts.

The Caspian Region (Plates 22G, 23A)

From Astara in the west to Hasan Qui Beg in the east the Caspian lowlands extend as a low-lying plain, some 640 km in length, but variable in width; in the extreme east it broadens out into the broad Turkmen lowlands. The Caspian sea is about 26 m below mean sealevel, although its extent depends upon the fluctuations of the inflow of the Volga River, which is subject to diversion for cultivation as well as to climatic fluctuation. Old shorelines are indicated by the presence of topographical flats and benches and marine erosion features. The coastline is characterized by a succession, first of coastal sand dunes, spits, and bars, then inland behind these a low-lying expanse of brackish lagoons and freshwater marsh, then a slightly higher and drier terrace zone, and finally, a piedmont zone at the base of the Alborz. The dunes carry a vegetation of coarse grasses and scrub, abundant in the west, where rainfall is much greater and thinning out in the east, to merge into the general steppe flora of the Turkmen plains. The marsh-lagoon area has a heavy vegetation of reeds, sedges, and aquatic plants, while the climax vegetation of the higher plain and foothills is the Hyrcanian forest. (Fisher, 1968.)

The Hyrcanian forest is a relic of the temperate, broad-leaved deciduous forest that covered large parts of Europe and northern Asia during the late Tertiary. It is characterized by: vigorous growth, tall trees, multi-storied structure, lianas, high regenerative capacity, and tree and shrub species diversity. This forest is much degraded, and largely destroyed in the lowlands. The whole forest area receives about 800 to 2000 mm precipitation per year. The maximum comes in autumn, but no season is without rain. (Bobek, 1968.)

The following lizard species have been recorded from within Iranian limits in the region of the Caspian Sea: *Trapelus agilis, Laudakia caucasia caucasia, Anguis fragilis colchicus, Ophisaurus apodus, Eremias velox velox, Lacerta chlorogaster, L. defilippii, L. praticola,* and *L. strigata. Lacerta chlorogaster* is a narrowly distributed endemic of the Hyrcanian forest within Iran, extending into the valley of the Atrek River in the Kopet Dagh; *Anguis fragilis* occurs in Iran only in the moist, lowland Caspian region; it is a widely distributed species of the western Palearctic, while *Ophisaurus apodus* is a species of the Balkans, Turkey, and Transcaucasia, which extends as far east as Afghanistan. *Laudakia caucasia* and *Lacerta defilippii* are mountain species which occur on the northern slopes of the Alborz. *Trapelus agilis* and *Eremias velox* extend into the southern Caspian region from the lowlands of Turkmenistan.

The Khuzestan Plain and the Persian Gulf Coast   (Plates 21E–G, 22E)

This is the largest expanse of true lowland within Iran. Most of the Mesopotamian valley in this area is covered by sediments brought down from the Zagros by the Karun and Karkheh rivers. Near the head of the Persian Gulf are freshwater and salt mangrove swamps. North and east of Ahvaz there are areas of sand dunes. Predominant vegetation is steppe-like ground cover, which tapers out into more desert-like formations with growing distance from the hills, toward the southwest. This lowland, with its vegetation floristically related to that of the Saharo-Arabian and Nubo-Sindian groups is termed *garmsir* ("warm land"). Annual precipitation is about 210–320 mm, coming mainly in the winter.

Geographically an extension of the Mesopotamian Plain, this region has a close faunal relationship to lowland Iraq and northern Arabia. The course of the Tigris River has apparently served as a barrier to the distribution of some forms, however. The fauna is not uniformly distributed, certain species being reported only from the more humid gulf coastal plain. The following lizards are known: *Phrynocephalus arabicus, Trapelus persicus, T. ruderatus, Laudakia nupta, Uromastyx loricatus, U. aegyptius, Asaccus elisae, Bunopus tuberculatus. Cyrtopodion gastrophole, C. scabrum, Hemidactylus flaviviridis, H. turcicus, Pristurus rupestris, Stenodactylus affinis, S. doriae, Acanthodactylus grandis, A. schmidti, A. boskianus, Mesalina brevirostris, M. watsonana, Ophisops elegans, Ablepharus pannonicus, Chalcides ocellatus, Eumeces schneideri princeps, Mabuya aurata septemtaeniata, Scincus scincus conirostris,* and *Varanus griseus.*

Most of these species occur widely in the North Arabian Desert, or more widely still. *Acanthodactylus boskianus, Scincus scincus,* and *Varanus griseus,* for example, extend across North Africa. Some species are more characteristic of the Iranian Plateau or the Zagros foothills, e.g., *Laudakia nupta, Trapelus ruderatus, Asaccus elisae Cyrtopodion scabrum,* and *Mesalina watsonana.*

Iranian Baluchistan and the Makran Coast   (Plates 20B–D, F–G, 21A)

In this region, the Zagros structure declines into two smaller formations, the first a system of ridges forming the coastal ranges and interior hills of Makran which continue with an east-west disposition into Pakistan, and second, a widely diverging fold structure, a high single anticline fronting the Qatar and Oman shores on the southern coasts of the Gulf of Oman. North of the Makran is an irregular upland dominated by the Kuh-e Taftan range. The narrow Kuh-e Basman range runs east-west, linking the uplands of eastern Iran to the Zagros and divides the central Lut from the Jaz Murian depression. Further east, the Kuh-e

Sultan chain serves as a link with the mountains of Baluchistan in Pakistan. North of Kuh-e Taftan is the Zahedan plateau, a variegated upland region broken by ridges.

The predominant vegetation of this region is that of scattered trees and shrubs with steppe-like ground cover, which becomes a more desert-like formation in the east (Makran). *Ziziphus spinachristi* and several species of *Acacia* are widespread. Annual precipitation is about 130–260 mm, from slight winter rains and irregular monsoon summer storms. The upper limit of this *garmsir* vegetation is about 1,400 m elevation.

Species recorded from this region within Iranian limits are: *Trapelus agilis, Laudakia nupta, Calotes versicolor, Phrynocephalus maculatus, P. scutellatus, Uromastyx asmussi, Agamura persica, Bunopus tuberculatus, Cyrtopodion brevipes, C. sagittifer, C. scabrum, Hemidactylus flaviviridis, H. persicus, H. turcicus, Stenodactylus doriae, Tropiocolotes persicus persicus, T.* cf. *steudneri, Teratoscincus scincus, Rhinogecko femoralis, Acanthodactylus blanfordi, A. micropholis, Eremias fasciatus, E. persica, Mesalina watsonana, Opisops elegans, Ablepharus pannonicus, Chalcides ocellatus, Eumeces schneideri zarudnyi, Ophiomorus blanfordi, O. brevipes, O. streeti, Varanus bengalensis*, and *V. griseus caspius*.

This lizard fauna is made up of two main elements: widely ranging plateau forms, most of which are confined here to the rugged, folded terrain of Baluchistan and Makran, and a low desert fauna, more or less restricted to the coastal plain of eastern Iran and western Pakistan. Many of these range no farther west than Bandar-e Lengeh, west of which the coastal environments are disrupted by the proximity of the mountains to the gulf; these include the species of *Phrynocephalus, Uromastyx asmussi, Cyrtopodion agamuroides, C. brevipes, Teratoscincus scincus, Acanthodactylus blanfordi, A. micropholis*, the species of *Ophiomorus*, and *Varanus bengalensis. Cyrtopodion brevipes, C. sagittifer*, and *Ophiomorus streeti* appear to be endemics of the Jaz Murian depression in Baluchistan. *Calotes versicolor*, widely distributed in Pakistan and eastward, has been reported only from a few date orchards in Iranian Baluchistan.

The Turkmen Steppe (Plate 25E)

Small portions of the lowland plains of Turkmenistan extend within Iranian borders in the northeastern corner of the country and in a narrow wedge east of the Caspian Sea, between the shore and the mountains. Species found in this drainage are: *Trapelus agilis, Laudakia caucasia caucasia, L. erythrogastra, Phrynocephalus helioscopus, P. mystaceus, Ophisaurus apodus, Crossobamon eversmanni, Cyrtopodion caspium, Teratoscincus scincus, Eremias grammica, E. intermedia, E. lineolata, E. nigrocellata, E. velox velox, Lacerta chlorogaster, Mesalina watsonana, Eumeces schneideri princeps, E. taeniolatus parthianicus, Mabuya aurata*, and *Varanus griseus caspius*, and possibly *Phrynocephalus interscapularis* and *Cyrtopodion longipes microlepis*.

Here on the fringes of the Turkmen steppe, many of these species are confined primarily to mountain slopes, the only species characteristic of the steppe itself are the species of *Phrynocephalus, Trapelus, Eremias, Crossobamon, Teratoscincus,* and *Varanus*, and most of these also occur on the Iranian Plateau, at least in Khorasan.

The Moghan Steppe (Plate 24D)

This region, drained by the Aras River, falls within Iranian limits only in the northernmost part of Persian Azarbaijan. This semiarid area receives less than 200 mm of precipitation annually and is covered with steppe vegetation; the surrounding mountains have been stripped of their natural forests and degraded steppe has been extended to their slopes (de Planhol, 1987). Lizard species in this region include: *Trapelus ruderatus ruderatus, Ophisaurus apodus, Cyrtopodion caspium, Eremias arguta, E. strauchi strauchi, Lacerta strigata, Ophisops elegans*, and probably *Eumeces schneideri princeps*. A few other species of the mountains of the Iranian Azarbaijan provinces probably should also be included within this drainage, such as *Lacerta raddei raddei* and *L. valentini valentini*.

The Zagros Mountains (Plates 22F, 24C, E, 25D)

This long mountain chain forms both a barrier between the plateau and the Mesopotamian lowlands and a corridor for the southward distribution of northern faunal elements. Geographically, the Zagros consist of two distinct subregions: a northwestern section extending from the northwestern frontier to a line extending through Qazvin-Hamadan-Kermanshah, forming a series of tablelands at an average of 1,500–2,000 m elevation, with greater heights in the north and west; the remainder of the Zagros, a series of domes or hogbacks formed of folds of parallel strike trending northwest-southeast to Bushire, becoming arcuate and curving eastward, extending as far as Bandar 'Abbas and Hormoz.

The Zagrosian forest, a semi-humid oak association, occurs on the outer slopes of the southern and southwestern margins of the Iranian Plateau, extending from the Turkish border through Iranian and Iraqi Kordestan and Lorestan into Fars. This is a dry, cold-resistant, and deciduous forest with summer-green deciduous oaks dominant, the ground cover receiving enough light to form a grassy and herbaceous steppe. Its lower limit lies at 700 to 800 m in the north, and at about 1,200 m in the south. The upper limit is 2,200 to 2,800 m elevation. Annual precipitation in this forest region is 500–750 mm, or more, mostly falling in winter and spring. (Bobek, 1968.)

Species known from the Zagros area are: *Trapelus agilis, T. ruderatus ruderatus, Laudakia nupta, Ophisaurus apodus, Cyrtopodion heterocercum, C. scabrum, Tropiocolotes helenae fasciatus, Eremias nigrolateralis, Lacerta princeps, L. strigata, L. zagrosica, Mesalina watsonana, Ophisops elegans, Ablepharus bivittatus, A. pannonicus, Eumeces schneideri princeps, Mabuya aurata, Ophiomorus persicus*, and *Varanus griseus*. Five taxa, *Asaccus kermanshahensis, Cyrtopodion heterocercum, Tropiocolotes helenae fasciatus, Lacerta princeps*, and *Ophiomorus persicus* are endemic to the Zagros and the contiguous mountains of Anatolia, but the extent of their distribution is undetermined. The known fauna is essentially that of the lower passes and consists mainly of wide-ranging Southwest Asian species.

The Western Foothills of the Zagros Mountains (Plates 21D, 22A–C, E)

The fauna of this foothill belt shares species with both the Zagros Mountains and the Mesopotamian lowlands, but also contains a number of species unique to this region. The species known from this region are: *Trapelus agilis, Laudakia nupta, Uromastyx loricatus, Eublepharis angramainyu, Asaccus elisae, A. griseonotus, Carinatogecko aspratilis, C. heteropholis, Cyrtopodion scabrum, Hemidactylus persicus, H. turcicus, Ptyodactylus* sp., *Tropiocolotes helenae helenae, T. persicus bakhtiari, Acanthodactylus nilsoni, Mesalina watsonana, Ophisops elegans, Ablepharus pannonicus, Eumeces schneideri princeps, Mabuya aurata septemtaeniata, Scincus scincus conirostris*, and *Varanus griseus*.

The endemic taxa of this list are of particular interest. *Eublepharis angramainyu* has its nearest relatives on the southeastern and northern borders of the Iranian Plateau, in Pakistan, Afghanistan, and Turkmenistan. The two species of *Tropiocolotes* have subspecies (or closely related species) in Persian Baluchistan and in Pakistan. *Asaccus griseonotus, A. kermanshahensis*, and their more broadly distributed congener, *A. elisae*, are related to species in the upland regions of Oman and the United Arab Emirates. *Carinatogecko* is a genus limited in its known extent to these foothills, although these species may eventually be treated as members of *Cyrtopodion* related to the *C. kotschyi* complex. Neither *Scincus scincus conirostris* nor *Uromastyx loricatus* can be said to be characteristic of the foothill belt; the former is found in eolian dunes caught in the fringes of the hills, while the latter occupies the lower alluvial fans and valleys. Conspicuously absent from this fauna are those species which are characteristic of the Central Plateau, such as the several species of *Phrynocephalus* and *Eremias*, which make up a large part of the diversity of the plateau fauna. Species of *Lacerta*, important elements of the northern and mountain faunas, do not appear in this foothill region.

The Alborz Mountains (Plates 24B, 25A–B)

The Alborz range, including the Talysh hills, extends in an arc from Astara at the border of the Azarbaijan Republic, around the southern end of the Caspian Sea to as far east as Jajarm, a distance of some 1,000 km. It includes Iran's highest peak, Mt. Damavand, 6,332 m elevation. The Talysh hills consist of a long, narrow hog's-back ridge rising to a maximum height over 3,000 m. In the central Zagros, two major chains of unequal height can be distinguished, separated by the valley of the Shahrud, the northern ridge being the higher in elevation. The highest elevations support icefields and glaciers. In the eastern part of the range there are three main ridges, with irregular plateau surfaces between. (Fisher, 1968.)

The lower elevations of the northern slopes, up to about 1,000 m, are covered by Hyrcanian forest (see above). The montane forest falls into two zones, the lower open and fully exposed flanks dominated by beech, while other associations are present on the more protected sun-facing slopes and in the ravines; upward of about 1,700 m beech is gradually replaced by a large oak (*Quercus macrantha*), accompanied by elm, ash, hornbeam, maple, wild pear, and many shrubs, including juniper. (Bobek, 1968.) Precipitation on the northern slope is 1,000–2,000 mm annually, distributed fairly evenly among seasons, with the maximum in autumn (Bazin, *et al.*, 1985). Juniper forest once covered the southern slopes of the Alborz chain, as many remnants demonstrate. The ground cover of this forest is a complete steppe complex. The upper limit of this dry forest approximates that of the Caspian montane forest; its lower limit seems to lie at 1,100–1,600 m. There are isolated patches of natural cypress forest in some valleys of the Alborz. (Bobek, 1968.) The southern slopes of the range are hardly less arid than the piedmont (150–300 mm) (Bazin, *et al.*, 1985).

The fauna of this range consists of two fairly well-defined segments: that of the dry southern slopes (these species having been included in the discussion of the plateau), and that of the much wetter, forested northern slopes (included in the section on the Caspian coast). A few species cross the passes; a few range along the lower crests. Known species of the Alborz are: *Trapelus agilis, T. ruderatus, Laudakia caucasia caucasia, Phrynocephalus persicus, P. scutellatus, Anguis fragilis, Ophisaurus apodus, Mabuya aurata transcaucasica, Cyrtopodion caspium, Eremias nigrocellata, E. persica, Lacerta defilippii, L. strigata, Ophisops elegans,* and *Eumeces schneideri princeps.*

Of all these species, only *Lacerta defilippii*, a montane species with Caucasian affinities, is known to occupy the mountain crests, occurring on both flanks of the range above tree line. Other species which at least cross the passes and occupy the mountain valleys are *Laudakia caucasia caucasia, Ophisaurus apodus, Eremias persica, Lacerta strigata,* and *Mabuya aurata transcaucasica. Cyrtopodion caspium* and *Eremias nigrocellata* are only known in the Alborz from the more arid eastern end of the range where it merges with the folds of the Kopet Dagh.

The Kopet Dagh (Plates 23B–D, 25F)

I use the term Kopet Dagh here to embrace the series of folds aligned northwest-southeast, including Gulul Dagh, Allahu Akbar, and Hazar Masjid, which are all part of the Caucasus system, and those folds which continue the structure of the Alborz into Khorasan, including Ala Dagh, Kuh-e Binalud, and Pusht-e Kuh. These two series of folds are separated by the faulting and downthrow trough containing the Atrak River. Technically, the Kopet Dagh range proper, as more narrowly defined by geographers, lies entirely within the borders of Turkmenistan. To the north and northwest, where the Kopet Dagh is open to penetration by moist winds, there is sufficient rainfall to support annual grasses, fairly thick scrub vegetation, and a scattering of trees such as alder, oak, juniper, and hornbeam (Fisher, 1968). Steppe with scattered almond and pistachio covers extensive areas of the northern foothills (Bobek, 1968).

The fauna of the more arid mountain folds stretching along the Iran-Turkmenistan border east of the Alborz has not been studied in any detail on the Iranian side, one reason being that routes crossing the border run to the west and east of these mountains. The relatively low Atrek Valley divides the two main folds of the range, and itself has been little travelled by zoological collectors. Much more investigation has taken place from the Turkmenistan side (see Ataev, *et al.*, 1994 for a summary of the herpetofauna). The known species within Iranian limits are: *Laudakia caucasia caucasia, Phrynocephalus helioscopus, Anguis fragilis, Ophisaurus apodus, Eublepharis turcmenicum, Cyrtopodion caspium, C. spinicauda, Eremias nigrocellata, E. strauchi kopetdaghica, E. velox velox, Lacerta chlorogaster, L. steineri, L. strigata, Ablepharus bivittatus,* and *A. pannonicus*.

In addition, Ataev, *et al.* (1994) list *Laudakia erythrogastra nurgeldievi, Trapelus sanguinolentus* [regarded here as synonymous with *T. agilis*], *Mesalina watsonana, Lacerta raddei raddei, Chalcides ocellatus, Eumeces taeniolatus parthianicus,* and *Varanus griseus caspius*. Most or all of these must occur within Iranian limits of the range as well.

Islands of the Persian Gulf (Plate 20E)

The islands along the Iranian side of the Persian Gulf represent Zagros anticlinal structures (Kassler, 1973). Almost nothing is known of the fauna of these islands, most of which lie close to the Iranian coast. A careful study of these islands may provide an answer to questions about earlier distributions along the gulf coasts. The few lizards known from these islands thus far are listed here, with the island from which they were taken in parentheses: *Bunopus tuberculatus* (Jazireh-ye Tanb-e Bozorg), *Pristurus rupestris* (Jazireh-ye Kharg; Qeshm Island), *Acanthodactylus micropholis* (Qeshm Island), *Mesalina brevirostris* (Qeshm Island; Jazireh-ye Tanb-e Bozorg), *M. guttulata* or *M. watsonana* (Jazireh-ye Hengam). I have not included in the Iranian fauna species from Bahrain Island, a politically separate entity. The amphibians and reptiles of this island have been listed by Gallagher (1971), including the following lizards: *Trapelus jayakari, Uromastyx aegyptius, Bunopus tuberculatus, Hemidactylus flaviviridis, H. persicus, Cyrtopodion scabrum, Pristurus rupestris, Stenodactylus arabicus, S. slevini, S. khobarensis, Acanthodactylus schmidti, Mesalina brevirostris, Scincus scincus conirostris, Mabuya aurata septemtaeniata* and a doubtful sight record for *U. thomasi*. Thus, the faunal relationships of this island are with the Arabian peninsula rather than with the Iranian Plateau.

For additional specific information on sympatry, consult the species distribution maps.

Ecological and Behavioral Influences on Lizard Distribution

Substrate Type

There is considerable correlation of substrate type and local distribution of lizards in Southwest Asia, as pointed out in my 1968 paper and by Szczerbak (1994) for Turkmenistan. Certain Iranian lizard species are adapted for life in or on eolian sand dunes, or at least on sandy soils, and are more or less restricted to such zones. Among these psammophiles are species of *Acanthodactylus, Scincus, Teratoscincus, Crossobamon eversmanni, Phrynocephalus interscapularis, P. mystaceus, Eremias acutirostris, E. grammica, E. lineolata, E. scripta*, and several species of *Ophiomorus*. The amphisbaenian *Diplometopon zarudnyi* is also a sanddweller. *Ophiomorus* and *Diplometopon* live beneath the surface of the sand, and their limbs have been greatly reduced, an adaptation facilitating the types of subsurface locomotion which these animals employ. The other species named above have the digits equipped with comb-like fringes of scales, an adaptation which has arisen independently in many groups in various sandy deserts throughout the world. *Scincus*, and to a lesser extent, species of *Phrynocephalus*, are adapted both for burrowing in soft sand and for sandrunning.

The species of *Laudakia* appear to be restricted to areas such as limestone outcrops and rocky cliff faces where both basking surfaces and deep crevices for retreat are provided. Species in this group include *L. caucasia, L. erythrogastra, L. melanura, L. microlepis*, and *L. nupta*. Such terrain is characteristic of the Iranian Plateau and the mountain and upland regions of its borders. The species of *Trapelus* occur on plains, valleys, and alluvial fans, on sandy, loam, clay, and gravel soils. Species specificity in regard to these different soil types has not been well established. These lizards are usually found in the vicinity of small rock piles, such as those erected by local inhabitants to mark the boundaries of grain fields. Such rock piles provide vantage points and basking areas upon which the lizards are able to orient themselves to sunlight for temperature control. They retreat into these piles for shelter. They also ascend low shrubs, and presence or absence of low bushes may be a factor in the distribution of some species. Such areas frequently interdigitate with the outcrops and boulder slides occupied by species of *Laudakia*, but the occupancy of the two environments is sharply defined.

Species of *Phrynocephalus* and *Eremias* show distinct preferences for particular soil types. *Phrynocephalus helioscopus, P. ornatus, P. persicus, P. raddei, P. scutellatus, Eremias fasciata, E. intermedia, E. nigrocellata*, and *Mesalina watsonana* prefer open clay and gravel plains, while *Phrynocephalus interscapularis, P. mystaceus, Eremias intermedia*, and *E. lineolata* are usually found on sandy plains and steppes. *Eremias pleskei* and *E. strauchi* prefer dry mountain slopes. *Phrynocephalus maculatus* is found on both sandy and clay surfaces of flatlands. Local distribution of many lacertid species may be determined by the availability of cracks and holes in clay and gravel soils, or burrows in plant-stabilized sandy soil. These crevices provide a retreat from predators and from temperature extremes.

A number of species are rock-inhabiting, able to negotiate the rough vertical surfaces of rock outcrops and montane habitats. These include the species of *Laudakia* and several species of small *Lacerta*, e.g., *L. cappadocica, L. defilippii, L. raddei, L. valentini*, and *L. zagrosica*. These lizards utilize the many angles and shadows of this environment for temperature regulation by basking and make use of the many crevices for retreat from predators and temperature extremes.

Among the geckos, *Teratoscincus* and *Crossobamon* have been mentioned as sanddwelling species. The various species of *Stenodactylus* and *Bunopus* also occur on sand, but to what extent they occupy other substrates is not known. *Agamura, Pristurus, Tropio-*

colotes, and *Cyrtopodion* usually are found on rocky slopes and cliff faces, in crevices and caverns, and in and about places of human habitation (particularly true of *Cyrtopodion scabrum*). *Asaccus elisae* is found in caverns in gypsum deposits and limestone, and occasionally as a house gecko. *Hemidactylus* and *Ptyodactylus* are similarly adapted to life on vertical surfaces. The discontinuous distribution of geckos of the genus *Eublepharis* should be investigated from an ecological standpoint. In Khuzistan, *E. angramainyu* was found only in the foothill areas where extensive gypsum deposits exist. It may be that these large geckos are dependent upon the cavernous areas in the gypsum where water persists throughout the year and a high relative humidity may be maintained.

Structures usually built of mud-brick provide additional habitat not only for the geckos mentioned above, but for rock-dwelling species of *Laudakia* as well. These lizards are often quite numerous on walls, houses, and monuments. There is usually an abundance of insect prey in such situations, attracted by the human inhabitants and their domestic animals and cultivated plants.

The Iranian species of *Uromastyx* are confined in their local distribution to well-drained alluvial soils wherein they are able to excavate their burrows. *Uromastyx asmussi* reportedly favors gravelly alluvium, whereas *U. loricatus* prefers silty-clay soils.

The most widely ranging forms in Southwest Asia are those occupying the greatest range of substrates, such as *Varanus griseus, Eremias persica, Trapelus agilis, Mabuya aurata,* and *Eumeces schneideri,* or those inhabiting the most continuously distributed substrates, such as *Mesalina watsonana*.

Evolution of the various lizard groups in Southwest Asia may be significantly correlated with this specific affinity for substrate type, and the discontinuous distribution of these substrates. Many populations or subpopulations may be effectively isolated genetically from one another over protracted periods of time. This has been discussed in the case of the obligate dune-dwelling species of *Ophiomorus* in a previous paper (S. Anderson and Leviton, 1966b). Even species physically and physiologically capable of crossing fairly narrow stretches of intervening unsuitable substrate may rarely do so.

The greatest number of individual lizards occupying the fringe areas of a population are often juveniles. It is these animals, unable to wrest already established territories from adults, that are most easily picked off by predators. It is among these juveniles also that one expects the greatest phenotypic variation, and from this peripheral group the occasional colonizers of unoccupied habitat separated by unfavorable terrain must usually be drawn. It is thus possible to visualize these patches of discontinuously distributed substrate types as analogous to islands, and the effects of "waif dispersal" and "founder effect" readily imaginable.

Dependence on substrate may also serve as a limiting factor on one or more fronts as populations extend and retract their ranges in response to climatic change. Species inhabiting low sandy plains, possibly physiologically able to negotiate the temperature and/or moisture gradient imposed by a bordering mountain range, may yet fail to cross the passes when confronted by the rocky cliffs, slides, and alluvial fans along the mountain front.

Vegetation

The distribution of lizards in relation to the flora is either a result of dependence upon common physical factors of the environment, substrate, precipitation, etc., or a consequence of certain physical requirements provided for the lizards by the vegetation. Certain types of shrubs, for instance, stabilize dune sands and provide suitable sites for burrow excavation among their roots. These burrows may be constructed by rodents, or even large arthropods, and thus the lizards are also dependent upon the presence of these animals as well as on the plants. In the arid areas of sparse vegetation, many lizards may depend for sustenance upon the insect species attracted to the vegetation, and consequently their local distribution depends upon the frequently narrowly restricted occurrence of certain plants.

The species of *Trapelus* climb into the branches of low steppe vegetation such as wormwood (*Artemisia*) and camel thorn (*Alhagi camelorum*) to orient themselves relative to the sun's rays and to escape the hot soil surface for temperature control. Such vantage points also enable them to survey their territories and to ambush prey. Vegetation may play a similar role among forest species in the mountains and along the Caspian coast, the patches of alternating light and shadow enabling efficient behavioral temperature control. A few species adapted to climbing tree trunks may extend their territories vertically as well as horizontally, and the ability to move quickly around the circumference of a tree trunk or to seek refuge under loose bark offers considerable protection from the larger predators. A few species, occurring in northern forested regions, prefer heavy vegetation (15% cover [Szczerbak, 1994]). These include *Anguis fragilis, Ophisaurus apodus, Lacerta chlorogaster, L. princeps*, and *L. strigata*. Skinks of the genera *Ablepharus, Eumeces*, and *Mabuya,* and probably *Ophiomorus persicus* also occur in such regions. These species are often found in riparian habitats.

While the relationship of Southwest Asian lizards to vegetation has been little studied, whatever dependence does exist is probably in relation to vegetation type rather than to particular species of plants. Certainly the role of vegetation in the creation and maintenance of soil and moisture conditions and other factors of the microclimate is obvious.

Temperature

It is well established that lizards are able to maintain their activity temperatures within a fairly narrow range. Although some physiological mechanisms, such as vasodilation and vasoconstriction, and changes in albedo are involved, these are generally accessory to the primary, behavioral means of temperature regulation. Thus the animal's body temperature at any given time is a product of its relation to ambient temperature, and reflected radiation from substrata. For this reason, the mean air temperature, either on a daily or a seasonal basis, is less meaningful than the extremes from which the animal is unable to escape. So, too, the number of hours and days during which the combination of climatic events enable the lizard to maintain a suitable activity temperature must be considered.

Temperature may be most critical to developmental stages, as the lizard must be able to place its eggs where they will be protected from lethal extremes as well as exposed to temperatures sufficiently high for development to proceed. Oviparous species must be developmentally labile to the extent that they are able to endure the inevitable fluctuations and inconsistencies in temperature and moisture characteristic of arid regions.

The behavioral means by which temperature is regulated differ considerably from group to group. Most lizards (at least diurnal species) bask, utilizing direct insolation to raise the body temperature to the activity range. Normal activity temperatures for agamid lizards in the foothills of Khuzistan were found to lie between 38° and 43°C, the larger *Uromastyx* foraging with cloacal temperatures as high as 44°C. A change in albedo correlated with body temperature was observed in these lizards, basking individuals being the darkest in appearance, those near or past the voluntary maximum being extremely light in color. Similar activity temperatures were recorded for lacertids and for *Varanus griseus* (S. Anderson, 1963).

Probably most species exhibit a seasonal shift in the daily hours of activity. In

Khuzistan, the lizards first become active in early spring for a few hours from midday when temperatures are highest. As the season wears on they begin activity progressively earlier in the day and remain active longer. Eventually, they seek cover during the hottest hours, and by mid-summer they are active only during the early morning and late afternoon, a reversal of this shift occurring in autumn.

Some species, such as the sand-dwelling *Ophiomorus* and *Scincus scincus conirostris*, have the opportunity to maintain fairly constant body temperatures, through their burrowing habits in eolian sand dunes. By moving onto the sand surface, or near the surface, during the warm hours, they come quickly to activity temperature, while both high and low extremes are readily avoided by burrowing a few centimeters below the surface.

The small lacertids are able to extend their activity periods into the hottest hours of the day by utilizing the small areas of shade provided by rock or bush, making brief forays into the sunlight to catch insects.

The agamids position themselves relative to the incident sunlight so that the maximum surface area is exposed during basking, the minimum during the hottest period. The small agamids ascend low bushes at midday and thus escape some of the heat re-radiated from the ground. A darkly pigmented peritoneum is characteristic of diurnal species in Southwest Asia, and is presumably related to the thermal environment affecting these animals.

One of the striking aspects of the lizard fauna of this desert region is the diversity of gecko species. These creatures are able to circumvent the problem of high daytime temperatures through exploitation of nocturnal activity. During the hottest season, when diurnal lizards are restricted to brief activity periods, nocturnal air temperatures remain high, due to the re-radiation from the heated ground surface. The activity of many insects and other arthropods is also largely confined to the night hours during this period. A few geckos apparently have become secondarily diurnal, or partially so. Such behavior is indicated for *Pristurus, Agamura*, and some *Cyrtopodion (C. agamuroides, C. gastrophole)*, all of which have darkly pigmented peritoneum. The habits of these lizards have not been studied, however, and only a few observations have been recorded.

Only the most cursory temperature observations have been recorded for lizards in Southwest Asia (S. Anderson, 1963), but considerable study of thermal problems in regard to lizards of the deserts of the southwestern United States, southern Africa, and Australia has been undertaken (see particularly Avery, 1982; Gans and Pough, 1982; Huey, 1982; Pianka, 1986:38–47 for reviews).

Mountain ranges often limit the distributions of animals found in bordering lowlands for several reasons, but certainly one of the most important is their imposition of a vertical temperature gradient. For instance, if continental temperature zones move southward in response to general lowering of temperatures, lizards living on the low Aralo-Caspian steppes find an increasing temperature barrier to their ascent through the passes through the east-west ranges separating the low steppes from the Iranian Plateau. Conversely, north-south ranges, such as the Zagros, may provide corridors for the southward penetration of upland northern elements. With general increase of continental temperatures, and increased aridity, some plateau species may find refuge in the higher elevations of the mountain masses present on the Iranian Plateau, while in response to climatic cooling and increased precipitation, species isolated in such mountain areas may descend to the plateau to become more widely distributed.

Moisture

Reptiles as a group have various physiological adaptations which enable them to exploit arid environments. The more obvious of these are well known, namely, a relatively impermeable integument, excretion of nitrogenous wastes as uric acid, and high degree of resorption of water in the kidneys. Many other aspects of their osmoregulatory physiology remain relatively unstudied. Studies on evaporative water loss in reptiles have been reviewed by Mautz (1983), but these do not include Southwest Asian species, thus far.

Certainly the amount of available water must be a limiting factor for many species in arid regions, moreso, perhaps, than extremes in temperature. Undoubtedly, behavioral adaptations play an important role in conservation of water in desert species. Many lizards spend the hours when evaporative rates are highest in burrows, where the relative humidity remains considerably higher than that of the general environment.

It has often been noted that although many desert lizards in captivity will not drink water standing in a vessel, they will, however, lap water sprayed in droplets on rocks, leaves or other objects (including other individuals) in their cages. This suggests that atmospheric water condensing as dew may be important to the survival of such species. There is very little precise information regarding dew in Southwest Asia, although it is well known to desert travellers that heavy dew frequently forms during the late night and early morning hours, even in extremely arid areas. Most field studies of water relations have been carried out on North American iguanians; for a review of this literature, see Nagy (1982).

Feeding Strategies and Feeding Niches

It is interesting to note that the great majority of lizard species examined in this regard exhibit a fairly wide latitude in dietary items. Most of these lizards apparently eat any small arthropod with which they come in contact and are able to capture and overcome. This virtually guarantees that the bulk of prey species will be coleopterans and hymenopterans, apart from the seasonal availability of such insects as termites and grasshoppers. Vegetable matter also appears in the stomach contents of several insectivorous and carnivorous species, sometimes in such quantity that its ingestion incidental to capture of insects can be ruled out. I have watched species of *Laudakia* feed on vegetation and Szczerbak (1994) states that some species of *Phrynocephalus* and *Eremias* also feed on plants. Similar observations have been recorded for lizards in desert regions of North America (Banta, 1961). This non-specificity or limited specificity as to diet is obviously an important adaptation in arid regions where population densities of prey species are usually low, and the appearance of any one species of insect tends to be seasonal. Even the vegetarian *Uromastyx* are reported to take animal food in captivity, and may occasionally do so in nature as well.

Present data are insufficient to demonstrate dietary specificity in any Iranian lizard species, but it has certainly not been ruled out for some. Ants predominate in the smaller species of *Phrynocephalus*, and in *P. ornatus* and *P. luteoguttatus* no other food material was found. This may, of course, only indicate the seasonal availability of these insects. Only beetles were found in the digestive tracts of the few specimens of *Teratoscincus scincus* and *T. bedriagai* which I examined. In captivity, specimens of *T. scincus* regurgitated mealworm larvae (*Tenebrio molitor*) and crickets, and fed successfully only on adult beetles. However, I have observed *T. s. scincus* feeding on spiders in the field in Turkmenistan. *Anguis fragilis* reportedly shows a preference for slugs and snails, its dentition being particularly adapted to these dietary items. It also feeds on earthworms and soft-bodied insect larvae. Vertebrates are taken by most of the larger lizards (other than *Uromastyx*); *Ophisaurus, Laudakia*, and *Varanus* will consume virtually any animal that they can catch and overcome, within the limitations of their feeding apparatus.

Behavioral patterns of prey seeking and capture have traditionally been categorized in insectivorous and carnivorous species as ambush (sit-and-wait) and active foraging "strategies." Casual observation of Iranian species leads one to put most agamids and gekkonids into the former category and most lacertids, scincids, and varanids into the latter. However, these behavioral patterns need to be viewed within the overall context of the management of energy flux through the body and, probably, water conservation as well (Y. Werner, pers. commun., Sept. 1995). Thus, total activity time management studies are needed that avoid oversimplification of feeding strategies.

Predation on Lizards

Lizards are preyed upon principally by other vertebrates, although small ground-dwelling geckos and lacertids are occasionally taken by scorpions, solpugids, and large hunting spiders. Szczerbak (1994) states that the snakes *Psammophis lineolatus, Coluber rhodorhachis,* and *C. karelini* are specialized herpetophages in Turkmenistan, and that many other snakes readily eat reptiles, e.g., species of *Eryx, Coluber, Ptyas, Lycodon, Lytorhynchus, Naja, Echis, Boiga,* and *Agkistrodon*; *Varanus griseus, Ophisaurus apodus,* and species of *Laudakia* have already been mentioned as feeding on other lizards. Birds of prey, especially raptors and shrikes, are important predators of diurnal lizards. According to Szczerbak (1994), the list of herpetophagous birds in Turkmenistan includes 66 species. In the more geographically diverse Iran, the list probably would be still longer. Reptiles are also consumed by virtually all predatory mammals, varying in size from shrews to wolves and hyenas. The list of known mammalian predators on lizards in Turkmenistan includes 20 species (Szczerbak, 1994), and again, we would expect the Iranian list to be longer still.

Syntopy

The necessary first step in describing animal communities is to identify the species that have the potential to interact. Sympatry is the term used for the overlapping distributional ranges of two or more species and can be determined in an armchair fashion by comparing distribution maps. This is useful for interpreting historical biogeography, but tells us little about the makeup of biotic communities. Syntopy refers to species living together in the same or adjacent biotopes where they have the opportunity to interact in some ecologically meaningful way, such as sharing and competing for the same resources, or through predator/prey relationships, and so forth, i.e., in some way constraining the fundamental niches of one another.

Much of the pioneering work in defining lizard communities has been done by Pianka and his coworkers (see especially Pianka, 1986 for a review). A study not only of species diversity, but relative abundance, resource partitioning in time and space, microhabitats, etc., is necessary to begin to understand the structure of lizard communities. Further, one must also consider other animals which may compete with, prey on, and serve as food for lizards. These include insectivorous birds and large arthropods, such as scorpions, solpugids, and hunting spiders among potential competitors.

At this point it is possible only to make a bare beginning at describing syntopy among Iranian lizards. While the most abundant diurnal species in a particular biotope can usually be identified in a short period, the rarest species may go undetected even after intensive investigation. I lived in Masjed Soleyman in Khuzestan for nine months without personally encountering all lizard species known from that general area. Here I will attempt to enumerate syntopic species for only a few of the best-collected biotopes from representative geomorphological provinces, with the expectation that these lists will be extended with future investigations.

Masjed Soleyman (Plate 21A–C)

The herpetofauna of this region was reported by S. Anderson (1963). Collecting was done over a nine-month period in this area in the foothills of the Zagros Mountains in Khuzestan Province in 1958. Details of the geography of this region were provided by S. Anderson (1963:417–419, fig. 1), but in brief, the foothill environment is set in a series of anticlinal hills and synclinal valleys from 150–1500 m. The formations are limestones, marls, and gypsum, the principal substrates of the biotope are alluvial fill in the valleys and exposed limestone outcrops on the hillsides. Early rains begin in November, March usually being the end of the rainy season. Spring temperatures are mild, but summer temperatures become increasingly high, with little nighttime cooling. Substrate temperatures exceed 50°C before midday in summer and remain high until late in the afternoon. The vegetation has been much degraded by human activities and is characterized by little woody vegetation, other than occasional thorny shrubs and widely scattered small trees. All arable land is under cultivation for wheat and barley during the winter and spring, and is barren the rest of the year. Annual herbaceous plants of a Mediterranean character proliferate in the cultivated fields and along seasonal water courses during the spring. The lizards which occur syntopically throughout this widespread biotope are: *Laudakia nupta nupta, Trapelus agilis, Eublepharis angramainyu, Asaccus elisae, Cyrtopodion scabrum, Hemidactylus persicus, Tropiocolotes helenae helenae, T. persicus bakhtiari, Mesalina watsonana, Mabuya aurata septemtaeniata,* and *Varanus griseus. Uromastyx loricatus* occurs locally at low elevations. In addition, *Ophisops elegans* may occur syntopically with some of these species at some localities. Thus, the lizard community of this biotope consists of 11–12 species at any one locality, *Laudakia, Trapelus,* and *Mesalina* being the most abundant diurnal species. Half of the species are nocturnal.

Kupal Dunes (Plate 21F)

This area was visited twice, once in 1958 and once in 1975. It is an area of active, wind-blown sand dunes, 1 km east of Kupal along the Ahvaz-Behbehan road at 31°16′N, 49°10′E, 25–50 m elevation. It is representative of a biotope locally developed at several places on the Ahvaz plains of Khuzestan. A few shrubs grow on the dunes and on the sandy fringes; some grasses and thorny shrubs are found in the "blow-out" depressions, that is depressions hollowed out by winds. The climate is Mediterranean, with winter rains, subtropical, hot, and arid. Lizard species occurring syntopically here: *Trapelus persicus, Bunopus tuberculatus, Stenodactylus doriae, Acanthodactylus schmidti, Scincus scincus conirostris,* and *Varanus griseus*. The amphisbaenid, *Diplometopon zarudnyi,* also occurs here. *Phrynocephalus arabicus* and *Stenodactylus affinis* have been collected in similar habitat 21 km north of Ahvaz, and might be expected to occur in the Kupal dunes, as well. The only other sand-adapted lizard known for the Ahvaz Plain is *Acanthodactylus grandis*; these were abundant in a more stabilized sandy area near Bushire.

Shah 'Abbas Caravanserai (Plate 23E)

The Shah 'Abbas Caravanserai, 34°44′N, 52°10′E, about 1,000 m elevation, near the foot of Siah Kuh, has served as a headquarters for biological field investigations of the fauna of the Kavir Protected Region. It is located on the Central Plateau in Tehran Province. I collected here on 13 June 1975 and examined specimens brought to the museum in Tehran (MMTT) by other zoologists over several years. The substrate here is dry, gravelly alluvium on flats and slopes, the vegetation shrubby steppe (dominated by *Artemisia herba-alba*). Elsewhere within the near vicinity are other surfaces, sandy or salty to various degrees (see

Nilson and Andrén, 1981, for a more detailed description of this region). I collected only two species of lizards during my brief visit, *Trapelus agilis* and *Agamura persica*, but the following are also known from this biotope at or near this locality: *Phrynocephalus scutellatus, P. maculatus, Uromastyx asmussi, Eremias andersoni, E. persicus, Mesalina watsonana Ophiomorus nuchalis,* and *Varanus griseus caspius* (see also Nilson and Andrén, 1981). Not all of these species may be present on all substrates, with *Phrynocephalus maculatus* and *Eremias andersoni* preferring sandy surfaces, and *P. scutellatus* tolerating saltier soils than most other lizard species.

Abandoned Village in Sistan (Plate 21B)

At various locations on the flat alluvial plain in Sistan, there are abandoned villages swamped in fine-grained eolian loess deposits. Vegetation around the village consisted of scanty shrubs, probably *Salsola* sp., although some wet-rice cultivation was carried out at the edge of the village. In addition to the drifts of loose material, this biotope consists of the inner and outer walls of the mud-brick buildings, some of these surfaces still covered with plaster. I collected at one such locality at 31°03'N, 61°38'E, 450 m elevation on 27–28 April 1975. The following species are syntopic at this locality: *Trapelus agilis, Bunopus tuberculatus, Cyrtopodion longipes, Teratoscincus bedriagai, T. microlepis, T. scincus keyserlingii, Ophiomorus tridactylus, Eremias acutirostris,* and *E. lineolata*. In addition to these nine species all living in close proximity to one another (within about 50 m linear distance), *Phrynocephalus maculatus maculatus* was collected on the flat plains nearby. Particularly noteworthy is the occurrence of all three species of *Teratoscincus* living syntopically (S. Anderson, 1993).

Hyrcanian Forest (Plate 23A–C)

I collected in the broadleaf forest (described above in the section on the Caspian [see page 43]) 26 km south of Amol on the road to Tehran, 36°17'N, 52°21'E, 450 m elevation on 28 May 1975, in (former) Mohammad Reza Shah National Park at two localities on 30 May 1975, 37°21'N, 56°01'E, 450 m elevation and 37°21'N, 56°12'E, 800 m elevation, and at 39 km south of Chalus on the road to Karaj, 36°19'N, 51°16'E, 600 m elevation, on 20 June 1975, all in Mazanderan Province. Only *Anguis fragilis, Ophisaurus apodus, Lacerta chlorogaster,* and *L. strigata* are known to be syntopic in this forest biotope.

Without wishing to belabor the obvious, I want to point out that syntopic occurrence presents us with a hypothesis of some importance. This is that the fundamental niche of any one species in a lizard community is constrained by the presence of other members of this community. This is based on the notion that the realized niche is determined by resource partitioning through competition and predation. However, in practice it is extremely difficult to test this hypothesis. Comparing the autecology of a species as a component of two or more lizard communities of differing composition can raise suggestive questions, but does not test the hypothesis, because the resource matrix is bound to differ amongst the communities studied. Removal/exclusion/replacement manipulation of communities might provide a test of the hypothesis, but such experiments are difficult in lizard communities. Pianka (1986:75–90) discussed theoretical and practical aspects of this problem and reviewed the literature.

Distributional Patterns of Iranian Lizards in Southwest Asia

A number of taxa, primarily at the generic level, should prove particularly instructive in any attempt to understand the historical biogeography of Southwest Asia. The genera *Laudakia, Eublepharis, Cyrtopodion, Ablepharus,* and *Ophiomorus* have their primary diversity in the elevated region stretching from western Turkey to the Himalayas. *Uromastyx, Ptyodactylus, Stenodactylus, Acanthodactylus, Chalcides,* and *Scincus* are genera of the low deserts stretching across North Africa, Arabia, and into Pakistan; *Lacerta* is a temperate western-Palearctic genus, which extends into the elevated regions of northwestern Southwest Asia and, *sensu lato*, via the Zagros, into the mountains of Oman. *Phrynocephalus, Crossobamon, Teratoscincus,* and *Eremias* are equally Central Asian and arid central Iranian Plateau in their distribution and diversity, their phylogenetic history reflecting the geomorphic history of the isolated interior basins of these regions. *Tropiocolotes* is similarly diverse in the upland areas of Southwest Asia and in lowland North Africa and Arabia. *Trapelus, Bunopus, Mesalina,* and *Ophisops* have their greatest diversity in the southern lowlands of Southwest Asia, yet each has one or more species which are widely distributed through the uplands and even into Central Asia. Four species, *Trapelus agilis* (including *T. sanguinolenta*), *Mesalina guttulata/watsonana* complex, *Ophisops elegans* and *Varanus griseus* are the most widely distributed lizards in Southwest Asia. Indeed, *V. griseus* is probably the most widely naturally distributed lizard in the world, having a range through the entire Palearctic desert region. Its foraging behavior, as well as its range, is more mammal-like than lizard-like. *Pristurus* is a genus of the coastal fringes of the Arabian peninsula and the Red Sea coasts of Africa. *Asaccus* is unique among lizard genera in sharing its diversity between the Zagros foothills and the upland areas of Oman and the United Arab Emirates. *Agamura, Carinatogecko,* and *Rhinogecko* are nominal gecko genera of restricted distribution, but also of disputed taxonomic status. The following genera are either Holarctic or pan-tropical, some of them discontinuously distributed: *Ophisaurus* (*sensu lato*), *Hemidactylus, Eumeces,* and *Mabuya*.

Relationships of the Iranian Lizard Fauna to Neighboring Areas

Iran shares the following species with adjacent regions:

Transcaucasian Republics. *Trapelus ruderatus, Phrynocephalus persicus, Anguis fragilis, Ophisaurus apodus, Cyrtopodion caspium, Eremias arguta, E. pleskei, E. strauchi, E. velox, Lacerta brandti, L. chlorogaster, L. defilippii, L. media, L. praticola, L. raddei, L. strigata, L. valentini, Ophisops elegans, Ablepharus bivittatus, A. pannonicus, Eumeces schneideri, Mabuya aurata.*

Turkey. *Laudakia caucasia, Phrynocephalus persicus, Trapelus ruderatus, Anguis fragilis, Ophisaurus apodus, Asaccus elisae, Cyrtopodion heterocercum, C. kotschyi, Hemidactylus turcicus, Eremias pleskei, E. strauchi, Lacerta cappadocica, L. media, L. praticola, L. princeps, L. raddei, L. strigata, L. valentini, Ophisops elegans, Chalcides ocellatus, Eumeces schneideri, Mabuya aurata, M. vittata, Varanus griseus.*

Iraq and Kuwait. *Laudakia nupta, Phrynocephalus arabicus, Trapelus agilis, T. persicus, T. ruderatus, Uromastyx aegyptius, U. loricatus, Ophisaurus apodus, Eublepharis angramainyu, Asaccus elisae, A. griseonotus, Bunopus tuberculatus, Carinatogecko heteropholis, Cyrtopodion scabrum, C. heterocercum, C. kotschyi, Hemidactylus flaviviridis, H. persicus, H. turcicus, Stenodactylus affinis, S. doriae, Acanthodactylus grandis, A. schmidti, Lacerta cappadocica, L. media, L. princeps, Mesalina brevirostris, M. watsonana, Ophisops elegans, Ablepharus pannonicus, Eumeces schneideri, Mabuya aurata, M. vittata, Scincus scincus, Varanus griseus.*

Lowland Gulf Coastal Arabian Peninsula. *Trapelus persicus, Phrynocephalus arabicus, P. maculatus, Uromastyx aegyptius, Bunopus tuberculatus, Cyrtopodion scabrum, Hemidactylus flaviviridis, H. persicus, H. turcicus, Pristurus rupestris, Stenodactylus* *doriae, Acanthodactylus schmidti, Mesalina brevirostris, Chalcides ocellatus, Mabuya aurata, Scincus scincus, Varanus griseus.*

Oman and Mountains of United Arab Emirates (UAE). *Phrynocephalus maculatus, Bunopus tuberculatus, Cyrtopodion scabrum, Hemidactylus flaviviridis, H. persicus, H. turcicus, Teratoscincus scincus, Acanthodactylus blanfordi, Ablepharus pannonicus, Chalcides ocellatus, Mabuya aurata, Varanus griseus.* This limited faunal relationship with Iran is particularly interesting, as the presence of *Phrynocephalus maculatus, Teratoscincus scincus, Acanthodactylus blanfordi,* and *Ablepharus pannonicus*, as well as the presence of species of *Asaccus* and *Lacerta* (*sensu lato*) in the mountains of Oman signal periods of contact of the mountains and lowlands of Oman and the UAE with the Zagros Mountains and the plateau areas of eastern Iran across the area now occupied by the Persian Gulf.

Pakistan. *Calotes versicolor, Laudakia caucasia, L. nupta, Phrynocephalus maculatus, P. ornatus, P. scutellatus, Trapelus agilis, T. ruderatus, Uromastyx asmussi, Agamura persica, Bunopus tuberculatus, Cyrtopodion agamuroides, C. kachhense, C. scabrum, Hemidactylus flaviviridis, H. persicus, H. turcicus, Rhinogecko femoralis, Teratoscincus microlepis, T. scincus, Acanthodactylus blanfordi, A. micropholis, Eremias acutirostris, E. fasciata, E. persica, E. scripta, Mesalina brevirostris, M. watsonana, Ophisops elegans, Ablepharus pannonicus, Chalcides ocellatus, Eumeces schneideri, E. taeniolatus, Ophiomorus blanfordi, O. brevipes, O. tridactylus, Varanus bengalensis, V. griseus.*

Afghanistan. *Calotes versicolor, Laudakia caucasia, L. erythrogastra, L. nupta, Phrynocephalus maculatus, P. mystaceus, P. ornatus, P. scutellatus, Trapelis agilis, T. ruderata, Uromastyx asmussi, Ophisaurus apodus, Agamura persica, Bunopus tuberculatus, Crossobamon eversmanni, Cyrtopodion caspium, C. longipes, C. scabrum, Hemidactylus flaviviridis, Teratoscincus bedriagai, T. microlepis, T. scincus, Eremias acutirostris, E. fasciata, E. grammica, E. intermedia, E. lineolata, E. nigrocellata, E. persica, E. scripta, E. velox, Mesalina watsonana, Ablepharus bivittatus, A. pannonicus, Eumeces schneideri, E. taeniolatus, Mabuya aurata, Ophiomorus tridactylus, Varanus bengalensis, V. griseus.*

Turkmenistan. *Trapelus agilis, Laudakia caucasia, L. erythrogastra, Phrynocephalus helioscopus, P. interscapularis, P. mystaceus, P. raddei, Ophisaurus apodus, Eublepharis turcmenicus, Bunopus tuberculatus, Cyrtopodion caspium, C. longipes, C. spinicauda, C. russowii, Crossobamon eversmanni, Teratoscincus scincus, Eremias arguta, E. grammica, E. intermedia, E. lineolata, E. persicus, E. scripta, E. strauchi, E. velox, Mesalina watsonana, Ophisops elegans, Ablepharus pannonicus, Eumeces schneideri, E. taeniolatus, Mabuya aurata, Varanus griseus.*

Paleogeographical Influences on Lizard Distribution

The known tectonic history and paleogeography of Eurasia during the late Paleozoic (Permian) through the Mesozoic has been provided by Sengör (1984) and only a brief synopsis, as pertains to our immediate areas of interest, will be presented here. During the early and middle Mesozoic, the Tethys consisted of two ocean areas separated by a strip or a string of continent(s), called the Cimmerian Continent, which had begun separating from the northern and northeastern margin of Gondwanaland during the Triassic (rifting had begun earlier). Paleo-Tethys, the east-facing embayment of Permo-Triassic Pangaea, began shrinking as the Neo-Tethys was evolving south of the Cimmerian Continent. The double obliteration of this Tethyan realm from the Late Triassic to the present has produced the Alpine-Himalayan system of orogenic belts. This system was produced in two parts, the first orogen, the Cimmerides, resulted from the closure of the Paleo-Tethys, and was largely overprinted from the eastern Carpathians through Afghanistan by the second, or Alpides, the product of the obliteration of the Neo-Tethys.

The terminal suturing of the Ghaznian, or Southwest Asian, Cimmerides (the segment between Turkish and Pamir orogens) occurred from the Late Triassic to the Early Jurassic. In the northwest, the connection between the eastern Greater Caucasus and the Talysh Mountains is interrupted by a narrow, fault-controlled basin in the Khoura depression, marked today by the Araxes Valley. The Khoura depression is a westerly continuation of the South Caspian depression, probably an extensional basin of early Cenozoic age, floored by modified oceanic lithosphere. The Paleo-Tethyan history of the Alborz Mountains was characterized by deformation during the Middle Triassic to Early Jurassic, along with sparse Permo-Triassic magmatism. The Kopet Dagh-Paropamisus-Western Hindu Kush orogenic belt rimming the Turan area on the south was growing during the late Paleozoic. Turan (the southern Central Asian block) and central Iran were united along the Mashhad suture by the Early Jurassic.

The Ghaznian Cimmerides formed one continuous orogenic belt to the Paropamisus before the Cenozoic disruption forming the Herat basin and the Harirud fault intervened. During the Triassic, the Paropamisus and the western Hindu Kush were dominated by strong subsidence and continued volcanism. Along with orogenic events, marine sedimentation continued into the Triassic, ending in the Late Triassic-Early Jurassic and immediately succeeded by uplift. (Sengör, 1984:26–28).

The Gondwanaland fragments south of the Paleo-Tethyan suture are characterized by a Pan-African basement and uppermost Precambrian to Middle Triassic Platform/shelf sequences, and central Iran and central and southern Afghanistan were affected by compressional deformation and low-grade metamorphism during the Late Carboniferous and Permian. Thus, Iran as a whole belonged to Gondwanaland during the Late Permian, although there is disagreement as to just where. A remaining question concerning the latest Paleozoic to early Mesozoic history of these regions is in regards to the number of microcontinents that rifted from Gondwanaland to open a number of oceans that are now represented by the numerous ophiolitic sutures in the region. According to Sengör's analysis, the numerous blocks now forming central Iran and Afghanistan south of the Herat-Wanch-Akbaytal lineament have always remained close to one another and separated from Gondwanaland as a single unit, but perhaps with considerable internal mobility. The "Waser Ocean," or Farah Rud Zone, formed within the foreland of the Ghaznian Cimmerides during the Permo-Triassic and separated the Farah Block from the Helmand Block (all of Afghanistan and Pakistan south of the Waser suture) during much of the Mesozoic. Only the Waser Ocean and the Neo-Tethys (now represented by the Waziristan suture) existed as oceanic realms during the Triassic. The closure of the Waser Ocean took place before the Middle Cretaceous. (Sengör, 1984:30.)

Still controversial is the age of opening of the ocean areas that surrounded the central Iranian microcontinent (Lut Block of some authors). These include the East Iranian Flysch Trough. In Sengör's view, these oceans may have resulted from post-collisional disintegration of the Cimmerian Continent. The East Iranian Flysch Trough persisted until earliest Oligocene times. (Sengör, 1984:31, 58.)

Rage (1988) presented a brief synopsis of broad paleogeographic factors of importance to terrestrial vertebrate distribution during the Mesozoic and early Cenozoic. He concluded that free dispersal of terrestrial vertebrates was possible in Pangaea during the Triassic from the standpoint of terrestrial connections. Nonetheless, fossil evidence indicates that faunal provinciality developed in Gondwana. During the Jurassic, there continued to be land communication between Gondwana and Laurasia. A Gondwanan pattern appeared during the Cretaceous, but it has not been possible to demonstrate a global difference between Gondwanan and Laurasian faunas during the Cretaceous. During the Cretaceous/Eocene interval there were intermittent crossings of the Tethys (although evidence for interchanges between Eurasia and Africa is still poor), and by the Cretaceous/Paleocene transition, there was probably already a terrestrial route between Laurasia and the still-southern Indian Plate.

Factors of Late Cretaceous through Quaternary paleogeography of Southwest Asia of importance to the historical biogeography of this region have been summarized by Wolfart (1987) and, with particular reference to mammalian zoogeography, by Tchernov (1988, citing particularly Adams et al., 1983; Bernor, 1983; Garfunkel and Bartov, 1977; Gvirtzman and Buchbinder, 1977, 1978; Horowitz, 1979; Madden and Van Couvering, 1976; Por, 1978; Rögl and Steininger, 1983; Thomas, 1985). The geographic evolution of this region during this period was dominated by the northward drift of various parts of the Gondwana continental complex gradually narrowing the Tethys Sea which connected the region of the present Mediterranean Sea and the Alpine region with the Indian/Pacific Oceans via southern Turkey, Syria, Iraq, Afghanistan, and Pakistan. See Robertson (1987) for factors contributing to the formation of the Oman Mountains and the Gulf of Oman. Seventy million years ago (mya) the Tethys was up to several thousand kilometers wide and separated pieces of the former Gondwanaland continent, namely Africa, Nubo-Arabia, and India, from the continents to the north. Thus, a distinct biogeographical provincialism must have existed in the Gondwana fragments and the northern continents. (Wolfart, 1987.)

In the transition from the Cretaceous to the Tertiary there were orogenetic activities and widespread regression, yet the relative positions and isolation of land areas of the Paleocene were not much different from the Late Cretaceous. This geography continued during Eocene and Early Oligocene, although tectonic activities increased during these periods due to the progressive approach and subduction of the Nubo-Arabian and Indian Gondwanian plates towards the northern continents. Towards the end of the Oligocene, Southwest Asian paleogeography was radically altered: the Tethys had largely disappeared, fragments of the Gondwanaland continent had been welded to the northern continents, although subduction continues to the present time. The central basin platform zone, which today constitutes the greater part of the Iranian Plateau, consists of small stable terranes which were long ago detached from Gondwanaland. Afghanistan, south of the Paropamisus and the northern Pamir ranges, as well as the neighboring crustal segments of Iran and the southern Pamir, were part of Gondwanaland and separated from the northern continents by the Tethys until towards the Early Tertiary. Northern Afghanistan, and probably northernmost Iran, must be considered part of the Eurasian plate. (Wolfart, 1987.)

Following the Late Oligocene regression, Southwest Asia was dominated by a continental environment during the Miocene. Marine sedimentation continued in the Euphrates-Persian Gulf furrow, the Taurus, Zagros, and Makran troughs, and the shelf of the Nubo-Arabian Shield. Miocene continental sediments occur in the northwestern shelf area of the Indian Shield and in numerous closed basins of Afghanistan, Iran, and southern Turkey. The beginning of extensive faunal exchanges between Asia and Africa, as indicated by numerous localities of continental Miocene faunas, demonstrate that the Tethys must have been closed, at least locally, towards the Middle and Late Miocene. An initial land bridge may have been present in the area of western Arabia, Jordan, and western Syria. (Wolfart, 1987.)

Thus far, one of the few herpetological systematic studies attempting to correlate Nei distances and geochronology of Southwest Asia is that of Mayer and Benyr (1994). Based on albumin evolution data, they see a division of their subfamily Lacertinae some 23–24 mya into two clades, a *"Lacerta"* line and a second lineage containing the other taxa. This second clade gave rise to the radiations of principal interest for Southwest Asian biogeography about 21 mya, an "African lineage" and a "Eurasian lineage." These lineages then radiated within their respective continents, *Acanthodactylus* originating within the African radiation some 15–16 mya, the various other clades now inhabiting Southwest Asia having diverged 12–17 mya.

Continental conditions have continued from the Late Miocene until the present in most parts of Southwest Asia. Only small areas were inundated by marine transgressions during the Pliocene, although marine sedimentation continued in the Zagros trough and its eastern continuation in the Makran-Baluchistan coastal regions and in gulf coastal Arabia in the area of the present United Arab Emirates. (Wolfart, 1987.)

In the large, closed, interior basins that exist everywhere throughout Southwest Asia, lacustrine and fluviatile deposits were formed under the arid conditions of the Quaternary (Wolfart, 1987).

Undoubtedly, the climatic fluctuations of the Quaternary have had a profound influence on the makeup and distribution of present-day faunal communities. Reliable data for older climatic epochs of the Quaternary are scarce, especially for arid areas. However, it is certain that the Quaternary is characterized by climatic variations with temperature differences of several degrees, causing glaciations and deglaciations in higher latitudes and altitudes. In Southwest Asia, boundaries between climatic zones were also shifted both vertically and horizontally. The depression of the snow limit in the northern Hindu Kush has been assumed to be up to 1,000 m during the Würm glaciation and between 500 and 800 m in Lebanon, on the basis of the presence of terminal moraines. While there is little question as to the cooling of the areas to the south of the glacial regions, the amount of precipitation is still a matter of dispute. The European glacial phases, however, cannot be correlated with "pluvials" in Southwest Asia. Although there are strandlines of Pleistocene lakes within the internal basins, it is certain that stream erosion was never sufficient to cut outflow channels which would open these basins to external drainage. (Wolfart, 1987; Fisher, 1968.)

The Persian Gulf is a broad, shallow epicontinental sea with a maximum depth of 110 m and a length of nearly 1,000 km. It is a shallow tectonic depression formed late in the Tertiary in front of the rising Zagros front. The water is deepest near the steeper Iranian slope, the slope of the Arabian side being much gentler (Kassler, 1973). Sea level dropped about 100 m during the Würm Period, and consequently the Persian Gulf dried up and the shoreline was in the Gulf of Oman with an estuary in the area of Hormoz (Felber, *et al.*, 1978; Wolfart, 1987). River valleys were then eroded down the slopes and the sea cut a series of platforms at its level of maximum retreat and at times of relative standstill during its postglacial rise (Kassler, 1973). This may have been only the most recent of several evacuations of the gulf during the Quaternary. And while this structural trough may have continued to serve as the drainage for the Tigris-Euphrates system, it is by no means certain that this drainage always exited to the Arabian Sea via the Gulf of Oman. Had the river drainage been diverted at times south of the Oman mountains, habitats might have been continuous across the area of the present Strait of Hormoz.

Thus, of herpetogeographic interest is less the presence of a few faunal elements (*Phrynocephalus maculatus, P. arabicus, Asaccus* species, *Teratoscincus scincus, Lacerta* species) in the eastern Arabian Peninsula in common with the Zagros mountains and the interior Plateau of Iran, but rather the absence of a more general and extensive exchange. Climatic and substrate limitations must have served to filter such an exchange, which seems to have been two-way, *Uromastyx, Mesalina* species and *Acanthodactylus* species entering the plateau from Arabia.

Chapman (1978) has summarized the geomorphology of the Oman Mountains (Al Hajar) and the Musandam Peninsula. These are rugged mountains that reach a maximum elevation of 3,000 m in the central part of the range. Northwards, the range becomes narrower and lower, plunging beneath the Strait of Hormoz at the Musandam Peninsula. The steep continental margin on the northeast side of the Oman Mountains drops abruptly into the Gulf of Oman basin. It is believed that the Oman Mountains started to rise in the late Oligocene (about 30 mya) and that the Gulf of Oman started to sink at the same time. Considerable vertical movement has produced the difference between the highest peak at 3,000 m above sea level and the center of the Gulf of Oman, 4,000 m deep. The major valleys in the Oman Mountains were probably initiated in the Miocene and experienced their greatest erosion in the Pliocene. Small canyon-like valleys may be Pleistocene in age. The Musandam Peninsula is notable for its drowned valleys. Isthmuses, and narrow straits which must have been isthmuses before their rupture by the sea, characterize this peninsula. It is estimated that the subsidence of the north and east coasts of northern Oman has exceeded 60 m during the last 10,000 years.

Physiographic and Climatic Barriers to Distribution

In summary, then, of the present physiographic barriers to and/or corridors of distribution, the interior basins are the most ancient portions of Iran, Afghanistan, and Central Asia, having been terrestrial environments from at least the Late Cretaceous to the present and in contact with one another since the Miocene. The many ranges of the interior plateau are the result of orogeny resulting from the uplift of the Iranian Plateau in the broad sense as a consequence of the collision of the Gondwanian plates with the northern continent, and their extents and elevations must have fluctuated considerably during the Tertiary and Quaternary. The extent of flatland steppe and desert habitats, undoubtedly influencing the diversification of such genera as *Eremias* and *Phrynocephalus*, must have varied greatly during most of the period of existence of interior basins, particularly during the periods when lacustrine deposits accumulated. The Alborz chain, having a long and complex orogeny, has also probably had a continuous terrestrial existence since the Miocene, although its status as a series of mountain ranges high enough to affect climate and to serve as a barrier between faunal areas is much more recent, probably only since the Pliocene. The Zagros range and its foothill belts received marine, and, later, terrestrial sediments, until the Pliocene/Quaternary. Their present elevations and influence on climate date only from the Pleistocene. The Persian Gulf, long a barrier between the Arabian Shield and the eastern interior plateau, was evacuated by the sea one or more times during the Pleistocene, allowing faunal contact between the eastern Iranian Plateau and eastern Arabia.

SYSTEMATIC SECTION

KEY TO THE LIZARD FAMILIES OF IRAN

1a. Fore- and hindlimbs present 2
1b. Fore- and hindlimbs absent, or hindlimbs minute vestiges without digits ....Anguidae (p. 111)
2a. Eyelids present 3
2b. Eyelids absent 8
3a. Scales on snout and top of head large; scales on venter usually large 4
3b. Scales on top of head mostly small 5
4a. Ventral scales usually larger than most scales on dorsum and flanks, sometimes with single median ridge or keel; femoral pores present .... Lacertidae (p. 199)
4b. Ventral scales usually about the same size as scales on dorsum and flanks and usually highly polished; femoral pores absent .... Scincidae (p. 262)
5a. Large lizards, adults often exceeding 1 m in total length; snout elongate, nostrils nearer eye than tip of snout, directed backwards and downwards; tongue long, slender, and deeply forked at tip (Fig. A) .... Varanidae (p. 295)
5b. Head short, blunt; snout short, nostrils nearer tip of snout than eye; tongue broad, fleshy, not deeply forked (Fig. B) 6

Tongues of (A) *Varanus bengalensis* and (B) *Calotes versicolor* (From Smith, 1935:18)

6a. Skin soft, body covered by small, rounded tubercles .... Eublepharidae (p. 116)
6b. Skin rough, body covered by imbricate scales, mostly keeled and usually sharply pointed behind 7
7a. Femoral pores present; tail strongly depressed throughout most of its length, shorter than snout-vent length, covered above by whorls of large spinous tubercles which are rounded at their bases .... Uromastycidae (p. 289)
7b. Femoral pores absent; tail not strongly depressed, except sometimes at base, longer than snout-vent length unless broken, without whorls of large, spinous tubercles rounded at base (large, keeled, mucronate scales may be arranged in annuli, however) ....Agamidae (p. 67)
8a. Top of head covered by large shields 9
8b. Top of head covered by tubercles; ventral scales granular or cycloid and imbricate, never juxtaposed and rectangular in shape ....Gekkonidae (p. 125)
9a. Ventral scales quadrangular in shape .... Lacertidae (*Ophisops*) (p. 254)
9b. Ventral scales like scales on dorsum .... Scincidae (*Ablepharus*) (p. 263)

Family AGAMIDAE
Agamas, agamids

The most recent revision of this family is that of Moody (1980), who recognized six major groups within the family. *Uromastyx*, along with *Leiolepis*, constitutes Group I, *Calotes* with other tropical arboreal agamids of South Asia, belong to Group V, and the other genera found in Southwest Asia and Africa belong to Group VI, in which *Agama*, formerly a widely-inclusive genus, is divided into six genera. *Agama* itself is redefined and restricted to Africa. Various authors subsequently have followed Moody's nomenclature, some recognizing his genera, others using his genera as subgeneric taxa. Leviton, *et al*. (1992:12–13) introduced a modified scheme of classification, which incorporated the findings of Joger (1991), Ananjeva and Sokolova (1990), and Macey, *et al*. (abstract, 1995) in an attempt to maintain a phylogenetic classification. I follow that scheme in this work. I define the genera within the limits of the six major groups specified by Moody (1980:110–145).

Estes, *et al*. (1988) pointed out that Agamidae is a metataxon composed of all Acrodonta except chamaeleonids, which are derived from agamids. Therefore, Agamidae (Gray, 1827) is a paraphyletic taxon, unless it includes chameleons, in which case it should be designated Chamaeleonidae (Gray, 1825), or unless it is restricted to exclude leiolepids (including *Uromastyx*) as well as chameleons. In fact, Moody (1980:174) had recommended the recognition of the family Uromastycidae, including the genera *Uromastyx* and *Leiolepis*. However, see Barsuk-Białynicka and Moody (1984), in which *Uromastyx* and *Liolepis* are regarded as one of three subfamilies of the Agamidae. I have followed Moody's (1980) arrangement here.

Key to the Genera of Agamidae in Iran

1a. Tympanum concealed or absent .... *Phrynocephalus* (p. 81)
1b. Tympanum exposed 2
2a. Well-marked dorsal crest (at least a row of erect scales on neck) .... *Calotes* (p. 68)
2b. No dorsal crest 3
3a. Caudal scales obliquely arranged, not forming annuli; tympanum small, its diameter less than half that of orbit, more or less deeply sunk (Fig. A) .... *Trapelus* (p. 99)
3b. Caudal scales forming more or less distinct annuli; tympanum usually larger than eye, its diameter at least half that of orbit, superficial (Fig. B) .... *Laudakia* (p. 69)

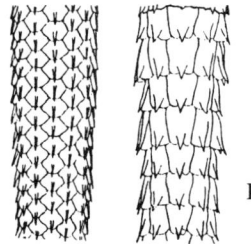

Dorsal surface of tail segments of
(A) *Trapelus agilis* and (B) *Laudakia caucasia*
(From Terentjev and Chernov, 1949, fig. 45)

Genus *Calotes* Cuvier, 1816
Variable lizards

Calotes Cuvier, 1817 (1816):35 (type species: *Lacerta calotes* Linnaeus, 1758, by absolute tautonymy).

Definition: Body usually compressed; dorsal scales regular, uniform in most species; a dorsal crest more or less developed, at least a row of erect scales on neck; gular sac usually present; an oblique fold or pit in front of the shoulder present or absent; tympanum naked. Tail long and slender, usually swollen and rounded at base in adult male, the scales on that part of it enlarged and thickened. No preanal or femoral pores. (Smith, 1935:180.)

Distribution: Southern Asia, from eastern Iran and Afghanistan to Sri Lanka and Sumatra, north to the Himalayan foothills, and east to South China. About 30 species (Wermuth, 1967). A single species in Iran.

Calotes versicolor (Daudin, 1802) (Plate 2A)
Indian garden lizard, variable agama

Agama versicolor Daudin, 1802:395, pl. 44 (Type locality: Pondicherry, India [restricted by Smith, 1935:192]; Holotype: said by Smith [1935:189] to be in MNHN, but not listed by Guibé [1954]).
Calotes versicolor: Fitzinger, 1826:49. — Gray, 1845:243–244. — Blanford, 1876:313–314. — Bedriaga, 1879:40. — Boulenger, 1885a:321–322. — Zarudny, 1903:10. — S. Anderson, 1963:475. — Wermuth, 1967:41. — S. Anderson, 1968:332; 1974:36, 42. — Schleich, 1977:127, 129. — Welch, 1983:18.

Diagnosis: Scales on sides of body pointing backwards and upwards (only slightly so in Afghan specimens examined); no fold or pit in front of shoulders; two separated spines above tympanum; 35–52 scales around middle of body.

Color pattern: Dorsum light sandy or olive through shades of brown to sooty gray; usually a series of light transverse bars and light dorsolateral stripes, more marked in juveniles; head and shoulders of males suffused with dull red or orange; throat and chest orange to red with black mottling; venter whitish with dark streaks (Minton, 1966:89, for specimens from Pakistan).

Size: Snout-vent length to 140 mm, tail to 350 mm in males (for Peninsular India) (females are usually slightly shorter than males) (*fide* Smith, 1935:191).

Habitat: Blanford found this lizard only on date palms in southeastern Iran. He felt that these animals may have persisted in date groves of historical antiquity, their presence dating from a period of favorable climate. Distribution to such areas through human agency, however, cannot be discounted. Such introductions might occur intermittently, populations neither long persisting nor becoming widely established. Zarudny also found this species in date groves in Baluchistan. To about 1000 m elevation.

Calotes versicolor

Distribution: In Iran, known only from eastern Baluchistan. This variable species ranges throughout the Indian subcontinent and Sri Lanka, the Andaman Islands, and through Southeast Asia, from Baluchistan and eastern Afghanistan in the west, Nepal in the north, to the northern Malay Peninsula and Sumatra in the south, and east to Hainan and Hong Kong. (Smith, 1935:191; Wermuth, 1967:42).

Remarks: Zarudny stated that both of his specimens (from Farra and Sarbaz) were lost. Comments on the habits of this species are found in Smith (1935:189–193) and Minton (1966:88–89). Differences between Afghan specimens and those from elsewhere in the range of *Calotes versicolor* have been discussed by Clark, *et al.* (1969:295). Auffenberg and Rehman (1993) have analyzed variation in some morphological characters, and have distinguished a population in eastern Afghanistan and northern Pakistan as a distinct subspecies, *C. versicolor nigrigularis* Auffenberg and Rehman, 1993.

Genus *Laudakia* Gray, 1845
Rock agamas, stellions

Laudakia Gray, 1845:254 (Type species: *Agama tuberculata* Hardwicke and Gray, 1827, by monotypy).

Definition: Body depressed; dorsal crest absent; gular sac absent; head depressed, more or less triangular; tympanum large, its diameter at least half that of the orbit, more or less superficial; fold in front of shoulder usually connecting with transverse gular fold; dorsal scales uniform or intermixed with larger ones; toes compressed, fifth extending beyond first; caudal scales forming more or less distinct annuli; males and some females with callous preanal and usually abdominal scales (Fig. 37). (Leviton, *et al.*, 1992:13.)

Distribution: From islands of the eastern Mediterranean and northern Egypt through southwestern and Middle Asia to northern India.

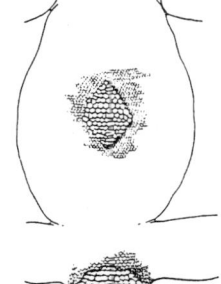

FIGURE 37. Callous preanal and abdominal scales in *Laudakia* sp. (After Smith, 1935).

Remarks: See Leviton, *et al.* (1992:12–13) for a discussion of the substitution of *Laudakia* for *Stellio* and restriction of the generic name to the Asian species previously placed in the genus (or subgenus) *Stellio*, and Henle (1995) for an alternative proposal. Baig and Böhme (1991), in a paper that appeared too late to be referred to by Leviton *et al.* (1992), first pointed out the nomenclatural problems associated with use of the name *Stellio*, and the fact that *Laudakia tuberculata* (the type species of the latter genus) has a number of autapomorphies which set it apart from the other mountain "stellions." Henle (1995), who provides a history of the use of the name "Stellio" in herpetology, proposes restricting the genus *Laudakia* to *L. tuberculata* and the use of the name *Plocederma* Blyth, 1854 (based on *Laudakia* [*Plocederma*] *melanura* Blyth, 1854) for the larger clade. I refrain here from one more taxonomic adventure with this group until systematic work currently underway by J. Robert Macey and others sheds more light on cladistic relationships within the Agamidae. Baig (1992), in his doctoral dissertation, advanced the hypothesis that the origin and subsequent evolution of the genus *Laudakia* centered on the region of the current Pamir Knot. Ananjeva and Tuniyev (1994) have published an hypothesis regarding the historical biogeography and differentiation within the genus *Laudakia*. Rastegar-Pouyani and Nilson (in preparation) are proposing a somewhat different interpretation; Macey, *et al.* (1997) present yet another view, emphasizing species related to *L. caucasia*.

Key to the Species of *Laudakia* in Iran

1a. Dorsal scales homogeneous; flanks without enlarged scales or tubercles; distal two-thirds or more of tail with segments composed of more than two annuli when viewed laterally (anterior portion of tail up to two or three head-widths posterior to vent may have only two annuli per segment), or segmentation indistinct 2
1b. Dorsal scales heterogeneous, back and usually flanks with scales of varying sizes intermixed 3
2a. Median dorsal scales in straight longitudinal series, 6–10 across middle of back, grading into dorsolateral scales; hemipenes of male unpigmented ....
....*Laudakia melanura lirata* (Blanford, 1874) (p. 75)
2b. Median dorsal scales in oblique longitudinal series, 16–20 across middle of back, clearly set off from dorsolateral scales; hemipenes of male black ....
....*Laudakia nupta* (De Filippi, 1843) (p. 78)
3a. One or two longitudinal rows of clusters of spiny tubercles on each side of body (Fig. A); 80–114 scales round middle of body; gular scales strongly keeled (weakly keeled in small juveniles), mucronate ....*Laudakia erythrogastra* (Nikolsky, 1896) (p. 73)
3b. Enlarged scales on flanks not arranged in longitudinal rows (Fig. B); 115 or more scales round middle of body; gular scales smooth, not mucronate 4

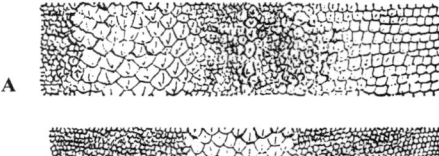

Dorsal scales of (A) *Laudakia erythrogastra*; (B) *L. caucasia* (From Bannikov, et al., 1977, figs. 28.4 [A] and 28.3 [B])

4a. Males with 115–188 scales around middle of body, females with 119–174 ....
.... *Laudakia caucasia* (Eichwald, 1831) (p. 70)
4b. Males with 177–235 scales around middle of body, females with 190–259 ....
.... *Laudakia microlepis* (Blanford, 1874) (p. 76)

Laudakia caucasia (Eichwald, 1831)
Caucasian agama, northern rock agama

Laudakia caucasia caucasia (Eichwald, 1831) (Plate 2B–C)
Caucasian agama, Northern rock agama

Lacerta muricata Pallas, 1814:20–21, pl. 4, fig. 1 (*nec* Shaw, 1801; Type locality: Moghan Steppe. Type(s): not located).
Stellio caucasius Eichwald, 1831:3:187 (Type locality: Tiflis and Baku "et alibii," Transcaucasia, here restricted to Baku, Azerbaijan; Type(s): not located).
Stellio caucasicus: De Filippi, 1865:352. — Blanford, 1876:322–326, pl. 20, fig. 1. — Bedriaga, 1879:38. — Nikolsky, 1899b:391. — Zarudny, 1903:11. — F. Werner, 1936:200. — Welch, *et al.*, 1990:58.

---

* Asterisks before names indicate extralimital species that may occur in Iran.

Stellio persicus J. Anderson, 1872:382–384, text-fig. 4 (Type locality: "Teheran," Iran; Holotype: ZSI 4830 [see Das, *et al.*, 1998]).
Agama caucasica: Boulenger, 1885a:367 (in part); 1889a:96 (in part). — F. Werner, 1903:340. — Nikolsky, 1907a:273; 1915:130–135, pl. 4. — Rostombekov, 1938:11–12. — Terentjev and Chernov, 1949:144–145. — Forcart, 1950:145 (in part). — Wettstein, 1951:267. — S. Anderson, 1963:475. — Clark, *et al.*, 1966:5. — S. Anderson, 1968:332. — Tuck, 1971a:54 (in part). — Ananjeva and Orlova, 1979:4–17, fig. 2. — Schleich, 1979:248–252, figs. 10–13. — Clark, 1991:37.
Agama microlepis: Schmidt, 1939:57 (in error). — Guibé, 1966a:98 (in part).
Agama caucasica caucasica: Wermuth, 1967:11. — S. Anderson, 1974:37, 42. — Bannikov, *et al.*, 1977:113. — Schleich, 1977:127, 129. — Andrén and Nilson, 1979:340.
Agama caucasia: Ananjeva and Orlova, 1979:4. — Orlova, 1981:136–148, figs 23–24. — Welch, 1983:16.
Stellio caucasius caucasius: Ananjeva and Ataev, 1984:5.
Laudakia caucasica caucasica: Leviton, *et al.*, 1992:12 (by implication). — Frynta, *et al.*, 1997:7–8.

Diagnosis: Tail divided into distinct segments, each composed of two whorls of scales; gular scales smooth; patch of enlarged scales on middle of flank; males with 115–188 (mean 145.1, $N = 43$) scales around widest part of body, females with 119–174 (mean 150.0, $N = 28$).

Color pattern: (Fig. 38.) In alcohol, initial fixation in formalin, basic ground color light olive to dark gray; head and tail often, but not always lighter than body, due in part to fact that head, and sometimes tail, lack dark markings; limbs without pattern, same as ground color of back, occasionally scattering of small dark spots on hind limbs; in some specimens indistinct light crossbars on forelimbs, particularly upper arms; tail often with distinct dark transverse bars, especially distinct in young, but many adults with uniformly colored tail; back with many dark-edged light ocelli, linked in transverse series across back, especially anteriorly, but fading out on vertebral line behind shoulders, black margins forming reticulum over dorsum; enlarged vertebral scales lighter than ground color of back in many, but not all individuals; enlarged scales scattered among small dorsolateral scales often light

FIGURE 38. *Laudakia caucasia*. (Blanford, 1876, pl. 20, fig. 1; G. H. Ford, artist.)

in color. Chest, abdomen, ventral surface of limbs, and often tail of mature males uniform dark gray, at least during breeding season, gular region reticulated dark gray and light yellowish; females and some males (not in breeding season?) with venters yellowish, uniform or mottled with gray.

There is considerable individual variation in color and pattern in these lizards. While this may prove to be geographically correlated, it probably also reflects seasonal variation, metachrosis, and other factors.

In life the ocelli are orange and there is frequently an orange suffusion on the venter.

Size: The largest Iranian male measures 153 mm snout-vent, tail 178 mm; largest Iranian female 152 mm snout-vent. Adults of this size rarely have complete tails. The smallest juvenile examined, collected in early October, is 38 mm snout-vent.

Habitat: (Plates 23C–D, 24B, D–E, 25D) These are mountain and upland lizards, invariably associated with rocky outcrops of sandstone, limestone, and basalt, scree-covered slopes, or with boulders in river beds. In cultivated areas they are seen on brick or rock fences and on abandoned buildings. In some lowland areas, as north of Gorgan, they are found very locally where large angular outcrops occur. Deep fissures or spaces beneath large boulders furnish retreats. In the Alborz Mountains they occur in the upper reaches of the forest and above on the north slope to at least 3600 m. They are present in the mountain ridges along the northwestern Iranian frontier. These mountains now lack forests, the slopes being covered primarily with herbaceous vegetation; deep gorges are present in this area, overgrown with shrubs at their edges. These lizards occur in suitable localities from sea level (locally, near the Caspian Sea) to about 4000 m in the Alborz and the mountains of the northern and eastern Central Plateau. Over most of their range in Iran the vegetation consists of sparse steppe shrubs reduced by overgrazing. See Orlova (1981:142–146) for a discussion of the ecology and natural history of *Laudakia c. caucasia* in the European portion of its range. Ananjeva and Tuniyev (1994:44) stated that *L. c. caucasia* has a semiarboreal mode of life where it occurs in forested regions, such as the gorges along the Kura River, the oak forests of northern Armenia, *Juniperus* and *Juniperus-Pistacia* forests of eastern Georgia and southern and eastern Armenia and the *Juniperus* and *Acer* forests of the Kopet Dag, the mountain range forming the border with Turkmenistan.

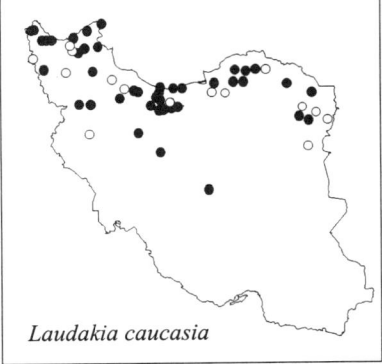

Laudakia caucasia

Distribution: *Laudakia caucasia* occupies almost the entire northern Iranian Plateau in suitable habitats, and is nearly confined to it. It has not been recorded on the western and southern slopes of the Zagros Mountains, where *L. nupta* occupies a similar habitat. In Turkey it is limited to a small area in the northeast bordering the Armenian and Iranian frontiers. The Clarks (1973:19) were able to document its westernmost limits east of Kars, Turkey, where the topography changes abruptly from rich grassland through steep, eroded, rocky hillsides down to open stony valleys. It does not overlap the distribution of *L. stellio*. It occurs in Transcaucasia west to the Suramask Ridge, and the mountainous part of Dagestan. East of the Caspian Sea, it is a mountain species in southern Turkmenistan, southern Uzbekistan, and southern Tajikistan. It occupies the mountainous regions of northern Iran (Alborz, Kopet Dagh), extending south in the inner Zagros regions at least as far as Kohrud in Esfahan Province. In northern

Afghanistan it occupies the area north of the Central Massif and extends south in northeastern Afghanistan to Paghman. It extends into the southern Central Asian republics.

All records of *L. caucasia* for Iraq appear to be in error, based on Schmidt's (1939:57) erroneous identification of specimens of *L. nupta* as this species. With these specimens as reference material in the Field Museum, specimens of *L. nupta* collected by Charles A. Reed and by Neal A. Weber in Iraq were also assigned erroneously to *L. caucasia* (Reed 1956; Reed and Marx 1959:98–99; Weber 1960:153). Nonetheless, it is likely to be discovered eventually in mountainous northeastern Iraq.

Remarks: Ananjeva and Ataev (1984) have described a narrowly distributed taxon, *Laudakia c. triannulata* from an isolated Meshed sand area near the village of Madau in southwestern Turkmenistan. It differs from both *L. c. caucasia* and *L. microlepis* in having three rings per tail segment, rather than two, 173–208 scales around middle of body in males, 167–202 in females. These lizards inhabit the slopes of sand ravines in stabilized ridged dunes. I examined six specimens from 38°10'N, 54°44'E, 14.2 km southwest of Madau (CAS 184640–2, 184650–2); midbody scale counts for three males: 146–186, for three females: 170–183. One male and one female have only two rings per tail segment; in the other four, the tail segments with three rings begin one to two head-lengths behind the vent. The third "ring" is present only on the sides of the tail, consisting of a short row of small scales inserted on the anterior part of the segment. Four other Turkmenistan specimens from 38°26'N, 56°18'E (CAS 179110–3) agree with the diagnosis for *L. c. caucasia* (males 164–174, females 151–180 midbody scales). Panov, *et al.* (1987) examined 218 specimens from seven localities along an 80 km distance between Karakala and Sharlouk in western Turkmenistan. They found populations from the western and central parts of this area to be more variable than those from other regions. They believe these populations represent a hybrid zone of secondary intergradation (I thank Michael Golubev for calling my attention to their paper).

Laudakia caucasia, *L. microlepis*, and *L. erythrogastra*, whose range overlaps the first two species, seem to form a related group. According to J. Robert Macey (pers. commun., 24 March 1997), who is studying the biochemical systematics of *Laudakia*, *L. erythrogastra* is the sister taxon of *L. caucasia*.

Laudakia erythrogastra (Nikolsky, 1896) (Plate 2D)
Khorasan agama, redbelly agama

Stellio erythrogaster Nikolsky, 1896:370–371; 1897:318–319, pl. 19, fig. 1 (Type locality: Kalender-Abad and Ferimun, eastern Iran; Syntypes: ZIL 8759, 8760); 1899:390–391. — Zarudny, 1903:10–11. — F. Werner, 1936:200.

Stellio erythrogaster var. *pallida* Nikolsky, 1897:319–321 (Type locality: Meshed, Iran; Holotype: ZIL 8761).

Agama erythrogastra: Nikolsky, 1915:119–121; 1916:320. — Terentjev and Chernov, 1949:145–147, text-fig. 60. — S. Anderson, 1963:475. — Clark, *et al.* 1966:5. — Wermuth, 1967:12–13. — S. Anderson, 1968:332: 1974:37, 42. — Bannikov, *et al.* 1977:114–115. — Schleich, 1977: 127, 129; 1979:244–249, figs. 8–9, 11–12. — Ananjeva and Orlova, 1979:14. — Welch, 1983:16. — Clark, 1991:37.

Agama erythrogastra pallida: Nikolsky, 1915:121.

Agama caucasica mucronata Guibé, 1957:137–138 (Type locality: Langarak, 60 km E Meched, on road to Sarakhs, Iran; Syntypes: MNHN). — Wermuth, 1967:11. — Schleich, 1977:127, 129.

Agama nupta: Guibé, 1957:137 (in error).

Laudakia erythrogastra: Leviton, *et al.*, 1992:12–13 (by implication).

Diagnosis: Tail segmented on proximal two-thirds, each segment composed of two whorls of scales; gular scales mucronate, strongly keeled in adults, weakly keeled in small juveniles; 80–114 scales round middle of body; one or two longitudinal rows of spiny tubercles on each side of body.

Color pattern: (Fig. 39.) In alcohol, initial preservation in formalin, predominantly olive-brown, with numerous irregular black-edged light marks on vertebral region, more regular light markings longitudinally and vertically arranged on flanks, these primarily confined to enlarged spiny tuberculate scales. Tail distinctly barred with alternate dark and light transverse markings, as are fingers and toes; forelimbs dark with narrow light crossbars. Venter light tan with irregular brown flecks, a dark marmoreal pattern on chin, neck, and chest; some males (in breeding coloration?) with venter of limbs, body, and tail black, throat and chest under shoulders yellow, with distinct black chevrons and other markings; venter suffused with orange in life, most intense on tail (perhaps breeding coloration). Small juveniles light gray with narrow black transverse marks across vertebral area; subadults with scattered black dots on vertebral area; small, distinct black spots on limbs and dorsolateral areas of back; throat mottled with gray.

Sexual dimorphism: According to Terentjev and Chernov (1949:147), adult males are black on the ventral surfaces, females orange.

Size: The largest male examined (from Afghanistan) is 147 mm snout-vent, tail 186 mm; largest female 111 and 133 mm, respectively.

Natural history: Digestive tracts examined contain beetles, ants, and grasshoppers, but no plant material.

Habitat: These lizards are found in areas of clayey and sandy-loamy soils, mainly in areas where there are colonies of the ground squirrel-like *Rhombomys*, the burrows of which serve as retreats for the lizards (Terentjev and Chernov 1949:147; S. Anderson and Leviton 1969:36–37). They have been found basking on small, man-made piles of rocks, under which they took cover when disturbed (Clark, *et al.* 1966:5). Unlike other members of the genus,

FIGURE 39. *Laudakia erythrogastra* from vicinity of Paghman, northern Afghanistan (FMNH 161187). (Photo by A. Leviton *in* Leviton and S. Anderson, 1969:36, fig. 3.)

they apparently avoid vertical slopes and rock outcrops. Where *Trapelus agilis* occurs in the same area, this smaller agamid occupies earth banks and shrubs. Clark (1991:37) dug them out of deep holes and crevices in earth banks and cliffs, where they hid in tunnels that were long and often branched into several passages. Ananjeva and Tuniyev (1994:44) stated that this species can occupy a wide spectrum of biotopes, but reaches its greatest density in gorges with pistachio (*Pistacia vera*) in Turkmenistan.

Laudakia erythrogastra

Distribution: Northeastern Iran, in the vicinity of Mashhad, southeastern Turkmenistan, northern Afghanistan north of the Central Massif and south through the mountain passes to Paghman; 900–2500 m elevation. Zarudny (1903:10–11) states that they occur on the plains and low hills between Mashhad and Torbat-e Jam, *Laudakia caucasia* being the species in the mountains to the northeast and southwest. In 1990, this species was reported from the eastern Kopet Dagh in Turkmenistan (Gorelov and Lukarevsky, 1990). This population was later described as *Stellio erythrogaster nurgeldievi*, by Tuniyev, Ataev, and Shammakov (1991).

Remarks: I have examined three specimens from Langarak identified as *Laudakia nupta* by Guibé (1957:137). These are very young specimens of *L. erythrogastra*. *Laudakia nupta* apparently does not occur this far north. As pointed out by Ananjeva and Orlova (1979:13–14), Guibé's *Agama caucasica mucronata* is a junior synonym of *L. erythrogastra*. Michael Golubev (pers. commun., 3 February 1995) has called my attention to an article on the possible hybridization of *L. caucasia* and *L. erythrogastra* (Zykova and Panov, 1990).

Laudakia melanura Blyth, 1854
Black rock agama

Laudakia (Plocederma) melanura Blyth, 1854:738 (Type locality: ?Salt Range, Punjab; Types: not located).

Laudakia melanura lirata (Blanford, 1874)

Stellio liratus Blanford, 1874a:453; 1876:320–322, pl. 20, fig. 2 (Type locality Saman, Dasht, Baluchistan, Pakistan; Holotype: ZSI 3410 [see Das, *et al*., 1998]).
Agama melanura: Boulenger, 1885a:363–364. — S. Anderson, 1963:475. — Wermuth, 1967:17. — S. Anderson, 1968:332. — Welch, 1983:17.
Agama melanura lirata: Mertens, 1969:34–35, text-figs. 8–9. — S. Anderson, 1974:36, 42.

Diagnosis: Caudal segments each composed of more than two whorls of scales, or annulation indistinct; median dorsal scales broader than long, in 6–7 straight longitudinal series, grading into dorsolateral scales; 120–130 scales around middle of body (Fig. 40); scales of back of head and middle of back strongly keeled.

Color pattern: Adult males dark brown with strongly contrasting sandy-yellow head and neck, and yellow mottling on shoulders; females dark brown with light dorsal spots, head dark like body, yellow on underside of thighs and anterior of tail.

Size: Males from the Mekran Coast of Pakistan to 135 mm snout-vent, tail 285 mm; females with corresponding measurements of 121 and 225 mm (Mertens, 1969:35).

Distribution: *Laudakia melanura* ranges from northwestern Punjab through Sind and

FIGURE 40. *Laudakia melanura.* (From Blanford, 1876, pl. 20, fig. 2.; G. H. Ford, artist.)

Baluchistan, at least to the vicinity of the Iranian frontier. There are no records within the borders of Iran, but the type locality of *L. lirata* is within a few miles of the Iranian border, and Shockley (1949:121) gives a sight record from Ras Jiunri, Pakistan, only 10 airline miles east of Iran. It may be expected to occur along much of the Makran Coast of Iran. Mertens (1969:34) revived Blanford's name to distinguish the Mekran population from the nominate form, in which the males are uniformly black during the breeding season. Curiously, he reports both subspecies from localities very close to one another on the right bank of the lower Hab River in Pakistan.

Remarks: See Smith (1935:218–219) and Minton (1966: 93) for notes on the natural history of *Laudakia melanura.* Apparently, the nearest relative of this species is *L. nupta,* although Ananjeva and Tuniyev (1994:46) consider it related to *L. erythrogastra.*

Laudakia microlepis (Blanford, 1874) (Plate 2E–F)
Small-scaled rock agama

Stellio microlepis Blanford, 1874a:453; 1876:326–327, pl. 19, fig. 2 (Type locality: Khan-i-Surkh pass north of Sarjan, between Kerman and Shiraz, 9000 ft [2745 m], and Kushkizard, between Shiraz and Esfahan, 8000 ft [2440 m]; Syntypes: BMNH 1946.8.28.74–77). — Bedriaga, 1879:38. — Nikolsky, 1897:317–318; 1899:391. — Zarudny, 1903:11. — F. Werner, 1936:200. — Ananjeva and Ataev, 1984:9.
Agama microlepis: Boulenger, 1885a:366. — F. Werner, 1895: 3–4; 1929:240; 1936:197. — Forcart, 1950:145. — Wettstein, 1951:435–436. — S. Anderson, 1963:475. — Guibé, 1966a:98 (in part). — Wermuth, 1967:17–18. — S. Anderson, 1968:332. — Schleich, 1977:127, 129. — Welch, 1983:17.
Agama caucasica: Forcart, 1950:145 (in part). — Tuck, 1971a:54 (in part).
Agama caucasica microlepis: S. Anderson, 1974:37, 42. — Bannikov, *et al.*, 1977:113.
Laudakia caucasica microlepis: Leviton, *et al.*, 1992:13 (by implication).

Diagnosis: Tail divided into distinct segments, each composed of two whorls of scales; gular scales smooth; patch of enlarged scales on middle of flank (sometimes absent); males with 177–235 (mean 197.4, N = 13) scales around widest part of body, females with 190–259 (mean 202.2, N = 16) (Fig. 41a).

Color pattern: As for *Laudakia caucasia* (see above). Males collected 27 km north of Birjand had nearly black shoulders and thorax, the head, abdomen, and anterior part of tail conspicuously contrasting light cream to yellow while they were displaying on rocks; they faded to nearly uniform grayish coloration when hiding, however. I find no consistent differences in color between *L. microlepis* and *L. caucasia*, nor any reliable means of distinguishing between the sexes of preserved specimens on the basis of color pattern.

Size: The largest Iranian male examined is 133 mm snout-vent, the largest female 149 mm; the smallest juvenile (from eastern Afghanistan), collected in mid-April is 40 mm snout-vent.

Habitat: The habitat of this species appears to be similar to that of *Laudakia caucasia* (see above), but niche differences in the area of overlap between the two species are yet to be determined.

Distribution: In the mountains of the southern, central, and eastern parts of the Iranian

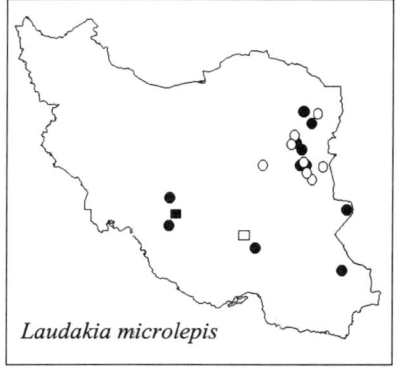
Laudakia microlepis

Plateau, from the area around Yazd-e Khvast and Kushk Zar, *Laudakia microlepis* replaces *L. caucasia*. In Afghanistan the former occurs south of the Kabul River Valley in the mountains of eastern Afghanistan, and in Waziristan and Baluchistan, Pakistan. It has also been taken west of the Central Massif, south of Herat, and in the mountains of eastern Iran along the Afghan frontier. The areas of overlap with *L. caucasia* are not clearly defined on the basis of material at hand, although both species have been collected in the mountains of northeastern Khorasan. It is noteworthy that there is no apparent overlap in the counts of scales around the body in females of the two taxa. The mean number of scales around the body in males of *L. caucasia* is 145.1, while for males of *L. microlepis* it is 197.4.

Remarks: Ananjeva and Ataev (1984:9) have made the case for recognizing *caucasia* and *microlepis* at the species level. In the region of Torbat-e Heydariyeh in northeastern Khorasan, the two apparently occur at localities very close together. I have seen only a few specimens from these localities, however, and cannot say if two distinct populations are actually represented. The narrow habitat restrictions on this species probably serve to maintain the genetic integrity of local populations. Zarudny (1903:11) states that the ridge of Kuh-e Kerat forms the boundary between the two forms in eastern Khorasan, while *Laudakia nupta* is found in the mountains in the vicinity of Chah-i-Ziru and Gyuishe, and at Bendun in Baluchistan.

Counts and measurements for syntypes of *L. microlepis* from Kushk Zar are as follows: BMNH 1946.8.28.74 female snout-vent 145 mm, tail 155 mm, scales around midbody 193. BMNH 1946.8.28.75 female snout-vent 132, tail broken, scales 199. BMNH 1946.8.29.77 female snout-vent 115, tail broken, scales 206.

In general, in animals of comparable size, the spines on neck and temporal region tend to be more pronounced in *L. caucasia* than in *L. microlepis*. This is a subjective observation, most evident when series of both are laid out for comparison. Most individuals of both species have enlarged scales scattered among very small dorsolateral scales. These often have a more or less obvious transverse and longitudinal arrangement. These enlarged scales are often within the light ocelli and the difference in size is visually exaggerated. In some specimens these enlarged scales are not as obvious as in others (relative size not as great). This is particularly true of those from northern Afghanistan. Specimens which I collected in eastern Iran between Torbat-e Heydariyeh and Birjand lack enlarged scales on the flank, and the scales of the back and flanks lack keels or mucros, flank scales being granular rather than sharply tubercular as in more northern specimens.

A male *Laudakia* of doubtful identity in the collection of the California Academy of Sciences (CAS 111931), received from Mahmoud Latifi of the Razi Institute, is said to come from Zabol, Sistan, which is within the range of *L. microlepis*. The count of scales around the body is 157, very like a specimen of *L. caucasia* from Babol (154). Zabol is an unlikely locality for *L. caucasia*, and I regard the reported provenance of this specimen with considerable doubt. This would be by far the lowest count for any specimen of *L. microlepis*.

Blanford (1876:327) suggested that his new taxon (*microlepis*) might prove to be a race of *L. caucasia*, and Wettstein (1953:267) also expressed this view. I follow Ananjeva and

Ataev (1984) and regard *microlepis* as a species distinct from *L. caucasia* on the basis of the nonoverlapping counts for females, in spite of the very close morphological resemblance of the two taxa. The biochemical systematic work being carried out by J. Robert Macey supports this view (see Remarks above under *L. caucasia*).

FIGURE 41. *Laudakia microlepis* (A) and *L. nupta* (B) (From Blanford, 1876, pl. 19, G. H. Ford, artist).

Laudakia nupta (De Filippi, 1843)
Large-scaled rock agama, yellow-headed agama

Agama nupta De Filippi, 1843:407 (Type locality: Persepolis, Iran; Holotype: Milan).

Diagnosis: Tail divided into more or less distinct segments, each composed of more than two whorls of scales when viewed from the side two or three head widths behind vent, or segmentation of tail indistinct; median dorsal scales broader than long, in oblique longitudinal rows (Fig. 41b).

Color pattern: Light brown, olive, or dark brown above, with many scattered black and yellow scales; tail with dark bars proximally, distal third to half black or dark brown, usually with narrow light rings; venter light tan; chin, throat, and chest of male dark blue, often mottled with white (yellow, pink, or orange in life); juveniles with dark transverse markings on back, these breaking up into reticulations with age. In males of *Laudakia n. fusca*, entire head yellow (at least in breeding season), body, limbs, and tail dark brown to sooty black. In the syntype I have examined, the head is pale brown, with darker flecks on temporal and occipital areas; throat yellow; enlarged scales of back each with lateral light marks, central dark streak; distal three quarters of tail black; venter dark brown, flecked with yellow on chest and sides.

Size: The largest Iranian male *Laudakia n. nupta* measures 166 mm snout-vent, the largest female 144 mm. The syntype of *L. n. fusca* is 164 mm snout-vent, and a male from the Karachi area of Pakistan measures 168 mm.

Natural history: During hottest hours of the day, individuals of *L. n. nupta* observed on walls of houses rested with head downward in shade and made occasional brief forays into the sunlight on the ground to capture insects. The area usually occupied by an individual seems to have a radius of about 15 m, while the basking area is usually not more than two to three meters diameter. Observations on the habits and temperature relationships of this species have been recorded in more detail in a previous paper (S. Anderson 1963:442–444).

Females collected in Fars Province in late December and early January contain ovarian eggs of various sizes, none very large. The smallest juvenile collected in Fars during this period has a snout-vent length of 46 mm. Blanford (1876:319) reports finding 10 eggs, each about two cm long, in the oviducts of a Kerman female in May. In Khuzestan, females with eggs in oviducts were collected in late August. Females collected in March and October had eggs up to 3.5 mm diameter in the ovaries. The smallest individuals (44 mm snout-vent) were collected on September, but half-grown specimens were seen from March through November.

This species is both herbivorous and carnivorous. It is known to eat scorpions, beetles, and other arthropods. It doubtless eats smaller lizards and anything it can catch and overpower. Stomachs examined primarily contained ants, beetles, beetle larvae, plant material, and occasionally orthopterans. A specimen from Fars contained a single date seed. An Afghan specimen had a lycosid spider and several rocks in the stomach.

Habitat: (Plate 22A–C, E) *Laudakia nupta* is the ecological counterpart of *L. caucasia* to the south of the latter's geographical range. They are abundant on and among large rocks of limestone and other outcrops having deep crevices. They are at home around human habitation, commonly seen on walls, mud-brick dwellings, and the tombs and monuments of graveyards. On the terrace of Persepolis, large individuals bask on the huge capitals and other ruins of the excavation. One was seen in a small gypsum cave, and they were seen on the rare trees of the foothill region. I have collected them on isolated boulders far from angular outcrops and on silty alluvium at the edge of a steep-sided ravine where there were erosion crevices for retreat. Their habitats include predominantly overgrazed land denuded of shrubbery, areas of scattered shrubs and trees with annual grasses and forbs, to the relatively dense scrub dominated by *Quercus brandtii* and *Crataegus* spp. in the Zagros Mountains west of Kermanshah.

Laudakia nupta

Distribution: The southern and western periphery of the Iranian Plateau, primarily in the outer ranges. *Laudakia caucasia* seems to be its ecological counterpart in the inner ranges of the plateau. It occurs in the foothills of the Zagros Mountains in eastern Iraq and western Iran, eastward through southern Iran into southern Afghanistan and Baluchistan, Pakistan. The ranges of *L. nupta* and *L. microlepis* overlap to some extent, and both were collected by Blanford at Kushk Zar. *Laudakia nupta* has been found in the mountains of eastern Khorasan, along the Afghan border, but Guibé's (1957:137) records for northeastern Iran are in error (see above under *L. erythrogastra*). Tuck (1979:98–99) found this species in the Turan Protected Region in the northern Dasht-e Kavir. This isolated occurrence is the northernmost record to date. There is a specimen (FMNH 74574), which I have examined, bearing the data: Syria: Habariyah, 6 km E Kheurbate al Ambachi, and ca 60 km E Damascus. Its occurrence in Syria requires confirmation, in my opinion. The extent of the distribution of *L. n. fusca* is not known. The only Iranian specimen I have seen which represents this form is the syntype from near Jalq, although the Binak specimen may belong to this subspecies. It is found in the hills near Karachi, Pakistan, and it is probably the representative of the species in outlying hilly country off the Iranian Plateau proper. The status of this latter taxon is in need of further investigation. From sea level to 2700 m elevation in Iran.

Remarks: The basal portion of the tail is not distinctly segmented in this species. The first distinct proximal segments contain only two annuli, and other segments often have but two rows of scales dorsally, three laterally. Tails are frequently broken; the lost portion is subsequently regenerated, the restored part being very distinct in appearance from the original.

Smith (1935:220) mentions that the enlarged dorsal scales are occasionally in straight longitudinal rows (specimens from Quraitu, on Iran-Iraq border, and another from Waziristan, Pakistan). This is the case in a juvenile specimen (CAS 86511) from Binak on the Persian Gulf; this specimen also lacks a distinct fold across the nape, and has 16 scales in the fifth caudal whorl. In specimens from Kermanshahan (CAS 141298) and from Diana, Iraq, the dorsal scales are in straighter than usual rows; all these specimens have the fold across the nape.

In the two specimens of *Laudakia n. fusca* I have examined (including a syntype), there are 18 scales in the fifth caudal whorl posterior to the postanal pocket. In those *L. n. nupta* in which I have made this count, there are 22–24 scales in specimens from Fars, including three topotypes from Persepolis. In Khuzestan lizards there are 24–28, 23–28 in those from Qom, 20–27 for Kerman, 18–23 in specimens from Afghanistan.

Laudakia melanura appears to be the nearest relative of *L. nupta*.

Laudakia nupta nupta (De Filippi, 1843) (Plate 2G)
Large-scaled rock agama

Agama nupta De Filippi, 1843:407 (Type locality; Persepolis, Iran; Holotype: Milan). — Boulenger, 1885a:365. — Nikolsky, 1907a; 273. — F. Werner, 1917:199–200. — Smith, 1935:219–220. — F. Werner, 1938:268. — Schmidt, 1939:57. — Wettstein, 1951:435. — Schmidt, 1955:203–204. — Hellmich, 1959:3–4. — S. Anderson, 1963:442–444. — Clark, et al., 1966:6. — Wermuth, 1967:20. — S. Anderson, 1968:332. — Tuck, 1971a:54, figs. 3–4. — Schleich, 1977:127, 129.
Stellio carinatus A. Duméril, 1851:107 (Type locality: "Persia"; Syntypes: MNHN 6953 [3]).
Stellio nuptus: De Filippi, 1865:352. — Blanford, 1876:317–319, pl. 19, fig. 1. — Bedriaga, 1879:38. — Blanford, 1881:676. — Nikolsky, 1899b:391–392. — Zarudny, 1903:12–13. — Werner, 1936:200. — Welch, et al., 1990:59.
Agama nupta nupta: Minton, 1966:91–92. — Wermuth, 1967:20. — Haas and Y. Werner, 1969:340. — S. Anderson, 1974:36, 42. — Tuck, 1979:98–99, fig. 2. — Welch, 1983:17.
Laudakia nupta nupta: Leviton, et al., 1992:13. — Frynta, et al., 1997:8.

Diagnosis: A prominent transverse fold across nape.

Laudakia nupta fusca (Blanford, 1876) (Plate 2H)
Yellow-headed agama

Stellio nuptus var. *fuscus* Blanford, 1876:319–320 (Type locality: Kalagan, Baluchistan, Iran, and near Jalk, Baluchistan, Iran; Syntypes [3]: BMNH 74.11.23.11 [not listed as a type in BMNH catalogue], also ZSI 6798 and MCZ 7220 [see Das, et al., 1998]). — Murray, 1884a:101. — Nikolsky, 1897:318.
Agama nupta var. *fusca*: Boulenger, 1885a:366. — Wermuth, 1967:20. — S. Anderson, 1974:36, 42.
Laudakia nupta fusca: Leviton, et al., 1992:13 (by implication).

Diagnosis: No fold of skin across nape.

Genus *Phrynocephalus* Kaup, 1825
Toad-headed agamas, toad agamas

Phrynocephalus Kaup, 1825:591 (Type species: *Lacerta caudivolvula* Pallas, 1811–1831 [= *Lacerta guttata* Gmelin, 1789], by subsequent designation of Fitzinger, 1843:18 [ICZN, 1964, art. 69a, iv]).

Definition: Body depressed; no dorsal crest; no gular sac; a transverse gular fold; dorsal scales uniform or intermixed with larger ones; tympanum rudimentary or absent, concealed beneath skin when present; tail rounded, depressed at base; no preanal nor femoral pores. Synapomorphy: 11 scleral ossicles (Moody, 1980:158).

Distribution: From southeastern Europe and Southwest Asia (including the Arabian peninsula and northern India) through Central and Middle Asia to East Asia (northern China and Mongolia). Approximately 40 species, seven species documented from Iran.

Remarks: Franz Werner (1936:200) lists *Phrynocephalus strauchi* (= *P. reticulatus strauchi*, according to Bannikov, *et al.* [1977:125]) from Iran, but I find no published locality records for this form. It occurs in the Fergana Valley, Uzbekistan. Richard and Erica Clark collected *P. raddei boettgeri* in northern Afghanistan, however, and it is possible that this latter taxon ranges into the northeastern corner of Iran. There are no records of *P. interscapularis* from Iran, although it has been listed in error (S. Anderson, 1968:332), and Terentjev and Chernov (1949:158) suggested that it probably occurs there. It has been collected in northern Afghanistan by the Clarks (Leviton and S. Anderson, 1970). *Phrynocephalus luteoguttatus* is also considered in the current work, although no reliable records exist for Iran. It inhabits southern Afghanistan and possibly will be found in Iranian Sistan. *Phrynocephalus clarkorum*, which inhabits southern Afghanistan and northern Baluchistan in Pakistan, has not yet been taken within Iranian limits. Because specimens of the subsequently named species, *P. clarkorum*, had been included in *P. ornatus* by Boulenger, I asked Ilya Darevsky to examine the Iranian specimens in the collections of the Zoological Institute of the Academy of Sciences, St. Petersburg, which had been identified as *P. ornatus* by Nikolsky. He informed me (pers. commun., 11 April 1970) that no specimens of *P. clarkorum* are among them.

Moody (1980:149, 156) hypothesizes that *Phrynocephalus* forms a sister group to the clade containing *Pseudotrapelus*, *Agama*, *Trapelus*, *Laudakia*, and *Xenagama*. Based on albumin immunological studies, Joger (1991) postulates that *Phrynocephalus* and *Laudakia* form a clade that is the sister group to *Agama*. This supports the conclusions of Ananjeva and Sokolova (1990), who found *Phrynocephalus* to be more closely allied to *Laudakia* than to *Trapelus* based on biochemical studies (they did not include other agamid genera in their cladogram).

Phrynocephalus laungwalensis Sharma, 1978 is a species from the Rajasthan Desert (India) at the southeastern periphery of the distribution of *Phrynocephalus*; it retains a number of plesiomorphic character states not shared with other members of the genus. Arnold (1992) has removed it from *Phrynocephalus* and reallocated it to his monotypic genus *Bufoniceps*. He comments on the probable origins of several of the derived character states that characterize *Phrynocephalus* and comments on the evolution of that genus. Thus, *Bufoniceps* is the sister taxon of *Phrynocephalus*.

Golubev and Sattorov (1992) believe that 10 species, mainly psammophilic, from Southwest and Middle Asia, form a compact, probably subgeneric group: *P. arabicus, P. clarkorum, P. euptilopus, P. interscapularis, P. laungualensis, P. luteoguttatus, P. maculatus, P. mystaceus, P. ornatus*, and *P. sogdianus*. This contrasts somewhat with the cladogram presented by Mezhzherin and Golubev (1989:73) based on an allozyme study of 11 species. However, Golubev (pers. commun., 3 February 1995) emphasizes that the principal importance of the 1989 paper was the demonstration of strong biochemical differences at the species level between *P. helioscopus* and *P. persicus*, and that they lacked adequate species representation to produce a phylogenetic hypothesis.

Macey, *et al.* (1995) presented a preliminary cladogram based on an allozyme study of 19 Central Asian and Chinese species (manuscript in preparation, *fide* Papenfuss, pers. commun., August 1998). Their study differed from previous studies in producing a tentative phylogeny in which clades conform to distinct geographic regions, and they provide an historical zoogeographic hypothesis to explain their phylogeny. Unfortunately, there is relatively little overlap in species studied by these authors with the study by Mezhzherin and Golubev; the two studies share *P. rossikowi*, *P. helioscopus*, *P. guttatus*, and *P. mystaceus*. There are significant differences in the placement of these species between the two studies (but see comment in above paragraph). Lichanova (1992) also presents a phylogenetic hypothesis based on biochemical data in an unpublished Ph.D. dissertation. The relationships within this genus are being studied actively by all of the above-mentioned authors, and there remains much unpublished data.

Key to the Species of *Phrynocephalus* in Iran

1a. Large fringed cutaneous fold at angle of mouth (Fig. A) ....
.... *Phrynocephalus mystaceus* (Pallas, 1776) (p. 92)
1b. No cutaneous fold at angle of mouth (Fig. B) 2

 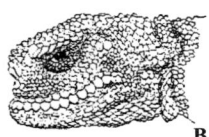

(A) Side of head of *Phrynocephalus mystaceus*; (B) side of head of *P. helioscopus* (After Terentjev and Chernov, 1949, fig. 65 [A]; after Bannikov, *et al.*, 1977:120, fig. 30 [B])

2a. Dorsal scales heterogeneous, small scales intermixed with strongly enlarged scales (Fig. C) 3
2b. Dorsal scales subequal, homogeneous (but in *P. raddei* clusters or single scales may appear to be of different size than the surrounding scales because they are swollen and tubercular with upraised posterior margins) 6

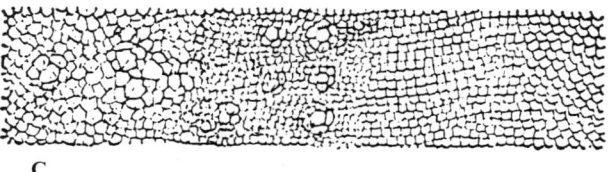

Dorsal scales of *Phrynocephalus helioscopus*
(After Bannikov, *et al.*, 1977, fig. 28.6)

3a. Enlarged dorsal scales flat, not tubercular, posterior border not sharply upturned; sides of back of head and neck with long, flat, upturned, fringe-like scales; both sides of 4th toe with long, well-developed fringes ....
.... *Phrynocephalus luteoguttatus* Boulenger, 1887 (p. 89)
3b. Some enlarged dorsal scales nail-like, often tubercular; large part of scale raised free of back; sides of back of head and neck without long, flat, upturned, fringe-like scales (but sometimes with short, spiny scales); one or both sides of 4th toe with short fringe..4
4a. Nasal shields in contact, or rarely separated by a single series of scales; crossbars on tail most intense (black) ventrally, though usually quite dark dorsally as well; crossbars always present ventrally .... *Phrynocephalus scutellatus* (Olivier, 1807) (p. 97)
4b. Nasals separated by 3–5 (exceptionally 1, usually 3) series of scales; crossbars on tail usually most intense dorsally, rarely absent, and much lighter or absent ventrally, sometimes interrupted dorsally, and seen as a series of spots along side of tail 5
5a. No longitudinal nuchal crest of mucronate scales; a distinct transverse fold of skin across back of neck; entire nostril not seen when viewed from side of head; width of space between nostrils considerably smaller than distance between nostril and preocular ridge .... *Phrynocephalus helioscopus* (Pallas, 1771) (p. 87)
5b. A longitudinal nuchal row of 3–8 mucronate tubercular scales; usually no transverse fold of skin across back of neck; entire nostril seen when viewed from side of head; width of space between nostrils equal to space between nostril and preocular ridge .... *Phrynocephalus persicus* De Filippi, 1863 (p. 94)
6a. Sides of head and neck with long, projecting, fringe-like scales; row of enlarged, upraised, tubercular scales on posterior margin of thigh and sides of tail forming short fringe (Fig. D); often a row of slightly enlarged scales along flank ....
.... *Phrynocephalus interscapularis* Lichtenstein, 1856 (p. 88)

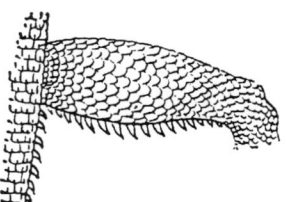

D
Fringe of scales on thigh and tail of
Phrynocephalus interscapularis
(After Bannikov, *et al.*, 1977, fig. 32)

6b. Sides of head and neck without projecting fringe-like scales; no fringe of scales on posterior margin of thigh and sides of tail; no enlarged scales along flank 7
7a. Scattered scales or clusters of scales on dorsum with upraised posterior margins, often swollen, tubercular; scales along midline of back prominently keeled ....
.... *Phrynocephalus raddei raddei* Boettger, 1890 (p. 96)
7b. No upraised swollen scales on dorsum; scales along midline of back smooth to indistinctly keeled 8
8a. Nasal shields separated by 1–3 series of scales 9
8b. Nasal shields in contact or partially separated 10
9a. Tail 130–160 percent of snout-vent length ....
.... *Phrynocephalus maculatus maculatus* Anderson, 1872 (p. 90)
9b. Tail 109–125 percent of snout-vent length ....
.... *Phrynocephalus arabicus* Anderson, 1894 (p. 84)

10a. Ventral surface of tail with indistinct dark crossbars or entire tip of tail black or gray......
.... *Phrynocephalus arabicus* Anderson, 1894 (p. 84)
10b. Tail with 4 or 5 jet black crossbars ventrally, tip of tail not black nor gray 11
11a. Distinct dark-margined light dorsolateral stripe from posterior angle of eye along body onto tail; single very elongate suborbital scale, two or three times as long as adjacent scales .... **Phrynocephalus clarkorum* Anderson and Leviton, 1967 (p. 85)
11b. No light stripe along sides of body; three suborbital scales of about equal size ....
.... *Phrynocephalus ornatus* Boulenger, 1887 (p. 93)

Phrynocephalus arabicus Anderson, 1894 (Plate 3A–B)
Arabian toad-headed agama

Phrynocephalus arabicus J. Anderson, 1894:377 (Type locality: Hadramut, southwestern Arabia; Syntypes: BMNH 97.3.11.51/1946.8.28.33 and 97.3.11.52/1946.8.28.34). — Leviton, *et al.*, 1992:16. — Welch, 1983:19.

Diagnosis: No cutaneous fold at angle of mouth; no fringe of scales on posterior border of thigh and sides of base of tail; sides of head and neck without projecting fringe-like scales; dorsal scales subequal, smooth to weakly keeled, homogeneous; no enlarged scales along flanks; nasals in contact, separated, or partially separated by a single scale; tail 106–125 percent of snout-vent length.

Color pattern: (Figs. 42a–b.) Dorsum reticulated with light and dark markings in preservative, probably yellow and brown in life; upper surface of limbs more or less distinctly barred, with speckling as on body; venter immaculate creamy white; tail speckled like body above, or with indication of faint bars; whitish below with posterior third black.

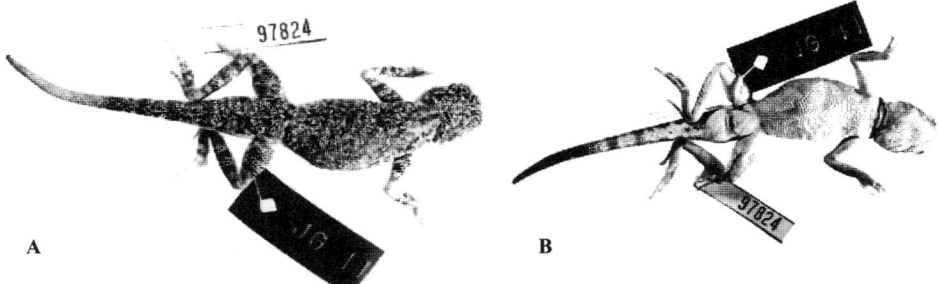

FIGURE 42. *Phrynocephalus arabicus* (from Abu Dhabi [CAS 97824]). (A) Dorsal view; (B) ventral view. (Photos by A. Leviton.)

Size: Largest male examined (Abu Dhabi): snout-vent 54.9 mm, tail 69.2 mm; largest female (Rub 'al Khali, Saudi Arabia): snout-vent 56.5 mm, tail 64.4 mm; smallest juvenile (Abu Dhabi): snout-vent 30.0 mm, tail 37.6 mm.

Habitat: No information has been recorded regarding habitat or natural history for this species in Iran, but see Gallagher and Arnold (1988:409) for their observations of *Phrynocephalus arabicus* in the Wahabi Sands, Oman. There, they occurred on fine, wind-blown sand on fairly open, level, or sloping sand between mounds of vegetation and between or at the foot of large, mobile dunes.

Distribution: Widely distributed on the Arabian peninsula, this species is known in Iran only from the Mesopotamian Plain in the vicinity of Ahvaz.

Remarks: A single specimen, collected by Douglas Lay, 29 June 1973, is the first record of this species for Iran, and the only record of the genus for Iran in the lowlands to the west

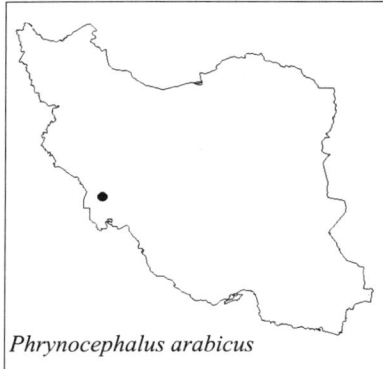
Phrynocephalus arabicus

of the Zagros Mountains. This represents an expected extension of the eastern Arabian sand desert fauna, which includes such species as *Trapelus persicus, Acanthodactylus schmidti, Scincus scincus conirostris, Diplometopon zarudnyi, Eryx jayakari*, and *Lytorhynchus gaddi*.

I examined the MMTT specimen in Tehran, without access to comparative material; however, it agrees generally with J. Anderson's original description and with Haas's (1957:68–69) description of *Phrynocephalus nejdensis* (= *P. arabicus, fide* Leviton and S. Anderson, 1967:159), considering the variation discussed by Leviton and S. Anderson (1967:159–164). There are 82 scales around middle of body (a specimen from Saudi Arabia, 20°05'N, 47°05'E, CAS 119252, has 96 scales around body; a series from Abu Dhabi, CAS 97814–25, has 79–106); pectoral and abdominal scales mucronate, some weakly keeled; adlabials reach mental on right, first labial on left; gulars mucronate; dorsal scales smooth, weakly keeled or smooth on limbs; nostrils directed forward, nasals separated by a single scale.

I examined the two syntypes of *P. arabicus* in the British Museum. In 1946.8.28.63 the nasals are in contact; the adlabials reach the second labial on both sides; the snout-vent length 35.3 mm, tail 40.6 mm (tail equals 115 percent of snout-vent), 74 scales around midbody. In 1946.8.28.64 the nasals are partly separated by a single scale; adlabials reach mental on left, first labial on right; snout-vent 36.1 mm, tail 38.3 mm, tail = 106 percent of snout-vent, 77 scales round midbody. The color pattern is as described above, but the back gives the impression of fine bright dots over darker, mottled brown ground; posterior third of tail is not black; there are three dark bars, the proximal least distinct.

See Leviton and S. Anderson (1967:159–164, figs. 3–4) for a discussion of the synonymy of this species.

On the basis of external morphology, this species appears to be most closely related to *P. maculatus*.

Phrynocephalus clarkorum Anderson and Leviton, 1967 (Plate 3C)
Clarks' toad-headed agama

Phrynocephalus ornatus Boulenger, 1887a:496 (in part), pl. 8, fig. 3c.
Phrynocephalus clarkorum S. Anderson and Leviton, 1967b:228–231, fig. 1 (Type locality: 20 mi [32 km] SE Kandahar, Afghanistan; Holotype: CAS 97989). — S. Anderson, 1974:35, 42. — Welch, 1983:19. — Welch, *et al.*, 1990:54.

Diagnosis: Dorsal scales subequal or increasing in size slightly from dorsolateral to middorsal line, homogeneous, keeled; nasal shields in contact; no spinose scales on neck or back of head; both sides of fourth toe and lateral aspect of third toe strongly fringed; suborbital scales elongate, 2–3 times as long as adjacent scales; lateral scales within broad dark lateral stripe smaller than adjacent scales; a distinct dark-margined light dorsolateral stripe from posterior angle of eye along body onto tail. Tail 124–153 percent of snout-vent length.

Color pattern: (Figs. 43a–b.) Light sandy gray above (in alcohol, initial preservation in formalin), with two rows of dark spots down back; a dorsolateral straight-edged, black-margined white stripe from behind eye down body onto tail; a straight-edged, black

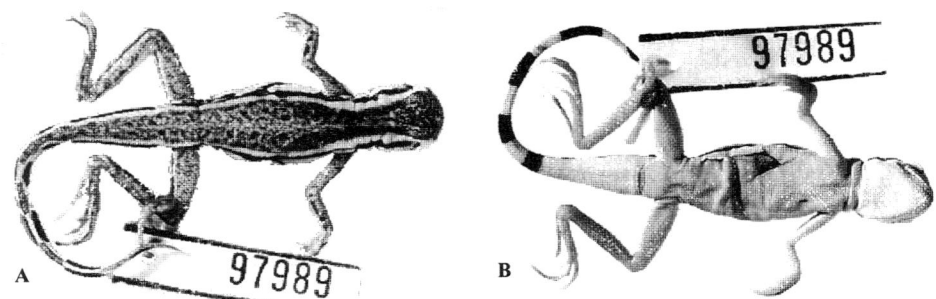

FIGURE 43. *Phrynocephalus clarkorum* (CAS 97989, holotype). (A) Dorsal view; (B) ventral view (From S. Anderson and Leviton, 1967, fig. 1) (Photos by A. Leviton.).

stripe of about equal width below this; another white stripe below that, followed by a thin black stripe; a dark streak on lower temporal region, and dark streak on hinder aspect of thigh; tail with a series of elongate oval brown spots down vertebral line, linked by light chain; venter white, no dark markings on chin or throat except for some black mottling below angle of jaw; tail below with five dark bars, tip of tail white. Light areas on underside of tail yellow in life (R. Clark, pers. commun.).

Size: The largest male examined measures 40 mm snout-vent, tail 60 mm; largest female 40 mm snout-vent, tail incomplete; a female with snout-vent length of 39 mm has a 58-mm tail.

Natural history: The scanty information about their habits indicates that the overall behavior of these lizards is similar to that of *Phrynocephalus ornatus*, but with significant differences. When pursued, they do not bury themselves in the sand and rarely run down holes (Clark, et al., 1969:296–297). See Clark (1992) for an account of behavior correlated with morphology and a comparison of behaviors of *P. clarkorum* and *P. ornatus*.

The oviducts of females collected in April and May in southern Afghanistan contain eggs, while there are developing ovarian eggs in specimens taken in March, August, and October. The smallest females with developing ovarian eggs measure 29–30 mm snout-vent; the oviducts are small in these animals, collected in early April.

Stomachs examined contain a preponderance of ants, but also small beetles and other insects.

See Clark (1992) for further discussion of these and other aspects of the natural history of this species.

Habitat: These lizards have been taken on the edges of dune areas and in sandy river beds (Clark, et al., 1969:296). Where they are syntopic with *Phrynocephalus ornatus*, there seems to be no difference in the general biotopes of the two species, *P. clarkorum* being more numerous on the course-grained sand of broad depressions amongst the dunes. Where they are syntopic, both occur in reduced numbers and roughly equal population densities and both appear to be more successful in areas where they do not occur together. (Clark, 1992:141).

Distribution: The Helmand and Argandab River systems of Afghanistan and Pakistan containing the main sand desert of the region. Its presence in Iran has not been confirmed. *Phrynocephalus clarkorum* appears to occur throughout much of the range of *P. ornatus*, and the presence of both in Sistan may be anticipated.

Remarks: *Phrynocephalus ornatus* appears to be the sister species of this taxon, based on external morphology.

Phrynocephalus helioscopus (Pallas, 1771)
Sunwatcher

Phrynocephalus helioscopus helioscopus (Pallas, 1771) (Plate 3D–E)
Sunwatcher

Lacerta helioscopa Pallas, 1771:1:457 (Type locality: Inderskija Gory, lower Ural River region [restricted by Mertens and Müller, 1928:26; see also Zhao and Adler, 1993:195]); Pallas, 1814:25–26, pl. 6, fig. 2.

Phrynocephalus helioscopus: Eichwald, 1831:3:185. — Bedriaga, 1879:39. — Nikolsky, 1915:147–157. — Guibé, 1957:138 (in part). — S. Anderson, 1963:475; 1968:332 (in part). — Tuck, 1971a: 54 (in part). — Schleich, 1977:127, 129. — Ananjeva, 1984:191–202, figs. 33–35.

Phrynocephalus helioscopus helioscopus: Mertens and Müller, 1928:26 (in part). — S. Anderson, 1974:35, 42. — Tuck, 1974a:62. — Bannikov, et al., 1977:121.

Diagnosis: Nasals separated by 3–5 (exceptionally 1) series of scales; width of space between nostrils more than one-half, but not equal to distance between nostril and preocular ridge; scales of back heterogeneous, enlarged scales nail-like, often tubercular, large part of scale raised free of back; sides of head and neck without long flat upturned fringe-like scales (but sometimes with short spiny scales); one or both sides of fourth toe with short fringe (Fig. 44); crossbars on tail usually most intense dorsally; nostril not visible in entirety when head viewed from side; no nuchal crest of mucronate, tubercular scales; transverse fold of skin across back of neck.

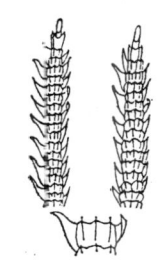

FIGURE 44. *Phrynocephalus helioscopus*. Ventral surface of digits (From Bannikov, et al., 1977, fig. 29.1–2).

Color pattern: (Fig. 45.) Dorsum light brown, light gray to dark gray (in alcohol, initial preservation in formalin), with dark irregularly shaped spots more or less regularly arranged transversely and longitudinally; hindlimbs with distinct transverse dark marks, enlarged tubercular scales within these dark marks; tail with dark dorsolateral spots separated on midline except on distal portion of tail; head with a few dark dots; venter dirty white to cream, chin, throat, and chest with dark gray spots, more pronounced in males; ventral surface of tail pale bluish gray, uniform, or with light gray transverse spots. In life there is considerable pink and blue in the dorsal pattern, as well as various shades of brown, and two kidney-shaped prescapular pink or red patches margined by light blue. According to Nikolsky (1915:153) the ventral surface of the tail tip is red in males (at least seasonally). There is great individual as well as geographic variation in color pattern. The above description is based on the preserved specimens from Iran which I have examined.

Sexual dimorphism: When both sexes are present in a series, males are easily distinguished by the pronounced swelling at base of tail, viewed from below.

Size: The largest Iranian male examined has a snout-vent length of 55 mm, tail 68 mm; largest female measures 57 mm snout-vent, tail 62 mm.

Natural history: According to Nikolsky (1915:156) the males bend the tail upward, exposing the red coloration of the under side. This he interprets as a means of attracting the female. Perhaps more likely it is an agonistic display between males. Other aspects of behavior are briefly summarized by Nikolsky (1915:155–157). Ananjeva (1981:198–202) summarizes work on the ecology and natural history of this species.

Females collected in Mazanderan in late October have small yolked follicles in the ovaries.

All digestive tracts examined contain ants, beetles, and other small insects.

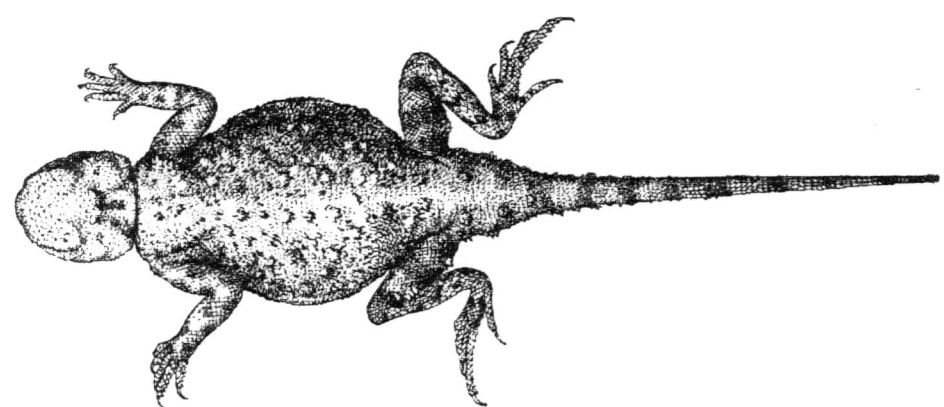

FIGURE 45. *Phrynocephalus helioscopus*. (From Terentjev and Chernov, 1949, fig. 62.)

Habitat: Terentjev and Chernov (1949:157) state that *Phrynocephalus helioscopus* lives on clay and stony deserts and semi-deserts, along dry stream beds, sometimes in sandy areas where there is a high admixture of pebbles. They prefer areas of sparse vegetation.

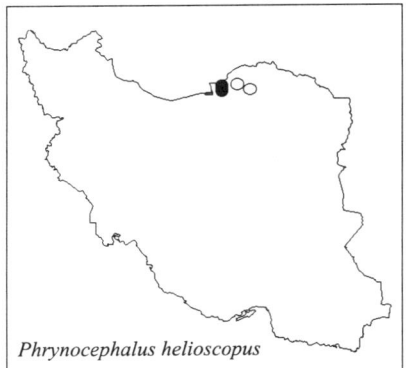

Phrynocephalus helioscopus

Distribution: *Phrynocephalus helioscopus* occurs in Iran in the lowlands of the Caspian Sea in the Gorgan region, but does not extend south of the Kopet Dagh. To what extent it may penetrate the valley of the Rud-e Atrak has not been determined. Beyond Iran, which lies at the southern fringe of its range, it occurs in the southeastern part of European Russia, Kazakhstan, Central Asian republics, and Mongolia. The subspecies *P. helioscopus saidalievi* Sattorov 1981 occurs in the Ferghana Valley, Uzbekistan.

Remarks: Mezhzherin and Golubev (1989:73) concluded that *Phrynocephalus helioscopus* is a sister taxon to a clade containing *P. moltschanovi*, *P. reticulatus*, *P. raddei*, *P. strauchi*, *P. rossikowi*, and *P. guttatus* on the basis of their allozyme studies (but see comments above under *Phrynocephalus* remarks). This is in general agreement with the cladogram of J. Robert Macey, *et al.* (pers. commun.), which places *P. helioscopus* as a sister species to the clade containing *P. raddei* and *P. rossikowi*.

**Phrynocephalus interscapularis* Lichtenstein, 1856 (Plate 3F)
Rose-shouldered toad agama

Phrynocephalus interscapularis Lichtenstein, 1856:12 (Type locality: Buchara). — S. Anderson, 1963:475; 1968:332; 1974:35, 42. — Bannikov, *et al.*, 1977:132.
Phrynocephalus interscapularis interscapularis: Welch, 1983:19.

Diagnosis: No cutaneous fold at angle of mouth; row of enlarged tubercular scales on posterior margin of thigh and sides of tail base form short fringe; sides of head and neck with long projecting fringe-like scales; supralabial scales separated from eye by two rows of scales; dorsal scales homogeneous; often a row of slightly enlarged scales along flanks; lower leg longer than head; toes long, third and fourth with fringe (Fig. 46).

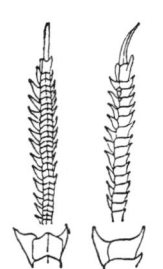

FIGURE 46. *Phrynocephalus interscapularis*. Ventral surface of digits. (From Bannikov, et al., 1977, fig. 29.4–5.)

Color pattern: Dorsum sandy, with whitish and blackish dots; rosy oval spot on back near shoulders; venter white; tip of tail black below; lower surface of tail with two to four black transverse bars, with white interspaces (sometimes red in life).

Size: Largest male examined: snout-vent length 38.2 mm, tail 42.5 mm; largest female examined: snout-vent 32.5 mm, tail 31.5 mm (specimens from Turkmenistan).

Habitat: Found only on sand. In the Karakum desert, Turkmenistan, I found them in white saxaul biotope.

Distribution: No records have been published as yet for Iran (although it is listed for Iran by Bannikov, et al., 1977:132), but it should be looked for in the Gorgan region near the Caspian Sea, and in the low elevations in the northeastern corner of the country. Along the eastern shores of the Caspian Sea it has been found close to the Iranian frontier at Chikishlyar, Turkmenistan (Nikolsky, 1915:183–184. See also Bannikov, et al., 1977:map 54).

Remarks: This is the smallest species in the genus. Nikolsky (1915:184–185) and Terentjev and Chernov (1949:158) briefly summarize the natural history of this species in Central Asia. *Phrynocephalus ornatus* and *P. clarkorum* appear to be the closest living relatives of this species.

Phrynocephalus luteoguttatus Boulenger, 1887 (Plate 3G)
Yellow-speckled toad agama

Phrynocephalus luteoguttatus Boulenger, 1887a:497–498 (Type locality: between Nushki and Helmand; Helmand, Afghanistan; Syntypes: BMNH 86.9.21.62–64/1946.8.28.44–49; 86.9.21.69–71/1946.8.28.36–38). — ?F. Werner, 1895:4. — S. Anderson, 1963:475. — Wermuth, 1967:83. — S. Anderson, 1968:332; 1974:35, 42. — Welch, 1983:19. — Welch, et al., 1990:55.

Diagnosis: About 30 scales from eye to eye across head; nasals in contact or partly separated; nostrils directed more or less directly forward; dorsal scales heterogeneous, enlarged scales strongly keeled, imbricate, flat, not tuberculate, posterior border not sharply upturned; sides of back of head and neck with long, flat, upturned fringe-like scales; both sides of fourth toe with long, well-developed fringes. Tail 85–102 percent of snout-vent length.

Color pattern: Dorsum yellowish brown to pale buff, with black dots and grayish reticulum enclosing pale yellow or golden spots; sides sometimes blackish, a blackish streak along outer side of leg present or absent; venter whitish (pink in life). Tip of tail with black crossbars or asymmetrical spots on lower surface (juveniles), or entire tip (more than distal third) black (adults). Boulenger (1889a, pl. 8, fig. 4) shows colors in life. In preserved specimens (from Afghanistan) a pink tinge is still visible in light areas on under side of tail.

Size: Snout-vent length to 46 mm, tail 48 mm in males; largest female 43 mm snout-vent, tail 38 mm. Smallest juvenile examined (late August) is 24 mm snout-vent.

Habitat: Inhabits sand dune areas sparse in vegetation.

Distribution: Afghanistan in the Helmand River basin and Pakistan in the desert basins of the Chagai and Nushki districts and western Las Bela. It probably enters Sistan in eastern Iran, but no reliable records exist for Iran. Franz Werner's (1895:4) record from between Kom (Qom) and Sultanabad (Arak) is at least 900–1000 km from the Helmand Basin, and is most likely a misidentification of *Phrynocephalus persicus* or a mistake in locality data.

Significantly, Werner did not include this species in his subsequent checklist of Persian reptiles (1936).

Remarks: Minton (1966:97) gives a brief account of this lizard's habits. A few observations are presented by Clark, *et al.* (1969:296). *Phrynocephalus euptilopus* appears to be closely related to this species.

I examined nine syntypes in the British Museum (Natural History). The posterior third of each dorsal scale has a mucronate upturned keel. This character is much more pronounced in the two adults (BMNH 1946.8.28.47–48) than in the young.

Phrynocephalus maculatus Anderson, 1872
Black-tailed toad agama

Phrynocephalus maculatus maculatus Anderson, 1872 (Plates 3H–I, 4A)
Black-tailed toad agama

Phrynocephalus maculatus J. Anderson, 1872:389–390, text-fig. 6 (Type locality: Awada [corrected to Abadeh, north of Shiraz, by Blanford, 1876]; Holotype: ZIS 4825 [see Das, *et al.*, 1998]). — Blanford, 1876:331–334. — Bedriaga, 1879:39–40. — Boulenger, 1885a:377–378. — Nikolsky, 1897:322. — Zarudny, 1903:14. — Annandale, 1906:197. — Nikolsky, 1907a:276; 1915:219–221. — Werner, 1936: 197. — Forcart, 1950:147. — Wettstein, 1951:437. — S. Anderson, 1963:475; 1968:332. — Schleich, 1977:127, 129. — Welch, *et al.*, 1990:55.

Phrynocephalus spiniventris Nikolsky, 1896:370; 1897:322–324, pl. 18, fig 3 (Type locality: Sistan, eastern Iran; Holotype: ZIL 8780). — F. Werner, 1936:200.

Phrynocephalus maculatus var. *spiniventris*: Nikolsky, 1900b:392. — Zarudny, 1903:14. — Nikolsky, 1915:221; 1916:322.

Phrynocephalus maculatus maculatus: Haas, 1957:68. — Y. Wermuth, 1967:83–84. — Bannikov, *et al.*, 1977:130. — Nilson and Andrén, 1981:136. — Welch, 1983:19.

Diagnosis: No cutaneous fold at angle of mouth; no fringe of scales on posterior border of thigh and sides of base of tail; sides of head and neck without projecting fringe-like scales (Fig. 47b); dorsal scales homogeneous; no enlarged scales along flanks; scales on vertebral region considerably larger than those on flanks; nasals separated by one to three scales; tail 140–158 percent of snout-vent length.

Color pattern: Dorsum light gray, flecked with lighter and darker pigment; four indistinct dark broad transverse marks on back, absent in large adults; limbs and toes barred with dark gray in young specimens; tail barred with dark gray dorsally in young, fading to indistinct spots in adults; barred with light gray on posterior half, coalescing with age to form more or less uniform gray ventral surface of distal half of tail, tip dark gray; entire ventral surface of distal half of tail black in mature adults; ventral surface of tail of young lizards of both sexes sometimes with considerable orange-red in life.

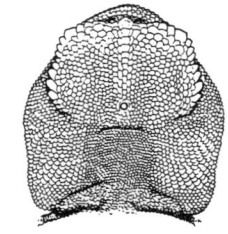
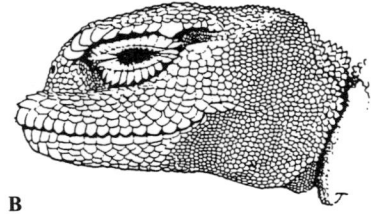

FIGURE 47. *Phrynocephalus maculatus*. (A) Dorsal view of head; (B) lateral view of head. (From Smith, 1935, fig. 60.)

Size: Snout-vent length to 91 mm, tail 128 in males.

Natural history: Like other species of *Phrynocephalus*, *P. maculatus* curls the tail upward when alarmed, reminiscent of iguanids such as *Callisaurus, Holbrookia,* and *Liocephalus*, which also have dark bars on the ventral surface of tail. In Sistan, I found them perched on almost every pile of dirt and gravel heaped at the edge of the road for road maintenance. From these perches they would run for up to 75 m across the hammada, freezing suddenly, their color rendering them invisible. I never saw them seek refuge in burrows.

Blanford (1876:334) reported finding two eggs (presumably in the oviducts) of each of two females collected near Bam in late April. In the only female I examined (snout-vent 65 mm), collected in southern Afghanistan in April, there are small eggs in the ovaries and the oviducts are enlarged.

In digestive tracts I examined, recognizable remains consist of ants, and Blanford (1876:334) chiefly found ants in specimens collected in Iran. Minton (1966:98) reports beetle and other insect remains, as well as seeds and other plant material.

Habitat: (Plate 20A) I found them to be lizards of flat desert, occurring on sandy flats with low, scattered shrubs in Kerman, on sandy, gravel-strewn hammada in Sistan, and on barren salt flats in Khorasan. Blanford (1876:333) and Minton (1966:98) found them in similar areas in both Iran and Pakistan.

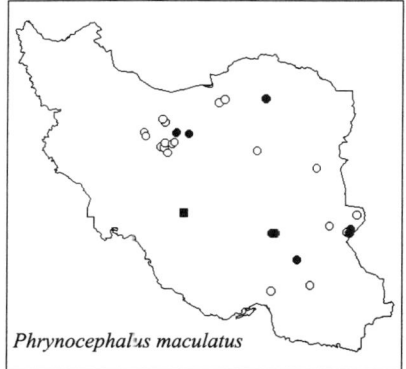

Phrynocephalus maculatus

Distribution: The Central Plateau of Iran, at elevations from 500 to 3000 m; east through southern Afghanistan and Baluchistan as far as Nushki, Pakistan. *Phrynocephalus maculatus* has been listed for Iraq by Smith (1935:234), Khalaf (1959:21), and Wermuth (1967:84); however, I find no published records for that country (see Leviton, *et al.*, 1992:17). *Phrynocephalus maculatus longicaudatus* Haas is found along the Persian Gulf coast of Saudi Arabia, and may extend into both Iraq and southwestern Iran. No species of *Phrynocephalus* crosses the Zagros Mountains, and there are no ecologically continuous areas between the present ranges of *P. m. maculatus* and *P. m. longicaudatus*. This latter taxon will probably prove to be specifically distinct from the nominal form, just as *Acanthodactylus schmidti*, with a distribution comparable to that of *P. m. longicaudatus*, once considered a subspecies of *A. cantoris*, has been elevated to species level. Definable taxa that are separated by the Gulf and the ecologically disjunct areas at the head of the Persian Gulf probably ought to be regarded as separate species, if one uses a phylogenetic species concept.

Remarks: Specimens from Sistan, Khorasan, and Kerman agree in general with Nikolsky's description of *spiniventris*, but there is considerable variation in the degree of carination of the ventral scales and those on the limbs. The marginal supraoculars vary in number from 9–12. I find no consistent characters by which to justify recognition of *spiniventris* as distinct from *maculatus*. Specimens from Saudi Arabia (*Phrynocephalus m. longicaudatus* Haas, 1957) are similar to specimens from Iran, Afghanistan, and Pakistan. In these Arabian specimens (CAS 84442–8, 84452, 84455, all paratypes) from Doha Dalum, the tail is 137–170 percent of the snout-vent length, whereas in the Iranian specimens it is 130–158 percent. In Iranian specimens I have counted there are 99–115 scales around middle of body, 106–119 in the Arabian specimens. I find no consistent differences between the Arabian lizards and Iranian specimens in the characters given in the diagnosis of *P. m. longicaudatus* (Haas, 1957:68), *viz.*, size and shape of supraocular scales, carination and mucronation of dorsal scales, length of tail, and coloration of tail.

Phrynocephalus mystaceus (Pallas, 1776) (Plate 4B–C)
Toad-headed agama

Lacerta mystacea Pallas, 1776:3:702, pl. 5, fig. 1 (Type locality: "Arenosis Naryn" and "deserti Comani," *fide* Pallas, but Naryn Steppe on north coast of Caspian Sea [restricted by Mertens and Müller, 1928; see also Zhao and Adler, 1993:196 for additional comments]; Type: not located).
Lacerta aurita: Pallas, 1814:21–23, pl. 5, fig. 1.
Phrynocephalus mystaceus: Nikolsky, 1900b:393–394. — Zarudny, 1903:15. — Nikolsky, 1915: 173–180. — F. Werner, 1936:197. — S. Anderson, 1963:475; 1968:332. — Schleich, 1977:127, 129.
Phrynocephalus mystaceus galli: Terentjev and Chernov, 1949:159. — Y. Wermuth, 1967:84. — S. Anderson, 1974:34, 42. — Tuck, 1979:99. — Bannikov, *et al.,* 1977:135. — Welch, 1983:19.

Diagnosis: A large, fringed, cutaneous fold at angle of mouth; well-developed lateral and medial fringes on digits (Fig. 48); tail equal to 92–114 percent of snout-vent length.

Color pattern: Dorsum sandy, with black and white dots and reticulations; a row of larger dark blotches on each side of vertebral line; venter white; faint dorsal crossbars (wider than interspaces) on tail, distal third gray; ventral aspect of tail tip (and more than distal third) black; usually a black spot on chest; chin and throat with gray reticulation.

FIGURE 48. *Phrynocephalus mystaceus.* Ventral surface of digits. (From Bannikov, *et al.,* 1977, fig. 29.7.)

Size: A male from Uzbekistan measures 122.7 mm snout-vent, tail 135.8. A female from the same locality: snout-vent 94.1 mm, tail 90.8 mm. The largest male I collected in Iran: snout-vent 77.7 mm, tail 71.0 mm; female: snout-vent 64.1 mm, tail 67.6 mm. The largest species in the genus.

Natural history: This species burrows in sand, usually in areas between dunes, digging a tunnel 70–80 cm long with wider chamber in level of moist sand. Seeks refuge from pursuit or digs into sand for the night by rapid lateral movements of body. Best known for its distinctive threatening posture, with anterior part of body raised on forelegs, mouth open wide, the flaps at sides exaggerating apparent size of mouth, while mucous membrane of mouth and cutaneous folds become red. Braced on widely spread hind limbs, the lizard hisses, whirls its tail in spirals and may jump toward its antagonist (Terentjev and Chernov, 1949:160). Photographs of this display are to be found in Mertens (1959b:pl. 42) and Vogel (1954:pls. 20–23). See Terentjev and Chernov (1949:159–160) for remarks on its natural history in the Central Asian republics. Ananjeva (1984:209–214) provides a summary natural history and ecology of this species.

Habitat: (Plate 24A) Sand dunes; I found them in association with *Tamarix* and other psammophilous shrubs and grasses in eastern Khorasan.

Distribution: Southeastern Russia, the Transcaspian Region, northern and eastern borders of the Caspian Sea, Turkmenistan, Uzbekistan, Kazakhstan, northeastern and eastern Iran and adjacent Afghanistan and Xinjiang, China. It occurs near the border of Iran

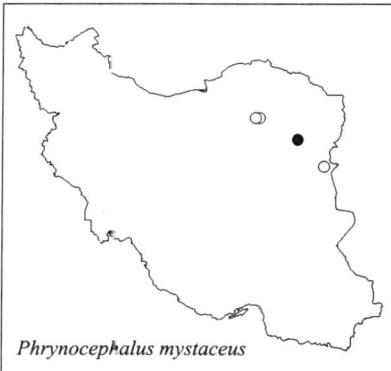
Phrynocephalus mystaceus

near Chikishlar, Turkmenistan, on the east coast of the Caspian Sea, and perhaps will be found within Iranian limits in the Gorgan region.

Remarks: In all specimens examined (seven from Khorasan, nine from northern Afghanistan, two from Repetek, Turkmenistan) there is a greatly enlarged scale (equal to about four adjacent scales) on each side of thorax behind axilla; occasionally an additional scale or two only slightly smaller adjacent to it. Such scales are absent in all other species of *Phrynocephalus* examined (all Southwest Asian species and several Central Asian forms). The Iranian specimens appear to be intermediate between the nominal taxon and *P. m. galli* Krassowsky (1932) in some characteristics. They agree with *galli* in color pattern and the fringes on the digits; they better fit the diagnosis of *mystaceus* in that the fold at the corners of the mouth does not reach the neck, and also that the tail is about equal to the snout-vent length. Only the keels of the breast scales are strongly developed, whereas keels of breast and belly are said to be strongly developed in *galli*, less distinct in *mystaceus*. The Afghanistan specimens differ from the Iranian specimens in lacking the larger dorsal spots and in some specimens having very light reticulations on chin and throat. Golubev and Sattorov (1992) regard *P. mystaceus* as a monotypic species based on study of morphological characters throughout the range.

Phrynocephalus ornatus Boulenger, 1887 (Plate 4D)
Striped toad agama

Phrynocephalus ornatus Boulenger, 1887a:496 (in part; Type locality: between Nushki and Helmand, Afghanistan [restricted by S. Anderson and Leviton, 1967:231]; Lectotype: BMNH 86.9.21.49/1946.8.28.20). — Nikolsky, 1897:324; 1899:393. — Zarudny, 1903:14. — F. Werner, 1936:200. — S. Anderson, 1963:475. — Wermuth, 1967:85. — S. Anderson, 1968:332. 1974:35, 42. — Welch, 1983:19. — Welch, et al., 1990:55

Diagnosis: Dorsal scales enlarge very gradually from flanks to mid-dorsal line, homogeneous; nasal shields in contact; no spinose scales on neck or back of head; both sides of fourth and outer aspect of third toes strongly fringed; three scales separate nasals from upper labials; two or three suborbital scales, none larger than adjacent scales; no dark-margined dorsolateral stripe between fore and hind limbs. Tail 119–132 percent of snout-vent length.

Color pattern: (Fig. 49.) Dorsum light sandy gray (in alcohol, initial preservation in formalin) with two rows of light orange or pinkish spots (partly ringed with dark, or each with small dark spots in center) down back; darker gray, festooned dorsolateral stripe, very faintly margined above with lighter color from eye down length of body onto tail; second gray stripe on lower temporal region; narrow light lateral stripe (much narrower than gray dorsolateral stripe), and below this a dark lateral stripe of still narrower width; gray streak along hinder part of thigh. Venter white, tail below with four or five dark brown to black bars, these faintly indicated in gray on dorsum of tail, each interrupted by elongate vertebral white spot; tip of tail white below. No dark markings on chin or throat. Tail lemon yellow in fresh specimens (Boulenger 1887a:496, 1889b:97, pl. 8, figs. 3, 3a). In life, brownish above, dark spots on back surrounded by orange (Clark, cited by S. Anderson and Leviton, 1967:233; Clark, 1990:141). My specimens had a deep rose spot bordered by blue on nape.

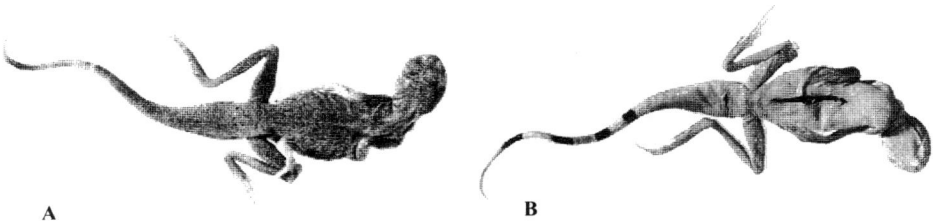

FIGURE 49. *Phrynocephalys ornatus*. Lectotype (BMNH 1946.8.28.20, male). (A) Dorsal view; (B) ventral view (From S. Anderson and Leviton, 1967, fig. 2). (Photos by A. Leviton.)

Size: Snout-vent to 45 mm, tail 58 in males; largest female 45 mm snout-vent, tail 55.

Natural history: Eggs (up to 12 mm long) are present in oviducts of females collected in Afghanistan in early April to May; all other females examined (March, July, August, September, October) have ovarian eggs, those of July specimens least developed (S. Anderson and Leviton, 1967:231). Recognizable stomach contents include small crickets and other orthopterans, ants, attid and other spiders. See Clark, *et al.*, (1969:297) for notes on behavior and environmental temperatures. Clark (1992) provides further information on the comparative natural history of *Phrynocephalus ornatus* and *P. clarkorum* in areas of syntopy.

Habitat: (Plate 24A) I found these lizards only on sand dunes and the sand of a dune-stabilization project in eastern Khorasan. They were found in the same habitat as *P. mystaceus*. The Clarks found them in sand dune areas and on bare stony terrain bordering the sand; Aitchison found them along the river on gravel plains, always near bushes. Clark (1992) records observations on the differences in behavior and habitat that distinguish this species from the morphologically similar *Phrynocephalus clarkorum*. The two species apparently overlap broadly in distribution, and were confused under the name *P. ornatus* until 1967 (see above under *P. clarkorum*). *Phrynocephalus ornatus* appears to be less restricted in habitat than *P. clarkorum* and is able to spread with the changing distribution of migratory sand accumulations (Clark, 1992:141).

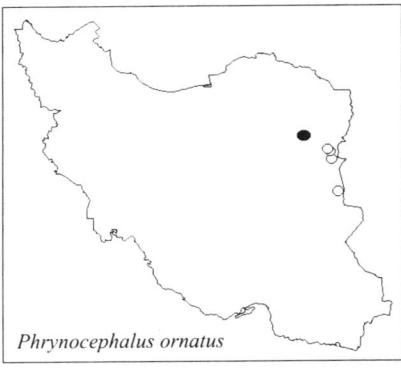

Phrynocephalus ornatus

Distribution: From the eastern deserts of Khorasan through the Helmand River basin of Afghanistan and desert basins of Baluchistan, Pakistan. Zarudny collected specimens identified as *Phrynocephalus ornatus* by Nikolsky (1897:324; 1899:393) in the Zirkuch region of eastern Iran. Ilya Darevsky (pers. commun.) has confirmed this identification, and states that this series does not include specimens of *P. clarkorum*.

Remarks: Golubev (pers. commun.) is describing the specimens from north of Gonabad and from the Zirkuch region as a new subspecies.

Phrynocephalus persicus De Filippi, 1863 (Plate 4E)
Persian toad agama

Phrynocephalus persicus De Filippi, 1863:387 (Type locality: along route between Armenia and Tehran, Iran; Syntypes: MSNG CE9597 [2 spec.], other syntypes in Turin); 1865:353–354. — S.

Anderson, 1872:385–389, text-fig. 5. — Blanford, 1876:329–331. — Bedriaga, 1879:39. — Nikolsky, 1907a:275–276; 1915:163–165. — F. Werner, 1936:200. — Forcart, 1950:147. — Cappocaccia, 1961:92. — S. Anderson, 1963:475. — Mezhzherin and Golubev, 1989:73 (by implication).

Phrynocephalus helioscopus: Boulenger, 1885a:371–372 (in part). — F. Werner, 1894:4. — Boulenger, 1899:378. — F. Werner, 1903:340. — Wettstein, 1951:437. — Guibé, 1957:138 (in part). — Clark, *et al*., 1966:6 — S. Anderson, 1968:332 (in part). — Tuck, 1971a:54 (in part). — Schleich, 1976: 189–193, figs. 1–3; 1977:127, 129.

Phrynocephalus horvathi: Rostombekov, 1938:12.

Phrynocephalus helioscopus persicus: Terentjev and Chernov, 1949:150–152. — Wermuth, 1967:81–82. — S. Anderson, 1974:35, 42. — Bannikov, *et al.,* 1977:121. — Welch, 1983:19.

Diagnosis: Nasals separated by 3–5 (exceptionally 1) series of scales; width of space between rostrils equal to distance between nostril and preocular ridge; scales of back heterogeneous, enlarged scales nail-like, often tubercular, large part of scale raised free of back; sides of head and neck without long flat upturned fringe-like scales (but sometimes with short spiny scales); one or both sides of fourth toe with short fringe; crossbars on tail usually most intense dorsally; entire nostril visible when head viewed from side; longitudinal nuchal crest of 3–8 mucronate, tubercular scales; no distinct transverse fold of skin across back of neck.

Color pattern: Dorsum light brown, light gray to dark gray (in alcohol, initial preservation in formalin), with three dark transverse (sometimes "butterfly-shaped") markings on back, interrupted on spine; hindlimbs with distinct transverse dark marks, enlarged tubercular scales within these dark marks; tail with dark dorsolateral spots, first two or three separated on midline; head with indistinct dark bar across front of hood; venter dirty white to cream, throat with a light gray mottling, more pronounced in males; ventral surface of tail pale bluish gray, uniform, or with light gray transverse spots. In life there is considerable pink and blue in the dorsal pattern, as well as various shades of brown, and two kidney-shaped prescapular pink or red patches margined by light blue. Blanford (1876:331) says that in some specimens the lower surface of the tail is pale green, red near the vent. There is great individual as well as geographic variation in color pattern. The above description is based on the preserved specimens from Iran which I have examined.

Sexual dimorphism: When both sexes are present in a series, males are easily distinguished by the pronounced swelling at base of tail, viewed from below.

Size: Largest male 46 mm snout-vent, tail 57 (a male from eastern Turkey is 52 mm snout-vent, tail 57 mm); largest female 59 mm snout-vent, tail 50.

Natural history: Apparently, the behavior is similar to that of *Phrynocephalus helioscopus*, but apart from Blanford's (1876:331) observations, nothing is known of its habits in Iran.

Females collected in Azarbaijan in late June have small yolked follicles in the ovaries. Specimens from northeastern Turkey collected in late May have eggs up to 11 mm in the oviducts. The smallest female examined (34 mm snout-vent, tail 34, from Turkey) has large ovarian eggs and enlarged oviducts.

All gut contents examined contain ants, beetles, and other small insects.

Habitat: Sobolevsky (1929) found *Phrynocephalus persicus* to be a characteristic species of the Diabar Valley, an arid region of semi-desert flora and fauna just north of the Iranian border in the Lenkoran District of the Azarbaijan Republic. Blanford found these lizards on open gravel or stony plains with scattered bushes on the Central plateau at 1300 to 2700 m elevation. Schleich (1976) collected it in West Azarbaijan (Iran) at about 1000

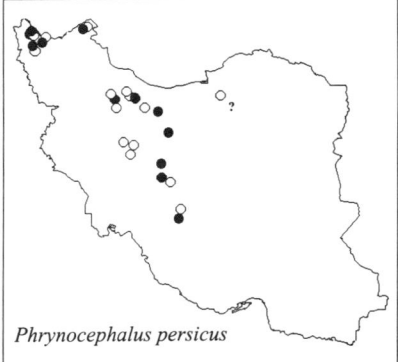
Phrynocephalus persicus

m in desert steppe of dry loamy soil admixed with small stones, cut by gullies, and having sparse vegetation, mostly legumes and grasses.

Distribution: *Phrynocephalus persicus* occupies the northwestern and western parts of the Central Plateau, north through Azarbaijan to northeastern Turkey and the Transcaucasian republics. *Phrynocephalus horvathi* Méhelÿ 1891 appears to be a synonym of *P. persicus*. *Phrynocephalus persicus* does not seem to occur in the eastern and central parts of the Central Plateau, nor does it penetrate west of the crest of the Zagros. Guibé's (1957:136) record from Tasuki in southeastern Iran requires confirmation.

Remarks: Mezhzherin and Golubev (1989:73), on the basis of allozyme studies, removed *Phrynocephalus persicus* from *P. helioscopus*. In addition to the characters given in the diagnosis, specimens from Pahlavi Dezh differ from specimens of *P. helioscopus* from Azarbaijan and eastern Turkey in having two very regular rows of scales on midline of hood distinctly larger than supraoculars, and in having larger scales on anterior part of hood. When *P. persicus* and *P. helioscopus* can be compared side-by-side, the former has the stouter appearance, broader head and body, and shorter neck.

Phrynocephalus raddei Boettger, 1890
Radde's toad agama

Phrynocephalus raddei raddei Boettger, 1890 (Plate 4F)
Radde's toad agama

Phrynocephalus raddei Boettger, 1890:262 (Type locality: Perevalnaya, Turkmenistan; Type: not located).
?*Phrynocephalus strauchi*: F. Werner, 1936:200.

Diagnosis: Upper surface of snout fairly steeply truncated, nostrils not visible or faintly noticeable when viewed from above; no cutaneous fold at angle of mouth; dorsal scales subequal, homogeneous; no transverse fold of skin across back of neck; sides of head and neck without projecting fringe-like scales; no fringe of scales on posterior margin of thigh and sides of base of tail; no enlarged scales along flank; no tubercular scales on dorsum; scales of upper surfaces of limbs and midline of back prominently keeled. A half-moon-shaped red patch on each scapular region, 21–24 lamellae under fourth toe; ridges on subdigital lamellae of third toe approximately uniformly developed in 3 rows.

Color pattern: Dorsum grayish or yellowish-sulfur with 3–5 pairs of dark transverse marks in similar or different degrees of expression. Living individuals characterized by the presence of a single rosy, rose-red, or dark red half-moon mark on each side of spine in region of scapula, bordered below by short blue stripe. Tail above with dark crossbars sometimes broken up into separated spots; ventral surface of tail with 4–5 dark bars, tip dark. Tail below pale blue in life, the base yellow or red.

Size: Snout-vent length to 56 mm in males, 58 mm in females; head and body 68–92 percent of tail length.

Habitat: (Plate 25E) Inhabits diverse kinds of clayey, sandy, and cobbly substrates, including stabilized sand and salt-flats, with sparse herbaceous vegetation.

Distribution: *Phrynocephalus raddei raddei* occurs in southern Turkmenistan and probably occurs in the valleys of the Kopet Dagh on the Iranian side of the border as well. *Phrynocephalus r. boettgeri* is found in southeastern Turkmenistan southern Uzbekistan, southwestern Tadjikistan, and northern Afghanistan (Bannikov, *et al.*, 1977:map 50; Leviton and S. Anderson, 1970:181).

Remarks: *Phrynocephalus reticulatus strauchi* Nikolsky is found in the Fergana Valley, Uzbekistan. It is listed for Iran by F. Werner (1936:200), although I find no documented records attributable to this species. If Werner's record is based on actual specimens, it is possible that they are *P. r. raddei*. On the other hand, M. Golubev (pers. commun., 28 July 1992) tells me that *P. scutellatus* closely resembles both *boettgeri* and *strauchi*, so Werner's record may be a misidentification of *scutellatus*, which is widespread in Iran, and morphologically variable (see below).

I examined Turkmenian specimens from the Ashkhabad region, 55 km N Ashkhabad (CAS 179772–3, 179775, 179777–8; 179881–2; 179896, 179898); and from 40 km E Tedzhen (CAS 172171). All those in which gender was determined were females (snout-vent 41–48 mm; tail 45–58 mm). In these specimens the most prominent swollen upraised scales are associated with the paired dark marks on either side of the spine; the dark markings make them seem particularly prominent; they are not really enlarged relative to the other scales of the back, or only slightly so; the largest scales of the back are less than twice the size of the smallest. CAS 179775, collected 21 May 1989, has the entire abdominal area almost filled with two oviducal eggs.

Phrynocephalus scutellatus (Olivier, 1807) (Plate 4G–I)
Gray toad agama

Agama scutellata Olivier, 1807:110, atlas pl. 42, fig. 1 (Type locality: Mt. Sophia, near Esfahan, Iran; Holotype: MNHN 6947).

Phrynocephalus olivieri Duméril and Bibron, 1837:517 (based on Olivier's type specimen of *Agama scutellata*). — De Filippi, 1865:354. — S. Anderson, 1872:386–387. — Blanford, 1876:327–329. — Bedriaga, 1879:39. — Boulenger, 1885a:370–371. — F. Werner, 1895:4. — Nikolsky, 1897:321–322; 1899:392. — Zarudny, 1903:13. — Nikolsky, 1907a:273; 1915:170–171. — F. Werner, 1929:240; 1936:197; 1938:268, 270.

Phrynocephalus olivieri var. *carinipes* Nikolsky, 1907a:273–274 (Type locality: Pudesck-kupa and Dschandak, Iran; Syntypes: ZIL 10236, 10240, 10241[2]); 1915:172. — F. Werner, 1936:200.

Phrynocephalus olivieri var. *brevipes* Nikolsky, 1907a:274–275 (Type locality: Naim-abad Damysan; Dasht-i-Kavir; Dshandak; Syntypes: ZIL 10235, 10362, 10344, 10239[2]); 1915:172–173. — Werner, 1936:200.

Phrynocephalus scutellatus: Smith, 1935:229–230. — Schmidt, 1939:59. — Forcart, 1950:146–147. — Wettstein, 1951:436–437. — Mertens, 1956:93. — Guibé, 1957:138. — S. Anderson, 1963: 475. — Clark, *et al.*, 1966:6. — Guibé, 1966a:98. — Tuck, 1971a:54–55, fig. 5. — S. Anderson, 1974:35, 42. — Schleich, 1977:127, 129. — Tuck, 1979:100. — Nilson and Andrén, 1981:136. — Welch, 1983:20. — Welch, *et al.*, 1990:55.

Diagnosis: Dorsal scales heterogeneous; enlarged scales nail-like, with free posterior margin often tubercular; more than 16 scales across head between eyes; width of space between nostrils equal to or less than half distance between nostril and preocular ridge; sides of back of head and neck without long, flat, upturned fringe-like scales (but sometimes with short, spiny scales); nasals large, in contact, or rarely separated by single series of scales; crossbars on tail most intense (black) and always present ventrally, though usually quite dark dorsally as well. Tail 118–157 percent of snout-vent length.

FIGURE 50. *Phrynocephalus scutellatus*. Dorsal pattern variation. (From Boulenger, 1889, pl. 8, figs. 2 & 2a.).

Color pattern: Dorsum grayish or brownish, with light and dark markings; often a large pale (lavender or rosy in life) area in the middle of the back surrounded by dark gray, two black irregular crossbars in front of it and two behind; limbs with two broad black crossbars dorsally; tail with black or brown annuli, most intense on ventral surface. Venter whitish. Some specimens speckled black and gray, or gray with small white spots, or brownish over entire dorsum. (Smith, 1935:230). When the limbs are folded in a resting position, the dark bar across the thighs lines up with another across the shank and one on the foot; there is a similar, if less perfect, alignment of bars on the forelimbs and the dark mark on back behind shoulders. Boulenger's color plate (1889a:pl. 8, fig. 2) shows the colors of a fresh specimen (see Fig. 50, which in grayscale shows pattern only).

Size: Snout-vent length 51 mm, tail 69 mm.

Natural history: Specimens I examined from the Central Plateau, collected in mid-May, have eggs in the oviducts; specimens from the Chagai District and Makran Coast, Pakistan, collected in February, also have oviducal eggs. Afghan specimens collected in August and Iranian specimens collected in mid-November have small, yolked follicles in the ovaries. The smallest juvenile examined (30 mm snout-vent) was collected in late August. Blanford's specimens collected in late February and March each contained four eggs about 12 mm long.

Digestive tracts examined contain ants primarily, occasionally beetles and flies. See Blanford (1876:327–329), Minton (1966:97), and Clark, *et al.* (1969:298) for field observations.

Habitat: Flat desert and steppe terrain of heterogeneous texture, especially sand and gravel hammada or gravelly alluvium on slight slopes and saline flats. They have not been found on sand dunes, clay soil, or homogeneous silty alluvium. Vegetation usually consists of sparse shrubs, e.g., *Artemisia, Zygophyllum, Alhagi, Tamarix*, etc. Substrate texture seems to be more important than vegetation, and their morphology, color pattern, and behavior allow them to bask undetected on such open terrain. From 150–2300 m elevation.

Distribution: In Iran, the whole of the Central Plateau bounded by the Zagros Mountains in the west and by the Alborz and Kopet Dagh in the north, and south through Baluchistan to the ranges of the Makran. It extends eastward along the border region of southern Afghanistan and northern Baluchistan, Pakistan.

Remarks: There is considerable individual variation in pholidosis and color pattern within populations of this species; this is true of the characters used by Nikolsky (1907a:273–

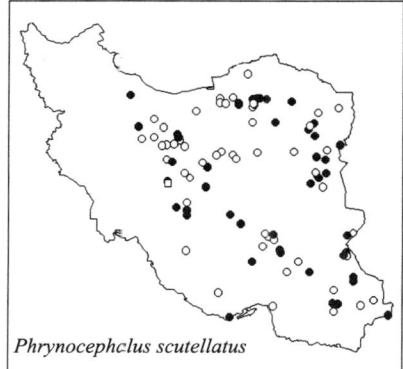
Phrynocephalus scutellatus

275) to define his varieties *carinipes* and *brevipes*. While some individuals have numerous, clearly enlarged, tubercular dorsal scales, both scattered and clustered, in others the enlarged scales are fewer, flatter, and smaller, and in a few specimens the dorsal pholidosis is nearly homogeneous. While variation in pholidosis may prove to be geographically correlated when large samples become available, extremes and intermediates in the character of the dorsal scales mentioned here are all to be found in the 55 specimens from Qom. In most of the 22 specimens examined from Pakistan (Nushki, Kharan, and Ormara; specimens in Royal Museum of Scotland, collected by J. A. Anderson between 1962 and 1968) the dorsal scales are almost homogeneous in appearance, the enlarged scales not tubercular nor standing out distinctly from the smaller scales. Usually the lower nasals are separated by a single scale or a single row of scales.

M. Golubev (pers. commun., 28 July 1992) suggests that *Phrynocephalus scutellatus* belongs to the "*ocellatus*" group (*ocellatus, strauchi, boettgeri, scutellatus, raddei*) and is nearest to *P. boettgeri* from southern Tajikistan and *P. strauchi* from Ferghana Valley. Based on my own observations and the data presented by J. Robert Macey, *et al.* (pers. commun., Sept. 1995), I would add *P. heliosopus* and *P. rossikowi* to this group. Recent unpublished biochemical studies (Golubev, pers. commun., 3 February 1995) further indicate that *ocellatus, strauchi,* and *raddei* are closer to one another than to *P. rossikovi*, and that *boettgeri* is, at best, only subspecific to *P. raddei*.

Genus *Trapelus* Cuvier, 1816
Agamas

Trapelus Cuvier, "1817" (1816):35 (Type species: "Le Changeant d'Egypte Geoffroy" [= *Agama mutabilis* Merrem, 1820, *fide* Wermuth, 1967:2], by monotypy).

Definition: Head rather high and short; tympanum small, its diameter less than half that of orbit, more or less deeply sunk; toes shorter, not compressed; fifth toe not extending as far as first; caudal scales not forming annuli; males with callose preanal scales only (Fig. 51). Synapomorphy: deep and narrow auditory tube (Moody, 1980:159).

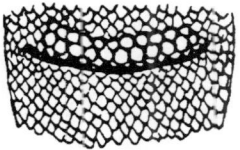
FIGURE 51. *Trapelus agilis.* Callous preanal scales. (From Nikolsky, 1915, fig. 23.)

Distribution: The deserts and steppes of North Africa, the Arabian Peninsula, Southwest Asia, and Central Asia. About eight species, three in Iran.

Remarks. Welch, *et al.* (1990:59) list *Trapelus rubrigularis* from Iran, but no citation is given for this listing.

Key to the Species of *Trapelus* in Iran

1a. Dorsal scales homogeneous, large scales of back grading into progressively smaller scales of flanks, no distinctly larger scales among them (Fig. A)............. .... *Trapelus agilis* (Olivier, 1804) (p. 100)

A *Trapelus agilis* (MNHN 5708, male [syntype])   B *Trapelus persicus* (BMNH 1946.8.11.39, male [syntype])
(Photos by A. Leviton)

1b. Dorsal scales heterogeneous, back and usually flanks with scales of varying sizes intermixed (Fig. B) 2

2a. Abdominal scales distinctly keeled; largest dorsal scales about twice width of adjacent small scales; at least anterior oval vertebral spots linked together to form undulating gray or lavender vertebral stripe on neck and back, bordered by brown (darker) stripes extending onto dorsal surface of head; males with distinct gular sac ....
.... *Trapelus persicus* (Blanford, 1881) (p. 104)

2b. Abdominal scales smooth (rarely faintly keeled); largest dorsal scales about three times width of adjacent small scales; oval vertebral spots often indistinct, contained within dark crossbars and not linked into longitudinal stripe; males without gular sac ....
.... *Trapelus ruderatus* (Olivier, 1807) (p. 107)

Trapelus agilis (Olivier, 1804) (Plate 5A–D)
Brilliant agama, steppe agama

Agama agilis Olivier, 1804:4:394 and atlas:pl. 29, fig. 2 (Type locality: neighborhood of Baghdad, Iraq; Syntypes: MNHN 5708[2]). — De Filippi, 1865:353. — J. Anderson, 1872:384. — Blanford, 1876:314–315. — Bedriaga, 1879:37. — Blanford, 1881:671. — Boulenger, 1887a:494. — Werner, 1895:3. — Nikolsky, 1897:317. — Zarudny, 1903:10. — Nikolsky, 1907a:272; 1915: 114–115. — Smith, 1935:221–223, text-fig. 58. — F. Werner, 1936:196–197; 1938:267, 270. — Schmidt, 1939:57. — Forcart, 1950:145–146. — Abel, 1952:125–133, text-figs. 4–5. — S. Anderson, 1963:444–451, fig. 11. — Clark, et al., 1966:4. — Wermuth, 1967:6. — S. Anderson, 1968:332. — Tuck, 1971a:53. — S. Anderson, 1974:36, 42. — Tuck, 1974a:61–62. — Schleich, 1979:59. — Tuck, 1979:97–98.

Lacerta sanguinolenta Pallas, 1814, vol. 3:23, pl. 4, fig. 2 (Type locality: Kum-Ankatar, on Terek River, Russia; Types: not located).

Agama sanguinolenta: Duméril and Duméril, 1851:102. — Boulenger, 1885a:1:343. — Nikolsky, 1915:102–112, fig. 23. — Lantz, 1918:14. — Terentjev and Chernov, 1949:140–141. — Bannikov, et al., 1977:105–108. — Orlova, 1981:149–160, figs. 25–26.

Agama isolepis Boulenger, 1885a:1:342–343 (Type locality: between Bampur and Magas, Iran [restricted by S. Anderson, 1966c:230–231]; Lectotype: BMNH 74.11.23.113). — Werner, 1903:340. — Annandale, 1906:197. — Nikolsky, 1915:112–114. — Werner, 1917:198.

Agama kirmanensis Nikolsky, 1900b:389–390 (Type locality: Kurin, Kerman Province, Iran; Holotype: ZIL 9321 [destroyed]). — Zarudny, 1903:10. — S. Anderson, 1963:475. — Wermuth, 1967:16. — S. Anderson, 1968:332.

Agama sanguinolenta var. *isolepis*: Nikolsky, 1907a:272. — F. Werner, 1936:200.

Agama kirmanensis var. *brevicauda* Nikolsky, 1907a:272–273 (Type locality: Kochrud, Irak-Adschemi, Iran; Holotype: ZIL 10335 [destroyed]).

Stellio kirmanensis: Werner, 1936:200.

Agama agilis agilis: Wettstein, 1951:433–434. — S. Anderson, 1966b:230. — Wermuth, 1967:6. —

Haas and Y. Werner, 1969:335–336. — Schleich, 1977:127, 129. — Nilson and Andrén, 1981: 134–135, fig. 4. — Welch, 1983:15.
Agama agilis isolepis: Wettstein, 1951:434–435. — Mertens, 1956:93. — Guibé, 1957:137; 1966a: 98. — S. Anderson, 1966a:230. — Wermuth, 1967:6. — Haas and Y. Werner, 1969:335–336. — Schleich, 1977:127, 129. — Welch, 1983:15.
Agama agilis sanguinolenta: S. Anderson, 1966a:230. — Wermuth, 1967:6–7. — Schleich, 1977: 127, 129.
Agama blanfordi blanfordi: Tuck, 1971a:53 (part; nec S. Anderson, 1966).
Trapelus agilis: Welch, et al., 1990:59. — Clark, 1991:36.
Trapelus agilis agilis: Leviton, et al., 1992:19–20, pl. 2G.

Diagnosis: Dorsal scales subequal, homogeneous, large scales of back grading into progressively smaller scales of flanks, no distinctly larger scales among them (but see remarks below); 1–3 rows of callose scales anterior to vent, more prominent in males than in females.

Color pattern: Dorsum gray or sandy, with more or less distinct dark brown or red crossbars containing a vertebral and one or two dorsolateral series of oval light spots, this pattern more distinct in females and young than in males, in which the back tends to be mottled with light-colored scales; venter cream, the throat and belly often streaked with brown, and in males with dark blue or lavender; a black patch in the shoulder fold.

Sexual dimorphism: In addition to color pattern differences mentioned above, the males are able to distend the gular region to display the blue color, although the gular "sac" is not as well developed as in *Trapelus persicus*. The "glandular" preanal scales which usually occupy two or three transverse rows in this species are more pronounced in males than in females. The dorsal crossbars of gravid females are usually dark orange or red in life.

Size: Snout-vent length (maximum reported): 115 mm; tail: 177 mm.

Habitat: This lizard is found on alluvial soils, on flat open plains of silt or gravel, hammadas, alluvial fans, and dry stream channels. It is associated with steppe and desert vegetation, e.g., *Artemisia herba-alba/Zygophyllum atricoides, Sueda* communities, *Artemisia/Alhagi, Euphorbia* spp., and shrub *Acacia*. In heavily grazed areas devoid of shrubby vegetation it is associated with small boulders and rock piles, such as those used to mark boundaries of wheat fields and gravel piles used in road construction and maintenance. In central Iran and in Baluchistan, it is also found on sandy flats and small dunes accumulating at the base of mountains. It utilizes shrubs, rocks, and rock piles (up to one meter high) as territorial observation posts and for orientation relative to the sun and heated surfaces for thermoregulation. It seeks refuge from predators and extreme temperatures beneath rocks and rock piles, and in sandy areas may burrow just beneath the sand surface. A "sit-and-wait" predator, it rarely ventures more than a very few meters from such refuge. Where shrubs, rocks, or rock piles are numerous, this species may be abundant, their observation posts less than 20 m apart. They often occur in close proximity to the large rock-inhabiting agamids, such as *Laudakia nupta*, and *L. caucasia* in foothill areas, but the microhabitats do not overlap, the larger species being confined to large rock outcrops with deep crevices.

Distribution: This species, in the broadest sense, extends from the western margins of the Iranian Plateau through Iran, Afghanistan, and Pakistan to western Punjab and north into the Asiatic steppes of the Central Asian republics and Russia to about 48°N. An isolated population occurs on the western side of the Caspian Sea in the Terek River region. In Iran, it extends from the foothills of the Zagros Mountains on the west, east through Khorasan, and north into Mazanderan. It extends south of the plateau to the Persian Gulf in Fars,

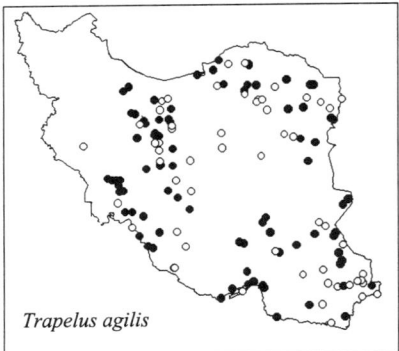

Trapelus agilis

Kerman, and Baluchistan. It is excluded only from the high mountains, Azarbaijan, the northern slopes of the Alborz Mountains, and the south coast of the Caspian Sea. Its presence on the Mesopotamian Plain is doubtful, in spite of the fact that the types were said to have come from the neighborhood of Baghdad; such localities must be given the broadest latitude in their interpretation. All Arabian records appear to be misidentifications of the closely related *Trapelus flavimaculatus* and *T. jayakari*. Three specimens from Yemen (FMNH 66219, 66434, 82295) which Schmidt (1953b:258) identified as *T. isolepis* (= *agilis*) I have redetermined as *Pseudotrapelus sinaitus*.

Remarks: There is considerable variation in pholidosis, color pattern, development of gular sac, and meristic data in this species. Much of this variation appears to be geographically correlated and the systematic status of several populations is currently under review. Specimens from northern Iran (Mazanderan and northern Khorasan Provinces), Turkmenistan, and northern Afghanistan have the most homogeneous dorsal pholidosis, with strong keels and long mucrones forming distinct ridges along back and flanks. The ventrals and gulars are also strongly keeled and mucronate. The epithet *sanguinolentus* has usually been applied by European writers to these populations, which occur on the northern edge of the Plateau of Iran and northward. It should be noted that the type locality for this taxon is on the Terek River, where there is an isolated population.

More variable are those populations found on the Central Plateau, south into Baluchistan, and eastward into Sistan, southern Afghanistan, and Pakistan. In general, the scales tend to be less strongly keeled, but still distinctly mucronate, while the dorsal scales are somewhat heterogeneous in size and carination. The name *isolepis* has been used for these lizards. However, the syntype of *Trapelus agilis*, which I examined (Fig. 52a,c), agrees more with these plateau populations than with either the northern or western lizards, although the type locality is given as Baghdad.

BMNH 1951.1.6.54 from Bandar-e Lengeh, collected by G. Popov, is an enigmatic specimen. It is a male, snout-vent 70 mm, tail 115 mm. The dorsal scales are heterogeneous, the enlarged scales on flanks more or less arranged in rows, not much larger than adjacent scales (a character that is variable in *Trapelus persicus*); the dorsal scales are less strongly keeled and mucronate than in specimens of *T. agilis* from Kerman Province. The ventral scales are weakly keeled, comparable in this respect to some *agilis*. The dorsal color pattern is rather indistinct, but the dorsal oval spots are not linked to form an undulating stripe and no dark pattern extends onto head. There is no pattern on venter, although the throat is dark gray (blue in life?). There are no dark marks anterior to shoulder as in most *agilis* and *persicus*. The callose preanal scales are in a single row (but this is true of some *agilis*). The gular sac is somewhat more pronounced than in most *agilis*, but not to the extent of most *persicus*. It lacks vertebral and lateral rows of enlarged, strongly mucronate scales on the neck. In my opinion, this specimen belongs with those populations I have identified with *T. agilis* rather than with *T. persicus*.

The most distinctive series examined is that from the western foothills of the Zagros Mountains (Khuzestan and Fars Provinces). They differ from all other populations in that there is reversal of the direction of imbrication of the scales on the anterior part of the neck, and the dividing line between forward and backward imbricating neck scales is often marked by a single large scale, all margins of which are free. The males of this series have the most heterogeneous dorsal scales, some approaching *T. persicus* in this character, and have the tail distinctly compressed laterally, differing in this respect from all specimens examined from elsewhere in the range, and from *T. persicus*. FMNH 170936 from Meshrageh, 53 miles [85 km] SE Ahvaz, has the most strongly compressed tail, and agrees with Nikolsky's description of *T. kirmanensis* in all respects, except that the ear opening is nearly as large as the eye opening, the callose preanal scales are in one row, and the tail is shorter (as in *T. kirmanensis brevicaudus*). Three of the paralectotypes of *Agama isolepis* Boulenger, BMNH

FIGURE 52. Syntypes of *Trapelus agilis* (MNHN 5708, male) (A, C) and *T. persicus* (BMNH 79.8.11.15.39/1946.8.11.39, male) (B, D). (A) Lateral view of head of *T. agilis*; (B) lateral view of head of *T. persicus*; (C) dorsal view of body of *T. agilis*; (D) dorsal view of body of *T. persicus*. (Photos by A. Leviton.)

79.8.15.10–12, from Dehbid, have the reversed imbrication on the neck, representative of the foothill population, although the male and young have only a single row reversed.

Pending a thorough analysis of meristic data for the many available specimens from Iran, Afghanistan, Pakistan, and Central Asia, I regard all of these populations as conspecific, and I do not believe use of subspecific names is warranted at his time (see Tsaruk [1985] for a different view). I am skeptical about the occurrence of this species in the immediate vicinity of Baghdad (the type locality) or elsewhere in lowland Iraq.

Trapelus agilis is the most widely distributed and easternmost representative of a species complex which includes *T. persicus* immediately to the southwest in the Mesopotamian Plain and North Arabian Desert, *T. jayakari* in the eastern Arabian Peninsula, *T. flavimaculatus* in western Arabia, and *T. savignii* west of the Jordan Rift in Israel and Egypt east of the Nile.

Trapelus persicus (Blanford, 1881)
Persian agama, Blanford's agama

Trapelus persicus fieldi (Haas and Werner, 1969)
Field's agama

Agama persica fieldi Haas and Y. Werner, 1969:337, fig. 3A, pls. 2–6 (Type locality: between Al-Gaisumah [=Al-Qaysumah] and Turaif, Saudi Arabia]; Holotype: MCZ 56866).
Trapelus persicus fieldi: Leviton, et al., 1992:21, pl. 3A–B.

Diagnosis: Dorsal scales heterogeneous, back and flanks with intermixed scales of varying sizes; ventral body scales distinctly keeled; conspicuous dorsal pattern of dark and light longitudinal stripes and a distinct, though fainter, vertebral pattern.

Distribution: Iraq, Kuwait, Jordan, northern Saudi Arabia, and probably the Ahvaz Plain in southwestern Iran.

Trapelus persicus persicus (Blanford, 1881) (Plate 5E–G)
Persian agama, Blanford's agama

Agama persica Blanford, 1881:674–676, pl. 59 (Type locality: Deh Bid and Kazerun, Iran; Syntypes: BMNH 79.8.15.43/1946.8.11.30, 79.8.15.39–42/1946.8.11.39–42). — Boulenger, 1885a:345. — Nikolsky, 1907:272. — F. Werner, 1936:200. — Hellmich, 1959:4. — S. Anderson, 1963:451–453, fig. 12.
Agama blanfordi Anderson, 1966b:230 (substitute name for *Agama persica* Blanford, 1881, preoccupied at that time by *Stellio persicus* J. Anderson, 1872 = *Stellio caucasius* Eichwald, 1831, then considered a species of *Agama*). — Wermuth, 1967:10.— S. Anderson, 1974:36, 42. — Schleich, 1977:127, 129.
Agama agilis: Anderson, 1968:309, 332 (part; not Olivier, 1807).
Agama blanfordi blanfordi: Tuck, 1971b:53 (part). — Welch, 1983:15.
Trapelus persicus persicus: Leviton, et al., 1992:21–22, pl. 3C.

Diagnosis: (Figs. 52b, d.) Dorsal scales heterogeneous, back and flanks with intermixed scales of varying sizes; ventral body scales distinctly keeled; at least anterior oval vertebral spots linked together to form undulating gray or lavender stripe on neck and back, bordered by darker brown stripes extending onto dorsal surface of head; male with distinct gular sac.

Color pattern: A dark stripe on side of neck extends across temporal region to eye, and a brown stripe crosses head at level of anterior part of the orbits. There is a tendency for the paravertebral light spots to link together to form longitudinal lines, these elements particularly noticeable in the young. In mature males the dorsal pattern is less distinct, and

FIGURE 53. *Trapelus persicus*. (From Blanford, 1881, pl. 59.)

back and flanks are flecked with white (Fig. 52d). Often these white flecks are single, slightly enlarged scales, and the effect is to make these scales look larger, the pholidosis more heterogeneous (Fig. B in key, page 100). All have a pattern of six crossbars containing vertebral and paravertebral light spots from shoulders to sacrum.

Sexual dimorphism: Males have a distinct gular sac, speckled or completely blue (dark gray to black in preservative). The "glandular" preanal scales are better developed in males than in females. In gravid females the dorsal crossbars are dark orange-red in life.

Size: The largest Iranian male measures 97 mm from snout to vent, tail 147. The largest Iranian female has a snout-vent length of 78 mm, tail 113. A male from Arabia is 110 mm snout-vent, tail 180 mm, while an Arabian female measures 95 and 152 mm, respectively.

Habitat: (Plates 21F–G) Almost all of the specimens I collected were found in sandy areas, active on the surface of dunes, most numerous on the fringes of dunes near low shrubs, into which they sought refuge when pursued. In the Karkheh National Park I found them on flat alluvial plain, where one was seen sitting on a small rock. It ran to one of many scattered small, thorny, green shrubs. I find no mention of the habitat of this species in the literature, other than my own previous brief remarks (S. Anderson, 1963:426, 451–452).

Distribution: The Mesopotamian Plain in lowland Iraq and southwestern Iran, the North Arabian Desert, including Jordan, and extending down the Persian Gulf coast of Saudi Arabia. It seems to be confined largely to the lowlands, but the extent of its penetration into the foothills and the Zagros Mountains is unknown; the types presumably came from these foothills and valleys of the Zagros (Deh Bid is approximately 2500 m above sea level and Kazerun at an elevation of about 1000 m). I did not find them at these localities during my brief travels through the Zagros in 1975, nor did I observe them in the lower foothills during nine months spent in that region, mostly at elevations of 200 to 1000 m. It is perhaps worth noting that the types of this species were among a collection made by a collector unknown, who traveled from Bushire to Esfahan via Shiraz, a route which would, indeed, have taken him through the alleged type localities; however, the bulk of the collection came from Bushire, and current knowledge of the distribution of the species makes it not wholly unlikely that there may have been a mixup

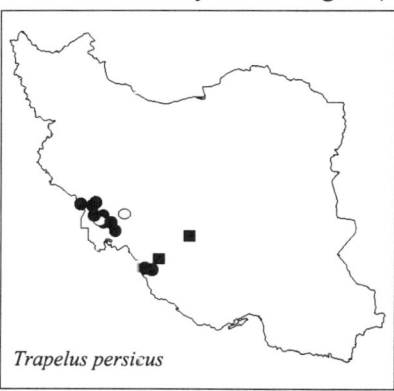

Trapelus persicus

in collection data, the road across the coastal plain from Bushire crossing typical habitat for *Trapelus persicus*.

Remarks: I have examined Blanford's types of *Agama persica* from Kazerun (BMNH 79.8.15.43/1946.8.11.30) and Deh Bid (BMNH 79.8.15.39–42/1946.8.11.39–42) (Figs. 52b, d; Fig. 53). They agree in all essentials of pholidosis and color pattern with specimens from the Ahvaz Plain of Khuzestan, Iran; Baghdad, Diala, and Sheikh Faleh's near Amara, Iraq; Qasr Burqa, Jordan; Dahran and Abqaiq, Saudi Arabia. In the specimen from Kazerun, an immature female, the ventral scales are not distinctly keeled, but the cornified layer has been lost; the dorsal color pattern is bleached; there is a distinct black bar in front of the shoulder. In BMNH 78.8.15.40, a male, and in 78.8.15.42, an immature male, the enlarged scales on the upper side of the neck tend to be in clumps, rather than in rows, but in the other male, 79.8.11.15.39, they are in rows. 79.8.11.15.41 is a gravid female.

The heterogeneity of some specimens of *Trapelus agilis* from the foothills of Khuzestan, as well as the fact that the types of *T. agilis* were said to have come from the vicinity of Baghdad (where specimens of *T. persicus* have been collected), and the types of *T. persicus* from Kazerun and Deh Bid, Iran (where *T. agilis* also occurs) led me to question the taxonomic status of *T. persicus* prior to examination of a syntype (S. Anderson, 1968:309). Were it not for the apparent sympatry of the two forms, I would hesitate to recognize them as more than subspecifically distinct on morphological grounds, considering the variation that exists in both taxa.

The two species are readily separable on the basis of the color pattern (see diagnosis). *Trapelus persicus* is further distinguished from *T. agilis* in having a short vertebral row of three or more enlarged, strongly mucronate scales on the anterior part of the neck, and another row of similar scales on each side of the neck between the dark longitudinal stripes. These scales are more obvious in juveniles and females than in males, where all scales are strongly mucronate and present less contrast. In some *T. agilis*, slightly enlarged scales can be seen in the same areas, but when this occurs, there is usually a single (or at most, two) enlarged vertebral scales, and a small cluster, rather than a row, on the side of the neck.

The development of the gular sac in males is variable, and the differences between *T. persicus* and *T. agilis* in this character are apparent only when a series of the two are available for comparison. Also of subjective interpretation is the heterogeneity of the dorsal scales. This is highly variable in *T. persicus*, but all specimens are less homogeneous than any specimens of *T. agilis* examined other than some foothill specimens. The latter are readily distinguished from *T. persicus* by color pattern, laterally compressed tail in males, and by the reversed imbrication of the anterior neck scales (see above under *T. agilis*), a condition never seen in *T. persicus*.

While the "glandular" preanal scales of *T. agilis* occupy from one to three rows, with a single exception (see below), none of the *T. persicus* examined has more than a single row. Typically, there is a clear disparity in size between the largest and smallest dorsal scales, the enlarged scales irregularly distributed. Groups of paravertebral scales have the mucrones sharply upturned, emphasizing the vertebral spots, which are usually composed of somewhat smaller, less mucronate scales. Of those examined, the Saudi Arabian specimens are the least heterogeneous in pholidosis.

In CAS 157138, a male, from Ataria, Iraq, 50 km east of Baghdad, the color pattern is indistinct, and the callose preanal scales are in two rows. The scale characteristics are extreme, the enlarged scales strongly mucronate and upturned; the enlarged scales of neck and back of head are very pronounced, pointed, and upturned.

In view of the possible confusion between *T. persicus* and foothill specimens of *T. agilis*, Nikolsky's (1907) specimens from Malamir (= Qal'eh-ye Tol) should be reexamined.

The relationships of this taxon seem to me to be with the more uniformly scaled species of the *T. agilis* complex, rather than with the *T. ruderatus – mutabilis* group, which has distinctly heterogeneous pholidosis.

Haas and Werner (1969:337–339) described *Agama persica fieldi*, based on material from between Al-Gusaimah and Turaif, Saudi Arabia. They stated that their form is also known from El Widien and from south of Tekrit in Iraq, and from Qa el Umari, northeastern Jordan, based on Steindachner's (1917:147–149, pl. 3) description and plate of a male and female from Mesopotamia.

Specimens I have examined from Iran and Saudi Arabia (CAS 84516, 84611, 84613–14, Dahran; CAS 84417, 84477, 84491–2, Qatif) do not have the distinct dorsal stripes of Haas and Y. Werner's holotype and female paratype (Haas and Y. Werner, 1969:378–380, pls. 4, 5a) although most show a tendency to fusion of elements of the dorsal pattern, most clearly on the neck (Fig. 52b), and several resemble the juvenile paratype (Haas and Y. Werner, 1969:380, pl. 5b). The Arabian specimens show this tendency more distinctly than do the Iranian lizards. Until a series is available from the Zagros Mountains in the vicinity of Kazerun and Deh Bid (if, indeed, this species actually occurs there), I withhold judgment about the subspecific status of the lowland Iranian populations. See the discussion by Haas and Y. Werner (1969:336–339) for further information.

Trapelus ruderatus (Olivier, 1804)
Olivier's agama

Trapelus ruderatus megalonyx Günther, 1864
Afghan ground agama

Trapelus megalonyx Günther, 1864:159 (Type locality: ?Afghanistan; Holotype: BMNH 60.3.19.1063/1946.8.11.34). — Welch, *et al.*, 1990:59.
Agama megalonyx: Boulenger, 1885a:347. — Alcock and Finn, 1896:566. — Annandale, 1905:88 — S. Anderson, 1963:475; 1968:332.
Agama microtympanum F. Werner, 1895:3, pl. 3, fig. 2 (Type locality: Persia; Holotype: NMW 19457); 1936:200. — S. Anderson, 1963:475. — Wermuth, 1967:18. — S. Anderson, 1968:332. — Welch, 1983:17.
Agama ruderata baluchiana Smith, 1935:223–224 (Type locality: Quetta District, Pakistan; Holotype: BMNH 1934.3.3.1/1946.8.11.20). — Schleich, 1977:127, 129.
Agama ruderata megalonyx: S. Anderson, 1974:36, 42.

Diagnosis: Nostril below canthus rostralis; dorsal scales heterogeneous, back with scales of varying sizes intermixed, enlarged scales not extending onto flanks; 68–88 scales around widest part of body; upper surface of thigh usually lacking distinctly enlarged scales, or with an area of large scales not intermixed with small scales; callose preanal scales in single row; none of neck scales with reversed imbrication; males without gular sac.

Remarks: See remarks under *Trapelus r. ruderatus*.

Trapelus ruderatus ruderatus (Olivier, 1804) (Plate 5H–I)
Horny-scaled agama, Olivier's agama

Agama ruderata Olivier, 1804:4:395, pl. 29, fig. 3 (Type locality: Persia and northern Arabia; Holotype: MNHN 2610); Olivier, 1807:429, pl. 29, fig. 3. — Bedriaga, 1879:37 — Murray, 1884b:370–371. — Boulenger, 1885a:348. — F. Werner, 1895:3; 1903:340. — Nikolsky, 1915:

115–117. — F. Werner, 1917:198–199; 1936:200. — Rostombekov, 1938:12. — Forcart, 1950: 146. — Wettstein, 1951:435. — Abel, 1952:125–133, figs 1, 2b. — Schmidt, 1955:203. — S. Anderson, 1963:453; 1968:332. — Welch, 1983:17.
Agama lessonae De Filippi, 1865:353 (Type locality: near Ispahan, Iran; Holotype: Turin Museum).
Trapelus ruderatus: J. Anderson, 1872:384–386.
Trapelus ruderata: Blanford, 1876:315–316.
Agama ruderata ruderata: Wermuth, 1967:22. — Tuck, 1971a:54. — S. Anderson, 1974:36, 42. — Bannikov, *et al.*, 1977:109. — Schleich, 1977:127, 129.
Trapelus ruderatus ruderatus: Leviton, *et al.*, 1992:22–23, pl. 3D.

Diagnosis: Nostril below canthus rostralis; dorsal scales heterogeneous, back and flanks with scales of varying sizes intermixed; 80–121 scales around widest part of body; upper surface of thigh with patch of enlarged scales among smaller scales; callose preanal scales in two or three rows; small patch of scales on neck just posterior to occiput in which direction of imbrication is reversed, i.e., with anterior margins imbricate; males without gular sac.

Color pattern: *Trapelus ruderatus ruderatus*: Dorsum sandy gray or grayish brown; five dark transverse bars on trunk, interrupted by vertebral series of light subquadrangular to ovoid spots, and one or two less distinct dorsolateral series on each side; tail also with dark bars interrupted by light vertebral spots; pattern sometimes indistinct in males; limbs indistinctly barred with brown; males with light blue cast on chin (at least seasonally), sometimes a rufous tinge to entire venter. Apparently considerable metachrosis is possible, as in *T. agilis*. *Trapelus r. megalonyx*: Dorsal pattern much like that of *T. r. ruderatus*, but vertebral light spots often clearly margined with very dark brown; up to three additional series of dark-edged spots on each side of the vertebral line, within the dark transverse bars, which are narrower than the interspaces; often dark bars enclosing light spots on limbs as well as back and tail; dark bar crosses head between eyes. In life transverse bars of females red (seasonal?), those of males dark brown or gray; throat pink in females; throat of males slatey, or with longitudinal gray stripes, venter flecked with gray (Clark, *et al.*, 1969:293).

Size: Largest Iranian *Trapelus r. ruderatus* male measures 64 mm from snout to vent, tail 94 mm. Largest Iranian female 81 mm snout-vent length, tail 94 mm. Specimens of *T. r. megalonyx* from Afghanistan: largest male 64 mm snout-vent, tail 94 mm; largest female 76 mm snout-vent, tail 94 mm. Terentjev and Chernov (1949:141) give 94 mm as the maximum snout vent length for *T. r. ruderatus*, while Blanford's (1876:316) largest specimen measured 85 mm snout to vent.

Sexual dimorphism: Males with distinct callose preanal scales.

Natural history: Blanford found this species in bushes more commonly than on the ground, but the Clarks seldom found *Trapelus r. megalonyx* near bushes in Afghanistan; when disturbed, the lizards ran a few meters then stopped motionless, relying on camouflage. In August they were active through the day, but tended to hide during the hottest hours, becoming more active by evening (Clark, *et al.*, 1969:294). Weber (1960:153) collected *T. r. ruderatus* in Iraq when soil temperatures were between 42.1°C and 49.6°C. Abel (1952: 125–133) recorded observations on this species in captivity.

All females measuring 42 mm or more in snout-vent length (January, May, August) have yolked ovarian follicles 1–3 mm diameter. Blanford (1876:316) found up to 13 eggs in oviducts of specimens collected in early June and later near Shiraz or Esfahan. The only specimens I have examined which have eggs in the oviducts are *T. r. megalonyx* collected in Afghanistan in May and August. The smallest juveniles examined (snout-vent length 27 mm, tail 32 mm) were collected in August. The limited data suggest that two clutches per year may be laid, one in late spring or early summer, and another in early fall.

Stomachs contain beetles, beetle larvae, ants, and spiders.

Habitat: I found this species on the terrace at Persepolis and Reed found it in Iraq in open stony places or on the walls of latrines and wash stands (Reed and Marx, 1959:98). I found it 10 km south of the Arax River in overgrazed *Artemisia*-steppe on rocky alluvium. Just north of the Iranian border it can be found in the arid and rugged Diabar Valley, which is crossed by barren ridges and cut by deep gorges (Sobolevsky, 1929). The Clarks found *Trapelus r. megalonyx* on a broad, open plain, most numerous near the road, in an area bare, dusty, stony in places, with sparse vegetation. The lizards were on exposed terrain with little cover. From near sea level to 2100 m elevation.

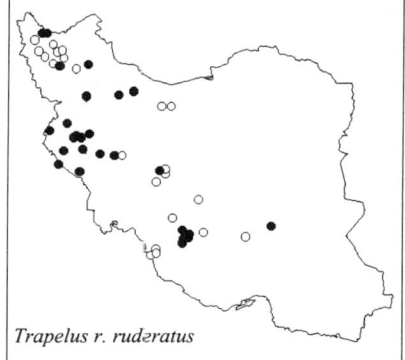

Trapelus r. ruderatus

Distribution: Essentially a species of the Iranian Plateau. In Iran, there are only three records from the eastern and central parts of the Central Plateau: two records for Kerman Province and a record of *Trapelus r. megalonyx* from the Perso-Baluch border. BMNH 95.5.29.1, a male, from Kerman (snout-vent 60 mm, tail 84 mm) has 91 scales at midbody and agrees with *T. r. ruderata* in every respect. There is a specimen in the University Museum, Copenhagen University (not seen by me) identified as *T. ruderatus* from Faisabad, Sistan, Afghanistan. All other Afghan records are from the eastern part of that country. The distributional limits of the two subspecies consequently have not been established, nor is it known whether or not there is a zone of intergradation. *Trapelus ruderatus* does occur south of the Iranian Plateau, at least locally, and I have examined specimens from northeastern Jordan, between pumping stations H4 and H5 southeast of Jebel Druze (FMNH 74567), "Moab, Syria," apparently the area east of the Dead Sea, now Jordan (USNM 56991; CAS 53990), and from approximately 32 km south of Damascus, Syria (FMNH 74566). These are all upland areas. Thus the range of *T. ruderatus* overlaps that of *T. mutabilis*. It barely extends north of the Iranian border into Azerbaijan. In Turkey it occurs in central and southeastern Anatolia, extending north up to the Lake Van region and as far west as Ankara (Başoğlu and Baran, 1977:202). All records for Arabia, Israel, and Sinai apparently refer to *T. mutabilis*.

Remarks: I have examined the type of *Agama microtympanum* F. Werner (from an unspecified locality in Iran), and find no significant difference between it and specimens of *T. ruderatus*. The cornified epidermal layer is lacking, perhaps accounting for the absence of keels on the scales. It is similar to other specimens of *T. ruderatus* in the size of the ear opening, which is about one-half the size of eye opening (about one-third diameter of eyeball). The color pattern is indistinct. The condition of the specimen suggests that it had dried out, perhaps prior to preservation, and this may account for loss of the cornified epidermal layer. There is considerable variation in degree of carination in this species, some individuals having very weakly keeled dorsal scales, and removal of the outer epidermal layer produces scales similar in appearance to those of *A. microtympanum*. The type of *A. microtympanum* has the callose preanal scales in a single row, and none of the scales of the nape are reversed in their imbrication; there are 83 scales around the widest part of the body. Thus it may be *T. r. megalonyx*. However, a juvenile from Shiraz also lacks reversed imbrication. Six male Iranian specimens have 80 to 102 scales around widest part of body, 12 females 93–121, and a small juvenile 105; a male from Syria has about 98. Specimens from Afghanistan and Pakistan have 68–88 scales around body. *Agama ruderata baluchiana* Smith has been considered a synonym of *T. r. megalonyx* and both have been regarded as synonymous with *T. ruderatus* (see discussions in Smith 1940:384, and Clark, *et al.*, 1969:292–4; Minton 1966:95–96 regards them as separate taxa, however).

I have examined the holotypes of *Trapelus megalonyx* Günther and *Agama ruderata baluchiana* Smith and could not see any obvious differences. The color pattern of the type of *megalonyx* is distinctive: eight light vertebral ocelli ringed with black, first on nape, eighth just behind pelvis. These are within gray crossbars of width similar to the ocelli, the crossbars alternating with light spaces that are slightly narrower; paravertebral rows of much smaller ocelli also within the dark crossbars. The snout-vent length of this specimen, probably a female, is 62 mm, tail 74 mm; there are 93 scales around midbody. The type of *baluchiana*, a female, is in bad condition and fragile, apparently having rotted prior to or during preservation. I did not attempt to count midbody scales. The color pattern cannot now be discerned.

Smith (1935:225) says of *Agama megalonyx* (which he does not consider a subspecies of *ruderata*): "The type, a female, probably came from Baluchistan. I have examined a second example (juvenile) from near Quetta. Annandale records two more from the Perso-Baluchistan border."

He says of *Agama ruderata baluchiana* (Smith, 1935:223–224): "Differs from the typical form in the dorsal scalation, the enlarged scales being less differentiated, less nail-like, more like those of *A. persica*, in the shorter limb, and in the coloration, there being no white streak along the back of the thigh. Only a single specimen is known." The two forms recognized by Smith are sympatric, in his view.

Trapelus mutabilis appears to be the species most closely related to *T. ruderatus*. It occupies desert country at low elevations from Iraq and northern Saudi Arabia westward through Syria, Jordan, Israel, Sinai, and North Africa to Morocco. *Trapelus mutabilis* has the nostrils on or slightly above the canthus rostralis, so that they are clearly visible when the head is viewed from above, while *T. ruderatus* has the nostrils below the canthus so that they are not seen from directly above. There are also color pattern differences, at least in the area of sympatry, *T. mutabilis* having the ground color light gray rather than tan and having but two crossbars on the back, one in the scapular area, the other across sacrum.

I am persuaded that *T. megalonyx* is distinct from *T. ruderatus*. My general subjective impression is that it may be closer to *T. persicus* than to *T. ruderatus*.

Family ANGUIDAE
Anguids, Lateral-fold lizards

In his revision of the family, based on the fossil and Recent species, Meszoely (1970) includes both *Anguis* and *Ophisaurus* in the subfamily Anguinae. The review by Petzold (1971), based on the living species, places *Anguis* alone in the Anguinae and assigns *Ophisaurus* to the Gerrhonotinae. The revision of the anguinomorph lizards by Rieppel (1980), based on an anatomical study of this entire clade, supports the placement of both genera in the Anguinae.

Key to the Anguidae of Iran

1a. A deep lateral fold from head to level of vent (A); teeth blunt, with conical crowns; vestiges of hind limbs appearing as elongate papillae at sides of vent (B) .... *Ophisaurus apodus* (Pallas, 1775) (p. 113)
1b. No lateral fold; teeth long and sharp; hind limbs lacking .... *Anguis fragilis colchicus* (Nordmann, 1840) (p. 111)

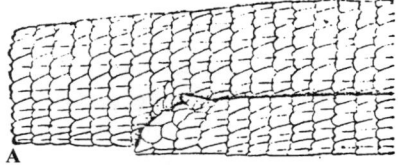

(A) Lateral fold and (B) Vestige of hind limb in *Ophisaurus apodus*
(From Bannikov, *et al.*, 1977, fig. 31)

Genus *Anguis* Linnaeus
Slow worms, blind worms

Anguis Linnaeus, 1758:227 (Type species: *Anguis fragilis* Linnaeus, 1758, by subsequent designation of Fitzinger, 1843:23).

Definition: No lateral fold. Scales rounded, in oblique rows on sides. Body serpentine. Limbs lacking. Teeth acutely pointed. Palatine bones toothless.

Distribution: Europe including Britain, western Asia, northwestern Africa. A single species.

Anguis fragilis Linnaeus, 1758
Slow worm, blind worm

Anguis fragilis Linnaeus, 1758:229 (Type Locality: Sweden [restricted by Mertens and L. Müller, 1928]).

Anguis fragilis colchicus (Nordmann, 1840) (Plate 6A–B)
Colchican slow worm

Anguis fragilis: Pallas, 1814:55–56. — De Filippi, 1865:355. — Bedriaga, 1879:26. — Nikolsky, 1907a:280; 1915:247–258, fig. 34. — F. Werner, 1936:200.
Otophis eryx var. *colchica* Nordmann *in* Demidoff, 1840:341 (*nomen conservandum*); Amph. Rept. (1842); pl. 3, figs. 1–3 (Type locality: Abasien [= Kuban region] and Mingrelien, Ukraine; Syntypes: not located).
Anguis orientalis J. Anderson, 1872:376, fig. 1 (Type locality: Rasht, Iran, on Caspian Sea; Holotype: ZIS 4829 [see Das, *et al.*, 1998]). — Blanford, 1876:394–395.

Anguis fragilis var. *colchica*: Boettger, 1888:902.
Anguis fragilis colchicus: Štěpanek, 1937:110. — Wettstein, 1951:437–438. — Mertens and Wermuth, 1960:87–88. — Guibé, 1966:98. — S. Anderson, 1963:475; 1968:332. — Tuck, 1971a:55. — Petzold, 1971:38. — S. Anderson, 1974:30. — Bannikov, *et al.*, 1977:142. — Schleich, 1977:127, 129. — Welch, 1983:57. — Dely, 1984:241–258, figs. 41–43.

Diagnosis: Ear opening distinct; 26–30 scale rows at midbody; internasal often in contact with frontal (Fig. 54a); dorsum of males frequently with blue spots.

Color pattern: Dorsum of adult brown or bronze, uniform over about 8–9 longitudinal scale rows, flanks sometimes darker brown to black, venter lighter buff or tan; back of males often with two irregular longitudinal rows of bright blue spots. Juveniles light tan on dorsum in preservative; silvery, copper, or golden in life, with a narrow dark brown to black middorsal stripe (actually two stripes, under magnification); sides, and sometimes venter, dark brown to black.

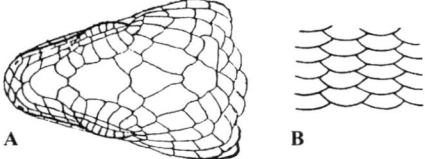

FIGURE 54. *Anguis fragilis*. (A) Dorsal view of head; (B) group of dorsal scales. (Head from Bannikov, *et al.*, 1977, fig. 35; dorsal scales from Terentjev and Chernov, 1949, fig. 42.)

Size: Terentjev and Chernov (1949:164) cite a snout-vent length of 265 mm. A female from Iran measures 185 mm from snout to vent, tail 195 mm, and a male measures 185 and 155+ mm, repectively. Dely (1984:250) gives 269 mm snout-vent and 270 mm tail for males, 290 mm snout-vent and 244 mm tail for females of this subspecies.

Natural History: Few natural history observations have been recorded for *Anguis* in Iran, although much is known about its biology, behavior, and ecology in the European parts of its range. An extensive synopsis of the natural history is provided by Petzold (1971:42–55; see also Smith, 1951:171–182) and by Dely (1984:251–255); Terentjev and Chernov (1949:165) give a brief account of habitat, behavior, and reproduction in the former U.S.S.R.). These lizards feed on slugs, snails, and earthworms, for which their long teeth are particularly adapted; they also take spiders and soft-bodied insects, particularly larvae. They are ovoviviparous, the young delivered in egg sacs which are ruptured immediately. The single Iranian female that I have examined, collected in late October, contains many yolked follicles of various sizes in the left ovary, the largest follicle about 2 mm in diameter.

Habitat: (Plate 23A) On 28 May 1975, I collected five specimens in dense broad-leaved summergreen Hyrcanian forest south of Amol at 450 m elevation. The adults were exposed on the leaf litter of the dark forest floor between 1340 and 1600 hours. Air temperature was 26°C. A juvenile was collected under a log. The composition and relationships of the Hyrcanian forest are discussed by Zohary (1963:22–30, 92–93) and Bobek (1968:284–285). Other reptiles and amphibians occurring in this habitat are *Ophisaurus apodus*, *Lacerta chlorogaster*, *Natrix tessellata*, *Rana ridibunda* (along streams and around ponds), and *Rana macrocnemis pseudodalmatina*.

Distribution: *Anguis fragilis* occurs from Britain and southern Scandinavia to northern Spain and east through Europe through Russia to the Urals. The subspecies *colchicus* occurs along the southern and eastern coasts of the Black Sea and south of the Caucasus to the Caspian. In Iran this species is restricted to the Caspian region, probably occurring not much farther east than the valley of the Rud-e Atrak. It is reported from Shahrud, on the south slope of the Kopet Dagh. De Filippi (1865:355) records specimens brought from Tehran by G. Doria, but these were probably collected on the Caspian coast. A map of the distribution of *A. fragilis* is provided by Arnold and Burton (1978:264, map 93) and its distribution in

Anguis fragilis

the former Soviet Union is mapped by Bannikov, et al. (1977:360, map 60). Petzold (1971:40–42) has presented a zoogeographic interpretation of the post-Pleistocene distribution of the subspecies of *Anguis fragilis*.

Remarks: Nikolsky (1915:252) and Terentjev and Chernov (1949:164–165) expressed the view that the lizards in the southeastern part of the species range were not distinct from the nominate subspecies, but Petzold (1971:38–42) reviewed a convincing body of evidence arguing for the validity of *colchicus* as a distinct taxon. Nonetheless, individuals cannot always be assigned reliably to subspecies on the basis of his characters. A male from 26 km south of Amol (CAS 141160) has the frontal and internasal broadly separated by the prefrontals (a characteristic of *A. f. fragilis*); this specimen has 26 scale rows. A female from the vicinity of Gorgan (FMNH 141672) has 30 scale rows at midbody.

Photographs of living *Anguis fragilis* are to be found in Petzold (1971), as well as in many other publications.

Genus *Ophisaurus* Daudin
Armored glass lizards

Ophisaurus Daudin, 1803:188 (Type species: *Anguis ventralis* Linnaeus, 1766, by monotypy. See note in Zhao and Adler [1993:199]).

Definition: A deep lateral fold from head to vent. Scales squarish-rhomboidal, in straight longitudinal and transverse series. Body serpentine. Limbs absent, or rudimentary hind limbs present. Teeth with conical or subspherical crowns. Pterygoids toothed; palatine and vomerine teeth present or absent.

Distribution: Southeastern Europe to eastern and southeastern Asia, western North America and Mexico. Eleven species recognized.

Remarks: Klembara (1981) resurrected the genus *Pseudopus* for *Lacerta apoda* Pallas, 1775 and two fossil species on the basis of his study of skull morphology. In his phylogenetic hypothesis, *Pseudopus* is the sister taxon to *Parapseudopodus* (Eocene), which gave rise to *Anguis* plus *Ophisaurus*. My continued use here of the name *Ophisaurus apodus* is not intended to imply any judgment on this issue.

Ophisaurus apodus (Pallas, 1775) (Plate 6C–D)
Sheltopusik, glass lizard

Lacerta apoda Pallas, 1775:435, pls. 9–10 (Type locality: Naryn Steppe, Russia, on north coast of Caspian Sea [but see Obst, 1978:136, who considers this locality unlikely and restricts the type locality to the region of the Terek river]; Holotype: not located); Pallas, 1814:33, pl. 6.
Pseudopus apoda: Blanford, 1876:387.
Ophisaurus apus: Nikolsky, 1897:325. — F. Werner, 1903:341. — Zarudny, 1903:22 — Nikolsky, 1907:280; 1915:240–246, figs. 32–33. — Morich, 1929:30. — F. Werner, 1936:200.
Ophisaurus apodus: Mertens and Müller, 1928:16. — Forcart, 1950:150. — Schmidt, 1955:204. — Mertens, 1957:120. — Guibé, 1957:139. — S. Anderson, 1963:475. — Clark, *et al.*, 1966:2. —

Guibé, 1966a:98. — S. Anderson, 1968:332. — Tuck, 1971a:55, fig. 6. — S. Anderson, 1974:30. — Bannikov, *et al.*, 1977:140. — Schleich, 1977:127, 129. — Welch, 1983:58.
Ophisaurus apodus apodus: Obst, 1978:129–140; 1984:259–274, figs. 44–46. — Clark, 1991:37–38.

Diagnosis: Ear opening distinct; vestiges of hind limbs appearing as elongate papillae at sides of vent; pterygoid teeth in two series; dorsal scales keeled. See Obst (1978:137) for distinctions between subspecies he recognizes.

Color pattern: Brown above, lighter below. Juveniles olive-gray with dark brown transverse bars. Clark (1991:38) describes his specimens in life as dull red-brown above, white-fawn below with red and gray patches.

Size: Terentjev and Chernov (1949:163) give 500 mm as maximum snout-vent length. Of Iranian specimens I have examined, a male measures 413 mm snout-vent, tail 584 mm; a female has a snout-vent length of 330 mm, tail 463 mm.

Natural history: Virtually nothing has been recorded about the natural history of *Ophisaurus* in Iran. Terentjev and Chernov (1949:163–164) have briefly summarized aspects of its biology in the southern republics of the former Soviet Union, and in general, these statements undoubtedly pertain to its natural history in the adjacent areas of Iran. It takes refuge in rodent burrows, under rocks, and among roots, emerging in early spring after hibernation. The two specimens I collected in Iran were found basking and sluggish at 0615 hours, 31 May, and 1030 hours, 20 June 1975. Up to 10 eggs are laid in early summer, hatching beginning in mid-summer. An Iranian specimen collected in November contains yolked follicles of various sizes in the ovaries, the largest about 3 mm diameter. The diet includes snails, slugs, beetles, grasshoppers, fledgling birds, lizards, small rodents, and eggs of birds and reptiles.

Habitat: (Plates 23D, 25A) In Iran, south of Chalus, I collected it in an open grassy area among riparian vegetation (*Alnus, Populus, Equisetum, Rubus*) along a swift stream in the Hyrcanian forest at 600 m elevation. Another specimen was collected in an irrigated field in a valley of the Kopet Dagh at an elevation of 1000–1300 m. Vegetation of the surrounding hillsides was primarily *Artemisia*. Other reptiles and amphibians in this latter locality were *Eremias velox, Laudakia caucasius, Mabuya aurata, Natrix tessellata, Bufo viridis*, and *Rana ridibunda*. Clark (1991:38) found them common in May in the Gulistan Forest.

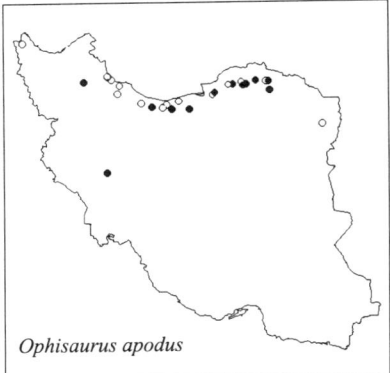

Ophisaurus apodus

Distribution: The southern Balkans, the Crimea and northeast coast of the Black Sea, the western drainages to the Caspian Sea and throughout Turkey; disjunct distribution in southern Turkmenistan, Uzbekistan, Tadjikistan, Kirghizistan, and northern Afghanistan. Thus far, it is known in Iran only from the Caspian region and from the mountainous areas of the northern Zagros, the Alborz, and the Kopet Dagh. Recorded from below sea level to at least 1,200 m elevation in Iran. Maps of the European distribution are available in Arnold and Burton (1978:264, map 94) and of the distribution in the former Soviet Union in Bannikov, *et al.* (1977:359, map 58). Its present distribution seems to be relictual, probably the remnant of a widespread Palearctic distribution. Its distribution is reminiscent of those of *Lacerta trilineata, Coluber najadum*, and *Telescopus fallax*, none of which extends as far east and northeast. Obst (1978, 1984) regards the populations of Europe and western Anatolia as the subspecies *O. a. thracius* Obst 1978; Iranian specimens belong to the typical subspecies of western Asia.

Remarks: For illustrations of living specimens see Petzold (1971), Leviton, *et al.* (1992: pl. 3G). My Iranian specimens have the head distinctly lighter in color than the body (see Plate 6C–D), a distinction supposed to be characteristic of *O. a. thracius* Obst.

Family EUBLEPHARIDAE

The only thorough review of this taxon is that of Grismer (1988). He follows Kluge (1987) in presenting evidence that the eublepharids are a monophyletic subgroup of the Gekkota on the basis of derived character states (medial contact of pares frontales, wide anterior section of nasal bones, and dorsal end of clavicle not projecting above dorsal end of scapulocoracoid). He recognizes two subfamilies, Aeluroscalabotinae and Eublepharinae, on the basis of his analysis. In his consideration of relationships among the outgroups of the Gekkota used in his analysis, Grismer (1988:372, fig. 4) utilizes Kluge's (1987:39–40) classification. He regards the Gekkota as composed of three families, the Eublepharidae, the Diplodactylidae (consisting of the Diplodactylinae and the Pygopodinae), and the Gekkonidae (Gekkoninae, Teratoscincinae, and Sphaerodactylinae). His hypothesis of the relationships among the Eublepharidae is summarized in his cladogram (Grismer, 1988:452, fig. 53).

Genus *Eublepharis* Gray, 1827
Leopard geckos, fat-tailed geckos

Eublepharis Gray, 1827:56 (Type species: *Eublepharis hardwickii* Gray, 1827, by monotypy).

Definition: Flat basioccipital bone; deep axial pockets (Grismer, 1988:441). Digits short, cylindrical, with transverse lamellae beneath, clawed, the claw partly concealed between two or four lateral scales and an upper scale; both eyelids well developed and movable; pupil vertical; males with preanal pores; dorsum with small juxtaposed scales and larger tubercles; tail shorter than head and body.

Distribution: A disjunct Southwest Asian distribution, including Iraq, Iran, Turkmenistan, Afghanistan, Pakistan, and northern India. At least four valid species, two known definitely from Iran. Börner (1981a) removed the East Asian species (*E. kuroiwai* and *E. lichtenfelderi*) from this genus, and Grismer (1988:371) concurred, placing both in the resurrected genus *Goniurosaurus* Barbour, 1908.

Remarks: See Kluge (1976:46–47) and Grismer (1988) for discussions of the relationships and zoogeography of this genus.

Eublepharis angramainyu and *E. turcmenicus* have been recorded from Iran, but *E.* cf. *macularius* occurs in eastern Afghanistan and eastern Baluchistan, Pakistan. Specimens were collected in eastern Khorasan by Zarudny (1903:9–10), but subsequently lost. Below, I list his localities tentatively under *E. macularius* (Appendix II). The distributions of the known populations of these closely related forms appear to be disjunct, and the observed morphological differences among populations is to be expected. Börner (1974, 1976, 1981a, 1981b) has seen fit to assign subspecific or specific names to each of the populations known from Pakistan, Afghanistan, and India. The distribution of these geckos as currently known extends primarily along the edges of the Iranian Plateau. This suggests that a once-continuous distribution has been fragmented by paleogeographic events antedating the development of the present fauna of Southwest Asia. Possibly the distributional discontinuity dates back to the uplifting of the Iranian Plateau during the Pliocene. In light of recent discoveries indicating previous faunal connections between Iran and the Arabian Peninsula (Arnold, 1972), *Eublepharis* should be looked for in the upland regions of Oman.

Closely related to the *macularius* group is *E. hardwickii*, known with certainty only from the hills of Chota Nagpur and Orissa and the adjacent districts, again a region of fairly long historical continuity as a continental area, but having been separated from areas now occupied by *E. macularius* by an arm of the Tethys Sea during the mid-Miocene, and since that time intermittently by climatic events, including increasing aridity in recent time (see discussion above, in section on paleogeography).

A curious structure in some eublepharid species is a deep axial pocket of unknown function. This is an invagination of the skin just behind the insertion of the forelimb. It is well developed in the Southwest Asian species of *Eublepharis*, but is absent in *Goniurosaurus*. It is not present in *Coleonyx* nor *Aeluroscalabotes*. This pocket has its greatest development in *Hemitheconyx caudicinctus* from Mauritania (CAS 55114), but is less developed in *H. caudicinctus* from Ghana (CAS 154299–302). Axillary pockets are not present in *Holodactylus africanus* that I have examined. I am confused by Grismer's (1988) remarks on axial pockets; on page 410 he lists character S92 as present = primitive, absent = derived, while in his matrix on page 468 he lists this character (C92 [*lapsus*?]) as derived in *Eublepharis*, variable in *Coleonyx* and *Hemitheconyx*. His fig. 31 (p. 409) shows *Eublepharis* as defined by the derived condition of character S92, and he includes the presence of deep axillary pockets in his diagnosis of *Eublepharis*. My observations of captive *E. macularius* have not shed any light on the function of these pockets. In preserved specimens they often contain many small mites. They bring to mind pockets in the neck and postanal regions of species of *Sceloporus*, which similarly harbor mites. Loveridge (1925) described a similar structure in *Gymnodactylus lawderanus* Stoliczka and termed it a "mite pocket." More recently, Arnold (1986) proposed that such pockets reduce the damage done to lizard hosts by ectoparasitic larval trombiculid mites, or chiggers, by concentrating them in restricted areas. Bauer *et al.* (1990, 1993) argued that such pockets result from phylogenetic and/or structural constraints on the skin of these lizards.

Key to the Species of *Eublepharis* in Iran

1a. 5–9 preanal pores (Fig. A); first labial not in contact with large, postmental chin shield (Fig. B) .... *Eublepharis turcmenicus* Darevsky, 1977 (p. 122)
1b. 11–17 preanal pores; first labial usually in contact with chin shield (Fig. C) 2

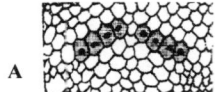 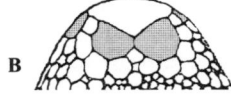 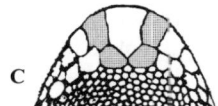

(A) Preanal pores in *Eublepharis turcmenicus*; (B) first lower labials not in contact with postmental chin shields in *E. turcmenicus;* (C) first lower labials in contact with postmental chin shields in *E. macularius*.

2a. Subdigital lamellae smooth (Fig. D) ....
.... *Eublepharis angramainyu* Anderson and Leviton, 1966 (p. 118)
2b. Subdigital lamellae each with several small tubercles (Fig. E)...
.... **Eublepharis macularius* (Blyth, 1854) (p. 120)

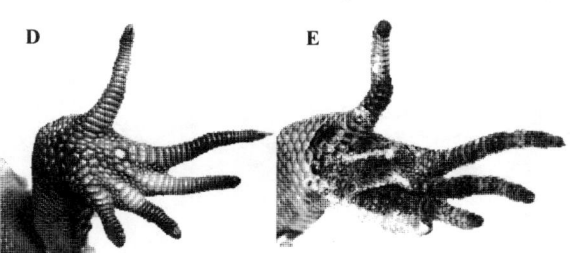

Ventral view of toes of *Eublepharis angramainyu* showing the smooth subdigital lamellae. (Photo by M. Giles.)

Ventral view of toes of *Eublepharis macularius* showing the tuberculate subdigital lamellae. (Photo by M. Giles.)

Eublepharis angramainyu Anderson and Leviton, 1966 (Plate 6E)
Western leopard gecko, Iranian fat-tailed gecko

Eublepharis macularius: Boulenger, 1885a:97 (in part; not Blyth, 1854). — F. Werner, 1917:197; 1936:200. — Smith, 1935:128. — S. Anderson, 1963:435–437, fig. 8.

Eublepharis angramainyu S. Anderson and Leviton, 1966a:1–5, figs. 1–2 (Type locality: between Masjed Soleyman and Batvand, Khuzestan Province, Iran; Holotype: CAS 86384). — S. Anderson, 1968:332. — J. J. Schmidtler and J. F. Schmidtler, 1970:239–241, figs. 1–2. — Tuck, 1971b:56. — S. Anderson, 1974:31, 43. — Schleich, 1977:127, 129. — Darevsky, 1978:207–208, fig. 3. — Welch, 1983:6. — Szczerbak and Golubev, 1986:29–30 [1996:29], fig. 7. — Grismer, 1988:441–452. — Kluge, 1991:8. — Leviton, *et al.*, 1992:36–37, col. pl. 4H. — Kluge, 1993:10.

Eublepharis ensafi Baloutch and Thireau, 1986:281–288, figs. 1–6 (Type locality: Fakke, ca 150 km N Ahvaz, Iran; Holotype: Musée Faculté des Sciences de Tehran 456).

Diagnosis: Subdigital lamellae smooth (Fig. D in key, page 117); mid-dorsal tubercles not as large as intertubercular spaces; chin shields in contact with first lower labials (Fig. C in key, page 117); ventral scales hexagonal; some elements of color pattern of head and body linearly arranged in adults (Fig. 55); males with uninterrupted series of 11–17 preanal pores, pores discernible in females. Grismer (1988:442–450; 1991:251–252) gives additional diagnostic characters: height of auditory meatus equal to distance between nostrils; mental scale shorter than wide; 41–48 eyelid fringe scales; widely spaced, pronounced and pointed dorsal tubercles, much smaller inter-tubercle granules; hexagonal ventral scales in 27–38 longitudinal rows; width of rostral 1½ times its height; undivided terminal lamellar scales; 2–3 transverse rows of ventral scales in each caudal whorl; dorsal scales of regenerated tail circular and slightly convex; supratemporal bone present; smooth basioccipital; clavicle extending above the scapulocoracoid and making broad contact with the suprascapula. Autapomorphies (Grismer, 1988): *pterygoid-palatine suture sharply V-shaped and posteriorly directed in its lateral margin*; the only species of the genus in which the *posterior margin of the coronoid shelf makes contact with the adductor fossa.*

Color pattern: (Fig. 55) Adults with a continuous light vertebral stripe, bordered on each side by a broken black stripe from occiput to base of tail; dark markings not confined primarily to tubercles and immediately surrounding scales (as in *Eublepharis macularius* from Afghanistan and Baluchistan) but confluent, linearly arranged along either side of vertebral stripe; dorsolateral dark markings also linearly arranged, confluent with transverse markings; head with a pattern of dark and light reticulations; no horseshoe-shaped mark, dark or light, on nape; limbs with numerous dark blotches; tail with numerous irregular dark

FIGURE 55. *Eublepharis angramainyu* (holotype, CAS 86384). (From Anderson and Leviton, 1966:2, fig. 1; photo by M. Giles.)

transverse markings, wider than the light interspaces; venter light tan. A juvenile has 3 dark transverse bars across dorsum, first on posterior part of neck and shoulders, second at midbody, third anterior to sacral region; middle bar largest, approximately equal to lighter interspaces, which are interspersed with dark tubercles; margins of dark bars darker than their central portions.

Size: Males 142–154 mm snout-vent length, tail 97–100 mm; females 126–127 mm snout-vent, tail 86–90 mm. Smith (1935:128) cites an Iranian specimen (as *Eublepharis macularius*) measuring 165 mm snout-vent.

Natural history: When alarmed, these lizards raise themselves high off the ground, even standing on their fingers and toes (S. Anderson, 1963:fig. 8). When captured, they give a long, loud, rattling squeak and attempt to bite. They often defecate, wrapping the short tail around their captor's hand. This may aid them in autotomizing the tail. In many ground-dwelling geckos caudal autotomy is basal, but in four specimens of *Eublepharis angramainyu* in which the tail has been broken and regenerated, the break is in the fifth caudal segment; in a single juvenile the break has occurred at the 10th segment. Two geckos were picked up dead on the road after being run over by automobiles, and in these the tails had been dropped at the base.

Two females, one collected May 22, the other August 20, have eggs in the oviducts; all other females, collected April 19, May 13, May 21, and September 5, have a single large egg (about 6–7 mm long) and several smaller eggs in each ovary. This suggests a rather long season of reproduction, probably several clutches of two eggs each being laid by each female.

Stomach contents include grasshoppers, scorpions, solpugids, large spiders, beetles, and other arthropods. Most of these creatures were seen on the roads at the same hours that the lizards were captured. Most likely, these large geckos will eat anything abroad at night that they can catch and overpower.

Foxes were observed feeding on individuals killed on the road, and probably prey regularly on them. Jackals, wolves, and owls are also abroad at night in the same region, as is *Telescopus tessellatus*, probably a lizard-eating snake.

Habitat: (Plate 22A–C) Specimens were collected on surfaced roads at night, and were fairly abundant on a few nights from mid-April until late May. A specimen was collected in late August as well. The area in which they were found is in the western foothills of the Zagros Mountains, in a region of extensive gypsum deposits. They were never found under stones in the area, and most likely they spend the day in the deep crevices and small caverns in the gypsum. In these retreats there is water for much or all of the year, and the relative humidity is fairly high. Other than the seasonal plantings of grains and the annual herbaceous plants in the spring, the hillsides where this animal lives are devoid of vegetation.

These geckos were collected when air temperatures were between 32°C and 34.4°C, road temperatures were 32.6°C to 36.4°C and at least 2°C higher than the surrounding soil temperature.

Distribution: Known from the western foothills of the Zagros Mountains, and the upper Tigris-Euphrates drainage in Iran, Iraq, and northeastern Syria. 300 to 1000 m elevation.

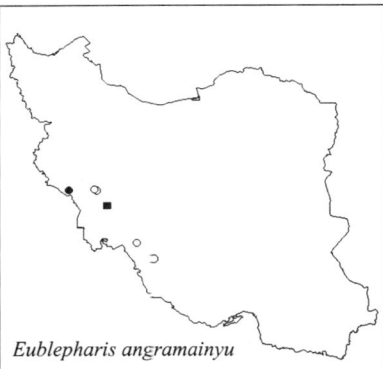

Eublepharis angramainyu

Remarks: A specimen from near the village of

Chalga, south of Chem-che, Kirkuk Liwa, Iraq (MCZ 51636) and another from Khanaquin, Iraq (CAS 157129) agree in all essentials with the Iranian series. Attempts to investigate the critical maximum temperatures for this species have been reported elsewhere (S. Anderson, 1963:436–437). The habits of this species are apparently similar to those of *Eublepharis macularius* as recorded by Minton (1966:73). The Schmidtlers (1970:239–241) have given an account of *E. angramainyu* maintained in captivity. Murray (1884) in *The Vertebrate Zoology of Sind* states: "There is, however, some risk attending the careless handling of these lizards, when killed or freshly preserved in spirit. The tubercles with which their bodies are studded contain a very irritant secretion, which, coming in contact with the naked skin of the back of the hand or other part of the body, occasions a numbness followed by a painful swelling of that part, and subsequently a species of *Herpes* which the natives in Sind cure by the application of a poultice made of chalk paste." I have handled several captive *E. macularius* as well as the numerous specimens of *E. angramainyu* that I captured in the field. Many of these animals were in an agitated state, yet never did I encounter any secretion from the tubercles. *Eublepharis macularius* has been kept in captivity frequently in recent years, but I have heard no reports of defensive secretions.

Grismer (1989) placed *Eublepharis ensafi* Baloutch and Thireau in the synonymy of *E. angramainyu*.

Eublepharis cf. *macularius* (Blyth, 1854)
Leopard gecko, fat-tailed gecko, panther gecko, spotted fat-tailed gecko

Cyrtodactylus macularius Blyth, 1854:737–738 (Type locality: Salt Range, Punjab; Holotype: ZSI).
Eublepharis macularius: Zarudny, 1903:9–10.
Eublepharis sp.: Szczerbak and Golubev, 1986:28, fig. 7.

Diagnosis: (Fig. 56) Subdigital lamellae each with several distinct small tubercles (Fig. E in key, page 117); mid-dorsal tubercles generally larger than intertubercular spaces; chin shields usually in contact with first lower labials (Fig. C in key, page 117); elements of dorsal color pattern not linearly arranged. Additional diagnostic characters (Grismer, 1988:442–450; 1991: 251–252): height of auditory meatus 1½ times internostril distance; rounded ventral scales in 21–30 rows; rostral twice as wide as high; 3 transverse rows of ventral scales in each caudal whorl; mental shorter than wide; 46–57 eyelid fringe scales; "cleft terminal subdigital lamellae" (see remarks below); widely spaced, pronounced and pointed dorsal tubercles, much smaller inter-tubercle granules; ventral scales rounded, in 21–30 transverse rows at midbody; dor-

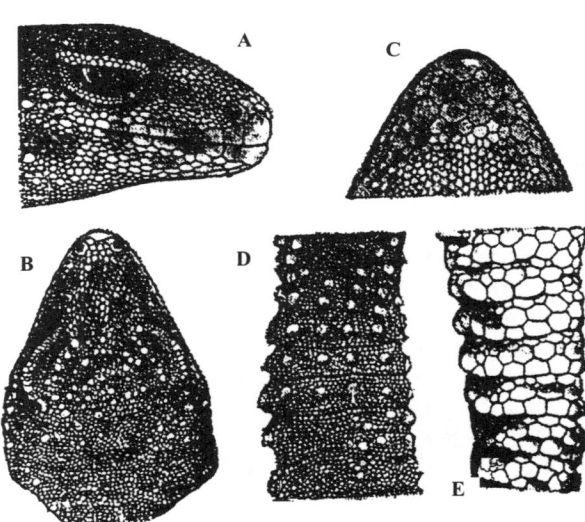

FIGURE 56. *Eublepharis macularius*. (A) Side of head; (B) top of head; (C) chin; (D) dorsal surface of tail; (E) ventral surface of tail. (After Szczerbak and Golubev, 1996:27, fig. 6.)

sal scales of regenerated tail circular and slightly convex; conspicuous supratemporal bone present; a straight pterygoid-palatine suture; a longitudinally-directed crest on ventral portion of basioccipital anterior to spheno-occipital tubercles, posterior section of basioccipital smoothly rounded; posterior margin of coronoid shelf does not contact the adductor fossa; clavicle extending above scapulocoracoid and making broad contact with the suprascapula.

Color pattern: Dorsal color of adults straw yellow to pale violaceous gray in life, often with tinge of pink; dorsum with blue-black spots, in some cases discrete and sparse, in others fusing into a reticulum; usually traces of dark juvenile bars remain visible. Juveniles dark brown to black dorsally, with two or three wide yellow bars across trunk, a white nape-band extending forward through ear onto lips. (Minton, 1966:73). See Szczerbak and Golubev (1986:27–28; 1996:27) for sexual dichromatism and age variation. Recently, captive breeders have selected color variants for the pet trade that exceed the variation observed in the field (see, for example, Tremper, 1997:16–17 for color photos). Apparently, the parental stock for the current pet trade came originally from Pakistan, probably Sind.

Size: Specimens from Pakistan measure to 158 mm snout-vent length. Males are larger than females.

Habitat: The specimens collected by Zarudny and identified as this species were found on hard clay soil strewn with sand where there were numerous bushes of *Zygophyllum*. The Clarks collected *Eublepharis macularius* in eastern Afghanistan on fairly open alluvial soil (Clark, et al., 1969:303). Minton (1966:73) found them in rock wall crevices in Pakistan.

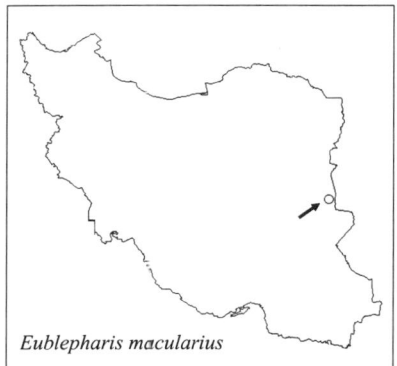

Eublepharis macularius

Distribution: From eastern Afghanistan south of the Hindu Kush and the Northwest Frontier Provinces and apparently generally through Pakistan south to Rajputana and the Khandesh District of India. In Iran, geckos tentatively assigned to this taxon are known only from Zarudny's record in eastern Khorasan.

Remarks: Zarudny collected two specimens which were lost before they could be deposited in the collections at St. Petersburg. He felt assured of his identification, due to the unique form of the tail, which, he said, could not be confused with that of any other lizard. Thus, there is little doubt that he collected *Eublepharis*, but in view of the recently recognized differences among the various populations, it is by no means certain that the geckos of eastern Khorasan are *E. macularius*. *Eublepharis turcmenicus* occurs in the mountains of northern Khorasan, *E. macularius montanus* is a name given by Börner (1976:9) to the westernmost population in Pakistan, while specimens in eastern Afghanistan have been named *E. afghanicus* by Börner (1976:10–12). Adding to the nomenclatural confusion is the fact that Börner (1974) described as a new species (*E. gracilis*) a live zoo specimen, locality unknown, but possibly from "inner or coastal [*sic*] Afghanistan" (Börner, 1976:12). Kluge (1991:8) lists *E. fasciolatus* Günther, 1864, *E. afghanicus* Börner, 1974, *E. montanus* Börner, 1976, *E. fuscus* Börner, 1981, and *E. smithi* Börner, 1981 as recognized subspecies of *E. macularius*, following Grismer's (1988:455) summary classification. See Szczerbak and Golubev (1986:6; 1996:5–6) for their comments on the nominal species and subspecies in the genus. They regard *E. fasciolatus* Günther as well as Börner's taxa, based largely on color pattern differences, as synonyms of *E. macularius*. They point out that changes in coloration among eublepharines depend on the physiological condition of the animal, particularly on hormonal factors.

Because it breeds easily in captivity, *Eublepharis macularius* is currently a popular animal in the pet trade (see, for example, Black, 1997:10–18), and is thus easily available for scientific research as well (e.g., Autumn and Denardo, 1995). To what extent conclusions based on captive animals of unknown provenance are transferable to populations in nature is problematical.

Eublepharis turcmenicus Darevsky, 1977 (Plate 6F)
Turkestan leopard gecko, Turkmenian fat-tailed gecko

Eublepharis turcmenicus Darevsky *in* Bannikov, *et al.*, 1977:83–84; 1978:204–209, figs 1–2 (Type locality: Bakharden, Kopet Dag, Turkmenistan. Holotype: ZIL 10103). — Welch, 1983:7. — Szczerbak and Golubev, 1986:30–33, figs. 7–9, col. pl. 1, fig. 3. — Grismer, 1988:441–452. — Kluge, 1991:8; 1993:10. — Grismer, 1991:251–253.

Diagnosis: (Fig. 57) Subdigital lamellae with weakly developed small tubercles; chin shields not in contact with first lower labials; ventral scales hexagonal; elements of dorsal color pattern not linearly confluent, 5–9 preanal pores, *interrupted medially by 1–4 scales lacking pores* (Fig. A in key, page 117); mental scale shorter than wide; 54–55 eyelid fringe scales; widely spaced, pronounced and pointed dorsal tubercles, much smaller inter-tubercle granules; hexagonal ventral scales in 20–22 longitudinal rows; width of rostral 1½ times its height; undivided terminal subdigital lamellar scales; 3–4 transverse rows of ventral scales in each caudal whorl; dorsal scales of regenerated tail circular and slightly convex; supratemporal bone present, but *smaller than in all other members of genus*; a straight pterygoid-palatine suture; a longitudinally-directed crest on ventral portion of basioccipital anterior to spheno-occipital tubercles, posterior section of basioccipital smoothly rounded; posterior margin of coronoid shelf does not contact the adductor fossa; clavicle extending above the scapulocoracoid and making broad contact with the suprascapula (Grismer, 1991:251–252).

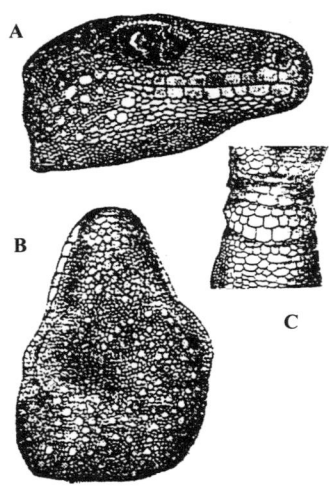

FIGURE 57. *Eublepharis turcmenicus*. A. Side of head; B. top of head; C. ventral surface of tail. (From Szczerbak and Golubev, 1966:27, fig. 8.)

Color pattern: (Fig. 58) Dorsum pinkish-white, with pattern of irregular elongate dark brown blotches on head, irregular roundish to squarish dark brown spots on back arranged to form more or less distinct broad transverse bars, these bars more distinct on tail; limbs with scattered dark spots; venter immaculate white.

Size: Snout-vent 130 mm, tail 80 mm. Kaverkin and Orlov (1996:99) give the overall length (tail included) of a hatchling female as 83.5 mm.

Habitat: (Plate 25F) Stony foothills of the Kopet Dagh, occupying hill slopes composed of crumbled schists and rock fragments and covered with *Artemisia* and *Ephedra* and sometimes trees, *Paliurus spinachristi*. They have been collected where burrows of *Microtus afghanus, Meriones persica*, or *Ochotona rufescens* were common. (Szczerbak and Golubev, 1986:30; 1996:31).

Natural history: See Szczerbak and Golubev (1986:31–33; 1996:31–32) and Rusta-

FIGURE 58. *Eublepharis turcmenicus* (holotype, ZIL 10108). (From Nikolsky, 1915:345, pl. 2.)

mov, et al. (1985). Rösler and Szczerbak (1993) recorded observations on growth, color pattern development and feeding for a young female specimen from Turkmenistan in captivity. Kaverkin and Orlov (1996:99) recorded their experiences with captive breeding of this species. Szczerbak and Golubev (1986:32; 1996:32) report food items from the feces of a single individual as including Heteroptera, Hymenoptera, Coleoptera, and a lacertid, probably *Eremias strauchi*. These authors also provide remarks on behavior of captive specimens.

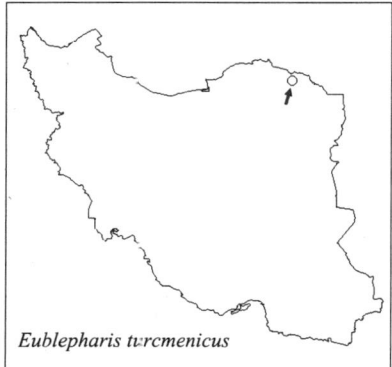

Eublepharis turcmenicus

Distribution: Known only from the valleys of the Kopet Dagh in northern Iran and adjacent southern Turkmenistan.

Remarks: Through the courtesy of Ilya Darevsky, I examined a single specimen of this species in 1966, and more recently, a second specimen (CAS 184771), also from Turkmenistan (39°07'N 55°08'E), collected by Theodore Papenfuss and Robert Macey. This taxon appears to be closest to *Eublepharis macularius*, but it is intermediate between *E. angramainyu* and *E. macularius* in certain characters. In *E. angramainyu* the subdigital lamellae are absolutely smooth, while in *E. macularius* they bear distinct small tubercles; in *E. turcmenicus* the lamellae are weakly but distinctly tuberculate. The dorsal color pattern is broken up into dark blotches, some of the blotches rather linearly arranged, but not longitudinally confluent as they are in *E. angramainyu*; remnants of two dorsal crossbars can just be distinguished. The mid-dorsal tubercles are subequal to the intertubercular spaces, more like *E. macularius* than *E. angramainyu*. In the specimens I examined there were 7 and 8 preanal pores, the series interrupted in two places by scales without pores in one specimen, and in the center in the other, differing in this respect from the other two species; the mental is followed by two large chin shields in contact with one another, or separated by a scale, but the chin is more like that of *E. macularius* in lacking rows of conspicuously enlarged scales between labials and gulars. The mental scale is shorter than wide, as in *E. macularius* and *E. angramainyu*. Grismer (1988:446) states that *E. macularius* is the only eublepharid species other than *Aeluroscalabotes* that has a cleft terminal subdigital lamella. In none of the specimens of *E. macularius, angramainyu*, or *turcmenicus* that I examined can I find a cleft subdigital lamella. CAS 184771 has three ventral caudal scales in the central row of each caudal whorl in the anterior third of tail, four in the posterior two-thirds, whereas specimens of *E. angramainyu* have two and three, respectively, while the *E. macularius* that I have examined have three throughout the length of the tail. I count 21 ventral scales across the midbody in CAS 184771, whereas Grismer gives 27–38 for *E. angramainyu* and 21–30 for *E. macularius*. Reluctant to do destructive dissection on the only available specimen of *E. turcmenicus*, I have not checked Grismer's osteological characters. Grismer (1988:442–450), without specimens of *E. turcmenicus*, and therefore based on the literature, felt that *E. turcmenicus* is virtually indistinguishable from *E. angramainyu*, and may be at best subspecifically distinct. Subsequently, he examined three specimens (Grismer, 1991), which he checked against the character analysis of his previous work. He concluded that *E. turcmenicus* is the sister taxon of *E. macularius*, and *E. angramainyu* the sister of that clade. The fourth species, *E. hardwickii*, is the sister taxon to all other species in the genus. The question of specific or subspecific differentiation of allopatric populations is, of course, a philosophical one, dependent upon what species concept one uses. All of the specimens I have examined are readily identifiable with one or another of the four taxa which Grismer recognizes tentatively at the species level. Grismer's (1991) conclusions are congruent with the prediction on biogeographical grounds that *E. turcmenicus* would be the sister taxon of *E. macularius*, considering the number of reptilian species common to Pakistan and Turkmenistan-Khorassan, but currently separated by the massif of the Hindu Kush.

Family GEKKONIDAE

Geckos

Of all the families of lizards that occur in the great Palearctic Desert, the Gekkonidae is represented by the greatest number of species and genera. The majority of lizard genera endemic to that region also belong to this family.

The systematics of this family pose the single greatest problem to our understanding of the lizard fauna of Southwest Asia. Kluge (1967, 1983, 1987) has elucidated many relationships at the subfamilial level, but the intergeneric relationships of the geckos of the Southwest Asian deserts are not yet well worked out. The conventional taxonomy (largely following Kluge 1967, 1983, and 1985) is used in the present treatise. A few introductory remarks are necessary, however, to outline the nature of the existing confusion.

Adaptation to the multitude of habitats within this vast desert, particularly to similar but geographically isolated habitats, has produced morphological parallels which have tended to obscure the phylogenetic relationships of these animals. Hence, differentiation at the generic level has been imperfectly perceived, and this is reflected in the taxonomy, which is based in large part on the most plastic of morphological features. For example, Bauer and Russell (1991) distinguished three types of pedal morphologies associated with dune-dwelling species: fringed toes (e.g., *Teratoscincus, Crossobamon, Stenodactylus*), spinous, swollen plantar surfaces (e.g., *Bunopus, Stenodactylus*), and webbed feet (*Stenodactylus arabicus*).

Kluge (1983) has begun to approach the problems of intergeneric relationships using a cladistic perspective. Within the gekkonines, Kluge has recognized the tribe Gekkonini, diagnosed by absence of the second ceratobranchial arch of the hyoid apparatus (a synapomorphy). The following Southwest Asian gekkonine genera are included in the Gekkonini: *Agamura, Alsophylax, Bunopus, Crossobamon, Hemidactylus, Stenodactylus, Cyrtopodion* (Kluge removes species having a second ceratobranchial — none of them Southwest Asian), and *Tropiocolotes* (as restricted by Kluge, 1983).

The following Southwest Asian genera are placed by Kluge (1983) provisionally in the "Ptyodactylini" (*incertae sedis* in the Gekkoninae, not defined by any synapomorphy) by virtue of their possessing a complete second ceratobranchial arch: *Asaccus, Microgecko, Pristurus* (*rupestris*), *Ptyodactylus*, and *Teratoscincus*.

Rhinogecko and *Carinatogecko* were not placed by Kluge (Kluge, 1991, 1993 follows Szczerbak and Golubev, 1986 in regarding *Rhinogecko* as a synonym of *Agamura*).

It is in Iran, centrally located in Southwest Asia, and most diverse ecologically, that the problems of generic recognition are most acute. Confusion regarding the type of *Alsophylax pipiens* resulted in perplexity as to the distinctions, if any, between the genera *Alsophylax* and *Bunopus*. This has been commented upon in previous papers (Leviton and S. Anderson, 1963:334–336; Mertens, 1965:2; Clark, *et al.*, 1969:298–299; Szczerbak and Golubev, 1977a:120–132; 1986:7–8; 1996:7–8).

In any case, a group of geckos having a heterogeneous dorsal pholidosis and carinate or tuberculate lamellae beneath the digits occurs in the arid areas throughout much of Southwest Asia. The articulations of the phalanges in these species do not give the impression of angularly bent digits seen in the genus *Cyrtopodion*, but this appears to be a matter of degree, and a complete morphological (but not necessarily phylogenetic) transition from "straight" to "angularly bent" digits probably exists. *Bunopus blanfordi, B. abudhabi*, and *B. crassicaudus* share digital characters and similarity of dorsal pholidosis with

"*Stenodactylus*" *lumsdeni* of Baluchistan, which in turn appears to be morphologically allied to *Crossobamon orientalis* and *C. maynardi*, which occur further to the east. *Stenodactylus lumsdeni* is here regarded as synonymous with *Bunopus tuberculatus*.

Golubev and Szczerbak (1981) erected the genus *Carinatogecko*, in which they placed *Bunopus aspratilis* and *Tropiocolotes heteropholis*, two species whose allocation had been doubtful. In the northeastern part of the Southwest Asian desert and in Central Asia, species with tuberculate pholidosis have been variously assigned to *Bunopus* and to *Alsophylax* (on the assumption that *Bunopus* was a synonym of *Alsophylax*; see Szczerbak and Golubev 1986:7–8; 1996:7–8 for details of this history). These species share digital characteristics with *Alsophylax pipiens*, and the dorsal tubercles are less pronounced than in *Bunopus*, and perhaps less constant a character. In this group, too, enlarged postmentals apparently may be present or absent, this character grading from enlarged, plate-like trapezoidal shields through slightly enlarged subcircular scales, to absent. *Tropiocolotes levitoni* of eastern Afghanistan ("*Alsophylax pipiens*" of Leviton and S. Anderson 1963:334–336; Clark, *et al.* 1969:298–300), appears to be allied to *T. depressus* in digital characters, color pattern, lack of dorsal tubercles, and lack of distinct postmentals. Golubev (1984:12) erected the subgenus *Asiocolotes* linking these two species, and Kluge (1993:4) recognizes this taxon at the generic level.

Terentjev and Chernov (1949:131) included northeastern Iran within the distribution of *Alsophylax pipiens* (which in their view included *A. microtus* and *A. laevis*), but I find no documented records for Iran. If *A. pipiens* (*sensu* Terentjev and Chernov) does occur in Iran, it can be distinguished from all other small geckos of that country by the possession of but a single small nasal and lack of spinose tubercles on the tail. Michael Golubev (pers. commun., 3 February 1995) stated that, despite Terentjev's and Chernov's (1949) statement, *Alsophylax pipiens* does not occur in Iran, but *A. laevis* might be found in Iran (or, probably, in northern Afghanistan). Kluge (1967:23) reassigned eastern species previously regarded as *Stenodactylus* to *Crossobamon*, with which they obviously belong. In many respects, these species resemble *Bunopus*.

John Anderson (1898:35) felt that a transition could be seen in digital characters from the carinate lamellae of *Stenodactylus* to the imbricate subdigital pholidosis of *Ceramodactylus*. Kluge (1967:24) regarded *Ceramodactylus* as a synonym of *Stenodactylus* and included the monotypic Arabian genera *Pseudoceramodactylus* and *Trigonodactylus* in *Stenodactylus* also. Arnold (1980b) followed this taxonomy.

Many species in Southwest Asia belong to monotypic genera, are assigned only provisionally to various genera, or have been shifted from one genus to another. These include *Trachydactylus spatulurus* (J. Anderson), *Trigonodactylus arabicus* Haas, *Pseudoceramodactylus khobarensis* Haas, *Cyrtodactylus amictopholis* Hoofien, *Alsophylax spinicauda* Strauch, and *Agamura femoralis* Smith. This situation reflects both our lack of understanding of the phylogeny of these geckos and the extent of morphological diversity in this family in Southwest Asia.

The several Southwest Asian species of *Cyrtopodion* appear to form a fairly well-defined natural group, with the above-mentioned exception of *C. amictopholis* as well as *C. chitralensis* and *C. stoliczkai*. The angular articulation of the phalanges is not as pronounced in the Southwest Asian *Cyrtopodion* as it is in many of the Southeast Asian members of the genus *Cyrtodactylus* (with which they were placed formerly), and there may prove to be a transition from the condition of the toes of *Bunopus* to that of *Cyrtopodion*. The digits of *Agamura persica* are also angularly bent.

Kluge (1985), through his designation of *Cyrtodactylus pulchellus* (Gray 1827) as the type species of *Gonydactylus* Kuhl and van Hasselt (1822), has made *Cyrtodactylus* Gray a junior objective synonym of *Gonydactylus*. Further complicating the taxonomy of this group is the fact that Szczerbak and Golubev (1984) separated the Palearctic species previously assigned to *Cyrtodactylus* (or to *Gymnodactylus* by several European authors) into a new genus, *Tenuidactylus* (Szczerbak and Golubev, 1984, type species *Gymnodactylus caspius* Eichwald, 1831), and further subdivided this latter genus into three subgenera (*Tenuidactylus, Mediodactylus, Mesodactylus*). But as Kluge (1985:98) has pointed out, *Cyrtopodion* Fitzinger (1843, type species *Stenodactylus scaber* Heyden, 1827) has priority for this assemblage. Thus, in the scheme proposed by Szczerbak and Golubev (1984), *Mesodactylus* Szczerbak and Golubev (1977, type species *Gymnodactylus kachhensis*) is a junior subjective synonym of *Cyrtopodion*, and has been so regarded in their later, comprehensive work (Szczerbak and Golubev, 1986; 1996).

In its habitus, its divided rostral, its basally autotomizing tail which is shorter than the snout-vent length, the absence of keeled tubercles or keeled imbricate scales, and the lack of enlarged plate-like postmentals, *Agamura persica* appears to form a well-defined monotypic genus. Smith felt that his species, *A. femoralis*, connected *Agamura* with *Cyrtodactylus*. S. Anderson and Leviton (1969:43–44) discussed the characters which seem to set *femoralis* apart from both genera as currently constituted. In 1973, de Witte described the genus *Rhinogecko* from Iran; Smith's *Agamura femoralis* clearly belongs to this genus. Szczerbak and Golubev (1986) regard *Rhinogecko* as a synonym of *Agamura* which in their scheme contains *femoralis* and *misonnei* as well as *persica*. They also include *Gymnodactylus gastropholis* Werner in *Agamura*. I follow these authors in retaining the combination *Cyrtopodion agamuroides*, contrary to Minton's (1966:80–81) view that this species belongs with *Agamura*.

The most thorough attempt to resolve and clarify most of the above confusion is that of Szczerbak and Golubev (1986; 1996). In general, I follow their taxonomic arrangement here, with some notable exceptions. Their work is based largely on external morphology (see their materials and methods section, pp. 13–14, and illustrations), and undoubtedly more detailed anatomical and biochemical studies will bring further changes to gekkonid taxonomy.

Key to the Genera of Gekkonidae in Iran

1a. Pupil of eye round .... *Pristurus* (p. 175)
1b. Pupil of eye vertically elliptical 2
2a. Digits strongly dilated (Figs. A-C) 3
2b. Digits not dilated (Figs. E-G) 5
3a. Each digit dilated at base, with double row of lamellae beneath, forming pads; terminal phalanges compressed (Fig. C) .... *Hemidactylus* (p.171)

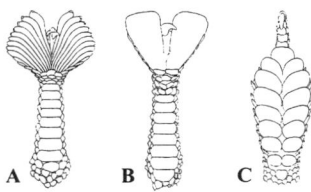

Toes from beneath showing adhesive pads: (A) *Ptyodactylus*; (B) *Asaccus*; (C) *Hemidactylus*
(From Arnold, 1986:402, fig. 17)

3b. Each digit dilated at apex, terminating in subtriangular expansion, claw lying in longitudinal groove dividing apical expansion (Figs. A–B) 4
4a. Apical expansion of digit with fine lamellae beneath, forming a fan-like pattern (Fig. A); postanal sacs present (Fig. D) .... *Ptyodactylus* (p. 177)
4b. Apical expansion of digit appears smooth beneath under low magnification (Fig. B) (covered with fine hair-like papillae seen under high magnification); postanal sacs absent .... *Asaccus* (p. 132)

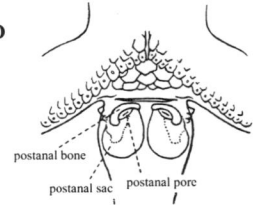

Preanal and femoral pores and postanal bones and sacs
in *Cyrtopodion* (From Smith, 1935:26, fig. 10)

5a. Digits with well defined lateral fringe of elongate, flexible pointed scales (Figs. E–F)..6
5b. Digits without fringe of elongate, flexible pointed scales, although scales may be denticulate (Fig. G) 8
6a. Small dorsal scales intermixed with rounded tubercles .... *Crossobamon* (p. 146)

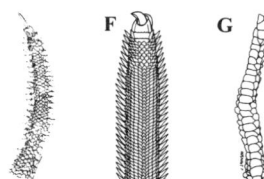

Ventral surface of fourth toe of (E) *Stenodactylus doriae*
(BMNH 87.9.23.3); (F) *Teratoscincus scincus*;
(G) *Cyrtodactylus heterocercum* (NMW 7286)
(Drawings by Jean Jahnson)

6b. Dorsal scales uniform, not intermixed with tubercles 7
7a. Dorsal scales small, not cycloid; dorsal scales of tail not large, not plate-like, and not strongly imbricate .... *Stenodactylus* (p. 181)
7b. Dorsal scales large, cycloid; tail covered above (at least on posterior two-thirds) by single row of large, plate-like strongly imbricate scales .... *Teratoscincus* (p. 185)
8a. Nostril at apex of prominent swollen or cylindrical caruncle formed by the nasal scales; rostral excluded from border of nostril .... *Rhinogecko* (p. 178)
8b. Nasal scales do not form cylindrical caruncle, although they may appear to be swollen around the nostril; rostral normally forms part of border of nostril 9
9a. Dorsal scales uniform, small, homogeneous .... *Tropiocolotes* (p. 190)
9b. Dorsal scales heterogeneous 10
10a. All scales of body and head, except labials and postmentals strongly keeled; small dorsal scales intermixed with larger scales or tubercles .... *Carinatogecko* (p. 143)
10b. At least some scales of head and body smooth, dorsal scales small, uniform, intermixed with larger tubercles 11
11a. Subdigital lamellae with a single transverse series of tubercles, particularly on the free

margin, seen under magnification (sometimes worn down in later part of epidermal cycle); distal phalanges not compressed .... *Bunopus* (p. 138)

11b. Subdigital lamellae smooth; distal phalanges compressed or not 12

12a. Postmentals (chin shields) present and well differentiated in size and shape from granular small scales of chin and throat (Fig. H) .... *Cyrtopodion* (part) (p. 149)

12b. Postmental shields absent (sometimes a short row of enlarged subcircular scales behind mental) (Fig. I) 13

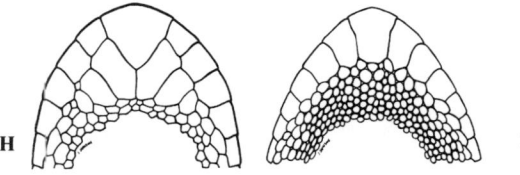

Ventral surface of head of
(H) *Cyrtopodion scabrum* (CAS 96175); (I) *Agamura persica* (FMNH 161054)
(Drawings by Jean Jahnson)

13a. Tail cylindrical, very slender, and of almost uniform diameter from base to tip (tip blunt), no mucronate tubercles or annuli (Fig. J) .... *Agamura* (p. 129)

13b. Tail tapering gradually (tip of original tail sharply pointed), two mucronate tubercles on either side of each annulus .... *Cyrtopodion* (part) (p. 149)

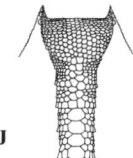

Ventral side of tail of *Agamura persica* (CAS 102505)
(Drawing by Jean Jahnson)

Genus *Agamura* Blanford, 1874
Spider geckos

Agamura Blanford, 1874a:455 (Type species: *Gymnodactylus persicus* A. Duméril, 1856, by subsequent designation of Smith, 1935:61).

Definition: Limbs slender; digits slender, clawed, cylindrical at base, with smooth transverse lamellae beneath, angularly bent, the distal phalanges slightly compressed, forming an angle with the basal portion of the digits, the claw between two enlarged scales; no enlarged plate-like postmental shields; dorsal scales small granules intermixed with larger tubercles; tail cylindrical, slender, diminishing suddenly in size after the basal portion (Fig. J in key, above), not longer than head and body; basal caudal autotomy only; pupil vertical.

Distribution: Iran, Pakistan, and Afghanistan. A single species (but see below).

Remarks: The relationships of the genus *Agamura* may be with *Cyrtopodion*, but close affinity with any living species within that genus is not apparent to me. The similarities between the two genera may be due to parallelism rather than to close common ancestry. Minton (1966) includes three species in this genus: *Gymnodactylus persicus* Duméril, 1856,

Agamura femoralis Smith, 1933, and *Gymnodactylus agamuroides* Nikolsky, 1899. Szczerbak and Golubev (1986:208–209; 1996:204) include four species within their concept of this genus, *viz. G. persicus* Duméril, 1856, *G. gastropholis* F. Werner, 1917, *A. femoralis* Smith, 1933, *Rhinogecko misonnei* de Witte, 1973, and are followed by Kluge (1991:2–3; 1993:3) in this interpretation. In this work, I regard *Agamura* as a monotypic genus.

Agamura persica (Duméril, 1856) (Plate 6G)
Blunt-tailed spider gecko, Persian spider gecko

Gymnodactylus persicus A. Duméril, 1856:481 (Type locality: Persia. Syntypes: MNHN 6761[3]).
Agamura cruralis Blanford, 1874a:455; 1876:356–358, pl. 23, figs. 3, 3a (Type locality: Bahu Kalat; Zamran; Nihing River; Askan, near Bampusht, Iran; and Ras Malan; Mand; Pakistan [restricted to Bahu Kalat and Askan, Baluchistan by Smith {1935:61}]. Syntypes: BMNH 74.11.23.52–54 [1946.8.25.33–35; MCZ 7136; ZMB 10234; ZSI 3487, 3501, 6811-6812 [*fide* Das, *et al.*, 1998]. ["We herein regard the BMNH type for which Boulenger {1885} provided measurements as the lectotype." {Bauer and Gûnther, 1991:281}. These authors also designate MCZ 7136 and ZMB 10234 as paralectotypes]). — Bedriaga, 1879:36. — Boulenger, 1885a:50–51. — F. Werner, 1938: 267. — Schleich, 1977:127, 129. — Welch, 1983:5.
Agamura persica: Blanford, 1874a:455; 1876:358–359, pl. 23, figs 4a, 4b. — Bedriaga, 1879:36. — Boulenger, 1885a:51. — Strauch, 1887:53. — F. Werner, 1895:2. — Nikolsky, 1897:316. — Zarudny, 1903:8. — Annandale, 1906:197. — Nikolsky, 1915:88–89. — F. Werner, 1936:196, 200. — Forcart, 1950:144–145. — Wettstein, 1951:431. — S. Anderson, 1963:474. — Wermuth, 1965:4. — Clark, *et al.*, 1966:6–7. — Guibé, 1966a:97. — S. Anderson, 1968:332. — Tuck, 1971a:5–56. — S. Anderson, 1974:34, 42. — Schleich, 1977:127, 129. — Tuck, 1979:100–101. — Nilson and Andrén, 1981:123–124, fig.3. — Szczerbak and Golubev, 1986:209–211; 1996: 204–207, figs. 94, 95. — Welch, 1983:5. — Welch, *et al.*, 1990:6. — Kluge, 1991:3; 1993:3. — Frynta, *et al.*, 1997:8.

Diagnosis: As for the genus.
Color pattern: (Fig. 59) Light grayish tan above, those in the western part of the range (most of Iran) with five distinct darker brown dorsal crossbars, first on nape, fifth on sacrum, nine to ten on tail; those from eastern part of range (extreme eastern Iran and Afghanistan) with three dark crossbars, first on nape, second behind shoulders, third in front of sacrum, a large area on middle of back without dark markings; lips, chin and flanks mottled with dark brown or gray; limbs with dark crossbars; venter cream, sometimes mottled with gray on lower surfaces of tail and limbs.
Size: Females 42–77 mm snout-vent, tail 34–59 mm. Males measure 35–65 mm snout-vent, tail 27–59 mm.
Sexual dimorphism: Males usually, but not invariably with two preanal pores, females with corresponding scales enlarged, with faint central depression. Males have large flat postanal bones, easily felt with a blunt probe inserted into postanal sac; females lack these bones.
Natural history: Blanford (1876:358) and Minton (1966:80) report these geckos active at midday, but state that they are primarily nocturnal. In spring (mid-April to mid-June), I found this species basking at 0625 hrs, air temperature 17.5°C, rock surface 15.5°C, active at 1340 hrs when surface temperature was 44°C, and exposed on a gravelly slope with a strong wind blowing at 1845 hrs. They are relatively slow-moving, easily captured by hand. They do not climb vertical surfaces.

A female collected in late April contains an egg 6 mm diameter in each oviduct. The only digestive tract examined containing food material had (lycosid?) spider remains.

FIGURE 59. *Agamura persica*. (From Blanford, 1876, pl. 23, fig. 3.)

Habitat: Barren plains, hillsides, rocky terraces, dry gravelly creekbeds, silty and gravelly alluvium, disintegrating shale and other rocks on slopes, and, according to Nilson and Andrén (1981:134), on sandy soils; associated with vegetation of widely scattered low steppe shrubs, such as *Artemisia*, and small herbaceous annuals. A series of eight specimens collected by the Street Expedition of 1962 at Galatappeh in December was found in torpor under a piece of paper in a hole dug as a sheep shelter. From 30 to 1900 m elevation.

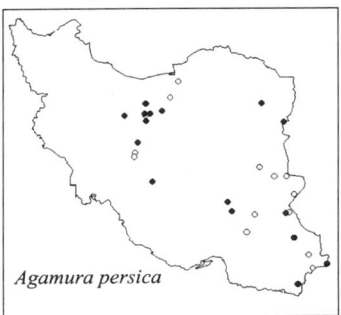

Agamura persica

Distribution: The Central Plateau from western Tehran, Esfahan, and Fars Provinces to Kerman and Baluchistan Provinces in Iran and contiguous similar areas of Afghanistan and Pakistan as far east as Sind and Waziristan.

Remarks: In all but four specimens examined, the rostral is completely divided, and tubercles are present on thigh and shank (rostral said to be only partly divided, tubercles lacking on hind limbs in *Agamura cruralis*). The tubercles on the back are rather flat, broader than long, usually weakly keeled, with a free posterior margin, or the posterior tubercles become trihedral. In Afghan specimens the dorsal tubercles are smaller, hence a greater distance between them. Each side of tail has a cluster of two or three enlarged tubercular scales. There is a median series of enlarged flat subcaudal scales, often longitudinally divided in part or all of tail. Tail distinctly segmented, each segment corresponding to one caudal vertebra. The dorsal surface of each segment is covered with three to five transverse rows of scales, the posterior row of each segment composed of enlarged scales, smooth or weakly keeled, with truncate posterior margins. In no case is the tail as long as head and body. Caudal autotomy appears to occur only at the base (see also Minton, 1966:80 in this regard). Mertens (1969:22) observed that the end of the tail is slightly flattened, and suggested the possibility that the terminal scales form an adhesive organ, as in *Lygodactylus*, for example. I find no indication that such is the case. FMNH 171249 has a short regenerated tail that is unlike the original tail in pholidosis, covered above and below with smooth, fairly homogeneous scales, not arranged in annuli, not segmented, and with no tubercles. The mental and adjoining first infralabials are elongate in these geckos, extending onto ventral surface of chin. No postmentals, but a row of enlarged scales borders the mental and first labials; these scales are somewhat larger in Iranian than in Afghan specimens.

Two preanal pores is the usual condition in males, but in the series from Qom, one has five, another six.

The peritoneum is darkly pigmented and the investiture of the intestine is nearly black, while stomach, liver, and gonads are not pigmented. Such pigmentation is characteristic of diurnal lizards and in contrast to strictly nocturnal species of geckos. The well-developed (but immovable) upper "eyelid" may also be an adaptation to diurnality, serving as a sunshade casting a shadow on the eye.

In the three syntypes of *Agamura cruralis* Blanford in the British Museum, the rostral is only partially divided; there are five dorsal crossbars in those in which the pattern can still be seen; tubercles are lacking on the hind limbs; the subcaudals are not divided into two longitudinal rows except at the flattened tip. I do not find any clear indication of preanal pores in these specimens. A specimen from Cheh Mossullum (FMNH 170934) agrees with Blanford's description of *A. cruralis* in having the rostral only partially divided and lacking tubercles on the limbs. It is a male with two preanal pores and I do not find other consistent differences between this and other specimens.

Pakistan specimens (Nushki; Ormara; Kila Safed; Upper Porali Ford) all have 5–7 crossbars. Males generally have only one (rarely two) preanal pore in an enlarged scale. Only one specimen (Ormara) has the rostral completely divided. Individuals from Porali Ford have an unusually distinct and contrasting color pattern, with distinct bars on limbs and tail. Most lack tubercles on the hind limbs, but many have enlarged scales.

Szczerbak and Golubev (1986:211; 1996:206) stated that in Iranian Baluchistan and to the north up to the Sistan basin and the Dasht-e Lut, enlarged chin shields appear in this species. They noted that these specimens had some divergence in other characters as well, and while a shortage of specimens did not permit final taxonomic resolution, they suggested that recognition of a subspecies, *A. persica cruralis*, was justified.

Genus *Asaccus* Dixon and Anderson
Southwest Asian leaf-toed geckos

Asaccus Dixon and S. Anderson, 1973:156–157, figs 1–3 (Type species: *Phyllodactylus elisae* Werner, 1895, by original designation).

Definition: (Fig. 60) Digits with paired terminal scansors that lack lamellae; no femoral pores; left oviduct absent, only one egg laid at a time; reduction of phalangeal formula of manus to 2.3.4.4.3; no transverse processes on autotomic caudal vertebrae, except sometimes the first (Arnold and Gardner, 1994). Cloacal sacs and postanal bones absent; second epibranchial arch of hyoid present; stapes perforate (stapedial foramen present); 28 amphicoelous sacral and presacral vertebrae; atlas paired; parietals paired; nasals paired, with long projection of premaxillary between nasals; anterior tip of mesoscapula lacking osseous or cartilaginous connection with precoracoid process; interclavicle shield-like; three pairs of sternal ribs, two pairs of mesosternal ribs; one large fenestra in clavicle; supratemporal absent; angular absent; frontal single; 9–10 premaxillary teeth, 50–60 total dentary teeth and 48–52 total maxillary teeth; 14 scleral ossicles; hypoischium cartilaginous, rod-like. (Dixon and S. Anderson, 1973:156–157.)

Distribution: *Asaccus* is known from southern Anatolia, eastern Syria, and eastern Iraq in the Euphrates drainage, and western Iran west of the Zagros Mountains. Seven species are recognized, three of them occurring in Iran. *Asaccus gallagheri* (Arnold, 1972), *A. montanus* Gardner, 1994, *A. platyrhynchus* Arnold and Gardner, 1994, and *A. caudivolvulus* Arnold and Gardner, 1994, occur in the mountains of northern Oman.

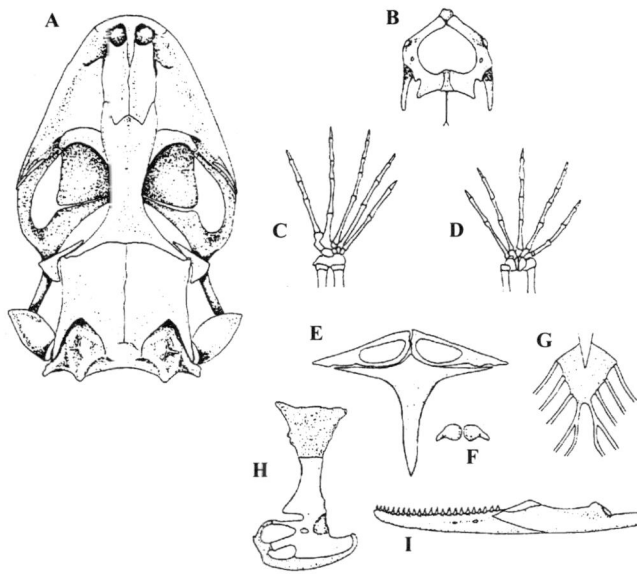

FIGURE 60. Some aspects of the osteology of the genus *Asaccus* (*A. elisae*): (A) dorsal view of skull; (B) ventral view of pelvic girdle; (C) dorsal view of pes; (D) dorsal view of manus; (E) ventral view of clavicles and interclavicle; (F) dorsal view of paired atlas; (G) ventral view of sternum and sternal ribs; (H) lateral view of the scapulocoracoid element; (I) lateral view of left dentary. (From Dixon and Anderson, 1973:156, fig. 1.)

Remarks: The work of Dixon and various coworkers has resulted in the reallocation of several species formerly assigned to the genus *Phyllodactylus*. The pholidosis of species now assigned to the genus *Asaccus* is similar to that of species of *Phyllodactylus* from the Western Hemisphere, and consists of enlarged tubercles arranged in longitudinal rows on the dorsum and tail, scattered randomly on head and limbs, and one large pair of terminal leaf-like lamellae on fingers and toes. The most striking character separating the two genera is the absence of cloacal sacs and postanal bones in *Asaccus*.

In *A. elisae*, *A. griseonotus*, *A. montanus*, and *A. caudivolvulus*, all specimens with original tails have the tip laterally compressed. The tip of the tail does not appear to possess an accessory adhesive structure, as occurs, for example, in *Lygodactylus*.

The cladogram provided by Arnold and Gardner (1994:439, fig. 5) suggests that *A. elisae* and *A. montanus* are sister species, while *A. griseonotus* is the sister taxon to the above clade. These authors state (p. 440): "Most species, including those comprising the two most basal branches of the phylogenetic tree, occur in the north Oman region of Arabia, and the two forms found in Iran and Iraq [now 3, since the discovery of *A. kermanshahensis*] are not a sister pair. The most parsimonious interpretation of this pattern is that the common ancestor of extant *Asaccus* initially speciated in Arabia and later invaded the north on two occasions." It seems to me equally likely (and no less parsimonious) that the current distribution represents a relictual vicariant pattern resulting from the geomorphic and associated climatic events that produced the elevated regions on either side of the present-day Persian Gulf. Further analysis of the relationships of *A. kermanshahensis* and additional collection in the Zagros ranges may help resolve this issue. Frynta, *et al.* (1997:8) report eight specimens collected at Choqa Zanbil as differing from both *A. elisae* and *A. kermanshahensis* in pholidotic characters. They plan to publish a detailed description. This material should be compared with specimens identified as *A. elisae* collected at Shush.

Key to the Species of *Asaccus* in Iran

1a. 4 pairs of postmentals bordered by 21–24 granules ....
......................................*Asaccus kermanshahensis* Rastegar-Pouyani, 1996 (p. 137)
1b. 2 pairs of postmentals bordered by 20 or fewer granules 2
2a. Largest dorsal tubercles more than one-half height of ear opening; tubercles extending onto occiput and temporal area, much larger than surrounding granules; whorls of caudal tubercles separated by 3–4 transverse rows of small scales (Figs. A–C) .... *Asaccus elisae* (Werner, 1895) (p. 134)
2b. Largest dorsal tubercles less than one-half height of ear opening; tubercles becoming much smaller on nape, usually not extending onto head, or if so, few in number, scarcely larger than surrounding granules; whorls of caudal tubercles separated by 6–7 transverse rows of small scales (Figs. D–F)...
.... *Asaccus griseonotus* Dixon and Anderson, 1973 (p. 136)

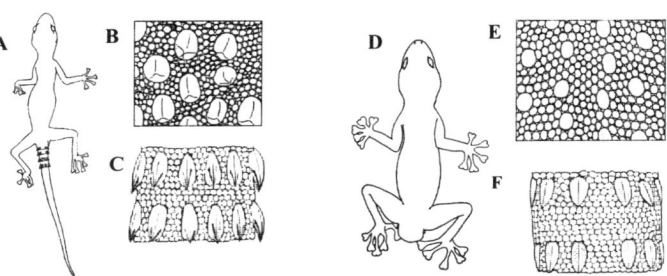

Asaccus elisae: (A) outline of animal; (B) dorsal scales of back; (C) dorsal surface of tail. *Asaccus griseonotus*: (D) outline of whole animal; (E) dorsal scales of back; (F) dorsal surface of tail. (After Eiselt 1973:175, [A–B] figs. 3-4, [D–E] figs. 1-2; Dixon and Anderson, 1973, [C–D] fig. 3.)

Asaccus elisae (Werner, 1895) (Plate 6H)
Werner's leaf-toed gecko

Phyllodactylus elisae F. Werner, 1895:14, pl. 3, figs. 1a-e (Type locality: ruins of Niniveh, near Mosul, Iraq; Syntypes: NMW [4 spec.]; BMNH 95.3.2.3[1946.8.24.39]); 1917:196–197; 1936:200. — Schmidt, 1939:56; 1955:202. — Mertens, 1957:126. — S. Anderson, 1963:442, 474. — Wermuth, 1965:135. — Anderson, 1968:333. — Haas and Werner, 1969:334. — Eiselt, 1973:175–178, figs. 3–11. — Schleich, 1977:127, 129. — Welch, 1983:10.
Phyllodactylus eugeniae Nikolsky, 1907a:268, pl. 1, fig. 1 (Type locality: Dizful and Abu-Garia, affluent to Karun River, Iran; Syntypes: ZIL 10261[6+], 10262[6+], 10263[4], 10270[3], 10349).
Asaccus elisae: Dixon and S. Anderson, 1973:157–158, figs. 1, 2 left. — Anderson, 1974:31, 43. — Kluge, 1991:3. — Leviton et al., 1992:28–29, col. pl. 4A–B. — Kluge, 1993:4. — Arnold and Gardner, 1994.

Diagnosis: (Figs. 61 and 62.) Two pairs of postmentals bordered by 18–20 granules (Fig. 61E); scales across supraorbital region coarse, as are those of snout (11–16 between postnasal scale and orbit, 14–19 across snout at level of third upper labials); tubercles of dorsum, limbs, and tail large, length of individual tubercle more than 64% (mean, 67.5%) of ear diameter; 8–14 longitudinal rows of enlarged dorsal tubercles; 10–12 large tubercles across rear of head between ears; 2–12 enlarged tubercles on upper arm above elbow; tail tubercles arranged in whorls, each whorl separated from other such tubercles by 2–3 granules (Fig. C in key, above); subtibial scales coarse; digital scansors not extending well beyond

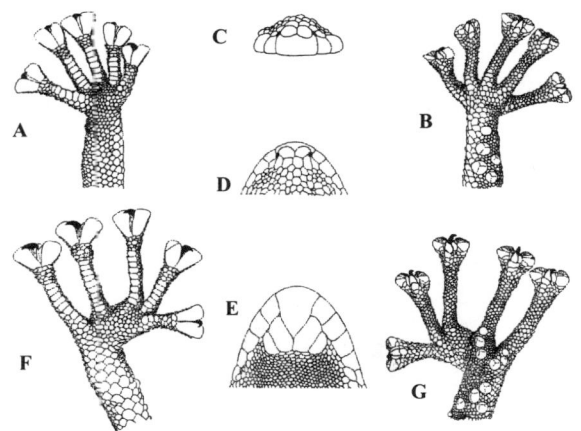

FIGURE 61. *Asaccus elisae*. (A) manus (ventral); (B) manus (dorsal); (C) snout from in front; (D) snout from above; (E) chin from below; (F) pes (ventral); (G) pes (dorsal). (From Eiselt, 1973:176, figs. 5-11.)

claws; phalanges in fourth toe reduced to 4; cloacal tubercle small; tail tip laterally compressed; subcaudal series of expanded scales not reaching vent area anteriorly; tail color not sexually dimorphic, with a series of dark transverse bars that extend ventrally (Dixon and S. Anderson, 1973:157; Arnold and Gardner, 1994).

Color pattern: Ground color usually dark brown with 0–5 darker brown dorsal crossbars, sometimes dorsum light tan with crossbars broken into dark spots; each tubercle usually tipped with white, giving freckled appearance to color pattern; tail with five brown to black bands and five white bands, each equal width; venter whitish. (Dixon and S. Anderson, 1973:158).

Size: Snout-vent length 57 mm, tail 67 mm.

Natural history: All females examined (collected June, July, August, late September) have ovarian follicles, largest 2 mm diameter. No immature specimens were collected. Stomach contents consisted primarily of spiders; winged ants and one beetle were also found in specimens from Khuzestan. Weber (1960:153) reports ants, mosquitos, roaches, and other arthropods from Iraq specimens.

Habitat: (Plate 21D) Specimens from Fars Province collected by the Street Expedition to Iran, 1962, were all found in caves; I collected a specimen in Lorestan under a large flake of exfoliating sandstone on a cliff face above a stream; most of those that I collected in the foothills of Khuzestan were found in a culvert under a road. Both H. Löffler (*in* Wettstein, 1951:433) and Weber (1960:153) report it as a house gecko in Iraq on the Mesopotamian

FIGURE 62. *Asaccus elisae*, from one of the syntypes. (From Werner, 1895, pl. 3.)

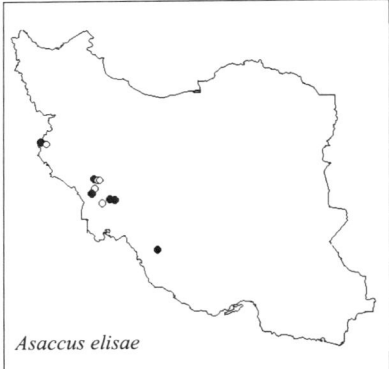

Asaccus elisae

plain, and I found it on the walls and ceilings of buildings at Shush on the Mesopotamian Plain in Iran. 150–1000 m elevation in Iran.

Distribution: Known from the Mesopotamian Plain and bordering foothills in southern Turkey (Bireçik), eastern Syria, Iraq, and southwestern Iran. Populations formerly assigned to this species from the mountains of the Sultanate of Oman and the United Arab Emirates (Arnold, 1977:100) have recently been described as two species, *Asaccus platyrhynchus* and *A. caudivolvulus* by Arnold and Gardner (1994).

Remarks: I have not examined the specimens from Mazu and from the vicinity of Shahbazan. In light of the description of the closely similar *Asaccus griseonotus* subsequent to identification and publication of these localities, these specimens should be reexamined.

Asaccus griseonotus Dixon and Anderson, 1973
Gray-spotted leaf-toed gecko

Asaccus griseonotus Dixon and S. Anderson, 1973:158–160, fig. 3 right (Type locality: 38.5 mi [61.6 km] from Shahabad, Kermanshahan, Iran; Holotype: FMNH 170824). — S. Anderson, 1974:31, 42. — Kluge, 1991:4. — Leviton, *et al.*, 1992:29. — Kluge, 1993:4.
Phyllodactylus ingae Eiselt, 1973:173–179, figs. 1–2, 12–20 (Type locality: 110 km by road SW Khorramabad, W junction with road to Malavi, Lorestan Province, Iran; Holotype: NMW 20452).

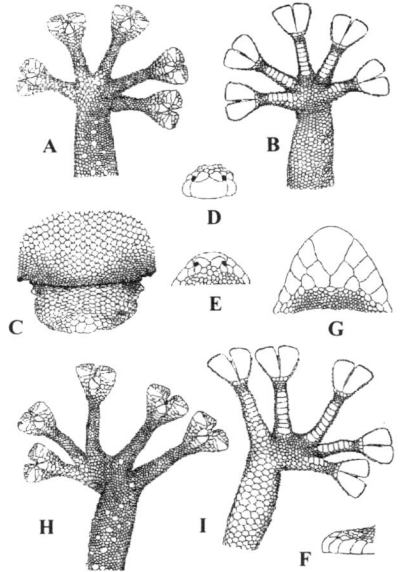

FIGURE 63. *Asaccus griseonotus* (NMW 20452; Holotype of *Phyllodactylus ingae*). (A) manus (dorsal); (B) manus (ventral); (C) anal region; (D) snout from front; (E) snout from above; (F) snout from left side; (G) chin from below; (H) pes (dorsal); (I) pes (ventral). (From Eiselt, 1973:177.)

Diagnosis: (Fig. 63) Two pairs of postmentals bordered by 15–18 granules; lacking enlarged tubercles across rear of head and on dorsal surface of upper arm; scales across supraorbital region coarse; tubercles of dorsum and limbs small, length of individual tubercle less than 46% (mean, 39.4%) of ear diameter, 10–13 longitudinal rows at midbody; 6–7 rows of granules separate enlarged tubercles of tail whorls; each tubercle of dorsum separated from its adjacent tubercle by 4–5 granules; subtibial scales moderate; digital scansors extending well beyond claws; phalanges in 4th toe reduced to 4; cloacal tubercle small; tail tip laterally compressed; subcaudal series of expanded scales reaching vent area anteriorly; tail color not sexually dimorphic; dorsal dark bars on tail do not extend ventrally (Dixon and S. Anderson, 1973: 158; Arnold and Gardner, 1994).

Color pattern: Dorsum ash gray with 20–30 small, dark gray spots scattered over dorsum; limbs with faint gray smudges on dorsal surfaces; ventral surfaces of limbs and body uniform dirty white; tail with seven gray to black bands, gray bands proximal, black distal, with white inter-

Asaccus griseonotus

spaces of equal width; tip of tail tan. (Dixon and S. Anderson, 1973:159).

Size: Snout-vent length to 71 mm (mean, 61.8 mm).

Distribution: Known only from Kermanshah and Lorestan Provinces in Iran and from Palegawra Cave, Sulaimaniyah Liwa, Iraq, all localities in the central Zagros foothills.

Remarks: In our original description of this species, the type locality was stated incorrectly to be in Kerman Province (*lapsus*) rather than in Kermanshah Province, although shown correctly on the accompanying map (Dixon and S. Anderson, 1973:158, fig. 2).

Nothing is known of the natural history of this species, other than the fact that Reed found these geckos in the cool, humid environment of the crevices in the limestone of Palegawra Cave in northeastern Iraq (Reed and Marx, 1959:97–98). Eiselt's (1973) specimen was found on a limestone cliff face close to a brook.

The name *A. griseonotus* Dixon and S. Anderson (November, 1973) antedates *P. ingae* Eiselt (December, 1973).

Asaccus kermanshahensis Rastegar-Pouyani, 1996
Kermanshah leaf-toed gecko

Asaccus kermanshahensis Rastegar-Pouyani, 1996:11–17, figs. 1–9 (Type locality: Mianrahan region, Zagros Mountains, 40 km NE city of Kermanshah, Kermanshahan Province, Iran, 1450 m; Holotype: TUZM 164R [field number] adult female).

Diagnosis: Four pairs of postmentals bordered by 21–24 granules; ear opening less than one-third diameter of orbit; enlarged dorsal tubercles smooth (pointed and weakly keeled on posterior portion of back), extending onto rear of head, tubercles lacking on dorsal surface of upper arm; diameter of individual tubercle greater than three-fourths ear diameter, about one-fourth eye diameter; 8–9 longitudinal rows of tubercles at midbody; 3–6 rows of granules separate enlarged tubercles of tail whorls; each tubercle of dorsum separated from its adjacent tubercle by 4–7 granules. (Rastegar-Pouyani, 1996:12–13).

Color pattern: Dorsum light grayish-cream with irregular, scattered dark brown patches, those of head and limbs much smaller than those on back and sides. No dark bands on tail (complete tail as yet unknown). (Rastegar-Pouyani, 1996:13)

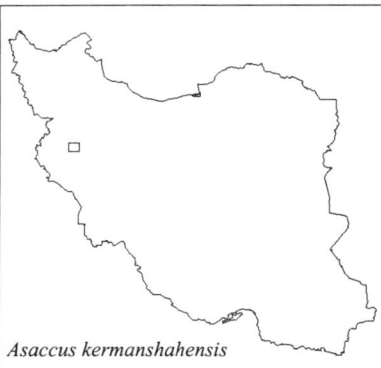

Asaccus kermanshahensis

Size: Maximum snout-vent length 55.7 mm (adult male). Complete tail unknown.

Natural History: Rastegar-Pouyani (1996:16–17) collected the three known specimens in a small cave when they were active at night at an air temperature of 35°C. During the day they hide in rock crevices within the cave.

Habitat: Known only from inside a small cave at the type locality. This cave is located in a narrow fault-valley through the Zagros mountains at 1450 m elevation. (Rastegar-Pouyani, 1996:16–17).

Distribution: Known only from the type locality, 40 km NE city of Kermanshah.

Remarks: Only three specimens are known, two males and one female, none of which has a complete, unregenerated tail. Thus, it is not known whether or not it shares the flattened tip of its two Iranian congeners. On the basis of the data presented, it is not possible to fit it within the key to the genus given by Arnold and Gardner (1994:425–426), nor into their proposed cladogram (Arnold and Gardner, 1994:439, fig. 5).

Genus *Bunopus* Blanford, 1874
Tuberculated geckos

Bunopus Blanford, 1874a:454 (Type species: *Bunopus tuberculatus* Blanford, 1874a, by monotypy).

Definition: Digits straight, not dilated, not angularly bent at any of the articulations (in contrast to *Cyrtopodion* and *Agamura*), clawed, with transverse subdigital lamellae which have projecting tubercles or pectinate anterior margins (examined under magnification; sometimes not evident if keratinized layer has been lost or worn excessively; can usually be seen at least on basal lamellae); dorsum with small juxtaposed scales intermixed with enlarged tubercles which are usually keeled, often trihedral; tail segmented, first segment of proximal portion of tail covered by 2–3 whorls of scales and one whorl of caudal tubercles, tubercles in each whorl in close contact with one another; pupil vertical; males with preanal pores; medium-size species, in which snout-vent length is usually 45 mm or more in mature adults. See Szczerbak and Golubev (1986:88–89; 1996:87) for additional characters.

Distribution: Israel; northern and eastern Arabia; Iran; Pakistan; Afghanistan; marginally into southern Turkmenistan. Four species, two in Iran.

Remarks: In my opinion, *Bunopus* is most closely related to *Crossobamon* and to *Carinatogecko*, and perhaps should include the latter. Szczerbak and Golubev (1986:88–100) recognize three species and place *B. blanfordi* Strauch, 1887 and *B. abudhabi* Leviton and S. Anderson, 1967 in the synonymy of *B. tuberculatus*, as does Arnold (1980a:278), followed by Leviton, *et al.* (1992:30–31).

Michael Golubev (pers. commun., February 1995) informed me that in his opinion, *Bunopus* is closely related to *Cyrtopodion (Cyrtopodion)*, whereas *Carinatogecko* is closely related to *Cyrtopodion (Mediodactylus)*. Thus, he interprets the relationships as representing two parallel branches derived from two close *Cyrtopodion* groups of species. If this is, indeed, the case, *Cyrtopodion*, as currently constituted, is paraphyletic.

Key to the Species of *Bunopus* in Iran

1a. Postmentals (chin shields) absent (Fig. A)...
.... *Bunopus tuberculatus* Blanford, 1874 (p.140)
1b. Postmentals present (Fig. B) .... *Bunopus crassicaudus* Nikolsky, 1907 (p. 139)

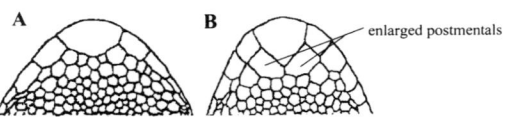

Chins of (A) *Bunopus tuberculatus* and (B) *B. cassicaudus*
(From Szczerbak and Golubev, 1996, figs. 38d, 44d, redrawn)

Bunopus crassicaudus Nikolsky, 1907
Thick-tailed tuberculated gecko

Bunopus crassicauda Nikolsky, 1907a:261–264, pl. 1, figs. 2, 2a (Type locality: Chara-Magommed-Abad, Irak-Adjemi Prov. [restricted by Szczerbak and Golubev, 1986:98]; Lectotype: ZIL 10233 [designated by Szczerbak and Golubev, 1986:98]). — F. Werner, 1936:200. — S. Anderson, 1968:332. — Minton, *et al.*, 1970:335–337, text-fig. 2. — S. Anderson, 1974:32, 42. — Szczerbak and Golubev, 1977a:131. — Szczerbak and Golubev, 1986:98–100 [1996:96–98], figs. 39, 44. — Kluge, 1991:4; 1993:5 . — Welch, 1983:6.
Alsophylax crassicauda: S. Anderson, 1963:474. — Wermuth, 1965:5. — Schleich, 1977:127, 129.

Diagnosis: A pair of postmentals in contact behind the pentagonal mental; ventrals smooth; posterior three-fourths of tail with enlarged subcaudal plates.

FIGURE 64. *Bunopus crassicaudus* (24 km W Rey, Iran). (Photo by A. Leviton, *in* Minton, *et al.*, 1970:336, fig. 2.)

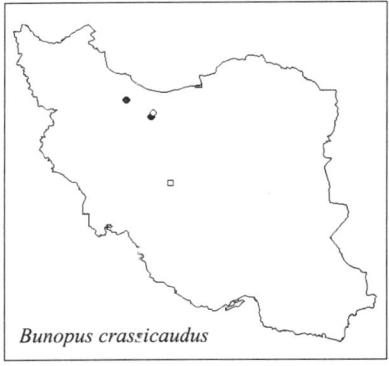

Bunopus crassicaudus

Color pattern: (Fig. 64) Brownish-gray above in life, with five darker transverse bars on body, first on nuchal region, fifth on sacral region; nine dark bars on tail, about equal to interspaces; side of head with dark longitudinal streak; venter white.

Size: Snout-vent length to 50 mm, tail 50 mm in males; snout-vent length to 53 mm in females.

Habitat: Two specimens collected by Minton were found at the edge of a cultivated alluvial plain.

Distribution: In Iran this species is known only from the northwestern part of the Central Plateau. Overlooked in a previous paper (Minton, *et al.*,

1970:335–337) was a record cited by Haas (1951:71) for Djebel Amri, northeastern Syria; this record requires verification, in my opinion.

Remarks: In the specimens I have examined, the series from Qazvin, the males have 5–7 preanal pores; the posterior three-fourths of the tail has enlarged ventral plates. The specimens agree with the description of the types and with the two specimens examined previously (Minton, *et al.*, 1970:335–337). See Szczerbak and Golubev (1986:98–99; 1996:96–97) for additional descriptive details.

Bunopus tuberculatus Blanford, 1874 (Plate 7A)
Baluch rock gecko, Arabian desert gecko, southern tuberculated gecko

Bunopus tuberculatus Blanford, 1874a:454; 1876:348–350, pl. 22, figs. 4, 4a, 4b (Type locality: near Bampur, Baluchistan, Iran [restricted by Szczerbak and Golubev, 1986]; Lectotype: BMNH 74.11.23.85/1946.8.22.84 [designated by Szczerbak and Golubev, 1986:89; see also Das, *et al.*, 1998]). — Bedriaga, 1879:36. — Nikolsky, 1897:316; 1899:387–388. — Zarudny, 1903:8–9. — Nikolsky, 1907a:261. — F. Werner, 1936:200. — Guibé, 1966a:97. — S. Anderson, 1963:439–440; 1968:332. Haas and Werner, 1969:332. — Tuck, 1971a:56. — S. Anderson, 1974:32, 42. — Schleich, 1977:127, 129. — Szczerbak and Golubev, 1977a:131, fig 8. — Szczerbak and Golubev, 1986:89–94 [1996:87–92], figs. 38–41, col. pl. 2, figs. 6–7. — Welch, 1983:6. — Welch, *et al.*, 1990:6. — Kluge, 1991:4. — Leviton, *et al.*, 1992:30–31, col. pl. 4C–D. — Kluge, 1993:5. — Frynta, *et al.*, 1997:9.

Alsophylax tuberculatus: Boulenger, 1885a:20–21. — Werner, 1917:194. — Forcart, 1950:144. — Wettstein, 1951:430–431. — Schmidt, 1955:201. — Wermuth, 1965:7. — Bannikov, *et al.*, 1977:93–94.

Stenodactylus lumsdeni Boulenger, 1887a:479 (Type locality: Afghanistan-Baluchistan frontier between Nushki and Helmand; Holotype: BMNH 86.9.21.8[1946.8.23.51]). — Nikolsky, 1899b:388. — Zarudny, 1903:9. — F. Werner, 1936:200. — S. Anderson, 1963:474. — Wermuth, 1965:176. — S. Anderson, 1968:332.

Gymnodactylus gabrielis F. Werner, 1936:195–196 (Type locality: Leb-e Kal and Kalwan, Iran; Syntypes: NMW 17303:1, 17303:2).

Bunopus biporus F. Werner, 1938:267 (Type locality: Ziarat, Baluchistan [Iran or Pakistan]; Holotype: NMW 15548).

Alsophylax (Bunopus) tuberculatus: Gorelov, *et al.*, 1974:33–35,fig. 1.

Crossobamon lumsdeni: Welch, 1983:6. — Welch, *et al.*, 1990:10.

Diagnosis: No postmental shields; belly covered with small, smooth, subcircular, subimbricate scales; subcaudal scales small, keeled.

Color pattern: (Fig. 66.) Sandy above, with five or six (rarely seven or eight) more or less distinct chocolate-brown transverse bars; a dark curved mark on occipital region more or less distinct; hind limbs crossbarred or mottled with brown, forelimbs mottled; venter white, under side of tail flecked with brown. In some adults the dark bars of the back tend to break up into spots.

Size: Males reach 54 mm snout-vent length, tail 64 mm; females with eggs in ovaries from 34–56 mm snout-vent.

Sexual dimorphism: Only the males have distinct preanal pores, although one juvenile male examined seemed to lack them.

Natural history: Iranian females collected between late August and early January contain ovarian eggs, usually five to seven in each ovary, none in oviducts. Small juveniles (20–24 mm snout-vent length) were collected throughout this period. Beetles were the only recognizable material in digestive tracts examined. (See Leptien, 1993:57 for observations on their natural history in the United Arab Emirates).

Habitat: (Plates 20G, 21B, F) I found this species most commonly on sandy soils of various degrees of compaction, from loose, blowing dune sand to hard silty surfaces, loose sand mixed with gravel and cobble of many sizes, less commonly on non-sandy surfaces, such as flat, silty alluvium. Blanford (1876:350) reported them in houses and under rocks on hillsides, and Minton (1966:74) found them among boulders along a stream in Pakistan. Szczerbak and Golubev (1986:93; 1996:91) describe the circumstances in which this species was collected in the Badkhyz, Turkmenistan. They appear unable to climb vertical surfaces. I found them associated with desert and steppe vegetation such as *Tamarix* and other psammophilous shrubs and bunch grasses, often where there were rodent burrows, and in a *Tamarix-Acacia* "woodland" in the Jaz Murian depression. From sea level to 1000 m in Iran.

Leptien (1993b:57) described the habitat of this species in the United Arab Emirates, where he found them only in sandy areas and among dunes. He found them in burrows, which he speculated that they dug themselves, and under discarded objects at the edges of roads.

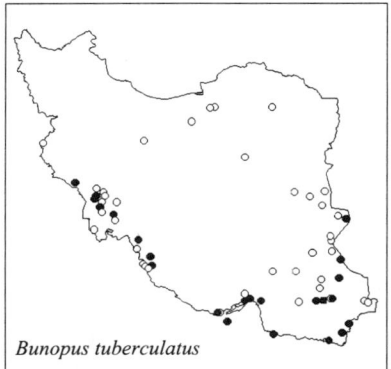

Bunopus tuberculatus

Distribution: Specimens identified as this species have been collected in Israel, Syria, Iraq, southwestern Iran, central and southeastern Iran, southern and eastern Afghanistan, southern Turkmenistan, and Baluchistan and Sind, Pakistan. Boulenger (1887a: 407) reported a specimen from Muscat, while Haas and Werner (1969:332) and Arnold (1977:107) record this species from Saudi Arabia. Arnold (1977:106) says that it occurs over most of Arabia, and that both *Bunopus blanfordi* and *B. abudhabi* should probably be considered synonyms of *B. tuberculatus*, although he offers no discussion or analysis in support of this assertion. *Bunopus tuberculatus* has recently been discovered in southern Turkmenistan (Szczerbak and Golubev, 1986:91; 1996:89; Szczerbak, 1994:310).

Remarks: I have examined 10 specimens from Blanford's syntypic series of 31. I have also examined the syntypes of *Gymnodactylus gabrielis* Werner and *Bunopus biporus* Werner. Blanford included the following remarks in his 1876 discussion:

"A variety of which I have specimens from Mand, Bahu Kalat, and Saman, in Baluchistan, differs so much in colour from the common form of the species that I was first inclined to consider it distinct. The ground colour is pale sandy, with the dark markings on the back almost confined to the enlarged tubercles, some of which, in patches, are brown, the patches having a tendency to form longitudinal rows. There is a dark mark from the nostril through the eye to above the shoulder; farther back it becomes broken up. The dorsal tubercles too in this form are small, and sometimes less distinctly trihedral. There appears, however, to be no constant distinction between the two varieties, which occur together."

Two of the types I have examined (BMNH 74.11.23.91[1946.8.20.43] from Mand; MCZ 7128 from "Baluchistan" [Fig. 65A]) are obviously the specimens referred to in the quotation above. The other specimens (Fig. 65B) agree with Blanford's figure (1876, pl. 22, fig. 4) (Fig. 66). I think that it is quite possible that the first two specimens mentioned represent a species distinct from the other syntypes I have examined.

I have examined, in addition to the type series, over 100 specimens from Iran, Afghanistan, and Pakistan, including one individual from Iranshahr, Iran, in the region of

FIGURE 65. *Bunopus tuberculatus.* (A) MCZ 7128, syntype; (B) BMNH 1946.8.22.86, syntype. (Photos by A. Leviton.)

Blanford's material. There is no difficulty in associating these specimens with the lectotype. All of these more recent specimens were fixed initially in formalin, while Blanford's specimens apparently were preserved directly in alcohol. In all the formalin-fixed specimens the carination of all the tubercles and enlarged caudal scales is pronounced, while the alcohol-preserved material is much softer and retains less of the keeled condition seen in the living animal.

I follow Wettstein (1951:430–431) in regarding *Gymnodactylus gabrielis* Werner and *Bunopus biporus* Werner as synonyms of *B. tuberculatus.* The syntypes of *G. gabrielis* differ from the lectotype of *B. tuberculatus* in having five, rather than six dark crossbars on the back, and six preanal pores; the tubercles and enlarged scales of tail annuli are more strongly keeled (preservation?). NMW 17303:1 has 12 longitudinal rows of dorsal tubercles; NMW 17303:2 has 14. The holotype of *B. biporus* is in poor condition due to its once having dried out. I count 10 very indistinct preanal pores in this specimen, which is evidently a female, the "pores" being only indentations. The subdigital lamellae are as in *B. tuberculatus.* The most striking characteristic of this animal is that the distal row of supracaudal scales of each annulus is not strongly enlarged, and in this feature it approaches MCZ 7128, also a female. This character is somewhat variable, however, in Blanford's series. In each annulus there is an increase of scale size distally.

In addition to the differences remarked upon by Blanford in the quotation above, BMNH 74.11.23.91[1946.8.20.43] and MCZ 7128 differ from the lectotype and other specimens in the following particulars: mental twice as broad as long, not projecting posteriorly beyond margin of adjacent labials, while in the lectotype of *B. tuberculatus* the mental is rounded behind, projecting beyond the posterior margin of the labials, and about 1½ times as broad as long; the tail is covered above with weakly keeled small scales, about five transverse rows to each segment; in the lectotypic series the posterior row of each segment is much larger than the preceding rows, and strongly keeled.

FIGURE 66. *Bunopus tuberculatus* Blanford. (From Blanford, 1876, pl. 22, fig. 4.)

The most striking differences are those noted by Blanford, *viz.*, the lack of distinct color pattern and the nature of the dorsal tubercles. In the two above-mentioned specimens, the tubercles are scarcely more than enlarged granules, and irregularly arranged, while in the other specimens they are distinctly trihedral and form regular longitudinal and oblique rows. There are no distinct tubercles on the back of the head as in typical specimens.

While I think it possible that BMNH 74.11.23.91/1946.8.20.43 and MCZ 7128 represent a taxon distinct from *B. tuberculatus* as here restricted, the species is variable, and these specimens may simply represent one extreme.

Specimens from Khuzestan have somewhat larger, more strongly keeled tubercles that are slightly more regular in arrangement than those of animals from southeastern Iran. The scales of the head are more swollen, conical, and rugose in the former than in the latter specimens. Two adult males examined from Khuzestan have 10 and 14 preanal pores, while those from southeastern Iran have 6–10. Geographically intermediate specimens from Fars are also intermediate in these characters, the males having 6–13 (mean 9.3) preanal pores. Specimens from Afghanistan have as few as 3 preanal pores. A female from Syria (FMNH 19739) and a male from Israel appear to belong to this taxon, and are in general agreement with Khuzestan specimens, except that the male has 7 preanal pores and the tubercles on the tail are less strongly developed. Much of the observed variation in color pattern appears to be due to the relative expansion and contraction of melanophores at time of preservation. There is also considerable variation, apparently not geographically correlated, in the degree of denticulation of the lateral digital scales.

I have examined the holotype of *Stenodactylus lumsdeni* Boulenger (BMNH 86.9.21.8/1946.8.23.51) and conclude that this nominal form also belongs in the synonymy of *B. tuberculatus*. It differs from the lectotype of *B. tuberculatus* in its more attenuate body and limbs, longer lateral denticulations on the digits, in having 7 rather than 6 dark crossbars on the body, 12 instead of 14 dorsal tubercles in the longest transverse series, and about 50 rather than 40 ventral and ventrolateral scales counted between the lowest row of tubercles on one side and that on the other (this latter difference due to fewer tubercles rather than scale size). I have compared the holotype with several other specimens of *B. tuberculatus*, and all of the above characters seem to vary independently. It differs from most other specimens examined only in having 7 rather than 5 or 6 transverse bars on the dorsum, although CAS 141098 from Sistan also shows this color pattern. The specimen is a female (contrary to Boulenger, 1887a:479) and lacks pores, but has 8 enlarged preanal scales, about 7 of which have faint depressions. The subcaudal scales are distinctly keeled. The dorsal tubercles are less strongly keeled than in most formalin-fixed specimens. The lateral denticulations are no longer than in several other specimens from various localities. The nostril is between the rostral, first labial, and two nasals, the third small "nasal" excluded by contact of the rostral with the second nasal. Most *B. tuberculatus* have three nasals, but in some specimens the uppermost is excluded on one or both sides.

Bunopus tuberculatus appears to be more closely related to *B. abudhabi* and *B. blanfordi* (considered synonyms of *B. tuberculatus* by Arnold [1980]) than to *B. crassicaudus*. It also seems to be related to the species currently assigned to *Crossobamon*.

Genus *Carinatogecko* Golubev and Szczerbak, 1981
Keel-scaled geckos

Carinatogecko Golubev and Szczerbak, 1981:34–41, figs. 1–3 (Type species: *Bunopus aspratilis* S. Anderson, 1973, by original designation).

Definition: (Fig. 67) All scales with exception of intermaxillaries, nasals, chin shields and upper and lower labials strongly keeled; 3 nasal scales contact nostril; digits weakly angularly bent, clawed, not dilated, not webbed nor ornamented, with keeled transverse subdigital lamellae; dorsal pholidosis heterogeneous, small juxtaposed scales intermixed with tubercles; pupil vertical; tail segmented, caudal tubercles with bases in middle of each segment, not in contact with one another, separated from posterior margin of segment by ring of scales (Fig. 67B). See Szczerbak and Golubev (1986:127; 1996:123) for additional characters.

Distribution: Western foothills of Zagros Mountains in Iran and Iraq. Two species.

Remarks: Golubev and Szczerbak (1986:17, [1996:17] fig. 1) show this genus as a sister taxon to *Cyrtopodion* + *Agamura*, according to their phylogenetic diagram. However, Golubev (pers. commun., 3 February 1995) informed me that my interpretation of this diagram leads to misunderstanding of their systematic hypothesis, and referred me to the diagram in Golubev and Szczerbak (1981:40, fig. 3), in which *Carinatogecko* is regarded as a branch derived from the subgenus *Cyrtopodion (Mediodactylus)* close to *C. heterocercum*. This hypothesis would, of course, make *Cyrtopodion* and *Mediodactylus* paraphyletic and necessitate placing the species included in *Carinatogecko* into *Cyrtopodion (Mediodactylus)*. For now, I retain the nomenclature of Golubev and Szczerbak (1986) without accepting their phylogenetic hypothesis until all of these generic taxa have been defined in terms of synapomorphies.

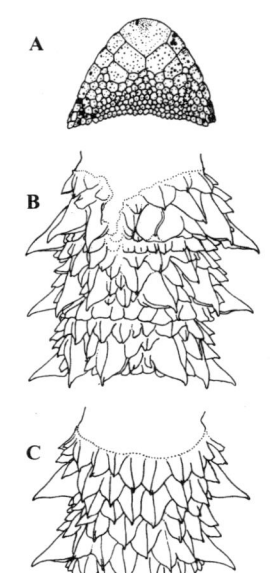

FIGURE 67. *Carinatogecko aspratilis* (USNM 193961, holotype). (A) chin from below; (B) tail from above; (C) tail from below. (From S. Anderson, 1973:357, fig. 2; drawn by Fanny Philips.)

Key to the Species of *Carinatogecko*

1a. Scales of middle of back distinctly larger than abdominals; caudal tubercles pointed, raised, with enlarged posterior facets; analogous dorsal tubercles present on forearms; 17–18 subdigital lamellae on 4th toe ....
....*Carinatogecko aspratilis* (Anderson, 1973) (p. 144)

1b. Scales of middle of back negligibly smaller or alike in size to abdominals; caudal tubercles not pointed, posterior facets not raised; no analogous tubercles on forelimbs; 15 subdigital lamellae on 4th toe ....
.............*Carinatogecko heteropholis* (Minton, Anderson, and Anderson, 1970) (p. 145)

Carinatogecko aspratilis (Anderson, 1973) (Plate 7B)
Iranian keel-scaled gecko

Bunopus sp.: Tuck, 1973:13.
Bunopus aspratilis S. Anderson, 1973:355–358, figs. 1–2 (Type locality: 35 km E Gach Saran [30°20′N, 50°48′E], Fars Province, Iran; Holotype: USNM 193961). — S. Anderson, 1974:32, 42. — Szczerbak and Golubev, 1977a:131. — Welch, 1983:6.
Carinatogecko aspratilis: Golubev and Szczerbak, 1981:35–37, fig. 1a–h. — Szczerbak and Golubev, 1986:127–130 [1996:124–125], fig. 58. — Kluge, 1991:4; 1993:5.

FIGURE 68. Holotype of *Bunopus aspratilis* Anderson (USNM 193961). (From S. Anderson, 1973, fig. 1.)

Diagnosis: Mid-dorsal scales distinctly larger than abdominal scales; caudal tubercles raised, pointed; scales of upper surface of forearm heterogeneous; lower row of distal lateral digital scales flat, keeled, not acuminate; 17–18 subdigital lamellae on fourth toe.

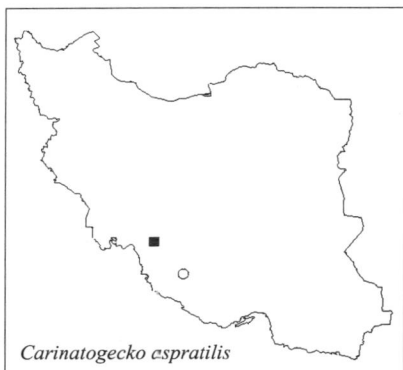

Carinatogecko aspratilis

Color pattern: (Fig. 68) Dorsum light gray-brown, back with six prominent, undulating, chocolate-brown, transverse bars narrower than interspaces, first one on nape, sixth across sacrum; top of head and lips mottled with dark brown; tail with 12 dark bars; venter light gray with irregular dark marks on throat and pectoral areas.

Size: Female: snout-vent length 27.4 mm, tail 29.6 mm. Males still unknown.

Habitat: The type and paratype were collected under small, flat stones adjacent to a dry watercourse in an area of sparse vegetation and few trees.

Distribution: Known only from the type locality in the southwestern Zagros foothills of Iran, and, recently, from Shiraz, Fars Province (H. Kami, pers. commun.; see Plate 7B).

Carinatogecko heteropholis (Minton, Anderson, and Anderson, 1970)
Iraqi keel-scaled gecko

Alsophylax persicus: Reed and Marx, 1959:97 (not Nikolsky, 1903).
Tropiocolotes heteropholis Minton, S. Anderson, and J. A. Anderson, 1970:357–358, fig. 12 (Type locality: Pirman Hotel, Salahedin, Salahedin Nahiya, Erbil Liwa, Iraq; Holotype: FMNH 74549). — S. Anderson, 1974:32, 43. — Welch, 1983:14.
Carinatogecko heteropholis: Golubev and Szczerbak, 1981:37–39, fig. 2a–h. — Szczerbak and Golubev, 1986:129–130 [124–127], fig. 59. — Kluge, 1991:4. — Leviton, *et al.*, 1992:31–32, pl. 33A. — Kluge, 1993:5.

Diagnosis: Mid-dorsal scales slightly larger than or subequal to abdominal scales; caudal tubercles distinctly swollen, but blunt, their posterior borders not upraised; scales of upper surface of forearm homogeneous; lower row of lateral subdigital scales acuminate; 15 subdigital lamellae on 4th toe.

Color pattern: Dorsum light gray-brown in preservative, with seven narrow dark chevrons, apices pointing caudad, bordered posteriorly with broader light crossbars; first chevron on nape, seventh across sacral region; similar marks on tail; limbs and digits faintly crossbarred with dark.

Size: Snout-vent 31 mm.

Habitat: I have no information regarding the habitat, other than that provided by Reed and Marx (1959:97) regarding collection of the holotype in Iraq. It was found under a bit of bark on the cement floor of the back porch of a hotel. Reed speculated that it may have come in with scrub-oak firewood. The Iranian specimen originates from an area that is also in the semihumid Zagrosian oak forest. Much of this area has been deforested, however, and I have no habitat information regarding the collection site.

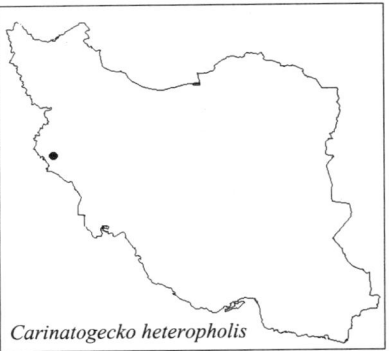
Carinatogecko heteropholis

Distribution: The two known specimens both come from the western Zagros foothills, the type from northeastern Iraq, the other from western Iran, at an elevation of about 1500 m, about 400 km southeast of the type locality.

Genus *Crossobamon* Boettger, 1888
Fringe-toed geckos

Crossobamon Boettger, 1888b:260 (Type species: *Gymnodactylus eversmanni* Wiegmann, 1834, by monotypy).

Definition: Digits straight (i.e., not distinctly angularly bent at any of the articulations as in *Cyrtopodion* and *Agamura*), not dilated, covered beneath with transverse lamellae which are pectinate, or have several keels; digits with comb-like fringe on each side; dorsum covered with small imbricate scales intermixed with larger tubercles Fig. 69B); nasal region not noticeably swollen (in the sense of *Stenodactylus* as redefined by Kluge, 1967:23–24); no postmental shields; pupil vertical; preanal pores present in males; scleral ossicles 14–15; angular absent; squamosal present. (In part after Kluge, 1967:23). See Szczerbak and Golubev (1986:46–47; 1996:46) for additional characters.

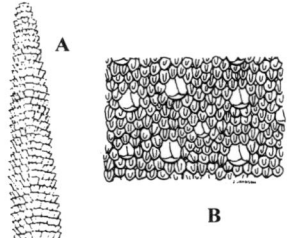
FIGURE 69. *Crossobamon eversmanni*. (A) Dorsal surface of tail (From Bannikov, *et al.*, 1977:84, fig. 22:2; (B) dorsal scales of back. (Drawn by Jean Jahnson.)

Distribution: Kazakhstan, Turkmenistan, eastern Iran, Afghanistan, and Pakistan. Three species recognized provisionally here (see below), one in Iran.

Remarks: Kluge (1967:23–24) redefined *Crossobamon* to include *"maynardi," "orientalis,"* and *"lumsdeni,"* as well as *C. eversmanni*, redefining the genus *Stenodactylus*, and restricting it to include only the western lowland species. I follow Kluge's concept of the two genera here, as have most recent workers, although I regard *Stenodactylus lumsdeni* Boulenger as conspecific with *Bunopus tuberculatus* Blanford (see above; Kluge, 1991:6 and 1993:7 lists *lumsdeni* as a synonym of *C. eversmanni*). *Crossobamon* appears to be closely related to *Bunopus* on the basis of external morphology, differing from that genus primarily in the form of the digits, which are adapted to sand-dwelling. In the phylogenetic hypothesis of Szczerbak and Golubev (1986:17[1996:17], fig. 1), *Crossobamon* is the sister taxon to *Stenodactylus*.

Crossobamon eversmanni (Wiegmann, 1834) (Plate 7C)
Eversmann's fringe-toed gecko

Gymnodactylus eversmanni Wiegmann, 1834:19, note 28 (Type locality: Kyzylkum Desert, Uzbekistan; Holotype: ZMB 435 [see Bauer and Günther, 1991:286]).
Crossobamon eversmanni: Boettger, 1888a:880; 1888b:260. — Nikolsky, 1897:315–316; 1900:388. — Zarudny, 1903:9. — Nikolsky, 1915:58–60, text-fig. 14. — F. Werner, 1936:200. — S. Anderson, 1963:474. — Wermuth, 1965:22. — S. Anderson, 1968:332; 1974:32, 42. — Bannikov, *et al.*, 1977:86–87. — Welch, 1983:6. — Welch, *et al.*, 1990:10. — Kluge, 1991:6.
Crossobamon eversmanni eversmanni: Szczerbak and Golubev, 1986:49–50 [1996:48–49], figs. 17–18. — Kluge, 1993:7.
Crossobamon eversmanni lumsdeni: Szczerbak and Golubev, 1986:50–54, figs. 20–21.

Diagnosis: Back with small dark dots, irregularly arranged or forming 10–12 more or less continuous, vermiculate rows; dark stripes that extend from nostrils over eye and above ear usually end on flanks in first third of body; tail with up to 30 transverse bars, which may coalesce to form longitudinal stripes on sides; 5–11 gular scales in contact with mental (Szczerbak and Golubev (1986:49; 1996:48).

Color pattern: (Fig. 70.) Light brownish yellow or sandy above, with many tiny dark brown dots, often largely confined to the dorsal tubercles, sometimes coalescing on vertebral line to form short crossbars in addition to the scattered dorsal dots; a dark line from rostral through eye to behind shoulder, sometimes extending length of flank; tail with dark crossbars, these fusing on distal third or more of tail; limbs and feet mottled and reticulated with dark brown; venter cream.

Sexual dimorphism: Males with 7–11 preanal pores, females with a similar series of enlarged preanal scales.

Size: The largest male examined measures 46 mm snout-vent, tail 65 mm; largest female snout-vent 50 mm, tail 72 mm.

FIGURE 70. *Crossobamon eversmanni*. (From Terentjev and Chernov, 1949:117, fig. 56.)

Habitat: (Plate 24A) In Iran, I found this species in areas of loose and compacted loess where there was fairly dense steppe shrub vegetation, e.g., *Convolvulus*, various legumes and composites, and in areas where psammophilous shrubs had been planted in regular rows in a dune-stabilization program, but where sand was still loose and blowing. About 900 m elevation. In Turkmenistan, I observed this species in typical white saxaul (*Haloxylon*) biotope in the Kara Kum.

Distribution: In Iran it is known only from the northeast and south to Sistan. In Afghanistan it is known from the low elevations to the north of the Central Massif. It should be looked for in the low coastal region of the southeastern corner of the Caspian Sea if suitable habitat occurs there, since it occurs in Turkmenistan.

Remarks: The natural history of this species has been summarized by Szczerbak and

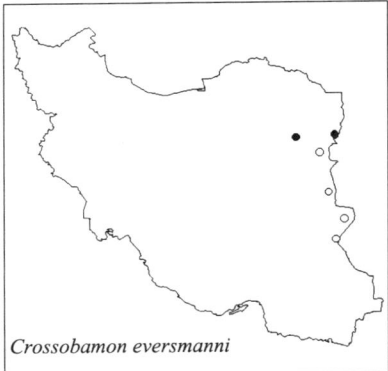

Crossobamon eversmanni

Golubev (1986:51–54; 1996:50–54). Until I collected it in 1975, this species had not been collected in Iran since Zarudny's expedition in 1903. Afghan specimens collected in mid-April have ovarian eggs; the smallest juvenile I have examined, collected at the same time, measures 26 mm snout-vent.

In CAS 141178, the only Iranian specimen compared directly with other material, while the toes are distinctly fringed, the fingers are denticulate, resembling those of *Bunopus tuberculatus*, whereas the fingers of specimens from Turkmenistan (CAS 94055–6) have comb-like fringes. In the Afghan specimens examined (CAS 120288–301, 121070), the fingers are more like those of the Iranian specimens than like those of the Turkmenistan geckos, although the lateral denticulation is somewhat more pronounced.

The ventral and subcaudal scales in this species are much like those of *Bunopus*, as are the subdigital lamellae. The tail, however, lacks the enlarged scales in the terminal row of each segment, and apparently it is not easily broken. Of 18 specimens examined, one has lost about two-thirds of the tail, another is missing the tip, and a third has regenerated a portion of the tail, about two-thirds of which has been lost. The regenerated part is quite obvious, being uniformly gray-brown, lacking transverse markings, with the scales morphologically distinct from those of the original. The long, thin, tapering tail is suggestive of the tails of lacertids, and probably represents an adaptation to rapid locomotion over loose sand.

This species is very similar to *Crossobamon maynardi*, which occurs in the Afghan-Baluch border region. The two differ principally in the color pattern, which consists of narrow longitudinal lines in *C. maynardi*. In my specimens from 35 km N Gonabad, the dorsal spots are fused to form similar stripes. I have examined one of the syntypes of *Stenodactylus maynardi* Smith (BMNH 1931.6.18.3[1946.8.23.36] and see little justification for recognizing it as a species distinct from *Crossobamon eversmanni*. I have not examined the type of *C. eversmanni*, however, nor had the opportunity to compare material from throughout the ranges of these two nominal forms. The fringes on the toes are longer in specimens identified as *C. maynardi* from Nushki, Pakistan (RSM spec.) than in *C. orientalis* I examined from Pakistan. In series from Nushki, Pakistan, individuals with lineolate pattern occur, as do individuals with color pattern as described for *C. eversmanni*.

Szczerbak and Golubev (1986:49–54; 1996:48–54) distinguish two subspecies, *C. e. eversmanni* of Turkmenistan and other Central Asian republics, northern Afghanistan, and eastern Iran, and *C. e. lumsdeni* of eastern Iranian Baluchistan (Hormak) and the Afghan-Pakistan border region. They place *C. maynardi* in the synonymy of *C. e. lumsdeni*. Apparently, they did not examine the type of *lumsdeni*. Since I regard *Stenodactylus lumsdeni* as a synonym of *Bunopus tuberculatus*, if the analysis of Szczerbak and Golubev is correct, the appropriate name for the southeastern taxon would be *Crossobamon eversmanni maynardi* Smith, 1933, in my opinion. Kluge (1991:6; 1993:7) lists *maynardi* as a subspecies of *C. eversmanni* and *lumsdeni* as a synonym of *C. e. eversmanni*.

Genus *Cyrtopodion* Fitzinger, 1843
Naked-fingered geckos, bent-toed geckos, thin-toed geckos

Cyrtopodion Fitzinger, 1843:18, 93 (Type species: *Stenodactylus scaber* Heyden *in* Rüppell, 1827, by original designation).

Definition: Digits clawed, slender, distal two or three phalanges make an angle with the proximal portion of the digits; a single series of smooth transverse subdigital lamellae; end of digit weakly or not at all laterally compressed; generally three or more rows of lateral scales on digits; no fringes or denticulations on lateral digital scales; longitudinal concavity in frontal region of head, as a rule, absent or weakly developed; usually not more than 30 scales between centers of top edges of eyes; pupil vertical with lobed anterior and posterior margins; usually two or three pairs of postmental shields, the first pair usually in contact behind the mental; males with preanal and/or femoral pores; segmentation of tail pronounced.

Distribution: Countries of the eastern Mediterranean (Egypt, the Balkans) east through Southwest Asia across the North Arabian Desert, through Pakistan to northern India and the flanks of the Himalayas. North to the southern republics of Central Asia.

Remarks: Annandale (1913) organized the Indian (in the larger sense of "British India") species of *Gymnodactylus* into five groups: Group I, based on *Stenodactylus scaber* Rüppell, including also *Gymnodactylus montiumsalsorum*, *G. brevipes*, *G. kachhensis*, and *G. elongatus*; Group II, monotypic, based on *Gymnodactylus stoliczkai*; Group III, based on *Puellula rubida* Blyth, including *G. gubernatoris*, *G. himalayicus*, and *G. khasiensis*; Group IV, based on *Gymnodactylus nebulosus* Beddome, including *G. oldhamii*, *G. triedrus*, *G. jeyporensis*, *G. deccanensis*, and *G. albofasciatus*; Group V, based on *Cyrtodactylus pulchellus* Gray, including *G. peguensis, G. consobrinoides, G. fasciolatus, G. frenatus, G. variegatus*, and *G. feae*.

Kluge (1985) regarded *Cyrtodactylus* as a junior objective synonym of *Gonydactylus* Kuhl and van Hasselt, 1822, type species *Cyrtodactylus pulchellus* Gray, 1827, by subsequent designation of Kluge (1985). Subsequently, however, he listed *Gonydactylus* as a *nomen nudum* (Kluge, 1993:7). The only Iranian species which Kluge (1993:8) listed as a species of *Cyrtodactylus* is *Gymnodactylus kirmanensis* Nikolsky, 1899 (see below), which he allocated without comment. *Tenuidactylus* is a junior subjective synonym of *Cyrtopodion* Fitzinger, 1843, type species *Stenodactylus scaber*, by monotypy (the footnote by Golubev *in* Szczerbak and Golubev [1996:126] notwithstanding).

In accord with the above statement, the subgeneric arrangement of Szczerbak and Golubev (1986; 1996) becomes as follows:

Cyrtopodion Fitzinger, 1843:93 (Type species: *Stenodactylus scaber* Heyden, 1827, by original designation). *Cyrtopodion agamuroides* (Nikolsky), *C. brevipes* (Blanford), *C. elongatus* (Blanford), *C. kachhense* (Stoliczka), *C. montiumsalsorum* (Annandale), *C. scabrum* (Heyden), *C. watsoni* (Murray). This subgenus thus corresponds generally with Annandale's Group I.

Mediodactylus Szczerbak and Golubev, 1977:130 (Type species: *Gymnodactylus kotschyi* Steindachner, 1870 by original designation). *Cyrtopodion amictopholis* (Hoofien), *C. heterocercum* (Blanford), *C. kotschyi* (Steindachner), *C. russowii* (Strauch), *C. sagittifer* (Nikolsky), *C. spinicauda* (Strauch).

Tenuidactylus Szczerbak and Golubev, 1984:53 (Type species: *Gymnodactylus caspius* Eichwald, 1831 by original designation). *Cyrtopodion caspium* (Eichwald), *C. fedtschenkoi* (Strauch), *C. longipes* (Nikolsky), *C. turcmenicum* (Szczerbak).

A group of species is unplaced as to subgenus, but collectively regarded as a "Tibeto-Himalayan" group:

Cyrtopodion chitralensis (Smith), *C. kirmanense* (Nikolsky), *C. mintoni* (Golubev and Szczerbak), *C. stoliczkai* (Steindachner), *C. tibetanus* (Boulenger).

In addition, a number of species have been described from northern Pakistan by Khan and others since Szczerbak and Golubev's book appeared, and so these are also unplaced, although Kluge (1991, 1993) included all of these, along with those of Szczerbak and Golubev's "Tibeto-Himalayan" group, in *Gonydactylus* Kuhl and van Hasselt, 1822, and subsequently in *Cyrtodactylus* (Kluge, 1993:8). Kluge (1991, 1993) recognized the above groupings (with slight rearrangements) at the generic level.

In this work, I have not arranged the species of *Cyrtopodion* according to subgenera, but rather, list them alphabetically. The subgeneric arrangement of Szczerbak and Golubev (1986:22, fig. 5) is not based clearly on synapomorphies. Obviously, the current arrangement of bent-toed geckos is unstable, for the group is assuredly not monophyletic, and, thus, much additional work remains to be done.

My own rather cursory observations of external morphology and pholidosis lead me to sort the species which I have had opportunity to examine and those which I can associate with my observations based on published descriptions into the four species groups below, differing slightly from the subgenera of Szczerbak and Golubev (1986). Apparent synapomorphies are in italics.

The *scabrum* group

Content: *Cyrtopodion scabrum, C. watsoni, C. kachhense, C. chitralensis, ?C. brevipes*.

A single row of enlarged, smooth subcaudals (except for base of tail, which is covered below with smaller smooth scales), or (*kachhense*) tail covered below throughout its length by small, smooth scales, three transverse rows to each caudal segment; terminal transverse row of each caudal annulus enlarged to form six strongly keeled "tubercles"; scales of lower surfaces of limbs smooth; four to nine preanal pores; no subfemoral tubercles; trihedral dorsal tubercles distinctly larger than interspaces; two or three pairs of plate-like postmental shields, the first in contact; peritoneum unpigmented; adults usually less than 55 mm snout-vent length.

The *agamuroides* group

Content: *Cyrtopodion agamuroides, C. gastrophole*.

A single row of enlarged, smooth subcaudals (except for base of tail, which is covered below with smaller smooth scales), or two longitudinal rows of enlarged scales; terminal transverse dorsal row of each caudal annulus consisting of six enlarged, strongly keeled scales; scales of lower surfaces of limbs smooth; limbs and tail long and slender compared with those of other groups in this assemblage; three to four preanal pores; no subfemoral tubercles; dorsal tubercles rounded, smooth or weakly keeled to subconical, 8–10 in longest transverse row, at least those of paravertebral rows distinctly smaller than interspaces; three to four pairs of plate-like postmentals, the anterior in contact; *peritoneum darkly pigmented*; adults usually less than 50 mm snout-vent length.

The *caspium* group

Content: *Cyrtopodion caspium, C. fedtschenkoi, C. montiumsalsorum, C. longipes, C. turcmenicum*.

A single row of enlarged, smooth subcaudals (except for base of tail, which is covered below with smaller smooth scales); terminal transverse dorsal row of each caudal annulus consisting of six enlarged, strongly keeled scales; scales of lower surfaces of limbs smooth; *both femoral and preanal pores present; one to six subfemoral tubercles usually present*; 12–16 rows of dorsal tubercles, distinctly keeled, distinctly larger than interspaces; two or three pairs of plate-like postmentals, the first in contact; peritoneum unpigmented; adults usually over 50 mm snout-vent length.

The *kotschyi* group

Content: *Cyrtopodion kotschyi, C. russowii, C. heterocercum, C. sagittifer, C. spinicauda*.

Caudal tubercles, six to each annulus, do not form terminal row, but are distributed around middle of each caudal segment; no subfemoral tubercles; dorsal tubercles strongly keeled, trihedral, larger than interspaces; peritoneum unpigmented; 13–23 lamellae under fourth toe; preanal pores only; adults usually less than 55 mm snout-vent length.

Of the species considered by Szczerbak and Golubev (1986), I have not seen the following: *Cyrtopodion tibetanus, C. mintoni, C. stoliczkai, C. kirmanense, C. amictopholis*, and *C. elongatus*. I have examined the type of *Gymnodactylus ingoldbyi* Procter (BMNH 1921.3.21.1/1946.8.23.23) from Ladha, Waziristan, placed in the synonymy of *Cyrtopodion watsoni* by Smith (1935:) who was followed by Szczerbak and Golubev (1986:188). In the original portion of the tail, the subcaudal plates form two longitudinal rows.

Key to the Species of *Cyrtopodion* in Iran

1a. No large chin shields .... *Cyrtopodion spinicauda* (Strauch, 1887) (p. 169)
1b. Chin shields (postmentals) present, usually in contact behind mental 2
2a. Subpostfemoral tubercles present among granules of lower surface of thigh, in short row of 2–6, often in contact with posterior row of large imbricate scales; males with continuous series of preanal and femoral pores 3
2b. No subpostfemoral tubercles; males with preanal pores only 6

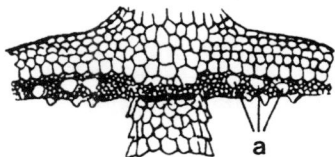

Ventral surface of thighs of *Cyrtopodion caspium* showing (a) subpostfemoral tubercles. (From Nikolsky, 1915:53, fig. 19.)

3a. 19–23 strongly keeled, mucronate tubercles in paravertebral row from occiput to level of vent; males with 22–31 (24–29 in Iranian and Afghan specimens examined) preanal and femoral pores (total both sides); 3 small, parallel stripes from front edge of orbit to edge of mouth; 115 or fewer ventral scales between postmentals and vent ....
....*Cyrtopodion caspium caspium* (Eichwald, 1831) (p. 155)
3b. 26–33 strongly keeled, nonmucronate trihedral or subtrihedral tubercles in paravertebral row from occiput to level of vent; males with 31–46 preanal and femoral pores (total of both sides); no such stripes on snout; 120 or more ventral scales between postmentals and vent 4
4a. 1st postmentals in contact, forming distinct suture (usually 10–40% of postmental length) ....*Cyrtopodion turcmenicum* (Szczerbak, 1978) (p. 170)
4b. 1st postmentals form short suture (less than 10% of postmental length) or separated by one or more gular scales 5

5a. 27–34 abdominal scales across middle of belly; 26–30 dorsal tubercles in paravertebral row from occiput to level of vent .... *Cyrtopodion longipes longipes* (Nikolsky, 1896) (p. 163)
5b. 35–39 abdominal scales across middle of belly; 31–33 dorsal tubercles in paravertebral row from occiput to level of vent .... **Cyrtopodion longipes microlepis* (Lantz, 1918) (p. 164)
6a. Subcaudal scales one head-width behind vent small, not enlarged, plate-like 7
6b. Subcaudal scales one head-width behind vent enlarged, plate-like, two serially arranged plates, or pairs of plates covering each caudal segment 10
7a. Subcaudal plates smooth 8
7b. Subcaudal plates distinctly keeled 9
8a. No scattered small tubercles among rows of enlarged dorsal tubercles; caudal tubercles form terminal rings of each annulus .... *Cyrtopodion kachhense* (Stoliczka, 1872) (p. 159)
8b. Scattered small keeled tubercles among the large trihedral dorsal tubercles which form fairly regular longitudinal rows; tubercles on tail arranged around middle of each segment, not in terminal scale row of each annulus .... *Cyrtopodion russowii zarudnyi* (Strauch, 1887) (p. 165)
9a. 23–30 abdominal scales across middle of belly; outermost row of caudal tubercles distinctly larger than other tubercles of each annulus .... *Cyrtopodion heterocercum heterocercum* (Blanford, 1874) (p. 158)
9b. 14–16 abdominal scales across middle of belly; outermost row of caudal tubercles not distinctly longer than other tubercles in each annulus .... *Cyrtopodion sagittifer* (Nikolsky, 1900) (p. 166)
10a. Subcaudal plates in two median series; dorsal tubercles distinctly smaller than interspaces; snout 2 to 2¼ times longer than diameter of eye .... *Cyrtopodion kirmanense* (Nikolsky, 1900) (p. 160)
10b. Subcaudal plates in a single median series; dorsal tubercles smaller or larger than interspaces; snout length less than twice diameter of eye 11
11a. Caudal tubercles arranged around middle of each caudal segment, not forming terminal ring of each segment .... **Cyrtopodion kotschyi* (Steindachner, 1870) (p. 161)
11b. Caudal tubercles or enlarged keeled scales forming terminal ring of each segment 12
12a. Dorsal tubercles distinctly smaller than interspaces, rounded, smooth or weakly keeled to subconical, but not distinctly trihedral; peritoneum and covering of some internal organs of abdominal cavity darkly pigmented; limbs and tail thin, attenuate 13
12b. Dorsal tubercles distinctly larger than interspaces, strongly keeled and trihedral; peritoneum and covering of organs of abdominal cavity without melanocytes; limbs and tail sturdy 14
13a. 24–28 abdominal scales across middle of belly (14–15 scales across belly in distance equal to length of snout); snout length less than 1½ times diameter of eye .... *Cyrtopodion agamuroides* (Nikolsky, 1900) (p. 153)
13b. 10–16 abdominal scales across middle of belly (6–8 scales across belly in distance equal to length of snout); snout length 1½ times diameter of eye .... *Cyrtopodion gastrophole* (F. Werner, 1917) (p. 156)
14a. 12–16 dorsal tubercles in longest transverse (chevron-shaped) series across back; width of dorsal tubercles distinctly smaller than greatest diameter of ear opening; 10–14 supralabials .... *Cyrtopodion scabrum* (Heyden, 1827) (p. 167)

14b. 10 dorsal tubercles in longest transverse series across back; width of dorsal tubercles nearly equal to greatest diameter of ear opening; 9 supralabials ....
....*Cyrtopodion brevipes* (Blanford, 1876) (p. 154)

Cyrtopodion agamuroides (Nikolsky, 1900) (Plate 7D)
Nikolsky's spider gecko, agamuroid thin-toed gecko

Gymnodactylus agamuroides Nikolsky, 1900:384–385 (Type locality: Neizar, Sistan, Iran and Duz-Abad and Pendsch-Sara, Kerman, Iran [restricted to Pendsch-Sara by Szczerbak and Golubev, 1986:196]; Syntypes: ZIL 9326–9328 [Lectotype: ZIL 9327, designated by Szczerbak and Golubev, 1986:196]). — Zarudny, 1903:7. — F. Werner, 1936:200. — Wettstein, 1951:432. — Welch, 1983:7.

Cyrtodactylus agamuroides: S. Anderson, 1963:438, fig. 9; 1968:332; 1974:34, 42. — Schleich, 1977:127, 129.

Gymnodactylus (Cyrtodactylus) agamuroides: Wermuth, 1965:47. — Szczerbak and Golubev, 1977a:130.

Tenuidactylus (Mesodactylus) agamuroides: Szczerbak and Golubev, 1984:54.

Tenuidactylus (Cyrtopodion) agamuroides: Szczerbak and Golubev, 1986:196–197, fig. 82, 87 (figs. 86 and 87 reversed); 1996:192–193, figs. 82, 87.

Cyrtopodion agamuroides: Kluge, 1991:6; 1993:9. — Frynta, et al., 1997:9.

Diagnosis: No subfemoral tubercles; males with preanal pores only, in scales significantly larger than surrounding ones; subcaudal plates one head-width behind vent enlarged, plate-like, in single median series; caudal tubercles forming terminal ring of each segment; dorsal tubercles distinctly smaller than interspaces, not distinctly trihedral; peritoneum darkly pigmented; limbs and tail thin and attenuate; 24–28 abdominal scales across middle of belly (14–17 scales across belly in distance equal to length of snout); snout length less than 1½ times diameter of eye.

Color pattern: Gray or tan above, with three longitudinal series of darker, subquadrangular spots and a less distinct lateral series; tail with 11–12 transverse dark bars; venter white.

Size: Snout-vent 49 mm, tail 63 mm.

Habitat: (Plate 21A) The only specimens that I personally collected were active at night among jumbled rocks on a rocky bluff above the Sarbaz River; none were found in or on the abandoned buildings present at the site. Another specimen was found by a boy beneath a rock near a village in a desert canyon (about 400 m elevation) during the day. Zarudny (1904) found them in abundance on the ruins of earthen buildings, in shady areas of clay, and on rock cliffs. He found them active in the shade during the day (Szczerbak and Golubev, 1986:196; 1996:192).

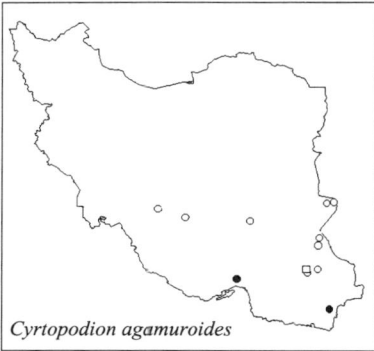
Cyrtopodion agamuroides

Distribution: Known from Kerman, Sistan, and Baluchistan Provinces in Iran; northwestern Las Bela in Pakistan.

Remarks: These geckos began calling at dusk, a "whit-whit" was audible for at least 100 m. Series varied from 9–20 chirps. The air temperature was 32°, rock surface 34°C. They were still calling at 0600 hrs the following morning, air 24.5°, rock 31°C. This species appears to be most closely related to *C. gastrophole* (Werner) (see above, under generic discussion).

Szczerbak and Golubev (1986:196; 1996:192)

claim that I (Anderson, 1963:438) misidentified my specimen from Kuhha-ye Genu (CAS 86370), as did Minton (1966:80) his, attributing them to *Agamura* (Minton placed this species in *Agamura*, whereas I regarded it as *Cyrtodactylus*). Because Szczerbak and Golubev (1986:196; 1996:192) place both the allocations of Anderson (1963) and Minton (1966) in the synonymy of their *Tenuidactylus*, it is not clear whether they mean that we misidentified our specimens as to species or to genus. I do not find my locality (Kuhha-ye Genu) plotted on any of their maps (figs. 82, 95). CAS 141075, the only specimen from Baluchistan that I had opportunity to examine carefully, differs from CAS 86370 from Kerman Province in having the dorsal tubercles less regularly disposed and in having a double row of enlarged subcaudal scales, rather than a single row of large plates. In the posterior half of the tail all scales are sharply keeled. This individual would fall out in the key at *C. kirmanense*, if it were not for its much shorter snout.

Cyrtopodion brevipes (Blanford, 1874)
Blanford's short-toed gecko

Gymnodactylus brevipes Blanford, 1874a:453; 1876:344–345, pl. 22, fig. 2 (Type locality: Aptan, near Bampur, 3000 ft, Iran [designated by Annandale, 1913b:315]; Holotype: ZSI 3465). — Bedriaga, 1879:35. — Murray, 1884b:102. — Boulenger, 1885a:28. — Welch, 1983:7.
Cyrtodactylus brevipes: S. Anderson, 1963:438, 474; 1968:332; 1974:34, 42.
Gymnodactylus (Cyrtodactylus) brevipes: Wermuth, 1965:49. — Szczerbak and Golubev, 1977a:130.
Mediodactylus brevipes: Kluge, 1991:20; 1993:22.

Diagnosis: (Fig. 71) No subfemoral tubercles; males with (4) preanal pores only; subcaudal scales one head-width behind vent enlarged, plate-like, in single median series; caudal tubercles forming terminal ring of each segment; dorsal tubercles distinctly larger than interspaces, strongly keeled and trihedral; limbs and tail sturdy; 10 dorsal tubercles in longest transverse series across back; width of dorsal tubercles nearly equal to greatest diameter of ear opening; 9 supralabials.

Color pattern: Gray above, with three indistinct longitudinal dusky streaks on back, formed of 8 dark violet-gray arrowhead-shaped marks; an indistinct dusky line from eye to shoulder; 12 dark transverse bars on tail; dorsal surface of head with dark marmoreal pattern; venter white.

Size: Snout-vent length 44 mm, tail 51 mm.

Habitat: Blanford's specimen came from an open sandy plain with scattered vegetation.

Distribution: Known with certainty only from the type locality in the Jaz Murian depression in Baluchistan, southeastern Iran. If Murray's identifications are correct (I am skeptical of all of Murray's identifications), it occurs along the Persian Gulf coast in Fars Province. He also records it (1892:68) from Nushki and Kirta, Baluchistan, Pakistan. However, Annandale, writing in 1913, states: "The type, which is in good condition and has apparently retained its natural colouration, still remains unique."

Remarks: In a list of the specimens in the museum of the Iran Department of Environment sent to me by Robert Tuck, there is a specimen from Iranshahr, near the type locality, identified as this species.

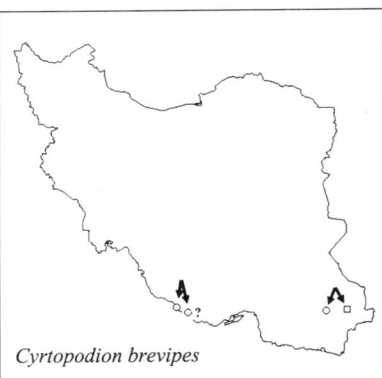
Cyrtopodion brevipes

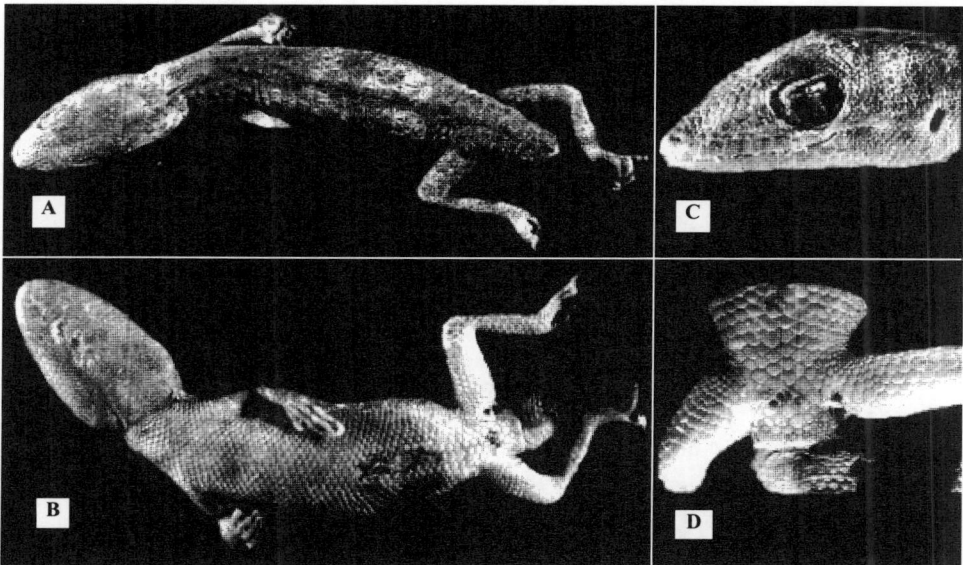

FIGURE 71. *Cyrtopodion brevipes* (holotype [ZSI 3465]): (A) dorsal view; (B) ventral view; (C) left side of head; (D) groin area, showing preanal pores. (Photos by B. K. Tikader, Calcutta, courtesy Michael Golubev.)

I have seen no specimens I can identify as belonging to this taxon. Minton (1966:77) placed it in the synonymy of *C. scabrum* without comment. The two appear to be closely related, but the described color pattern for *C. brevipes* differs markedly from that of *C. scabrum*, and as I have not seen the type of *C. brevipes*, I recognize them tentatively as distinct taxa.

Szczerbak and Golubev (1986:185, footnote, 1996:181) say that judging from photographs of the type specimen of *Gymnodactylus brevipes* received from Calcutta (Fig. 71), this specimen resembles *Cyrtopodion kachhense* in the small number of preanal pores (4) and small dorsal tubercles and *C. scabrum* by large ventral scales (20–22). They believe that, in any case, it belongs to their subgenus *Cyrtopodion*.

Cyrtopodion caspium (Eichwald, 1831)
Caspian bent-toed gecko

Cyrtopodion caspium caspium (Eichwald, 1831) (Plate 7E)
Caspian bent-toed gecko, Caspian thin-toed gecko

Gymnodactylus caspius Eichwald, 1831:181 (Type locality: Baku, Azerbaijan, on the Caspian Sea; Syntypes: ZIL 3181–2 [Lectotype: ZIL 3182, designated by Szczerbak and Golubev, 1986:132]). — De Filippi, 1865:352 (in part). — Blanford, 1876:347–348. — Boulenger, 1885a:26–27. — Strauch, 1887:45. — Nikolsky, 1897:312; 1899:378. — Zarudny, 1903:4. — Nikolsky, 1907a:260; 1915:71–76, text-figs. 17–19. — Lantz, 1918:13–14, pl. 1. — F. Werner, 1936:200. — Bannikov, *et al.*, 1977:100. — Welch, 1983:7.
Cyrtodactylus caspius: Underwood, 1954:475. — S. Anderson, 1963:438, 474; 1968:332. — Tuck, 1971a:56. — S. Anderson, 1974:33, 42. — Schleich, 1977:127, 129. — Tuck, 1979:101.
Cyrtodactylus scaber: Guibé, 1966a:97 (in part; not *Stenodactylus scaber* Heyden, 1827:15).
Gymnodactylus (Cyrtodactylus) caspius: Szczerbak and Golubev, 1977:130.
Tenuidactylus (Tenuidactylus) caspius: Szczerbak and Golubev, 1984:53.

Tenuidactylus (Tenuidactylus) caspius caspius: Szczerbak and Golubev, 1986:132–141 [1996:133–138], figs. 60–62, col. pl. 3, fig. 1.
Tenuidactylus caspius: Kluge, 1991:34. — Clark, 1991:36. — Kluge, 1993:37.

Diagnosis: Subfemoral tubercles present; males with continuous series of 23–31 femoral and preanal pores (total both sides); 19–23 strongly keeled, mucronate dorsal tubercles in paravertebral row from occiput to level of vent; first pair of postmentals in broad contact.

Color pattern: Dorsum gray or brown with dark transverse bars, often indistinct or broken into spots, five of these bars on body, first on neck, 11–12 on tail; upper part of head brownish with indistinct dark spots; snout often with three dark streaks on each side, running from eye to lip; venter white.

Size: The largest male examined measures 60.7 mm snout-vent; largest female, 59.8 mm snout-vent; both specimens from northern Afghanistan. Szczerbak and Golubev (1986:132; 1996:129) report male snout-vent lengths to 72.0 mm, females to 68.5 mm.

Natural history: Nothing has been published on the natural history of this species in Iran. A female collected in northern Afghanistan in mid-April has an oviducal egg about 5 mm long. The smallest juvenile examined (snout-vent 20.7 mm) was collected in late August. Only beetle remains can be identified in the digestive tracts of six specimens I examined. See Terentjev and Chernov (1949:137), Szczerbak (1984:49–52), and Szczerbak and Golubev (1986:138–142; 1996:134–138) for summaries of the natural history in regions of the former Soviet Union.

Habitat: This gecko inhabits steep cliff faces, living in rock crevices; found among ruins and as a house gecko. Terentjev and Chernov (1949:137) report that it is found in burrows of mammals and turtles, clinging to the walls during the day. Sobolevsky (1929) records it from the Moghan Steppe, a flat, uniform plain covered with scanty herbaceous vegetation to the north of the Iranian border in the Lenkoran District of Azerbaijan. Szczerbak and Golubev (1986:137–138; 1996:134–135) have given details of its several habitats throughout the former Soviet Union.

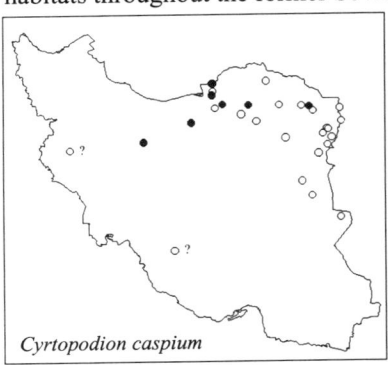

Cyrtopodion caspium

Distribution: Azerbaijan; Turkmenistan; Uzbekistan; southern Tadjikistan; southwestern Kazakhstan; northern Afghanistan; Iran. The northern border of its range is a line from Komsomolets Bay on the northeastern shore of the Caspian Sea to the northern coast of the Aral Sea and Syr Darya. In Iran it is known from the Gorgan region of Mazanderan, from northern and eastern Khorasan, extending south to Sistan. Sea level to 1700 m. A second subspecies, *C. c. insularis* (Akhmedov and Szczerbak, 1978), has been described from an island in the Caspian Sea, and is known only from the type locality.

Cyrtopodion gastrophole (Werner, 1917) (Plate 7F)
Werner's bent-toed gecko, Farsian spider-gecko

Gymnodactylus gastropholis F. Werner, 1917:194–196 (Type locality: Fars Province, Iran; Holotype: ZFMK 27095 [formerly University Göttingen Zoological Museum, Andreas field no. 74]); 1936:200.
Cyrtodactylus gastropholis: S. Anderson, 1968:332; 1974:34, 42.

Agamura gastropholis: Szczerbak and Golubev, 1986:211–213 [1996:207], figs. 95–96. — Kluge, 1991:3; 1993:3.
Cyrtopodion (Cyrtopodion) gastropholis: Leviton, *et al.*, 1992:33–34, col. pl. 4E. — Frynta, *et al.*, 1997:9.

Diagnosis: Anterior pair of enlarged postmentals in contact; no subfemoral tubercles; males with 4 preanal pores; subcaudal scales one head-width behind vent enlarged, plate-like, in single median series; snout-length one and one-half times diameter of eye; dorsal tubercles distinctly smaller than interspaces, rounded; peritoneum darkly pigmented; limbs and tail thin, attenuate; 10–16 abdominal scales across middle of belly (6–8 scales across belly in distance equal to length of snout).

Color pattern: Light sandy gray above with darker brown spots transversely and longitudinally arranged, sometimes coalescing to form 6 distinct crossbars, first on neck, sixth on sacrum. Limbs with indistinct crossbars; tail with distinct dark brown crossbars.

Size: Snout-vent length 43–47 mm; none of the specimens examined has a complete tail.

Habitat: (Plate 22D) I collected a single specimen in a mud-brick building constructed as a shelter over a well; it was exposed on the wall during the midday hours in a darkened passage with steps leading down to the well water. The building was at the margin of the coastal plain where there is an abrupt transition to the first range of foothills of the Zagros Mountains. 100 m elevation.

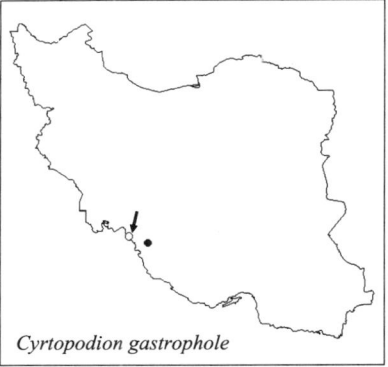

Cyrtopodion gastrophole

Distribution: The coastal plain of the Persian Gulf in Fars Province, southern Iran; known only from the holotype and five paratypes, for which no specific locality was cited, and from the single additional record cited below (see also Appendix II).

Remarks: This species is closely allied to *Cyrtopodion agamuroides*, but differs in the very large abdominal scales which are much larger than the dorsal tubercles, and approximately as wide as the greatest diameter of the ear opening. The abdominal scales of *C. agamuroides* are smaller than the dorsal tubercles, much smaller than the ear opening. The dorsal tubercles of *C. gastrophole* are subcircular, with no tendency toward becoming subtrihedral as there is in those of *C. agamuroides*. The tubercles are rounded and smooth to weakly keeled, lower and less strongly keeled than in *C. agamuroides*. Werner gives 10 as the number of abdominal scales at midbody in the holotype, but I count 14 distinctly larger than the lateral scales. There are three pairs of postmentals, the third much the smallest. On the dorsal surface of the tail each annulus consists of four transverse rows of scales, the distal two rows of each annulus enlarged and keeled. The regenerated tail is much altered in appearance, the dorsal scales being more uniform, while there are no distinct annuli; the ventral caudal scales are transversely enlarged, forming subcaudal lamellae.

In a single specimen (CAS 100472) there are 10 rather than 8 longitudinal rows of dorsal tubercles at midbody. There are 26–30 dorsal tubercles in a paravertebral row in these specimens.

The specimens of the original series are in poor state of preservation and the color pattern has faded almost completely. Werner stated that there were four indistinct crossbars on the dorsum; I count six, but cannot tell if these are actually transverse bars or discrete spots as in my specimen (MMTT 1049) and in *C. agamuroides*.

Szczerbak and Golubev (1986:211) select ZFMK 27095 as lectotype. However, Werner (1917:194) clearly designated Universität Göttingen Zoologische Museum register number 74 as the type. The type series, along with most of the Göttingen collection, has been redeposited in Bonn, and ZFMK 27095 does correspond to number 74, which was a field number, rather than a register number (W. Böhme, pers. commun.).

I am not persuaded that this species should be transferred to the genus *Agamura*, at least as I define that genus (above).

Nothing is known of the habits of this species, but the fact that the peritoneum and the investiture of some of the viscera are darkly pigmented, in contrast to most species in the genus, suggests at least a partially diurnal life style.

Cyrtopodion heterocercum (Blanford, 1874)
Blanford's rough-scaled gecko, Asia Minor thin-toed gecko

Cyrtopodion heterocercum heterocercum (Blanford, 1874)
Asia Minor thin-toed gecko

Gymnodactylus caspius: De Filippi, 1865:352 (in part; not Eichwald, 1831).
Gymnodactylus heterocercus Blanford, 1874a:453–454 (Type locality: Hamadan, Iran; Holotype: MSNTO [lectotype TZM R2532, adult female, "Persia dono Doria", designated by Szczerbak and Golubev (1986:183)]); 1876:345–347, pl. 22, figs. 3, 3a. — Bedriaga, 1879:34. — Boulenger, 1885a:30. — F. Werner, 1936:200. — Stepanek, 1937b:279.
Gymnodactylus heterocercus heterocercus: Wettstein, 1951:431–432. — Welch, 1983:7.
Cyrtodactylus heterocercus: S. Anderson, 1963:438, 474; 1968:332. — Minton, *et al.*, 1970:359–361, fig. 13. — Schleich, 1977:127, 129.
Cyrtodactylus heterocercus heterocercus: S. Anderson, 1974:33, 42.
Gymnodactylus (*Cyrtodactylus*) *heterocercus*: Szczerbak and Golubev, 1977:130.
Tenuidactylus (*Mediodactylus*) *heterocercus*: Szczerbak and Golubev, 1984:54.
Tenuidactylus (*Mediodactylus*) *heterocercus heterocercus*: Szczerbak and Golubev, 1986:182–184 [1996:180], figs. 79–80.
Mediodactylus heterocercus heterocercus: Kluge, 1991:20; 1993:22.
Cyrtopodion (*Mediodactylus*) *heterocercus heterocercus*: Leviton, *et al.*, 1992:35 (by implication).

Diagnosis: Dorsum with enlarged trihedral tubercles in 12–14 longitudinal rows; no enlarged subcaudal plates, tail covered below with strongly keeled imbricate scales in 5–7 longitudinal rows; sides of tail with greatly enlarged sharply keeled mucronate scales (Fig. 72C).

Color pattern: About 8 indistinct dark angular transverse bars on dorsum, apices pointing caudad, first bar on neck, 8th across sacral region; these tend to fuse and coalesce; 13 dark bars on tail.

Size: Snout-vent 45 mm, tail 50 mm.

Distribution: The typical subspecies is apparently known from 4 specimens, all from Hamadan, Kermanshahan, and two questionable records from Persepolis, Fars Province. The subspecies *C. heterocercum mardinense* (Mertens) is known from Gaziantep, Nisi, Mardin, and Siirt, all in southeastern Turkey. A specimen from Shawklawah, Irbil Prov., northern Iraq (CAS 178913) agrees in all particulars with Mertens's (1924) description of *C. h. mardinense* and differs in significant details of pholidosis and proportions from specimens from Hamadan. A lowland Iraq record for Al Basrah (Nader and Jawdat, 1976:20,

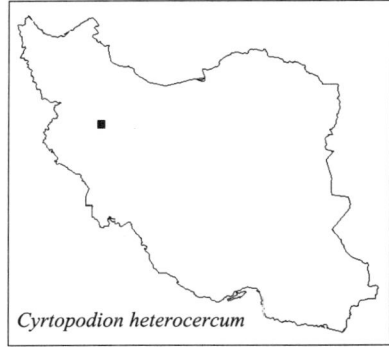

Cyrtopodion heterocercum

map 3) requires confirmation since elsewhere this species occurs in foothills and at elevations to about 1000 m (Leviton, et al., 1992:35).

Remarks: The only specimen of this subspecies that I have examined (NMW 7286) has been described in an earlier paper (Minton, et al., 1970). The strongly depressed habitus of this species is one of its most striking distinctions. Judging from published descriptions, this species may be related to *Cyrtopodion kotschyi colchicus* (Nikolsky), and possibly to *C. kotschyi danilewskii* (Strauch). After examining Mertens's photograph (1924:pl. 12, fig. 1) of *C. h. mardinense*, I am doubtful of the close relationship of that taxon to *C. heterocercum*. Mertens (1952:52–53), suggested that *C. h. mardinense* might be a subspecies of *C. kotschyi*. *Cyrtopodion heterocercum mardinense*, previously known only from northern Iraq and southeastern Turkey, has recently been reported from Halab, Syria (Moravec and Modry, 1994:53–54, fig. 1).

In the phylogenetic scheme presented by Szczerbak and Golubev (1986:22; 1996:21), *C. heterocercum* is the sister species of *C. sagittifer*.

Little is known of the habits or habitat of this species. The specimens reported by Wettstein were collected on screens of buildings at the American Mission to Hamadan.

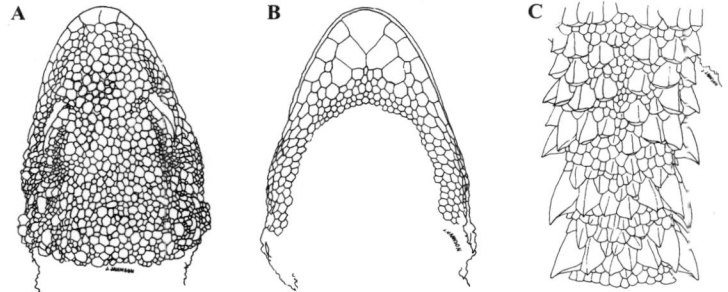

FIGURE 72. *Cyrtopodion heterocercum* (NMW 7286): (A) dorsal view of head; (B) ventral view of chin shields; (C) dorsal view of tail. (Drawn by Jean Jahnson.)

Cyrtopodion kachhense (Stoliczka, 1872) (Plate 7G)
Warty rock gecko, Kachhi thin-toed gecko

Gymnodactylus kachhensis Stoliczka, 1872a:79–81 (Type locality: Kachh [Pakistan?]; Lectotype: ZSI 5162 [designated by Annandale, 1913:315]). — Annandale, 1913b:315.
Cyrtodactylus kachhensis: Underwood, 1954:475. — S. Anderson, 1974:33, 42.
Tenuidactylus (Mesodactylus) kachhensis: Szczerbak and Golubev, 1984:54.
Tenuidactylus (Cyrtopodion) kachhensis: Szczerbak and Golubev, 1986:185–188 [1996:181–183], figs. 81–82.
Cyrtopodion kachhensis: Kluge, 1991:7; 1993:9.

Diagnosis: Anterior pair of enlarged postmental shields in contact behind mental; no subfemoral tubercles; males with preanal pores only; subcaudal scales one head-width behind vent small, smooth, not enlarged and plate-like; no scattered small tubercles among the rows of enlarged dorsal tubercles (Fig. 73); caudal tubercles form terminal ring of each annulus; investiture of lower viscera darkly pigmented, usually visible as dark spot through skin of belly.

Color pattern: Dorsum light brown or gray with 5 irregular longitudinal rows of small dark spots, sometimes more or less confluent into crossbars; venter white.

Size: Snout-vent length of males 34–43 mm, females 35–42 mm (Minton, 1966:83).

FIGURE 73. *Cyrtopodion kachhense*. (From Smith, 1935:44, fig. 15.)

Habitat: In Pakistan, it inhabits dry, rocky places, hiding under rocks and euphorb shrubs; it is rarely encountered as a house gecko (Minton, 1966:78).

Distribution: The only Iranian record is that of Annandale (1913b:315) for Bushire in Fars Province. It is distributed throughout much of Kutch, Sind and coastal Las Bela, Pakistan (Minton, 1966:78).

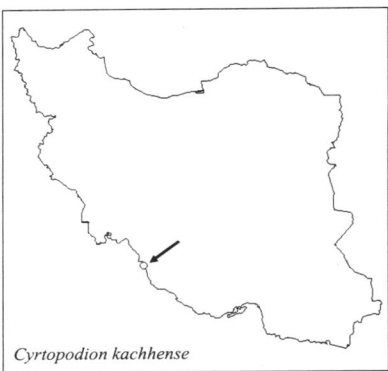

Cyrtopodion kachhense

Remarks: Considering the difficulties in correctly identifying species of *Cyrtopodion* without comparative material of all of the species of Southwest Asia, I am skeptical of the occurrence of *C. kachhense* in Iran, and Szczerbak and Golubev (1986:188; 1996:183) do not include Iran in its range. These authors (Szczerbak and Golubev, 1986:185 [1996:181], footnote) have called attention to the resemblance between *C. kachhense* and *C. brevipes*. Annandale presumably compared the Bushire specimens with the type of *C. kachhense*; however, he also had the type of *C. brevipes*. Of course, it is possible that the species has been carried to any of the Persian Gulf ports. Szczerbak and Golubev (1986:186) select NMW 17383:5 as the lectotype, but Annandale (1913) had already designated a lectotype.

Minton (1966:78) summarizes the natural history of this species.

Cyrtopodion kirmanense (Nikolsky, 1900)
Kerman bent-toed gecko, Kirman thin-toed gecko

Gymnodactylus kirmanensis Nikolsky, 1900:381–383 (Type locality: Kuh-e Taftan, Sargad, Iran, and eastern Kerman, Iran; Syntypes: ZIL 9329[6], 9330[6] [Lectotype ZIL 9330B, designated by Szczerbak and Golubev, 1986:207]). — Zarudny, 1903:5–7. — F. Werner, 1936:200. — Schmidt, 1939:55. — Welch, 1983:7.

Cyrtodactylus kirmanensis: S. Anderson, 1963:438, 474. — Kluge, 1967:27. — S. Anderson, 1968: 332; 1974:33, 43. — Schleich, 1977:127, 129. — Kluge, 1993:8.

Gymnodactylus (Cyrtodactylus) kirmanensis: Wermuth, 1965:56. — Szczerbak and Golubev, 1977a:130.

Tenuidactylus kirmanensis: Szczerbak and Golubev, 1984:55; 1986:206–208 [1996:201–203], figs. 89, 93.

Gonydactylus kirmanensis: Kluge, 1991:12.

Diagnosis: Anterior pair of enlarged postmental shields in contact; no subfemoral tubercles(?); males with 4 preanal pores; subcaudal scales one head-width behind vent enlarged, plate-like, smooth, in 2 longitudinal series of large and small scales; dorsal tubercles distinctly smaller than interspaces; snout 2 to 2¼ times longer than diameter of eye. See Szczerbak and Golubev (1986:207; 1996:203) for additional characters.

Color pattern: Dorsum gray, with 5 longitudinal series of black quadrangular spots; tail with 13–16 black transverse bars; venter white.

Size: Snout-vent length 48 mm, tail 57 mm.

Habitat: "This species is encountered in abundance in the mountainous terrain between the Sistan and Bampur Depressions. It has been found on the plains and in the mountains. It inhabits sheer rocky cliffs in the mountains; river banks and dry channels; shady terraces, cracks, niches; occasionally, on loose fragments of rock boulders. It is most frequently found on granites and, less often, on conglomerates and other rocks." (Szczerbak and Golubev, 1996:203, based on Zarudny, 1904).

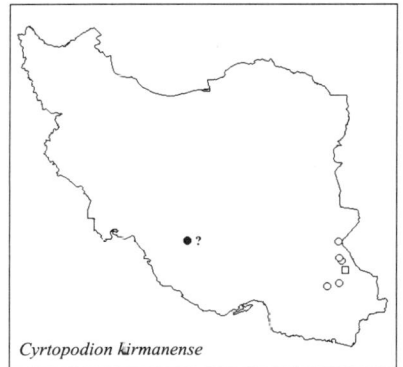

Cyrtopodion kirmanense

Distribution: Known with certainty only from the types, which came from the southern part of the Central Plateau. Zarudny (1903:5–7) subsequently collected specimens which he identified as this species from several localities between Sistan and Bampur. If still extant, these specimens should be compared with the original description.

Remarks: I am uncertain as to the identification of FMNH 210007, which Schmidt (1939:55) identified as *Cyrtopodion kirmanense*. The first pair of chin shields are separated by a tiny scale behind the mental; the snout is only about one and one-third to one and one-half the diameter of the eye; the subcaudals are smooth, in two longitudinal rows, 2 pairs per tail segment; the keeled caudal tubercles are 6 in number, in the terminal ring of each segment; there are three pairs of postmentals, the dorsal tubercles are oval, bluntly keeled, not trihedral, and those of the paravertebral rows smaller than adjacent interspaces, most longer than broad, and all distinctly smaller than ear opening; no subfemoral tubercles; specimen apparently a female, with 4 enlarged preanal scales, no enlarged scales on femur; about 28 scales across abdomen (14 in space equal to length of snout), but skin of belly wrinkled, and accurate count hard to obtain; 12 rows of dorsal tubercles, outermost very small, about 29 in a paravertebral row from occiput to vent, but irregularly arranged on neck and back of head. Six indistinct black crossbars on neck and body, 10 on tail, limbs barred. The peritoneum has at least some dark pigment, and at least part of the digestive tract is very dark. It would appear to be in the *agamuroides-gastrophole* species group, but habitus not so attenuate as in those forms.

The specimen does not fit the description of any known species very well. Many of the Southwest Asian species are known from so few specimens, however, that we have little idea of the variation that may exist. Compare the above description with Szczerbak and Golubev's (1986:206 [1996:202], fig. 93) illustration of the lectotype, to which Schmidt's specimen seems to bear little resemblance.

Szczerbak and Golubev (1986; 1996) place this species in their "Tibeto-Himalayan" group of the genus without assigning a subgeneric name.

Cyrtopodion kotschyi (Steindachner, 1870) (*sensu lato*)
Kotschy's gecko

Gymnodactylus kotschyi Steindachner, 1870:329, pl. 1, figs. 1–2 (Type locality: Syros Island, Cyclades

[restricted by Mertens and Müller, 1928:24]; Syntypes: NMW [Lectotype: BMW 10868, designated by Szczerbak and Golubev, 1986:160]). — Boettger, 1880:192. — F. Werner, 1936:200.
Cyrtodactylus kotschyi: Underwood, 1954:475. — S. Anderson, 1963:438, 474; 1968:332; 1974:33, 43.
Gymnodactylus (Cyrtodactylus) kotschyi syriacus: Wermuth, 1965:59.
Mediodactylus kotschyi syriacus: Welch, 1983:10.
Tenuidactylus (Mediodactylus) kotschyi: Szczerbak and Golubev, 1984:54; 1986:160–163 [1996:156–163].
Mediodactylus kotschyi: Kluge, 1991:20; 1993:22.
Cyrtopodion (Mediodactylus) kotschyi: Leviton, et al., 1992:35.

Diagnosis: Anterior pair of enlarged postmentals in contact; no subfemoral tubercles; males with 4 preanal pores; subcaudal plates one head-width behind vent enlarged, plate-like, in single median series; snout length less than twice diameter of eye; caudal tubercles arranged around middle of each caudal segment, not forming terminal ring of each segment.

Color pattern: Dorsum gray with darker transverse bars; venter white (Fig. 74).

Size: Snout-vent length to 42 mm, tail 52 mm.

Distribution: Southern Europe along Mediterranean, from southern Italy eastward to Israel, Egypt, Syria, Turkey, northern Iraq, and probably northwestern Iran. Twenty subspecies recognized by Wermuth (1965:56–59), 25 recognized by Szczerbak and Golubev (1986:161–163 [1996:156–163] *q.v.* for details of distribution), 28 by Kluge (1991:20–21; 1993:22).

Remarks: In his original description of this species, Steindachner indicated that among the 42 specimens upon which he based the taxon were specimens from Persia. Through the kindness of Franz Tiedemann, I recently received a list of Iranian lizards in the Vienna collection; NMW 17323 is listed as a specimen of *C. kotschyi* from "Persien" collected by Kotschy, 5 January 1845, and this specimen must have been among of the type series. No specific locality is cited, although Boettger (1880:192) gives Shiraz as a locality, citing Steindachner (but not giving the date) as his source. This is the only published locality record of this species for Iran, as far as I have been

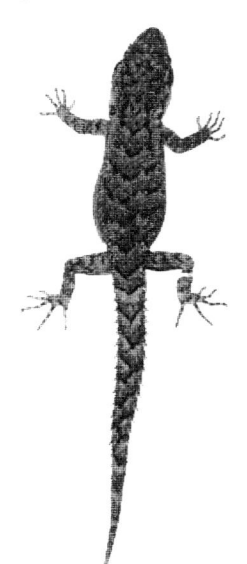

FIGURE 74. *Cyrtopodion kotschyi.* (CAS 159049).

able to determine. All subsequent listings of Iran within the range of *C. kotschyi* appear to derive from Steindachner's original paper. It should be looked for in the oak forests of the lower elevations of the Zagros. *Cyrtopodion kotschyi syriacus* (Štěpánek, 1937) occurs in Syria and Anatolian Turkey, and is probably the form of northern Iraq as well. Wermuth includes Iran in its distribution, but I do not know if this is on the basis of specimens examined. Also recognized from Anatolia are *C. k. danilewskii* (Strauch, 1887), and *C. k. ciliciense* (Baran and Gruber, 1982), *C. k. beutleri* (Baran and Gruber, 1981), *C. kotschyi karabagi* (Baran and Gruber, 1982), *C. k. ponticus* (Baran and Gruber, 1982), while *C. k. colchicus* (Nikolsky, 1902) is the form from northeastern Turkey in the vicinity of Artvin. Any specimens from Iran identified as *C. kotschyi* in collections should be compared with *C. heterocercum*.

See Beutler (1981b:64–71) for discussion of the ecology and natural history of this species in Europe, Szczerbak and Golubev (1986:165–167, 1996:161–163) for *C. k. danilewskii* in Turkey and the Crimea.

Cyrtopodion longipes (Nikolsky, 1896)
Nikolsky's long-toed gecko

Cyrtopodion longipes longipes (Nikolsky, 1896) (Plate 7H)
Nikolsky's long-toed gecko, long-legged thin-toed gecko

Gymnodactylus longipes Nikolsky, 1896:369; 1897:313–315, pl. 19, fig. 2 (Type locality: Neh [Nehbandan], Iran; Syntypes: ZIL 8809[6], 8810[3], 8811 [Lectotype: ZIL 8810, designated by Szczerbak and Golubev, 1986:153]); 1900:379. — Zarudny, 1903:4–5. — F. Werner, 1936:200. — Welch, 1983:8.
Gymnodactylus fedtschenkoi: Forcart, 1950:144 (not Strauch, 1887).
Cyrtodactylus fedtschenkoi: S. Anderson, 1963:438, 474 (in part; not *Gymnodactylus fedtschenkoi* Strauch, 1887); 1968:332; 1974:33, 42. — Schleich, 1977:127, 129.
Cyrtodactylus longipes: S. Anderson, 1963:438, 474; 1968:332.
Gymnodactylus longipes longipes: Gorelov, Darevsky, and Szczerbak, 1974:37–38, figs. 1, 3v. — Bannikov, et al., 1977:103.
Gymnodactylus (*Cyrtodactylus*) *longipes*: Wermuth, 1965:60. — Szczerbak and Golubev, 1977b:130.
Cyrtodactylus longipes longipes: Leviton and S. Anderson, 1984:272, fig. 2.
Tenuidactylus (*Tenuidactylus*) *longipes*: Szczerbak and Golubev, 1984:53. 1986:153–156, figs. 64, 68, col. pl. 3, fig.4.
Tenuidactylus (*Tenuidactylus*) *longipes longipes*: Szczerbak and Golubev, 1986:153–156 [1996:152], figs. 64, 68, col. pl. 3, fig.4.
Tenuidactylus longipes longipes: Kluge, 1991:35; 1993:37.

Diagnosis: Enlarged dorsal tubercles separated by small scales; 27–36 abdominal scales across middle of belly; 26–30 dorsal tubercles in paravertebral row from occiput to level of vent; 32–33 preanal and femoral pores (total both sides), series not separated by scales lacking pores, females with corresponding depressions in scales; 12–16 interorbital scales; length of forelimb 38–45% of snout-vent length; first pair of postmentals usually in relatively broad or point contact, or separated by one or two scales.

Color pattern: Light gray above; dorsum with 5–6 dark crossbars, tail with 8–10 dark bars, limbs with 6–9 dark bars; venter white without dark markings.

Size: Snout-vent length 60.5 mm, tail 73+ mm.

Habitat: (Plate 21B) Found in caves, on walls of mud-brick buildings, and between layers of flaking rock on bare, rocky hills in dry, open country.

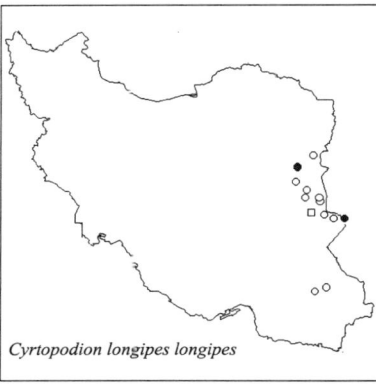
Cyrtopodion longipes longipes

Distribution: Extreme eastern Khorasan and Baluchistan, Iran, southwestern Afghanistan. To the north in Turkmenistan it is apparently replaced by *C. l. microlepis* (but see Orlov, 1981).

Remarks: I follow Szczerbak and Golubev (1986, 1996) in assigning CAS 141095 from Sistan to this taxon, rather than to *Cyrtopodion* [*longipes*] *voraginosis* as in the paper describing the latter taxon (Leviton and S. Anderson, 1984), and I here use their reduction of the taxa *voraginosus* and *microlepis* to subspecific status under *C. longipes*. I quote their reasoning from the English version of their book (Szczerbak and Golubev, 1996:152):

"A confused situation has arisen in connection with the description of a new species from Afghanistan ... The types of this form from the Chicago museum that were available to us convinced us that geckos from the southeastern foothills of Hindukush can be distinguished under this name [*C. voraginosus*]. However, they can only be given the subspecific status. But the authors of the description included a specimen from eastern Iran (Sistan Province) in the supplementary materials. We disagree with that, because it is the nominate form that is found in the foothill system of the eastern Iranian crests of Kayen and Pelenghan. In addition, as it follows from the table in the original description, it is this very specimen that possesses the characters, which prevent its unequivocal assignment to the new subspecies. Possibly, the absence of significant physical barriers could have led to the formation of a wide intergradation zone between these forms in the plains portion of southwestern Afghanistan and the contiguous regions of eastern Iran in the recent times."

It should be noted, however, that these authors had not examined the specimen in question.

Cyrtopodion longipes microlepis (Lantz, 1918)
Small-scaled long-toed gecko, small-scaled thin-toed gecko

Gymnodactylus microlepis Lantz, 1918:11–13, pl. 1, fig. 1 (Type locality: River Tajan [Harirud], Perso-Afghan-Turkmen frontiers; Lectotype: MMSU 151a [designated by Gorelov, et al., 1974:37]).
Cyrtodactylus fedtschenkoi: S. Anderson, 1963:438, 474 (in part; not *Gymnodactylus fedtschenkoi* Strauch 1887); 1968:332; 1974:33, 42.
Gymnodactylus (Cyrtodactylus) longipes microlepis: Gorelov, et al., 1974:35–38, figs. 1–3. — Bannikov, et al., 1977:103.
Cyrtodactylus longipes microlepis: Leviton and S. Anderson, 1984:272, fig. 2.
Tenuidactylus (Tenuidactylus) longipes microlepis: Szczerbak and Golubev, 1986:156 [1996:152–153].
Tenuidactylus longipes microlepis: Kluge, 1991:35; 1993:37.

Diagnosis: Dorsal tubercles separated by small scales; 31–33 tubercles in paravertebral row from occiput to level of groin; 35–39 abdominal scales across middle of belly; total of 27–35 preanal and femoral pores (seen as slight depressions in the scales in females), the series not separated by scales lacking pores; 12–14 interorbital scales.

Color pattern: Similar to *Cyrtopodion l. longipes*.

Size: Snout-vent length 63.5 mm, tail 82 mm.

Natural history: See Szczerbak and Golubev (1986:157–158; 1996:153–155).

Habitat: See Szczerbak and Golubev (1986:157; 1996:153–154).

Distribution: Known only from the Tedzhen Valley, of the western Badghyz Plateau, Turkmenistan on the northeastern border of Iran, but may occur in adjacent Iran and Afghanistan. Replaced to the south in eastern Iran by *Cyrtopodion l. longipes* and in Afghanistan south of the Hindu Kush by *C. l. voraginosus* (Leviton and S. Anderson, 1984).

Remarks: I follow Szczerbak and Golubev (1986:156–157) in listing *C. voraginosus* and *C. microlepis* as subspecies of *Cyrtopodion longipes*. These taxa are closely related and are also related to *C. fedtschenkoi*. See Leviton and S. Anderson (1984:274–275) for a discussion of relationships and zoogeography. Szczerbak and Golubev (1986:157) restrict the distribution of *C. voraginosus* to southern Afghanistan and regard specimens from Sistan as *C. l. longipes*. I find no indication that they had examined any of the Sistan specimens, however. Michael Golubev (pers. commun., 3 February 1995) has examined Zarudny's specimens from Chah-i-Ziru (= Chah-e Ziru), Neh (= Nehbandan) (type locality for *C. l.*

longipes), Siahkoh, and Neizar, all near the Sistan basin, and identifies all of them as *C. l. longipes*. He expressed the opinion that *C. l. voraginosus* is probably restricted to the foothills of the Hindu Kush in Afghanistan. See Leviton and S. Anderson (1984) for discussion of relationships and zoogeography.

Cyrtopodion russowii (Strauch, 1887) (Plate 8A)
Transcaspian bent-toed gecko, gray thin-toed gecko

Gymnodactylus russowii Strauch, 1887:49–51, figs. 10–12 (Type locality: Nowo Alexandrowsk; Chodschent; Mangyschlak; Mursa-Robat; Mohol-Tau; Tschimkent; Brunnen Abadchir; Tschinas; Saamin; Golodnaja Steppe; Utsch-Kurgan at Naryn; Chark-Usjur [all localities in the Central Asian republics of the former Soviet Union; type locality restricted to Nowo Aleksandrowsk by Mertens and Wermuth {1960:78} and further restricted and clarified as 30 km east of Fort Shevchenko {formerly Fort Aleksandrovsk}, Kazakhstan, on the Mangyschlak Peninsula by Szczerbak and Golubev {1986:167}, but see Zhao and Adler {1993:179} for additional comments]; Syntypes [37]: ZIL 3658[2], 3659, 3660, 3700[3], 3701[2], 4192, 4193[6], 4194, 4195[5], 4310[2], 5037, 5197, 5218, 5224, 5800[2]; CAS 94050–52; USNM 69891–3 {Lectotype: ZIL 3658 adult male, designated by Szczerbak and Golubev, 1986:167}).

Cyrtopodion russowii zarudnyi (Nikolsky, 1900)
Zarudny's bent-toed gecko, Zarudny's gray thin-toed gecko

Gymnodactylus zarudnyi Nikolsky, 1900:385–387 (Type locality: Neizar, Sistan, Iran; Syntypes [5]: ZIL 9334[6] [Lectotype: ZIL 9334a, designated by Szczerbak and Golubev, 1986:172]). — Zarudny, 1903:7–8. — F. Werner, 1936:200.
Gymnodactylus russowii: Terentjev and Chernov, 1949:132–133 (in part). — Bannikov, *et al.*, 1977: 98–99.
Cyrtodactylus russowii: Underwood, 1954:475. — S. Anderson, 1963:474; 1968:332; 1974:33,43. — Szczerbak, 1984:75–83.
Cyrtodactylus zarudnyi: S. Anderson, 1963:438, 474; 1968:332.
Gymnodactylus (*Cyrtodactylus*) *russowii*: Wermuth, 1965:66 (in part).
Mediodactylus russowii: Welch, 1983:10.
Tenuidactylus (*Mediodactylus*) *russowii zarudnyi*: Szczerbak and Golubev, 1986:172 [1996:167–171], figs. 73–74.
Cyrtopodion russowii: Welch, *et al.*, 1990:17.
Mediodactylus russowii zarudnyi: Kluge, 1991:21; 1993:22.

Diagnosis: Anterior pair of postmentals separated by one or two scales; no subfemoral tubercles; males with 2–4 preanal pores; subcaudal scales one head-width behind vent small, smooth, not enlarged and plate-like; scattered small keeled tubercles among the large trihedral dorsal tubercles which form fairly regular longitudinal rows; tubercles on tail arranged around middle of each caudal segment, not in terminal row of each segment.
Color pattern: Dorsum light gray, uniform, or with 6 indistinct dark transverse bars; venter white.
Size: Snout-vent length to 51 mm.
Habitat: Szczerbak (1981) and Szczerbak and Golubev (1986:172–175; 1996:168–171) present details of the habitat and natural history of *Cyrtopodion r. russowii* in the former Soviet Union. It seems to be nearly restricted to flatland desert areas. Where it occurs in sand deserts and semi deserts (Turkmenistan, southern Tajikistan) it lives on the trunks of saxaul (*Haloxylon*) and other shrubs and trees. In clay deserts and loess foothills (in the northerly part of its range), it is common on cliffs in river and ravine plains, under rocks and on walls of inhabited and abandoned buildings.

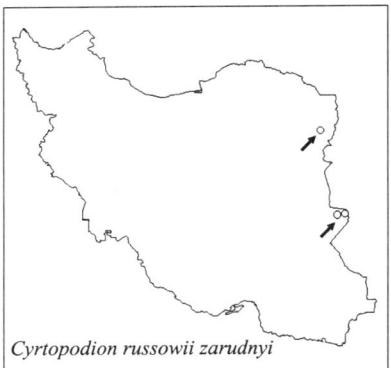
Cyrtopodion russowii zarudnyi

Distribution: The inclusion of *Cyrtopodion russowii* in the Iranian fauna was based on Chernov's (1959) assignment of *Gymnodactylus zarudnyi* Nikolsky to the synonymy of this widely distributed Central Asian gecko. Bannikov, *et al.* (1977:99) included northeastern and eastern Iran in its distribution. Szczerbak and Golubev (1986:172) recognize *zarudnyi* as a subspecies of *C. russowii*. *Cyrtopodion r. russowii* occurs close to the Iranian border in Turkmenistan and *C. r. zarudnyi* is known only from eastern Khorasan and Sistan in Iran. The distribution of the species as a whole extends across the Central Asian Republics, north to the latitude of the northern extent of the Aral Sea and east to northwestern China. There is a single locality in European Russia on the west coast of the Caspian Sea.

Remarks: I have examined six of the syntypes of *Gymnodactylus russowii*, all from Chinaz, Uzbekistan. In these specimens the postmentals are smaller than in other species of Southwest Asian *Cyrtopodion*, and in two specimens the anterior pair is separated by a scale of nearly the same size. The dorsal tubercles are less regularly arranged in rows than in most species. The six large keeled scales of each caudal segment are separated from one another by a single scale, and are disposed around the middle of each segment, rather than forming the terminal row. In the distal two-thirds of the tail there is a single median series of enlarged subcaudal plates, but these are not as broad as in most species.

I have seen neither the types of *G. zarudnyi*, nor the 1959 paper by Chernov, who had access to the types of both nominal forms. *Gymnodactylus zarudnyi* was said to differ from *C. russowii* in having 24–26 abdominal scales across the belly rather than 30. In the six syntypes of *C. russowii* that I have examined, I count 25–29. Nikolsky also stated that his form differed in having the anterior postmentals separated, but this character is variable in the types of *C. russowii*.

Cyrtopodion sagittifer (Nikolsky, 1900) (Plate 8B)
Jaz Murian bent-toed gecko, Bampur thin-toed gecko

Gymnodactylus sagittifer Nikolsky, 1900(1899):379–381 (Type locality: Bampur and Farra, southeastern Iran [restricted to Bampur by Szczerbak and Golubev, 1986:180]; Syntypes [3]: ZIL 9331–9333 [Lectotype: ZIL 9331, designated by Szczerbak and Golubev, 1986:180]). — Zarudny, 1903:5. — F. Werner, 1936:200.
Cyrtodactylus saggitifer (*sic*): S. Anderson, 1974:33.
Cyrtodactylus sagittifer: S. Anderson, 1974:43.
Gymnodactylus (*Mediodactylus*) *sagittifer*: Szczerbak and Golubev, 1977b:130; 1984:54; 1986:180–181, fig. 78.
Mediodactylus sagittifer: Welch, 1983:10. — Kluge, 1991:21; 1993:22.

Diagnosis: Dorsal tubercles oval, keeled; diameter of ear opening smaller than half longitudinal diameter of eye; anterior pair of enlarged postmentals in contact; no subfemoral tubercles; males with 2–4 preanal pores; subcaudal scales one head-width behind vent small, keeled, not forming large plates; 14–16 abdominal scales across middle of belly (9–12 scales in a distance across belly equal to length of snout).

Color pattern: Dorsum gray with gray-brown spots arranged in 5 longitudinal rows, the spots of the central row arrowhead-shaped, the spots of the outer row confluent to form longitudinal stripe from eye. In life the pattern of dark marks forms seven M-shaped marks emphasized by white posterior margins. Limbs and tail barred with dark gray and white; venter white.

Size: Snout-vent length 33 mm, tail 32 mm. A male with a complete tail measures 26 mm snout-vent, tail 37 mm.

Habitat: (Plate 20G) Zarudny (1903:5) collected two specimens on trunks and branches of *Acacia arabica* and one on a wall of an old underground building. I collected them on the trunks of *Tamarix* in a *Tamarix-Acacia* "woodland" west of Bampur. The geckos began calling from the trees at dusk, apparently one from each tree; air temperature 33°C, trunk surface 33°C at 1805 hours. The call was a series of high-pitched "tsk" notes which could be heard up to 50 meters away. A male and female were caught on the same tree. I suspect that one tree constitutes the territory of one pair. About 500 meters elevation.

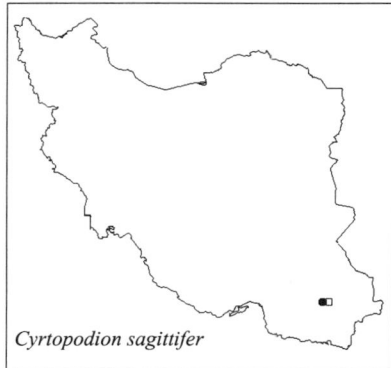

Cyrtopodion sagittifer

Distribution: Known only from the Jaz Murian depression in Baluchistan.

Remarks: My specimens represent a rediscovery of this species, which had not been collected since the topotypes were collected by Zarudny in 1903. The dorsal pattern, like that of *C. heterocercum*, appears to be cryptic for lizards living on tree bark. Wettstein (1951:431–432) placed this species in the synonymy of *C. h. heterocercum*, apparently on the basis of the original description. Both appear to be in the *kotschyi* group, and are perhaps one another's nearest relatives, as in Szczerbak and Golubev's (1986:22; 1996:21) phylogenetic hypothesis. *Cyrtopodion sagittifer* differs from *C. heterocercum* in its larger abdominal scales and in the smaller, more uniformly-sized caudal tubercles.

Cyrtopodion scabrum (Heyden, 1827) (Plate 8C)
Keeled rock gecko, rough-tail gecko, rough thin-toed gecko

Stenodactylus scaber Heyden *in* Rüppell, 1827:15, pl. 4, fig. 2 (Type locality: Tor, Sinai, Egypt, and Abyssinian Coast; Lectotype: SMF 8180, male [designated by Mertens, 1967:60]).
Gymnodactylus geckoides: Blanford, 1876:348 (not Spix, 1825). — Bedriaga, 1879:34.
Gymnodactylus scaber: Murray, 1884b:102–103. — Boulenger, 1885a:27–28. — F. Werner, 1895:1–2. — Annandale, 1913b:315. — F. Werner, 1936:200. — Schmidt, 1955:201. — Mertens, 1956: 91; 1957:126. — Welch, 1983:8.
Cyrtodactylus scaber: Underwood, 1954:475. — S. Anderson, 1963:437, 474. — Guibé, 1966a:97 (in part). — S. Anderson, 1968:332. — Haas and Y. Werner, 1969:333. — Tuck, 1971a:56; 1974d: 107; 1974a:62. — S. Anderson, 1974:34, 43. — Schleich, 1977:127, 129.
Gymnodactylus (*Cyrtodactylus*) *scaber*: — Wermuth, 1965:66–67. — Szczerbak and Golubev, 1977b:130.
Tenuidactylus (*Mesodactylus*) *scaber*: Szczerbak and Golubev, 1984:54.
Tenuidactylus (*Cyrtopodion*) *scaber*: Szczerbak and Golubev, 1986:190–192 [1996:186–188], figs. 83–84.
Cyrtopodion scaber: Welch, *et al.*, 1990:17. — Kluge, 1991:7; 1993:9.
Cyrtopodion (*Cyrtopodion*) *scaber*: Kluge, 1985:98. — Leviton, *et al.*, 1992:34, col. pl. 4F–G. — Frynta, *et al.*, 1997:9.

Diagnosis: Anterior pair of enlarged postmentals in contact; no subfemoral tubercles; males with 4–7 preanal pores; subcaudal scales one head-width behind vent enlarged, plate-like, in single median series; snout length less than twice diameter of eye; caudal tubercles forming terminal ring of each tail segment; dorsal tubercles distinctly larger than interspaces, strongly keeled and trihedral, 12–16 in longest transverse series across back, width of tubercles distinctly less than greatest diameter of ear opening; 10–14 supralabials.

Color pattern: Dorsum sandy, back with brown spots arranged in regular longitudinal series; limbs and tail with narrow dark transverse bars; venter white.

Size: Largest male examined 51 mm snout-vent; largest female 55 mm snout-vent.

Natural history: In the foothills of Khuzestan this is the most common house gecko, rarely seen away from human habitation. These geckos become active during late daylight hours in summer. At night they wait just beyond the circle of light thrown on a wall by a light fixture, darting forward to capture insects attracted to the light.

All females examined contain ovarian eggs, the largest (3 mm diameter) in mid-August. Judging from small juveniles collected and from condition of reproductive organs, eggs are laid in spring and probably through summer in clutches of two, hatching at least through early autumn. I noted egg laying in Khuzestan in late August, eggs measuring 7×10.5 mm. Hatchlings were common in mid-August.

Stomach contents include flies, ants, and beetles.

Habitat: I know this species only as a house gecko in Iran, and I find no literature records to indicate that it has been found in other situations. I have found it on both inside and outside walls of inhabited and abandoned buildings and gardens. I have found them on the same buildings with *Asaccus elisae, Hemidactylus turcicus,* and *Pristurus rupestris.* I do not recall seeing them on ceilings, whereas both *Asaccus* and *Hemidactylus* move easily on plastered ceilings. Elsewhere in its distribution it occurs in other habitat situations (Minton, 1966:77; Loveridge, 1947:65). Sea level to 1800 m elevation.

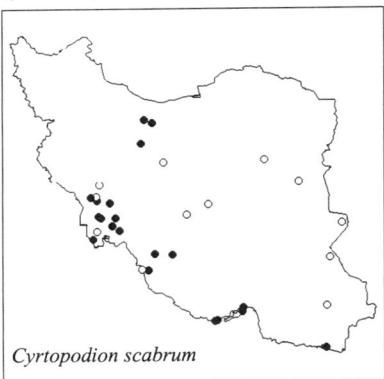

Cyrtopodion scabrum

Distribution: From Egypt south along the coast of the Red Sea to Ethiopia (port cities), the Persian Gulf coast of the Arabian Peninsula, and east across the arid regions of Southwest Asia (southern Anatolia, eastern Syria, Iraq, and Iran) to southern Afghanistan, Pakistan, and northwestern India. In Iran it is known from the plains and foothills of Khuzestan, the coast of the Persian Gulf, and Sistan. It appears to occur spottily on the Central Plateau as an urban gecko. The single record for Gorgan (Guibé, 1966a:97) is a specimen of *Cyrtopodion caspium* (see above). The specimen reported from between Herat and Islam Qala, northern Afghanistan (Clark, *et al.,* 1969:302) also proves to be a juvenile *C. caspium* on reexamination.

Remarks: The dorsal tubercles of this species are very large, strongly keeled, and trihedral, arranged in regular longitudinal and transverse rows, those of the transverse rows in contact or nearly so. There is a vertebral row of much smaller keeled tubercles, occasionally absent, and not corresponding in position to the transverse rows. There are usually three pairs of postmental shields, the posterior pair much smaller than the other two; a number of specimens have only two pairs, however. I find no obvious differences between the Iranian specimens and those from elsewhere in the range, although those from Afghanistan and Pakistan tend to have slightly smaller abdominal scales than those from Iran and Arabia.

Kluge (1991:7) lists *Cyrtodactylus basoglui* Baran and Gruber from Turkey as a synonym, following Szczerbak and Golubev (1983). A number of authors (Selcer and Bloom, 1994; Bloom, *et al.*, 1986; Klawinski, *et al.*, 1994; Vaughan, *et al.*, 1996) have studied the natural history of *Cyrtopodion scabrum* as a recent introduction to the fauna of southern United States at the port of Galveston, Texas, where it appears to be displacing ecologically the previously introduced *Hemidactylus turcicus*.

Cyrtopodion spinicauda (Strauch, 1887) (Plate 8D)
Kopet Dagh bent-toed gecko, spiny-tailed thin-toed gecko

Alsophylax spinicauda Strauch, 1887:58–59, figs. 15–16 (Type locality: Schahrud, Iran; Holotype: ZIL 4047). — Nikolsky, 1915:65–66. — F. Werner, 1936:200. — Wermuth, 1965:6. — S. Anderson, 1968:332; 1974:34, 43. — Bannikov, *et al.*, 1977:93.
Gymnodactylus (*Mediodactylus*) *spinicauda*: Szczerbak and Golubev, 1977b:130.
Mediodactylus spinicauda: Welch, 1983:10.
Tenuidactylus (*Mediodactylus*) *spinicauda*: Szczerbak and Golubev, 1984:54; 1986:175–179, figs. 74, 76–77, col. pl. 3, fig. 8.
Mediodactylus spinicaudus: Kluge, 1991:21; 1993:22.

Diagnosis: Dorsum covered by rounded tuberculate scales, as large as scales of snout, intermixed with enlarged rounded tubercles irregularly arranged; caudal scales arranged in annuli with three mucronate tubercles on each side of annulus; supralabials 9–11, anterior 5 or 6 largest; infralabials 7–9; nostril between rostral, first labial, and two or three nasals; no distinct chin shields. Males with 2–4 preanal pores.

Color pattern: Pale gray above with 7 narrow undulating dark transverse bars; tail with 7 dark bars; venter white.

Size: Largest male, from Turkmenistan portion of Kopet Dagh, measures 44.4 mm snout-vent, tail broken; largest female 48.2 mm snout-vent length.

Habitat: (Plate 25F) Those found in the Kopet Dagh in Turkmenistan were collected in rock crevices and under stones, one in a rocky area, the other on a clay slope in an area of sparse vegetation (wormwood, low shrubs, and annuals) and a few isolated small rocks. Other reptiles encountered in the same area include *Testudo* (*Agrionemys*) *horsfieldi*, *Cyrtopodion caspium*, *Eumeces schneideri*, *E. taeniolatus*, *Leptotyphlops vermicularis*, and *Coluber rhodorachis*. (Ataev, Bogdanov, and Shammakov, 1968:1421). Theodore Papenfuss and Robert Macey (Papenfuss, pers. commun.) collected their specimens in Turkmenistan under rocks on a treeless slope in the Kopet Dagh. Szczerbak and Golubev (1986:177–179; 1996:173–174) have summarized what is known of the natural history of this species.

Distribution: Known only from the Kopet Dagh, in northern Iran and southern Turkmenistan, from about 500–1500 m elevation.

Remarks: In Iran, this species is known only from the holotype, but is much better known from the Turkmenistan portion of the Kopet Dagh. Ataev, *et al.* (1968:1420) give a brief description of their two specimens from Meimili Spring and from between

Cyrtopodion spinicauda

Ashkhabad and the Kalinsky settlement, which agree in all essentials with the original description. No mention is made by these authors or in the previous descriptions of femoral or preanal pores, nor of characters critical to placing this species within *Cyrtopodion*. The illustration in Szczerbak and Golubev (1986:177 [1996:172], fig. 76g) clearly shows 4 preanal pores in ZIL SR 364. Szczerbak and Golubev (1977a:130–131; 1984; 1986) have redefined *Alsophylax*, removed this species from that genus and reassigned it to their subgenus *Mediodactylus*, thus aligning it with *C. amictopholis, C. kotschyi, C. heterocercum, C. russowii*, and *C. sagittifer*. This species is the only Southwest Asian member of the genus which lacks well-defined chin shields; however, the scales bordering the mental and sublabials are considerably larger than other scales of the chin (see Szczerbak and Golubev, 1986:176 [1996:172], fig. 76).

I have examined 14 specimens from the northern foothills of the Kopet Dagh, Turkmenistan, collected by Theodore Papenfuss and Robert Macey in May 1992. All males have two preanal pores; there are three nasal scales in these specimens; there are no enlarged plates on the ventral side of tail, which is covered with small, smooth scales; all specimens have broken or regenerated tails; the tail may break at any postpygal vertebra, apparently. The viscera are unpigmented, but the innermost layer of the skin of the venter is darkly pigmented. There are often two enlarged scales, always separated by a small scale, touching the mental and first infralabials. These have 11 supralabials and 7–9 infralabials.

Cyrtopodion turcmenicum (Szczerbak, 1978), new comb. (Plate 8E)
Turkmenian thin-toed gecko

Gymnodactylus turcmenicus Szczerbak, 1978:39 (Type locality: Agashly locality near Kushka, Badghyz, Turkmenistan; Holotype: ZIK Re No. 10, adult male).

Diagnosis: Both femoral and preanal pores present; one to six subfemoral tubercles present; 30 or more preanal-femoral pores in males; usually more than 115 (up to 133) ventral scales from postmentals to vent; 25–32 ventral scales across midbody; 8–12 scales across head; dorsal tubercles triangular roundish or trihedral, usually large, 26–30 in paravertebral row from occiput to level of vent, usually one to three additional tubercles present; 2–3 (usually 3) pairs of postmentals, the scales of first pair broadly in contact; usually two rows of scales between row of femoral pores and vent; finger tips of adpressed forelimb reach tip of snout; length of forelimb 33–38% of body length). (Szczerbak and Golubev, 1986:141 [1996:138].)

Size: Adult males: 41.4–80.0 mm snout-vent length; adult females: 40.6–71.0 mm snout-vent (Szczerbak and Golubev, 1986:142 [1996:138]).

Color pattern: Dorsum ocher with 5 poorly defined brownish transvers bars on body and about 10 such dark bars bars on tail. Venter without spots.

Natural history: Szczerbak and Golubev (1986:144–146 [1996:140–142]) present the only study of the ecology, behavior, and life history of this gecko. It has not been studied in Iran or Afghanistan.

Habitat: In Badghyz they occur on sandstone cliffs in hill country with grass cover and pistachio trees. At Karabil they were found only on vertical limestone cliffs, being absent from talus slopes and

Cyrtopodion turcmenicum

individual boulders. Along the Kushka River, these geckos were found on conglomerate rock cliffs. (Szczerbak and Golubev (1986:144 [1996:140–141], fig. 65).

Distribution: Southeastern Turkmenistan and northern Afghanistan from the divide between the Kushka and Murghab Rivers and the southern extreme of the Karabil upland eastward through the Paropamisus to the western hills of the Hindu Kush. There is a single specimen from Gorgan, Mazanderan Province, Iran (NMW 19724; not seen by me) in the Vienna museum, collected by Franz Steiner 12 August 1968 and presumably identified by Josef Eiselt. 570–720 m elevation.

Remarks: I add this species to the Iranian fauna only tentatively, based on the single specimen identified as such cited above and on the photograph of a lizard collected by a recent Czech expedition to Iran (see Plate 8E). There is no published record for Iran, although its presence is not unexpected, since it occurs near to the northeastern border of Iran.

Genus *Hemidactylus* Oken
House geckos, leaf-toed geckos

Hemidactylus Oken, 1817:1183 (based on Cuvier's Hemidactyles, 1816 [1817], Régne Anim. 2:v, 47 [Type species: *Gecko tuberculosus* Daudin = *Hemidactylus mabouia* {Moreau de Jonnès} by subsequent designation of Stejneger {1907:172}, but see also Wermuth, (1965:69) and Zhao and Adler {1993:183}]).

Definition: Digits free or partly webbed, the distal phalanges short or long, rising angularly from within a dilated basal portion, covered above with scales, not denticulate laterally, covered below on undilated portion by scales or lamellae, on dilated portion by a double series of lamellae; all digits clawed; pupil vertical; dorsal scales granular, subimbricate, uniform, or with enlarged tubercles intermixed.

Distribution: Africa, southern Europe, southern Asia; Oceania; tropical America. Seventy-five species recognized by Kluge (1993). Three species known from Iran.

Remarks: Blanford's (1876:342) record of *Hemidactylus maculatus* from Gwadar on the Makran coast just east of the Iranian border probably refers to *H. brookii*. In any case, the animal was probably present adventitiously through human agency. The specimen from Dizak, southeastern Iran, designated "*H.* sp." and figured (pl. 22, fig. 1) (Fig. 75) by Blanford (1876:343–344) appears to differ from *H. persicus* and *H. turcicus* in color pattern and dorsal pholidosis. It is also unlike *H. flaviviridis* in having numerous trihedral dorsal tubercles and fewer labials.

FIGURE 75. *Hemidactylus* sp. from Iran (From Blanford, 1876, pl. 22, fig. 1.)

In the key which follows, I have included the widely distributed *H. brookii*, although it has not been recorded from Iran. It becomes established so readily near ports that it would be surprising if it does not show up in storehouses and habitations at places along the Persian Gulf.

In a previous paper (S. Anderson, 1974:31, 43) I listed *Hemidactylus garnotii* Duméril and Bibron from Iran, based on information received from I. Darevsky (pers. commun.) regarding Zarudny's collections from Chah Bahar. In a more recent conversation, Darevsky informs me that these specimens are *H. flaviviridis*.

Key to the Species of *Hemidactylus* in Iran

1a. No enlarged dorsal tubercles, or if tubercles present, these are rounded, feebly keeled, not regularly arranged (tubercles not present in Iranian, Afghan, Pakistan, or northern Indian specimens examined); males with femoral pores only ....
....*Hemidactylus flaviviridis* Rüppell, 1840 (p. 172)
1b. Enlarged dorsal tubercles numerous, strongly keeled, arranged in more or less regular longitudinal series; males with preanal pores only, or with both femoral and preanal pores 2
2a. Males with 15–27 femoral and preanal pores; 6–10 lamellae under 4th toe ....
....*Hemidactylus brookii* Gray, 1845
2b. Males with preanal pores only; 8–14 lamellae under 4th toe 3
3a. 8–11 lamellae and pairs of lamellae under basal expanded portion of 4th toe; 7–10 supralabials and 7–9 infralabials; males with 2–10 preanal pores ....
.... *Hemidactylus turcicus* (Linnaeus, 1758) (p. 174)
3b. 12–14 lamellae and pairs of lamellae under basal expanded portion of 4th toe; 10–12 supralabials and 8–10 infralabials; males with 9–13 preanal pores ....
....*Hemidactylus persicus* Anderson, 1872 (p. 173)

Hemidactylus flaviviridis Rüppell, 1840 (Plate 8F)
Yellow-bellied house gecko

Hemidactylus flaviviridis Rüppell, 1840:18, pl. 6, fig. 2 [see Leviton, *et al.*, 1992:38 for comment on date of publication] (Type locality: Massawa Island, Eritrea; Lectotype: SMF 8772 [designated by Mertens, 1967:55]). — J. Anderson, 1896:26. — Forcart, 1950:145. — Schmidt, 1955:203. — S. Anderson, 1963:474. — Wermuth, 1965:74. — S. Anderson, 1968:332; 1974:31, 43. — Schleich, 1977:127, 129. — Welch, 1983:8. — Welch, *et al.*, 1990:23. — Kluge, 1991:14. — Leviton, *et al.*, 1992:38, col. pl. 5A. — Kluge, 1993:15.
Hemidactylus cocteaui: Murray, 1884b:101.
Hemidactylus coctaei: F. Werner, 1929:240.

Diagnosis: Tail without sharp, denticulated lateral edge (although there is a ventrolateral row of small, pointed tubercles); outer postmentals in contact with labials; no enlarged dorsal tubercles (at least none in Southwest Asian specimens examined); males with 5–14 femoral pores on each side; 11–14 lamellae under fourth toe.

Color pattern: Gray to pinkish tan dorsally, uniform or with 5 transverse, undulating darker bands edged with white on posterior margins, first on neck, fifth on sacral region; tail barred above; a pale band through eye onto temporal region; considerable metachromatism.

Size: Snout-vent to 90 mm; tail 90 mm.

Natural History: Eggs were laid in our laboratory at the California Academy of Sciences between April and mid-June by captive specimens from Pakistan. Specimens collected in March near Karachi have a single egg with a calcareous shell in each oviduct. Females from Minab in November have small yolked ovarian follicles. Breeding behavior has been described for this species by Prashad (1916:834–838) and Mahendra (1936:250–281).

Habitat: Apparently this species is strictly a house gecko in Iran, found on both outside and inside walls and ceilings of inhabited and abandoned buildings on or near the coast of the Persian Gulf. Near sea level.

Distribution: From the northeast African and Arabian shores of the Red Sea and around the coasts of Arabia and Iran, across Pakistan, eastern Afghanistan and northern India to

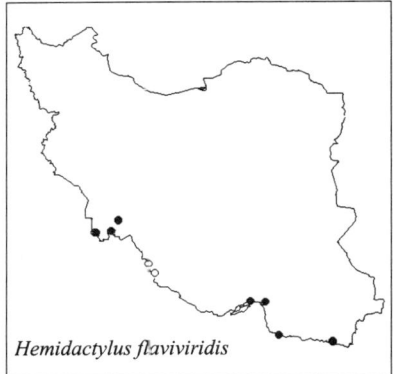
Hemidactylus flaviviridis

West Bengal and south to the vicinity of Bombay. Coastal towns and villages of southern Baluchistan, Kerman, Fars, and Khuzestan Provinces in Iran. The fact that they are house geckos in ports in the western part of their range and appear to be most closely related to *Hemidactylus leschenaulti* of peninsular India and Sri Lanka suggest that they are native to central India and have become distributed westward along trade routes through human agency.

Remarks: I have examined specimens from Baghdad and Tell Asmar, Iraq, which seem not to differ from Iranian, Afghan, and Pakistan specimens. Although an inland locality, Baghdad is on the Tigris River, and is a trade center of long standing; Tell Asmar is near Baghdad, on the Diyala River.

Specimens collected in October and November in Iran (Khuzestan and Minab) have subcutaneous fat deposits in the area of the lower abdomen. These deposits are much more pronounced in females than in males, and are not to be confused with fat bodies lying within the body cavity. I do not recall ever having seen such subcutaneous fat in any other lizard species. However, it is present in *Rhotropus* and some species of *Phelsuma* (A. M. Bauer, pers. commun.).

Hemidactylus persicus Anderson, 1872 (Plate 8G)
Persian gecko

Hemidactylus persicus J. Anderson, 1872:378–379, fig. 2 (Type locality: Iran, no exact locality given [probably near Bushire, *fide* Blanford, 1876:342, Shiraz according to Smith, 1935:87]; Holotype: ZSI 5961 [see Das, *et al.*, 1998]). — Blanford, 1876:342–343. — Nikolsky, 1907a:268. — F. Werner, 1917:196; 1936:200. — Mertens, 1956:92. — S. Anderson, 1963:441, 474. — Wermuth, 1965:82. — S. Anderson, 1968:333; 1974:31, 43. — Schleich, 1977:127, 129. — Welch, 1983:8. — Welch, *et al.*, 1990:24. — Kluge, 1991:15. — Leviton, *et al.*, 1992:38, col. pl. 5B–D. — Kluge, 1993:16. — Frynta, *et al.*, 1997:9.

Diagnosis: Tail without sharp, denticulated lateral edge; numerous enlarged, strongly keeled dorsal tubercles arranged in 14–16 more or less regular longitudinal series; 12–14 lamellae and pairs of lamellae under basal expanded portion of fourth toe; 10–12 supralabials and 8–10 infralabials; males with 9–13 preanal pores.

Color pattern: Light brown or grayish dorsally, with individual tubercles white, dark brown or black in an irregularly arranged pattern of small spots; usually a dark streak on side of head; venter whitish.

Size: Snout-vent length 71 mm for largest male examined, 67 mm for largest female.

Natural history: Females collected in May, June, and mid-August have ovarian eggs (largest in mid-June), while a female collected May 12 has an egg in each oviduct (all specimens from foothills of Khuzestan). One stomach examined contains primarily grasshopper remains.

Habitat: I collected this gecko at night on surfaced roads in the foothill region of Khuzestan, when air temperatures were between 31.4°C and 36.4°C, the road surfaces 35.4°C to 38.0°C, usually 3–5°C higher than the surrounding soil. *Hemidactylus persicus* is also found occasionally in houses in Khuzestan, but the common house gecko there is

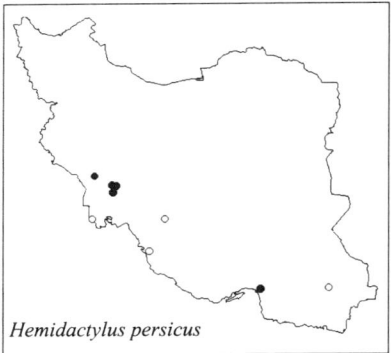
Hemidactylus persicus

Cyrtopodion scabrum. These geckos were collected only at night during the period when the rocky hills are almost completely denuded of vegetation and the seasonal streams long since dry, although some water remains in sinkholes and small caverns in gypsum formations.

Distribution: Coastal eastern Arabia north to southern Iran and Iraq, east to Sind and Waziristan, Pakistan. In Iran it is known from Khuzestan, Fars, Kerman, and Baluchistan Provinces. Sea level to 1000 meters in Iran.

Hemidactylus turcicus (Linnaeus, 1758) (*sensu lato*) (Plate 8H)
Mediterranean gecko, Turkish gecko, Turkish warty gecko

Lacerta turcica Linnaeus, 1758:202 (Type locality: "in Oriente" [restricted to Asiatic Turkey by Schmidt, 1953b]. Holotype: not located).
Hemidactylus robustus Heyden *in* Rüppell, 1827:19 (Type locality: Egypt, Arabia, and Abyssinia [= Ethiopia] [restricted to Abyssinia by lectotype designation by Mertens, 1967:55]; Lectotype: SMF 8720 [designated by Mertens, *loc. cit.*])
Hemidactylus turcicus: Boettger, 1876:57. — F. Werner, 1936:200. — Schmidt, 1955:203. — S. Anderson, 1963:474. — Guibé, 1966a:97. — S. Anderson, 1968:333. — Tuck, 1971a:56. — Schleich, 1977:127, 129. — Leviton, *et al.*, 1992:39, col. pl. 5E–F.
Hemidactylus karachiensis Murray, 1884b:361 (Type locality: Sind; Holotype: not located).
Hemidactylus turcicus turcicus: Mertens, 1925:60. — Wettstein, 1951:433. — Guibé, 1957:137. — S. Anderson, 1974:31, 43. — Welch, 1983:8. — Welch, *et al.*, 1990:25. — Kluge, 1993:17.
Hemidactylus parkeri Loveridge, 1936:59 (Type locality: "Zanzibar"; Holotype: not located).
Hemidactylus turcicus parkeri: Arnold, 1980a:283. — Kluge, 1991:16.

Diagnosis: Tail without sharp denticulated lateral edge; numerous enlarged, strongly keeled dorsal tubercles arranged in 14–16 more or less regular longitudinal rows; 8–11 lamellae and pairs of lamellae under basal expanded portion of fourth toe; 7–10 supralabials and 7–9 infralabials; anterior postmentals often in contact with second infralabials; males with 2–10 preanal pores.

Color pattern: (Fig. 76) Pinkish brown, light brown, sandy gray, or sandy yellow dorsally; an indistinct dark streak from nostril to temporal region; back mottled with darker brown, but many of the tubercles white; venter white.

FIGURE 76. *Hemidactylus turcicus* (CAS 159044).

Size: Snout-vent 57 mm, tail 60 mm.

Habitat: Apparently known only as a house gecko in Iran, found on walls and ceilings in both inhabited and abandoned buildings. Salvador (1984:99–102) summarizes the ecology and natural history of this species in Europe. Gallagher and Arnold (1988:406–407) observed them as tree geckos associated with *Prosopis cineraria*, calling on the trees at night and in the early morning, foraging on the ground at night, in the Wahiba Sands, Oman.

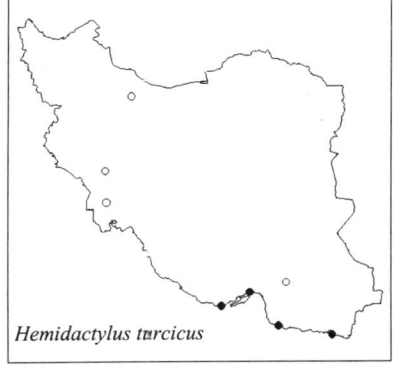

Hemidactylus turcicus

Distribution: From Morocco to Egypt and south through Somalia to northern Kenya; the coastal areas of the Mediterranean and Red Seas, through Southwest Asia to Sind, Pakistan; the Canary Islands; the islands and coastal regions of the Caribbean and the southeastern U.S.; isolated populations in Virginia and Maryland. They have been collected primarily in port towns of the Persian Gulf in Iran, although there are scattered inland records (Shahbazan, Qazvin, Rig Mati) as is also the case in Turkey, Jordan, and Iraq, but all of these localities lie along trade routes. There is a single record for Turkmenistan (Obst, 1984), but this was rejected as an established population by Szczerbak and Golubev (1986:13; 1996:13).

Remarks: See Arnold (1986c:419) for a discussion of the nomenclatural uncertainties regarding populations of this species. The populations that occur in Iran have been variously assigned to *H. t. turcicus* or separated as a distinct taxon by various workers (Loveridge, 1936:59; Lanza, 1978; Arnold, 1980b:283). *Hemidactylus robustus* Heyden is the oldest available name if these lizards are regarded as distinct from the Mediterranean populations. Such simple taxonomic treatments are probably not appropriate for animals introduced along trade routes, perhaps from several sites of origin.

Genus *Pristurus* Rüppell
Semaphore geckos

Pristurus Rüppell, 1835:16 (Type species: *Pristurus flavipunctatus* Rüppell, 1840, by monotypy).

Definition: Splenial bone absent; tendency for extensive contact of palatal portions of maxillae behind premaxilla; failure of lateral downgrowths of frontal bone to meet on midline and form tube (Kluge 1983:472); digits slender, clawed, cylindrical at base; distal phalanges compressed, forming angle with basal portion of digits, lower surface of which covered by row of plates; body not depressed, covered with uniform granules; tail compressed and keeled in males; pupil circular; no preanal nor femoral pores; no cloacal bones nor sacs.

Distribution: Islands and coastal regions of eastern Africa, Arabia, and Southwest Asia bordering the Red Sea, Gulf of Aden, Gulf of Oman, and Persian Gulf. Kluge (1993:29–30) recognized 18 species.

Remarks: *Pristurus*, a gekkonine genus, has many adaptations in common with the New World sphaerodactyline geckos, *viz.*, diurnal habits, sexual dichromatism, round pupil. "eyelid" formed by circumorbital ring of tissue, simple, undilated digits as in *Gonatodes*. small adult size, lack of preanal pores, lack of cloacal sacs and bones, and loss of splenial (Kluge 1967:33; 1983:472).

Pristurus rupestris Blanford, 1874 (Plates 8-I, 9A)
Blanford's semaphore gecko

Pristurus rupestris Blanford, 1874a:454 (Type locality: Muscat, Oman [restricted by Schmidt, 1952a:2]; Syntypes: ZSI [Annandale, 1906:91]); 1876:350–352, pl. 23, figs. 1, 1a. — Murray, 1884b:103; 1884b:365–366. — Boulenger, 1885a:53–54. — F. Werner, 1917:197. — S. Anderson, 1963:474; 1968:333; 1974:31, 43. — Schleich, 1977:127, 129. — Leviton, *et al.*, 1992:40, col. pl. 5G–H.
Pristiurus (sic) *rupestris*: Bedriaga, 1879:36.
Pristurus rupestris iranicus Schmidt, 1952:2 (Type locality: Bushire, Iran; Holotype: ZMUC 3476); 1955:201–202. — Wermuth, 1965:151. — Tuck, 1971a:56. — Schleich, 1977:127, 129. — Welch, 1983:11. — Kluge, 1991:28; 1993:30.

Diagnosis: Rostral enters nostril; hind limb reaches ear opening; 16–24 lamellae under fourth toe; tail of male strongly compressed, with feeble crest or row of enlarged scales in males only, never extending onto body.

Color pattern: Olive-gray dorsally, with or without pale vertebral stripe; back and sides with rufous spots arranged in longitudinal lines, those on back larger than those on sides, and with white posterior border (spots absent in preservative); dark mark from nostril through eye to temporal region; sides of head, neck, chin, and throat spotted with black; tail barred with brown or gray.

Sexual dimorphism: Tail more strongly compressed in males and bearing dorsal crest consisting of row of elongated scales, ventral crest of two to three rows.

FIGURE 77. *Pristurus rupestris*. (From Blanford, 1876, pl. 23, fig. 1.)

Size: Snout-vent length 32 mm, tail 53 mm. Largest of 43 Iranian specimens measured has snout-vent length of 28 mm, tail 27 mm.

Natural history: Blanford found them active on the surface of rocks in late morning; they took refuge in crevices in limestone only when approached, and were difficult to capture due to their activity and the numerous cracks and fissures.

I find no oviducal eggs in specimens collected in late November; ovaries appear regressed. Small juveniles (snout-vent length 16 mm) were collected then.

Habitat: (Plate 20C–E) Blanford (1876:352) found these geckos in houses as well as on limestone rocks, and Schmidt (1955:202) reported specimens collected under rocks in cultivated fields. I found them on trunks of palm trees in a hotel garden in Minab. Each palm seemed to have one male gecko. They moved easily up the tree to avoid capture, but seemed reluctant to go higher than about 3 meters up the trunk. Gallagher and Arnold (1988:407) also found them on trees (*Prosopis*) in the Wahiba Sands, Oman. I found them active on the surface of stones in a dry wash, on walls of abandoned buildings on a beach and among rocks near Bandar-e Lengeh, and on faces of sandstone bluffs above the town of Qeshm on Qeshm Island in the Persian Gulf. They were active at midday in full sun (air temperature 30°C, surface of sandstone 38°C) but near cracks or shade where they could unload heat. They were also active at dusk (air 26°C, wall surface 22°C). For an account of the habitat and natural history of this species in the United Arab Emirates, see Leptien (1993a:34–36).

Distribution: Persian Gulf coast of Iran, gulf islands, and the coasts of the Arabian Peninsula; southwestern Jordan; northern Somalia. Murray (1892:69) recorded this species

Pristurus rupestris

near the Afghan border in Baluchistan, Pakistan, but as is the case with many of Murray's records, subsequent collectors have failed to verify these localities. Sea level to 300 m elevation.

Remarks: I am unable to recognize Schmidt's subspecies on the basis of the characters he cited. I have examined one of his paratypes (FMNH 69296) and find that there is a narrow light vertebral stripe, Schmidt's comments to the contrary notwithstanding. Six specimens from southern Arabia (FMNH 18220; Schmidt, 1939:56) also have black spots on the chin, although they are less conspicuous than in Iranian specimens. These southern Arabian specimens seem to lack the vertebral line. They work out to *Pristurus flavipunctatus* in Loveridge's key (1947:42), and I believe them to be closer to that species than to *P. rupestris*. Specimens from Yemen (FMNH 66337–97, 66426–32) are characterized by very prominent dark markings on chin and throat, these very elongate, rather than discrete dots; in some, the black marks form irregular stripes crossing the throat. These specimens also have a prominent black spot just above the shoulder, lacking in Iranian specimens. The paratype of *P. r. iranicus* has relatively larger, flatter, less granular scales than those from Aden, but resembles the Yemen specimens in this respect. Specimens from Cheh Mossulum and Minab agree with Schmidt's specimen. Most specimens lack a light dorsal stripe, but such is occasionally present, although not prominent. The dorsal and ventral crests on tails of males are less well developed in Iranian specimens than in Aden males. In Aden specimens the elongate scales of the dorsal crest begin just posterior to level of vent. Characteristic of diurnal lizards, the peritoneum and lower viscera are darkly pigmented, as are the fat bodies; the top of the skull, particularly in the frontal and parietal areas, is black.

Genus *Ptyodactylus* Goldfuss, 1820
Fan-footed geckos

Styodactylus (= *Ptyodactylus*) Goldfuss, 1820:158 (spelling error due to a *lapsus*, spelled *Ptyodactylus* elsewhere in text and index [Type species: *Gecko lobatus* Geoffroy St. Hilaire, 1827 (= *Lacerta hasselquistii* Donndorff, 1798), by subsequent designation of Fitzinger, 1843:18 (ICZN, 1964, 67[g])]).

Definition: Digits free, slender at base, strongly dilated at apex, covered above with scales, not denticulate laterally; digits with transverse shields inferiorly on basal part, dilated portion with two diverging series of lamellae, fan-like, a small median fissure between the two series; all digits clawed, the claw retractile between scales in the anterior notch in the apical expansion; pupil vertical; eyelid an immovable circumorbital ring; dorsum covered with small, smooth, juxtaposed granules, uniform, or mixed with larger tubercles; subimbricate scales on venter; tail tapering, subcylindrical.

Distribution: Heimes (1987) recognizes six taxa in this genus, distributed across northern Africa (south to Ghana and Cameroon), the Nile Valley, scattered, mostly coastal, localities on the Arabian Peninsula, the North Arabian Desert (Israel, Jordan, Syria, Iraq) and one species (*Ptyodactylus homolepis* Blanford) isolated in Sind, Pakistan. Heimes makes no mention of Schmidt's (1955:203) record for Iran.

Remarks: I have not examined the specimen (ZMUC 3447, female) recorded as

Ptyodactylus hasselquistii (Donndorff) by Schmidt (1955:203), subsequently cited by S. Anderson (1963:474; 1968:333; 1974:31, 43) and Schleich (1977:127, 129). Jens Rasmussen (pers. commun., 12 August 1993) informs me that this specimen cannot be located. Smith (1935:79) also includes Persia in his statement of the range of this species. In view of Heimes' (1987) revision of the genus, it would seem likely that this record would prove to be *P. puiseuxi* Boutan, 1893, but the presence of the genus in Iran awaits confirmation, in my opinion. This species is known from Israel, Jordan, Syria, and western Iraq, whereas in the northern part of its range, *P. hasselquistii hasselquistii* seems not to occur east of the Rift (Heimes, 1987:fig.9, but see also Arnold, 1986d:421 and Leviton, *et al.*, 1992:41). *Ptyodactylus guttatus* Heyden, 1827 overlaps the ranges of *P. h. hasselquitii* and *P. puiseuxi* and is known as far east as central Syria.

Deferring judgment as to the specific identity of the specimen recorded from Iran (Schmidt, 1955:203), I present the following key, adapted from Heimes (1987:221–222), to the species of *Ptyodactylus* known from Southwest Asia.

Key to the Southwest Asian Species of *Ptyodactylus*

1a. Back covered with small, homogeneous granular scales ....
  .... *Ptyodactylus homolepis* Blanford, 1876
1b. Back covered with small granules and longitudinal rows of enlarged tubercles 2
2a. Tubercles conical or rounded, without keels; large tubercles in front of the ear ....
  .... *Ptyodactylus puiseuxi* Boutan, 1893
2b. Tubercles more or less oval, keeled; no tubercles in front of the ear 3
3a. Tail distinctly shorter than head and body; dorsum predominantly spotted ....
  .... *Ptyodactylus guttatus* Heyden, 1827
3b. Tail about as long as head and body or slightly shorter; dorsum with 5 crossbars as ground-pattern .... *Ptyodactylus hasselquistii hasselquistii* (Donndorff, 1798)

Ptyodactylus incertae sedis

Ptyodactylus hasselquistii: Schmidt, 1955:203. — S. Anderson, 1963:474; 1968:333; 1974:31, 43. — Schleich, 1977:127, 129.
Ptyodactylus hasselquistii hasselquistii: Welch, 1983:11.

Genus *Rhinogecko* de Witte, 1973
Swollen-nose geckos

Rhinogecko de Witte, 1973:1–6, figs. 1–6 (Type species *Rhinogecko misonnei* de Witte, 1973, by monotypy).

Definition: Nostril at apex of a prominent swollen or cylindrical caruncle formed by 3 nasals or 3 nasals and the first supralabial (Fig. 78B, C [on p. 181]); digits cylindrical, more or less compressed, slightly depressed at base, covered below with a series of transverse lamellae (Fig. 78E); a row of enlarged scales on lower surface of thigh (Fig. 78D); tail thin, cylindrical, ending in a point, swollen at base; males with preanal pores.

Distribution: The Dasht-e Lut, eastern Iran, and Kharan and Chagai Districts, northwestern Pakistan. Two species recognized here.

Remarks: I have examined the holotype of *Agamura femoralis* Smith (BMNH 12.3.26.12/1946.8.20.48, male, Kharan, Baluchistan, Pakistan), and 10 specimens from "Koh-i-Taftan," Chagai District, Pakistan (AMNH 92681–2; RSM 1964–58–13–20) (J. A.

Anderson [who collected the RSM specimens], pers. commun., tells me that this latter locality is Taftan, a railway station, not Koh-i-Taftan). In my opinion, *Rhinogecko misonnei* de Witte is doubtfully distinct from *Agamura femoralis* Smith. I also believe that these geckos are not particularly closely related to *Agamura persica* (Duméril), and that de Witte's genus is justified, at least until relationships among these Southwest Asian geckos are better understood. A preliminary discussion of the generic relationships of the Pakistan specimens was presented in a previous paper (S. Anderson and Leviton, 1969:43–44).

The above specimens lack the basal constriction of the tail characteristic of *Agamura persica*, have mental shields, and toes not so distinctly angularly bent as in *Agamura*. There is no dark pigment in the peritoneum as in *Agamura*, and they lack a pronounced "eyeshade." The type of *femoralis* has 6 preanal pores; tail segments of 5 transverse scale rows dorsally, the fifth enlarged, more strongly keeled, but not tuberculate, as in many species of *Cyrtopodion*; scales of underside of tail not clearly set off from those of dorsal side; distal 3 segments of tail compressed, rest of tail slightly depressed; scales of dorsum juxtaposed, tubercles subcircular, weakly keeled; 5 dark crossbars, first on neck, fifth on sacrum; the nostril is between the first labial and 3 nasals. De Witte's illustration and description show the first labial excluded from the nasal caruncle, the difference being that in the type of *femoralis* the first labial and a nasal are fused. Measurements of the holotype of *femoralis*: snout-vent 52, tail 58, head length 14, head width 10, forelimb 22, hindlimb 30 mm.

In 9 specimens from "Koh-i-Taftan," the nasal caruncle is formed by 3 nasals and the first labial; in one, the supralabial is transversely divided so that the nostril is surrounded by 4 nasals, the first labial being excluded. In de Witte's specimens the rostral and first labial are excluded from the border of the nostril, which sits upon a prominent nasal caruncle formed by 3 nasal scales. Four males have 6 preanal pores, one has 4, as does de Witte's male paratype. Seven of the "Koh-i-Taftan" specimens have 5 dorsal crossbars, like the holotypes of both *femoralis* and *misonnei*, while one has 6. A specimen from 56 km E Nak Kundi (= Nok Kunde), Chagai District, Pakistan (USNM 158550) has the nasal caruncle formed of three nasals only, as in de Witte's Iranian specimens, and has 8 preanal pores. In this specimen the nasal caruncles are very long and reflexed backward, so that the nostrils point up and back. At the time I examined this specimen, I had no other material for direct comparison, but my impression was that the shape of the head was different from the Taftan specimens, with the snout shorter and head higher.

Szczerbak and Golubev (1986:213–216; 1996:207–211) regard these specimens as representing two distinct species, both of which occur in eastern Iran and western Pakistan. I follow their judgment here, somewhat tentatively. They include both species in the genus *Agamura*. The following key is adapted from their key to that genus.

Key to the Species of *Rhinogecko* in Iran

1a. Nasal shields noticeably swollen, 17–22 scales across abdomen .... *Rhinogecko femoralis* (Smith 1933) (p. 179)
1b. Nasal shields sharply swollen and erect; 26–28 scales across abdomen .... *Rhinogecko misonnei* de Witte 1973 (p. 180)

Rhinogecko femoralis (Smith, 1933)
Sharp-tailed spider gecko, Kharan spider gecko

Agamura femoralis Smith, 1933:17 (Type locality: Kharan, Baluchistan; Holotype: BMNH

12.3.26.12/1946.8.20.48). — Szczerbak and Golubev, 1986:215–216 [1996:210–211], fig. 95, 98. — Welch, *et al.*, 1990:5. — Kluge, 1991:2; 1993:3.

Rhinogecko femoralis: de Witte, 1980:2 (by implication).

Diagnosis: 17–21 scales across abdomen; nasal shields noticeably swollen; a row of enlarged scales on lower surface of thigh; 5–6 distinct pores; tail slightly longer than body. Snout-vent length 52–60 mm. (Szczerbak and Golubev, 1986:215; 1996:210).

Color pattern: Dorsum gray, tinged with brown, 5 broad dark brown bars across body, 8–10 on tail; pattern more vivid in juveniles. (Minton, 1966:80).

Size: 52–60 mm snout-vent length.

Habitat: Pakistan specimens have been collected near rocky outcrops in sandy country. They are nocturnal and terrestrial. (Minton, 1966:80).

Distribution: Probably extreme eastern Baluchistan in Iran; Kharan and Chagai districts in Pakistan. Szczerbak and Golubev (1986:210, fig. 95, p. 216) list Kuh-e Taftan and Mirjawa specimens as from Iran. However, both of these localities are in Pakistan (see Minton, 1966 and de Witte, 1980; see above comment regarding confusion about the name Koh-i-Taftan). Mirjawa is at the border with Iran, however, and there is a Mirjaveh (or Mir Javeh), 29°01'N, 61°28'E, on the railroad, just on the Iranian side of the border; the Rud-Mirjawa (Pakistan) — Rud-e Mirjaveh (Iran) is a stream which crosses the border. There is also a Kuh-e Taftan, 28°36'N, 61°06'E, in Iranian Baluchistan. Occurrence of this form in Iran is almost certain, but not confirmed.

Rhinogecko misonnei de Witte, 1973
Misonne's swollen-nose gecko, Misonne's spider gecko

Rhinogecko misonnei de Witte, 1973:1–6, figs. 1–6 (Type locality: Dasht-i-Lut [30°13'N 58°47'E], Iran; Holotype: IRSNB 2514). — S. Anderson, 1974:33. — Schleich, 1977:127, 129. — de Witte, 1980:1–3. — Welch, 1983:12.

Agamura misonnei: Szczerbak and Golubev, 1986:213–214 [1996:207–210], figs. 95, 97. — Welch, *et al.*, 1990:5. — Kluge, 1991:3; 1993:3.

Diagnosis: 26–28 scales across abdomen; nasal shields sharply swollen and erect, forming a short tube-like structure (Fig. 78B,C); a row of 9–12 enlarged scales on lower surface of thigh (Fig. 78D); 4–8 very faintly expressed pores; tail slightly longer than body.

Color pattern: Dorsum gray, tinged with brownish, with 5 or 6 broad dark brown crossbars, 7–10 on tail, limbs with broad brown bars less dark than those of body and tail, lips flecked with dark brown, venter whitish.

Rhinogecko misonnei

Size: Snout-vent length to 81 mm, tail 73 mm.

Habitat: The holotype of *R. misonnei* was found in the Dasht-e Lut in an area devoid of vegetation for 100 km in all directions; the paratype was found on a gravel desert with very meager vegetation. These geckos apparently feed on insects blown into the desert by the violent winds of this region (de Witte, 1973:6).

Distribution: Known from the Dasht-e Lut, which encompasses portions of Kerman, Khorasan, and Baluchistan Provinces in Iran and from Baluchistan, Pakistan, in the vicinity of Nak Kunde (or Nok Kunde).

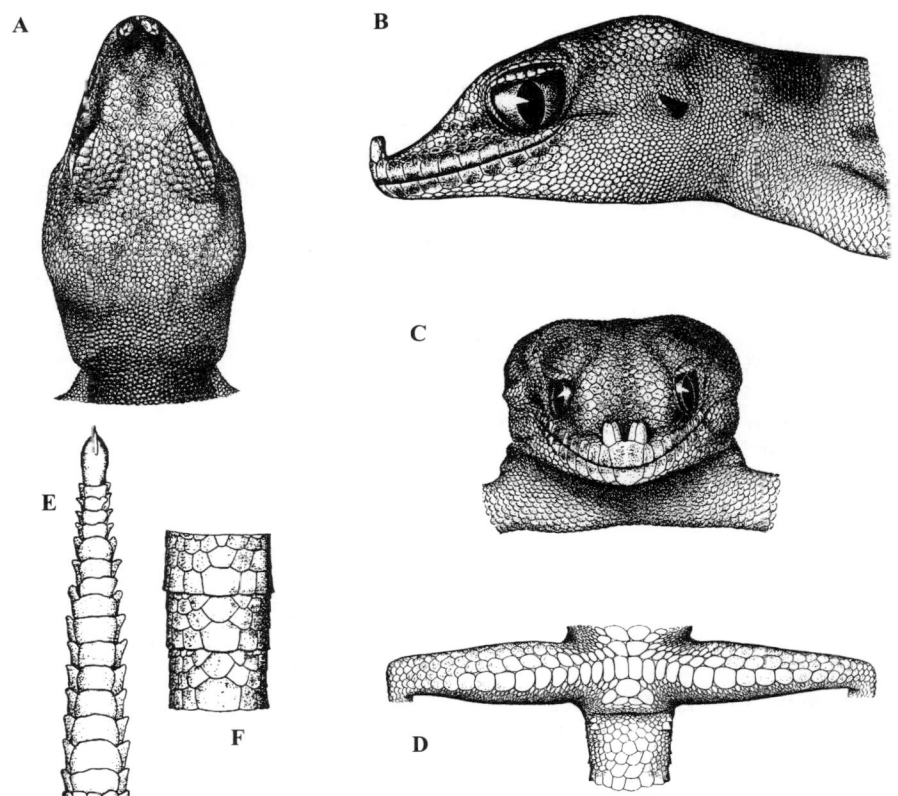

FIGURE 78. *Rhinogecko misonnei* (holotype): (A) top of head; (B) side of head; (C) head-on view of head; (D) femoral and preanal scales; (E) ventral surface of digit; (F) ventral surface of tail. (From de Witte, 1973, figs. 3, 1, 2, 4, 5 and 6, respectively.) Reprinted with permission Royal Belgian Institute of Natural Sciences.

Remarks: Subsequent to his original description of *Rhinogecko misonnei*, de Witte (1980) examined additional material, including the specimens which I examined, cited above, in addition to two specimens from "Koh-i-Taftan" in Minton's personal collection and an additional specimen from Nok Kunde (SMF 63457). He regards the specimens from Koh-i-Taftan as *Agamura femoralis* (which he, too, would assign to *Rhinogecko*) and the two male specimens from the vicinity of Nok Kunde, Chagai District, Pakistan as *R. misonnei*. It is the form of the nasal caruncle, reflexed and tubular in *R. misonnei* as opposed to rounded and swollen in *R. femoralis*, which most clearly sets these two groups of specimens apart.

Genus *Stenodactylus* Fitzinger, 1826
Comb-fingered geckos

Stenodactylus Fitzinger, 1826:13, 47 (Type species: *Stenodactylus elegans* Fitzinger, 1826 [= *Ascalabotes sthenodactylus* Lichtenstein, 1823], by subsequent designation of Fitzinger [1843:18]).

Definition: "The only gekkonid genus in which the phalangeal formula is reduced to 2 3 3 4 3 on both manus and pes. Premaxilla and frontal single, nasal and parietal bones paired, splenial present, supratemporal and angular bones absent, scleral ossicles 20–28, second ceratobranchial element of hyoid lacking, postcranial calcified endolymphatic sacs present, vertebrae amphicoelous, cloacal sacs present in both sexes and cloacal bones present in males, eye covered by a spectacle, pupil vertically elliptic, scaling more or less homogeneous, preanal pores reduced to one on each side or absent, toes without complex musculature or adhesive pads, with 1–15 longitudinal rows of scales beneath and sometimes a lateral fringe of pointed scales. Eggs with calcified shells, produced in clutches of 1 or 2; voice present." (Arnold, 1980b:373).

Distribution: From western part of the Saharo-Sindian desert region across North Africa and south to northern Kenya east through the Sinai and Arabian peninsulas, Israel, Jordan, southeastern Turkey, Syria, Iraq, into southeastern Iran.

Remarks: This genus has been reviewed by Arnold (1980b). He recognizes 11 species, following Kluge (1967) in excluding the eastern species *S. orientalis*, *S. maynardi*, and *S. lumsdeni*, now assigned to *Crossobamon* (see above), and including *Ceramodactylus* Blanford, 1874a, *Pseudoceramodactylus* Haas, 1957, and *Trigonodactylus* Haas, 1957. Two species occur in Iran.

Key to the Species of *Stenodactylus* in Iran

1a. Back with 4 dark crescentic crossbars; 10–11 supralabials; forelimb does not reach beyond tip of snout .... *Stenodactylus affinis* (Murray, 1884) (p. 182)
1b. No dark crossbars on back; 12–15 supralabials; forelimb reaches beyond tip of snout .... *Stenodactylus doriae* (Blanford, 1874) (p. 183)

Stenodactylus affinis (Murray, 1884) (Plate 9B)
Murray's comb-fingered gecko

Ceramodactylus affinis Murray, 1884b:103–104 (Type locality: Tanjistan, Iran; Syntypes: BMNH 84.7.25.1/1946.8.23.33, 87.9.22.2/1946.8.23.60). — Boulenger, 1885a:14. — F. Werner, 1936: 200. — Schmidt, 1955:200–201. — S. Anderson, 1963:474; 1968:333. — Schleich, 1977:127, 129.
Stenodactylus affinis: Wermuth, 1965:175. — S. Anderson, 1974:32, 43. — Arnold, 1980a:391–393, fig. 9. — Welch, 1983:12. — Kluge, 1991:33. — Leviton, et al., 1992:43, col. pl. 6D. — Kluge, 1993:35.

Diagnosis: "Medium sized, up to at least 60 mm from snout to vent; toes not depressed, at most slightly fringed, three rows of scales beneath; rostral and sometimes upper labial scales reach nostril which is directed forwards and sometimes upwards and somewhat outwards; preanal pores often present, cloacal tubercles in a single row; autotomy restricted to tail-base; prefrontal projection absent or weak; upper border of preotic sloping downwards to some extent; epipterygoid separated from skull roof in protracted skull; 25 presacral vertebrae; 6 nuchal ribs; 4–6 basal caudal vertebrae; clavicle expanded; tail not dark, but with 6 to 8 dark bands." (Arnold, 1980a:391).

Color pattern: Sandy gray or brown above, with darker flecks, light dots, and 4 dark crescentic transverse bars, anterior on occiput, posterior on sacral region; 6–8 dark bars on tail; lips with dark vertical bars; venter whitish.

Size: Snout-vent length 60 mm, tail 29 mm.

Habitat: The specimens recorded by Schmidt (1955:200–201) were collected under stones in a cultivated field. My specimens were collected at night on an unpaved road running through flat terrain of silty alluvium.

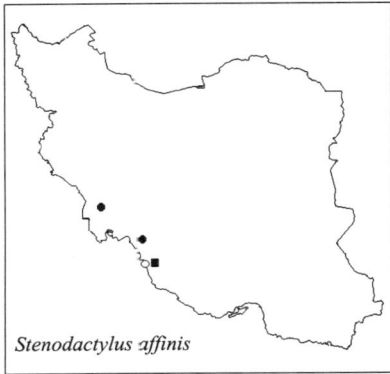
Stenodactylus affinis

Distribution: Known only from southeastern Khuzestan and southern Fars Provinces, Iran.

Remarks: I have examined the two syntypes. In these female specimens there are 2 enlarged preanal scales separated by 7 smaller scales. The nostril is between rostral, first labial, and 3 nasals, uppermost nasals in contact behind rostral. Scales rugose, particularly ventral and caudal scales, head scales and those of back less so; about 157 scales around middle of body. Nasal region surrounding nostrils slightly swollen, much less so than in other Southwest Asian species in the genus. Tail, which is apparently original and complete, tapers gradually from base, but is thicker and blunter at end than in other members of the genus; it is distinctly compressed in its distal one-fourth. Limbs, tail, and digits shorter and stouter than in specimens from Arabia identified as *Stenodactylus doriae*. Fringes on digits sharp, and shorter than in specimens of *S. doriae* and Blanford's illustration (1876:pl. 23, fig. 2a) of a digit of the latter. Toes very similar to those of *S. slevini* Haas, except that while the median row of subdigital scales is somewhat enlarged, these scales do not form such distinct lamellae as they do in *S. slevini*; fingers of both species have distinct lamellae, however. About 30 scales on midline of under side of fourth toe, while in *S. slevini* about 38, 43 in *S. doriae* from Dhahran, 38 in *S. doriae* from Tangistan, 68 in *S. major* from Abu Dhabi. Color pattern is faded in these specimens, although the crossbars can be seen (see Arnold, 1980b:pl. 9); no indication of reticulate pattern such as characteristic of *S. doriae*. Diameter of eye equal to distance from corner of orbit to nostril; In Arabian and Iranian specimens of *S. doriae* the eye is less than this distance. MMTT 61, collected by Douglas M. Lay in 1973, is also *S. affinis* according to Arnold (1980a:370). My notes on this last specimen, written in Tehran in 1975, indicate that it differs from the syntypes examined in having nasals separated behind the rostral by 2 scales, no preanal pores or enlarged scales, distinct lamellae under toes as well as fingers, forelimb reaching beyond tip of snout. My impression while in Tehran, with no access to comparative material, was that the specimen might be *S. slevini*, although it differed from Arabian specimens in color pattern and in shorter toes.

Stenodactylus affinis is the sister species to a clade containing *S. slevini*, *S. leptocosymbotes*, and *S. doriae* according to the phylogenetic hypothesis put forward by Arnold (1980b:401).

Stenodactylus doriae (Blanford, 1874) (Plate 9C)
Doria's comb-fingered gecko

Stenodactylus guttatus: De Filippi, 1865:352 (not Cuvier, 1829:58).
Ceramodactylus doriae Blanford, 1874a:454–455 (Type locality: Bandar 'Abbas, Iran; Holotype: MZUT); 1876:353–354, pl. 23, figs. 2, 2a (Type locality corrected to: one [-day's] march from Bandar 'Abbas on road to Kerman). — Murray, 1884b:103. — Boulenger, 1885a:13–14, pl. 2, fig. 4. — F Werner, 1917:b93; 1936:200. — Guibé, 1957:137. — S. Anderson, 1963:474; 1968: 333. — Schleich, 1977:127, 129.
Stenodactylus doriae: Wermuth, 1965:175–176. — Arnold, 1980a:397–399, figs. 1e (col. pl.), 12–13. — Welch, 1983:12.—Kluge, 1991:33. — Leviton, *et al.*, 1992:44, col. pl. 6E–F, 7A. — Kluge, 1993:35.

Diagnosis: "Medium to large-sized, up to 83 mm from snout to vent; toes depressed with a distinct lateral fringe of pointed, toothed scales and 5–13 rows of scales beneath; rostral and often first upper labial scale reach nostril; preanal pores nearly always present, cloacal tubercles usually in 2 rows; autotomy restricted to tail base; prefrontal projection strong; epipterygoid well separated from skull roof in protracted skull; usually 24 presacral vertebrae, 6 nuchal ribs and 6 basal caudal vertebrae; clavicle expanded." (Arnold, 1980a:397).

FIGURE 79. *Stenodactylus doriae*. (From Blanford, 1876, pl. 23, fig. 2.)

Color pattern: (Fig. 79) Pale buff dorsally, with brown reticulation enclosing round whitish spots; tail with brown and white crossbars; venter white.

Size: The holotype measured 63 mm snout-vent, tail 50 mm (Blanford, 1876:354). Largest Arabian specimen examined has snout-vent length of 74 mm, tail missing. A specimen 65 mm snout-vent has a tail 59 mm. According to Arnold (1980a:399), males reach 59 mm snout-vent, females 83 mm.

Habitat: (Plate 21F–G) Specimens from 12 km east of Shaqu were collected on loose sand among scattered low desert shrubs, the largest of which were *Tamarix*. Other species collected at the same time were *Bunopus tuberculatus*, *Teratoscincus scincus*, and *Lytorhynchus ridgewayi*. The locality near Kupal consists of a small dune field with small shrubs and grasses in blow-out areas. The sand in depressions among dunes was quite moist. Other species collected or seen at this locality included *Bufo viridis*, *Diplometopon zarudnyi*, *Trapelus persicus*, *Acanthodactylus schmidti*, *Scincus scincus conirostris*, *Varanus griseus*, and *Eryx jayakari*. The *Stenodactylus* and the *Eryx* were collected at night. The Alchangi locality was an area where compacted silt formed the low hills, with a loose sand blow-on. Low shrubs were scattered over the hills. The *Stenodactylus* were collected at 1915 h and shortly thereafter, air temperature 31°C, sand 33°C. Other species collected near this locality during the day were *Trapelus persicus*, *Acanthodactylus grandis*, and *Mabuya aurata*. See Gallagher and Arnold (1988:408) for notes on habitat and behavior in the Wahiba Sands, Oman.

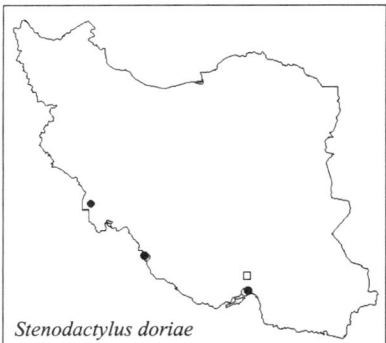

Stenodactylus doriae

Distribution: The coastal plain of Fars and Kerman Provinces and the lower Mesopotamian Plain in Khuzestan, Iran; known from throughout the Arabian Peninsula, north to Syria, west to Israel.

Remarks: Arnold (1980a:397) placed the widespread *Ceramodactylus major* Parker, 1930 in the synonymy of *Stenodactylus doriae*. According to the phylogenetic hypothesis of Arnold (1980a:401), *S. doriae* is most closely related to *S. leptocosymbotes* Leviton and S. Anderson, 1967.

Genus *Teratoscincus* Strauch, 1863
Plate-tailed geckos, skink geckos

Teratoscincus Strauch, 1863:col. 480 (Type species: *Teratoscincus keyserlingii* Strauch, 1863 [= *Stenodactylus scincus* Schlegel, 1858], by monotypy).

Definition: Digits straight, not angularly bent at any of the articulations, not dilated, all clawed, with lateral fringe of long pointed scales, and minute granular scales below; body covered with uniform, cycloid, imbricate scales; tail with large transverse plates above; pupil vertical; males without preanal or femoral pores; angular bone present. Supraorbital bones present (Bauer and Russell, 1988).

Distribution: Central Asia, from southern Mongolia and western China through Turkmenistan and other Transcaucasian republics to eastern Iran; through Afghanistan to Baluchistan, Pakistan; eastern United Arab Emirates. Four species recognized, three of which occur in eastern Iran.

Remarks: The distribution of this genus is unlike that of any other gekkonid genus. The greatest morphological diversity is found in the desert area of the southeastern Iranian Plateau. Its distribution is complementary to, and nearly entirely allopatric with that of *Stenodactylus*, the gekkonine genus to which it is probably most closely related, according to Kluge (1967:23), although the two genera do occur together on the gulf coast of Iran and the Arabian Peninsula. It is the only genus in the Gekkoninae to retain the angular bone, and Kluge (1967:33) suggested that this may indicate derivation from eublepharine stock. It is now regarded by some workers as belonging to its own subfamily, Teratoscincinae (Kluge, 1987:40; Grismer, 1988:372).

Werner (1967) has shown that the regenerated tail in this genus is nearly indistinguishable from the original tail in its external morphology. This is undoubtedly related to its unique ability to produce a clearly audible sound by the rubbing together of the large imbricate caudal scales.

Among the peculiarities of pholidosis in these geckos is the fact that the first upper labial never forms part of the margin of the nostril. In *Teratoscincus microlepis* and *T. bedriagai* the lowest of the four nasals has a lunate upper margin, usually excluding the rostral as well as the first labial from the border of the nostril. In *T. scincus* there are also four nasals, but in addition, there is a narrow, rim-like crescentic scale bordering the nostril, usually completely excluding the four lowest nasals as well as the rostral from the nostril.

In my opinion, based on the less derived nature of the scales, *T. microlepis* is the most primitive member of the genus, while *T. scincus* appears to be the most specialized structurally.

Key to the Species of *Teratoscincus* in Iran

1a. Large cycloid scales of dorsum extend forward to occiput (Fig. A) ....
....*Teratoscincus scincus* (Schlegel, 1858) (p. 183)
1b. Large cycloid scales not extending forward beyond shoulders (Fig. B) 2

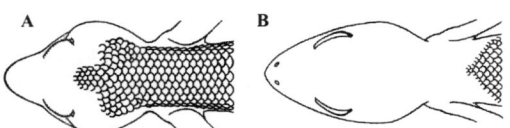

(A) *Teratoscincus scincus*; (B) *Teratoscincus bedriagai* (Drawn by Jean Jahnson)

2a. Not more than 60 scales round middle of body ....
.... *Teratoscincus bedriagai* Nikolsky, 1899 (p.186)
2b. About 100 scales round middle of body ....
.... *Teratoscincus microlepis* Nikolsky, 1899 (p. 187)

Teratoscincus bedriagai Nikolsky, 1899 (Plate 9E–F)
Bedriaga's plate-tailed gecko, Bedriaga's skink gecko

Teratoscincus bedriagai Nikolsky, 1899:146–147 (Type locality: Zirkuch and Sistan, eastern Iran [restricted to Chadschi-du-i Tschaghi by Szczerbak and Golubev, 1986:45]; Syntypes: ZIL 9157, 9158[2], 9159[3], 9160, 9161–3; BMNH 99.7.25.3/1946.8.23.39 [Lectotype: ZIL 9161, designated by Szczerbak and Golubev, 1986:45]); 1899c:378–379. — Zarudny, 1903:3–4. — F. Werner, 1936:200. — S. Anderson, 1963:474. — Wermuth, 1965:181–2. — S. Anderson, 1968:333. — Tuck, 1971a:57. — S. Anderson, 1974:32, 43. — Schleich, 1977:127, 129. — Tuck, 1979:102. — Welch, 1983:13. — Szczerbak and Golubev, 1986:45–46 [1996:45], figs. 10, 16. — Kluge, 1991:35. — S. Anderson, 1993:8–9. — Kluge, 1993:37.

Diagnosis: Enlarged scales of dorsum do not extend forward beyond the shoulder, neck covered by small granular scales; ventral scales slightly smaller than or subequal to dorsals; 36–52 scales round the middle of body; 9–11 supralabials; 9–10 infralabials.

Color pattern: Dorsum light sandy or cream, head with a brown crescentic mark from eyes onto occiput, dark vertical bars on snout, below eye, and on temporal region, back with four or five brown caudally-pointing chevrons, lighter and broken up in adults; tail with two or three brown crossbars, becoming lighter and indistinct in adults; limbs without dark pattern; venter white.

Size: Snout-vent length 72 mm, tail 35 mm (literature record, sex unknown).

Natural history: The only Iranian specimen for which stomach contents were examined had various kinds of beetles in the stomach. Arachnids, including solpugids, are also eaten (S. Anderson and Leviton, 1969:47).

Habitat: (Plate 21B) I collected this species in and around an abandoned village in Sistan at night on silty eolian soil. The village was being "drowned" in loose wind-blown alluvium, and vegetation around the village consisted of weedy "tumbleweed" shrubby plants. The geckos had just emerged from holes in the soil at 1840 h, and some retreated back into them when approached. Air temperature was 22°C, surface 30°C. All three species of *Teratoscincus* were collected together within a 25-meter radius, *T. scincus* being the most abundant, *T. microlepis* the least numerous. Other species collected at this locality included *Trapelus agilis, Bunopus tuberculatus, Cyrtopodion longipes, Eremias acutirostris, E. lineolata, Ophiomorus tridactylus,* and *Eryx t. tataricus. Teratoscincus bedriagai* was also taken along with *T. scincus* in an area of alluvial hills strewn with gravel and where such sand as was present was compacted and mixed with gravel. The vegetation in this locality was scanty, mostly herbaceous. (S. Anderson, 1993:8–9). Szczerbak and Golubev (1996:45) quote from Zarudny (1904): ". . . found on gravel soil occasionally sprinkled with small road metal and covered in places with a thin layer of sand. Several individuals were observed . . . on highly saline, loose soil covered with a thin salt crust. Very common in the Sistan Depression, where it occurs in the untilled

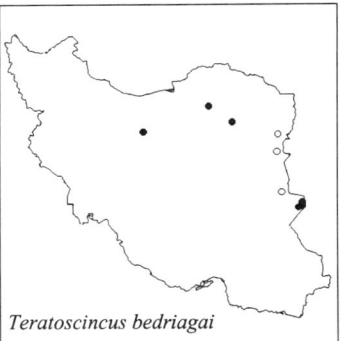
Teratoscincus bedriagai

agricultural areas of this region, as well as, and especially outside of it, further south in the desert, overgrown with tamarisk bushes. Here it stayed on clay and loamy soils, covered in places with small sand mounds."

Distribution: The northern and eastern desert basins of the Central Plateau of Iran, Sistan, and the desert regions of southern Afghanistan as far east as Kandahar.

Remarks: This form is most closely allied to *Teratoscincus przewalskii* Strauch, differing, according to published descriptions, in its large head scales and smaller ventral scales proportionate to the dorsals, and in the color pattern.

Iranian specimens are rare in collections, apparently; the species had not been collected in Iran since Zarudny collected the types (April through October 1898) until the 1962–63 Street Expedition collected a single specimen near Zabol and Ranck and Herman of the mammal survey of the National Museum of Natural History (Washington) collected two specimens in 1962 (Tuck, 1971a:54). Robert Tuck and I collected 10 in 1975.

Teratocscincus microlepis Nikolsky, 1899 (Plate 9G)
Baluch plate-tailed gecko, small-scaled skink gecko

Teratoscincus microlepis Nikolsky, 1899:145–146 (Type locality: Duz-Ab, eastern Iran; Holotype: ZIL 9164); 1899c:376–378. — Zarudny, 1903:3. — F. Werner, 1936:200. — Guibé, 1957:136. — S. Anderson, 1963:474. — Wermuth, 1965:182. — S. Anderson, 1968:333; 1974:32, 43. — Schleich, 1977:127, 129. — Welch, 1983:13. — Szczerbak and Golubev, 1986:43–45 [1996:44–45], figs. 14, 16. — Welch, *et al.*, 1990:32. — Kluge, 1991:35. — S. Anderson, 1993:8–9. — Kluge, 1993:37.

Diagnosis: Cycloid scales on back feebly imbricate, not extending beyond shoulders; about 100 scales around middle of body.

Color pattern: Light tan above, with narrow dark brown or black oblique or V-shaped bars, becoming indistinct or disappearing with age; a dark U-shaped mark on occiput; light spots on flanks; tail with 5–8 dark transverse bars.

Size: The largest female examined measures 77 mm snout-vent, tail incomplete; largest male 73 mm snout-vent, tail 50 mm.

Habitat: I collected this species only in the habitat described above under *T. bedriagai*. Minton (1966:76) found it in sandy areas of Pakistan together with, but less numerous than *T. scincus*, and Guibé (1957:136) stated that it was a burrower in dune sand. Zarudny found it on highly saline-saturated soils covered with a salt crust.

Distribution: Recorded in Iran only from Baluchistan and Sistan; adjacent southern Afghanistan and Baluchistan, Pakistan, along Afghan border.

Remarks: In CAS 141096, the large cycloid imbricate plates are present only on the distal third of the tail, the proximal two-thirds covered dorsally by transverse rows of much smaller imbricate scales. In 41 specimens from Pakistan (RSM specimens) the large plates are present on the distal one-third to one-half of the tail. A female collected in southern Afghanistan in early April has large ovarian eggs.

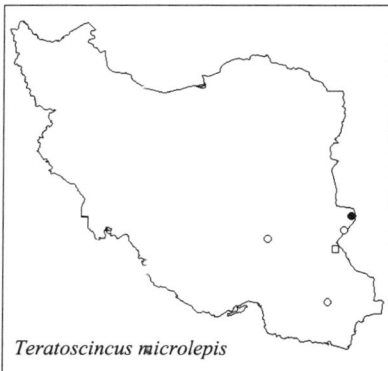
Teratoscincus microlepis

Teratoscincus scincus (Schlegel, 1858)
Turkestan plate-tailed gecko, common skink gecko

Stenodactylus scincus Schlegel, 1858:16 (Type locality: Ili River, eastern Turkestan [= southern Xinjiang Uygur Autonomous Region], China [*fide* Zhao and Adler, 1993:186]; Holotype: RMNH).

Teratoscincus scincus keyserlingii Strauch, 1863 (Plate 9H)
Keyserling's plate-tailed gecko, Keyserling's skink gecko

Teratoscincus keyserlingii Strauch, 1863:480 (Type locality: Seri-Tschah, eastern Iran [probably Sar-i-Chah, Khorasan, *fide* Blanford, 1876:355]; Syntypes: ZIL 2395–6, ZMB 6872 [ZIL 2396 designated as lectotype by Szczerbak and Golubev, 1986:38]). — Blanford, 1876:354–355. — Bedriaga, 1879:36. — Strauch, 1887:68–71. — Kulagin, 1888:15.

Teratoscincus scincus: Boulenger, 1885a:12–13, pl. 2, fig. 3. — Nikolsky, 1900:176. — Zarudny, 1903:3. — Annandale, 1906:197. — Nikolsky, 1915:51–54, text-figs. 12–13. — F. Werner, 1936:200. — Forcart, 1950:144. — S. Anderson, 1963:474. — Wermuth, 1965:182. — Anderson, 1968:333; 1974:32, 43. — Tuck, 1979:102. — Leviton, *et al.*, 1992:46, col. pl. 7F–G.

Teratoscincus zarudnyi Nikolsky, 1896:370; 1897:309–312, pl. 18, fig. 1 (Type locality: Rum, eastern Iran; Holotype: ZIL 8804 [destroyed, *fide* I. S.Darevsky, pers. commun.]); 1900:176; 1915:54–56. — Werner, 1936:200.

Teratoscincus scincus keyserlingii: Mertens, 1956:92. — Schleich, 1977:127, 129. — Bannikov, *et al.*, 1977:85. — Welch, 1983:13. — Szczerbak and Golubev, 1986:38, figs. 10, 12. — Welch, *et al.*, 1990:33. — Kluge, 1991:35; 1993:37.

Teratoscincus scincus scincus: Guibé, 1957:136. — Schleich, 1977:127, 129. — Welch, 1983:13.

Diagnosis: Cycloid scales of back strongly imbricate, extending onto posterior part of head; 28–34 scales around middle of body. Adults with longitudinal dark stripes, rather than transverse bars (Fig. 80).

FIGURE 80. *Teratoscincus scincus keyserlingii*. (From Boulenger, 1889, pl. 8, fig. 1.)

Color pattern: Dorsal ground color mottled with light gray, yellow, orange, various shades of brown (subject to metachrosis); two wide longitudinal dark brown stripes down body, but not onto tail, which lacks distinct markings; sides and belly pinkish to white. Young dark yellow to light orange, with 4–5 sooty to black transverse bars on body, similar numbers on tail (Minton, 1966:76). Yellows and oranges are lost in preserved specimens, which are dirty white with dark brown markings as described. In specimens I have examined there are four to six narrow longitudinal dark stripes, sometimes interrupted, resulting from the breakup of the transverse bars of the young, rather than two broad stripes.

Size: Largest male examined measures 116 mm snout-vent, tail 76 mm; a large female 98 mm snout-vent, tail 65 (male from Nushki, Pakistan, female from southern Afghanistan).

Natural history: My field observations are in accord with those of Minton (1966:76). In areas where they occur, these geckos are fairly abundant and easily collected at night. When a headlamp is used, the reflected bright orange eye-shine is readily seen for about 40 meters. Females collected in southern Afghanistan in mid-March, early April, and mid-November have eggs up to 4 mm in the ovaries, no oviducal eggs. A small juvenile (50 mm snout-vent) was collected near Iranshahr in late November by the 1962 Street Expedition. Digestive tracts contain mostly beetles; termites were found in specimens from southern Afghanistan (S. Anderson and Leviton, 1969:47), and Minton found cricket and small vertebrate remains in Pakistan specimens. Cherlin, *et al.* (1983) reported on the thermobiology of *Teratoscincus s. scincus* in the Kara Kum.

Habitat: (Plates 20G, 21B, 24A) In addition to the habitats described above under *Teratoscincus bedriagai*, I found this species on loose sand as well as on sandy surfaces of varying degrees of compaction, on salt-encrusted sand, and sand mixed with gravel. Vegetation varied from *Tamarix-Acacia* "woodland," sand-hummocks with rodent burrows at the base of *Tamarix* and bunch grasses, to scanty, low, psammophilous shrubs of various species. Minton (1966:76) found that they lived in burrows 25–40 cm deep and blocked the entrances with sand. My observations of *T. s. scincus* in white saxaul (*Haloxylon*)vegetation in the Kara Kum, Turkmenistan, are generally in accord with the above comments.

Teratoscincus scincus keyserlingii

Distribution: Szczerbak and Golubev (1986: 38), the most recent reviewers of this genus, assign all of the Iranian specimens to *Teratoscincus scincus keyserlingii*. In Iran this gecko occurs on the Central Plateau, west as far as the steppe region near Argavani south of Tehran, north to Chahar Deh in Mazanderan, south to the vicinity of Bandar 'Abbas, east through Sistan. The typical subspecies should be looked for in suitable habitats near the border with Turkmenistan. *Teratoscincus scincus* has long been known from Central Asia, Afghanistan, and Pakistan, but only recently has been found on the Arabian Peninsula in the eastern United Arab Emirates (Arnold, 1977:83). The typical subspecies occurs in the southern republics of Central Asia and in Afghanistan north of the Hindu Kush, while *T. s. keyserlingii* occupies eastern Iran, southern Afghanistan and western Pakistan. *Teratoscincus scincus rustamovi* is the subspecies of the Fergana Valley (Szczerbak and Golubev, 1986: 35–42).

Remarks: I concur with Szczerbak and Golubev (1986:38; 1996:37) that *Teratoscincus zarudnyi* is a synonym of *T. s. keyserlingii*. The former does not appear recognizable on the basis of those scale characters in which it is said to differ from the latter, *viz.*, number of scales across head between eyes, rows of enlarged scales on occiput, and scale rows at midbody. I have examined 35 specimens from Baluchistan, Pakistan, southern Afghanistan, and central and southeastern Iran; there are 28–37 scale rows at midbody, 32–53 scales across head between centers of eyes (not counting scales of "eyelid"), and 4–15 enlarged scales across occiput in these specimens. All of these geckos have longitudinal stripes, even the smallest juveniles (snout-vent 51 mm) having the beginnings of paravertebral stripes linking the prominent transverse bars. There are 4–6 transverse bars or paravertebral lines broken into four dashes on the body and one on the neck in those specimens in which they are still visible (faintly visible in most large adults). This is in contrast to two specimens (one small juvenile 47 mm snout-vent, one female 80 mm snout-vent) from 20 km S Andkhoy, northern Afghanistan, in which there is no trace of longitudinal stripes; these specimens have 6–7 crossbars on body and 2 on neck, and these are more undulating in shape than those of the southern specimens. In these two specimens there are 29–32 scales around the body, 39–41 scales across head, 8–9 enlarged scales on occiput. I am unable to find consistent differences between these and the southern geckos in other characters. Mertens (1956:92) initially recognized *keyserlingii* as a subspecies of *T. scincus*, but his series from Pakistan (Mertens, 1969:30) convinced him that the two nominal forms were not separable. Szczerbak and Golubev (1986:37) distinguish *T. s. keyserlingii* from *T. s. scincus* on the basis of color pattern (see diagnosis above).

I have examined 55 specimens from Turkmenistan collected in 1992 by Theodore Papenfuss and Robert Macey. These are patterned like those from northern Afghanistan, with a basic pattern of seven dark crossbars breaking up on the vertebral line in adults. These fit the diagnosis for *T. s. scincus* provided by Szczerbak and Golubev (1986:37; 1996:37). The smallest juvenile (45 mm snout-vent length) was collected in early May. After observing *T. s. scincus* in white saxaul biotope in the Kara Kum of Turkmenistan (August-September, 1995), I am fully persuaded that the two taxa are distinct (Plate 10B–C).

The skin of *T. scincus* is notoriously delicate, and large areas of skin, including the dermal layer, are often lost during capture. Many museum specimens are so badly damaged in this regard that scale counts cannot be made. The large imbricate scales of the body are unusually thick, while the interconnecting skin is very thin and easily torn, so that even individual scales can be torn loose. The easily torn skin in these and other gecko species is apparently an adaptation facilitating escape from the grasp of predators. Perhaps the loosely attached large imbricate scales make possible the loss of a minimum amount of skin, while the thickness of the scales themselves provides a relatively impermeable integumentary covering in the arid environment inhabited by these geckos. This species appears to be more specialized in this regard than either *T. bedriagai* or *T. microlepis*. Bauer and coworkers (Bauer, *et al.*, 1989; Bauer and Russell, 1992; Bauer, *et al.*, 1993) have commented extensively on the adaptive significance and the physical basis of integumentary loss in geckos, and specifically in *T. scincus* (Bauer, *et al.*, 1993).

The buzzing sound that these geckos and the other species of *Teratoscincus* make (by moving the tail so that the large imbricate plates rub together) is very faint, and I could hear it only by holding the animal close to my ear. Aaron Bauer (pers. commun.) made a similar observation in specimens of *T. s. scincus* that he collected in Kazakhstan.

Genus *Tropiocolotes* Peters, 1880
Dwarf geckos

Tropiocolotes W. Peters, 1880:306, pl. unnumbered, fig. 1 (Type species: *Tropiocolotes tripolitanus* W. Peters, 1880, by monotypy).

Definition: A gekkonine genus (in the sense of Kluge, 1967) rarely exceeding 35 mm snout-vent length, digits slightly angularly bent, not dilated, not fringed, not webbed, nor ornamented, covered below with a single series of transverse lamellae; pupil vertical; dorsal scales uniform, small, homogeneous, imbricate to subimbricate; postanal sacs present; preanal and femoral pores usually absent (two preanal pores in male *T. steudneri* according to J. Anderson, 1898:49, and in *T. depressus*).

Distribution: From western Sahara across North Africa to Israel, Sinai, the Arabian Peninsula, Iran, eastern Afghanistan, Pakistan. Four species in Iran.

Remarks: See the following for discussion of the systematic problems associated with this genus Guibé (1966b), Minton, *et al.* (1970), Kluge (1983), Golubev (1984), Szczerbak and Golubev (1986; 1996). Kluge (1983) split the genus *Tropiocolotes* as recognized by Minton, *et al.* (1970) and by S. Anderson and Leviton (1972), resurrecting the genus *Microgecko* for *helenae, latifi,* and *persicus*, which he found to have fused nasal bones and a second ceratobranchial. He did not consider the species *levitoni* from eastern Afghanistan. Golubev (1984:12) and Szczerbak and Golubev (1986:101) divide *Tropiocolotes* into three subgenera on the basis of minor differences in pholidosis. They place *tripolitanus, steudneri,* and *scorteccii* in the typical subgenus, *helenae, persicus,* and *latifi* in the subgenus *Microgecko*, and *depressus* and *levitoni* in the subgenus *Asiocolotes* Golubev, 1984. Kluge (1991, 1993) follows this arrangement, recognizing the three taxa, *Tropiocolotes, Microgecko,* and *Asiocolotes* at the generic level. For the present work, I retain *Tropiocolotes* for all of these species, without making a judgment here as to the monophyly of this assemblage.

Key to the Species of *Tropiocolotes* in Iran

1a. Dorsal scales weakly keeled; subdigital lamellae distinctly tricarinate ....
.... *Tropiocolotes steudneri* (Peters, 1869) (p. 196)
1b. Dorsal scales smooth; subdigital lamellae smooth, not distinctly tricarinate 2
2a. No postmentals (chin shields) ....
.... *Tropiocolotes latifi* (Leviton and Anderson, 1972) (p. 193)
2b. Postmentals present 3
3a. A single pair of postmentals 4
3b. Two pairs of postmentals 5
4a. 65–84 dorsal scales between axilla and groin; 0–6 indistinct dark dorsal crossbars with white posterior margins, sometimes two dorsolateral series of spots ....
.... *Tropiocolotes helenae helenae* Nikolsky, 1907 (p. 191)
4b. 80–92 dorsal scales between axilla and groin, 5 distinct crossbars with white posterior margins .... *Tropiocolotes helenae fasciatus* (Schmidtler and Schmidtler, 1972) (p. 193)
5a. Dark dorsal crossbars of body and tail broader than interspaces ....
..... *Tropiocolotes persicus bakhtiari* (Minton, Anderson, and Anderson, 1970) (p. 195)
5b. Dark dorsal crossbars less than one-half width of interspaces ....
.... *Tropiocolotes persicus persicus* (Nikolsky, 1903) (p. 195)

Tropiocolotes helenae (Nikolsky, 1907)
Banded dwarf gecko

Microgecko helenae Nikolsky, 1907a:265–268, pl. 1, figs. 4, 4a (Type locality: Alchorschir; Aguljaschker; Isfagan; Bidezar, Khuzestan Province, Iran [restricted to Bidezar (= Bid Zard) by J. J. Schmidtler and J. F. Schmidtler 1972:60; Alchorschir by designation of lectotype by Szczerbak and Golubev, 1986:111, 113]; Syntypes: ZIL, all destroyed; Lectotype: ZIL 10242 [designated by Szczerbak and Golubev, 1986:111]).

Tropiocolotes helenae helenae (Nikolsky, 1907) (Plate 10D–E)
Helen's banded dwarf gecko, Khuzestan dwarf gecko

Microgecko helenae Nikolsky, 1907a:265–268, pl. 1, figs. 4, 4a. — F. Werner, 1936:200. — Tuck, 1971b:477–482, figs. 1–4. — Kluge, 1983:472.
Tropiocolotes persicus helenae: S. Anderson, 1968:333 (in part).
Tropiocolotes helenae: Minton, *et al.*, 1970:345–349, figs. 6A–B, 8, 9. — Tuck, 1971a:59. — Frynta, *et al.*, 1997:9.

Tropiocolotes helenae helenae: J. J. Schmidtler and J. F. Schmidtler, 1972:60–62, fig. 1 (left). — S. Anderson, 1974:32, 43. — Schleich, 1977:127, 129. — Welch, 1983:14.
Tropiocolotes (*Microgecko*) *helenae*: Szczerbak and Golubev, 1986:111–115 [1996:109–111], figs. 51–52.
Microgecko helenae helenae: Kluge, 1991:21; 1993:22.

Diagnosis: Single pair of postmental shields, not in contact with one another (Fig. 81b); 65–84 dorsal scales between axilla and groin.

Color pattern: Back tan, gray, or brown, with or without indistinct, undulating dark transverse bars, narrower than interspaces and bordered posteriorly with white, back sometimes uniform, sometimes with two dorsolateral series of white spots; a brown line from rostral through eye to ear or continuing to shoulder; venter white; limbs uniform grayish tan, tail with 8–12 dark transverse bars narrower than interspaces and bordered posteriorly with white; regenerated portion of tail uniform black (Fig. 82).

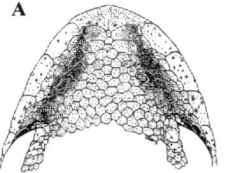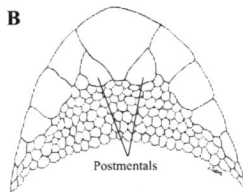

FIGURE 81. *Tropiocolotes helenae*: (A) dorsal surface of snout; (B) ventral view of head showing chin shields. (Drawings by Lynnette Sabre.)

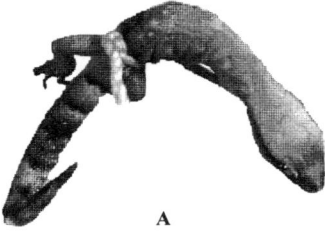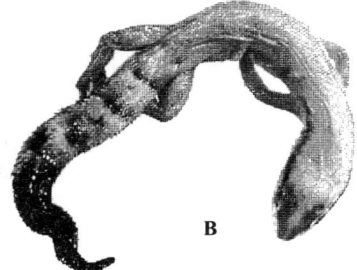

FIGURE 82. *Tropiocolotes helenae*. (A) USNM 153693, with original tail; (B) USNM 153699, with regenerated tail. (Photos by A. Leviton, in Minton, et al., 1970.)

Size: Snout-vent length 25 mm (literature record, sex unknown).

Habitat: Tuck (1971b:481) described the habitat in which his series was collected. They were taken under small stones along a dry stream bed. The terrain was rolling foothills with sparse vegetation of grasses, thorny shrubs, euphorbs, and scattered oaks. His series of 20 was collected in February. Nikolsky's specimens were collected in December and March. The Schmidtlers collected their specimens at Mehkuh during rainy weather in April on a bare hill, and their Mian Kotal specimens during clear April weather under small flat stones in shaded areas where there were pistachio and oak trees.

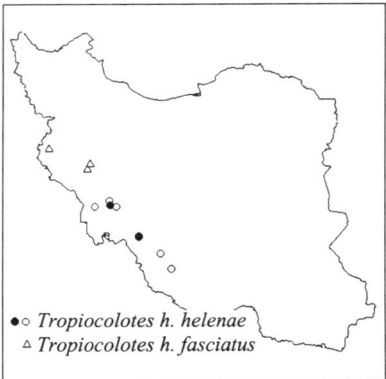

● ○ *Tropiocolotes h. helenae*
△ *Tropiocolotes h. fasciatus*

Distribution: All known specimens appear to have been collected in the foothills of the Zagros Mountains of southwestern Iran. I have not been able to locate all of the localities cited by Nikolsky, but they are given as "Arabistan" and "Kuchistan," both within the modern province of Khuzestan. "Isfagan" may refer to Esfahan, but is cited as being in western "Kuchistan." All references to this taxon other than those in the above synonymy appear to refer to other species (see below).

Tropiocolotes helenae fasciatus (Schmidtler and Schmidtler, 1972) (Plate 10F)
Schmidtlers' dwarf gecko

Tropiocolotes helenae fasciatus J. J. Schmidtler and J. F. Schmidtler, 1972:62–65, fig. 1 right (Type locality: Sorkh-e Dize, 125 km W Kermanshah, on road to Baghdad, Kermanshahan Province, Iran; Holotype: ZSM 500/68). — S. Anderson, 1974:32. — Schleich, 1977:127, 129. — Welch, 1983:14.
Microgecko helenae fasciatus: Kluge, 1991:21; 1993:22.
Tropiocolotes (*Microgecko*) *helenae fasciatus*: Leviton, et al., 1992:47, col. pl. 7E.

Diagnosis: A single pair of postmental shields, in contact with one another at a point; 80–92 dorsal scales between axilla and groin.

Color pattern: Dorsum brownish gray in life, with five small dark crossbars about two to three scale rows wide, fifth at level of vent, these followed by five equally narrow whitish crossbars; unregenerated tail with six black, yellow-margined crossbars; indistinct dark gray temporal band, not passing through ear opening; limbs uniform light brownish gray; venter whitish.

Size: Snout-vent 27 mm, tail 23 mm (literature record, sex unknown).

Habitat: The type locality lies in a valley, about 1500 m elevation, on the western slope of the Zagros Mountains. The climate is cooler and wetter (about 700 mm precipitation) than that in which *Tropiocolotes h. helenae* is found. Open oak forest characterizes this locality, with willows and poplars along the stream bank. The three paratypes were collected in a similar habitat nearly 200 km to the south; all were collected under flat rocks in late April and late May. (J. J. Schmidtler and J. F. Schmidtler, 1972:64–65).

Distribution: Known from only two localities on the western slopes of the Zagros Mountains in Kermanshahan and Lorestan Provinces, Iran; 1000–1500 meters elevation.

Remarks: Szczerbak and Golubev (1986:113; 1996:109–110) decline to recognize *Tropiocolotes h. fasciatus* as a distinct taxon, believing that only clinal variation in color pattern and scalation is involved. I reserve judgment at this point.

Tropiocolotes latifi Leviton and Anderson, 1972 (Plate 10G)
Latifi's dwarf gecko

Tropiocolotes latifi Leviton and S. Anderson, 1972:1–7, figs. 1–4 (Type locality: Kerman, Kerman Province, Iran; Holotype: CAS 134365). — S. Anderson, 1974:32, 43. — Schleich, 1977:127, 129. — Welch, 1983:14. — Moravec and Černy, 1994:88. — Kami and Vakilpoure, 1996:153. — Frynta, et al., 1997:9.
Tropiocolotes (*Microgecko*) *latifi*: Szczerbak and Golubev, 1986:119–120 [1996:116–117], fig. 53.
Microgecko latifi: Kluge, 1983:472; 1991:21; 1993:22.

Diagnosis: Internasals enlarged, in contact, followed by an additional pair of enlarged shields twice as large as succeeding scales (Fig. 83A); no postmentals (Fig. 83B), mental rounded behind, bordered by 8 granules; about 75 dorsal scales from axilla to groin; nostril surrounded by four scales.

Color pattern: (Fig. 84) Dorsum sandy (in preservative), each dorsal scale with fine reticulum of black when viewed through microscope; a middorsal dark spot between forelimbs, two dorsolateral spots just anterior to hind limbs; eight dark crossbars on tail, narrower than interspaces; granules encircling eye dark; dark line extending from posterior

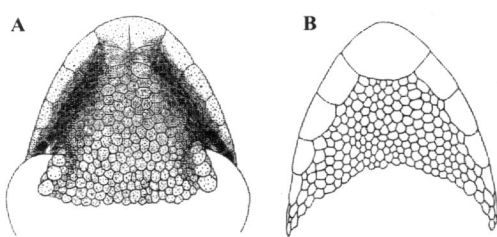

FIGURE 83. *Tropiocolotes latifi*. (A) Dorsal view of snout showing enlarged internasals followed by a pair of enlarged shields; (B) ventral view of chin showing absence of enlarged postmentals. (Drawings by Lynnette Sabre.)

margin of eye across temporal region and above axilla for short distance onto side of body; no dark line on snout, which is slightly darker than upper surface of head; venter cream.

Size: Holotype: snout-vent 17 mm, tail 16 mm. The holotype specimen is a juvenile, probably a female. Moravec and Černy's four specimens had snout-vent lengths of 21.5 to about 26 mm.

Habitat: The specimens collected by Černy were found under stones, one in a "bushed wadi," one on the shore of a small stream near a village, and two on a mountain ridge (Moravec and Černy, 1994:88).

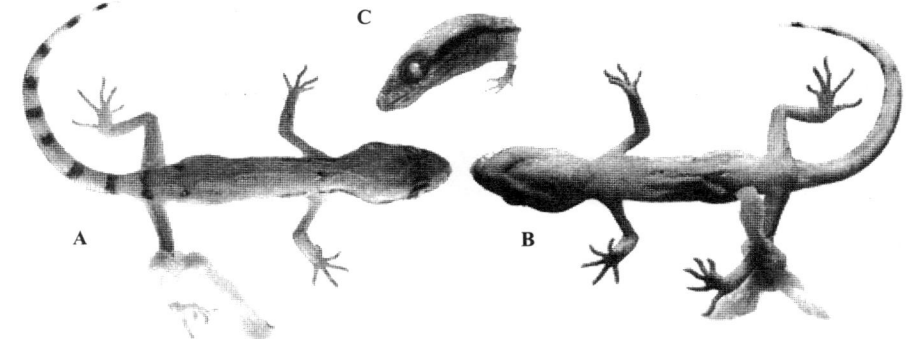

FIGURE 84. *Tropiocolotes latifi* (Holotype, CAS 132365). (A) Dorsal view of body; (B) ventral view of body; (C) lateral view of head. (Photos by A. Leviton.)

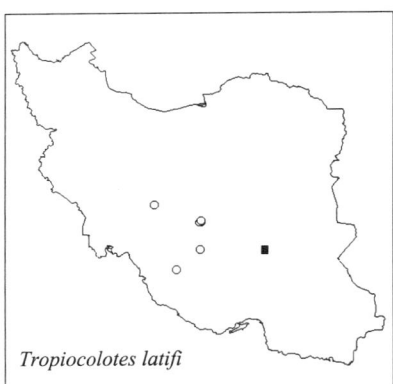

Tropiocolotes latifi

Distribution: Known from Kerman, the type locality, on the Central Plateau of Iran. The specimens identified by Moravec and Černy (1994) as this species come from a site about 340 km NW of Kerman, while that of Frynta, et al. (1997:9) are from Fars Province to the west. Thus, it may prove to be broadly distributed on the Central Plateau.

Remarks: The specimens identified as this species by Moravec and Černy (Plate 10G) differ in color pattern from the holotype: "(NMP6V 34799): A dark lateral stripe extends from the nostril across the eye and temporal region above the exila [sic]. Five dark crossbars on dorsum, the first one lies between forelimbs and joins lateral lines. A trace of the sixth additional crossbar between third and fourth ones. A dark crossbar also on the rest of the broken tail. Living specimens: The same pattern of the dark crossbars on a yellowish background and the lateral dark stripe extending from the nostril backwards above the axilla .... The specimen from Mazraeh has two additional uncompleted crossbars on the dorsum. Tail when original with 8 distinct crossbars..., when regenerated uniformly yellow." The analysis of these authors was based upon comparison with the original description, rather than comparison with the type specimen.

Tropiocolotes persicus (Nikolsky, 1903)
Persian dwarf gecko

Alsophylax persicus Nikolsky, 1903:95 (Type locality: Degak [Dehak] in the region of Dizak, Iran; Holotype: ZIL 10005).

Tropiocolotes persicus persicus (Nikolsky, 1903)
Persian dwarf gecko

Alsophylax persicus Nikolsky, 1903:95. — F. Werner, 1936:200.
?*Tropiocolotes helenae*: Mertens, 1956:92–93.
Bunopus persicus: S. Anderson, 1963:474.
Microgecko helenae: Guibé, 1966a:98 (not *Microgecko helenae* Nikolsky, 1907a).
Tropiocolotes persicus persicus: S. Anderson, 1968:333. — Minton, *et al*. 1970:348–350, fig. 3d. — S. Anderson, 1974:32, 43. — Schleich, 1977:127, 129. — Welch, 1983:14. — Welch, *et al.*, 1990:33. — Frynta, *et al.*, 1997:9.
Microgecko persicus: Kluge, l983:472.
Tropiocolotes (*Microgecko*) *persicus persicus*: Szczerbak and Golubev, 1986:115–117 [1996:114–115], figs. 53–54.
Microgecko persicus persicus: Kluge, 1991:21; 1993:22.
Tropiocolotes persicus: Kluge, 1991:35 (*lapsus*?).

Diagnosis: Two pairs of postmentals; dark transverse bars of body narrow, less than half width of interspaces, and with distinct light margins; tail bars narrower than interspaces.

Color pattern: Dorsal ground color lemon-yellow in life; five dark transverse bars on body, first on neck, fifth just anterior to hind limbs; nine dark bars on tail, extreme tip of tail black; dark stripe from tip of snout through eye and along side of head and neck to level of second dorsal bar; limbs without dark markings; venter white.

Size: Snout-vent 30.7 mm, tail 32.6 mm (literature record, sex unknown).

Distribution: Only three specimens are known with certainty, the type, the specimen recorded by Guibé (1966a:98) as *Microgecko helenae* from 100 km N Iranshahr, and the specimen from Mirjawa, Pakistan, at the border of Iran, described by Minton, *et al*. (1970:348–350). An additional specimen from the same region of Iranian Baluchistan may belong to this taxon: Mertens's (1956:92–93) specimen from 20 km SW Pip. The specimen listed from Chahak by Frynta, *et al.* (1997:9) would extend the range of this subspecies considerably, across the ecological barrier posed by the Zagros Mountains. This serves to illustrate that the relationships within this genus remain to be adequately investigated.

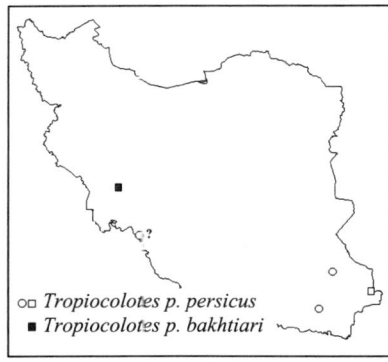

○ *Tropiocolotes p. persicus*
■ *Tropiocolotes p. bakhtiari*

Tropiocolotes persicus bakhtiari Minton, Anderson, and Anderson, 1970 (Plate 10H)
Bakhtiari dwarf gecko

Tropiocolotes helenae: S. Anderson, 1961:287–289 (not *Microgecko helenae* Nikolsky, 1907a).
Microgecko helenae: S. Anderson, 1963:440–441, fig. 10 (not Nikolsky, 1907a).
Tropiocolotes persicus helenae: S. Anderson, 1968:333 (in part; not Nikolsky, 1907a).
Tropiocolotes persicus bakhtiari Minton, S. Anderson, and J. A. Anderson, 1970:351–353, figs. 4A–B, 10 (Type locality: between Masjed Soleyman and Sar-i-Gach, Khuzestan Province, Iran; Holotype: CAS 86408). — S. Anderson, 1974:32, 43. — Schleich, 1977:127, 129. — Welch, 1983:14.

Alsophylax persicus: Tuck, 1971b:481 (not Nikolsky, 1907a).
Microgecko persicus: Kluge, 1983:472.
Tropiocolotes (*Microgecko*) *persicus bakhtiari*: Szczerbak and Golubev, 1986:117–118 [1996:115], figs. 53–54.
Microgecko persicus bakhtiari: Kluge, 1991:21; 1993:22.

Diagnosis: Two pairs of postmentals; dark transverse bars of body and tail broader than interspaces; anterior pair of postmentals in contact (Fig. 85B).

Color pattern: Dorsal surfaces cream in life, with four chocolate-brown transverse bars between shoulders and pelvis, darkest on posterior margin of each bar; dark bar on nape, posterior corners of which extend back and down to meet next crossbar, anterior corners extending forward, meeting dark marking which lies just posterior to eye and over ear opening; 11 dark crossbars on tail; limbs, snout, and labials lightly dusted with brown; venter immaculate creamy white.

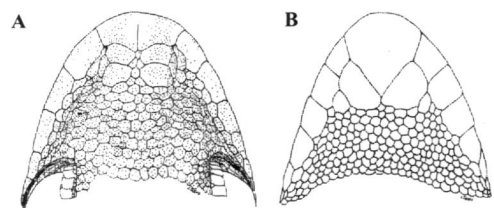

FIGURE 85. *Tropiocolotes persicus bakhtiari* (Holotype, CAS 86408). (A) Dorsal view of snout; (B) ventral view of chin shields. (Drawings by Lynnette Sabre.)

Size: Snout-vent 28 mm, tail 28.5 mm (literature record, sex unknown).

Natural history: The single known specimen was collected on a surfaced road at 2000 h in mid-May. The road surface was 37.6°C, the air 34.2°C. The gecko was very agile, able to jump several centimeters.

Distribution: Known only from the type locality in the lower foothills (600–700 m elevation) of the western Zagros Mountains.

Tropiocolotes cf. *steudneri* (Peters, 1869) (Plate 10-I)
Steudner's dwarf gecko

Gymnodactylus steudneri W. Peters, 1869:788 (Type locality: Sennar, Sudan; Holotype: ZMB 5476 [not located by Bauer and Günther, 1991:302]).
Tropiocolotes steudneri: Boulenger, 1891:98. — Guibé:1966a:98; 1966b:337–338, figs. 3a, 4, 5a–d. — Schleich, 1977:127, 129. — Welch, 1983:14. — Kluge, 1991:35; 1993:37.
Tropiocolotes (*Tropiocolotes*) *steudneri*: Szczerbak and Golubev, 1986:106–108 [1996:103–106], figs. 48–49.

Diagnosis: Dorsal scales homogeneous, imbricate, weakly keeled (under magnification) 51 around middle of body; subdigital lamellae distinctly tricarinate; first pair of chin shields (postmentals) in broad contact, second pair small, separated by three scales; hind limb reaches nearly to axilla. (Diagnosis based on Iranian specimens only).

Color pattern: Dorsum light sandy beige with 6 or 7 light brown crossbars sometimes broken and irregular, first on nape, last across sacrum; a dark mark from snout through eye to first crossbar; a dark bar across snout in front of eyes; tail with 7–9 dark crossbars narrower than interspaces; limbs with indistinct dark bars; venter white.

Size: Snout-vent 26 mm; tail 27 mm (literature record, sex unknown).

Habitat: (Plate 20D) My specimen (MMTT 1048) was found in the morning under a rock in a wash running through gravelly alluvium, vegetation consisting of low shrubs about 35 cm high and sparse dry grass; a few small trees grew in low drainage courses. Other species found at this locality were *Bufo surdus*, *Pristurus rupestris*, *Mesalina watsonana*, and *Echis carinatus sochurecki*.

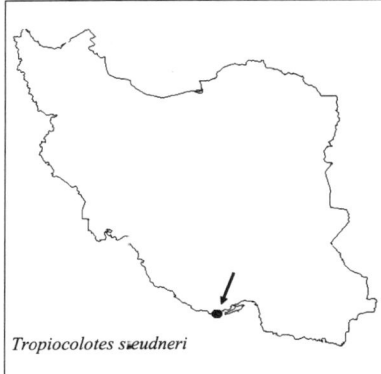
Tropiocolotes steudneri

Distribution: Known in Iran only from the vicinity of Bandar-e Lengeh on the Persian Gulf coast. 0–50 m elevation.

Remarks: Only two specimens have been collected in Iran, both from about the same locality. I follow Guibé (1966a) in assigning these Iranian geckos to *Tropiocolotes steudneri*, but I suspect that this population represents an undescribed taxon, especially since *T. steudneri* has not been found in the vast region between Israel and Bandar-e Lengeh. MNHNP 1966.18 may be an immature male; MMTT 1048, a mature female, was not available for direct comparison with material of *T. steudneri* at time of writing.

The following description is of Guibé's (1966a) specimen, MNHNP 1966.18: Rostral twice as broad as deep, three-fourths divided; nostril between rostral, first labial, and two small nasals; two supranasals in contact behind rostral, about twice as large as surrounding scales of snout; scales of top of head and snout uniform, subimbricate, rugose; 12 scales across head between "eyelids"; 9/10 upper labials (8/7 to center of eye); 7 lower labials. Mental pentagonal, pointed behind. Anterior pair of postmentals in broad contact behind mental, in contact with first lower labials only; second pair of postmentals small, separated by three scales; scales of occipital region granular, polygonal, juxtaposed, rugose.

Ear opening vertically oval, about one-fourth the diameter of eye; snout length greater than distance from ear to posterior border of eye; diameter of eye slightly less than distance to ear.

Dorsal scales homogeneous, imbricate, weakly obtusely keeled; ventrals similar to dorsals; 51 scales around midbody; 45 ventrals from level of axilla to level of groin; 55 dorsals from level of shoulders to level of vent. Tail covered above and below with uniform strongly keeled imbricate scales.

Limbs covered with scales similar to dorsals; digits angularly bent, compressed distal to angle, covered below with strongly tricarinate lamellae; third toe longest; lamellae on right foot 8–16–20–18 (skin stripped from 5th toe, 18 on 5th toe of left foot); fingers less strongly bent than toes, lamellae similar.

Forelimb reaches in front of eye, does not reach groin; hind limb just fails to reach axilla.

Snout-vent 25.5 mm, tail 26.5 mm; head length (tip of snout to angle of jaw) 8 mm; head width 3.5 mm; forelimb 11 mm, hind limb 11.5 mm.

MMTT 1048, mature female, is in overall agreement with the above description.

The Iranian specimens differ from *T. steudneri* from Israel, Egypt, and Algeria, and from *T. nattereri, T. depressus*, and *T. levitoni* in having the dorsal and ventral body scales keeled, although not as strongly so as those of *T. tripolitanus*; all scales of tail strongly keeled. However, Guibé (1966b:342) states regarding the dorsal pholidosis of *T. steudneri*: "... chez les femelles est toujours constituée par des écailles plane, subarrondies, imbriquées, par contre chez les mâles les écailles sont un peu plus allongées, de form plus losangiqueleur partie médiane est parfois un peu saillante, simulant une car ne difficile à distinguer." The hind limbs are shorter than those of *T. nattereri*; dorsal scales are larger than those of *T. depressus*, and Iranian specimens differ from all of these in having distinctly tricarinate subdigital lamellae; *T. scorteccii, T. latifi*, and *T. depressus* lack enlarged postmentals in contact behind the mental; *T. levitoni* has smooth granular dorsal scales, whereas the present specimens have keeled imbricate scales and differ from *T. tripolitanus tripolitanus, T. t. somalicus, T. t. algericus,* and *T. t. occidentalis* in having smaller dorsal scales.

This population is closely related to *T. steudneri, T. nattereri, T. tripolitanus,* and *T. scortecci* and not closely related to any of the other species which occur in Iran. There is, however, a large hiatus in the known distribution of this group. *T. scortecci* is known only from the Hadramaut and Dhofar of southern Arabia, *T. steudneri* distributed from Israel west through North Africa to Algerian Sahara, *T. nattereri* (considered synonymous with *T. steudneri* by Szczerbak and Golubev, 1986:106; 1996:103) in the Sinai Peninsula, possibly west into Libya, and *T. tripolitanus* from Egypt west across the whole of North Africa to Spanish Sahara.

At the time I collected MMTT 1048 a second specimen was seen at the same locality, but it escaped among rock rubble. Both geckos had a cluster of bright red mites in each axillary region. MMTT 1048 laid a single egg the day following capture.

Family LACERTIDAE

Lacertids, typical lizards

A recent revision of this family based on morphology is that of Arnold (1989), building upon his earlier morphological studies (Arnold, 1973, 1982, 1984, 1986c). He presented cladograms (Arnold, 1989:237–239, figs. 23–25) summarizing his analysis of relationships within the family and offered a biogeographic scenario for the evolution of the lacertids. His findings necessitated some nomenclatural changes and will require further alterations if a truly phylogenetic classification is to be developed. Apart from *Lacerta*, which is a paraphyletic taxon, the genera which occur in Iran belong to a xeric-adapted Ethiopian and Saharo-Eurasian clade of advanced taxa, apparently derived from a common ancestor with *L. jayakari*.

However, Mayer and Bischoff (1996) amended Arnold's scheme on the basis of morphology and karyotype, recognizing the genus *Zootoca* Wagler, 1830 (type species: *Lacerta vivipara* Jacquin, 1787) as the sister taxon of *Lacerta* (*sensu stricto*), the genus *Omanosaura* Lutz, Bischoff, and Mayer, 1986 (type species: *Lacerta jayakari* Boulenger, 1887; also including *L. cyanura* Arnold, 1973, both species from the Oman mountains), *Timon* Tschudi, 1836 (type species: *Lacerta lepida* Daudin, 1802; also including *Lacerta princeps* Blanford, 1874 and *L. pater* Lataste, 1880), and *Teira* Gray, 1838 (type species: *Lacerta dugesii* Milne–Edwards, 1829). Their phylogenetic tree shows *Omanosaura* as the sister taxon to a clade containing *Eremias, Mesalina,* and perhaps *Ophisops*. *Timon* is possibly sister to a clade containing *Lacerta* (*sensu lato*) subgenera *Archaeolacerta* and *Apathya* and the *L. saxicola* complex.

Mayer and Benyr (1994) published a phylogenetic hypothesis for the lacertids based on albumin evolution. They established two subfamilies, Gallotinae (*Gallotia* + *Psammodromus*) and Lacertinae, this initial division having occurred between the Oligocene and Lower Miocene. Within the Lacertinae, an "African lineage" includes *Acanthodactylus* as the sister taxon to *Adolfus*, the split having occurred some 15–16 million years ago. A "Eurasian lineage" includes *Ophisops* as sister to the clade which includes *Omanosaura*, which is sister to *Eremias* + *Mesalina*. *Timon* is questionably sister to the clade containing "*Archaeolacerta,*" which is sister to the *L. cappadocica* + *L. saxicola* complex. All of these taxa represent an Early to Middle Miocene radiation, according to these authors.

Thus far, two centuries of increasingly detailed and sophisticated analysis of the relationships within the Lacertidae has failed to produce a widely accepted consensus, and still more analysis can be expected.

Key to the Genera of Lacertidae in Iran

1a. Eyelids immovable, eye covered by a transparent shield (spectacle) (Fig. B); collar absent or weakly developed .... *Ophisops* (p. 254)
1b. Eyelids movable (Fig. A); collar well developed 2

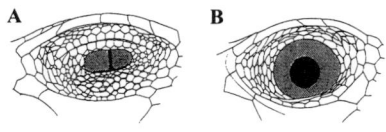

Eyelid of (A) *Mesalina*; (B) *Ophisops*
(After Arnold, 1989, fig. 15)

2a. Nostril separated from first supralabial by lower nasal shield 3
2b. Nostril in contact with first supralabial or separated from it by very narrow brim 4
3a. Ventral plates in straight longitudinal series (Fig. C); lower nasal resting on first supralabial .... *Mesalina* (p. 249)
3b. Ventral plates in tessellated or oblique longitudinal series (Fig. D), converging posteriorly; lower nasal resting on two or three supralabials .... *Eremias* (p. 207)

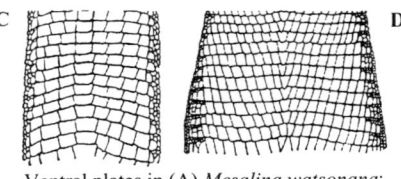

Ventral plates in (A) *Mesalina watsonana*;
(B) *Eremias grammica*
(From Bannikov, et al., 1977, fig. 41)

4a. Digits with lateral fringes .... *Acanthodactylus* (p. 200 and 256)
4b. Digits without lateral fringes .... *Lacerta* (p. 228 and 259)

Genus *Acanthodactylus* Fitzinger, 1834
Spiny-toed lizards, fringe-toed lizards, sand spine-foot lizards

Acanthodactylus Fitzinger *in* Wiegmann, 1834:10 (Type species: *Lacerta boskiana* Lichtenstein, 1823 [= *Lacerta boskiana* Daudin, 1802], by monotypy).

Definition: Head shields normal, but occipital absent; nostril between 2 nasals and first upper labial; lower eyelid scaly; collar distinct; dorsal scales small and juxtaposed or large and imbricate; ventral plates subquadrangular, smooth, imbricate; digits subcylindrical, with keeled lamellae below, and lateral denticulation, at least on outer side of toes; femoral pores present (Smith 1935:370; Salvador 1982:8). Hemipenes with armature present (Arnold, 1986a:1234–1235).

Distribution: Iberian Peninsula and northern Africa, through mainly lowland Southwest Asia to northwestern India. Does not penetrate Iranian Plateau to any great extent, but enters basins of Afghanistan and Baluchistan. Thirty species, six recorded from Iran. The widely distributed *Acanthodactylus opheodurus* Arnold (1980b:296) may be expected in lowland southwestern Iran. Rastegar-Pouyani (pers. commun., 23 October 1996 and in press) informs me that he has identified *A. boskianus* (Daudin, 1802) from Iran and has collected an as-yet-undescribed species from Qasr-e Shirin, Kermanshahan Province near the Iraq border.

Remarks: This genus has been revised by Salvador (1982) and by Arnold (1983), who recognize nine species groups. Of the Iranian species, *Acanthodactylus micropholis* and *A. grandis* belong to monotypic groups; *A. blanfordi* and *A. schmidti* belong to the *cantoris* group of seven species (*A. arabicus, A. cantoris, A. gongrorhynchatus, A. haasi*, and *A. tilburyi* also included). (See Addendum to the Lacertidae, p. 255ff., for updates on Iranian *Acanthodactylus*.)

The genus is Saharo-Sindian in its distribution. The center of diversity is the Arabian Peninsula and contiguous Mesopotamian lowlands and Syrian Desert. Only one species reaches southwestern Europe, and this is a western North African species. Of the nine species groups, three are primarily North African. Only one species (*A. cantoris*) reaches the deserts of the northwestern Indian Peninsula.

Arnold (1986e) regards *Acanthodactylus* as the sister taxon of *Mesalina* plus *Ophisops*.

See Arnold's paper for discussion of osteological, hemipenial, and external characters and their distribution and variation within the genus. In the phylogenetic scheme of Mayer and Benyr (1994), *Acanthodactylus* is the sister taxon to *Adolfus* in their African lineage.

Key to the Species of *Acanthodactylus* in Iran
(See Addendum to the Lacertidae [p. 256] for a revised key to the Iranian *Acanthodactylus*)

1a. 3 scales around fingers but only two visible from medial side view (Fig. A); ventrals usually 10 in longest transverse row across belly 2
1b. 4 scales around fingers, but only three visible from medial side view (Fig. B); ventrals 13–18 in longest transverse row across belly 4

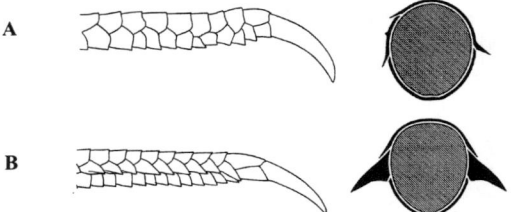

Third fingers of *Acanthodactylus*, viewed from the side and in cross section. As viewed from the side (A) only two rows visible, (B) three complete rows visible. (From Arnold, 1986, fig. 22)

2a. Temporal scales minute; dorsal scales feebly keeled; 49 or more scales across middle of body .... *Acanthodactylus micropholis* Blanford, 1874 (p. 205)
2b. Temporal scales medium-sized; dorsal scales strongly keeled, 19–55 across middle of body 3
3a. Eyelid barely pectinate; 4th toe strongly pectinate ....
.... *Acanthodactylus boskianus* (Daudin, 1802) (p. 257)
3b. Eyelid strongly pectinate; 4th toe scarcely pectinate ....
.... *Acanthodactylus opheodurus* Arnold, 1980 (p. 258)
4a. Ventral scales in oblique or irregular longitudinal series, not forming straight longitudinal rows; 18–22 dorsal scales in transverse series between hind limbs ....
.... *Acanthodactylus grandis* Boulenger, 1909 (p. 203)
4b. Ventral scales in straight longitudinal rows down middle of venter, outer series may be somewhat oblique; 10–16 dorsal scales in transverse series between hind limb 5
5a. Dorsal color pattern reticulate, not lineate even in young specimens, indistinct in large adults; scales on sides of dorsum double the size of those on central dorsum, 38–54 dorsal scales across middle of back; 13–18 ventral plates in longest transverse series .... *Acanthodactylus schmidti* Haas, 1957 (p. 206)
5b. Dorsal color pattern lineate, young specimens with 6 dorsal and one lateral light longitudinal streaks, with or without round white spots between them; some adults nearly uniform, no distinct pattern; scales on sides of dorsum equal to those on central dorsum; 34–46 dorsal scales across middle of back; 12–16 ventral plates in longest transverse series .... *Acanthodactylus blanfordi* Boulenger, 1918 (p. 201)

Acanthodactylus blanfordi Boulenger, 1918 (Plate 11A)
Blanford's fringe-toed lizard

Acanthodactylus cantoris: Blanford, 1876:381–383, pl. 26, figs. 3, 3a–b (not Günther, 1864:73). —

Bedriaga, 1879:31. — Boulenger, 1887a:60–61. — Werner, 1895:4–5. — Nikolsky, 1900:395. — ?Zarudny, 1903:18.

Acanthodactylus cantoris var. *blanfordi* Boulenger, 1918c:154 (Type locality: Bam and Jask, Iran; Dasht and Mand, Pakistan; Syntypes: BMNH 1946.9.3.54–55, 1946.9.8.33–34, 1946.9.8.43–44); 1921:94–95. — Smith, 1935:372–373. — Werner, 1936:201; 1938:268. — Forcart, 1950:148. — Wettstein, 1951:440. — Mertens, 1956:93–94. — S. Anderson, 1963:456–459; 1968:333. — Tuck, 1971a:57. — S. Anderson, 1974:40, 43. — Schleich, 1977:127, 129. — Welch, 1983:30.

Acanthodactylus cantoris cantoris: Guibé, 1966a:98 (not Günther, 1864).

Acanthodactylus blanfordi: Salvador, 1982:151–155, fig. 103–105, map 31. — Arnold, 1983:311–315. — Welch, *et al.*, 1990:114.

Diagnosis: (Fig. 86) Four supraoculars; one row of granules between supraoculars and superciliaries; usually 5 supralabials anterior to subocular; subocular separated from lip; temporals sharply keeled; ventrals in 12–16 straight longitudinal series; 34–46 dorsal scales across middle of back, no marked gradation between median dorsal and dorsolateral scales; 27–38 gular scales on median line; 4 series of scales around fingers.

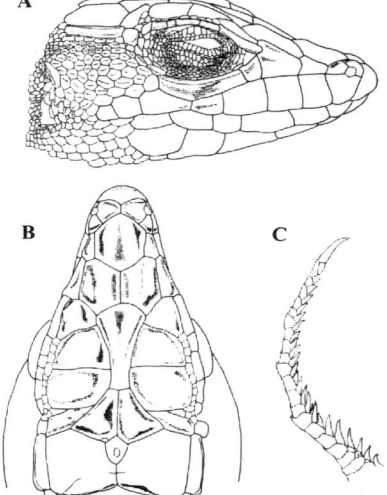

FIGURE 86. *Acanthodactylus blanfordi* (BMNH 1946.9.8.44, syntype). (A) Side of head; (B) top of head; (C) side of fourth toe. (From Salvador, 1982, figs. 103–105.)

Color pattern: Six or 7 dorsal and one lateral light longitudinal streaks (cream in life), sometimes broken into series of white spots, on a buff to golden brown ground, the stripes of this ground often with single lines of black spots, especially in juveniles; some large adults nearly uniform light brown; hind limbs with dark-ringed light spots; venter uniform cream to grayish white.

Size: Snout-vent length 66 mm (to 75 mm, according to Arnold 1983:312), tail 131 mm.

Natural history: Blanford (1876:383) surmised that eggs were probably laid in summer and hatched in the fall, due to the fact that he found numerous young in November, but only adults or near-adults in January and February. Minton (1966:107) found females with large eggs from late March through July, with the young appearing at the end of June, becoming more numerous through October. I found no small juveniles in late October, 1958, and females had yolked ovarian follicles at that time. Five of 14 females collected by the Street Expedition in late November or early December had eggs in the oviducts, the largest 8 mm long; all others had large, yolked follicles (up to 2 mm). There were no small juveniles in this series. Minton's (1966) observations indicated that sexual maturity is reached in one year.

Stomachs examined contain termites, ants of several species, beetles, hemipterans, larval insects of various kinds, all primarily ground–dwelling forms. I observed one eating a grasshopper.

Habitat: (Plate 20G) I found this lizard only on sand where sparse vegetation (*Acacia, Tamarix*, salt-tolerant shrubs and grass) was growing, both on dunes and sandy flats. Blanford (1876:383) recorded similar observations in Iran, while in Pakistan, Minton (1966:7) also found them in clay and gravel desert adjacent to sandy tracts. They retreat to

FAMILY LACERTIDAE

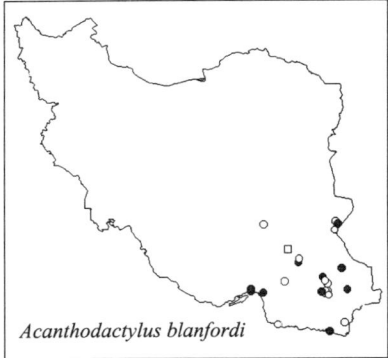

Acanthodactylus blanfordi

holes, usually among roots. They are absent from the higher elevations of the Plateau of Iran, the highest record being 1400 m.

Distribution: Occurs in southeastern Iran (Baluchistan and Kerman Provinces), southwestern Pakistan, southern Afghanistan (Helmand drainage and northern border area of Pakistan according to Arnold, 1983:313), northern coast of Sultanate of Oman (Arnold, 1980a:295; Salvador, 1982:151); I believe F. Werner's (1895:4–5) record between Tehran and Qom to be in error.

Remarks: Arnold (1983:313–314) regards *Acanthodactylus schmidti* as the probable sister species of *A. blanfordi*, which is also closely related to the similar *A. arabicus* and its relatives.

Acanthodactylus grandis Boulenger, 1909 (Plate 11B)
Giant fringe-toed lizard, Mesopotamian fringe-toed lizard

Acanthodactylus grandis Boulenger, 1909a:189 (type locality: Jerud and Ataibé, E Damascus, and Khan Agach, between Damascus and Kutaife, Syria; syntypes: BMNH 1909.4.20.27–29/1946.9.2.69–70; MNHNP 23.8–11). — Salvador, 1982:110–113; figs. 66–68, map 22. — Arnold, 1983:316–317. — Leviton, *et al.*, 1992:52–53, col. pl. 8B. — Welch, 1983:30.

Acanthodactylus fraseri Boulenger, 1918c:373 (Type locality: Zobeya [=Az Zubayr, 30°23'N, 47°43'E], Shariba, Iraq. Syntypes [BMNH 1917.6.18.3/1946.8.7.40] and BNHM). — S. Anderson, 1968:333; 1974:40, 43. — Welch, 1983:30.

Diagnosis: (Fig. 87) Four supraoculars (4th broken up in all specimens I have seen); one row of granules between supraoculars and superciliaries; usually 4 supralabials to below center of eye; temporals granular, not keeled; ventral plates not forming straight longitudinal series, 14–16 in longest transverse row; 18–22 dorsal scales in transverse series between hind limbs; 4 series of scales around fingers; lateral fringe on toes scant (well-developed on Iranian specimens checked by Salvador, 1982; the fringe is never really absent, in as much as the extra row of scales that forms it is always there).

Color pattern: (Fig. 88) Gray above in alcohol, tan or light brown in life, with 4 longitudinal lines of conspicuous small black spots superimposed on a pattern of pale bluish-gray spots (creamy white in life), these spots narrower than spaces between lines of spots; dark spots of dorsum sometimes coalesce transversely or longitudinally to form reticulations; flanks with less regularly arranged dark markings; venter cream to grayish white (based on Iranian specimens).

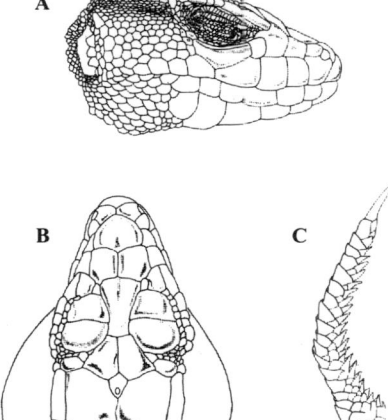

FIGURE 87. *Acanthodactylus grandis* (BMNH 1909.4.20.29, syntype). (A) Side of head; (B) top of head; (C) side of fourth toe. (From Salvador, 1982, figs. 66–68.)

Size: Largest Iranian specimen examined: snout-vent length 74 mm, tail 122 mm.

Natural history: Females collected January 6–11 and 17–20, 1963 by Street Expedition

FIGURE 88. Dorsal view of *Acanthodactylus grandis* showing typical color pattern (CAS 102535). (Photo by A. Leviton.)

contain yolked ovarian follicles, the largest 3 mm diameter. Stomachs of these specimens contain ants, ant larvae, wasps, and beetles. They were active when the air temperature was 38°C, sand surface 42.5°C.

Habitat: (Plate 21G) I found this species only on active dunes on which low thorny shrubs were growing. Some of these areas were under cultivation for cucumbers.

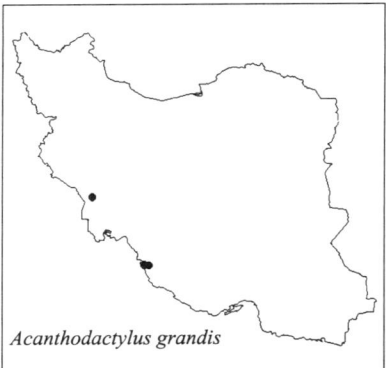

Acanthodactylus grandis

Distribution: Eastern Lebanon, Jordan, Syria, Iraq, northern Saudi Arabia, and the Mesopotamian lowlands of Khuzestan and Fars in southwestern Iran.

Remarks: Variation across Iraq appears to connect *Acanthodactylus fraseri* with the Syrian specimens described as *A. grandis* (Arnold, 1983:317). The dorsal scales of Iranian specimens and the types of *A. fraseri* from southeastern Iraq are larger than those of the Syrian animals and strongly keeled, rather than smooth. Salvador (1982:110) has placed *A. fraseri* in the synonymy of *A. grandis*, and has commented on the variation correlated with geography. The specimens collected by the Street Expedition to Iran in 1962– 63 were the first records of the species from Iran. Haas's (1957:72) specimens identified as *A. fraseri* from Arabia were redetermined and described by Leviton and S. Anderson (1967:171–178) as new species (*A. gongrorhynchatus* and *A. haasi*), while a specimen from Badanah, northern Saudi Arabia, identified by Haas (1960) as *A. scutellatus*, was identified as *A. grandis* by Salvador (1982:110). Schmidt's (1941:162) specimen of *A. fraseri* from north–central Saudi Arabia appears to be a juvenile *A. schmidti*.

Salvador (1982:112) cites a specimen (MNHN 1966.45) from Khash, 28°14′N, 61°14′E as this species, but this locality is in Baluchistan, far to the east of the next nearest locality, and beyond his stated limits of distribution. This locality is not plotted on his map.

Arnold (1983) regarded both nominal species as part of what he termed the *A. grandis* complex:

"It is possible that the *A. grandis* complex is best regarded as a single species but available samples are too small and scattered to be certain about this. The irregular variation of populations intermediate between typical *A. grandis* and *A. fraseri* may reflect the geography of Mesopotamia, for here the comparatively arid country favored by *Acanthodactylus* is divided up by the Tigris and Euphrates rivers and their tributaries which flood seasonally, so populations may be substantially discontinuous. Other *Acanthodactylus* species, especially *A. boskianus*, also show considerable variation in this area."

Subsequently, however (Arnold, 1986c:424), he accepted the synonymy of the two names without further comment.

Acanthodactylus grandis appears to be the sister taxon to *A. schreiberi* plus *A. boskianus* (Arnold, 1983:333).

Acanthodactylus micropholis Blanford, 1874 (Plate 11C)
Yellow-tailed lizard, Persian fringe-toed lizard

Acanthodactylus micropholis Blanford, 1874b:33 (Type locality: Magas, Iran [designated by Smith, 1935:373]; Syntypes: BMNH 1946.9.3.71–72; ZMB 9333; ZSI 5301 [*fide* Das, *et al.*, 1998]); 1876:383, pl. 26, fig. 2. — Bedriaga, 1879:31. — Lataste, 1885:503. — Boulenger, 1887a:63; 1890b:171. — Nikolsky, 1899b:394–395. — Zarudny, 1903:17–18. — Boulenger, 1918b:147; 1921:76–78. — Smith, 1935:373. — F. Werner, 1936:201. — Forcart, 1950:149. — S. Anderson, 1963:475; 1968:333; 1974:40, 43. — Schleich, 1977:127, 129. — Salvador, 1982:19–22, figs. 1–3, map 1. — Arnold, 1983:311. — Welch, 1983:30. — Welch, *et al.*, 1990:114.

Diagnosis: (Fig. 89) Only 2 entire supraoculars; one or 2 rows of granules separating supraoculars and superciliaries; 4 supralabials anterior to subocular; minute smooth temporals; ventrals in 10 straight longitudinal series; dorsals feebly keeled, 48 or more across middle of body; 3 series of scales around fingers.

Color pattern: (Fig. 90) Gray above in alcohol, light brown in life, usually with 7 white or cream (sometimes bright yellow in life) longitudinal streaks on body, sometimes with light spots in dorsolateral dark stripe; interspaces between light streaks often blackish, with a series of small, round white dots; vertebral stripe bifurcates on nape; limbs with light spots on light brown ground; venter white or cream; in life, tail is sometimes (perhaps seasonally) bright yellow.

Size: Snout-vent length 63 mm, tail 125 mm.

Habitat: According to Blanford (1876:383) this species occurs in situations similar to those in which "*A*[*canthodactylus*]. *cantoris*" (*A. blanfordi*) is found, but at higher elevations, being less abundant than the latter near the coast, replacing it completely over 3000 ft (915 m). Minton found that in Pakistan it prefers sandy stream beds and canyons; *A. blanfordi* inhabits sandy alluvium between hills and sea coast, the two species meeting only at the mouths of canyons. I found it at only one locality along the coast, on ground at the base of a shrub growing on roadside rubble. The soil in the area was gravelly, the vegetation scattered low desert shrubs. There was no sandy soil in the vicinity.

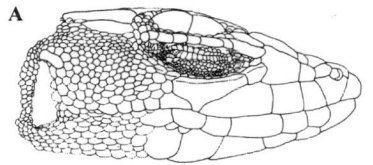
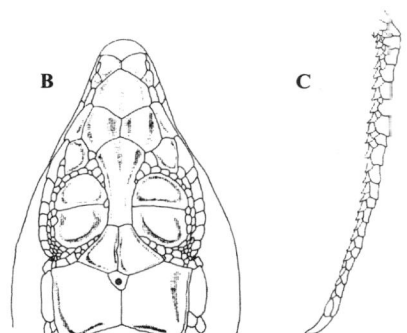

FIGURE 89. *Acanthodactylus micropholis* (BMNH 74.11.23.78, syntype). (A) Side of head; (B) top of head; (C) side of fourth toe. (From Salvador, 1982, figs. 1–3.)

Distribution: Southeastern Iran (Kerman and Baluchistan), Qeshm Island in the Persian Gulf, southern Pakistan east to Las Bela (Minton, 1966:108; Salvador, 1982:19–20).

Remarks: In his description of this species, Boulenger (1921:76–78) says, "56 to 63

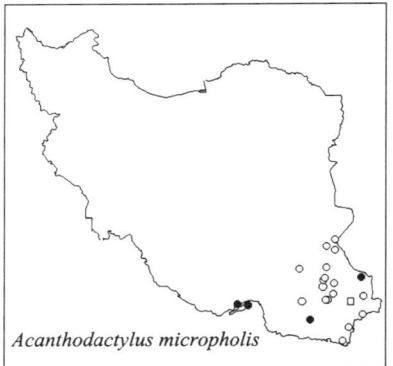

Acanthodactylus micropholis

scales across the middle of the body . . .," while according to his tabular data there are 48–52; Blanford (1876:383) gives the range as 49–53. In the two specimens from Qeshm Island there are 50 and 52 dorsal scales at midbody, 10 longitudinal series of ventrals, the outermost series very small, 27 and 28 transverse series of ventrals (29–31, Boulenger), 30 and 26 gulars (28–33, Boulenger), 21–22 femoral pores (21–25, Boulenger), and 4 upper labials anterior to subocular which borders mouth. The outer one or two longitudinal white streaks are broken into dots. These are the only specimens known from Islands of the Persian Gulf. Arnold (1983:333, cladogram) regards *Acanthodactylus micropholis* as a sister taxon of the *cantoris*-group (see also Arnold, 1983:311).

FIGURE 90. *Acanthodactylus micropholis*. (From Blanford, 1976, pl. 26, fig. 2; G. H. Ford, artist.)

Acanthodactylus schmidti Haas, 1957 (Plate 11D)
Schmidt's fringe-toed lizard

?*Acanthodactylus cantoris*: Werner, 1917:201–202.
Acanthodactylus cantoris schmidti Haas, 1957:72–73 (Type locality: Dhahran, Saudi Arabia; Holotype: CAS 84599). — S. Anderson, 1963:456–459; 1968:333; 1974:40, 43. — Schleich, 1977: 127, 129. — Welch, 1983:30.
Acanthodactylus schmidti: Salvador, 1982:146–151, figs. 98–102, map 30. — Arnold, 1983:311–315. — Leviton, et al.,1992:55, col. pl. 8E.

Diagnosis: (Fig. 91) Three large supraoculars (1st, 2nd, 3rd; 4th usually divided in two); 5 supralabials anterior to subocular; subocular not bordering mouth; usually 2 keeled supratemporals; temporals sharply keeled; ventral plates in 12–16 oblique longitudinal series, outer rows consisting of pointed scales; 32–54 dorsals across middle of back, posterior dorsolateral scales double the size of the middorsals; dorsal color pattern reticulate, not lineate even in young specimens.

Color pattern: Dorsum covered with a reticulum of tan or light brown, enclosing light areas which form spots more or less regularly arranged in longitudinal rows; limbs and tail similarly patterned; large adults almost uniform tan on back, retaining pattern on extremities; young and adults similar, never with distinct longitudinal stripes; venter cream to grayish white.

Size: Largest Iranian specimen examined: snout-vent 85 mm; some Arabian specimens reach 105 mm.

Natural history: Females collected in mid-August and mid-September have yolked

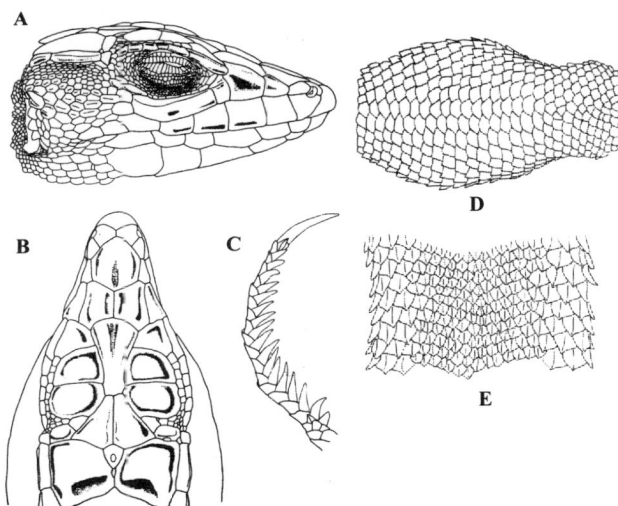

FIGURE 91. *Acanthodactylus schmidti* (BMNH 1953.1.8.55). (A) Side of head; (B) top of head; (C) side of fourth toe; (D) ventral scales; (E) dorsal scales. (From Salvador, 1982, figs. 98–102.)

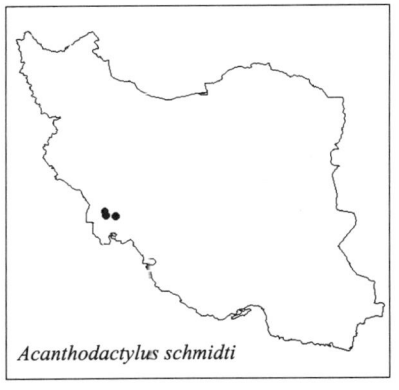

Acanthodactylus schmidti

follicles; no small juveniles were collected in August and September. Ants, termites, spiders, beetles, and blossoms were in stomachs I examined. See S. Anderson (1963:457) for observations on summer activity and behavior.

Habitat: (Plate 21F) I found this lizard only on active dunes where sparse shrubs provide shade and shelter. The sand was moist a few centimeters beneath the surface. In Oman, Gallagher and Arnold (1988:410) recorded it primarily as a species of wind-blown sand with scrubby vegetation, where it digs burrows at the base of shrubs. See also Arnold (1984) for additional habitat information for the eastern United Arab Emirates.

Distribution: Over much of the Arabian Peninsula, extending into southern Jordan and Iraq, entering lowland southwestern Iran in Khuzestan and Fars (see Salvador, 1982:146, map 30).

Remarks: A photograph of the holotype has been published by Leviton and S. Anderson (1967, fig. 8).

Salvador (1982) recognizes *schmidti* and other populations previously considered subspecies of *Acanthodactylus cantoris* as full species. He restricts the distribution of *A. arabicus* Boulenger to South Yemen, and finds no specimens considered as intermediate between *arabicus* and *schmidti* or between *schmidti* and *blanfordi* on the Arabian Peninsula. He also states: "In Iran there is an enormous and as yet uninvestigated area separating *schmidti* from *blanfordi* making it impossible, at this time, to know the relationship between these two species in this zone." This region is not entirely "uninvestigated," as I have traversed some of it, and in my judgment it is a region lacking in suitable habitat for *Acanthodactylus*. Thus the populations are geographically separated by hundreds of kilometers.

Acanthodactylus blanfordi is the sister species of *A. schmidti* (Arnold, 1983:313–314, 329–330).

Genus *Eremias* Fitzinger, 1834
Racerunners, desert lacertas

Eremias Fitzinger *in* Wiegmann, 1834:9 (Type species: *Lacerta variabilis* Pallas, "1811–1831" (1827) [= *E. arguta* (Pallas), 1773], by subsequent designation of Fitzinger, 1843:21).

Diagnosis: Head shields normal, but occipital often vestigial or absent; nostril between 3 or 4 nasals, not touching labial; lower eyelid scaly; collar complete or nearly so; dorsal scales small or granular, subimbricate or juxtaposed; ventral plates subquadrangular, imbricate, smooth, in converging longitudinal rows; digits with or without lateral fringes; tail cylindrical; femoral pores present (except in *Eremias aporosceles*, considered a synonym of *E. acutirostris* by Szczerbak [1974]).

Key to the Species of *Eremias* in Iran

1a. Subocular bordering mouth 2
1b. Subocular not bordering mouth 12
2a. A complete row of lateral scales of 4th toe forming distinct fringe or comb in its entire length (Fig. B) 3
2b. Lateral scales of 4th toe not forming distinct fringe (Fig. A) 4

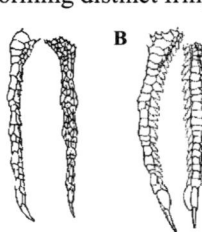

Lateral (left) and ventral (right) sides of fourth toes of *Eremias*: (A) *E. fasciatus* and (B) *E. lineolata*. (From Lantz, 1928, pl. 1, fig. 1.)

3a. Row of femoral pores reaches well short of knee (Fig. C); 4th toe with 2 complete rows of subdigital scales, i.e., a total of 4 scales counted around toe (except that an extra scale may be present at a joint); supracaudal scales keeled, but not pointed behind (Fig. E); broad dark dorsolateral stripe from nostril through eye, along body and side of tail, one or two additional narrower dark stripes medial to these on each side, the remainder of the dark dorsal stripes interrupted and anastomosing to form reticulate pattern ....
.... *Eremias scripta* (Strauch, 1867) (p. 224)
3b. Row of femoral pores reaches knee (Fig. D); 4th toe with single row of subdigital scales, i.e., a total of 3 scales counted around toe (except an extra scale may be present at a joint); supracaudal scales strongly keeled and acuminate (Fig. F); dorsal pattern of 7 dark stripes, outer dorsolateral stripe broadest ....
.... *Eremias lineolata* (Nikolsky, 1896) (p. 218)

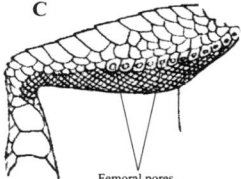 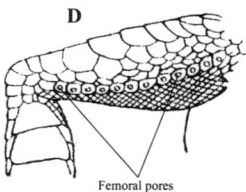  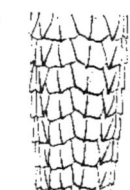

Femoral pores of (C) *Eremias scripta* and (D) *E. lineolata*. (From Terentjev and Chernov, 1949, fig. 54)

Supracaudal scales of (E) *Eremias scripta* and (F) *E. lineolata*. (From Bannikov, et al., 1977, figs. 45:2)

4a. The 2 series of femoral pores broadly separated, space between series at least ⅓ length of each .... *Eremias pleskei* Bedriaga, 1907 (p. 223)
4b. The 2 series of femoral pores meeting, or separated by space not greater than ⅓ length of each 5

5a. Back with 5–11 dark stripes, broader than interspaces, none of the stripes containing light ocelli or spots; stripes persistent in adults, but sometimes indistinct so that back appears almost uniform sandy; usually only single median collar scale distinctly larger than adjacent gulars 6
5b. Light ocelli or spots on upper flanks (rare exceptions), dark stripes of juvenile breaking up in adults to form spots or broken lines; usually several collar scales distinctly larger than adjacent gulars 7
6a. Frontal and supraocular scales separated by complete row of granules; 4th toe with 2 complete rows of subdigital scales and a complete row of sharply pointed lateral scales, i.e., total of 4 scales counted around penultimate phalanx ....
.... *Eremias fasciata* Blanford, 1874 (p. 213)
6b. Frontal and supraocular scales not separated by complete row of granules; 4th toe lacking complete row of distinctly pointed lateral scales; i.e., total of 3 scales counted around penultimate phalanx..
.... *Eremias andersoni* Darevsky and Szczerbak, 1978 (p. 211)
7a. Four submaxillary shields, smaller shields lateral and posterior to 4th submaxillary
.... *Eremias lalezharica* Moravec, 1994 (p. 217)
7b. Five submaxillary shields... 8
8a. Adults with dark interrupted dorsolateral black stripe forming ocelli with white spots, this dorsolateral pattern not contrasting strongly with interrupted dark stripes and spots of dorsum; juveniles with 3 dark stripes on dorsum between white-spotted dorsolateral stripes, vertebral stripe black, bifurcated on nape (dark stripes breaking up into several irregular rows of dark spots with age); ventral surface of tail carmine red in juveniles (in life) .... *Eremias velox velox* (Pallas, 1771) (p. 227)
8b. Adults usually with black dorsolateral stripe more or less continuous for at least major portion of its length 9
9a. Black dorsolateral stripe uniform, without white spots; 41–42 gular scales between submaxillary shields and collar; (juvenile color pattern not yet known) ....
.... *Eremias nigrolateralis* Rastegar-Pouyani and Nilson, 1997 (p. 220)
9b. Black dorsolateral stripe containing white spots, black stripe contrasting strongly with dorsal color pattern; less than 40 gular scales; juvenile with 4 dark stripes on dorsum between dorsolateral white-spotted stripes, vertebral stripe light colored (dark stripes breaking up into 4 more or less regular rows of dark spots with age); ventral surface of tail not red in juveniles 10
10a. Adults with more or less distinct rows of dark spots on dorsum between dorsolateral dark stripes, the latter usually with white spots in a single row within each stripe; distal portion of tail bluish in juveniles (in life)..........*Eremias persica* Blanford, 1875 (p. 221)
10b. Adults usually without dark stripes or spots on middorsum; dorsolateral region with alternate light and dark spots, often fusing longitudinally, forming 2–4 longitudinal stripes, often broken, the impression being 3–4 rows of light spots on flanks; ventral surface of tail yellow in juveniles (in life) 11
11a. 23–33 gulars; 56–68 dorsals; 24–35 scales in 9–10th caudal annulus; West and East Azarbaijan, Iran .... *Eremias strauchi strauchi* Kessler, 1878 (p. 225)
11b. 19–28 gulars; 48–59 dorsals; 20–26 scales in 9–10th caudal annulus; eastern Mazanderan, northern Khorasan ...... *Eremias strauchi kopetdaghica* Szczerbak, 1972 (p. 226)
12a. 4th toe with distinct fringe on both lateral and medial sides, formed by complete row

of sharply pointed lateral scales and complete row of similar medial scales; ungual lamellae of fingers and toes with prominent flat lateral expansions 13

12b. 4th toe without distinct fringe; ungual lamellae without prominent lateral expansion 14

13a. Scales of flanks not larger than those of back; broad plates on lower surface of tibia more than twice as broad as adjacent scales ....*Eremias acutirostris* (Boulenger, 1887) (p. 211)

13b. Scales of flanks distinctly larger than those of back; plates on lower surface of tibia not twice as broad as adjacent scales ...*Eremias grammica* (Lichtenstein, 1823) (p. 215)

14a. 5th toe with 2 complete rows of subdigital scales and incomplete row of small lateral scales; 2nd supraocular (1st of 2 large undivided supraoculars) as long as or shorter than its distance from 2nd loreal .... *Eremias arguta* (Pallas, 1773) (p. 212)

14b. 5th toe with single complete row of subdigital scales and a few scattered lateral scales not forming complete row; 2nd supraocular (1st of 2 large undivided supraoculars) longer than its distance from 2nd loreal 15

15a. 4th toe with single row of subdigital scales (Fig. H) ....*Eremias intermedia* (Strauch, 1876) (p. 216)

15b. 4th toe with 2 rows of subdigital scales (Fig. G), internal much the larger ....*Eremias nigrocellata* (Nikolsky, 1896) (p. 219)

Ventral surface of toes of *Eremias*: (G) *E. nigrocellata* and (H) *E. intermedia*. (From Bannikov, et al., 1977, fig. 48.)

Distribution: Some 27 species recognized, from northern China, Mongolia, Korea, Central Asia, to southeastern Europe, south through the Iranian Plateau to Baluchistan. Thirteen species occur in Iran.

Remarks: *Eremias* has been revised by Szczerbak (1974), who separated those species which range across North Africa and lowland Southwest Asia into the genus *Mesalina*.

Szczerbak recognizes five subgenera; the Iranian species fit within this classification as follows:

Eremias (*Eremias*)
   lalezharica
   velox
   strauchi
   nigrolateralis
   persica
Eremias (*Ommateremias*)
   arguta
   intermedia
   nigrocellata

Eremias (*Rhabderemias*)
   lineolata
   scripta
   pleskei
   fasciata
   andersoni
Eremias (*Scapteira*)
   grammica
   acutirostris

Thus, except for the subgenus *Pareremias* (a Central and East Asian group), all of the major species groups of the genus are represented on the Iranian Plateau. I have not arranged species by Szczerbak's subgenera here. While the groupings seem intuitively justified, based on the Iranian species I have seen, they are not all defined by synapomorphies. In general, Szczerbak's subgenera are supported by Arnold's (1986:1236–1237) study of the hemipenes. Arnold's (1989) phylogenetic hypothesis regards *Eremias* as the sister taxon of the clade containing *Acanthodactylus, Mesalina*, and *Ophisops*. According to the hypothesis of Mayer and Benyr (1994), *Mesalina* and *Eremias* are sister taxa in a larger clade containing *Omanosaura* and *Ophisops* and not at all closely related to *Acanthodactylus*.

Szczerbak's monograph summarizes information on geographic variation, distribution, natural history, and behavior of the species covered; it includes photographs of all species.

Frynta, *et al.* (1997:9–10) reported 8 specimens of an undetermined species related to *E. persica* from the Zagros Mountains in Esfahan Province. They plan to publish a description of these specimens.

Eremias acutirostris (Boulenger, 1887) (Plate 11E)
Pointed-snouted racerunner, reticulate desert lacerta

Scapteira acutirostris Boulenger, 1887a:114–115 (Type locality: between Nushki and Helmand, Afghan-Baluch border region; Holotype: BMNH 86.9.21.88/1946.8.7.46).
Eremias (*Scapteira*) *acutirostris*: Lantz, 1928a:41; 1928c:136. — S. Anderson, 1978:22.

Diagnosis: Subocular not bordering mouth (Fig. 92); fourth toe with distinct fringe on both lateral and medial sides; ungual lamellae of fingers and toes with prominent, flat, lateral expansions; scales of flanks not larger than those of back; series of broad plates on lower surface of tibia more than twice as broad as adjacent scales.

Color pattern: A fine reticulum of reddish brown over dorsum of back and limbs enclosing pale tan or cream spots; top of head light brown with scattered small dark brown spots and marks; venter pale yellow or creamy white.

Size: Snout-vent 70 mm.

Habitat: (Plate 21C) I found this species active on loose drifting loess in an abandoned village in Sistan. Minton (1966:109) found them on dune slopes and blowouts in Pakistan where they ran rapidly over loose surfaces and burrowed into the sand and tunnelled beneath the surface.

FIGURE 92. Head of *Eremias acutirostris*. (From Szczerbak, 1974, fig. 81.)

Distribution: The Sistan Basin of Iran and Afghanistan and the sandy tracts along the Afghan-Pakistan border.

Remarks: Szczerbak (1974:261–266) considers *Eremias aporosceles* (Alcock and Finn, 1896), which occurs in the same general region of the Afghan-Pakistan border, to be synonymous with this species.

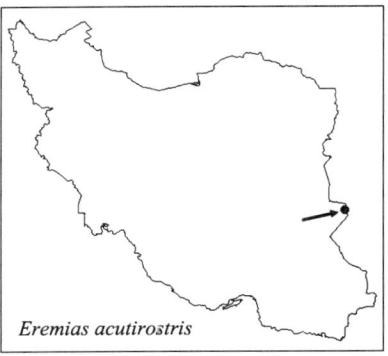
Eremias acutirostris

Eremias andersoni Darevsky and Szczerbak, 1978
Anderson's racerunner

Eremias andersoni Darevsky and Szczerbak, 1978:13–15, figs. 1–2 (Type locality: 34°30′N, 52°40′E,

40–45 km E Daryacheh-ye Namak, Dasht-e Kavir, Iran; Holotype: MMTT 1671). — Nilson and Andrén, 1981:138.
Rhabderemias andersoni: Welch, 1983:55.

Diagnosis: Subocular in contact with edge of mouth; frontal and supraoculars not separated by a complete row of granules; nostril situated among 3 nasals and widely separated from supralabials (Fig. 93); ventrals in 13–14 oblique longitudinal series; series of femoral pores separated by space not greater than one-fourth of one row; supracaudals weakly keeled; fourth toe without long flat projecting scales.

Color pattern: Sandy gray dorsally with 9 dark brown longitudinal stripes, medial stripes broken into separate, more or less wavy segments, some passing onto tail, one extending from eye along flank to anterior half of tail; limbs above with light circular or oval spots on dark brown background. Venter white.

Size: Snout-vent 40 mm, tail 90 mm.

Habitat: Isolated areas of semistabilized sands in stony deserts. They were numerous in low areas between dunes and on sands with grassy and subshrubby vegetation. They climb into low bushes.

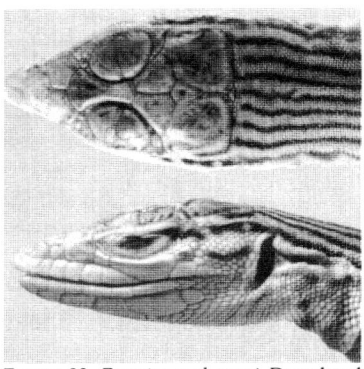

FIGURE 93. *Eremias andersoni*. Dorsal and lateral views of head of holotype. (From Darevsky and Szczerbak, 1978, fig. 2.) With permission Journal of Herpetology.

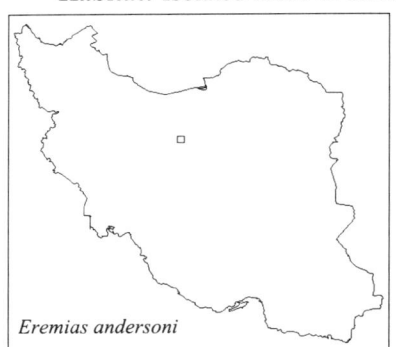

Eremias andersoni

Distribution: Known only from the type locality.

Remarks: This species is known only from the holotype and two paratypes. The latter are in the Zoological Institute of the Russian Academy of Sciences in St. Petersburg. The holotype is in the natural history museum in Tehran. I have not examined these specimens. According to the authors of this species, it is morphologically closest to *Eremias fasciata*, but its subgeneric position is not yet clear.

Eremias arguta (Pallas, 1773) (Plate 11F)
Steppe-runner, racerunner, arguta

Lacerta arguta Pallas, 1773:718 (Type locality: "Irtin australiorem . . . M[are] Caspium" [between the Ural and Emba rivers, a region along north shore of Caspian Sea, Kazakhstan, *fide* Szczerbak, 1974:167; see Zhao and Adler, 1993:202 and Bauer and Günther, 1995:56 for additional comments]; Neotype: ZIL 13205 [designated by Szczerbak 1974:149]).
Eremias arguta: Gray, 1845:39. — ? F. Werner, 1936:198, 201. — Terentjev and Chernov, 1949:207–208. — S. Anderson, 1963:476; 1968:333; 1974:39, 43. — Schleich, 1977:127, 129. — Clark, 1991:40.
?*Podarces* (*Eremias*) *arguta*: Bedriaga, 1879:33.
Eremias (*Ommateremias*) *arguta*: Szczerbak, 1974:147–167, figs 44, 50–52.
Eremias arguta deserti-transcaucasica: Szczerbak, 1974:170–171, fig. 53 (map, p. 166).
Ommateremias arguta deserti: Welch, 1983:55.
Ommateremias arguta transcaucasica: Welch, 1983:55.

Diagnosis: (Fig. 94) Three nasals, lower in contact with 2 or 3 anterior supralabials; 2 large supraoculars, first usually shorter than second, as long as, or shorter than its distance from second loreal; frontonasal single; subocular not bordering mouth; the 2 series of femoral pores separated by a space at least one-half the length of each; fifth toe with 2 complete rows of subdigital scales and one incomplete row of small lateral scales.

Color pattern: Gray dorsally, young with white ocelli edged with black, these sometimes confluent into transverse bands, rarely into 6 or 8 longitudinal streaks; ocelli persist in adults or replaced by black marbling or irregular transverse bars; dark spots or blotches sometimes present on head; venter white. (Boulenger 1921:342–347). Clark (1991:40) describes his specimen as having two longitudinal lines of dark-edged ocelli on either side of the spine on a gray ground (see also Ovenden's illustration in Arnold and Burton, 1978, col. pl. 17, fig. 4).

Size: Snout-vent length 100 mm.

Habitat: Sobolevsky (1929) records it from the Lenkoran District north of Iran in areas of scanty herbaceous vegetation, from the low Diabar Valley with its broken terrain to the Moghan Steppe, a flat uniform plain, and the unforested ridges along the Persian frontier. Clark's specimen (in SMF) was taken on a stone wall bordering a field in hilly, partly cultivated land.

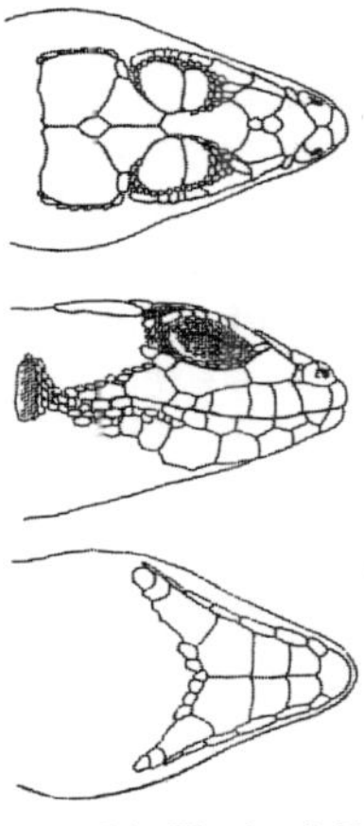

FIGURE 94. Head of *Eremias arguta*. (From Szczerbak, 1974, fig. 44.)

Distribution: The species as a whole extends from the eastern Balkans across Central Asia to Siberia and Mongolia south of about 52°N. Known with certainty in Iran only from the western slopes of the Talysh Mountains and Clark's locality southeast of Tabriz in East Azarbaijan.

Remarks: I have seen no Iranian specimens, but Szczerbak examined specimens from the Talysh Mountains near the border of the Azerbaijan republic and regards this population as intergrades between *Eremias a. deserti* (Gmelin, 1789; type locality: Yaichki steppe between the Volga and the Urals) and *E. a. transcaucasia* Darevsky, 1953 (type locality: Mech-Mazra near Sevan, Armenia). Guibé's (1957:139) record for Mahneh in Khorasan refers to *E. nigrocellata* while his specimens from Sarakhs are *E. intermedia* (see below under these species). Werner's and Bedriaga's records for northeastern Iran also require confirmation.

Eremias arguta

Eremias fasciata Blanford, 1874 (Plate 11G)
Sistan racerunner

Eremias fasciata Blanford 1874b:32 (Type locality: Saidabad, southwest of Kerman [restricted by Smith, 1935:386]. Syntypes BMNH 74.11.23.35–37/1946.8.7.57–59, 1917.3.6.25–26/ 1946.8.7.34–35 [Szczerbak, 1974:242 designated ZMB 9329 from Kerman as lectotype]); 1876: 374–377, pl. 25, fig. 3. — Boulenger, 1887a:99–100. — Zarudny, 1897:356. — Nikolsky, 1897: 329; 1900:399. — Zarudny, 1903:21. — Nikolsky, 1915:431–433. — Boulenger, 1921:318–320. — F. Werner, 1936:198. — S. Anderson, 1963:476; 1968:333; 1974:39, 43.

Podarces (Eremias) fasciata: Bedriaga, 1879:32.

Eremias (Rhabderemias) fasciata: Lantz, 1928b:90–95, pl. 1, figs. 3a–b. — Szczerbak, 1974:241–247, figs. 75–77.

Rhabderemias fasciata: Welch, 1983:55. — Welch, et al., 1990:118.

Diagnosis: Three nasals, lower in contact with 2 or 3 anterior supralabials (Fig. 95); fourth toe with 2 complete rows of subdigital scales, 28–30 in a row, and complete row of sharply pointed lateral scales, i.e., total of 4 scales counted around penultimate phalanx; 21–30 gular scales in straight median series; 2 series of femoral pores separated by space not greater than one-fourth length of each.

Color pattern: (Fig. 96) Dorsum with alternating light and dark lines, the dark lines varying from pale tan to chocolate brown, light lines creamy white to buff; adult with 5–8 dark stripes, no light spots within dark stripes; head uniform brown.

Size: Snout-vent 63 mm, tail 111+ mm.

Natural history: Szczerbak (1974:246, table 30) gives an analysis of the stomach contents of 22 Iranian specimens. Afghan specimens collected mid-May have yolked follicles up to 2 mm.

Habitat: (Plate 21B) I found these lizards on plains having a variety of soil surface types — sandy flats, gravelly hamada, salt-encrusted silt, drifting loess, and silty alluvium. Associated vegetation was low scattered steppe shrubs (e.g., *Artemisia, Alhagi, Tamarix, Acacia*).

Distribution: Eastern Iran on the plateau, southern Afghanistan in the Helmand River basin, and Baluchistan, Pakistan. Iranian localities lie between 450 and 1700 m elevation. Morich (1929:31) records this species from Bagir and Chardzhou in Turkmenistan,

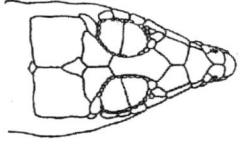
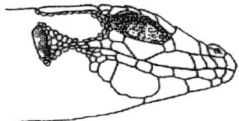
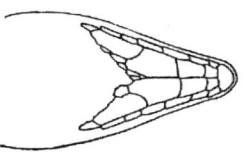

FIGURE 95. Head of *Eremias fasciata*. (From Szczerbak, 1974, fig. 78.)

but subsequent authors (Szczerbak 1974) have not cited these records.

Remarks: CAS 141067 agrees in every respect with Blanford's original description and with the plate, except that the scales of the foot are weakly keeled, rather than smooth (this could be a difference in the optical systems used). There are six dark stripes on the body, all but the middle two broader than the interspaces. All run the length of the body, the two dorsal-most pairs fusing on tail; there is an additional very light and indistinct stripe on flank. The tail was bluish in life.

Eremias fasciata

Blanford made no mention of a blue tail in life. His description is a composite of 12 specimens. He described the pattern as 6–8 dark brown stripes, each pair uniting or terminating on anterior portion of tail, or occasionally in middle of back. His plate (1876, pl. 25) (Fig. 96) shows a vertebral stripe (or a pair?) terminating in middle of back. He stated that specimens from Narmashir and Baluchistan are paler.

Boulenger's description indicates more variation, but in addition to the types, he includes specimens from Helmand, the Perso-Baluch frontier, and Kheran. His description of color pattern indicates 9 dark stripes as wide as interspaces, or 11 on nape, 10 on middle of body, 7 on posterior part of body; tail bluish toward the end. I count 7–9 dark stripes in Blanford's specimens in the British Museum, including the faint lateral stripes.

CAS 141215 and 141218 from Khorasan and CAS 141100 from Sistan appear to be *E. fasciata*, but differ from CAS 141067 in having 8–9 stripes on back and in having yellow, rather than blue tails in life.

FIGURE 96. *Eremias fasciata*. (From Blanford, 1876, pl. 25, fig. 3; G. H. Ford, artist.)

Specimens from southern Afghanistan tentatively identified as this species differ from the original description in having 9–11 dark stripes on dorsum as well as an additional faint lateral stripe on each side.

Eremias grammica (Lichtenstein, 1823) (Plate 11H)
Reticulate racerunner

Lacerta grammica Lichtenstein *in* Eversmann, 1823:140 (Type locality: Karakum, Turkmenistan; Lectotype: ZMB 1095 [designated by Szczerbak, 1974:248]).
Scapteira persica Nikolsky, 1900:395–397 (Type locality: Tscharachs, Zirkuch, Iran. Syntypes: ZIL 9322[3], 9323[5]; not *Eremias persica* Blanford, 1874). — Zarudny, 1903:18 — Boulenger, 1921:371. — Werner, 1936:201.
Eremias (Scapteira) grammica: Lantz, 1928c:117–122, pl. 1, fig. 5, pl. 2, fig. 4. — Terentjev and Chernov, 1949:210–211. — S. Anderson, 1968:333; 1974:39, 43. — Szczerbak, 1974:247–261, figs. 78–80, photo 2. — Bannikov, *et al.* 1977:187–188.
Eremias (Scapteira) zarudnyi Lantz, 1928c:123–127 (substitute name for *Scapteira persica* Nikolsky, 1899, preoccupied by *Eremias persica* Blanford, 1874).
Scapteira grammica: Welch, 1983:56. — Welch, *et al.*, 1990:118.

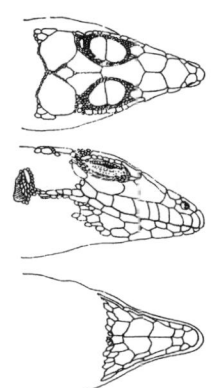

FIGURE 97. Head of *Eremias grammica*. (From Szczerbak, 1974, fig. 78.)

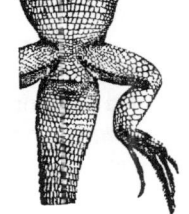

FIGURE 98. Ventral aspect of abdomen, hindlimb, and tail of a male *Eremias grammica*. (From Lantz, 1928, pl. 2, fig. 4).

Diagnosis: (Figs. 97 and 98) Lower nasal resting on 2 or 3 supralabials; subocular not bordering mouth; subdigital lamellae smooth, 20–22 under fourth toe, which has distinct fringe on lateral and medial sides, formed by complete row of sharply pointed scales; ungual lamellae of fingers and toes with prominent, flat lateral expansions; scales of flanks distinctly larger than those of back; broad plates on lower surface of lower leg not twice as broad as adjacent scales.

Color pattern: Fine reticulum of dark brown or reddish brown over back and limbs enclosing cream or tan spots; top of head light brown with scattered small dark brown spots and marks; venter creamy white.

Size: Snout-vent length 100 mm.

Natural History: Tsellarius (1977) studied the ecology of this species in the eastern Kara Kum. It has not been studied in Iran.

Habitat: North of Gonabad, I found them on the still-loose sands of dune stabilization areas. *Tamarix*, thorny psammophilous shrubs, and grasses constituted the vegetation. This agrees with observations in the former Soviet Union (Terentjev and Chernov, 1949:211).

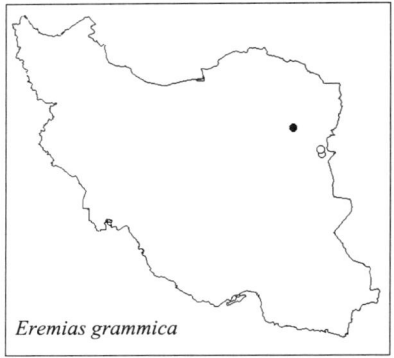
Eremias grammica

Distribution: In Iran known only from eastern Khorasan (Plate 24A); occurs in adjacent lowland regions of Afghanistan north of the Hindu Kush and in the adjoining Central Asian republics, Turkmenistan, Uzbekistan, north into Kazakhstan and east into western Xinjiang, China.

Remarks: The two Iranian specimens (CAS 141198, 141202) which I have compared with Afghan material differ from the latter in having the first supraocular broken up into small scales, whereas the Afghan specimens have a distinct first preocular separated from the second by a series of granules. In the Iranian specimens the size difference between lateral and middorsal scales is less pronounced than in the Afghan specimens.

Eremias intermedia (Strauch, 1876) (Plate 11-I)
Aralo-Caspian racerunner

Podarces (*Eremias*) *intermedia* Strauch, 1876:28 (Type locality: Kizil Kum, Aralo-Caspian Desert; Lectotype: ZIL 3664 [designated by Szczerbak 1974:181] see also Bauer and Günther, 1995:52–53).
Eremias intermedia: Boulenger, 1887a:100–101. — Lantz, 1918:14; 1928:111–117, pl. 1, fig. 6. — F. Werner, 1936:201. — Terentjev and Chernov, 1949:205–206. — S. Anderson, 1968:333; 1974:39, 43.
Eremias arguta: Guibé, 1957:139 (in part; not *Lacerta arguta* Pallas, 1773:718).

Diagnosis: Three nasals, lower in contact with 2 or 3 supralabials; subocular not bordering mouth; fourth toe without distinct fringe, with single row of subdigital scales; space separating series of femoral pores at least one-third length of each; tympanic shield usually distinct; fourth supraocular usually distinct.

Color pattern: Dorsum gray with brown tinge; longitudinal rows of small round or oval whitish spots edged with dark brown or black on body; a few scattered dark, usually elongated spots or narrow stripe along middorsum; limbs with more or less ocellate spots above; venter white.

Size: Snout-vent length 69 mm. Largest Iranian specimen: snout-vent 54 mm, tail 91 mm.

Habitat: Stabilized sands covered with semi-shrubby vegetation; uses rodent burrows as refuge. By contrast, closely similar *Eremias nigrocellata* is reported to inhabit hard, mainly loess soils. (Terentjev and Chernov, 1949:206).

Distribution: Southern Central Asia and northern Afghanistan; it has been taken in the valley of the Tajan River at the point where the borders of Iran, Afghanistan, and Turkmenistan come together. Apparently, no previous specimens within the borders of Iran have been documented, although Terentjev and Chernov (1949:206) said that it occurs in northeastern Iran. Szczerbak (1974:182–183, fig. 56) listed no Iranian records.

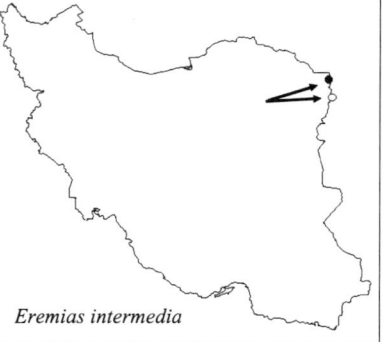
Eremias intermedia

Remarks: The only Iranian specimens I assign to this species are those from Sarakhs, identified by Guibé (1957:139) as *Eremias arguta*. They differ from Iranian specimens identified as *E. nigrocellata* in the following characters: snout-vent length/tail length (SV/T) = 0.53–0.60; supraoculars completely separated from frontal by row of granules; fourth toe with single row of subdigital scales and complete row of much smaller lateral scales; 16–18 (mean 16.8) ventrals in longest transverse row; ocelli on dorsum and limbs small, indistinct in alcohol. I find no obvious difference from *E. nigrocellata* in the tympanic shield (small or indistinct) or fourth supraocular (absent or indistinct). I have examined two specimens (BMNH 85.11.30.1–2/1946.8.7.47–48, Desert of Kizilkum) from the series on which Strauch based his species. In both, the fourth supraoculars are completely broken up into small scales, the SV/T ratio is 0.82 in the adult and 0.63 in the immature specimen. In the latter specimen the supraoculars are completely separated from the frontal and the frontoparietals by a ring of small scales, while in the adult, the supraoculars are in contact with the frontal.

Eremias lalezharica Moravec, 1994
Lalehzar racerunner

Eremias (Eremias) lalezharica Moravec, 1994:61–66, figs. 1–2 (Type locality: "Lalezhar, 29°31′N, 56°51′E, N foot of Mt. Lalezhar" [= Lalehzar *fide* U.S. Board of Geographical Names, Gazetter #19, Iran], Kerman Province, Iran, 2800–3100 m elevation. Holotype: NMP6V 34555/3).

Diagnosis: "A species of subgenus *Eremias* (subocular bordering mouth, only one frontonasal, two supraoculars, femoral pore series separated by a very short space and reach the knee . . .). It differs from all other known species by several [3–5 (8)] smaller shields situated laterally and posterior to each fourth submaxillary shield, instead of a distinct individual fifth submaxillary. The fifth [*sic*; fourth] submaxillary is exceptionally well developed and the smaller shields are located laterally and posteriorly to it. The smaller shields can also surround the fourth submaxillary and together with it border the third submaxillary. Higher rate of separation of fourth (41.7%) and third (16.7%) submaxillary shields from lower labials, as consequence of above described arrangement of the chin shields, is another characteristic feature of the new species . . ." (Moravec, 1994:61).

Color pattern: (Fig. 99) (In preservative, initial fixation in formalin) two dorsal and two dorsolateral light longitudinal stripes more or less interrupted into small whitish spots;

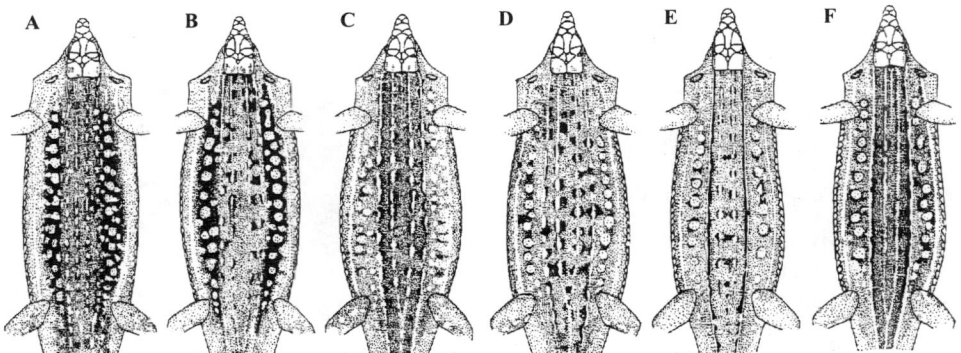

FIGURE 99. Dorsal pattern variation in *Eremias lalezharica*. (A-B) NMP6V 34555/1-2, male paratypes; (C) NMP6V 34555/3, holotype; (D-E) NMP6V 34555/4-5, female paratypes; (F) ZFMK 54840. (From Moravec, 1994, fig. 2).

lateral and median edges of these spots bordered with black; on flank a light lateral stripe from posterior edge of ear to groin; between dorsolateral and lateral stripes a row of formerly blue ocelli extending from in front of forelimbs to above insertion of hindlimbs; ocelli sporadically associated, above or below, with other smaller spots; upper labials dark with white markings; dorsal surface of limbs dark with lighter spots; venter white, lateral ventrals marked with black. Males have more conspicuous ocelli with better developed black margins fusing to form conspicuous dark lateral stripe; females have less interrupted dorsolateral and dorsal light stripes. (Moravec, 1994:64; fig. 2).

Size: Snout-vent to 71 mm, tail to 116 mm.

Habitat: The type locality is a mountain plateau in the vicinity of the village of Lalehzar [= Laleh Zar]. The vegetation consists of degraded steppe dominated by *Artemisia herba-alba*, *Astragalus* sp., *Zygophyllum* sp., *Salvia* sp., and *Ferula* sp;. gardens and fields containing *Ligurus sativus* and poplar; and wet meadows with *Orchis* sp., *Pedicularis* sp., *Carex* sp., *Juncus* sp., *Mentha* sp., *Eleocharis pauciflora*. All specimens were collected in open fields of soil and stones washed down the slopes of Mt. Lalehzar [= Laleh Zar]. Large solitary stones and scanty vegetation were characteristic. The lizards were not rare and were frequently seen along the banks of a drainage ditch and in the vicinity of irrigated gardens. (Moravec, 1994:64–66).

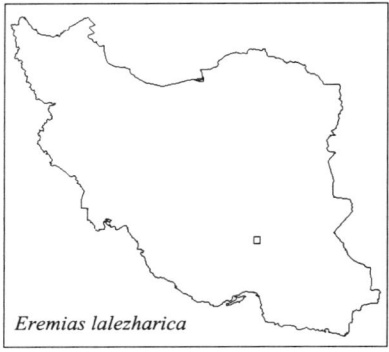
Eremias lalezharica

Distribution: Known only from the type locality in the mountains of south-central Iran.

Remarks: Gravid females were collected in late May. (Moravec, 1994:66).

Eremias lineolata (Nikolsky, 1896) (Plate 12A)
Striped racerunner

Scapteira lineolata Nikolsky, 1896:371 (Type locality: between Feizabad and Nusi, eastern Iran; Lectotype: ZIL 8801 [designated by Szczerbak, 1974:202]); 1897:330–332, pl. 18, fig. 4. — Zarudny, 1897:358. — Nikolsky, 1900:395. — Zarudny, 1903:18. — Boulenger, 1921: 363–365. — F. Werner, 1936:201.

Eremias (Rhabderemias) lineolata: Lantz, 1928c:79–84, pl. 1, figs. 7a–b, pl. 2, fig. 1, pl. 3, fig. 2. — Terentjev and Chernov, 1949:202–203, text-fig. 79. — S. Anderson, 1963:476; 1968: 333. — Tuck, 1971a:57. — S. Anderson, 1974:39, 43. — Szczerbak, 1974:202–212, figs. 62–64. — Bannikov, et al., 1977:183. — Schleich, 1977:127, 129.

Rhabderemias lineolata: Welch, 1983:55.

Diagnosis: (Figs. 100 and 101) Subocular bordering mouth; lateral scales of fourth toe in complete row length of toe, forming distinct fringe; single row of subdigital scales, i.e., total of 3 scales counted around fourth toe (except an extra scale may be

FIGURE 100. Ventral aspect of abdomen, hindlimb, and tail of *Eremias lineolata*. (From Terentjev and Chernov, 1949, fig. 79.)

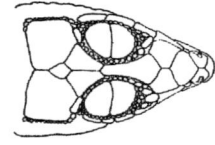
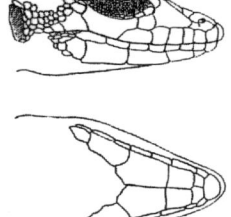
FIGURE 101. Head of *Eremias lineolata*. (From Szczerbak, 1974, fig. 62.)

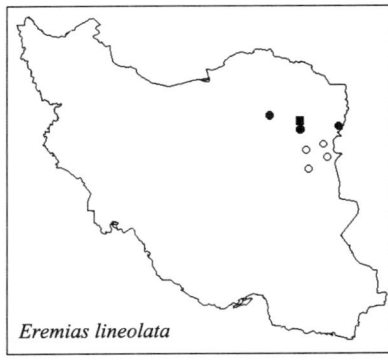
Eremias lineolata

present at joints); tibial scales more than twice as large as adjacent scales.

Color pattern: Dorsum sandy cream with 7 or 8 brown stripes, outer dorsolateral stripes broadest; head uniform light brown or with small dark spots; limbs with light rounded or oval spots on dark brown background; sides and underside of tail yellow in life; venter white. Adults and juveniles similarly marked.

Size: Snout-vent length 55 mm, tail 116 mm.

Habitat: (Plate 24A) Sandy, loess, and clay soils where there is scanty shrubby vegetation.

Distribution: Southern Central Asian republics, northeastern Iran (Khorasan), and northern lowland Afghanistan.

Remarks: *Eremias scripta* appears to be the sister taxon for this species.

Eremias nigrocellata Nikolsky, 1896
Black-ocellated racerunner

Eremias arguta: Kessler, 1878:171 (in part; not *Lacerta arguta* Pallas, 1773:718). — Guibé, 1957:139 (in part).
Eremias nigrocellata Nikolsky, 1896:371 (Type locality: Sistan, Iran [restricted by Szczerbak 1974:191]; Lectotype: ZIL 8800 [designated by Szczerbak 1974:191]); 1897:326–328, pl. 18, fig. 2. — Zarudny, 1897:357–358. — Lantz, 1928c:106–111. — Terentjev and Chernov,1949:206–207. — S. Anderson, 1968:333. — Tuck, 1971a:58. — Anderson, 1979:39, 43. — Bannikov, *et al.*, 1977:182. — Schleich, 1977:127, 129.
Eremias intermedia var. *nigrocellata*: Nikolsky, 1899b:399. — Zarudny, 1903:21–22. — Morich, 1929:31. — F. Werner, 1936:201. — S. Anderson, 1963:476.
Eremias intermedia: Nikolsky, 1907a:282 (in part; not *Podarces intermedia* Strauch 1876:28); 1915:442–446. — Boulenger, 1921:333–336.
Eremias (*Ommateremias*) *nigrocellata*: Szczerbak, 1974:190–199, figs. 57–59.
Ommateremias nigrocellata: Welch, 1983:55.

Diagnosis: (Fig. 102) Three nasals, lower in contact with 2 or 3 anterior supralabials; subocular not bordering mouth; fourth toe without distinct fringe, with 2 rows of subdigital scales, internal much the larger; tympanic scale usually small or indistinct; fourth supraocular usually indistinct.

Color pattern: Dorsum gray with fairly large light gray or nearly white circular spots bordered with black in 6–10 more or less regular longitudinal rows, ocelli faint or absent on middorsum; venter white. Young and adults similarly marked, young most prominently (Fig. 105a).

Size: Snout-vent length 83 mm, tail 100 mm.

Habitat: Lives in hard, mainly loess soils, sometimes with admixture of sand or pebbles, covered with ephemeral vegetation, according to Terentjev and Chernov (1949). In contrast, the closely similar *Eremias intermedia* is said to prefer stabilized sand.

Distribution: Apparently discontinuously distributed — northern and eastern Iran; southeastern Turkmenistan, southern Uzbekistan, southwestern Tajikistan, and northern Afghanistan; a large hiatus between the Iranian localities and the Central Asian and Afghanistan distribution (see Szczerbak, 1974:193, fig. 59, map). Iranian localities are between 1300 and 1700 m, while known Afghan localities lie below 700 m.

Remarks: Guibé's (1957:139) specimens from Mahneh, which he identified as *Ere-*

mias arguta, are *E. nigrocellata*. They differ most notably from Iranian specimens of their nearest relative, *E. intermedia* (see above) in the following characters: snout-vent length/tail length = 0.70–0.85; supraoculars in contact with frontal; fourth toe with row of subdigital scales and row of ventrolateral, somewhat smaller scales, sometimes incomplete, and complete or nearly complete row of dorsolateral scales; 17–20 (mean 19.2) ventrals in longest transverse row; ocelli on dorsum, limbs, and tail distinct, larger than in *E. intermedia*. In all Iranian specimens I have examined (and in 2 from Tajikistan) the tympanic shield, while small, is distinct, and the fourth supraocular is broken up.

Subsequent to his original description of the form, Nikolsky came to regard *nigrocellata* as a subspecies of *E. intermedia* and Boulenger regarded the two as synonymous. It is not possible, without examining the specimens, to be sure that the material examined by Boulenger from Astrabad belongs to *E. nigrocellata* rather than to *E. intermedia*. The character of the subdigital scales is difficult to use and subject to interpretation. Although he regarded *E. nigrocellata* a subspecies of *E. intermedia*, the types of *nigrocellata* are not among the material listed by Nikolsky under *intermedia* (1915:443–444). I have examined one specimen (BMNH 99.4.20.4/1946.8.7.44) from Nikolsky's type series. The supraoculars are in contact with the frontal and the toes are as described for the specimens from Mahneh. The tympanic scale is small, but distinct, and the supraocular is broken up.

I am not persuaded, on the basis of the very limited material that I have seen, that *E. nigrocellata* is distinct from *E. intermedia*.

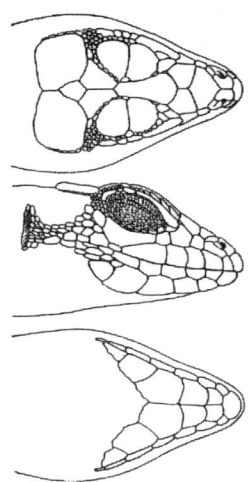

FIGURE 102. Head of *Eremias nigrocellata*. (From Szczerbak, 1974, fig. 57.)

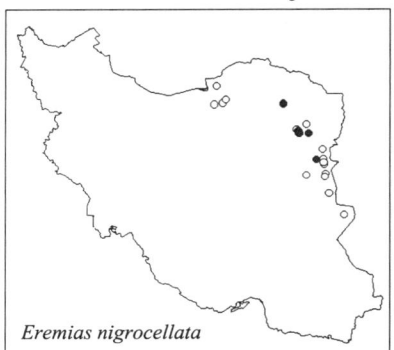

Eremias nigrocellata

Eremias nigrolateralis Rastegar-Pouyani and Nilson, 1998
Black-sided racerunner

Eremias nigrolateralis Rastegar-Pouyani and Nilson, 1998:94–101, figs. 1–8 (Type locality: 30°52'N, 53°09'E, 150 km northeast of Shiraz, Fars Province, Iran, about 1800 m; Holotype: GNHM Re. Ex. 5147).

Diagnosis: Lower nasal resting on 2 or 3 supralabials; subocular bordering mouth; lateral scales of fourth toe not forming distinct fringe; single frontonasal; two supraoculars; femoral pores separated by a short space, row of pores not reaching knee; 5 pairs of submaxillary shields, third pair separated by a single longitudinal row of narrow scales; small, rudimentary tympanic shield, more than 125 temporal scales; 41–42 gulars between submaxillaries and collar; 64–69 dorsal scales counted across back; upper caudal scales not distinctly keeled; a wide, uniformly black dorsolateral stripe strongly contrasting with dorsum; no ocelli on body and limbs. (Rastegar-Pouyani and Nilson, 1997:95–97).

Color Pattern: In life, dorsum light tan suffused with pale greenish-brown with two or four longitudinal series of dark spots; a wide uniformly black stripe without light ocelli, running from upper temporal region, becoming less intense on tail, disappearing on distal third of tail; upper surface of tail sandy gray; upper surface of forelimbs uniformly sandy grayish-tan; hind limbs similar in color, uniform or with a few scattered small, dark spots; no light ocelli on limbs or body; upper surface of head olive-brown, two dark brown blotches on parietal region, dark blotches on temporal and labial regions; ventrolateral region bluish-gray with a longitudinal series of 5–9 dark spots; venter bluish-white cream; lower surface of tail and gular region whitish cream (Rastegar-Pouyani and Nilson, 1997:99).

Eremias nigrolateralis

Size: Holotype (female): snout-vent length 78 mm; tail 122 mm; paratype (male): s-v 84 mm, tail 127 mm (Rastegar-Pouyani and Nilson, 1997:99).

Habitat: Wide, open silt and gravel plain with luxuriant steppe vegetation of *Artemisia herba-alba, Astragalus, Zygophyllum,* and *Euphorbia*. The two known specimens were collected syntopically with *Eremias persica, Trapelus agilis,* and *Phrynocephalus scutellatus.*

Distribution: Known only from the type locality in Fars Province, south-central Iran.

Eremias persica Blanford, 1875 (Plate 12B)
Persian racerunner, steppe lacerta

Eremias variabilis: De Filippi, 1865:354 (in part; not *Lacerta variabilis* Pallas, 1811).
Eremias persica Blanford, 1875:31 (Type locality: near Esfahan, Iran [restricted by Smith, 1935:383]; Lectotype: BMNH [designated by Szczerbak, 1974:124, but register number not given; see Bauer and Günther, 1995:40]); 1876:370–373, pl. 26, fig. 1. — Nikolsky, 1897:330; 1900:398. — Zarudny, 1903:20–21. — Nikolsky, 1907a:282; 1915:429–431. — Terentjev and Chernov, 1949:198. — S. Anderson, 1974:39, 43. — Bannikov, et al., 1977:175. — Tuck, 1979:103. — Nilson and Andrén, 1981:138, fig. 7. — Welch, 1983:53. — Clark, 1991:40–41. — Welch, et al., 1990:115. — Leviton, et al., 1992:57, col. pl. 8G. — Frynta, et al., 1997:9.
Podarces (Eremias) persica: Bedriaga, 1879:32.
Eremias velox: Boulenger, 1887a:97–99 (in part; not *Lacerta velox* Pallas, 1771). — F. Werner, 1895:4; 1903:342–343. — Annandale, 1906:197.
Eremias velox var. *persica*: Boulenger, 1921:312–314. — Smith, 1935:383–385. — F. Werner, 1936:201; 1938:268, 270. — Schmidt, 1939:66. — Forcart, 1950:149. — S. Anderson, 1963:476. — Clark, et al., 1966:2. — Guibé, 1966a:98. — S. Anderson, 1968:333. — Tuck, 1971a:58. — Schleich, 1977:127, 129.
Eremias velox velox-persica: Wettstein, 1951:440.
Eremias (Eremias) persica: Lantz, 1928b:53–60. — Szczerbak 1974:123–131, figs. 33–36.

Diagnosis: (Fig. 103) Lower nasal resting on 2 or 3 supralabials; subocular bordering mouth; lateral scales of fourth toe not forming distinct fringe; fourth toe with single complete row of subdigital scales, a complete row of somewhat smaller ventrolateral scales, and a few scattered, much smaller dorsolateral scales not forming complete row (total of 3 scales counted around penultimate phalanx of fourth toe); the 2 series of femoral pores separated by space not greater than one-fourth length of each; usually several collar scales distinctly larger than adjacent gulars; 28–39 gulars; 56–70 dorsals; 23–35 scales in ninth or tenth caudal annulus; broad lateral dark stripe enclosing one or 2 rows of white spots.

Color pattern: (Figs. 104a-b and 105b) Juveniles pale gray or tan dorsally, with 4 dark brown longitudinal stripes on back between dorsolateral white-spotted brown to black stripes, vertebral stripe being light; in adults dark stripes break up into 4 more or less regular rows of black spots on light ground; broad dorsolateral stripe remains more or less continuous for at least major portion of its length, containing single or double row of white ocelli and contrasting strongly with sandy-brown ground color; venter white; distal part of tail bluish to greenish blue in juveniles in life.

Size: Snout-vent length to 93 mm in males, 98 mm in females; tail to 168 mm.

Habitat: (Plates 20B, 21C, 24C) Common throughout the Central Plateau on open plains and slopes. I found them on a variety of surfaces including gravel alluvium, silt and gravel, sand and gravel, dry loose and compacted loess, gravel and rock, hamada, and disintegrating shale. Vegetation was almost always open steppe, such as *Artemisia herba–alba/Zygophyllum atriplicoides* association. They walk around the base of shrubs when pursued, breaking cover when hard-pressed and running directly for the nearest large shrub. I found them active at air temperatures from 15.5°C to 28.5°C and surface temperatures 20°–32°C.

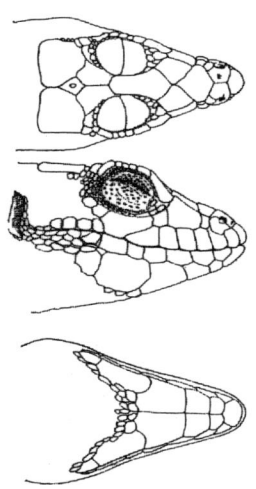

FIGURE 103. Head of *Eremias persica*. (From Szczerbak, 1974, fig. 33.)

Distribution: Central and eastern portion of the Iranian Plateau at elevations from 600–3050 m. In Iran it extends west to Qazvin and to the eastern front of the Zagros Mountains. In the north it reaches southern Turkmenistan, extending east and south through southern Afghanistan and Baluchistan to Waziristan, Pakistan. There is a record for Kuretu [= Quraitu], Iraq at the Iranian border (Procter, 1921:240 [see Leviton, et al., 1992, for comment]). Rastegar-Pouyani informs me (pers. commun., 23 October 1996) that he has collected specimens of this species from Kermanshah, Hamadan, and Avaj, and thus the Quraitu locality now seems more probable.

Remarks: *Eremias persica* is sympatric with *E. velox* in scattered localities on the northern and western margins of the Central Plateau in Iran and in southern Turkmenistan near the Iranian border; it is sympatric with *E. strauchi* in northeastern Khorasan and perhaps in the Kopet Dagh. The distributions of these three forms are complementary for the most part, but *E. persica* occupies the entire Central Plateau of Iran and the southern basins of

FIGURE 104. *Eremias persica*; (A) adult and (B) juvenile. (From Blanford, 1876, pl. 26, figs. 1 and 3, respectively.)

Eremias persica

Afghanistan and the basins of Baluchistan, Pakistan, while *E. v. velox* occurs in the lower, but more northern arid basins of southern Central Asia and northern Afghanistan. *Eremias strauchi* is the form of the Transcaucasian regions and the Kopet Dagh, extending into the northern mountains of Iran.

Color patterns of adults and juveniles appear to be sufficiently constant to distinguish the three forms from one another. In juveniles of *Eremias persica* the end of the tail is blue, while in those of *E. strauchi* the lower surface of the tail is yellow, and in those of *E. v. velox* the lower surface of the tail is red (G. Peters, 1964:463). Scale characters employed by Nikolsky (1915), Lantz (1928), and Boulenger (1921) to separate the three taxa are not sufficiently constant as diagnostic features. Although color pattern is variable and ontogenetic changes can be seen, all specimens I identify as *E. persica* have light ocelli enclosed in a black lateral stripe. The length of the suture between the supranasals is variable, but generally less than one-third length of frontonasal; amount of contact of infranasal with rostral is highly variable, from just touching (rarely not in contact) to a distinct suture; occasionally an interprefrontal is present in specimens from Sistan; length of interparietal greater than (rarely equal to) length of suture between parietals; dorsal caudal scales keeled in all specimens, but not as strongly so as in specimens of *E. v. velox*; basal median subcaudals smooth, becoming keeled in more posterior scales.

Szczerbak (1974:123) designated the lectotype as being in the British Museum and from near Esfahan, but did not give a register number. There are two specimens from "near Isfahan" in that collection, BMNH 74.11.23.30–31/1946.8.7.32–33. BMNH 1946.8.7.32, a male, comes closest to Szczerbak's counts and measurements for his lectotype.

Eremias pleskei Bedriaga, 1907
Pleske's racerunner

Eremias pleskei Bedriaga, 1907:238 (Type locality: Nachitschewan, Erivan, Transcaucasia [Nakhichevan]; Lectotype: ZIL 6724 [designated by Szczerbak, 1974:234]). — Lantz, 1928c:84–90, pl. 2, fig. 3, pl. 3, fig. 1. — Terentjev and Chernov, 1949:201–202. — Clark, *et al.* 1966:7. — S.

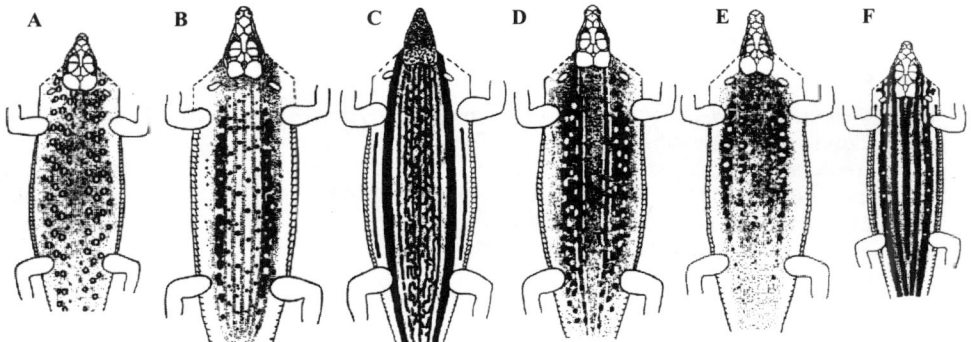

FIGURE 105. Color patterns of (A) juvenile *Eremias nigrocellata*, (B) adult *E. persica*, (C) adult *E. scripta*, (D) adult *E. strauchi*, (E) adult *E. velox*, and (F) juvenile *E. velox*. (After Terentjev and Chernov, 1949, figs. 81, 78, 80, 77, 76 and 75, respectively.)

Anderson, 1968:333; 1974:38, 43. — Bannikov, et al., 1977:186–187. — Schleich, 1977:127, 129. — Clark, 1991:40.
Eremias fasciata pleskei: Nikolsky, 1915:433–434.
Eremias (Rhabderemias) pleskei: Szczerbak, 1974:234–241, fig. 72–74.
Rhabderemias pleskei: Welch, 1983:55.

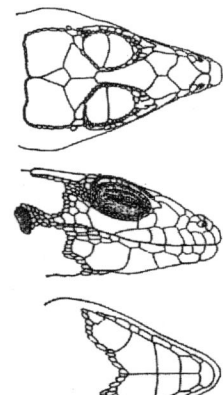

Diagnosis: Lower nasal resting on 2 or 3 supralabials; subocular bordering mouth (Fig. 107); lateral scales of fourth toe not forming distinct fringe; series of femoral pores separated by space at least one-third length of each (Fig. 106).

Color pattern: Light brown above, with 7 brown longitudinal lines broader than interspaces, vertebral stripe bifurcating on nape, stripe on flank faint; limbs with light ocelli; venter cream; lower surface of tail yellow and sides of body yellowish in life.

FIGURE 106. Ventral aspect of abdomen, hindlimb, and tail of male *Eremias pleskei*. (From Lantz, 1928, pl. 2, fig. 3.)

Size: Snout-vent length 60 mm, tail 106 mm.

Habitat: Semidesert sandy or stony soils of *Artemisia* steppes.

Distribution: Very narrowly distributed, restricted to the Armenian Plateau of Armenia, Nakhichevan, and Azerbaidzjan, the extreme northwest corner of Iran, and in northeastern Turkey near the borders of Armenia and Iran.

FIGURE 107. *Eremias pleskei*. (From Szczerbak, 1974, fig. 72.)

Remarks: The single Iranian specimen examined (CAS 96263) differs from 2 northern extralimital specimens examined in having 5 rather than 6 supralabials anterior to subocular, and in having fifth chin shield in contact with the infralabials. In 19 specimens from Turkey, fifth chin shield is separated from infralabials, and usually 6 supralabials (rarely 5 or 7) anterior to subocular.

Nikolsky (1915:434) speculated that De Filippi's (1865:354) record of *Eremias pardalis* in the Yerevan steppe refers to this species.

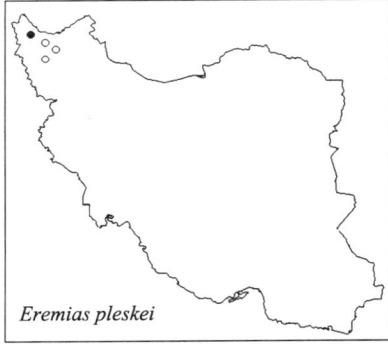

Eremias pleskei

Eremias scripta (Strauch, 1867) (Plate 12C–D)
Sand racerunner

Podarces (Scapteira) scripta Strauch, 1867:424 (Type locality: Aralo–Caspian desert. Lectotype: ZIL 3669 [designated by Szczerbak, 1974:213]); 1868a: cols. 327–328.
Eremias (Rhabderemias) scripta: Lantz, 1928c:73–79, pl. 2, fig. 2, pl. 3, fig. 3. — Terentjev and Chernov, 1949:204–205. — S. Anderson, 1963:476; 1968:333; 1974: 38, 43. — Szczerbak, 1974:212–229, figs. 65–67.
Eremias scripta scripta: Bannikov, et al., 1977:185.
Rhabderemias scripta: Welch, 1983:55.

Diagnosis: (Fig. 108) Subocular borders mouth; lateral scales of fourth toe in complete row for entire length of toe, forming distinct fringe, 2 complete rows of subdigital scales, i.e., total of 4 scales counted around fourth toe (except extra scale may be present at joints).

Color pattern: (Fig. 105c) Dorsum sandy gray, broad dark dorsolateral stripe from nostril through eye, along body and sides of tail, one or 2 additional dark stripes mediad to these on each side, remainder of dark dorsal stripes interrupted and anastomosing to form reticulate pattern; limbs with light ocelli or oval spots on dark brown background; venter white. Adults and young similarly patterned.

Size: Snout-vent length 66 mm, tail 140 mm.

Habitat: In Pakistan these lizards inhabit flat sandy terrain with relatively numerous shrubs and clumps of grass (Minton 1966:109); Terentjev and Chernov (1949:205) reported that in the Central Asian republics they prefer the area between dunes, as well as sands with vegetation.

Distribution: Central Asian republics; southern desert region of Afghanistan; Chagai District of Baluchistan, Pakistan; Terentjev and Chernov (1949:204), Bannikov, et al. (1977:185), and Szczerbak (1974:215) stated that it is found in eastern Iran, but I find no published locality records for Iran, nor specimens cited. If it occurs in Iran, it is to be expected in northern Khorasan, and if the southern population is indeed the same species as that of Central Asia, it may occur along the Iran-Afghan and Iran-Pakistan borders in Baluchistan. There is a large hiatus between the Central Asian populations and that south of the Hindu Kush (see Szczerbak 1974:215, fig. 67, map).

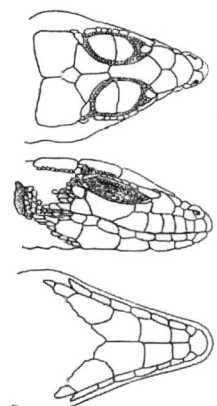

FIGURE 108. *Eremias scripta*. (From Szczerbak, 1974, fig. 65.)

Eremias strauchi Kessler, 1878
Strauch's racerunner

Eremias strauchi strauchi Kessler, 1878 (Plate 12E)
Strauch's racerunner

?*Eremias variabilis*: De Filippi, 1865:354 (in part; not *Lacerta variabilis* Pallas, 1811).
Eremias strauchi Kessler, 1878 (suppl.):166, pl. 2 (Type locality: Etshmiadzin, Armenia; Lectotype: MLSU 166 [designated by Szczerbak 1974:114]). — Lantz, 1928c:60–67, pl. 3, figs. 4–5 (in part). — Rostombekov, 1938:15. — Terentjev and Chernov, 1949:196–197, fig. 77 (in part). — S. Anderson, 1974:39, 43. — Clark, 1991:41.
Eremias velox var. *persica*: Boettger *in* Radde, 1886:50 (*fide* Lantz, 1928; not *Eremias persica* Blanford 1874). — Boulenger, 1921:312–314 (in part).
Eremias velox strauchi: Nikolsky, 1915:427–428 (in part). — S. Anderson, 1963:476 (in part); 1968:333 (in part). — Schleich, 1977:127, 129 (in part).
Eremias strauchi strauchi: Szczerbak, 1971:83–86. — Bannikov, et al., 1977:173–174. — Welch, 1983:53.
Eremias (*Eremias*) *strauchi strauchi*: Szczerbak, 1974:117–118.

Diagnosis: (Fig. 109) Lower nasal resting on 2 or 3 supralabials; subocular bordering mouth; lateral scales of fourth toe not forming distinct fringe; series of femoral pores separated by space not greater than one-fourth length of each; fourth toe with single row of subdigital scales, complete row of somewhat smaller ventrolateral scales, and a few scattered, much smaller dorsolateral scales not forming complete row (total of 3 scales around penultimate phalange of fourth toe); usually several collar scales distinctly larger than adjacent gulars; 23–33 gulars; 56–68 dorsals; 24–35 scales in ninth or tenth caudal annulus; vertebral zone of back uniform in adults, usually without dark or light spots; ventral surface of tail yellow in juveniles.

Color pattern: (Fig. 105d) Ground color yellow-gray, olive-gray, or brownish-gray; vertebral region uniform, without spots in adults; broad dark dorsolateral and flank area contains up to 4 complete or partial rows of light spots, dorsolateral dark stripe frequently breaking up in places, leaving dark-margined light ocelli; spots and ocelli continue onto neck and onto anterior part of tail; lower row of spots or ocelli bright blue or yellow-green (in life); upper labials with alternate light and dark marks; juveniles similar to adults, but with more distinct black spots, tending to fuse longitudinally into alternating light and dark lines; lower surfaces of thighs and tail bright yellow.

Size: Snout-vent length to 68 mm, tail 122 mm.

Habitat: I found them on flats and gentle slopes, primarily on silty soils, but also on steep rubbly slopes of red sandstone and on flat areas of a basalt flow. Vegetation was sparse overgrazed steppe shrub.

●○ *Eremias s. strauchi*
▲△ *Eremias s. kopetdaghica*

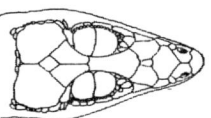
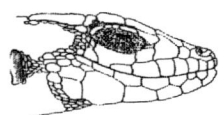
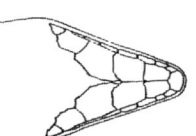

FIGURE 109. *Eremias strauchi*. (From Szczerbak, 1974, fig. 29.)

Distribution: The Azarbaijan Provinces of northwestern Iran, and extending onto the high Armenian Plateau of Azerbaijan and Armenia; to about 3500 m.

Remarks: Guibé (1957:138) assigned his specimens from the vicinity of Gonbad-e Kavus, northeastern Mazanderan, to "*E*[*remias*]. *velox strauchi*" on the basis of the presence of an internasal shield. He stated that the interparietal was longer than the suture between the parietals. Both these characters are found in some specimens of *E. persica*. In the high number of dorsal scales (68–70), they agree better with *E. persica* than with *E. strauchi kopetdaghica*. I have not seen these specimens. I identified three specimens in the Iran National Museum of Natural History in Tehran as *E. strauchi* but did not compare with other material or with Szczerbak's description: MMTT 1413, 1423, Kerman (Ostan 9): N Baft, N slope Kuh-e Shah; MMTT 1421, Mazanderan (Ostan 12): 41 km N Semnan to Firuzkuh.

Eremias strauchi kopetdaghica Szczerbak, 1972 (Plate 12F)
Kopet Dagh racerunner

Eremias velox: Nikolsky, 1897:329–330 (in part; not *Lacerta velox* Pallas, 1771); 1915:418 (in part).
Eremias velox strauchi: Nikolsky, 1915:427–428 (in part). — ?Guibé, 1957:138. — S. Anderson, 1963:476 (in part). — 1968:333 (in part). — Schleich, 1977:127, 129 (in part).
Eremias strauchi: Lantz, 1928c:60–67, pl. 3, figs. 4–5 (in part). — Terentjev and Chernov, 1949:196–197, fig. 77 (in part). — S. Anderson, 1974:39, 43 (in part).
Eremias strauchi kopetdaghica Szczerbak, 1972:83–86, fig. B (Type locality: vicinity of Ai-Dere-Tuzli-Tepe, Kara-Kalinskii, Turkmenistan; Holotype: ZIK 4); — Bannikov, *et al.*, 1977:174. — Welch, 1983:53.
Eremias (Eremias) strauchi kopetdaghica: Szczerbak, 1974:118–123, fig. 31 lower.

Diagnosis: Differs from nominal subspecies in having larger scales: 19–28 gulars, 48–59 dorsals, 20–26 scales in ninth or tenth caudal annulus.

Color pattern: In life, dorsum sandy brown, vertebral area without stripe or spots; 2 prominent black dorsolateral stripes on each side of back, broader than interspace, margined with and sometimes enclosing small white and bright yellow spots; posterior of thighs with black reticulum; dark-margined white spots extend onto tail; head brown with indistinct darker markings; venter creamy tan, lower surfaces of tail and thighs bright yellow in adults.

Size: Snout-vent length to 76 mm; tail 122 mm.

Habitat: I found them near Robat-e Sang on the slopes and ridges of fracturing and eroding sandstone mountains, and in narrow alluvium-filled valleys; vegetation consisted of sparse, overgrazed shrubs. They were unlike *Eremias persica* in behavior, in that they went up steep slopes among rocks.

Distribution: (see map p. 226) The Kopet Dag of northern Khorasan and eastern Mazanderan, into southern Turkmenistan, descending to 150 m or below on the northern flanks; the Qal'eh Manar Mountains of Khorasan, and may occur in the isolated mountain regions of the Iranian Plateau.

Eremias velox (Pallas, 1771)
Rapid fringe-toed lizard; Central Asian racerunner

Eremias velox velox (Pallas) (Plate 12G)
Rapid fringe-toed lizard; Central Asian racerunner

Lacerta velox Pallas, 1771:406, 457 (Type locality: "Inderskiensem lacum," [see Zhao and Adler, 1993:204 for note on type locality]; Neotype: ZIL 16233 [designated by Szczerbak 1974:84]; see Bauer and Günther, 1995:57 for additional remarks).

Eremias velox Wiegmann, 1834:9. — ?Blanford, 1876:374. — Boulenger, 1887a:97–99 (in part). — Nikolsky, 1915:414–427 (in part). — Boulenger, 1921:308–312. — Terentjev and Chernov, 1949:194–196, figs. 75–76.

Podarces (Eremias) velox: Bedriaga 1879:32.

Eremias velox velox: Lantz, 1918:14; 1928:44–52, pl. 1, figs. 2a–b. — S. Anderson, 1968:333. — Tuck, 1971a:58. — S. Anderson, 1974:39, 43. — Bannikov, et al. 1977:172. — Schleich, 1977:127, 129. — Welch, 1983:54. — Clark, 1991:41.

Eremias (Eremias) velox velox: Szczerbak, 1974:98–99.

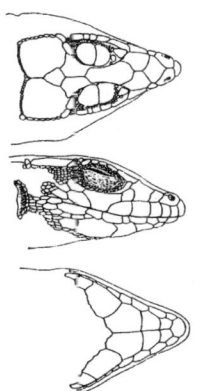

FIGURE 110. *Eremias velox*. (From Szczerbak, 1974, fig. 24.)

Diagnosis: (Fig. 110) Lower nasal resting on 2 or 3 supralabials; subocular bordering mouth; lateral scales of fourth toe not forming distinct fringe; fourth toe with single row of subdigital scales, complete row of somewhat smaller ventrolateral scales, and a few scattered, much smaller dorsolateral scales not forming complete row (total of three scales counted around penultimate phalanx of fourth toe); usually several collar scales distinctly larger than adjacent gulars; 23–25 gulars; 46–56 dorsal scales; series of femoral pores separated by space not greater than one-fourth length of each; adults with dark interrupted dorsolateral black stripe forming ocelli with white spots, middle of back with dark spots or dark-margined ocelli; young with alternating dark and light stripes, vertebral stripe dark, bifurcating on nape; lower surface of tail carmine red in young.

Color pattern: (Fig. 105e-f) Gray, brownish, or dark brown on dorsum, adults with irregular longitudinal rows of dark spots and sometimes dark-margined white ocelli; flanks with light ocelli (sometimes bright blue in life) resulting from break-up of dark dorsolateral line present in juveniles; venter white, lower surfaces of throat, shoulders, and groin sometimes yellowish in life, at least in males; young with 3 dark stripes on dorsum between dorsolateral stripes which contain light spots; vertebral stripe dark, bifurcating on nape; ventral surface of tail and sometimes backs of thighs carmine red in juveniles.

Size: Snout-vent length to 77 mm, tail to 144 mm.

Habitat: (Plate 23D) I found them on silty alluvium of flats and low hills in irrigated cultivated fields and in areas of steppe vegetation (*Artemisia, Tamarix, Acacia*). South of 'Abbasabad they occur on dry, salt-encrusted silt.

Distribution: In Iran it occurs on the southern coast of the Caspian Sea (Blanford 1876:374), valleys of the Kopet Dagh, and scattered localities on the northern and western margins of the Central Plateau, south at least to southern Kerman Province, and west to Arak and Kharakan. In Central Asia it extends from the Volga to western Mongolia and into China as far east as Gansu and Nei Mongol (see Szczerbak, 1974: 92, fig. 28, map; Zhao and Adler, 1993:204).

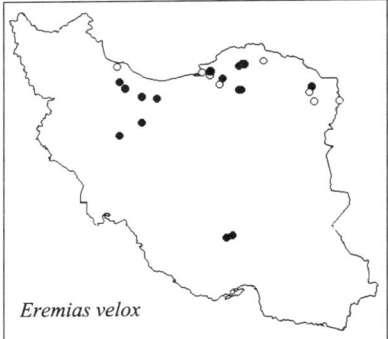

Eremias velox

Genus *Lacerta* Linnaeus, 1758
Lacertas

Lacerta Linnaeus, 1758:200 (Type species: *Lacerta agilis* Linnaeus, 1758, by subsequent designation of Fitzinger, 1843:20).

Definition: Parietal foramen present; frontal bones paired; clavicle strongly expanded medially; interclavicle cruciform, lateral arms not strongly directed forwards; sternal fontanelle almost always present, nearly always roughly oval; sexual variation in number of presacral vertebrae; all presacral vertebrae with ribs, except first 3 cervicals; free ribs divided into 2 series, an anterior one of long ribs and posterior one of short ribs. Hemipenes symmetrically bilobed; typically armature absent and lobes not complexly folded; lobes usually plicate, microornamentation variable; apical regions of lobes usually short with small sulcal lips; no large conical papillae at lobe tips.

Head shields normal; nostril usually in contact or close to the upper labial, bordered posteriorly by one, 2, or rarely 3 postnasals; lower eyelid usually scaly, although a small transparent window may be present; anteriorly parietals typically not extending to outer margin of postorbital bone; first supratemporal often large; masseteric often present. Dorsal body scales small or moderate (smaller than proximal caudals); collar well marked; ventral scales smooth, truncate, strongly overlapping or not, in 6–10 longitudinal rows. Toes cylindrical or compressed, usually tubercular beneath (occasionally strongly keeled); femoral pores present. Tail long, unmodified. (Arnold, 1973:330).

Key to the Species of *Lacerta* in Iran

1a. Lower eyelid with 5–7 transparent shields edged with black; subdigital lamellae keeled .... *Lacerta cappadocica urmiana* (Lantz and Suchow, 1934) (p. 232)
1b. Lower eyelid without transparent shields; subdigital lamellae smooth or tuberculate.. 2
2a. Ventral plates more or less rectangular with rectilinear or nearly rectilinear posterior margins (Fig. A), juxtaposed to subimbricate 3
2b. Ventral plates shaped like inclined parallelograms (Fig. B) with notches between longitudinal rows, posterior and lateral edges strongly overlapping 10

3a. Dorsal scales distinctly keeled; collar serrated (Fig. C) 4
3b. Dorsal scales smooth, granular; collar not serrated (Fig. D) 6

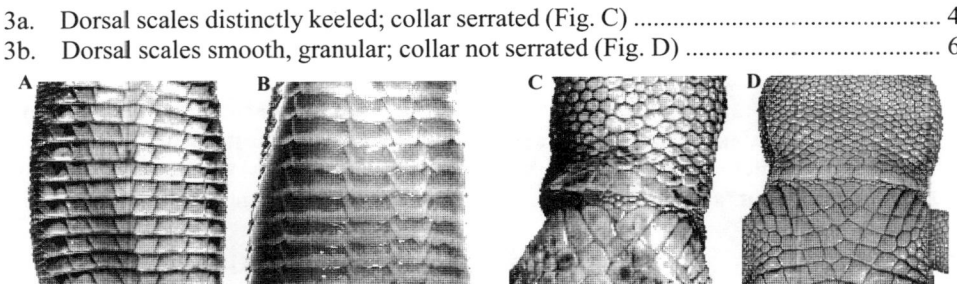

Generalized patterns of ventral scutes

Generalized patterns of collars (serrated or notched between scales) (Fig. C), and smooth edge (Fig. B)

4a. Venter gray or black, never greenish or yellowish ....
.... *Lacerta mostoufi* Baloutch, 1976 (p. 238)
4b. Venter greenish or yellowish in life.. 5
5a. Rostral touches nostril, or barely separated from it; 43–49 dorsals at midbody; 27–32 lamellae under 4th toe .... *Lacerta chlorogaster* Boulenger, 1908 (p. 234)
5b. Rostral does not touch nostril; 32–43 dorsals at midbody; 20–25 lamellae under 4th toe.. *Lacerta praticola* Eversmann, 1834 (p. 239)
6a. 5–6 (usually 5, rarely 4) supralabials anterior to subocular; normally 2 superposed postnasals (sometimes fused on one or both sides of head); pterygoid teeth strongly developed; outer ventrals with small black spots ....
.... *Lacerta brandtii* De Filippi, 1863 (p. 231)
6b. 3–4 (usually 4, rarely 5) supralabials anterior to subocular; normally single postnasal; no pterygoid teeth; outer ventrals without black spots (turquoise spots present in males)
.... 7
7a. Row of granules between supraoculars and supraciliaries often interrupted; dorsum green or greenish-yellow in life .... *Lacerta valentini valentini* Boettger, 1892 (p. 248)
7b. Supraciliary scales invariably separated from supraoculars by complete row of 6–18 granules; dorsum brown, brownish, gray or black tones, not green in life 8
8a. Dorsum with a vertebral stripe formed of close-set black dots, breaking into distinct, larger spots on tail; a row of black dots down either side of back; flanks dark with light ocelli in a single row .... *Lacerta steineri* Eiselt, 1995 (p. 245)
8b. Dorsal pattern not as above, back with small dark spots and mottlings, sometimes more or less linearly arranged 9
9a. Venter in life brick red; Alborz and Kopet Dagh ranges ....
.... *Lacerta defilippii* Camerano, 1877 (p. 236)
9b. Venter in life yellowish, bluish, or greenish-white; Talysh Mountains ....
.... *Lacerta raddei* Boettger, 1892 (p. 243)
10a. Ventral plates in 10 longitudinal series; 34–37 dorsals at midbody 11
10b. Ventral plates in 6 or 8 longitudinal series; 38 or more dorsals at midbody 12
11a. Outer row of ventrals (marginals) smooth; 20–22 gulars; 13–17 femoral pores on each side; lower edge of subocular ½ or less than ½ maximal length of shield ....
.... *Lacerta princeps princeps* Blanford, 1874 (p. 240)
11b. Outer row of ventrals (marginals) keeled; 17–19 gulars; 16–21 femoral pores on each side; lower edge of subocular ½ or more than ½ maximal length of shield ....
.... *Lacerta princeps kurdistanica* Suchow, 1936 (p. 241)

12a. 17–21 femoral pores, row of pores reaches knee (Fig. E); usually less than 20 temporal scales; 5th chin shield always well developed; juveniles usually with uninterrupted lateral line in addition to vertebral and dorsolateral lines .... *Lacerta strigata* Eichwald, 1831 (p. 246)

12b. 12–16 femoral pores, row of pores does not reach knee (Fig. F); usually more than 20 temporal scales; 5th chin shield small or absent; juveniles with lateral light line interrupted in its anterior half ..... *Lacerta media media* Lantz and Cyrén, 1920 (p. 237)

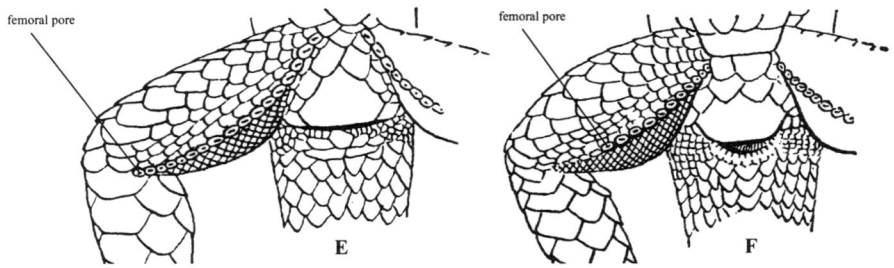

Row of femoral pores reaches knee (Fig. E) and does not reach knee (Fig. F)
(From Bannikov, *et al.*, 1977, fig. 56)

Distribution: As redefined by Arnold (1973), the genus *Lacerta* (*sensu lato*) encompasses about 35 species, centering on the least xeric regions of the Mediterranean and extending through the northern and western mountain ranges of Southwest Asia. Only two species (*L. vivipara* and *L. agilis*) occur beyond this region, extending the range of the genus across northern Eurasia to about 63°N in eastern Siberia and to Japan. In Southwest Asia, the genus is confined to the Mediterranean coast, Turkey, and north-south trending mountain ranges, such as the Lebanon and Anti-Lebanon and the Zagros Mountains; in the northern mountains it extends east into the western Kopet Dagh. The genus *Lacerta* (*sensu lato*) is also found on the Arabian Peninsula in the mountains of Oman. *Lacerta mostoufi* appears to be a disjunct outlier on the Central Plateau of Iran. Twelve species are recorded for Iran.

Remarks: In his review of the genus, Arnold (1973) removed the tropical and southern African species from *Lacerta*, and raised the subgenera *Podarcis* and *Gallotia* to generic level. Arnold's *Lacerta* part I consists of large, robust species having strongly imbricate ventrals and serrated collars. These species usually inhabit areas of dense, shrubby vegetation. *Lacerta princeps*, *L. strigata*, and *L. media* are the species of this group found in Iran.

Arnold's *Lacerta* part II consists of smaller species (usually less than 90 mm snout-vent length) and almost all have brightly colored venters, at least in breeding males. They occupy a wide variety of structural niches, but many are adapted to living on and around rock faces. All of the remaining Iranian species belong to this group.

Lacerta part I is a continuously distributed group, apparently of recent expansion, while the larger and more diversified group, *Lacerta* part II, is discontinuously distributed, its range presumably broken and restricted by post-Pleistocene climatic events (Arnold, 1973:340, 355). Arnold's (1989:fig. 23) analysis of the Lacertidae recognizes a more complex condition of paraphyletic relationships within this nominal genus. Among the Iranian species, *L. cappadocica*, *L. princeps*, *L. media*, and *L. strigata* belong to one of the two main branches of his tree that includes *Takydromus* and the Ethiopian and Saharo-European genera, as well as *L. jayakari*, *L. vivipara*, *L. agilis*, and *L. lepida*. All of the other species are on the other branch, which also includes the genera *Podarces*, *Algyroides*, *Psammodromus*, and *Gallotia*.

In this scheme, *L. brandtii* is the primitive member of a clade which includes *Gallotia* and *Psammodromus* in addition to *L. parva*. The relationships of the rock lizards to other taxa remain to be clarified, but the *L. saxicola* complex, centered on the Transcaucasian region, including the Iranian species *L. defilippi, L. raddei,* and *L. valentini*, appears to form a clade (the more recently described *L. steineri* Eiselt, 1995 would also join this clade) and *L. chlorogaster* and *L. praticola* may also be related to this group (Arnold, 1989:235). Rastegar-Pouyani and Nilson (1998) have described yet another species of this group, *Lacerta zagrosica*, from the central Zagros Mountains of Iran (see Addendum to the Lacertidae, p. 255).

Arnold (1989:246) points out that the primitive species of *Lacerta* (*sensu lato*) represent a paraphyletic taxon, but that many relationships remain to be resolved: "It seems better, for the present, to accept *Lacerta* s. lat. as an admittedly paraphyletic assemblage confined to the Palearctic region and employ subgenera within it if supposed relationship is to be formally emphasized." I adhere to this practice in this book, although Mayer and Bischoff (1996) have raised *Zootoca, Omanosaura, Timon,* and *Teira* to the generic level and E. N. Arnold (pers. commun.) agrees that, with respect to the Southwest Asian taxa, that *Timon* and *Omanosaura* appear to represent clades that will stand up. Recognizing the generic taxa proposed by Mayer and Bischoff still leaves *Lacerta* paraphyletic, as does the hypothesis of Mayer and Benyr (1994).

Wettstein (1951:438) cites a juvenile specimen from Khorramabad, Mazanderan [not to be confused with the much better known Khorram-Abad, the capital of Lorestan Province] as "*L. danfordi*?" This identification seems most unlikely, but I have not examined the specimen. It had 52 dorsal scales at midbody, 23 subdigital lamellae, 2 postnasals, a large, round masseteric, small round upper temporals, and lacked pterygoid teeth. Schleich (1977:127, 129) also lists *L. danfordi*, presumably on Wettstein's record; Welch (1983:33) lists it, citing Schleich.

Frynta, *et al.* (1997:10) list an undetermined specimen from Khalkhal resembling *L. raddei*. A description of this specimen is anticipated. It should be compared with specimens I collected near this same locality.

Lacerta brandtii De Filippi, 1863 (Plates 12H–I, 13A)
Brandt's Persian lizard

Lacerta brandtii De Filippi, 1863:387 (Type locality: Basminsk [Basmenj], between Tabriz and Tehran, Iran; Syntypes: MSNTO); 1865:354 — Blanford, 1876:362, pl. 25, fig. 1. — Bedriaga, 1879:29. — Boulenger, 1887a:38–39. — Nikolsky, 1915:386–388; 1916:324–325. — Boulenger, 1920:299–301. — F. Werner, 1936:200. — Rostombekov, 1938:12–13. — Lantz and Cyrén, 1939:228–237, pl. 1, figs. 1–2. — Terentjev and Chernov, 1949:189–190. — Wettstein, 1951:439 — S. Anderson, 1963:476. — Clark, *et al.*, 1966:8. — S. Anderson, 1968:333; 1974:40, 43. — Bannikov, *et al.*, 1977:202–203. — Schleich, 1977:127, 129. — Welch, 1983:33. — Clark, 1991:38–39.

Diagnosis: Transverse series of ventral plates with rectilinear or nearly rectilinear posterior margins, in 8 longitudinal rows; dorsal scales smooth; 2 superposed postnasals; pterygoid teeth present; masseteric shield present.

Color pattern: (Fig. 111) Pale brown or olive-gray above, with small black spots tending to form reticulation; sides of head and body bronze with numerous whitish, dark-edged ocelli, largest above axil often bright blue; males with whitish-yellow throat and abdomen, outer abdominal scales greenish with small black and bright blue dots; in females, dorsolateral stripe white, breaking up into spots posteriorly, throat and abdomen light bluish, head greenish during breeding, anal region, thighs, and lower base of tail orange. Young gray above with 2 series of black markings along back; sides with black reticulate network enclosing round white spots; 2 to a few large, blue, black-edged ocelli above axil.

Size: Snout-vent length to 75 mm.

Habitat: (Plate 24C) At the Maragheh paleontological site, I found these lizards active in gullies and on the slopes of redeposited volcanic tuff, taking refuge under and among boulders, but soon emerging into sunlight. Vegetation was overgrazed steppe shrub, with *Euphorbia* predominant. The Clarks found them numerous in a dry stream gully on a hillside above a lake, running swiftly across open fields from bush to bush and hiding beneath small boulders (Clark, *et al.*, 1966:8).

Lacerta brandti

Distribution: Northwestern Iran in East Azarbaijan Province and contiguous regions of the Azerbaidzhan Republic (Talysh and Zuvand), and the vicinity of Kuh Rang in Esfahan Province; distribution appears to be discontinuous; 500–2600 m in Iran.

Remarks: The Street Expedition to Iran of 1968 extended the distribution of this species significantly by collecting seven specimens in Esfahan Province at an elevation of nearly 2600 m. These should be compared directly with material from East Azarbaijan.

The relationships of this species remain problematic. Several researchers (e.g., G. Peters, 1962:460; Arnold, 1973:339–340, 1989:210) proposed that it is probably most closely related to *Lacerta fraasii* of Lebanon and *L. parva* of Turkey, but is intermediate between these two distinctive species and the more typical species of *Lacerta* part II. All three inhabit arid steppe-like environments. Böhme (1993: 141–142) compared the microornamentation of their hemipenes and claimed to have falsified this hypothesis, but he did not provide a satisfactory alternative. As E. N. Arnold pointed out to me several years ago (pers. commun., 25 November 1995), Böhme had shown that one possible synapomorphy, a particular derived pattern of microornamentation, found in *L. parva* and *L. fraasi*, does not occur in *L. brandtii*, a circumstance that might be expected if *L. parva* and *L. fraasi* are one another's closest relatives. A detailed description of *L. brandtii* and comments on habitat and behavior are given by Lantz and Cyrén (1939:228–237).

FIGURE 111. *Lacerta brandti* De Filippi. (From Blanford, 1876, pl. 25, fig. 1.)

Lacerta cappadocica (Werner)
Anatolian rock lizard

Lacerta cappadocica urmiana (Lantz and Suchow, 1934)   (Plate 13B–C)
Lake Urmia rock lizard

Apathya cappadocica urmiana Lantz and Suchow, 1934:294–299, fig. (Type locality: Berdesur River Gorge, Kherra, Iran, about 20 km SW of Rezaiyeh; Holotype: ZIL 12657). — S. Anderson, 1963:475; 1968:333; 1974:40, 43. — Schleich, 1977:127, 129.
Lacerta cappadocica urmiana: Eiselt, 1979:417–418, pl. 6. — Welch, 1983:33. — Leviton, *et al.*, 1992:59, pl. 8H.

Diagnosis: Lower eyelid scaly, a transparent disc formed of 5–7 scales edged with black; nostril surrounded by 3–6 scales including first supralabial; first supraocular undivided, not in contact with frontal; 7–8 (rarely 6) superciliaries; collar distinct; dorsal scales smaller than caudals, juxtaposed, 52–60 across middle of body; 22–27 femoral pores; 8 longitudinal rows of ventrals; digits feebly compressed, keeled lamellae below.

Color pattern: Dorsal surface of head olive-brown, uniform; dorsum with more or less distinct light dots arranged in 2 or 4 longitudinal rows with irregular dark edges, having tendency to form black transverse bars; dorsolateral light stripes indistinct on neck, bluish-white on body, continuing with more or less emarginate borders, sharply pronounced and meeting on base of tail; tail gray or bluish-white, often bright blue on posterior three-fourths; row of large blue ocelli on sides of body, above and below, with more or less regular rows of whitish flecks; venter bluish-white, often with reddish tinge on throat and anal region (breeding males?); ventral surface of tail blue.

Size: Snout-vent length to 67 mm, tail 150 mm; a juvenile collected in mid-September measures 33 mm snout-vent.

Natural history: Two stomachs examined contained a small species of grasshopper.

Habitat: (Plate 25G) West of Lake Reza'iyeh I found this species on rocky outcrops near a series of step-like waterfalls, basking on ledges where they could move easily from sunlight to shadow and could take refuge in crevices. This was a narrow gorge with steep, loose, rocky slopes, with overgrazed steppe vegetation on hillsides, and olives, *Juglans, Pistacia, Crataegus* along the watercourse, around pools, and on ledges above each fall. Reed found them in northeastern Iraq (Reed and Marx, 1959:100) and Clark and Clark (1973:24) in similar habitats in Turkey.

Lacerta cappadocica

Distribution: This subspecies is known from West Azarbaijan Province in gorges on the west slope of the Reza'iyeh basin. It is known from Byelyaki, Kurdistan, on the Iran-Turkey border and from northeastern Iraq. In Turkey it occurs east from Siirt and Cizre. The species occurs outside Iran in southern and eastern Turkey and northeastern Iraq. Four additional subspecies are currently recognized (Eiselt, 1979): *Lacerta c. cappadocica* F. Werner from south-central Anatolia, *L. c. wolteri* (Bird) from southeastern Anatolia and northern Iraq, *L. c. muhtari* Eiselt from east of the Euphrates to Lake Van in Turkey and south to northern Iraq, and *L. c. schmidtlerorum* Eiselt, from the vicinity of Diyarbakir and Viransehir, Turkey.

Remarks: The Street Expedition to Iran of 1962–63 collected the first specimens of this taxon since its description, and Eiselt collected two specimens in 1968. In his revision of *Lacerta cappadocica*, Eiselt (1979:407) gives the following counts for *L. c. urmiana*: superciliaries 6–10; superciliary granules 9–27; transparent scales in lower eyelid 4–10; dorsals 51–68; gulars 25–34; femoral pores 20–27; fourth toe lamellae 24–30.

Arnold (1973:337–338) placed this species in a group with *L. cyanura* and *L. jayakari* of Oman and suggested that it may also be related to *L. danfordi*, a geographic neighbor. However, he informs me (pers. commun., 25 November 1995) that he now considers the relationship with *L. jayakari* and *L. cyanura* less likely, as reflected in a more recent paper (Arnold, 1989: fig. 23).

Lacerta chlorogaster Boulenger, 1909 (Plate 13D–E)
Green-bellied lizard

Lacerta boettgeri Méhelÿ, 1907:88 (*nomen nudum*); 1909:583; 1910:593–594. — Nikolsky, 1910: 497. — F. Werner, 1913:15.
Lacerta muralis: Nikolsky, 1907a:281 (in part; not *Seps muralis* Laurenti, 1768). — F. Werner, 1936:200 (in part).
Lacerta chlorogaster Boulenger, 1909b:934–936, pl. 67 (Type locality: Enzeli [Bandar-e Pahlavi, now Bandar-e Anzali], Iran, south coast of Caspian Sea; Syntypes: BMNH 1908.8.7.14–15/ 1946.9.1.87–88, 1908.8.7.29–34/1946.9.2.28–33). — Nikolsky, 1915:394–396. — Boulen-ger, 1920:292–295. — Terentjev and Chernov, 1949:189. — Mertens, 1957:122–123. — S. Anderson, 1963:476; 1968:333. — Tuck, 1971a:58; 1974:62–63, fig. — S. Anderson, 1974:40, 43. — Bannikov, *et al.*, 1977:211. — Schleich, 1977:127, 129. — Welch, 1983:33. — Eiselt, 1995:60–62, 66–67, 69–70, figs. 1–2, tab. 1, pl. 2, figs. 1–4. — Frynta, *et al.*, 1997:10.
Lacerta muralis: Nikolsky, 1907a:281 (in part; not *Seps muralis* Laurenti, 1768). — F. Werner, 1936:200 (in part).
Lacerta saxicola defilippi: Forcart, 1950:148 (in part; not *Podarcis defilippii* Camerano, 1877).

Diagnosis: Transverse series of ventral plates with rectilinear or nearly rectilinear posterior margins, in 6 longitudinal series; 43–53 dorsal keeled scales; single postnasal; rostral touches nostril or just separated from it; parietal foramen present; collar serrated; 20–25 gulars on midline of throat from third pair of chin shields to collar (Fig. 112); 27–35 lamellae under fourth toe; no pterygoid teeth. The following from Eiselt (1995:62, table 1): Pileus length/breadth 1.8–2.1; number of small femorals across 3–5; number of small tibials across 14–23; temporals between massetericum and tympanicum 1–4; temporals between massetericum and 1st supratemporal 1–4; marginals along 10 rows of ventrals 13–28.

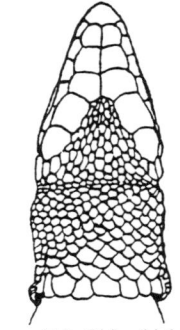

FIGURE 112. Chin shields and scalation of throat in *Lacerta chlorogaster*. (After Nikolsky, 1915, fig. 48.)

Color pattern: Head and back grayish olive, greenish, or bronze in male, with bright green on sides of head, with or without small black spots or vermiculation; sides and limbs yellow-green or bronze with black reticulum, often enclosing large, whitish ocelli, or black with small yellowish-gray spots; a few turquoise spots sometimes present behind shoulder; female pale or golden brown above, with small blackish spots and dark brown lateral stripe with undulating margins; occasionally dorsolateral and/or ventrolateral series of small whitish spots; markings of body usually continued on anterior part of tail; venter of male light green or yellow-green with turquoise and black spots on outer ventrals; throat blue or bluish-green; anal region and ventral surface of hind limbs often yellow. (Modified from Boulenger, 1920).

Size: Snout-vent length 72 mm.

Natural history: Females collected in late October have yolked follicles of various sizes. Stomachs examined primarily contain grasshoppers; spiders, beetles, and other insects were also present.

Habitat: (Plate 23B) This is a forest species, easily ascending tree trunks and rock walls to exploit patches of sunlight in shadowy areas they inhabit (Mertens, 1957:123). Sobolevsky (1929) found it a characteristic species of forests of lowlands and low mountain belt to 600–700 m elevation. The vegetation of the area was characterized by lianas, grapes, ivy, giant ferns, and trees such as *Parrotia persica, Quercus castaneaefolia, Acer insigne, Ficus carica, Gleditschia caspia,* and *Albizzia julibrissin.* I found them in Hyrcanian forest (mixed deciduous forest) of similar description on the lower slopes of the Alborz Mountains and in the Kopet Dagh. They were on logs and tree trunks and in the grass of the forest floor. *Lacerta chlorogaster* has been found sympatrically (but probably not syntopically) with *L. defilippii* in much of its range, and with *L. steineri* at the type locality of the latter species (see below).

Distribution: Southern and southwestern coasts of the Caspian Sea in Iran and Azerbaidjan Republic. Occurs east as far as the Atrak River of northern Khorasan, from slightly below sea level on the Caspian coast to 900–1500 m on the Atrak. Franz Werner's (1913) record of this species from Tehran is surely in error. Nikolsky's (1915:394) record from "Robat-i-Satid," if referring to the locality in eastern Khorasan (Robat-e Safid), certainly requires reexamination of the three specimens (ZIL 10192). There is a specimen in the Iran National Museum of Natural History with the locality "Qom," some 225 km south of the expected distribution of this species; apparently, it was collected in a park within the city, and thus it may have come there adventitiously with vegetation from the Caspian area.

Lacerta chlorogaster

Remarks: Arnold (1973:336) places this species among the more typical members of his *Lacerta* part II, which includes the species placed in the subgenus *Archaeolacerta* by Mertens and others. In his 1989 paper, he further places *L. chlorogaster* firmly among the *L. saxicola* assemblage. The history of this restricted endemic is undoubtedly closely tied to that of the relictual Hyrcanian forest of the southern Caspian coast. This Euro-Siberian flora is of Arcto-Tertiary origin and is part of the Euxino-Hyrcanian floral province, which includes the southern coastal forests of the Black and Caspian Seas (Zohary, 1963:22). According to the analysis of Eiselt (1995:67, fig. 2), it is the sister taxon of a clade composed of *L. raddei, L. defilippii,* and *L. steineri* (*L. praticola* was not included in this analysis). Eiselt further points out the presumed ecological similarity of *L. chlorogaster* to *L. clarkorum* Darevsky and Vedmederja, 1977, which inhabits a somewhat similar biotope in eastern Turkey. *Lacerta laevis* also inhabits moist-forest biotopes and shares some morphological similarities with *L. chlorogaster*. Whether the resemblances of these species are synapomorphic as a result of shared ancestry or parallel adaptations to similar habitats remains to be determined.

Eiselt (1995:70) speculates that recognition of two subspecies, distinguishing the Talysh-Lenkoran population from that of the Alborz, may be justified once sufficient material is available from the whole of the Alborz part of the distribution.

Lacerta defilippii (Camerano, 1877) (Plate 13F–G)
Alborz lizard

Lacerta muralis: De Filippi, 1865:354 (not *Seps muralis* Laurenti, 1768). — Blanford, 1876:361–362. — Boulenger, 1887a:29–30 (in part). — Boettger, 1893:83. — F. Werner, 1936:200 (in part).
Podarcis defilippii Camerano, 1877:90, pl. 3, figs. 1–3 (Type locality: Lar Valley, northwest of Tehran, Iran; Syntypes: Turin?).
Lacerta muralis var. *fusca*: Bedriaga, 1879:29.
Lacerta defilippi: Bedriaga, 1879:30. — Darevsky, et al., 1984:102–108. — Eiselt, 1995:60–62, 65, 67, 69, figs. 1–2, tab. 1, pl. 2, figs. 5–8. Frynta, et al., 1997:10.
Lacerta muralis var. *defilippii*: Boettger, 1886:44 (in part). — Boulenger, 1904:337; 1913:195, pl. 23, fig. 2; 1920:271, 288–290.
Lacerta muralis fusca var. *persica* Bedriaga, 1886:199 (Type locality: Persia; Holotype: BMNH 1928.12.8.870/1946.9.1.75).
?*Lacerta depressus*: F. Werner, 1903:341–342 (not *Podarcis depressus* Camerano, 1878 [= in part *Lacerta saxicola* Eversmann, 1834; in part *L. rudis* Bedriaga, 1886]).
Lacerta saxicola var. *defilippii*: Méhelÿ, 1909:519; 1910:592–593. — Nikolsky, 1915:370–373 (in part); 1916:323 (in part). — Moritz, 1929:31. — Lantz and Cyrén, 1936:164 (in part). — Terentjev and Chernov, 1949:187–189 (in part). — Forcart, 1950:148 (in part). — Wettstein, 1951: 438. — S. Anderson, 1963:476. — Clark, et al., 1966:8. — Darevsky, 1967:60–63 [1978:63–65], text-fig. 20, photo 11. — S. Anderson, 1968:333 (in part); 1974:40, 43. — Schleich, 1977:127, 129.
Lacerta raddei defilippii: Bannikov, et al., 1977:218.
Lacerta defillipi: Welch, 1983:34.
Lacerta saxicola de Filippi: Clark, 1991:38–39.

Diagnosis: Ventral plates more or less rectangular in shape, with rectilinear or nearly rectilinear posterior margins; collar not serrate; dorsals smooth, 41–56 across middle of back; usually single postnasal; complete row of granules separates supraoculars from supraciliaries; pterygoid teeth absent; 22–27 gulars; 14–20 femoral pores on each side; 14–18 scales round middle of shank. The following from Eiselt (1995:62, table 1): Pileus length/breadth 2.0–2.3; number of small femorals across 3–6; number of small tibials across 16–21; lamellar scales beneath fourth toe 27–32; temporals between massetericum and tympanicum 2–6; temporals between massetericum and first supratemporal 1–3 (Fig. 113); marginals along 10 rows of ventrals 13–31.

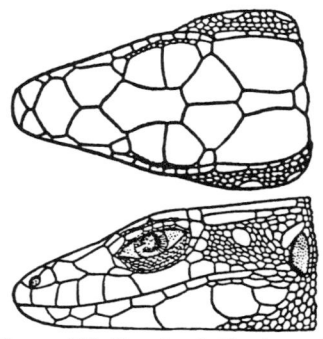

FIGURE 113. Dorsal and side views of head of *Lacerta defilippii*. (After Darevsky, 1967, fig. 20.)

Color pattern: Dorsum cinnamon, olive-yellow, to brownish or olive-gray, occipital stripe (middorsal area of back) finely and irregularly spotted with dark brown to black, often forming reticulum; sides (temporal stripe) darker than back, set with longitudinally arranged light ocelli, a row of ocelli marking upper margin of lateral dark stripe. Venter often brick red in life, outermost ventrals with pale blue spots on exterior edge.

Size: Snout-vent length to 58 mm.

Habitat: (Plate 24B) I found them in the Alborz range on both north and south slopes with loose scree and on rocky outcrops. Vegetation in these localities was shrubby, up to about 2 m in height. On the southern slope I found them at the upper margin of the Hyrcanian forest region. They were active on loose jumbled rock, road edge near shrubs, and on branches and leaves of shrubs at air temperature 18.5°C, surface 14°C. Recorded localities are from below sea level near Chalus to 3355 m, a remarkable elevational range for any lizard species. Eiselt (1995:65) studied two populations from quite distinct Biotopes: "Polur" consists of a grassy alpine landscape with relatively few scattered rocky parcels (about 2,000–3,000 m) in the central Alborz highlands near Damavand (close to the type locality of the species), "Chalus" characterized by a rocky, "gorge-like" river valley (about 800 m), 50 km south of the Caspian coast.

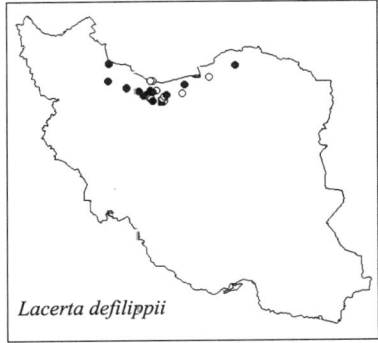

Lacerta defilippii

Distribution: The Alborz Mountains of Iran and the Kopet Dagh of Iran and Turkmenistan.

Remarks: In their most recent analysis of the Iranian populations of the *saxicola*-group, Darevsky, *et al.* (1984:102–108) removed *defilippii* from *L. raddei* and raised it to full species status.

The population from Mazanderan in the valley of the Rud-e Atrak of the Kopet Dagh is probably well isolated from the Alborz population, and may prove to be a distinct taxon. It should be compared with the recently described *L. steineri* Eiselt (see below).

Lacerta media Lantz and Cyrén, 1920
Three-lined lizard, giant green lizard

Lacerta media media Lantz and Cyrén, 1920 (Plate 13H–I)
Three-lined lizard, giant green lizard

Lacerta trilineata media Lantz and Cyrén, 1920:33 (Type locality: Tiflis [Tbilisi], Georgia [restricted by Mertens and Müller, 1940:44]; Syntype: BMNH 1960.1.4.38). — Cyrén. 1924:1–82. — Rostombekov, 1938:13–14. — Müller, 1939:12. — Forcart, 1950:148. — G. Peters, 1964a:243, figs. 17–19. — Clark, *et al.*, 1966:8–9. — S. Anderson, 1968:333. — Tuck, 1971a:58–59; 1974b:107. — S. Anderson, 1974:40, 44. Bannikov, *et al.*, 1977:197. — Schleich, 1977:127, 129. — Welch, 1983:37.
Lacerta viridis var. *major*: Boulenger, 1920:82–88 (in part; not *L. v. major* Boulenger, 1887a [= *L. t. trilineata* Bedriaga, 1840]).
Lacerta media: Terentjev and Chernov, 1949:179, text-fig. 69. — S. Anderson, 1963:476.
Lacerta trilineata: G. Peters, 1962:127, 152, fig. 3c–d.
Lacerta media media: J. F. Schmidtler, 1986b:127. — Leviton, *et al.*, 1992:59–60, col. pl. 9A–B.

Diagnosis: Ventral plates trapezoidal, with notches between plates, in 6 longitudinal rows; enlarged marginals lacking or few in number; collar strongly serrated; usually 2 superposed postnasals; one or two preoculars (Fig. 114); usually 20–26 temporal scales; fifth submaxillary small; 12–16 femoral pores, row of pores not reaching knee.

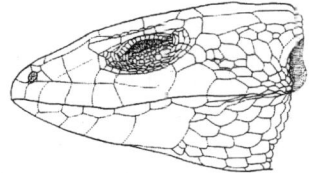

FIGURE 114. Side of head of *Lacerta media*. (After Haas and Y. Werner, 1969, fig. 4A.)

Color pattern: Young dark olive-brown, with 5 light lines on dorsum, anterior half of lateral streak broken up into spots, often additional row of light spots in temporal region; adults without pattern of light lines or spots; dorsum predominantly green to olive-brown, uniform or speckled, mottled, or flecked with black. All adults with blue on sides of neck. Venter greenish-white to greenish-yellow. Three-year-old and older males with blue flecks or bands on flanks.

Size: Snout-vent length to at least 116 mm in Iranian specimens, tail 263 mm.

Natural history: Females I examined, collected in late July, have 2 mm ovarian follicles. Stomachs of all specimens I examined primarily contained grasshoppers; beetles, ants, and insect larvae also were found.

Habitat: I found this species only in the grass understory of a poplar grove along a stream near a small village 42 km west of Kermanshah. According to G. Peters (1964a:243), it inhabits brushy or sparsely forested areas at various elevations, depending upon the mean (or extreme?) temperatures of the warm summer months. It rarely occurs on distinctly dry slopes in rocky steppes above 2000 m with a boreal climate, descending to 600 or 700 m in brush-strewn river valleys. See J. F. Schmidtler (1986a) for comments on habitat in Turkey.

Distribution: The species ranges from mid-Asia Minor (Turkey) and the eastern shores of the Mediterranean (south to Israel) to northern and western Iran. *Lacerta m. media* occurs in northern and eastern Turkey, northeastern Iraq, the Caucasus, and western Iran; 1000–1500 m elevation in Iran. J. F. Schmidtler (1986b) recognizes five populations of this subspecies, which he thinks may eventually be worthy of taxonomic recognition. He refers to those he examined from Iran (Ostan 5) as "*L. m. media*-Zagrosform."

Lacerta mostoufi Baloutch, 1976
Dasht-e Lut lacerta

Lacerta mostoufi Baloutch, 1976:1379–1384, figs. 1–2 (Type locality: Deh Salm, 31°12′N, 59°19′E, about 200 km N Malek Mohammad Mountains; Holotype: MMTT 1582).

Diagnosis: (Fig. 115) Transverse series of ventral plates with nearly rectilinear posterior margins, in 6 longitudinal series; dorsal scales keeled, 46 across midbody; single postnasal; collar serrated; 17–23 gulars on midline of throat from third pair of chin shields to collar; 25 lamellae under fourth toe; venter grayish or black.

Color pattern: In alcohol, dorsum entirely black, gular region and lower surfaces of feet and digits whitish, venter brilliant black with colorless borders. In life, dorsum gray or olive, lower surface of femur pale yellow, venter grayish or blackish. (Baloutch, 1976:1381–1383).

Size: Snout-vent 65 mm, tail 123 mm.

Habitat: Baloutch's notes on the habitat do not indicate the substrate on which these lizards occur, but the region of the Malek Mohammad Mountains is in the Dasht-e Lut, where precipitation and humidity are very low, the soil salty, and the scanty vegetation

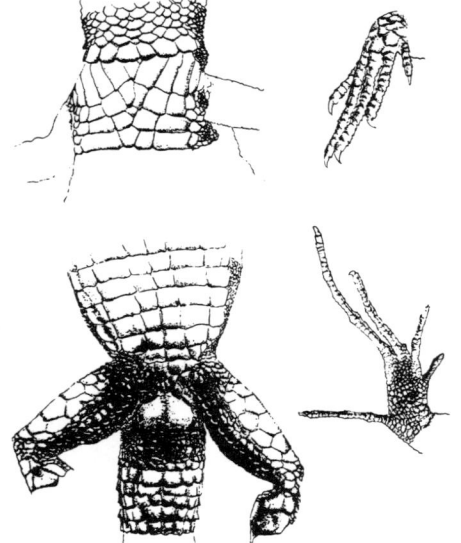

FIGURE 115. *Lacerta mostoufi*, holotype. (From Baloutch, 1976, fig. 2.) Reprinted with permission Muséum National d'Histoire Naturelle.

includes *Calligonum, Seidlitzia*, and *Artemisia herba-alba*. Such rain as occurs falls from October to May. The village of Deh Salm has date palms and other vegetation, the vegetation decreasing toward the mountains, where there are few shrubs. The soil is sandy, precipitation is greater than in the Malek Mohammad Mountains, and shrubby vegetation consists of *Tamarix, Seidlitzia, Calligonum*, and *Halo-xylon*. (Baloutch, 1976:1383–1384).

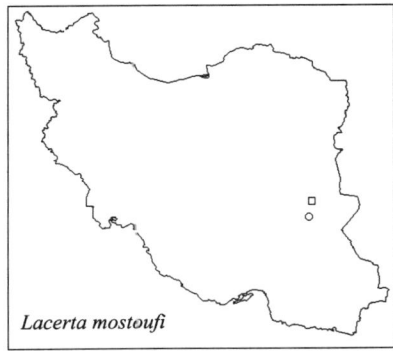
Lacerta mostoufi

Distribution: Known only from the Dasht-e Lut in Baluchistan, 800–1000 m elevation.

Remarks: This species is known only from the male holotype and two juvenile paratypes. The original description is not sufficiently detailed to place it with certainty within one of Arnold's sections of the genus, or even to place it in the key, except tentatively. Its small adult size, the implication that it has a single postnasal, the apparent character of the ventral plates, and the author's statement that it is related to *Lacerta saxicola* argue for its position in Arnold's *Lacerta* part II. If 6 pairs of chin shields is the normal condition for this species, it shares this character with *L. chlorogaster* (in which the third pair is small); other Iranian members of this genus have 5 (normally 4, or the fifth small in *L. media*). Apparently, it is strongly melanistic, again unique for Iranian species, and this might be an adaptation to dark surfaces such as basalt outcrops, although the substrate at site of capture was not stated. The description is not detailed as to the condition of the ventral plates and the illustration suggests that it is intermediate between those species having the very rectilinear juxtaposed plates of saxicolous members of part II of the genus and the trapezoidal imbricate plates with notched margins characteristic of part I. This and the serrated collar would argue for a ground-living species in part II, according to Arnold (1973:318–320). It resembles *L. chlorogaster* and *L. praticola* in the keeled dorsals and serrated collar, but differs from both in the greater number (12) of plates in the collar (7–9 in *L. chlorogaster* and *L. praticola*); it further differs from *L. chlorogaster* in having fewer lamellae under the 4th toe and from *L. praticola* in the smaller dorsal scales.

Herman A. J. in den Bosch (pers. commun.) has examined a paratype (MNHP 1976.81, formerly MMTT 1584) and informs me that he regards it as a specimen of *L. praticola*. Arnold (pers. commun., 25 November 1995) concurs. If this identification is correct, it strongly suggests that there has been a mix-up of specimens and/or data. Baloutch showed me his three specimens in 1975 before he published his description. These bore no tags and were in a jar with specimens of other species of lizards. I did not examine them under magnification, nor compare them with other material or published information, and my recollections of the animals are far too faint to make further comment. Eiselt (1995:70) suggests that the description of the color may indicate formalin-darkening.

The biogeographic implications of the discovery of a species of *Lacerta* some 500 km east-northeast of its nearest congener in part I (*L. princeps*) and some 850 km southeast of its closest part II neighbor (*L. defilippii*) make it of considerable interest, whether it is a valid species, or an isolated population of a previously described species.

Lacerta praticola Eversmann, 1834 (Plate 14A–B)
Meadow lizard

Lacerta praticola praticola Eversmann, 1834
Meadow lizard

Lacerta praticola Eversmann, 1834:345, pl. 30, fig. 2 (Type locality: Narzan).
Lacerta praticola praticola: Terentjev and Chernov, 1949:185–186, fig. 6a, map 17. — Bannikov, et al., 1977:204. — Orlova, 1978:204–215. — Stugren (1984:323).

Diagnosis: 24–28 transverse series of ventral plates with nearly rectilinear posterior margins, in 6 longitudinal series; dorsal scales keeled, 29–49 across middle of back; single postnasal; rostral not touching nostril (Fig. 116); 16–30 temporals; collar serrated; 5 pairs of chin shields, the first 2 pairs in contact on midline; 15–17 gulars on midline of throat from second pair of chin shields to collar; 11–13 femoral pores on each side; 24–31 lamellae under 4th toe.

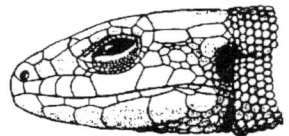

FIGURE 116. *Lacerta praticola*, view of side of head. (From Terentjev and Chernov, 1949, fig. 73a.)

Color pattern: Light brown above with green or olive tinge. Narrow dark brown occipital stripe along crest, and wide brown or dark brown stripe on sides of body. Usually a small number of small brown spots between lateral and occipital stripes. Young (and sometimes also adults) unicolor. Throat white. Abdomen of males greenish, of females yellow. (Terentjev and Chernov, 1949:186).

Size: Snout-vent 43–54 mm in males, 46–64 mm in females.

Habitat: Nothing is recorded of its habitat in Iran. In the Caucasus it lives in broadleaf forests, slopes of river canyons, in meadows and grassy brush woods (Terentjev and Chernov, 1949:186). See Stugren (1984:328–330) for comments on ecology and natural history of this species in its European distribution.

Distribution: The Caucasus in southern Russia and in the Transcaucasian republics. Terentjev and Chernov (1949:186), Bannikov, *et al.* (1977:209) and I. S. Darevsky (pers. commun., 20 January 1994) state that this Caucasian lizard occurs in northwestern Iran. The only published locality records for Iran appear to be those of Orlova (1978), cited again by Stugren (1984:323). A second subspecies, *L. praticola pontica* Lantz and Cyrén, occurs in the Balkans and on the Black Sea coast of the Caucasus, south to Krasnodar, Russia (Bannikov, *et al.*, 1977:209). See also Stugren (1984: 319–323, figs. 35–36).

Lacerta praticola

Lacerta princeps Blanford, 1874
Zagrosian lizard

Lacerta princeps princeps Blanford, 1874 (Plate 14C–D)
Zagrosian lizard

Lacerta princeps Blanford, 1874b:31 (Type locality: mountain pass near Niriz [Neyriz], Iran, 7000 ft [2333 m]; Holotype: ZSI 3351 [*fide* Das, *et al.*, 1998]); 1876:364–367, pl. 24. — Bedriaga, 1879:28 — Boulenger, 1887a:18–19. — Méhelÿ, 1910:594–596. — Boulenger, 1920:95–97. — Fejérváry, 1936:1–21. — F. Werner, 1936:200. — S. Anderson, 1963:476; 1968:333.
Lacerta princeps princeps: Eiselt, 1968b:411–412, pl. 2, figs. 1–5; 1969:209–22, pl. 1, figs. 1–7; 1970:109–114, text-figs. 1–3. — S. Anderson, 1974:40, 43. — Schleich, 1977:127, 129. — Welch, 1983:35. — Frynta, *et al.*, 1997:10.
Timon princeps princeps: Mayer and Bishoff, 1996:169. — Muldur, 1998:75–79. — Bosch. 1998:80–87, figs. 1–3.

Diagnosis: Transverse series of ventral plates with notches between plates, in 10 longitudinal rows; collar strongly serrated; 2 superposed postnasals; rostral bordered by 5 shields including rostral; 34–37 strongly keeled, subimbricate dorsal scales at midbody;

outer row of ventrals (marginals) not keeled, adjacent 1–3 rows of flank scales smooth or weakly keeled; neck scales smooth; 20–22 gulars; lower edge of subocular less than half maximal length of shield; 13–17 femoral pores on each side.

Color pattern: (Fig. 117) In adult males dorsum uniform beige with pale greenish cast in life, top of head olive-gray; olive-gray zone extends over supratemporals and temporals, coalescing with light gray-blue, which covers maxillary and mandibular region; gulars between maxillaries white, throat and collar light straw yellow; venter white; axillary region and sides of thorax with prominent black-ringed, intense blue ocelli, often connected in superposed pairs or triads, posteriorly small ocelli or black spots on flanks in longitudinal row extending to insertion of hind limb, but not onto base of tail. Adult females (snout-vent over 133 mm) like males, but ground color somewhat darker and showing faint indication of spotted pattern of juveniles; fewer ocelli on flanks. (Eiselt, 1969:213–215 and col. pl.). Boulenger (1920:97) described the ventral surface of the tail in the female he examined as red. This may be seasonal coloration. Juveniles (snout-vent length less than 75.5 mm) and half-grown females (snout-vent 117 mm) various shades of light olive-gray to yellow-gray to brown dorsally in life, often with light greenish cast; top of head greenish-gray or olive to olive-brown; maxillary and mandibular regions pale bluish, throat pale yellow; dorsum and flanks very irregularly flecked with brownish black in small animals, dark brown in older individuals, dark color forming transverse or oblique bars; dark median stripe about 5 mm long on nape; 2–4 longitudinal rows of more or less distinct round spots on body, pale blue in axillary region and thorax, becoming whitish posteriorly; upper sides of extremities, especially hind limbs, white-spotted.

Size: The holotype is the largest known specimen: snout-vent 145 mm. Eiselt's (1969) largest male was 143 mm snout-vent, tail 252 mm; his largest female was 141 mm snout-vent, tail 289 mm.

Habitat: (Plate 22F) The climatic zone at Shiraz and Neyriz is xerothermomediterranean (150–200 physiologically dry days in the course of the annual dry period). Sarchun and the mountains north of Neyriz are cold dry steppe (5–8 months of frost, part of which is during the dry period). The vegetation of these areas is impoverished *Pistacia-Amygdalus* shrub-steppe. Specimens were collected on dry knolls where numerous white-blooming dogroses were interspersed among the isolated shrubs. (Eiselt, 1968:410–411; 1969:210).

● ○ ▫ *Lacerta p. princeps*
△ ◪ *Lacerta p. kurdistanica*

Distribution: The nominate subspecies is known from Neyriz, about 160 km east of Shiraz to Sarchun, about 125 km southwest of Esfahan. There is a hiatus of about 500 km between the known range of this subspecies and the known range of *Lacerta p. kurdistanica*. The distribution may prove to coincide more or less completely with the remaining almond-pistachio and deciduous oak climax forests of the Zagros Mountains.

Lacerta princeps kurdistanica Suchow, 1936 (Plate 14E)
Kurdistan lizard

Lacerta princeps kurdistanica Suchow, 1936:303–308, pl. 2 (Type locality: Kurdistan, Iran [restricted to Biare = Beydarvaz by Eiselt, 1968:412]. Holotype: ZIL 11441:b). — Eiselt, 1968:412–433,

pls. 1, 3–5; 1970:109–114. — S. Anderson, 1974:40, 43. — Welch, 1983:35. — Leviton, *et al.*, 1992: 60.
Lacerta princeps: Mertens, 1952b:353–355.
Timon princeps kurdistanica: Rykena and Bischoff, 1977:41–43. — Rykena, *et al.*, 1977:174–184.—
Mayer and Bischoff, 1996:169 (by implication). — Mulder, 1998:75–79. — Bosch, 1998:80–87.

Diagnosis: Outer row of ventrals (marginals) keeled, as are all flank scales; neck scales keeled; lower edge of subocular half or greater than half maximum length of shield; 17–19 gulars; 16–21 femoral pores on each side.

Color pattern: In life, dorsum light yellow- to olive-brown, often olive-gray, crossed by irregular network of dark brown, forming irregular crossbars on back and broad longitudinal zone on sides, this pattern distinct to scarcely noticeable; dark brown flecks on top of head; longitudinal rows of whitish spots contained within dark network, largest and most sharply delineated anteriorly, smaller and irregular caudad and ventrad, often occupying only one or two scales; especially in axillary region, ocelli formed by dark brown (never black) pigment surrounding light spots, in other areas dark vertical bars each enclose a row of light spots; light spots always extend over sides of sacral region onto sides of tail base; tail often checkered in appearance; extremities, especially hind limbs, spotted with whitish; venter white to light grayish-yellow. Juveniles (less than 57 mm snout-vent) appear uniform at first glance due to indistinct pattern. Top of head, supratemporals, and upper temporals olive to chestnut brown, indistinctly dark-flecked, remainder of head (from behind tympanum, below mental, all infralabials, chin shields, and anterior gulars) intensive brownish- to bluish-black; throat and collar light orange-red, with pale orange to reddish-brown on side of neck; longitudinal rows of light spots greenish to intense yellowish- to bluish-green (orange color disappears in alcohol). Old adult females show wash of bluish-gray on part of head which is black in males. (Eiselt, 1968:423–425.)

Size: The largest male (holotype) measures 148 mm snout-vent, tail 303 mm; largest female, snout-vent 131 mm.

Habitat: The localities lie within the thermomediterranean climatic zone (100–150 dry days) (Mardin and Siirt Vilayets in Turkey, and northern Iraq), and in cold, dry steppe (5–8 months of frost, partly combined with dry period) (northeastern Iraq, northwestern Iran, Hakkari Vilayet in Turkey). All are within the Irano-Turanian xerophilous and summer-

FIGURE 117. *Lacerta princeps*. (From Blanford, 1876, pl. 24.)

green Zagros oak forest climax vegetation type characterized by the presence of *Quercus brandtii*. (Eiselt, 1968:410–411). Rykena, *et al.* (1977) observed these lizards in the field in southeastern Turkey (Mardin hills) where they preferred oak-shrub habitat of the slopes of valleys. These authors present an account of the ecology and behavior of this subspecies in the field and in captivity.

Distribution: (see map, p. 241) The region of the Zagros oak forest in northwestern Iran, northeastern Iraq, and southeastern Turkey (35°–38°08'N, 40°16'–46°15'E).

Remarks: Debate over the relationships of this species continues. G. Peters (1962a: 454) excluded *Lacerta princeps* from close relationship to the subgenus *Lacerta*, whereas Arnold (1973:332) felt strongly that it belonged to his *Lacerta* part I (the equivalent of subgenus *Lacerta* [*sensu stricto*]). Similarly, Eiselt (1968:428–431), based on a study of the skull, pholidosis, and color pattern, argued for its close relationship to *L. viridis*, both probably being derived from the *L. strigata–agilis* complex. More recently, Mayer and Bischoff (1996) have placed it in the genus *Timon*, probably related to *L. cappadocica* + *L. saxicola* complex.

Observations on the natural history of both subspecies are to be found in Eiselt (1968:413–414), Clark and Clark (1973:30), and Mulder (1998). Bosch (1998) recorded observations on their growth, feeding, and hibernation in captivity.

Lacerta raddei Boettger, 1892
Azerbaidzhan lizard

Lacerta raddei raddei Boettger, 1892 (Plate 14F–G)
Azerbaidzhan lizard

Lacerta muralis var. *raddei* Boettger, 1892:142 (Type locality: Nieuvadi, Arax Valley, Armenia; Lectotype: SMF 12054). — Mertens, 1922b:173.
Lacerta muralis var. *defilippi*: Boulenger, 1904:337 (in part; not *Podarcis defilippii* Camerano, 1877); 1920:288–290 (in part).
Lacerta saxicola var. *defilippi*: Méhely, 1909:519, pl. 18, figs. 1, 3. — Nikolsky, 1915:370–373 (in part). — Lantz and Cyrén 1936:164 (in part). — Terentjev and Chernov, 1949:187–189 (in part), — Mertens, 1957:82. — S. Anderson, 1968:333 (in part).
Lacerta saxicola raddei: Darevsky, 1967:83–89 [1978:91–97], text-figs. 34–35, photo 12. — S. Anderson, 1974:44.
Lacerta raddei raddei: Bannikov, *et al.*, 1977:218. — Darevsky, *et al.*, 1984 :102–108. — Welch, 1983:36. — Eiselt, *et al.*, 1993:68–70, figs. 2–7; 1995:60–62, 66–67, figs. 1–2, tab. 1.

Diagnosis: Ventral plates more or less rectangular in shape with rectilinear or nearly rectilinear posterior margins; collar not serrated; dorsals smooth, 41–62 across middle of body; usually single postnasal; complete row of granules separates supraoculars from supraciliaries (Fig. 118); pterygoid teeth absent; 20–29 gulars; 13–23 femoral pores on each side; 15–22 scales round middle of shank. The following from Eiselt (1995:62, tab. 1): Pileus length/breadth 1.9–2.3; number of small femorals across 4–8; number of small tibials across 17–21; lamellar scales beneath fourth toe 25–32; temporals between massetericum and tympanicum 2–5; temporals between massetericum and first supratemporal 2–5; marginals along 10 rows of ventrals 14–31.

Color pattern: Dorsum various shades of light brown, from cinnamon to yellowish brown; numerous dark spots irregularly arranged along spine (occipital stripe), sometimes forming two parallel rows; 3–4 longitudinal rows of light ocelli along sides within darker brown band (temporal stripe), these ocelli light blue in axillary region in life; irregular upper margin of lateral dark band usually marked by distinct row of whitish ocelli; comparatively large, dark-edged ocelli usually restricted to lower margin of dark (temporal) stripe. Venter whitish, bluish, greenish-white, or bright yellow. During breeding season most light ocelli on sides of male become vivid light blue, as do outer ventrals. (Darevsky, 1967:85).

Size: Snout-vent length to 69 mm in males, 67 mm in females (Darevsky, 1967 [1978 translation]:92).

Habitat: (Plate 24E) In the Lenkoran District of Azerbaidzhan to the north of the border of Persian Azarbaijan, *Lacerta r. raddei* ranges from the barren rocks within the high altitude forest zone at 1600–1800 m elevation to the high ridges without forest cover along the Iranian border, and down in the Diabar Valley, an area of deep gorges, barren rocks, and low ridges devoid of vegetation. (Sobolevsky, 1929).

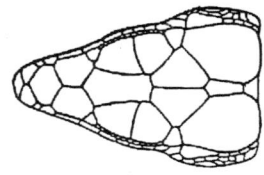
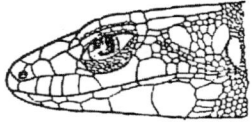

FIGURE 118. *Lacerta raddei.* Dorsal (A) and lateral (B) views of head. (From Darevsky, 1967, fig. 34.)

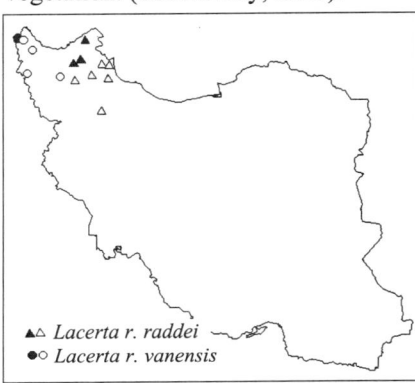

▲△ *Lacerta r. raddei*
●○ *Lacerta r. vanensis*

Distribution: Armenia and Azerbaidzhan, Lenkoran Region, Talysh Mountains, Karabagh, and the northern parts of East Azarbaijan Province of Iran, passing over the bordering Aras River and the bordering range of the Qareh Dagh (relatively little material; southernmost locality: 31 km S Ahar, 67 km SW Ardebil and perhaps near Khalkhal on the Boghrovdagh-chain). (Eiselt, 1995:66).

Remarks: My specimens of *Lacerta defilippii* (until recently, considered a subspecies of *L. raddei*) from the Alborz between Chalus and Karaj had bright orange-red outer ventrals with blue spots, throat, breast, and midbelly suffused with lighter orange-red. My specimens of *L. raddei* from 31 km S Ahar had bright yellow venters, including entire chin, throat, chest, and hind limbs, blue spots on outer ventrals. Eiselt, *et al.* (1993:66–70) reexamined CAS 141273 along with other specimens from northwestern Iran and considered them intermediate between *L. r. raddei* and their new subspecies, *L. raddei vanensis* Eiselt, J. F. Schmidtler, and Darevsky 1993 from Turkey (but see below).

Lacerta raddei vanensis **Eiselt, Schmidtler, and Darevsky, 1993** (Plates 14H, 15A)
Lake Van lacerta

Lacerta raddei vanensis Eiselt, *et al.*, 1993:65–70, figs. 1–6 (Type locality: Burgberg (castle mountain) in the city of Van, eastern Turkey, about 1720 m. Holotype: NMW 32999). — Schmidtler, *et al.*, 1994:61, fig. 6.

Diagnosis: A medium-sized *Lacerta raddei*, distinguished from *L. r. raddei* Boettger and *L. r. nairensis* Darevsky by numerous small, characteristic, clearly differentiated character states, of which however only a few are conspicuous, whereby the discrimination from *raddei* appears to be rather less distinct in contrast to *nairensis*. In the case of *vanensis* there is a single central preanal (instead of two); much more frequently the massetericum is considerably smaller, the number of supraciliary granules is clearly greater; this is true also for the number of temporals between the massetericum and tympanale, while the number of small scales counted transversely across the throat is clearly less. The dorsal pattern is relatively undistinguished from that of *raddei*, extremely distinct from *nairensis* and varies from almost unspotted to marked with double longitudinal rows of coarse spots. (Eiselt, *et al.*, 1993:66). The following from Eiselt (1995:62, table 1): number of dorsals 43–53; pileus length/breadth 1.9–2.2; number of small femorals across 5–7; number of small tibials across 16–21; lamellar scales beneath fourth toe 28–33; temporals between massetericum and tympanicum 3–7; temporals between massetericum and 1st supratemporal 1–3; marginals along 10 rows of ventrals 21–30.

Color pattern: Dorsum light grayish olive, back very loosely and finely spotted; a longitudinal row of larger rounder lighter spots is disposed in each supraciliary stripe; the dark temporal bands contain similar spots resembling ocelli, of which ¾ are prominent as light blue axillary spots; flanks in part finely blue-spotted; venter light yellowish white, femoral pore region in life distinctly yellow, outer longitudinal row of ventrals gray-blue. (Eiselt, *et al.*, 1993:68.)

Size: Holotype (male): snout-vent length 63 mm, tail 126 mm (largest known specimen).

Distribution: (see map, p. 244) Easternmost Turkey, east of Lake Van and West Azarbaijan Province of Iran, west and north of Lake Reza'iyeh, to Sero and Khoy (Eiselt, 1995:66).

Remarks: My specimens of *Lacerta raddei* from Qaraqachid and 'Arab-e Dizehsi have a pale blue venter, bright blue spots on outer ventrals. Eiselt, *et al.* (1993) regarded specimens from West Azarbaijan as intermediate between *L. r. raddei* and *L. r. vanensis*. Subsequently, however (Schmidtler, *et al.*, 1994:61, fig. 6), they showed *L. r. vanensis* as occurring at Maku in West Azarbaijan Province. Based on their findings, I include this subspecies here as occurring in Iran and tentatively assign specimens from west and north of Lake Reza'iyeh to this taxon. Nonetheless, on the basis of my examination of the specimens I collected, I find no differences so distinct as to be used as key characters for identifying individuals as to subspecies.

Lacerta steineri Eiselt, 1995 (Plate 15B)
Steiner's lacerta

Lacerta (Archaeolacerta) steineri Eiselt, 1995:63–65, figs. 1 (map) –2, pl. 1 (Type locality: Gole-Loweh near Minou-dasht [33°11'N, 65°21'E — in error in original publication; correctly: 37°11'N, 55°21'E], SE Gonbad-e Gavous, NE Iran; Holotype: NMW 33715, male).

Diagnosis: Ventral plates more or less rectangular in shape with nearly rectilinear posterior margins; collar not serrated; dorsals smooth, 53–60 across middle of body; single postnasal; row of granules separating supraoculars and supraciliaries unbroken; 25 gulars; 16 femoral pores on each side. A rather flat-headed *Lacerta* Part II (sensu Arnold [1973]; *Archaeolacerta* sensu Méhelÿ [1909:424]) of medium size. Distinguished from all similar *Lacerta*-taxa (*chlorogaster, defilippi, raddei, valentini*) by its characteristic, nearly unvarying pattern-type, its relatively longer pileus length (length/breadth ratio: 2.2–2.4), much smaller massetericum, greater number of dorsals (53–60), smaller tibial scales (21–26), as well as smaller temporals between the massetericum and the tympanicum (3–5), or, respectively, the first supratemporal. *L. steineri* is separated from *defilippi* and *raddei* by its lower number of small femoral scales and marginals (19–26). The smooth, only slightly swollen dorsals and slender body-form readily allows *steineri* to be distinguished from the stouter *chlorogaster* and *praticola* with their keeled dorsals. (Eiselt, 1995:62–63, table 1.)

Color pattern: Ground color of the back brownish-olive, pattern black. Pileus lightly lineated, the dorsum traversed by a vertebral row of small flecks, which, longitudinally connected, form a narrow, jagged stripe; on either side, paralleling the vertebral stripe, a series of small flecks; the two supraciliary stripes are only indicated by an indistinct longitudinal row of somewhat lighter markings, which on either side is delimited by the blackish margins of the dark reddish-brown temporal band; the latter run along the flanks and enclose large whitish ocelli and are continued on the sides of the tail as a series of dark spots, as also the dorsal longitudinal stripe along the tail is broken up into similar flecks. (Eiselt, 1993:64, 68, pl. 1). Eiselt made no mention of the ventral color, but the specimens had been in preservative since their collection in 1968.

Size: Holotype: snout-vent 68 mm, tail 126 mm; largest specimen (male?): snout-vent 71 mm.

Habitat: The syntopic occurrence of *Lacerta steineri* and *L. chlorogaster* has been documented, as Steiner also collected a specimen of *L. chlorogaster* (NMW 33893:2) from the type locality of *L. steineri*, as had Franz Ressl, a distinguished observer and collector of a 1974 trip to "Gole-Lowe" (NMW 33893:2), who captured a further example of *chlorogaster*. The latter noted of this region that it was a thick-foliaged virgin forest, but that the lizards come, not out of this forest, but from the northern edge (in part steep-edged) with luxuriant herbaceous vegetation and interspersed stony blocks. Eiselt (1995:65) speculated that just as with the close proximity of *chlorogaster* and *defilippii*, where inhabitants of two closely adjacent ecological niches occur, the morphology of *steineri* with its *muralis*-like habitus, its slender head, its incompletely ossified supraocular lamellae, and its unkeeled, scarcely swollen dorsals, clearly an adaptation to life in a rocky, fissure-rich habitat, is consistent with a niche similar to that of *defilippii*, while in the case of *chlorogaster* the arched head with its primitive, almost or fully ossified supraocular lamellae and clearly keeled dorsals points to a life on the forest floor, tree trunks and forest edges.

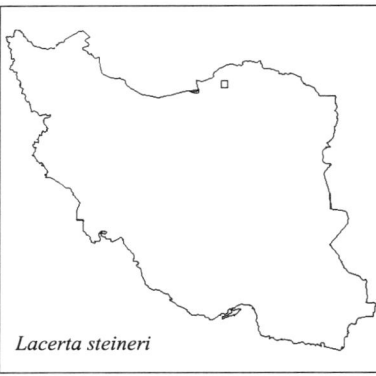
Lacerta steineri

Distribution: Known only from the type locality at the edge of the subtropical moist forest of the western Kopet Dag.

Remarks: Paratypes: 6 males, one female, NMW 33716:1–7, same data as holotype. Eiselt's (1995:67, fig 2) dendrogram represents *Lacerta steineri* as the sister taxon of *L. defilippii* in the larger clade that also contains *L. raddei*.

Lacerta strigata Eichwald, 1831 (Plate 15C)
Five-streaked lizard, Caspian green lizard

Lacerta strigata Eichwald, 1831:189 (Type locality: Krasnowodsk, northern Caucasus, Russia [restricted by Mertens and Muller, 1928:41; see also Darevsky, 1984:89]). — J. Anderson, 1872:372–373. — Blanford, 1876:364. Bedriaga, 1879:29. — Terentjev and Chernov, 1949:176–177, text-fig. 67. — Mertens, 1957:120–122. — Guibé, 1957:139. — G. Peters, 1962:127–152. — S. Anderson, 1963:476; 1968:333. — Tuck, 1971a:58. — S. Anderson, 1974:40, 44. — Bannikov, *et al.*, 1977:194–196. — Darevsky, 1984:82–99, figs. 7–9. — Clark, 1991:39. — Frynta, *et al.*, 1997:10.

?*Lacerta viridis* var. *astrabadensis* Eichwald, 1841:83 (Type locality: Astrabad [Gorgan], Iran).
Lacerta viridis: Blanford, 1876:364 (not *Seps viridis* Laurenti, 1768). — Bedriaga, 1879:28. — F. Werner, 1936:200. — S. Anderson, 1963:476.

Lacerta viridis var. *vaillanti* Bedriaga, 1886:95 (Type locality: Persia).
Lacerta viridis var. *strigata*: Boulenger, 1887a:17–18; 1899:378. — F. Werner, 1903:341. — Nikolsky, 1907a:281; 1915:286–292. — Boulenger, 1920:77–82. — Morich, 1929:31. — Rostombekov, 1938:14. — Guibé, 1966a:98. — Schleich, 1977:127, 129.
Lacerta viridis var. *woosnami* Boulenger, 1917:277 (Type locality: south coast of Caspian Sea and Bash Nurashin, Iran; Syntypes: BMNH 1908.8.7.6–13/1946.8.5.27–33); 1920:88–90. — Welch, 1983:38.
Lacerta strigata strigata: Wettstein, 1951:438. — Schleich, 1977:127, 129. — Welch, 1983:36.

Diagnosis: Transverse series of ventral plates with notches between plates, in 6 (rarely 8) longitudinal series; collar strongly serrated; usually 2 superposed postnasals, 25–29 subdigital lamellae under fourth toe; 37–49 dorsals at midbody; tympanic shield present; 17–21 femoral pores, row of pores complete up to knee; rarely over 20 temporal scales; fifth submaxillary well developed.

Color pattern: Young light olive-brown, almost always with light line from supramaxillary region to hind limb in addition to vertebral and dorsolateral light lines (total of 5 lines on body) (Fig. 119). Adults without light lines or spots; dorsum green (in various shades), uniform, or speckled with black dots. In lowland populations there is no blue on underside of head, throat, neck, or flanks; hind limbs and tail (often sacral region as well) light to gray-brown; remnants of three dorsal lines still in evidence, dorsum finely sprinkled to coarsely flecked with black. Old males in mountain populations have blue coloration on chin, throat, and sides of neck, varying in extent and intensity. (G. Peters 1962:139–142).

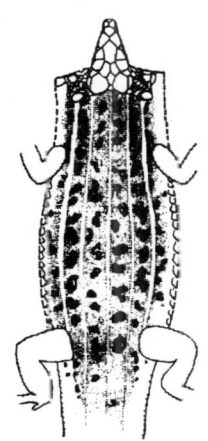

FIGURE 119. *Lacerta strigata*, dorsal pattern of subadult female. (From Terentjev and Chernov, 1949, fig. 67.)

Size: Snout-vent length to at least 102 mm in Iranian specimens, tail 202 mm.

Habitat: (Plate 22G) I found this species active among rushes in a depression in the back dunes of the coastal strand of the Caspian Sea and at the edge of an alder stand on the Mordab west of Bandar-e Pahlavi. They run rapidly and crash into vegetation. According to Nikolsky (1915:291), this lizard lives in brushy areas, avoiding both bare steppes and dense forests. It prefers hills on the outskirts of forests where there is vegetation for cover, rather than flat areas. Mertens (1957:122) reported that Shuz found it to be most common in pasture land converted from brushland, where shrubs and bushes were interspersed with grassy meadows. Small individuals were found also among rushes (*Juncus maritimus*) along the strand of the Caspian Coast. Darevsky (1984:93–97) gives an account of its ecology and natural history in Transcaucasia.

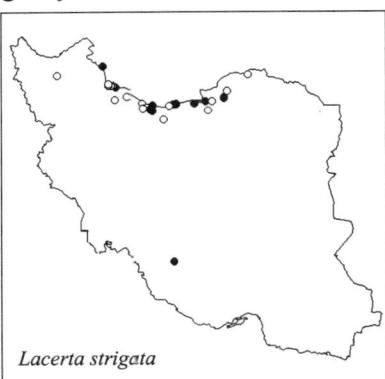

Lacerta strigata

Distribution: In Iran it extends south to Shiraz and north of the Alborz it extends east to Bandar Shah on the east coast of the Caspian Sea and north across the border into Turkmenistan. It extends north to Kaspiskij, north of the Terek River, and Armawir in Russia; eastern Turkey and possibly northeastern Iraq (but see Darevsky, 1984:86–87). The Shiraz record is surprising, although there is a specimen of record in the British Museum (both Arnold and I have confirmed the identity; this specimen appears to have been obtained by Major St. John, a reliable collector). This is the only record south of the Alborz. From below sea level to about 1800 m elevation.

Remarks: Specimens from near Babol Sar appear to have had blue on the sides of the throat and on the chin in life, contrary to G. Peters's statement regarding lowland populations of this species. See J. F. Schmidtler (1986b) and Clark (1991:39) for remarks on the status of this species.

Haas and Y. Werner (1969:341–343) record specimens of *Lacerta media* as well as a somewhat problematic juvenile specimen of *L.* cf. *strigata* from near Ruwanduz Iraq, but possible habitat differences were not recorded by the collector, apparently. Their specimen tentatively assigned to *L. strigata* lacked the typical juvenile color pattern and had 8 (rather than 6) rows of ventral plates.

Lacerta valentini Boettger, 1892
Valentin's lizard

Lacerta valentini valentini Boettger, 1892   (Plate 15D–E)
Valentin's lizard

Lacerta muralis var. *valentini* Boettger, 1892b:145 (Type locality: Bazarkent, northeastern Armenia. Holotype: SMF 12064).
Lacerta valentini valentini: Bannikov, et al., 1977:222.
Lacerta valentini: Darevsky, et al., 1984:102–108.

Diagnosis: Ventral plates more or less rectangular in shape with nearly rectilinear posterior margins; collar not serrated; dorsals smooth, 41–53 across middle of body; single postnasal; row of granules separating supraoculars and supraciliaries usually incomplete, rarely complete or absent (Fig. 120); 21–29 gulars; 14–22 femoral pores on each side. The following from Eiselt (1995:62, table 1): Pileus length/breadth 1.9–2.1; number of small femorals across 4–5; number of small tibials across 16–18; lamellar scales beneath fourth toe 24–28; temporals between massetericum and tympanicum 1–4; temporals between massetericum and first supratemporal 0–2; marginals along 10 rows of ventrals 16–24.

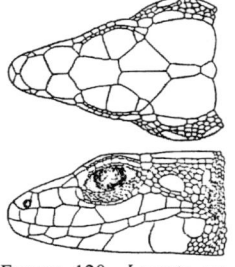

FIGURE 120. *Lacerta valentini*. Top and side of head. (From Darevsky, 1969, fig. 39.)

Color pattern: Dorsal ground color olive-yellow, yellowish green, greenish yellow, bright green, or greenish brown; occipital stripe consists of large irregularly spaced black or dark brown blotches, some covering up to 30–40 body scales, in some cases reduced in size, forming continuous reticulum; narrow parietal stripes of ground color bordered by central occipital stripe extend onto dorsal third of tail; broad black or dark brown temporal stripes bear 1–3 rows of rounded bright (blue in postaxillary zone) spots forming centers of fused dark ocelli; head with irregular black blotches and spots. During breeding season venter including throat and head of both sexes bright yellowish orange, extreme lateral rows of ventrals in males acquire bright or light blue color. (Darevsky 1967:104–105).

Size: Snout-vent to 70 mm in males, 78 mm in females.

Habitat: Nothing known of habitat in Iran. In the Caucasus it is a hill steppe form occurring on outcrops of solid rocks and (rarely) argillaceous scarps, on individual stone blocks and piles of large boulders in hill-meadow, and in subalpine and alpine zones.

Darevsky (1967:197–236) discusses aspects of ecology of this and other rock lizards in the former USSR.

Distribution: Occurs in northwestern Iran according to Ilya Darevsky (pers. commun.), but I do not have information as to precise localities.

Genus *Mesalina* Gray, 1838
Sand lizards, mesalinas

Mesalina Gray, 1838a:282 (Type species: *Mesalina lichtensteinii* Gray, 1838 [= *Lacerta rubropunctata* Lichtenstein, 1823).

Definition: Head shields normal; occipital shield usually present; lower nasal in contact with first supralabial only; nostril between 3 nasals and widely separated from supralabials; sometimes 2 or more transparent shields in lower eyelid; abdominal plates in parallel longitudinal rows.

Distribution: North Africa and Southwest Asia, from Morocco to Pakistan, Somalia to Turkmenistan. 13 species, 2 in Iran.

Key to the Species of *Mesalina* in Iran

1a. Occipital in contact with interparietal (Fig. 122), or separated from it by a small shield; large transparent scales of lower eyelid edged with black; ventral plates in 10 longitudinal series .... *Mesalina watsonana* (Stoliczka, 1872) (p. 251)
1b. Occipital absent or minute, not in contact with interparietal (Fig. 121); transparent shields of lower eyelid not edged with black; ventral plates usually in 12 (rarely 10 or 14) longitudinal series .... *Mesalina brevirostris* Blanford, 1874 (p. 249)

Remarks: In his review of the Palearctic genus *Eremias*, Szczerbak (1974) resurrected Gray's *Mesalina* for the distinctive group of small species centering on North Africa and the Saharo-Sindian region of Southwest Asia. Arnold's (1986:1255) study of the hemipenes of lacertid lizards supports the holophyly of this genus. Nonetheless, recent interpretations of phylogeny have considered *Eremias* and *Mesalina* sister taxa (Mayer and Benyr, 1994; Mayer and Bischoff, 1996) or at least closely related (Arnold, 1989).

Mesalina brevirostris Blanford, 1874
Blanford's short-nosed desert lizard

Mesalina brevirostris brevirostris Blanford, 1874 (Plate 15G–H)
Blanford's short-nosed desert lizard

Mesalina brevirostris Blanford, 1874b:32 (Type locality: Tumb Island [Persian Gulf] and Kalabagh, Punjab [restricted by Schmidt, 1939:66]; Syntypes: BMNH 80.11.10.40/1946.8.6.34 and ZSI 3474 [*fide* Das, et al., 1998]); 1876:379–380.
Mesalina pardalis Blanford, 1876:377–379 (in part; not *Lacerta pardalis* Lichtenstein, 1823:99).
Eremias brevirostris: Boulenger, 1887a:89; 1921:273–276. — F. Werner, 1936:201. — Schmidt, 1955:204. — Hellmich, 1959:5. — S. Anderson, 1968:333; 1974:38, 43. — Schleich, 1977:127, 129.
Eremias brevirostris forma *typica*: Angel, 1936:111–112.
Eremias brevirostris brevirostris: Haas and Y. Werner, 1969:352–356, figs. 8B, 9B. — Leviton, et al., 1992:61–62, pl. 9D.

Diagnosis: Three nasals, lower in contact with rostral and first supralabial; ventral plates in 12 (rarely 10) straight longitudinal series; occipital absent or minute, not in contact with interparietal (Fig. 121); collar curved or angular, free; head not strongly depressed, 1⅕ to 1⅓ as long as broad; 34–50 scales across middle of back; 19–28 lamellae under fourth toe.

Color pattern: Gray or grayish brown above, usually with numerous large whitish ocelli edged with black, pattern highly variable; sometimes small dark brown or rusty spots on back and larger ones on sides; tail with dark lateral spots; venter white.

Size: Snout-vent length 56 mm, tail 102 mm.

Mesalina brevirostris fieldi (Haas and Werner, 1969)
Field's short-nosed desert lizard

Eremias brevirostris fieldi Haas and Y. Werner, 1969:356–359, figs. 8A, 9A, 10–12, pls. 16–17, 18A–B (Type locality: Mahor Birinji, southwestern Iran; Holotype: MCZ 56617). — Welch, 1983:54.

Diagnosis: 30–39 (usually 33–35) dorsal scales across middle of back; 16–20 lamellae under fourth toe.

Color pattern: Light ocelli on back arranged in more or less regular longitudinal rows, some of dark margins confluent, forming dark crossbars; on tail pattern gradually changes to alternating light and dark half-rings.

Natural history: The following remarks pertain to Iranian *Mesalina brevirostris* generally, without respect to subspecies. Specimens from Ahram, collected in early January, contain yolked follicles up to 3 mm diameter in ovaries; none had eggs in the oviducts, whereas 3 or 4 females of *M. watsonana* from the same locality contain oviducal eggs. Females from Meshrageh, collected in mid-October have much smaller ovarian follicles. Digestive tracts contained species of ants, lycosid spiders, wasps, and beetles. Weber

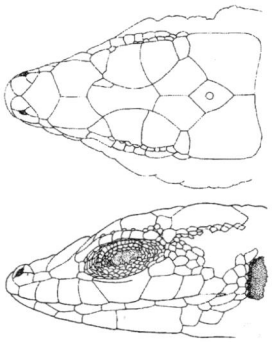

FIGURE 121. Top and side views of head of holotype of *Mesalina brevirostris fieldi*. (From Haas and Y. Werner, 1969, figs. 8A and 9A.)

(1960) listed several species of ants, moths, locusts, hemipterans, weevils and other beetles, lepidopteran larvae, mantids, orthopteran nymphs, roaches, spiders, solpugids, and scorpions as dietary items in Iraq specimens.

Habitat: At Ahram in coastal Fars, both *Mesalina brevirostris* and *M. watsonana* have been collected together. Both apparently occur in the upper Mesopotamian Plain, and the habitats probably are in contact where foothills and plain come together. Minton (1966:110) found both species near the coast in Las Bela, Pakistan, *M. brevirostris* on sandy or silty soil, *M. watsonana* in rocky situations along dry stream beds.

Distribution: The species as a whole is distributed, probably discontinuously, from Sinai (the southern tip, and Tiran Island off the southern tip of the peninsula according to Hoofien, 1957), northern Saudi Arabia, Syria, Jordan, Iraq, southwestern Iran and islands in the Persian Gulf, Pakistan (Las Bela and the Iranian Plateau according to Minton, 1966:110) to Punjab, northern India. In Iran it is known only from the Mesopotamian Plain and the Persian Gulf coast from Bushire northward. It has not been collected in southeastern Iran, even though it is present on islands off that coast.

Undoubtedly several populations will prove recognizable; Angel (1936:111–112) has described a small-scaled subspecies, *Mesalina b. microlepis*, from Syria, between Homs and Qariatein. However, he assigned specimens from Palmyra northward to the typical subspe-

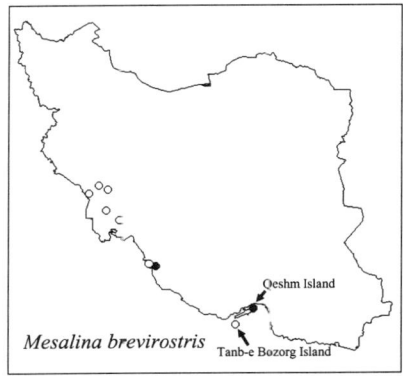

cies, and Haas and Y. Werner (1969:356) regarded specimens from eastern Syria, Iraq, and Jordan as well as Pakistan to be *M. b. brevirostris*. They felt that Arabian material may represent a distinct subspecies. They described *M. b. fieldi* from specimens coming from Mahor Birinji and Gotwand (= Gatvand), localities close together in southwestern Iran. These localities are close to Schmidt's (1955:204) locality, Saleh Abad (= Andimeshk). It seems likely that Iranian specimens from various localities are assignable to both *M. b. brevirostris* and *M. b. fieldi* as defined by Haas and Werner. Arnold (1986e:1253–1254) suggested on the basis of hemipenial differences that a population from the Jerrahi River in southwestern Iran might deserve taxonomic recognition. This would seem to be further justification for recognizing *M. b. fieldi*.

Remarks: In the series from Ahram and Meshrageh, the subocular borders the mouth. These specimens and those of Boulenger (1921:276) from Bushire have counts which agree with those of Haas and Y. Werner (1969) for Mahor Birinji and Gatvand. Boulenger's specimens from the Jarrahi River have considerably higher dorsal counts. In 11 specimens from Dammam, Saudi Arabia, the subocular is uniformly excluded from the labial border by a small scale. Haas and Y. Werner (1969:352–359) found this to be variable in specimens from Syria and Iraq, but did not mention it in regard to their Arabian specimens. In their 22 Iranian specimens the subocular apparently borders the mouth. Three specimens counted from Rutba, Iraq (FMNH 11358) have 45–50 dorsals, and others checked from Iraq and Saudi Arabia have 37–46, all within the range for *Mesalina b. brevirostris*. One of Boulenger's (1921:276) specimens from the Jarrahi River near the head of the Persian Gulf had 50 dorsals, while his lowest count, 36 dorsals, within the range given for *M. b. fieldi*, came from Jerud, Syria, between Homs and Damascus. Four unusually large specimens (snout-vent 64 mm) from 5 mi [8 km] N Zarqa, Jordan, north of Amman (FMNH 74535) have high dorsal counts (54–60). Clearly the distinctions among populations require further study and the application of subspecific designations at this time has no zoogeographic significance; however, the importance of geographically or, perhaps, ecologically correlated variation in morphology should not be overlooked, and the application of subspecific names serves to call attention to this phenomenon until such time as it is studied adequately.

Mesalina watsonana (Stoliczka, 1872) (Plate 16B)
Persian long-tailed desert lizard

?*Eremias pardalis*: De Filippi, 1865:354 (not *Lacerta pardalis* Lichtenstein, 1823:99).
Eremias guttulata watsonana Stoliczka, 1872b:86 (Type locality: Sind, between Karachi and Sukkur; Syntypes BMNH 74.4.29.1436/1946.8.7.75, ZSI 4929, 5050, 5223–5225, and NMW). — Smith, 1935:389–390. — Forcart, 1950:149–150. — Wettstein, 1951:440. — Schmidt, 1955:204. — Mertens, 1956:94. — Guibé, 1957:139. — S. Anderson, 1963:460–462. — Guibé, 1966a:98. — Clark, *et al*., 1966:7. — S. Anderson, 1968:333. — Haas and Y. Werner, 1969:350–351. — Tuck, 1971a:57, fig. 7. — Schleich, 1977:127, 129. — Welch, 1983:54.
Mesalina pardaloides Blanford, 1874b:32 (Type locality: Henjam Island, Persian Gulf; Holotype: ZSI 3381); 1876:381.
Mesalina pardalis: Blanford, 1876:377–379 (in part; not *Lacerta pardalis* Lichtenstein).

Podarces (*Eremias*) *watsonana*: Bedriaga, 1879:33.
Podarces (*Eremias*) *pardalis*: Bedriaga, 1879:33.
Podarces (*Eremias*) *pardaloides*: Bedriaga, 1879:33.
Eremias pardaloides: Boulenger, 1887a:87.
Eremias guttulata: Boulenger, 1887a:87–89 (in part). — F. Werner, 1895:4. — Nikolsky, 1897:328; 1899c:398. — Zarudny, 1903:19– 20. — Nikolsky, 1907a:281–282; 1915:411–414, text-fig. 51. — F. Werner, 1917:202. — Lantz, 1918:15. — Boulenger, 1921:258–261. — Morich, 1929:31. — F. Werner, 1929:240; 1936:198. — Terentjev and Chernov, 1949:193–194. — S. Anderson, 1974: 38, 43.
Eremias guttulata guttulata: S. Anderson, 1963:476.
Mesalina guttulata watsonana: Szczerbak, 1974:275–277, fig. 86. — Bannikov, *et al.*, 1977:163. — Tuck, 1979:103–104. — Nilson, and Andrén 1981:138. — Welch, 1983:54.
Mesalina guttulata: Leviton, *et al.*, 1992:62–63 (in part), col. pl. 9E.
Mesalina watsonana: Arnold, 1986a:1254. Frynta, *et al.*, 1997:10.
Mesalina watsonnana: Clark, 1991:41.

Diagnosis: Three nasals, lower in contact with rostral and first supralabial; ventral plates in 10 (rarely 8) straight longitudinal series; small occipital present (Fig. 122); larger transparent scales of lower eyelid edged with black. Collar complete or nearly so, its scales distinctly enlarged (Smith, 1935:390). Hemipenis relatively long, the basal parts of the lobes not folded, although the apical sections are; armature very elongate and cleft for most of its length with narrow clavulae; lips of sulcus strongly developed (Arnold, 1986:1243–1244).

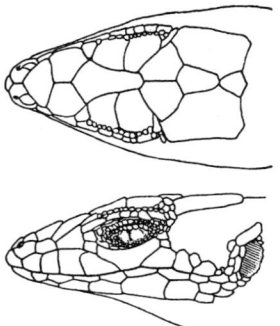

FIGURE 122. Dorsal and side views of head of *Mesalina watsonana*. (From Szczerbak, 1974, fig. 86.)

Color pattern: Grayish or olivaceous above; dorsum with longitudinal series of small white spots, edged or accompanied by black spots, small white spots edged with black on flanks; often light dorsolateral stripe running to eye; limbs marbled with black and white; black streak on posterior aspect of thigh; venter whitish, some individuals from some localities having pale yellow throat and chest.

Size: Snout-vent length 56 mm, tail 114 mm.

Natural history: I found oviducal eggs in specimens collected in early March in Khuzestan; specimens collected in early March, early April, and mid-August have yolked follicles up to 3 mm diameter. Hatchlings were seen from late April through November, half-grown individuals not appearing until late May or June. Specimens from Ahram (Fars Province) collected in early January have oviducal eggs, while one from Galatappeh, Esfahan Province, collected in late December has an egg in the oviduct; other females from these localities have large yolked ovarian follicles. A specimen collected in late April between Tehran and Karaj at 1158 m has oviducal eggs. In southern and eastern Afghanistan, specimens collected in late February, early March, and mid-April contain large eggs, while specimens collected in mid-July through August have regressed testes and ovaries. Hatchlings were found in early August. This is in accord with Minton's (1966:110) data for Pakistan, which may indicate early spring breeding, the first young appearing about May, and reaching maturity by winter. In Turkmenistan, egg-laying occurs in May and June, with the young appearing in July (Terentjev and Chernov 1949:194). The implications of these incomplete data are not clear. One may speculate that in the parts of the range where winters are mild and springs warm, egg-laying occurs throughout the winter and spring, and perhaps in late summer or early autumn, with a period of gonadal regression during the summer. More than one clutch is probably laid, and oviposition may vary with annual fluctuation in weather conditions. In high elevations and the northern parts of the range, egg-laying is probably limited to late spring. Maturity appears to be reached in the first year, at least in the southern regions.

Spiders, crickets, beetles, ants and ant larvae and other small insects make up the diet. See S. Anderson (1963:460–462) for observations on behavior. Tsellarius and Cherlin (1981) studied the ecology of this species in the Badghyz, Turkmenistan.

Habitat: (Plates 20A–B, 21C, 22A–C, F) This lizard is abundant on hard soils of plains and alluvial fans throughout much of Iran and is found on hillsides, valleys, and along stream courses. I have found it on gravelly and silty alluvium, sand and gravel steppe, disintegrating shale slopes, and saline hamadas. It avoids solid rock substrate and is rarely found on sandy soils, although I found it in a sandy wash near Kerman and on and around sandy hummocks in a *Suaeda* association northeast of Bandar 'Abbas. Vegetation in areas where it occurs is usually scanty desert or steppe shrub, or areas stripped bare of perennial vegetation. It has been found together with the closely related *Mesalina brevirostris* at Ahram in Fars, Mahor Birinji in Khuzestan, and coastal Las Bela, Pakistan as previously noted (see p. 250).

Mesalina watsonana

Distribution: Occurs widely on the Iranian Plateau and extends as far north as southern Turkmenistan. It occurs throughout Iran and Afghanistan at elevations below about 2500 m. In Iran it is absent only in the high mountains, along the Caspian coast, and in the Azarbaijan provinces. It appears to be absent from Turkey. Its distribution is largely complementary to that of *Ophisops elegans*, which appears to be its ecological counterpart to the northwest. The two species are sympatric over a large area of Iran, however.

Remarks: Arnold (1986a:1253–1254) found that *Mesalina guttulata* and *M. watsonana* differ radically in the size and form of hemipenial morphology and recognized them as distinct at the species level. The lips of the sulcus of the hemipenis are strongly developed in *M. watsonana*, the hemipenis is relatively long and the basal parts of the lobes are not folded, although the apical sections are; the armature is very elongate and cleft for most of its length with quite narrow clavulae (Arnold, 1986:1243–1244). "Compared with *M. g. watsonana*, *M. g. guttulata* has an extremely small, slender hemipenis with tiny unfolded lobes that are asymmetrical, the median one being larger" (Arnold, 1986:1244). He concluded that they are allopatric or nearly so, *M. guttulata* (Plate 16A) occurring in North Africa, Arabia, Israel, Syria, Iraq, and probably southwestern Iran, while *M. watsonana* is found in much of the rest of Iran, Turkmenistan, Afghanistan, Pakistan, and northwest India. Here, I have included all of the Iranian specimens under *M. watsonana*. I examined the hemipenes of specimens from the Masjed Soleyman area of southwestern Iran (CAS 86249, 86430, 86440), the vicinity of Bandar-e Lengeh (CAS 141048) as well as specimens from the Central Plateau of Iran, Baluchistan, Afghanistan, and Turkmenistan. In none of these specimens did I find the notably small, slender hemipenis described for *M. guttulata*. Thus, while the Zagros Mountains might seem a reasonable geographic barrier between the two species, such does not seem to be the case. However, if it becomes feasible to recognize the two species on the basis of external morphology, some other geographic correlation with distribution may become apparent. Smith (1935:390) says: "In the Indian form [*M. guttulata watsonana*] the collar is complete or nearly so, and its scales are distinctly enlarged; in the African form [*M. g. guttulata*] it is free only at the sides, and its scales are not or but slightly enlarged." In all of the Iranian specimens, the collar is as described for *watsonana*.

Genus *Ophisops* Ménétriés, 1832
Snake-eyed lacertas

Ophisops Ménétriés, 1832:63 (Type species: *Ophisops elegans* Ménétriés, 1832, by monotypy).

Definition: Lower eyelid fused with upper, with large transparent disc (Fig. 123); ventral plates smooth; collar weakly defined, or absent in middle; subdigital lamellae keeled; dorsal scales rhombic, imbricate, strongly keeled; femoral pores present.

Distribution: Southeastern Europe (from Balkan Peninsula eastward), North Africa (eastern Morocco and western Algeria eastward), Southwest Asia (north to Transcaucasia, south to southern Iran in Zagros Mountains), and India. The distribution of the genus is disjunct, with a gap of some 1300 km separating the easternmost records of *O. elegans* and the westernmost localities of *O. jerdoni*. At least eight currently recognized species, including those that were previously assigned to *Cabrita* (see Arnold, 1989 for synonymy of *Cabrita* with *Ophisops*); a ninth, *O. meizolepis*, known only from the type, collected in the Punjab, has been regarded as synonymous with *O. elegans* (Smith, 1935:379); however, this locality is over 1600 km east of the known range of *O. elegans*. A single species in Iran.

Remarks: Boulenger (1921) speculated that *Ophisops* and the related Indian genus, *Cabrita*, were derived from a *Nucras*-like stock; *Nucras* is a genus at present confined to southern Africa. Arnold (1989) thinks that *Ophisops* is probably the sister taxon of *Mesalina*, the two together being the sister clade to *Acanthodactylus*, certainly a more coherent relationship biogeographically. In the scheme of Mayer and Benyr (1994) it is the sister taxon to the clade containing *Omanosaura* as sister to *Mesalina* + *Eremias*.

Ophisops elegans Ménétriés, 1832 (Plate 16C–D)
Snake-eyed lizard

Ophisops elegans Ménétriés, 1832:63 (Type locality: vicinity of Baku, Azerbaijan; Syntypes: MNHN 544, USNM 21396). — S. Anderson, 1974:38, 44. — Tuck, 1974a:63, fig. — Welch, 1983:39. — Welch, et al., 1990:116. — Leviton, et al., 1992:63–64, col. pl. 9G–H. — Frynta, et al., 1997:10.
Ophiops elegans: De Filippi, 1865:354. — J. Anderson, 1872:374–375. — Blanford, 1876:367–369. — Bedriaga, 1879:31. — Boulenger, 1887a:75–77. — Boettger, 1893:91. — Boulenger, 1899: 378–379. — Nikolsky, 1900:397. — F. Werner, 1903:343. — Zarudny, 1903:19. — Nikolsky, 1907a:282; 1915:403–407, text- figs. 49–50. — F. Werner, 1917:201.
Ophiops elegans var. *persicus* Boulenger, 1918b:160 (Type locality: Superghan [Sopurghan], Lake Urmi [Daryacheh-ye Reza'iyeh], Ispahan [Esfahan], Shiraz, and Kerman, all Iran. Syntypes: BMNH 99.9.30.22–24/1946.8.4.43–45, 99.9.30.5–8/1946.8.4.70–73, 99.9.30.19–21/ 1946.8.4.74–76, 74.11.23.99/1946.9.4.2–3, 99.9.30.11/1946.9.4.4); 1921:215–216.
Ophiops elegans forma *typica*: Boulenger, 1921:211–214.
Ophisops elegans elegans: Lantz 1931:31–42. — Smith, 1935:379–380. — Rostombekov, 1938:14. — Schmidt, 1939:63–64. — Terentjev and Chernov, 1949:192–193. — Forcart, 1950:149. — Wettstein, 1951:439. — Schmidt, 1955:205. — Guibé, 1957:139. — S. Anderson, 1963:462. — Clark, et al., 1966:9. — S. Anderson, 1968:333. — Tuck, 1971a:59. — Bannikov, et al. 1977:234. — Schleich, 1977:127, 129. — Clark, 1991:39–40.
Ophisops elegans ehrenbergi: F. Werner, 1938:270. — S. Anderson, 1963:476.

Ophisops blanfordi Schmidt, 1939:64–65 (Type locality: Halfaya, 20 mi [32 km] E Amara, Iraq. Holotype: FMNH 19721); 1955:205. — Hellmich, 1959:5. — S. Anderson, 1963:476.
Ophisops elegans blanfordi: S. Anderson, 1968:333. — Tuck, 1971a:59. — Schleich, 1977:127, 129.
Ophisops elegans sspp.: Haas and Y. Werner, 1969:346–349.

Diagnosis: Upper head shields smooth, 27–28 scales and plates round body; snout shorter than breadth of head across eyes; scales on nape small and granular; supraoculars separated from superciliaries by a series of small granules.

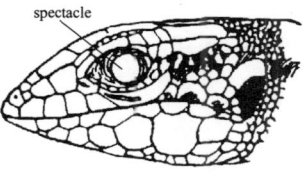

FIGURE 123. *Ophisops elegans*. Note absence of eyelids; eye covered by a spectacle. (From Nikolsky, 1915, fig. 50.)

Color pattern: Olive-greenish or brownish above, with 2 light dorsolateral stripes, upper running from supraciliary margin to tail; lower running from below eye through ear, along flank to hind limb; upper margin of dorsolateral stripe spotted with black; light stripes occasionally absent in adults; black vertebral streak often present, at least on neck; a series of small vertebral spots occasionally present; supralabials, sides of neck, and interval between light stripes spotted with black; venter greenish white; distal half of tail often reddish. Hatchlings with contrasting chocolate brown and bright yellow stripes.

Size: Snout-vent length to 62 mm.

Natural history: Specimens collected at Esfahan in late April have eggs in the oviducts, while specimens collected at Khvoy in northwestern Iran have oviducal eggs in late June. All other females examined (collected in March, April, August, and October) have yolked ovarian follicles; ovarian eggs can be seen in two specimens only 29 mm snout-vent (one collected in early October, the other in late January). The smallest juvenile, collected in mid-October, measures 20 mm snout-vent. Stomachs I examined contained a wide variety of ground-living arthropods, including beetles, beetle larvae, spiders, ants, solpugids, and orthopterans.

Habitat: (Plate 25G) Common on stony plains and hillsides; it does not occur in sandy desert regions, and is rarely found on flat hamadas and silty plains. On the Central Plateau it occurs primarily over 1000 m, but in northern regions and on the Mesopotamian Plain it is found at lower elevations. I found it on rocky alluvial slopes and hilltops, sloping silt and gravel fans, igneous outcrops with loose rubble, and rolling hills of silty alluvium. Vegetation varied from nearly barren areas to cultivated and fallow fields, overgrazed steppe, open areas in riparian woodland, fairly dense *Quercus brandtii* and *Crataegus* scrub, to oak forest in the Zagros Mountains. In many areas it occupies habitats apparently similar to those in which the sympatric *Mesalina watsonana* is found, but I never found the two species to be syntopic.

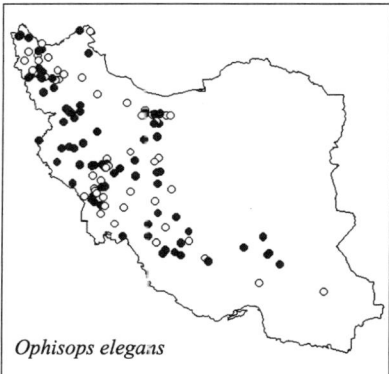

Ophisops elegans

Distribution: *Ophisops elegans*, with its variously differentiated populations, extends from the Bosporus through Southwest Asia through Iran, extending south to the Sinai Peninsula and Red Sea coast of northern Egypt, Jordan, Iraq, and north to the Transcaucasian republics. In Iran it occurs east through the western part of the Central Plateau, and in the south as far east as Kerman. It is not known from the eastern part of the Iranian Plateau (eastern Iran, Afghanistan, and Baluchistan, Pakistan) but should be looked for at higher elevations; to 3000 m elevation.

Remarks: On the basis of the material I have examined, I conclude that the application of subspecific names to various populations of this species does not, at present, help to clarify systematic and zoogeographic problems (but see Darevsky and Beutler, 1981:470–471). There is now sufficient material in museums that a review of the whole complex could be undertaken profitably. As Haas and Y. Werner (1969:349) pointed out, the characters usually employed in the diagnoses of subspecies may prove to vary independently and clinally. It may be that *O. e. blanfordi* will prove a valid taxon of the Mesopotamian Plain, intergrading with one or more subspecies (*O. e. elegans, O. e. ehrenbergi,* and *O. e. persicus* are existing names) over a broad zone in the Zagros foothills. Worthy of note is the large number of localities in common for *O. elegans* and *Mesalina watsonana*. Both species occupy similar habitats and are similar in size and behavior.

ADDENDUM TO SECTION ON FAMILY LACERTIDAE

During the time that this book was in proof, three lacertid species were added to the fauna of Iran, necessitating modification of two keys and the addition of three species accounts. I have also added an account for *Acanthodactylus opheodurus*, a widely distributed species not yet recorded from Iran, but likely to occur in lowland southwestern Khuzestan.

Revised Key to the Species of *Acanthodactylus* in Iran

1a. 3 scales around fingers, but only two visible from medial side view (Fig. A, p. 199)......2
1b. 4 scales around fingers, but only 3 visible from medial side view (Fig. B, p. 199) 5
2a. Ventrals 10 in longest transverse row across belly...3
2b. Ventrals 8 in longest transverse row across belly..
....*Acanthodactylus nilsoni* Rastegar-Pouyani, 1998 (p. 258)
3a. 2 entire supraoculars...................*Acanthodactylus micropholis* Blanford, 1874 (p. 205)
3b. 4 entire supraoculars...4
4a. Eyelid barely pectinate; 4th toe strongly pectinate..
....*Acanthodactylus boskianus* Daudin, 1802 (p. 257)
4b. Eyelid strongly pectinate; 4th toe scarcely pectinate..
....*Acanthodactylus opheodurus* Arnold, 1980 (p. 258)
5a. Ventral scales in oblique or irregular longitudinal series, not forming straight longitudinal rows; 18–22 dorsal scales in transverse series between hind limbs..........................
....*Acanthodactylus grandis* Boulenger, 1909 (p. 203)
5b. Ventral scales in straight longitudinal rows, at least down middle of venter, outer series may be somewhat oblique; 10–16 dorsal scales in transverse series between hind limbs
....6
6a. Dorsal color pattern reticulate, not lineate even in young specimens, indistinct in large adults; scales on sides of dorsum double the size of those on central dorsum, 38–54 dorsal scales across middle of back; 13–18 ventral plates in longest transverse series
....*Acanthodactylus schmidti* Haas, 1957 (p. 206)
6b. Dorsal color pattern lineate, young specimens with 6 dorsal and one lateral light longitudinal streaks, with or without round white spots between them; some adults nearly uniform, no distinct pattern; scales on sides of dorsum equal to those on central dorsum; 34–46 dorsal scales across middle of back; 12–16 ventral plates in longest transverse series........................... *Acanthodactylus blanfordi* Boulenger, 1918 (p. 201)

Acanthodactylus boskianus (Daudin, 1802)
Bosc's fringe-toed lizard

Lacerta boskiana Daudin, 1802, 3:188, pl. 36, fig. 2 (Type locality: "L'Île Saint-Domingue" (in error), probably Mediterranean Egypt; Holotype: MNHN 2762).
Acanthodactylus boskianus var. *euphraticus* Boulenger, 1919:550 (Type locality: Ramadieh on the Euphrates; Syntypes: BMNH 1946.8.4.83-90).
Acanthodactylus boskianus: Rastegar-Pouyani (in press).

Diagnosis: Medium to large size species. Usually four entire supraoculars but, at times, the first one divided. Pectinate anterior border of ear opening. Keeled temporals. Slightly denticulated eyelids. Conspicuous gular fold. Three series of scales on fingers. Ventrals arranged in 10 straight longitudinal rows. Very large, keeled, imbricate dorsals. Granular scalation on sides of the body. Moderate to intense pectination of fourth toe. Large, imbricate, and sharply keeled scales on the upper surface of the tail. (Salvador, 1982:23).

Color pattern: "In the youngest specimens (snout-vent length 35 mm) there are seven black bands. Except for at its beginning point, the vertebral band is continuous. At the SVL of 39 mm, the vertebral has already begun to separate in two and continues to do so until the process is completed at the SVL of 44 mm. At this age, eight black bands are present. These remain until the SVL of 55 mm, at which time the pattern may begin to reticulate.

"Abundant white ocelli cover the limbs. The undersides are chalky-white. In adults the bands gradually disappear and the body takes on an overall darkish gray coloring. In some adults, however, remains of the bands are conserved." (Salvador, 1982:26).

"In the male [Iranian] specimen, dorsum sandy grey with two light dorsolateral stripes on each side enclosing a broad brown band with light reticulations, also a weakly visible dark-brown vertebral stripe present; base of tail with two light lateral stripes, distal $4/5$ of tail uniformly grey dorsally, upper surface of limbs greyish-brown with numerous light spots; upper surface of head olive-brown; all of the ventral surfaces whitish. In the female specimen, dorsum is light brown with 7 narrow, light stripes, the two dorsolateral ones on each side being lighter and in strong contrast with the dark-brown pattern of back, the three vertebral and paravertebral ones duller, proximal $1/4$ ventral part of tail whitish, distal $3/4$ pink or bright-red, other ventral surfaces whitish. In the juvenile specimen, upper surface of head is light olive, dorsum dark-brownish-black with 6 strongly contrasting light lines, vertebral stripe whitish on neck, disappears towards the posterior part of back." (Rastegar-Pouyani, in press; quoted from manuscript).

Size: The largest Iranian specimen measured 65.5 mm SV, tail 134 mm.

Habitat: The Iranian specimens were collected on sandy hills covered with various species of *Astragalus*.

Distribution: Mauritania, Rio de Oro, Morocco, Algeria, Mali, Niger, Tunisia, Libya, Chad, Nigeria, Sudan, Ethiopia, Egypt, Israel, Jordan, Iraq, Syria, Turkey, Saudi Arabia, South Yemen, Oman, United Arab Emirates (Salvador, 1982:23, map 2). A single locality is known for Iran, 2 km W Harsin (34°17′N, 47°24′E), Kermanshah Province (Rastegar-Pouyani, in press).

Acanthodactylus boskianus

Acanthodactylus nilsoni Rastegar-Pouyani, 1998
Nilson's spiny-toed lizard

Acanthodactylus nilsoni Rastegar-Pouyani, 1998:257–265 (Type locality: 34°30'N, 45°33'E, 5 km S Qasr-e Shirin, Kermanshah Province, Iran, about 7 km from Iran-Iraq border, 285 m; Holotype: GNHM Re. ex. 5145, adult male).

Diagnosis: Eight longitudinal and 26 transverse roes of ventral plates; 48–52 imbricate, keeled scales across dorsum; toes covered with three rows of scales; 9/7 supralabials, 4 anterior to subocular; subocular not bordering mouth; temporals keeled, more than 100; gular fold distinct; 27–27 gular scales; 10–12 scales in collar; 23–24 lamellae under fourth toe which is moderately denticulated laterally; eyelids with weak pectination. (Rastegar-Pouyani, 1998:258).

Color pattern: Upper surface of head olive brown; temporal region and upper surface of forelegs light brown; upper surface of hindlimbs with many large whitish spots bordered by dark brown circles; dorsum grayish brown, with 6–8 interrupted longitudinal dark stripes, the dorsolateral ones having a tendency to form a reticulation; vertebral region without dark stripe; upper surface of tail light brownish gray; ventral surfaces whitish. (Rastegar-Pouyani, 1998:260).

Size: Snout-vent length 73.4 mm; tail 143 mm (based on the larger of the two known adult male specimens).

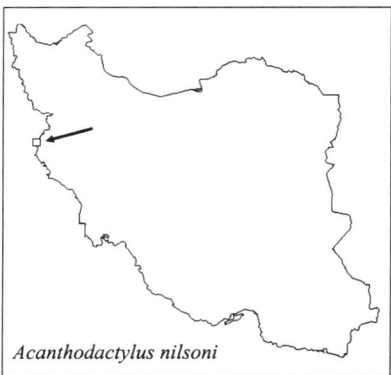

Acanthodactylus nilsoni

Habitat: Stony hills and soft-sandy alluvial substrate covered by desert-adapted vegetation including *Tamarix, Astragalus, Zygophyllum, Artemisia, Euphorbia,* and *Alhagi* as well as grasses. The habitat lies in a lowland area at the foot of the western Zagros Mountains. Other lizard species found in this area include *Laudakia n. nupta, Trapelus r. ruderatus, Ophisops elegans, Mabuya aurata, Uromastyx loricatus,* and *Varanus g. griseus.* (Rastegar-Pouyani, 1998:261, 263, fig. 9).

Distribution: Known only from the type locality, although the biotope apparently extends south, southeast, and southwest into Iraq.

Remarks: Rastegar-Pouyani considers his species a member of Salvador's (1982) *Acanthodactylus boskianus–schreiberi* group.

**Acanthodactylus opheodurus* Arnold, 1980
Arnold's fringe-toed lizard

Acanthodactylus opheodurus Arnold, 1980b:296, pls. C–D and col. pl. 17 (Type locality: Jazir coast [18°40'N, 16°40'E], Sultanate of Oman; Holotype: BMNH 1969.314).

Diagnosis: Small species with a wide head and a short snout. Rounded and protruding nasals. Four supraoculars. One row of granules between the superciliaries and the supraoculars. Very conspicuous subocular keel. Keeled temporal scales. Large ear opening bordered anteriorly by 3 or 4 scales. Intensely pectinate eyelids. Small tympanic present. Subocular separated from the lip and wedged between the fourth and fifth supralabial. Ventrals arranged in 10 straight longitudinal rows. Flat, keeled dorsals much larger than laterals, numbering from 29 to 36 across the middle of the body. Intense pectination of toes. Three rows of scales on fingers. Large, keeled scales on the upperside of the tail. Color pattern with three dark dorsal bands.

Color pattern: This species is characterized by an overall sandy-gray coloring with three very conspicuous dorsal bands and one very reticulate band along the sides (Salvador, 1982:47; 1982:45). Arnold (1980b:300) reported that a Dhofar female and juveniles had the distal part and underside of tail bright red in life.

Size: Males to 62 mm SV, females to 60 mm (Arnold, 1980b:300).

Habitat: Specimens from Dhofar were collected in a flat area with relatively hard sand surface and small dense shrubs on hummocks (Arnold, 1980:300).

Natural history: Arnold's (1980:300) Dhofar specimens sought refuge at the base of shrubs when pursued and wriggled or waved their brightly colored tails.

Distribution: This species occurs over a large area which includes Saudi Arabia, Oman, South Yemen, Kuwait, Iraq, Jordan, and Israel (Salvador, 1982:45–46, map 8). It has not been recorded from Iran, although it occurs in Kuwait and Iraq in habitats which are similar to those found in Iran; it should be looked for in southwestern Khuzestan.

Remarks: Salvador (1982:46) believes this species is closely related to *Acanthodactylus yemenicus* Salvador, 1982 and *A. felicis* Arnold, 1980 of the southern Arabian peninsula. He places these species in his *yemenicus*-group, which also includes *A. masirae* Arnold, 1980. Prior to Arnold's recognition of this species, it had been identified in collections as the sympatric *A. boskianus*.

Revised Key to the Species of *Lacerta* in Iran

1a. Lower eyelid with 5–7 transparent shields edged with black; subdigital lamellae keeled ....*Lacerta cappadocica urmiana* (Lantz and Suchow, 1934) (p. 232)
1b. Lower eyelid without transparent shields; subdigital lamellae smooth or tuberculate...2
2a. Ventral plates more or less rectangular with rectilinear or nearly rectilinear posterior margins, juxtaposed to subimbricate..3
2b. Ventral plates shaped like inclined parallelograms with notches between longitudinal rows, posterior and lateral edges strongly overlapping 11
3a. Dorsal scales distinctly keeled; collar serrated 4
3b. Dorsal scales smooth, granular; collar not serrated 6
4a. Venter gray or black, never greenish or yellowish ....
....*Lacerta mostoufi* Baloutch, 1976 (p. 238)
4b. Venter greenish or yellowish in life..5
5a. Rostral touches nostril, or barely separated from it; 43–49 dorsals at midbody; 27–32 lamellae under 4th toe............................*Lacerta chlorogaster* Boulenger, 1908 (p. 234)
5b. Rostral does not touch nostril; 32–43 dorsals at midbody; 20–25 lamellae under 4th toe ....*Lacerta praticola* Eversmann, 1834 (p. 239)
6a. 5–6 (usually 5, rarely 4) supralabials anterior to subocular; pterygoid teeth strongly developed; at least outer ventrals with black spots in males 7
6b. 3–4 (usually 4, rarely 5) supralabials anterior to subocular; normally single postnasal; no pterygoid teeth; outer ventrals without black spots (turquoise spots present in males)...8
7a. Normally 2 superposed postnasals (sometimes fused on one or both sides of head); 8 longitudinal rows of ventrals; outer ventrals with small black spots.................................
....*Lacerta brandtii* De Filippi, 1863 (p. 231)
7b. 5 supralabials anterior to subocular; a single postnasal; 10 longitudinal rows of

ventrals; pterygoid teeth strongly developed; all ventrals with distinct black spots in males .... *Lacerta zagrosica* Rastegar-Pouyani and Nilson, 1998 (p. 260)
8a. Row of granules between supraoculars and supraciliaries often interrupted; dorsum green or greenish-yellow in life .... *Lacerta valentini valentini* Boettger, 1892 (p. 248)
8b. Supraciliary scales invariably separated from supraoculars by complete row of 6–18 granules; dorsum brown, brownish, gray or black tones, not green in life 9
9a. Dorsum with a vertebral stripe formed of close-set black dots, breaking into distinct, larger spots on tail; a row of black dots down either side of back; flanks dark with light ocelli in a single row..*Lacerta steineri* Eiselt, 1995 (p. 245)
9b. Dorsal pattern not as above, back with small dark spots and mottlings, sometimes more or less linearly arranged 10
10a. Venter in life brick red; Alborz and Kopet Dagh ranges..
....*Lacerta defilippii* Camerano, 1877 (p. 236)
10b. Venter in life yellowish, bluish, or greenish-white; Talesh Mountains........................
.... *Lacerta raddei* Boettger, 1892 (p. 243)
11a. Ventral plates in 10 longitudinal series; 34–37 dorsals at midbody 12
11b. Ventral plates in 6 or 8 longitudinal series; 38 or more dorsals at midbody................ 13
12a. Outer row of ventrals (marginals) smooth; 20–22 gulars; 13–17 femoral pores on each side; lower edge of subocular ½ or less than ½ maximal length of shield........................
....*Lacerta princeps princeps* Blanford, 1874 (p. 240)
12b. Outer row of ventrals (marginals) keeled; 17–19 gulars; 16–21 femoral pores on each side; lower edge of subocular ½ or more than ½ maximal length of shield......................
....*Lacerta princeps kurdistanica* Suchow, 1936 (p. 241)
13a. 17–21 femoral pores, row of pores reaches knee; usually less than 20 temporal scales; 5th chin shield always well developed; young specimens usually with uninterrupted lateral line in addition to vertebral and dorsolateral lines ....
....*Lacerta strigata* Eichwald, 1831 (p. 246)
13b. 12–16 femoral pores, row of pores does not reach knee; usually more than 20 temporal scales; 5th chin shield small or absent; young specimens with lateral light line interrupted in its anterior half.......*Lacerta media media* Lantz and Cyrén, 1920 (p. 237)

Lacerta zagrosica Rastegar-Pouyani and Nilson, 1998
Zagros Mountains lacerta

Lacerta zagrosica Rastegar-Pouyani and Nilson, 1998:267–277, figs. 1–7 (Type locality: 32°58′N, 50°40′E, 3 km NW Fereydin Shahr, about 140 km NW Esfahan, Zagros Mountains, Esfahan Province, Iran, 2450 m; Holotype: GNHM Re. ex. 5149, adult male).

Diagnosis: 10 longitudinal rows of ventral plates, 58–61 smooth, granular scales across dorsum, single postnasal, 7 pairs of submaxillary scales, complete row of granules between supraciliaries and supraoculars, obtusely keeled subdigital lamellae, masseteric shield very small or absent, dorsum greenish or olive brown with numerous dark spots, venter blue with black spots on ventrolateral region. (Rastegar-Pouyani and Nilson, 1998:268).

Color pattern: In life, dorsal surface of head olive brown green with irregular dark spots and dots, ground color of dorsum green, with numerous dark spots on sides having tendency to form a reticulation, encompassing light green spaces; these dark dots and spots less numerous on the vertebral region, which looks almost uniformly light green; dorsal surface of tail also light green with irregular dark spots scattered throughout; upper surface of limbs reticulated, dark oval and round ocelli encompassing light green spaces; all of ventral surfaces dark blue, mixed with black spots and dots, these black markings most numerous on the ventrolateral and gular regions; postanal region, corresponding four or five transverse scale rows, strongly contrasting in color with all other parts of body, being yellowish cream; ventral surface of tail uniformly light turquoise blue. (Rastegar-Pouyani and Nilson, 1998:272).

Size: Male: snout-vent length 67.8 mm; tail 114 mm; female: snout-vent length 57.8, tail 97 mm.

Sexual dimorphism: (Based on the male holotype and female paratype, both adults). The venter of the female is more or less whitish blue, rather than the dark blue of the male, lower surfaces of limbs and tail light blue, pectoral and gular regions turquoise blue; dorsum is olive brown with light ocelli in the female, not showing the light green of the male. (See Rastegar-Pouyani and Nilson, 1998:271, fig. 4 for color photograph of female paratype).

Habitat: The habitat, in the central Zagros Mountains, is an area with large, fragmented rocks and boulders and scant, low vegetation. *Lacerta zagrosica* is a saxicolous species living in the same habitat that is typical for *Laudakia caucasia*, foraging, basking on, and seeking refuge under large rocks. *Mabuya aurata* was collected in the same habitat. A color photograph of the habitat is presented by Rastegar-Pouyani and Nilson (1998:275, fig. 7).

Distribution: Known only from the type locality.

Lacerta zagrosica

Remarks: *Lacerta zagrosica* is a member of Arnold's *Lacerta* Part II, but within that large group, its relationships remain to be evaluated.

Family SCINCIDAE

Skinks

The cladistic analysis of the scincomorphan lizards by Presch (1988) places the Scincidae and the Xantusiidae as sister taxa within the scincomorphan clade that includes cordylids, lacertids, teiids, and gymnophthalmids. The last review of the subfamilial classification of skinks was that of Greer (1970), who recognized four subfamilies. The species that occur in Iran and elsewhere in Southwest Asia belong to two subfamilies: the morphologically most primitive Scincinae, which includes *Chalcides, Eumeces, Ophiomorus*, and *Scincus*, and the most derived subfamily, Lygosominae, including *Ablepharus* and *Mabuya*.

Key to the Genera of Scincidae of Iran

1a. Eyelids immovable (spectacle) (Fig. A); small species (adults less than 65 mm from snout to vent); limbs well developed .... *Ablepharus* (p. 263)
1b. Eyelids movable; adults more than 65 mm from snout to vent; limbs well developed or reduced 2

Ablepharus bivittatus showing spectacle
(From Terentjev and Chernov, 1949, fig. 47)

2a. Digits fringed laterally .... *Scincus* (p. 287)
2b. Digits not fringed 3
3a. Limbs greatly reduced, with less than five digits; body elongate, serpentine ....
.... *Ophiomorus* (p. 278)
3b. Limbs well developed, with five digits; body robust 4
4a. Lower eyelid with transparent shield (Fig. C) 5
4b. Lower eyelid without transparent shield (Fig. B) .... *Eumeces* (p. 269)

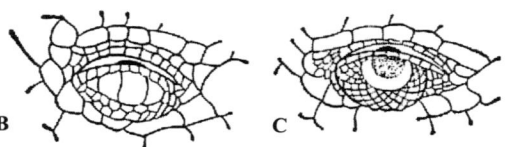

(B) *Eumeces schneideri* showing opaque scales of lower eyelid
(C) *Mabuya aurata* showing transparent window in lower eyelid
(From Bannikov, et al., 1977, fig. 37)

5a. Nostril between nasal and rostral, in emargination of latter; scales smooth; back with numerous dark-margined light ocelli irregularly transversely arranged ....
.... *Chalcides* (p. 267)
5b. Nostril in nasal shield; dorsal scales usually distinctly, but weakly bi- or tricarinate; back without ocelli .... *Mabuya* (p. 274)

Genus *Ablepharus* Fitzinger, 1823
Lidless skinks, snake-eyed skinks

Ablepharus Fitzinger *in* Lichtenstein *in* Eversmann, 1823:145 (Type species: *Ablepharus pannonicus* Fitzinger *in* Lichtenstein *in* Eversmann, 1823, by monotypy).

Definition: Supranasals absent; no movable eyelids, lower eyelid fused with upper, forming transparent spectacle covering eye (see Fig. A in key, p. 262) ; pentadactyl limbs relatively weakly developed; usually no strongly enlarged shields on outer side of forearms and shins; outer ear opening small, its length contained more than five times in diameter of eye or ear opening completely absent; no posterior projecting process of palatines separating pterygoids; palatine processes of pterygoids not meeting; no recurved process of pterygoids; nine pleurodont teeth on premaxillary bone; 18–26 rows of smooth scales around middle of body. Oviparous. (Fühn, 1969a:24; Eremchenko and Szczerbak, 1986:64).

Distribution: Southeastern Europe (southeastward from Czech Republic and Hungary to European Turkey and the islands of the eastern Mediterranean), Southwest Asia (Turkey and the eastern shores of the Mediterranean through the Transcaucasian republics, Iran, Afghanistan, and Pakistan), and Central Asian republics (north to the Aral Sea, east to 80°E).

Remarks: Fühn (1969a, 1969b) redefined the genus *Ablepharus*, restricting it to the Eurasian species. In his opinion the relationships of the genus are with *Leiolopisma* rather than with African and Australian "ablepharine" species groups. Eremchenko and Szczerbak (1986) revised the genus subsequently, referring *Ablepharus alaicus* Elpatjevsky 1901 to their genus *Asymblepharus* Eremchenko and Szczerbak 1980, along with a Pleistocene species, *A. borealis* Darevsky and Tschumakov, 1962, known from the Altai in Khazakhstan. In their phylogenetic hypothesis, *Ablepharus* is derived from a common ancestor with *Scincella*. Seven species recognized, of which two are known from Iran.

Key to the Species of *Ablepharus* in Iran

1a. Prefrontals usually forming a median suture; two frontoparietals (Fig. A) .... *Ablepharus bivattatus* (Ménétriés, 1832) (p. 263)
1b. Prefrontals separated; usually a single frontoparietal (Fig. B)... .... *Ablepharus pannonicus* (Lichtenstein, 1823) (p. 264)

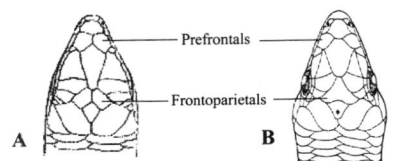

(A) *Ablepharus bivittatus* (From Blanford, 1876, pl. 27, fig. 2a)
(B) *Ablepharus pannonicus* (From Smith, 1935, fig. 70)

Ablepharus bivittatus (Ménétriés, 1832) (Plate 16E–F)
Two-streaked snake-eyed skink

Scincus bivittatus Ménétriés, 1832:64 (Type locality: Perimbal, Talysch Mountains, Azerbaijan; Syntypes: ZIL 563–565 [Lectotype: ZIL 565, designated by Eremchenko and Szczerbak, 1986:71]).
Ablepharus Menetriesii Duméril and Bibron, 1839:5:811 (Type locality: Perimbal, Talysh Mountains, Azerbaijan). — De Filippi, 1865:355.
Ablepharus bivittatus: Gray, 1844:64. — Strauch, 1868b:563;1868c:col. 366. — Blanford, 1876:

390–391, pl 27, figs. 2, 2a. — Bedriaga, 1879:26. — Boulenger, 1887a:353. — Nikolsky, 1899a: 176; 1915:315–316. — F. Werner, 1936:201. — Rostombekov, 1938:15. — Terentjev and Chernov, 1949:173–174. — S. Anderson, 1963:476; 1968:333. — Eremchenko and Szczerbak, 1986: 69–86, figs. 17–21.

Ablepharus bivittatus bivittatus: Wettstein, 1960:62. — Fühn, 1969b:71; 1969b:27–29, text-figs, 3f., 4–5. — S. Anderson, 1974:37, 44. — Bannikov, *et al*., 1977:155. — Welch, 1983:26.

Diagnosis: Four longitudinal rows of scales on dorsum; 22–25 scales around midbody; palpebral circle complete, with three enlarged granules on upper border, separating supraoculars from eye; three supraoculars, two in contact with frontal; three temporal plates in first row behind eye; prefrontals large, usually forming median suture and separating frontal from frontonasal; two frontoparietals; nasal semidivided; four upper labials in front of subocular. (Fühn, 1969a:27; Eremchenko and Szczerbak, 1986:70).

Color pattern: Body and tail bronze-olive above, in preservative, cinnamon brown in life, with darker spots with light central shaft, or each scale of back with two to three short dark marks; dark stripe, edged above and below with white, along entire body, through eye, from nostril; limbs same ground color as body with darker markings tending to form stripes; venter greenish white.

Size: Snout–vent length of females to 60 mm, males to 50 mm.

Natural history: A specimen collected near Divandarreh in early August has ovarian follicles up to 1 mm in diameter. See Eremchenko and Szczerbak (1986) for a synopsis of the natural history of this species, as known for the Transcaucasian portion of its range.

Habitat: (Plate 25B) I encountered this species only twice, on and near steep loose rocky slopes where the lizards sought refuge in small spiny legume shrubs. Blanford (1876:391) found it common on open dry level ground scattered with very small thorny bushes. The skinks hid among roots of these shrubs when pursued.

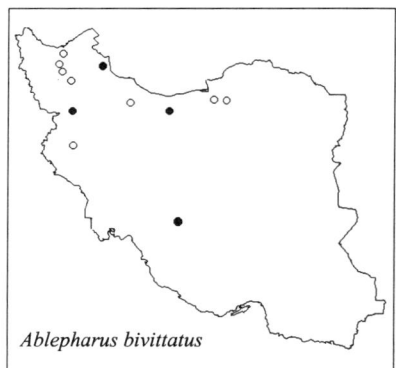

Ablepharus bivittatus

Distribution: Armenian and Azerbaijan republics; northern Iran (to 3600 m) and the Zagros Mountains (2500 m in western Iran); Turkmenistan in the Kopet Dag.

Remarks: In the three specimens from Lar there is a distinct light lateral line from snout, below eye, to axillary region, fading out on the body, the broader dark line is distinct along entire length of body and onto tail. The scales of the back each have two to three short dark marks. All three specimens have 22 scales around the middle of body.

Ablepharus lindbergi Wettstein from upland Afghanistan and Punjab, and *A. alaicus* Elpatjevsky from Khirgiz, northeastern Tadzhikistan, southeastern Kazakhstan, and western Xinjiang, China, have been regarded as subspecies of *A. bivittatus* by Wettstein (1960), Fühn (1969a, 1969b) and subsequent authors, but Eremchenko and Szczerbak (1986), in their thorough review of the genus, regarded *A. lindbergi* as a distinct species and placed *A. alaicus* in their genus *Asymblepharus*.

Ablepharus pannonicus Fitzinger, 1823 (Plate 16G)
Asian snake-eyed skink

Ablepharus pannonicus Fitzinger *in* Lichtenstein *in* Eversmann, 1823:145 (Type locality: Bokhara;

Syntypes: ZIL [lost]; ZMB [lost]). — Smith, 1935:310–311. — Forcart, 1950:148. — Hellmich, 1959:3. — S. Anderson, 1963:476; 1968:333. — Tuck, 1971a:59. — S. Anderson, 1974:37, 44. — Schleich, 1977:127, 129. — Eremchenko and Szczerbak, 1986:119, fig. 29, map (fig. 30). — Welch, et al., 1990:62. — Leviton, et al., 1992:66, col. pl. 10B. — Frynta, et al. 1997:10.

Ablepharus brandtii Strauch, 1868c:565; 1868b:col. 368 (Type locality: Samarkand, Turkestan; Syntype: BMNH 72.5.30.15/1946.8.18.47). — Blanford, 1876:391–394, pl. 27, figs. 1, 1a. — Elpatjevsky, 1901:1. — Nikolsky, 1907a:283; 1915:490–492. — Morich, 1929:32. — F. Werner, 1936:198. — Terentjev and Chernov, 1949:170–171. — Guibé, 1996a:98. — Schleich, 1977: 127, 129.

Ablepharis pusillus Blanford, 1874b:33 (Type locality: Bussora [= Basra], Iraq; Syntypes: BMNH 74.11.23.27/1946.8.18.48; ZSI); 1976:461.

Ablepharus grayanus: Nikolsky, 1900:401 (not *Blepharosteres grayanus* Stoliczka, 1872). — Zarudny, 1903:23. — F. Werner, 1936:201. — Terentjev and Chernov, 1949:171. — S. Anderson, 1963:476; 1968:333; 1974:37, 44. — Welch, et al., 1990:62.

Ablepharus brandti var. *brevipes* Nikolsky, 1907a:283 (Type locality: Dech-i-Diz and Karun River, Iran; Syntypes: ZIL 10188–10189).

Ablepharus persicus Nikolsky, 1907a:283–285, pl. 1, fig. 5 (Type locality: Schachrud [= Shahrud], Iran; Holotype: ZIL 10342); 1915:502. — F. Werner, 1916:203; 1936:201. — S. Anderson, 1963: 476; 1968:333.

Ablepharus pannonicus pannonicus: Fühn, 1969b:71; 1969a:35–38, text-figs. 3d, 12–13. — Bannikov et al., 1977:158. — Welch, 1983:27.

Ablepharus pannonicus grayanus: Fühn, 1969b:71; 1969a:38–40, text-figs. 3a, 14–15 (in part; not *Blepharosteres grayanus* Stoliczka, 1872). — Bannikov, et al., 1977:158. — Welch, 1983:27.

Diagnosis: Body slender, legs not meeting when adpressed; four longitudinal rows of scales on dorsum; two upper scales of periocular circle much enlarged; frontoparietal usually single (exceptionally, paired); prefrontals separated; four anterior supralabials; ear opening small but distinct; nasal semidivided; scale rows at midbody usually 20, rarely 18 or 22. (Fühn, 1969a:36–37; Eremchenko and Szczerbak, 1986:120).

Color pattern: (Fig. 124) Olive or brownish above in preservative, with metallic gloss; dark brown dorsolateral stripe with whitish edge above; flanks with less distinct dark longitudinal lines; whitish upper lips; sides of tail with small spots regularly arranged; limbs above with light and dark longitudinal lines; whitish or brownish below. In life, ventral surface of tail and hind legs brick–red or orange during breeding.

FIGURE 124. *Ablepharus pannonicus*. (From Blanford, 1876, pl. 27, fig. 1.)

Size: Snout-vent length to 50 mm (usually 35, tail 60 mm).

Natural history: Murray (1884a:354–355) found this species to be diurnal in Karachi, feeding on red ants. Terentjev and Chernov (1949:171) state that it hides beneath stones, shrubs, or in cracks in the soil, entering by serpentine movements with limbs pressed to the body.

Two specimens from western Afghanistan, collected the first half of April, have yolked ovarian follicles. Smith (1935:311) found three large eggs with no trace of an embryo in one specimen. Terentjev and Chernov (1949:171) report oviposition from late April to end of May at 800 m in Tajikistan, a month to a month and a half later at 2200–2300 m; usually three to four eggs per clutch, hatchlings 20 mm, first appearing in July.

Small beetles, acridids, and ants are recorded as food items.

Habitat: Occurs in grassy areas, up to 2500 m in mountains; often found in damp areas near irrigation ditches, never ranging far from water (Terentjev and Chernov, 1949:171; Fühn, 1969a:37). Minton (1966:103) found them on rocky hillsides sparsely forested with juniper at 7000 feet (2300 m) in Pakistan. They were found under rocks or creeping among leaves on forest floor.

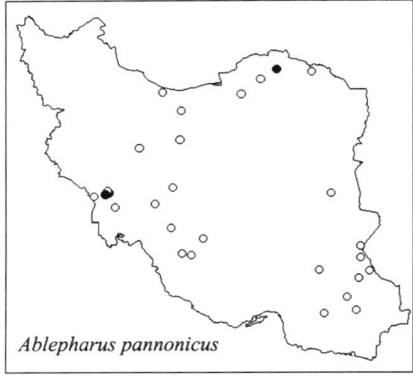

Ablepharus pannonicus

Distribution: Syria (a single specimen from Aleppo in MCZ [Eremchenko and Szczerbak, 1986]); SW Saudi Arabia, N Yemen, SE Kuwait, N Oman, Iraq; northern and western Iran; Central Asian republics (Kopet Dagh, ranges of Pamiro-Altai and their northern slopes to Leninabad; east to Darvaz); Afghanistan; upland Pakistan; Punjab, Pakistan.

Remarks: In two specimens from south of Kabul, Afghanistan, the ring of small scales encircling the spectacle is composed entirely of scales from the lower lid, the upper scales of the ring being hidden by the two larger scales of the rudimentary upper lid, the lower scales of the ring are hidden by the fifth upper labial. In a Khorasan specimen (FMNH 141477), the two scales of the rudimentary upper lid form the superior portion of the ring, the upper scales of the lower lid having disappeared. Fusion of the lids is total in all specimens. The specimens from eastern Afghanistan differ from the Iranian specimens and a specimen from south of Herat, northwestern Afghanistan, in having the two middorsal scale rows margined with dark brown, forming three dark longitudinal lines continued onto tail. None of the specimens of *Ablepharus grayanus* from Pakistan show such a pattern.

There are 21 scale rows at midbody in a specimen from Khuzestan and a specimen from Iraq (FMNH 83533), 20 in specimens from Iraq (BMNH 74.11.23.27/1946.8.18.48, type of *A. pusillus*), Khorasan, Tajikistan (BMNH 72.5.30.15/1946.8.18.47, type of *A. brandtii*), eastern Afghanistan, and Pakistan, 22 in specimens from eastern Afghanistan and Tajikistan, 18 in specimens from Baghdad, Iraq (FMNH 28309), Shalgahi, Iran (BMNH 1969.1531 [labelled *A. brandtii festae* in the collection, but does not agree with the type of that taxon, which has shorter limbs, frontoparietal divided, and lacks two enlarged scales at top of palpebral circle]), and northwestern Afghanistan.

Fühn (1969a) placed *Ablepharus persicus* in the synonymy of *A. pannonicus*, but apparently did not examine the type. The former species was said to be characterized by a single frontoparietal and the prefrontals in contact; there were 18 scale rows in the type. Hellmich (1959:3) had eight specimens from Harmalah with 18 scale rows; the character of the frontoparietal was reported as variable, being undivided or half or entirely divided. These specimens agree in the number of longitudinal scale rows with *A. kitaibelii* and *A. persicus*. They are intermediate between *A. kitaibelii*, which has the frontoparietal divided, and *A. persicus* and *A. pannonicus* which have a single frontoparietal. Hellmich did not state whether or not the prefrontals are in contact.

Khalaf (1959:45) uses the name *Ablepharus kitaibelii* (Bibron and Bory) for the Iraqi form, placing *A. pannonicus* Fitzinger, *A. brandti* Strauch, and *A. pusillus* Blanford in the synonymy without giving his reasons for doing so. In his diagnosis, he states, "frontoparietal single," whereas *A. kitaibelii* has the frontoparietal divided. Of eight specimens from

Baghdad (Khalaf (1960:14), one had the prefrontals in contact. Schmidt (1939:67) and Weber (1960:154) used the name *A. brandtii festae* Peracca for specimens from Baghdad. Fühn (1969a) made no mention of Peracca's name in his brief revision of the genus. Eremchenko and Szczerbak (1986:128) placed *A. festae* in the synonymy of *A. kitaibelii kitaibeli*.

Ablepharus grayanus (Stoliczka, 1872) appears to be very closely related to *A. pannonicus*, differing primarily in the absence of an external ear opening and in the lack of a red tinge on the undersurface of tail and legs. Fühn (1969a) regarded the two taxa as only subspecifically distinct, the populations of Afghanistan and Pakistan probably being intergrades, explaining the sympatric occurrences in some localities. Because Fühn examined only four specimens of *A. pannonicus* and three of *A. grayanus*, none of these specimens of questionable identity, apparently, and as I have examined specimens of *A. pannonicus* from the supposed area of intergradation, as well as a large series of *A. grayanus* from Pakistan without seeing evidence of intergradation, I am unwilling to accept Fühn's view, which does not seem logical, in any case. Eremchenko and Szczerbak (1986), in their revision of the genus, regarded the two as distinct species, their map (fig. 29) and lists of localities (p. 113) showing *A. grayanus* as restricted to Pakistan east of the Iranian Plateau in the Indus drainage. I follow these authors, as I have seen no specimens from Iran which I attribute to *A. grayanus*. However, neither they nor I have examined the specimens from Kerman, Baluchistan, and southern Khorasan cited by Nikolsky (1900, 1915), Zarudny (1903), and followed by subsequent authors as *A. grayanus*. If Nikolsky identified them on the basis of the lack of an external ear opening, the question of the distribution of *A. grayanus* needs reexamination.

In a series of 25 specimens of *A. grayanus* from Pakistan there is no external ear opening, but the position of the ear is clearly evident as a small depression. All specimens have 18 scales around the midbody except for a single specimen with 19. A single specimen has the frontoparietal divided and in two other specimens it is partially divided. In none of the specimens are the prefrontals in contact. In CAS 99997 the frontal is completely fused with both prefrontals, and frontoparietal is partially fused with interparietal. The circumorbital ring in these specimens is composed of scales entirely derived from the lower lid, the uppermost scales of the ring hidden beneath one or two scales of the rudimentary upper lid, or these marginal scales of the lower lid may be absent, the superior portion of ring formed by scales of the upper lid. The fifth supralabial partially conceals the lower scales of the ring. Fusion of the lids is complete in all specimens.

Genus *Chalcides* Laurenti, 1768
Cylindrical skinks

Chalcides Laurenti, 1768:64 (Type species: *Chalcides tridactyla* Laurenti, 1768 [= *Lacerta chalcides* Linnaeus, 1758], by subsequent designation of Smith, 1935:349).

Definition: Palatine bones not meeting in midline of palate, which is toothless; teeth subconical; nostril between nasal and rostral, in emargination of latter; supranasals present; prefrontals and frontoparietals absent; lower eyelid with undivided transparent disc; body more or less elongate; limbs short or vestigial.

Distribution: From Spain east through the Mediterranean countries of southern Europe; islands of the Mediterranean; from Canary Islands east across North Africa, through lowland Southwest Asia to Sind, Pakistan. About eight species currently recognized, one occurring in Iran.

Chalcides ocellatus (Forsskål, 1775)
Ocellated skink

Chalcides ocellatus ocellatus (Forsskål, 1775) (Plate 17A)
Ocellated skink

Lacerta ocellata Forsskål, 1775:13 (Type locality: Egypt).
Gongylus ocellatus: J. Anderson, 1872:377. — Bedriaga, 1879:26.
Seps (Gongylus) ocellatus: Blanford, 1876:395.
Chalcides ocellatus: Boulenger, 1887a:400–401. — J. Anderson, 1898:210–219. — F. Werner, 1917: 202; 1936:201. — Welch, 1983:24. — Leviton, *et al.*, 1992:67–68, col. pl. 10C–D.
Chalcides ocellatus ocellatus: Mertens, 1921:118. — Schmidt, 1955:205. — S. Anderson, 1963:476; 1968:333. — Tuck, 1971a:59. — S. Anderson, 1974:37, 44. — Schleich, 1977:127, 129.

Diagnosis: Snout conical, end of snout scarcely projecting beyond mouth; ear opening much larger than nostril; limbs pentadactyl; nostril pierced just above suture between rostral and first supralabial; usually fifth labial enters orbit (Fig. 125); 28–32 scales around body.

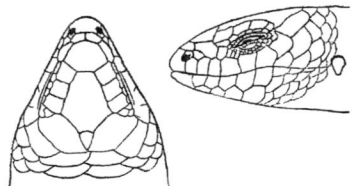

Color pattern: Light brown above, shading to pale yellow or cream on sides and belly; black spots transversely arranged or confluent, forming irregular crossbars, each black spot with a central cream spot or longitudinal bar; labials cream, margined with black; top and sides of head speckled with dark.

FIGURE 125. *Chalcides ocellatus*. (From Smith, 1935, fig. 79.)

Size: Snout-vent length to 122 mm, tail 84 mm.

Natural history: Reported to be live–bearing (Minton, 1966:102–103).

Habitat: In Pakistan and Turkey it is found in both sandy and rocky terrain, below 165 m (Minton, 1966:102; Clark and Clark, 1973:42). J. Anderson (1898:210–219) found it one of the most prevalent lizards in Egypt, present on alluvium, edges of the desert, oases, and from towns and villages as well as open country.

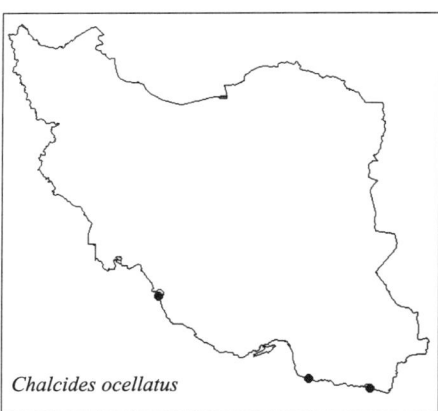

Chalcides ocellatus

Distribution: From the Algerian Sahara through Egypt and Sinai, and from Greece through southern Turkey and lowland Southwest Asia and along the shores of the Persian Gulf to the Mekran Coast of Pakistan. The distribution, at least in Asia, appears to be primarily littoral and along major river systems. The furthest inland localities are reported from Israel, Syria, and Jordan; In Iran and Pakistan it has been recorded only from coastal port localities. It may be that the eastern part of the range is due to its transport through human agency, perhaps in sand ballast of the dhows that have plied the Persian Gulf for centuries.

Genus *Eumeces* Wiegmann, 1834
Long-legged skinks, opaque-lidded skinks

Eumeces Wiegmann, 1834:36 (Type species: *Scincus pavimentatus* Geoffroy St.-Hilaire, 1827 [= *Scincus schneiderii* Daudin, 1802] by subsequent designation of Wiegmann, 1835:288).

Definition: Palatine bones not meeting on midline of palate; pterygoids separated on median line; maxillary teeth conical or with rounded, spheroidal crowns; pterygoid teeth present; eyelids well developed; nostril pierced in nasal, which may be single, partly or completely divided; supranasals present; prefrontals, frontoparietal, and interparietal distinct (Fig. 126); limbs well developed, pentadactyl, all digits clawed.

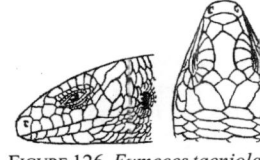

FIGURE 126. *Eumeces taeniolatus*. (From Nikolsky, 1916, figs. 62–63.)

Distribution: North and Central America from Canada to Nicaragua; Bermuda; Africa north of the Sahara; Cyprus; Southwest Asia from Mediterranean to northwestern India, and from the Transcaucasian and Transcaspian republics to northern Arabia and the Persian Gulf; Southeast and East Asia. Known fossil in the Pliocene (Romer, 1956:549). About 45 species recognized; two species, one with two or three subspecies, occur in Iran.

Key to the Species of *Eumeces* in Iran

1a. Two median rows of dorsal scales united into single row of broad scales (Fig. A); postnasal shield present .... *Eumeces taeniolatus parthianicus* Szczerbak, 1990 (p. 272)
1b. Two median rows of dorsal scales broader than those on flanks (Fig. B); no postnasal shield 2

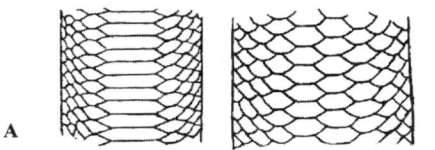

Dorsal scales of (A) *Eumeces taeniolatus*; (B) *Eumeces schneideri*
(From Nikolsky, 1916, figs. 64 and 67, respectively)

2a. Base of tail reddish in life (preserved individuals in which color has faded cannot be identified to the subspecies level, as far as I have been able to determine) ....
....*Eumeces schneiderii zarudnyi* Nikolsky, 1899 (p. 271)
2b. Base of tail not reddish in life, dorsum with or without orange or reddish flecks
.... *Eumeces schneiderii princeps* (Eichwald, 1839) (p. 269)

Eumeces schneiderii (Daudin, 1802)
Schneider's skink, gold skink

Scincus schneiderii Daudin, 1802, 4:291–292 (Type locality: not stated; probably Egypt or Sinai [*fide* Taylor, 1935:123], or Cyprus [*fide* Mertens, 1946a:55]; Holotype: originally in MNHN, probably lost).
Eumeces schneiderii: Boulenger, 1887a:383–384.

Eumeces schneiderii princeps (Eichwald, 1839)   (Plate 17B–D)
Red-marked skink

Euprepis princeps Eichwald, 1839:303 (Type locality: Talysch Mountains, Azerbaijan; Holotype: ?Moscow).

Mabouia aurata: Günther, 1864:82 (*nec Lacerta aurata* Linnaeus, 1758).
Plestiodon Aldrovandi: De Filippi, 1865:354 (not Duméril and Bibron, 1839).
?*Eumeces pavimentatus*: Blanford, 1876:354 (not *Scincus pavimentatus* Geoffroy St.–Hilaire, 1827). — Bedriaga, 1879:27.
Eumeces schneiderii: Boulenger, 1887a:383–384 (in part). — Nikolsky, 1897:333; 1900:399. — Boulenger, 1899:379–380. — Zarudny, 1903:22. — Nikolsky, 1907a:285; 1915:511–515, figs. 65–68 (in part). — F. Werner, 1917:202–203. — Lantz, 1918:15. — F. Werner, 1936:201. — Rostombekov, 1938:16. — S. Anderson, 1968:333 (in part). — Tuck, 1971a:59–60, fig. 8.
Eumeces schneiderii princeps: Mertens, 1920:179. — Eiselt, 1940:218, figs. 3b, 3c, 3d (in part). — Mertens, 1946a:58–59. — Terentjev and Chernov, 1949:168–169. — S. Anderson, 1974:38, 44. — Bannikov et al., 1977:148. — Nilson and Andrén, 1981:138–139. — Darevsky, 1981:360, fig. 63 (p. 357). — Welch, 1983:24. — Leviton, et al., 1992:68. — Frynta, et al., 1997:11.
Eumeces schneiderii schneiderii: Mertens, 1924c:182.
Eumeces princeps: Taylor, 1935:138–141, pl. 3, fig. 3, text-figs. 10, 14.
Eumeces schneiderii variegatus Schmidt, 1939:68–69 (Type locality: Persepolis, Iran; Holotype: FMNH 21008). — S. Anderson, 1963:476; 1974:38, 44. — Bannikov et al., 1977:148 — Schleich, 1977:127, 129. — Welch, 1983:24.

Diagnosis: Postnasal absent; palpebral scales separated from superciliaries; median preanal scales overlap outer; 2 azygous postmentals; nasal completely divided; plates on lower eyelid large, much higher than wide; ear with 3–4 acute lobules; 26–28 scales around midbody; tail not red at base in life.

Color pattern: Dorsum nearly uniform brownish slate to lavender, some scattered gray flecks; indistinct, narrow lateral cream line from posterior labials through ear along sides to groin; below this line, grayish, becoming lighter below; tail lighter above than body; limbs lighter and browner than body. (Taylor, 1935:138). In life, often with yellow-orange to brick-red spots, which may be more or less regularly arranged in longitudinal or transverse rows, light lateral stripe orange to red, becoming lighter on tail; venter yellowish to yellowish-orange. (Terentjev and Chernov, 1949:168).

Size: Snout-vent length to 165 mm.

Natural history: Eggs laid not earlier than end of July in Asian republics north of Iran. A wide variety of food items recorded, including acridids, beetles, dipterans, and other insects as well as spiders, mollusks, and other lizards. (Terentjev and Chernov, 1949:169). Darevsky (1981:361–364) presents a synopsis of the published information on the ecology and natural history of this species.

Habitat: (Plate 25C) Foothills and lowlands below 1400–1500 m on loess and clay soils, stony areas where there is grassy or shrubby vegetation, seeking refuge under stones, in burrows of other animals, or its own shallow burrows 40–100 cm long. (Terentjev and Chernov, 1949:169). Reed collected this skink in Iraq in late April in a field of barley and in tents, but not in mud-brick rubble nor under rocks, where *Mabuya aurata* was found (Reed and Marx, 1959:104). The Clarks, on the other hand, found them seeking refuge under piles of stones and in holes under rocks in Turkey (Clark and Clark, 1973:42). They stated that this skink requires relatively high temperature for activity, finding them active at air temperatures of 25–30°C, surface temperatures of 33.5–44°C.

Distribution. This subspecies occurs in Iraq, eastern Turkey, western and northern Iran, Transcaucasian republics west to eastern Georgia, Transcaspian republics in the foothills of the Kopet Dagh, Nura Tau Mountains, and northern foothills of the Turkestan Ridge, east to the foothills of the Pamiro Alai. The eastern extent of its range in Iran is not yet known, nor have areas of intergradation with *Eumeces schneiderii zarudnyi* been determined. The species as a whole extends across North Africa east from Morocco through western Asia to

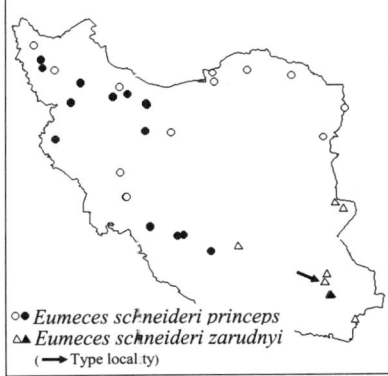

○● *Eumeces schneideri princeps*
△▲ *Eumeces schneideri zarudnyi*
(➝ Type locality)

Baluchistan and the Makran coast of Pakistan. The extent to which it may extend south into the Arabian Peninsula is not known. A mandible of this skink has been identified from Layer B of Shanidar Cave, Iraq, and if not intrusive, would be approximately 12,000 years old (Reed and Marx, 1959:104).

Remarks: Here I place *Eumeces schneideri variegatus* in the synonymy of *E. s. princeps*. Schmidt (1939:69) described the color pattern of *E. s. variegatus* as follows: "Back brown, with vermiculate darker markings, some of which may be arranged in vertical rows, while the mid-dorsal spots tend to form longitudinal lines; under surfaces paler brown." Schmidt's holotype and only specimen is immature. I believe that this specimen is simply an unusually marked *E. s. princeps*. The pattern is not particularly striking. Mertens (1946a:59) records a juvenile color pattern for a Transcaucasian specimen of *E. s. princeps*, in which dark and light spots gave "salt and pepper" appearance to the back, the vertebral zone patternless, no lateral stripe present. The holotype of *E. s. variegatus* has the nasal completely divided, three or four large scales on the lower eyelid, characters supposed to distinguish *E. s. princeps* from *E. s. schneiderii*. The dark markings of the dorsum are confined largely to the margins of the scales, except on the two mid-dorsal rows, in which many of the scales have a median longitudinal dark bar. USNM 153595 from 38 km N of Shiraz, almost exactly the same locality as Schmidt's, has a color pattern typical of *E. s. princeps*, as do Blanford's specimens from Neyriz (BMNH 74.11.23.22–23), about 100 km southeast of Persepolis.

Eumeces schneiderii zarudnyi Nikolsky, 1900
Zarudny's skink

Eumeces pavimentatus: Blanford, 1876:387–388 (in part; not *Scincus pavimentatus* Geoffroy St.-Hillaire, 1827). — Bedriaga, 1879:27.

Eumeces zarudnyi Nikolsky, 1900:399–401 (Type locality: Bazman, Iran [restricted by Taylor, 1935:142]; Lectotype: ZIL 9339 [designated by Taylor, 1935:142]). — Zarudny, 1903:22. — Taylor, 1935:142–143, fig. 10. — F. Werner, 1936:201; 1938:270–271. — Forcart, 1950:148. — S. Anderson, 1963:476. — Schleich, 1977:127, 129.

Eumeces schneiderii princeps: Eiselt, 1940:218 (in part; not *Euprepis princeps* Eichwald, 1839).

Eumeces schneiderii zarudnyi: Mertens, 1946:58. — S. Anderson, 1974:38, 44. — Darevsky, 1981:360. — Welch, 1983:24. — Welch, et al., 1990:77.

Diagnosis: Postnasal absent; palpebral scales separated from superciliaries; median preanal scales overlap outer; two azygous postmentals; nasal completely divided; ear with 5–6 acute lobules; limbs overlap when adpressed; scales in 26–28 rows at midbody; base of tail red in life.

Color pattern: Dorsum brownish gray in preservative, venter creamy white; a white lateral stripe runs from below or behind eye, through ear to thigh, this stripe bordered above by stripe darker than dorsum, most pronounced in juveniles; two or three pinkish red stripes on posterior one-third to two-thirds of dorsum in freshly preserved specimens (vanishing eventually in alcohol), interrupted anteriorly, extending onto basal portion of tail; occasionally scattered pinkish marks on lateral body scales. (Based on 9 specimens from 11 miles west of Iranshahr, collected by Street Expedition to Iran, 1962–63). In life dorsum blue-gray to lead gray, venter porcelain whiten; light lateral stripe orange to cinnabar red; somewhat duller orange-red flecks on upper side of tail (Mertens, 1969:47, for specimens from Astola Island off the Mekran Coast of Pakistan).

Size: Snout-vent to 120 mm, tail 202 mm (Mertens, 1969:48). The largest Iranian male measures 117 mm snout-vent.

Natural history: Specimens collected in late November have well developed yolked ovarian follicles. Digestive tracts contain beetles and other insect remains; one specimen contains a seed in the intestine, and in two specimens the lower intestine is packed with sand.

Distribution: (see map p. 271) The southeastern Iranian Plateau in Kerman, Baluchistan, and Sistan Provinces in Iran, the Helmand Basin and southern desert regions of Afghanistan, Baluchistan and Mekran Coast of Pakistan.

Remarks: BMNH 1904.12.7.3 from Ormara, Makran Coast, Pakistan, is lighter on the dorsum than specimens of *Eumeces schneiderii princeps*, and also has three faint but distinct darker stripes from occiput onto base of tail, a darker dorsolateral stripe on each side, bordered by a cream stripe, and a faint lateral dark stripe between legs. Minton (1966:102) also commented on a specimen from Ormara, and concluded tentatively that *E. blythianus* (J. Anderson) is at best subspecifically distinct from *E. s. zarudnyi*, and that the two taxa intergrade along the Makran Coast of Pakistan. Mertens (1969:46) also regarded *blythianus* as a subspecies of *E. schneiderii*.

Eumeces taeniolatus (Blyth, 1854)
Ribbon–sided skink

Eurylepis taeniolatus Blyth, 1854:739–740 (Type locality: Alpine Punjab [?= Salt Range, Punjab], India; Holotype: ZSI 2382 [*fide* Das, *et al.*, 1998]).

Eumeces taeniolatus parthianicus Szczerbak, 1990 (Plate 17E-G)
Parthian skink

Eumeces taeniolatus: Stoliczka, 1872a:75–76. — Taylor, 1935:111–119, pls. 4,5, text-figs. 9–10. — Terentjev and Chernov, 1949:169–170. — S. Anderson, 1963:476; 1968:333; 1974:38, 44. — Bannikov, *et al.*, 1977:150. — Welch, 1983:24. — Welch, *et al.*, 1990:77. — Leviton, *et al.*, 1992:69.
Eumeces scutatus: Lantz, 1918:15. — Morich, 1929:32.
Eumeces taeniolatus parthianicus Szczerbak, 1990:33–40, fig. 1b, fig. 2b (Type locality: Northern slope of central Kopet Dag, Chuli, 25 km west of Ashkhabad, Turkmenistan; Holotype: ZIK Re 18 no. 17660 adult male).

Diagnosis: A single median dorsal series of greatly widened scales (Fig. A in key, p. 269), scales around body in odd numbers; usually four or five pairs of nuchals followed by paired median scales for short distance; postnasal present. Usually 21 scale rows around middle of body; dorsal median dark stripe broken into series of transverse spots.

Color pattern: Pale brown above in alcohol, young with three dark brown longitudinal stripes formed of closely connected spots; vertebral stripe occupies greater part of vertebral scales, two others upon upper half of neck and flank, freely spotted with white; in later life stripes, particularly vertebral, become more or less broken up into spots; venter deep saffron in life, fading to white in alcohol. Terentjev and Chernov (1949:170) state that the lower jaw and limbs are light bluish in life, the tail sky blue, light blue-green or blue-green. Neither Minton (1966) nor Mertens (1969) mention this color for Pakistan specimens of *Eumeces* *taeniolatus* (one was kept by Mertens in captivity for nine years), nor was it present in two specimens (male and female) from Pakistan that I kept in captivity for more than a year.

Size: Snout-vent length to 150 mm in specimens from Turkmenistan (Terentjev and Chernov, 1949:168); tail may be 1½ times snout-vent length. Blanford (1875b:195) records a specimen 175 mm snout-vent from Kashmir.

Natural history: Nothing seems to be recorded of reproduction in this skink. Minton (1966:101) collected a small juvenile (*Eumeces t. taeniolatus*) in early September. Carabid beetles, orthopterans, and spiders are the recorded food items.

Habitat: Ingoldby found the nominate subspecies abundant in Waziristan in the burrows of the gerbil, *Tatara indica*, in the sandy patches around the roots of bushes on the stony plain bordering the foothills near the Afghan border; Murray found it among hedges and in gardens in Sind (Smith, 1935), while Minton (1966:101) collected the species in sparse grassland with loose clay soil and on rocky hillsides up to 6800 ft (2090 m) elevation. Specimens of *Eumeces t. taeniolatus* in my laboratory spent most of the time buried in the loose gravel of the terrarium floor.

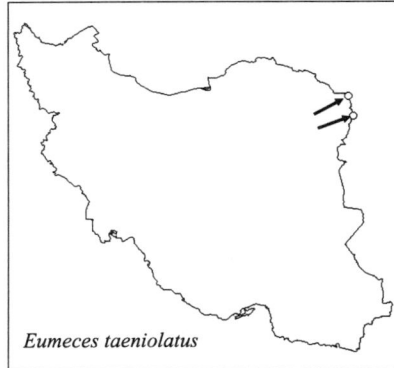

Eumeces taeniolatus

Distribution: This subspecies occurs in southern Turkmenistan, eastern Iran, and northern Afghanistan; the typical subspecies is found in Pakistan, Kashmir, and probably southern Afghanistan. While Terentjev and Chernov (1949:170) state that it is found in northern and northeastern Iran, the only records I find are those of Sarakhs and the River Tajan on the Iran/Afghan/Turkmenistan border. In Afghanistan the only known locality other than the River Tajan is Pandjvai near Kandahar. (Leviton and S. Anderson, 1970:192). *Eumeces taeniolatus arabicus* Szczerbak, 1990 is an apparently widely disjunct population in northwestern and western Saudi Arabia, and Yemen (Aden area).

Remarks. This rather specialized species appears to have no close relatives, but Taylor (1935:110) regarded it closest to *Eumeces schneiderii* of the living species. Its distribution appears to be widely disjunct. I have made only cursory examinations of specimens and without comparing materials from throughout the range. I examined three specimens from the Arabian peninsula, CAS 149442 from Wadi Libah, 30 km NE Al Jumum, BMNH 1903.6.26.17 from El Kubar, Amiri Country and BMNH 1978.2268 from 16 km from Taif on Taif Abha road, Saudi Arabia. These Arabian specimens are very distinctive in having three wide, dark longitudinal stripes on the head and neck from snout to level of forelimbs, where they break up into small spots (see Leviton, *et al.* [1992: col. pl. 10G]). These were subsequently described by Szczerbak (1990) as *E. t. arabicus*. In a series from Pul-i Hatan, Turkmenistan (BMNH 91.10.6.15–23) and a large series from southern Turkmenistan collected by Theodore Papenfuss and Robert Macey from along the border with Iran these stripes are much less distinct and tend to be broken up on the neck. Szczerbak (1990:33–40) carried out a brief review of the species, breaking it into three taxa on the basis of these color pattern differences and the number of scales around the middle of the body. The color pattern differences are clearly shown in his photographs (figs. 1 and 2). His 50 specimens of *E. t. parthianicus* were consistent in having 21 scales around the body, whereas two of 12 specimens of *E. t. taeniolatus* had 20, the rest 21. Two of three specimens of *E. t. arabicus* had 19, one had 21. Minton (1966:101) and Mertens (1969:48) consider *Plestiodon scutatus* Theobald, 1868 a probable junior objective synonym of *Eurylepis taeniolatus* Blyth, 1854. Annandale (1906:148), on the other hand, compared specimens which he identified as *E. scutatus* from Sind, Rajputana, northern Kashmir, Chitral, and Afghanistan with Blyth's types and considered the two taxa as quite distinct, without providing the data for his conclusion. The possible existence of two distinct taxa in Pakistan and eastern Afghanistan may warrant further analysis.

Genus *Mabuya* Fitzinger, 1826
Mabouyas

Mabuya Fitzinger, 1826:23, 52 (Type species: *Mabuya dominicensis* Fitzinger, 1826 [= *Lacertus mabouya* Lacepède, 1788; see comment in Leviton, *et al.*, 1992:69 and note in Zhao and Adler, 1993:211]).

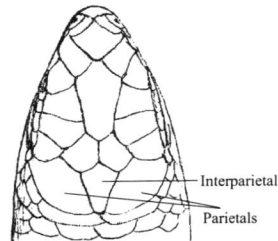

FIGURE 127. Dorsal view of head of *Mabuya aurata*. (From Blanford, 1876, pl. 27, fig. 3.)

Definition. Palatine bones in contact mesially; palatal notch entirely separating pterygoids, extending forward to between centers of eyes; pterygoid teeth minute or absent. Maxillary teeth conical or bicuspid. Eyelids movable, lower with or without more or less transparent disc. Ear distinct, tympanum more or less deeply sunk. Nostril pierced in single nasal; supranasals present; prefrontals present; frontoparietals sometimes united in single shield; interparietal sometimes united with parietals (Fig. 127). Limbs well developed, pentadactyl. Digits subcylindrical or compressed, with transverse lamellae inferiorly.

Distribution. Africa and Madagascar; southern Asia; the East Indies; Central America; the West Indies; South America. About 85 species recognized; two occur in Iran.

Key to the Species of *Mabuya* in Iran

1a. Parietal scales usually in contact behind interparietal; nuchals and postnuchals with three strongly developed keels; often a distinct light vertebral stripe, usually dark–margined and clearly set off from ground color ...... *Mabuya vittata* (Olivier, 1804) (p. 277)
1b. Parietal scales not in contact (Fig. 127), separated by interparietal; nuchals smooth, postnuchals smooth or very weakly keeled; no light vertebral stripe 2
2a. 60–62 gulars plus ventrals counted from mental shield to vent ....
.... *Mabuya aurata septemtaeniata* (Reuss, 1833) (p. 275)
2b. 65–72 gulars plus ventrals .... *Mabuya aurata transcaucasica* Chernov, 1926 (p. 275)

Mabuya aurata (Linnaeus, 1758)
Golden grass skink

Lacerta aurata Linnaeus, 1758:209 (Type locality: "Jersea Anglorum," Cyprus; Holotype: NHRM [see Andersson, 1900:14]).
Euprepis affinis De Filippi, 1865:354 (*nec Tiliqua affinis* Gray, 1838:289 [= *Mabuia affinis*, Boulenger, 1889:166]; Type locality: Kazvin [Qazvin], Iran; Syntypes: Genoa and Turin Museums).
Mabuya aurata: Andersson, 1900:14.

Diagnosis: Lower eyelid with undivided, more or less transparent disc; 32–38 scales

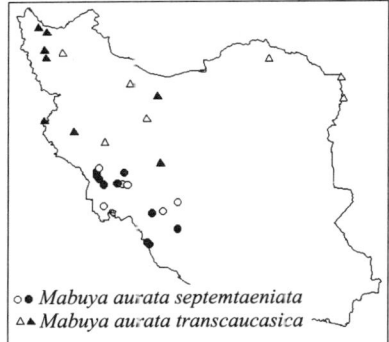

○● *Mabuya aurata septemtaeniata*
△▲ *Mabuya aurata transcaucasica*

around body, dorsals feebly tricarinate to smooth; prefrontals not in contact; 16–22 lamellae under 4th toe; parietals not in contact (Fig. 127); no light vertebral stripe.

Size: Snout-vent length 105 mm.

Distribution: The species is found in Eritrea, Arabia, islands of the northeastern Mediterranean, Turkey, Lebanon, Syria, Jordan, Iraq, western and northern Iran, southern regions of Armenia, Nakhichevan, southern Turkmenistan, and Uzbekistan. It is of doubtful occurrence in Afghanistan, except, perhaps, in the northern lowlands. Records for Sind and Baluchistan, Pakistan date from Murray and have not been confirmed. Two subspecies (see below) are recognized from Iran.

Mabuya aurata septemtaeniata (Reuss, 1834) (Plate 17H)
Southern grass skink

Euprepis septemtaeniatus Reuss, 1834:47 (Type locality: Massawa, Eritrea).
Euprepis septemtaeniata: Blanford, 1876:388–390. — Boettger, 1880:188. — Nikolsky, 1897:335.
Euprepis septemtaeniatus: Bedriaga, 1879:27.
Mabuia septemtaeniata: Boulenger, 1887a:177–178 (in part); 1899:379. — Nikolsky, 1907a:285; 1915:487–489 (in part). — F. Werner, 1917:202. — F. Werner, 1936:201 (in part).
Mabuya aurata: Andersson, 1900:14. — S. Anderson, 1968:333 (in part); 1974:38, 44 (in part). — Schleich, 1977:127, 129 (in part). — Welch, 1983:27 (in part). — Welch, et al., 1990:85.
Mabuya aurata septemtaeniata: Schmidt, 1939:66 (in part). — Terentjev and Chernov, 1949:166–167. — S. Anderson, 1963:463–464. — Haas and Y. Werner, 1969:360. — Schleich, 1977:127, 129 (in part). — Akhmedov, 1983:84–85. — Akhmedov and Szczerbak, 1987:20–24. — Leviton, et al., 1992:70–71, col. pl. 11B–C.

Diagnosis: 60–62 gulars plus ventrals.

Color pattern: Light brown above, with four longitudinal dark brown stripes beginning at occiput, distinct on nape, breaking up into spots or disappearing on posterior part of back; broad, dark stripe, spotted with white, from nostril, on anterior half of flank, passing along upper half of flank, usually bordered above and below with light stripe, often breaking up and becoming indistinct on posterior half of flank; venter white.

Habitat: (Plates 21G, 22A–C) In Khuzestan this species inhabits areas of the foothill region where rock crevices provide a retreat (S. Anderson, 1963:464).

Distribution: (see map above) Southern Iran, lowland Iraq, northeastern Saudi Arabia (Al Hasa south to Hofuf), Bahrain, northern Oman (Muscat); Eritrea. Zones of possible intergradation with *Mabuya a. transcaucasica* in central Iran have not been identified.

Mabuya aurata transcaucasica Chernov, 1926 (Plate 18A)
Transcaucasian grass skink

Euprepis affinis De Filippi, 1863:25 (not *Tiliqua affinis* Gray, 1838).
Mabuia septemtaeniata, Boulenger, 1887a:177–178 (in part); 1899:379. — Zarudny, 1903:22–23. — Nikolsky, 1915:487–489 (in part). — Lantz, 1918:15. — Morich, 1929:32. — F. Werner, 1936:201 (in part).
Mabuya aurata: Andersson, 1900:14 (in part). — Rostombekov, 1938:15–16. — S. Anderson, 1968:

333 (in part); 1974:38, 44 (in part). — Schleich, 1977:127, 129 (in part). — Welch, 1983:27 (in part). — Frynta, *et al.,* 1997:11.

Mabuya aurata septemtaeniata: Mertens, 1924a:377 (in part, not Reuss, 1834). — Schmidt, 1939:66 (in part). — Terentjev and Chernov, 1949:166–167 (in part). — S. Anderson, 1963:463–464. — Bannikov, *et al.*, 1977:145. — Schleich, 1977:127, 129 (in part). — Clark, 1991:38.

Mabuya aurata transcaucasica Chernov, 1926:64 (Type locality: Migri and Ordubat, Armenia; Syntypes: ZIK)

Mabuya aurata affinis: Wettstein, 1951:440 (*nec* Gray, 1838). — Mertens, 1957:126–127. — Tuck, 1971a:60. — Schleich, 1977:127, 129. — Akhmedov, 1983:84–85. — Akhmedov and Szczerbak, 1987:20–24.

Diagnosis: 65–72 gulars plus ventrals.

Color pattern: Olive-brown above, with four longitudinal dark brown stripes on head, breaking up into spots on nape, disappearing on posterior part of back; broad, dark stripe, spotted with white, from nostril, passing along upper half of entire flank, usually bordered above and below with white stripe, continuing onto tail; limbs brown, with white speckles; venter white.

Natural history: FMNH 21016 from Esfahan, collected in mid-August, contains what appear to be full-term fetuses in the oviducts. Digestive tracts examined contain spiders, orthopterans, and other arthropod remains.

Habitat: (Plates 23D, 25D) According to Terentjev and Chernov (1949:167), it is found in areas with shrubs or high grassy vegetation, on canyon slopes, banks of irrigation ditches, and stony fences.

Distribution: (see map, p. 275) Armenia and Nakhichevan, close to the Iranian border, southern Turkmenistan, northern and central Iran, possibly northwestern Afghanistan.

Remarks: The most recent analysis of the populations of *Mabuya aurata* is that of Akhmedov and Szczerbak (1987). They recognized three population groups based on dorsal color pattern and scale counts: *Mabuya aurata aurata* in western Asia Minor, *M. a. septemtaeniata* in southern Iran, Baluchistan (Pakistan), and the lowland countries of Southwest Asia, and *M. a. affinis* (De Filippi, 1863) in the Transcaucasian republics, northern and central Iran, Turkmenistan, and Afghanistan. They pointed out that the population from the Kopet Dagh seems to differ from neighboring populations (assigned to *affinis*), but they did not raise it to subspecific rank. They placed *Euprepis fellowsii* Gray, 1845 in the synonymy of the typical subspecies, although they apparently did not examine material from Cyprus, the stated type locality for *Mabuya aurata*. I examined the type series of *Euprepis fellowsii* from Xanthus (BMNH 1946.8.17.90–96; see fig.1, p. 21 in Akhmedov and Szczerbak, 1987). They fit the diagnosis of *M. aurata* given above, except that there is a light vertebral line, at least represented by the disruption of the darker dorsal pattern. The dorsal pattern is distinctive, the longitudinal stripes broken up into more or less rectangular dark spots, those of the paravertebral stripes larger than or subequal to the interspaces, the quadrangular spots of the lateral lines interrupted by narrow light interspaces. They do not agree in color pattern with *M. a. septemtaeniata* (see Akhmedov and Szczerbak, 1987:fig.2, p. 21, which is of a specimen from Saati, near the type locality for *septemtaeniata*, according to Arnold [1986:428], and not from Iran as indicated by Akhmedov and Szczerbak). The type locality for *Mabouia septemtaeniata* Boulenger is Massawa, Eritrea, and while it seems unlikely on biogeographic grounds that Iranian specimens represent the same taxon, Arnold (1986d:428) suggested that this skink may have been introduced accidentally both in Eritrea and Muscat.

The name *Euprepis affinis* De Filippi, 1863 is preoccupied by *Tiliqua affinis* Gray, 1838, which refers to a specimen of *Mabuya* in the British Museum, type locality unknown (Boulenger, 1889:166). *Mabuya transcaucasica* Chernov, 1926 appears to be an available name that applies to the populations designated by Akhmedov and Szczerbak as *affinis*. *Mabuya vittata* of the countries of the eastern Mediterranean to western Iran and *M. dissimilis* of Pakistan, Afghanistan, and northern India appear to be the closest living relatives of *M. aurata*.

Mabuya vittata (Olivier, 1804) (Plate 18B)
Bridled skink

Scincus vittatus Olivier, 1804:103 (Type locality: sands west of Rosetta [Rashid], Egypt [see Leviton, et al., 1992:72 for comment on citation]; Holotype: MNHN 197).
Mabuia vittata: Boulenger, 1887a:176–177.
Mabuya vittata: Wettstein, 1928b:783 — J. J. Schmidtler and J. F. Schmidtler, 1972:65. — S. Anderson, 1974:37, 44. — Schleich, 1977:127, 129. — Leviton, et al., 1993:72.

Diagnosis: Lower eyelid with undivided, more or less transparent disc; dorsal scales strongly tricarinate; prefrontals usually not in contact; parietals usually in contact behind interparietal; limbs not, or but slightly overlapping when adpressed; a distinct light vertebral stripe, usually dark-margined, usually present.

Color pattern: Dark olive-brown with a lighter vertebral stripe edged with black (often black margins broken into dots); very narrow light lateral line on each side of body, sometimes white, more or less distinctly edged with black, and another, somewhat broader, from in front of and below eye through ear and along flank, edged above and below with black; plates of head, and often dorsal scales, edged with dark brown; these stripes barely discernible in some specimens, the lizards being almost uniform in appearance. Venter yellowish or greenish white.

Size: Snout-vent length to about 90 mm.

Natural history: Ovoviviparous. Specimens collected in late April to mid-May and early June in Tunisia contained eggs with embryos in middle stages of development (Mosauer, 1934:56). Twenty-one of 24 females examined (collected in Turkey mid-April to mid-May) contained eggs, the smallest 4 mm in diameter, the largest 9 × 6 mm (Clark and Clark, 1973:45).

Habitat: Schmidtler and Schmidtler (1972:65) found this skink for the first time in Iran on a dry eastern slope, nearly devoid of vegetation near the northern limit of palm growth. They collected *Laudakia nupta*, *Trapelus ruderatus*, *Asaccus elisae*, *Coluber rhodorhachis*, *Pseudocyclophis persica*, *Natrix tessellata*, and *Leptotyphlops hamulirostris* (?) in the same habitat. In Tunisia, *Mabuya vittata* is a typical inhabitant of oases, and unlike *Chalcides ocellatus*, does not live on clay walls, but chiefly in the humid grass and herbaceous vegetation growing along ditches and brooks. It does not hesitate to jump into the water to escape pursuit (Mosauer, 1934:56). In Turkey, it occupied similar habitats and was occasionally abundant, particularly around clumps of vegetation and hedges bordering sand. Unlike *C. ocellatus* of the same areas, it was never seen on the open sand, and was very shy, quickly vanishing into the underbrush. It was active at an air temperature of 16°–20°C, and never found out of hiding below 14°C. In some areas

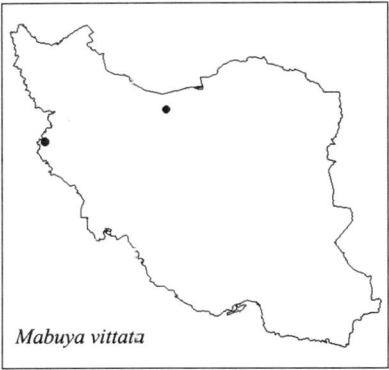
Mabuya vittata

it occurred with *Mabuya aurata*, but the microhabitats did not overlap, the latter preferring rocks and crevices. (Clark and Clark, 1973:44).

Distribution: Mediterranean coasts of North Africa, from Algeria through Egypt; Cyprus and Rhodes; Turkey, Lebanon, Israel, Syria, Jordan, Iraq, and western Iran.

Remarks: The Schmidtlers' (1972) specimen came from not far east of the Iraq border, collected on April 26, 1968. The California Academy of Sciences received a specimen from Lar, near Mt. Damavand in the Alborz Mountains, through the courtesy of Mahmoud Latifi of the Institut d'Etat des Serums et Vaccins Razi. This specimen lacks a distinct light vertebral line, but the light dorsolateral stripes are distinct. In cases where the identification of Iranian *Mabuya* are in doubt, the strong keels on the nuchals of *M. vittata* should distinguish this species from *M. aurata*, in which the nuchals always seem to be smooth.

Genus *Ophiomorus* Duméril and Bibron, 1839
Sand skinks, snake skinks, ophiomores

Ophiomorus Duméril and Bibron, 1839:799 (Type species: *Ophiomorus miliaris* Duméril and Bibron, 1839 [junior objective synonym of *Anguis punctatissimus* Bibron and Bory de St. Vincent, 1833], by monotypy).

Definition: Palatine bones not meeting on midline of palate; pterygoid teeth usually present; teeth conical or with obtuse crowns. Eye small; lower lid with undivided transparent disc. Ear opening absent or hidden. Nostril in suture between nasal and supranasal, close to rostral; prefrontals usually distinct; frontoparietal and parietal distinct. Body elongate. Limbs small or absent.

Distribution: From Greece through Southwest Asia to northwestern India.

Remarks: This genus was reviewed by S. Anderson and Leviton (1966b). It is composed of ten species, six of which inhabit Iran. The species fall into two groups from the standpoint of habitat selection as well as morphology. The eastern forms (*Ophiomorus chernovi, O. brevipes, O. blanfordi, O. nuchalis, O. streeti, O. tridactylus, O. raithmai*), with the exception of *O. nuchalis*, are sand burrowers, and their morphological adaptation to this mode of life is at once apparent. The three western species (*O. persicus, O. latastii,* and *O. punctatissimus*) have been collected under rocks and do not live in areas of wind-blown sand.

Consideration of the morphology and distribution of the genus invites the speculation that the ancestral form may have inhabited the area which is now the Iranian Plateau at a time when this area had a climate supporting a more or less continuously distributed biome, perhaps of savannah or grassland vegetation type.

With the breakup in the continuity of this habitat, possibly coincident with increased orogeny in the western portion of the region, one line of specialization resulted in the legless western species occupying an upland, under-rock habitat. With increasing aridity and further fragmentation of the environment of the Central Plateau of Iran, a line of specialization adapted to life in the wind-blown sand developed. The discovery of *O. nuchalis* by Nilson and Andrén (1978) in nonsandy habitat in the western Dasht-e Kavir on the Central Plateau appears to support these speculations as it appears to link the eastern and western groups of the genus both morphologically and ecologically. In several respects it resembles the hypothetical ancestor of the eastern species proposed by S. Anderson and Leviton (1966b:530–531).

As pointed out by S. Anderson and Leviton, the morphological specialization of the eastern species has been directional in a geographic sense, from north to south and west to east. The least specialized of the four-fingered forms is the most western (*O. nuchalis*), while the southeasternmost species (*O. blanfordi*) is the most modified. Of the three-fingered forms, *O. streeti*, the least specialized with respect to head scalation and color pattern, is found furthest west; the other two species occur to the east of the Central Plateau, and there is a suggestion that even more highly specialized populations may exist to the east of these.

Because these animals are obligate sand-dwellers, they are dependent on the progressive movement of the sand for their distribution. It is particularly noteworthy that the winds of such force as to be responsible for dune migration blow from the high pressure areas to the north of Iran south across the Central Plateau, and in summer, winds blow from the west toward the low pressure center of India and Pakistan.

The sand-dwelling populations of *Ophiomorus* are apparently isolated from one another by physical barriers to the distribution of the dunes in which they live. These populations are thus somewhat analogous to island populations in their separation. Contact can be reestablished only as the agency of wind in conjunction with changing landforms brings existing dune areas into contact. Thus, while geographic isolation may be relatively transitory, immediate genetic isolation can be assumed to be complete.

Ophiomorus and the related genera, *Chalcides* and *Scincus*, may have arisen from a *Eumeces*-like ancestor (S. Anderson and Leviton, 1966b).

Key to the Species of *Ophiomorus* in Iran

1a. Fingers four, toes three 2
1b. Fingers three, toes two or three 4
2a. Scale rows 20 at midbody (counts must be made exactly midway between snout and vent) .... *Ophiomorus blanfordi* Boulenger, 1887 (p. 279)
2b. Scale rows 22 at midbody 3
3a. Nuchals equal to or about 1½ times size of dorsals ....
.... *Ophiomorus brevipes* (Blanford, 1874) (p. 280)
3b. Nuchals about 2½ times size of dorsals ....
.... *Ophiomorus nuchalis* Nilson and Andrén, 1978 (p. 282)
4a. Toes two .... *Ophiomorus persicus* (Steindachner, 1867) (p. 283)
4b. Toes three 5
5a. Parietals in contact posteriorly; prefrontals not in contact with supralabials (20 scale rows at midbody) .... *Ophiomorus streeti* Anderson and Leviton, 1966 (p. 283)
5b. Parietals not in contact posteriorly; prefrontals in contact with supralabials (usually 22, occasionally 20 scale rows at midbody)..*Ophiomorus tridactylus* (Blyth, 1853) (p. 285)

Ophiomorus blanfordi Boulenger, 1887
Blanford's snake skink

Zygnidopsis brevipes: Blanford, 1879:128 (*nec Zygnopsis brevipes* Blanford, 1874).
Ophiomorus blanfordi Boulenger, 1887a:395, pl. 33, fig. 1 (Type locality: southern Persia or Baluchistan, restricted to Chah Bahar, Iran by S. Anderson and Leviton [1966b:511]; Holotype: BMNH 80.11.10.188); 1887b:523–525). — F. Werner, 1917:203. — Smith, 1935:347–348. — F. Werner, 1936:201. — S. Anderson, 1963:476. — S. Anderson and Leviton, 1966b:511–512, figs 2e,2f, 3c. — S. Anderson, 1968:333. — Tuck, 1971a:60. — S. Anderson, 1974:37, 44 — Schleich, 1977: 127, 129. — Welch, 1983:25. — Welch, *et al.*, 1990:89.

Diagnosis: Fingers four, toes three, 20 scales round middle of body.
Color pattern: (Fig. 129) Cream or pale brown above in preservative, the two median

dorsal scale rows with scattered dark spots, forming more or less distinct longitudinal lines; scale rows immediately lateral to these median dorsal rows without dark markings; third and fourth rows from vertebral line with dark brown dots forming well marked lines which extend forward along sides of head. Top of head with or without central dark streak.

Size: The largest specimen examined, a female, has a snout-vent length of 96 mm.

Ophiomorus blanfordi

Distribution: Coastal sand dunes of southern Iran and Pakistan (Ras Jiunri [Jiwani], 16 air km east of Iranian border). Extent of distribution unknown; no precise locality is given by F. Werner (1917:203) for his specimens, hence a single Iranian locality and a single Pakistan locality a few miles away are known.

Remarks: Apart from the fact that it lives in coastal dune sands, nothing of its natural history is known. BMNH 80.11.10.88 is the specimen described by Boulenger. A second specimen lacks a tag and a second label on the bottle says "91.9.14.5 holotype." However, the register number indicates that this specimen was acquired by the British Museum after the species was described, although it was collected by Blanford at the same time as the type.

Ophiomorus brevipes (Blanford, 1874) (Plate 18C)
Short-legged snake skink

Zygnopsis brevipes Blanford, 1874b:33 (Type locality: Saadatabad, southwest of Kerman, Iran; Holotype: ZSI 3464 [*fide* Das, *et al.*, 1998]); 1876:397–399, pl. 27, figs. 4, 4a. — Bedriaga, 1879: 28.
Ophiomorus brevipes: Boulenger, 1887a:395–396; 1887b:525–526. — Nikolsky, 1900:401–402. — Zarudny, 1903:24–25. — Nikolsky, 1915:515–516. — Smith, 1935:348. — F. Werner, 1936: 201. — Forcart, 1950:148. — Wettstein, 1951:441. — S. Anderson, 1963:464–465, text fig. 13. — Guibé, 1966a:98. — S. Anderson and Leviton, 1966b:510–511, text figs. 2c, 2d, 3b. — S. Anderson, 1968:333; 1974:37, 44. — Schleich, 1977:127, 129. — Welch, 1983:25.

Diagnosis. Fingers four; toes three; 22 scales round middle of body; interparietal broader than long; nuchals equal to or about 1½ times size of dorsals.

Color pattern. (Figs. 128, 130) Cream or pale brown above in life and in preservative, each scale of the two median dorsal rows with central brown spot, these forming two distinct lines down length of body and tail. Scale rows immediately lateral to these median rows lack dark markings; next three lateral rows each with dark marking on each scale, forming

FIGURE 128. *Ophiomorus brevipes.* (From Blanford, 1876, pl. 27, fig. 4.)

FAMILY SCINCIDAE

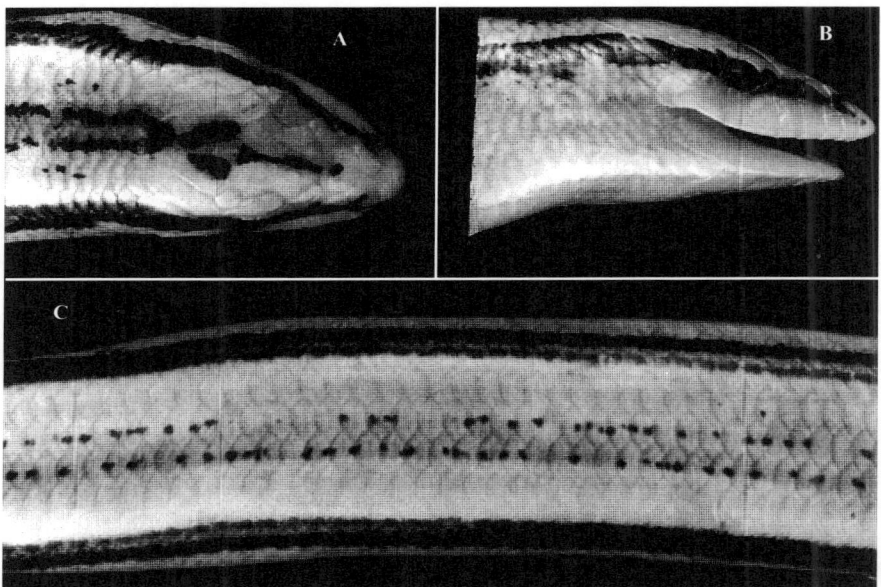

FIGURE 129. *Ophiomorus blanfordi.* (A) Dorsal view of head; (B) lateral view of head; (C) dorsal view of body. (From S. Anderson and Leviton, 1966, figs. 2e-f and 3c, respectively; photos by A. Leviton.)

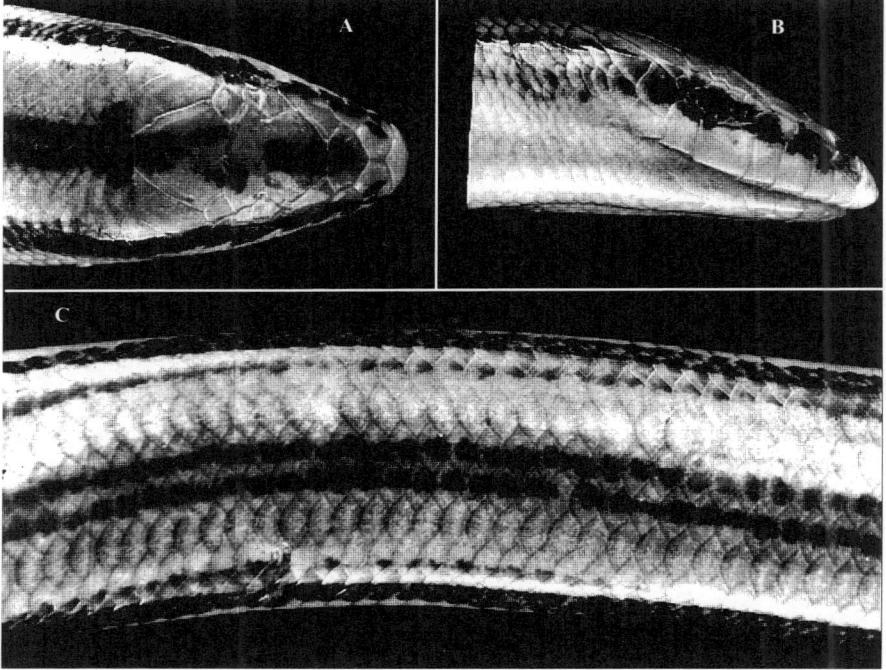

FIGURE 130. *Ophiomorus brevipes.* (A) Dorsal view of head; (B) lateral view of head; (C) dorsal view of body. (From S. Anderson and Leviton, 1966, figs. 2c-d and 3b, respectively; photos by A. Leviton.)

either three distinct lateral stripes, or single broad stripe length of body and tail. Dark line from nostril through eye and temporal region, contacting lateral stripes on body; single line on frontonasal and frontal, divided on posterior portion of frontal, reuniting on posterior portion of interparietal. Venter immaculate white.

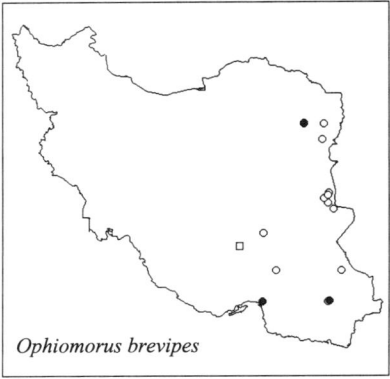
Ophiomorus brevipes

Size: The largest female examined is 91 mm snout-vent; the largest male measures 82 mm.

Natural history: Females collected in late November to early December contain yolked ovarian follicles, the largest slightly less than 2 mm diameter.

Habitat: The single specimen I collected was active just below the sand surface of a small active dune against a ridge near Minab. Vegetation consisted of scattered small shrubs.

Distribution: Sandy regions of eastern Iran (Kerman, Khorasan, and Baluchistan).

Remarks. Nothing seems to be known about the biology of this skink.

Ophiomorus nuchalis Nilson and Andrén, 1978 (Plate 18D)
Plateau snake skink

Ophiomorus nuchalis Nilson and Andrén, 1978:559–564, figs. 1a–c (Type locality: Siah Kuh, 34°44′N, 52°11′E, Kavir Protected region, about 150 km S Tehran, Iran; Holotype: NHMG Re. ex. 4418); 1981:139. — Welch, 1983:25.

Diagnosis: Fingers four, toes three, 22 scale rows at midbody, nuchals about 2½ times larger than dorsals.

Color pattern: "Cream or pale brown above, each scale of the two median dorsal rows with a central longitudinal black line, forming two distinct lines along body and tail. Scattered dark spots on some of the scales on the scale rows immediately lateral to the median rows. Next three lateral rows with a dark marking on each scale forming three distinct lateral stripes along body and tail. White below, with small black spots immediately posterior to vent. A dark line from nostril through eye and temporal region, in contact with lateral stripes on the body. A single line on frontonasal and frontal, divided on posterior portion of frontal but reuniting on posterior portion of interparietal." (Nilson and Andrén, 1978:559).

Size: Adult female: snout-vent to 98 mm, tail 81+ mm.

Habitat: The two known specimens were collected under stones on almost bare gravel ground in the lower hills around Cheshmeh Shah at the border of the north-facing slopes of Siah Kuh. Vegetation was sparse and patchy in distribution, *Artemisia herba-alba* the dominant species. The area is one of broken rocky mountains and alluvial plains. There is no loose sand in the vicinity of the type locality. (Nilson and Andrén, 1978:563 and fig. 1d, p. 560).

Distribution: Known only from the type locality on the Dasht-e Kavir in the western portion of the Central Plateau of Iran.

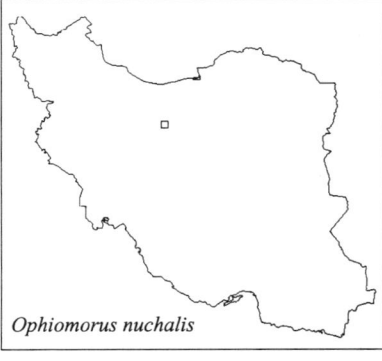
Ophiomorus nuchalis

Remarks. While most closely related to *Ophiomorus brevipes* and *O. chernovi*, this species is the least specialized for a fossorial existence of all of the eastern species of the genus. The snout is more conical, with a less sharp edge, the frontonasal is shorter, and there is no developed ventrolateral edge along the body. It is known only from the holotype and one paratype, both adult females. (Nilson and Andrén, 1978).

Ophiomorus persicus (Steindachner, 1867) (Plate 18E)
Persian snake skink

Hemipodion persicum Steindachner, 1867d:265 (Type locality: Persia; restricted to 5 km southeast of Pol-i-Abgineh, Fars Province, Iran, approximately 29°33'N 51°46'E [S. Anderson and Leviton, 1966b:528]; Syntypes: NMW 10398:1, 2 and 10399:1, 2). — Bedriaga, 1879:27.
Hemipodium persicum: Blanford, 1876:394.
Ophiomorus persicus: Boulenger, 1887a:396–397; 1887b:526–528. — Nikolsky, 1915:517. — F. Werner, 1936:201. — S. Anderson, 1963:476. — S. Anderson and Leviton, 1966b:528–529, text-figs. 7e–f, 8c. — S. Anderson, 1968:333; 1974:37, 44. — Schleich, 1977:127, 129. — Welch, 1983:25. — Frynta, et al., 1997:11.

Diagnosis: Fingers three; toes two.

Color pattern: (Fig. 131) Cream or light brown above in preservative, each dorsal and lateral scale with a brown spot, these more or less confluent to form lines the length of body; these dark lines are distinct on sides, four median lines being much paler; dots not confluent on tail. All scales of tail with dark spot. Brown line from nostril through eye, onto temporal region; upper surface of head mottled with brown. Venter cream or tan.

Size: Largest male 69 mm snout-vent length; largest female 82 mm.

Natural history: Specimens collected in early January contain yolked ovarian follicles up to nearly 2 mm diameter.

Distribution: (see map, below) Known from two localities on the western slope of the Zagros Mountains.

Remarks: This secretive lizard is probably spread much more widely in the Zagros Mountains than the limited material would indicate. Its biology is unknown.

Ophiomorus streeti Anderson and Leviton, 1966
Street's snake skink

Ophiomorus streeti S. Anderson and Leviton, 1966b:512–577, text– figs. 4a–b, 5a (Type locality: 11 mi [17.6 km] west of Iranshahr, Iran; Holotype: FMNH 141551). — S. Anderson, 1968:333; 1974:37, 44. — Schleich, 1977:127, 129. — Welch, 1983:25.

Ophiomorus persicus

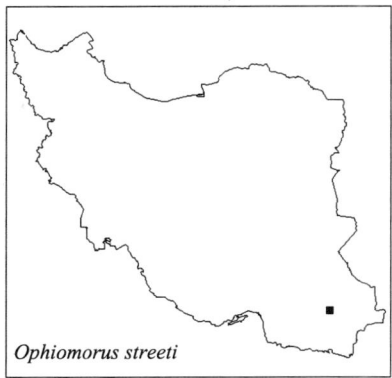

Ophiomorus streeti

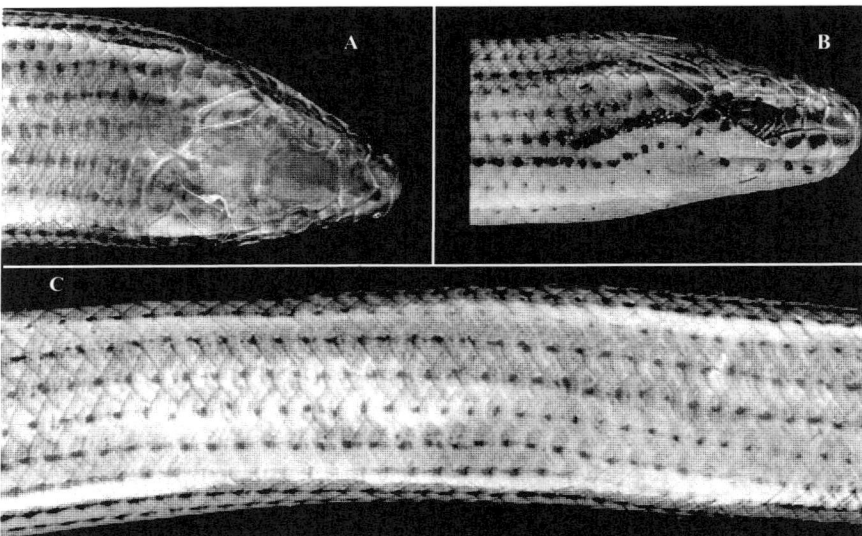

FIGURE 131. *Ophiomorus persicus*. (A) Dorsal view of head; (B) lateral view of head; (C) dorsal view of body. (From S. Anderson and Leviton, 1966, figs. 7e-f and 8c, respectively; photos by A. Leviton.)

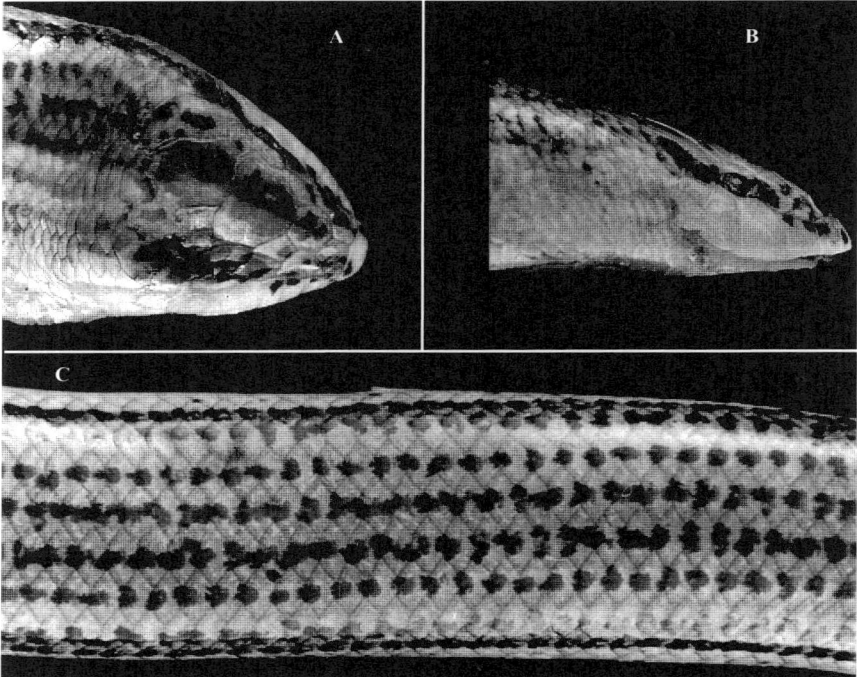

FIGURE 132. *Ophiomorus streeti*. (A) Dorsal view of head; (B) lateral view of head; (C) dorsal view of body. (From S. Anderson and Leviton, 1966, figs. 4a-b and 5a, respectively; photos by A. Leviton.)

Diagnosis: Fingers three; toes three; 20 scales round middle of body; prefrontals not in contact with upper labials.

Color pattern: (Fig. 132) Cream above in preservative, each scale of eight dorsal rows with dark brown spot, these forming eight longitudinal lines extending from posterior head shields to level of insertion of hind limbs; posteriorly six lines continue onto tail. Venter and sides immaculate white.

Size: Snout-vent length 91 mm.

Distribution: (see map, p. 283) Known only from the type locality, which lies within the enclosed internal drainage basin of the Jaz Murian depression of southern Baluchistan.

Remarks: Nothing is known of the biology of this form. Particularly noteworthy is the fact that specimens of *Ophiomorus brevipes* were collected in the same general area at the same time, late November to early December by the Street Expedition to Iran 1962–63.

Ophiomorus tridactylus (Blyth, 1853) (Plate 18F–G)
Three-toed sand skink

Sphenocephalus tridactylus Blyth, 1853:654 (Type locality: Afghanistan; Syntypes: ZSI 2526–2529, 2531–2532 [*fide* Das, *et al.*, 1998]). — Blanford, 1876:395. — Bedriaga, 1879:28.
Ophiomorus tridactylus: Boulenger, 1887a:394–395 (in part); 1887b:520–523 (in part); 1890a:222, text-fig. 59 (in part) — Nikolsky, 1899b:402. — Zarudny, 1903:25–27. — Annandale, 1906:197. — Smith, 1935:346, text-fig. 78 (in part). — F. Werner, 1936:201. — S. Anderson, 1963: 476. — S. Anderson and Leviton, 1966b:517–519, text-figs. 4c, 4d, 5b. — S. Anderson, 1968:333; 1974: 37, 44. — Schleich, 1977:127, 129. — Welch, 1983:25. — Welch, *et al.*, 1990:90.

Diagnosis: Fingers three; toes three; 22 scales around middle of body; parietal in contact with anterior temporal; prefrontals in contact with upper labials.

Color pattern: (Fig. 133) Cream or pale brown in preservative, sandy-beige in life, uniform, or with dorsolateral brown line on either side, from nostril through eye, on body and extending onto tail, composed of more-or-less confluent dots on one or two rows of scales; occasionally a few scattered dots on dorsum; brown dots arranged in lines on dorsal surface of tail. Some Iranian specimens with six more or less distinct longitudinal dark lines. Dorsal surface of hind limbs with brown dots. A few dark marks on head shields present or absent. Venter immaculate cream.

Size: Males examined measure up to 91 mm snout-vent, females to 95 mm.

Natural history: A specimen collected in late November contains yolked ovarian follicles.

Distribution: The sandy areas of the Helmand Basin and adjacent regions of Afghanistan, Iran, and northern Baluchistan, Pakistan.

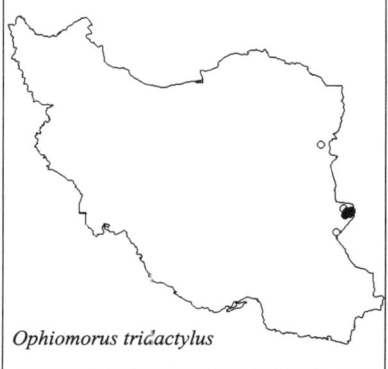
Ophiomorus tridactylus

Remarks: Apart from the fact that they are "sand-swimmers" in active dune areas, little has been recorded about the biology of this skink. Minton's (1966:106) remarks on the natural history refer to *Ophiomorus raithmai* S. Anderson and Leviton.

Local residents of Sistan refer to this species as "mar-rig," literally "sand-snake." They seemed to have no fear of them, but considered *Eryx tataricus*, found in the same locality, to be dangerously venomous.

Specimens labelled as from "Zabol" have six indistinct dark dorsal lines composed of dots on indi-

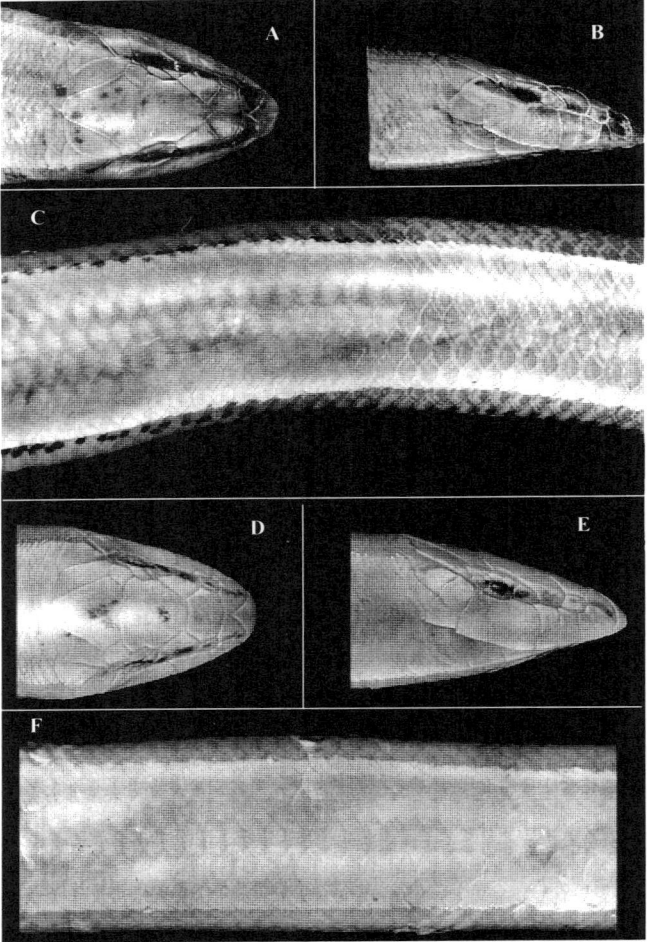

FIGURE 133. *Ophiomorus tridactylus*, from Seistan Basin, Iran: (A) Dorsal view of head; (B) lateral view of head; (C) dorsal view of body; from Punjab, India: (D) dorsal view of head; (E) lateral view of head; (F) dorsal view of body. (From S. Anderson and Leviton, 1966, figs. 4c-d and 5b and figs. 6a-c, respectively; photos by A. Leviton.)

vidual scales of the paravertebral rows, the adjacent row, and fourth row on either side of vertebral line; more distinct and less broken up in some specimens than in others. There are dark dots on the hind limbs. However, specimens I collected east of Zabol in Sistan had only a faint lateral stripe from nostril onto tail. S. Anderson and Leviton (1966b:518–519) commented on specimens without precise locality data from "Baluchistan" and "Punjab" which were heavier bodied than Afghan and Iranian specimens, and lacked the characteristic dorsolateral dark lines on body and tail (Fig. 133D–F). Since that time, the California Academy of Sciences received, through the courtesy of Jeromie A. Anderson, similar specimens from Nushki, near the Afghan border in Pakistan. These specimens were a beautiful saffron-yellow in life and lived in the laboratory for several years in a gallon jar filled with fine sand, feeding exclusively on small meal worms (*Tenebrio*). This, again, raises the question of the true identity of Blyth's type, since no mention of dorsolateral stripes was made in the original description. I am of the opinion that at least two distinct taxa are still included under the name *O. tridactylus*.

Genus *Scincus* Laurenti, 1768
Skinks, sandfishes

Scincus Laurenti, 1768:55 (not Gronovius, 1763[1]; type species: *Scincus officinalis* Laurenti, 1768 [= *Lacerta scincus* Linnaeus, 1758, by reason of the fact that they are objective synonyms, *S. officinalis* having been proposed to avoid tautonomy in the nominal genus *Scincus*], by subsequent designation of Fitzinger, 1826:23).

Definition: Palatine bones not meeting on midline of palate; pterygoids toothed; single premaxillary prolonged forward into a sharp rostrum; postorbitals and prefrontals completely fused. Maxillary teeth conical or with obtuse tubercular crowns. Eyelids well developed, scaly. Ear opening more or less completely hidden under scales. Snout depressed, cuneiform, mouth inferior; nostril between an upper and lower nasal. Supranasals, prefrontals, frontoparietals, and interparietal distinct. Limbs well developed, pentadactyl; digits flattened, fringed laterally, with transverse lamellae inferiorly; tail shorter than body.

Distribution: North Africa, the Arabian Peninsula, southern Israel, Jordan, Iraq, and southwestern Iran. Four species currently recognized, only one of which occurs in Iran. This is a strictly Saharo-Sindian genus with its center of diversity in the Arabian peninsula. Interestingly, it is complementary in its distribution with that of *Ophiomorus*, a genus primarily Iranian in its distribution. The two genera do not overlap, as far as known (Murray's *Scincus arenarius*, supposedly from the Hab River area of Sind, though searched for by several experienced collectors, has never been rediscovered, although there are recent rumors of such a skink collected by Jogis, a caste of professional animal collectors, according to J. A. Anderson (pers. commun.). Arnold and Leviton (1977:224, 229–30) placed it in synonymy with *S. mitranus* and provided a convincing explanation of its Arabian origin.

Scincus scincus (Linnaeus, 1758)
Common skink, sandfish

Lacerta stincus (*lapsus calami* for *scincus*) Linnaeus, 1758:205.
Scincus stincus: Flower, 1933:788.
Scincus scincus: El-Toubi, 1938:5 — K. P. Schmidt, 1939:69.

Scincus scincus conirostris Blanford, 1881 (Plate 18H)
Iranian skink, Blanford's sandfish

Scincus conirostris Blanford, 1881:677, fig. 1 (Type locality: Tangyak, 7 mi [11 km] south[east] of Bushire, Iran; Syntypes: BMNH 79.8.15.1–3/1946.8.20.55–57). — Boulenger, 1887a:391. — Werner, 1936:201. — S. Anderson, 1963:462–463; 1968:333; 1974:37, 44. — Schleich, 1977:127, 129.
Scincus scincus conirostris: Khalaf, 1960:15. — Arnold and Leviton, 1977:221–224, pl. 2a, pl. 3b. — Welch, 1983:26. — Leviton, *et al.*, 1992:74, col. pl. 11D–F.

Diagnosis: External ear orifice relatively large, its upper margin reaching continuation of line made by lower edges of upper labial scales and typically covered by two serrated scales. Six supraoculars, anterior not in contact with frontal; supranasals in contact behind

---

[1] Gronovius' work, *Zoophylacium Gronovianum*, published in three volumes between 1763–1781, in which the name *Scincus* first appears, has been rejected for purposes of zoological nomenclature by the International Commission on Zoological Nomenclature (Opinion 261 [published 10 August 1954] in the *Opinions and Declarations* of the Commission [see also Hemming and Noakes, 1958, *Official Index of Rejected and Invalid Works in Zoological Nomenclature*, p. 31]).

rostral; frontoparietals usually in extensive contact; parietals 1.5–2.5 times length of adjoining nuchals; rostrals and loreals usually separated from frontonasal (Fig. 134); 26–28 scale rows at midbody. No dark vertical bars or spots on flanks.

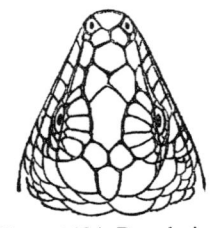

FIGURE 134. Dorsal view of head of *Scincus scincus conirostris*. (From Blanford, 1881, fig. 1.)

Color pattern: Light brown or cream above, each dorsal scale edged with light brown, dark spot near posterior margin of each scale, white mark on either side of dark marking. In life there are (at least in adults) 9–10 vertical orange-yellow to brick-red bars on sides of body, extending onto dorsum, these fading completely in alcohol.

Size: Males to 112 mm snout-vent, females to 98 mm (Arnold and Leviton, 1977:223). The largest specimen I examined (BMNH 79.8.15/1946.8.20.55, male) measures 93 mm snout-vent, tail 57 mm.

Habitat: (Plate 21F) Like the other members of the genus, this species inhabits dunes of windblown sand. They are able to run with some alacrity over the dune surface, the tail leaving a broad, undulating track in the sand. They burrow very rapidly into the loose sand, their wedge-shaped snout and short forelimbs, which are folded back along the sides of the body when the animal burrows into the sand, particularly adapting them to this mode of progression. Although measurements of body temperature during normal activity were not taken, they appear to have a high heat tolerance, the critical thermal maximum being around 50°C. (S. Anderson, 1963:463). Unlike many of the lizard species that share this habitat and are active early in the morning, *Scincus* becomes active in the late afternoons, when sand surface temperatures are still high.

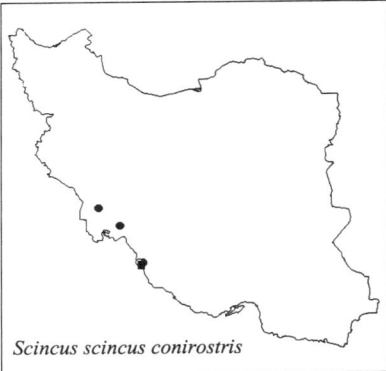

Scincus scincus conirostris

Distribution: Low sandy deserts of southeastern Iraq and Kuwait, Persian Gulf coast of northern Saudi arabia, eastern and central Arabia south to Hadramaut; southern Yemen, Bahrain Island, eastern United Arab Emirates, and Khuzestan and coastal Fars Provinces of southwestern Iran.

Remarks: Aside from the above observations, nothing seems to be known about the biology of this form, which appears to be most closely related to *Scincus s. meccensis* Wiegmann, which occurs in the western Arabian Peninsula. Arnold and Leviton (1977:221) placed *S. gasperettii* Haas in synonymy with *S. s. conirostris*.

Two specimens from Iraq appear somewhat darker in color than the Iranian specimens examined, due to the fact that the brown spots on the scales of the former are somewhat larger, possibly due to conditions of preservation.

In this genus, as in other genera inhabiting small areas of windblown sand, the characters which separate the recognized taxa are often slight and may represent minor genetic differences. That these characters exhibit constancy may be due to the small size of the populations living in these geographically separated areas. Reestablishment of contact between populations is dependent upon fortuitous movement of the dunes, hence our current recognition in nomenclature of individual populations may have brief temporal significance and reflect little of the biological realities of the relationships of these populations.

Family UROMASTYCIDAE[2]

Spiny-tailed lizards and granular-scaled lizards

Moody (1980:142, 174) recommended recognition of the family Uromastycidae encompassing the genera *Uromastyx* and *Leiolepis*. In his subsequent study of *Uromastyx* (Moody, 1987), he stated that *Uromastyx* and *Leiolepis* form a monophyletic group that is a sister group to all other agamid genera (see also Frost and Etheridge [1989] and de Queiroz and Gauthier [1994]).

Genus *Uromastyx* Merrem 1820
Spiny-tailed lizards, thorny-tailed lizards

Uromastyx Merrem, 1820:56 (Type species: *Stellio spinipes* Daudin, 1802 [= *Uromastyx aeqyptius* (Forsskål, 1775)], by subsequent designation of Fitzinger, 1843:18).

Definition: Body depressed, without crest; dorsal scales small, uniform, or intermixed with larger ones; no gular sac; a transverse gular fold; tympanum distinct. Tail short, depressed, covered with whorls of spinous scales; preanal and femoral pores present in most species. Upper central incisors replaced in adult by downward projection of premaxilla, forming large cutting edge with lateral incisors. Synapomorphies: angular foramen absent; rounded vertebral centra; posterior coracoid fenestra (Moody, 1980:159).

Distribution: From northwestern India throughout southwestern Asia and the Arabian Peninsula to Saharan North Africa and arid regions of Ethiopia and Somalia. Moody (1987) recognizes 15 species. Three species in Iran.

Remarks: Welch, *et al.* (1990:60) include Iran in the distribution of *Uromastyx hardwickii* but cite no source, and this is clearly in error.

Key to the Species of *Uromastyx* in Iran

1a. Whorls of spinous scales on upper surface of tail not separated by small scales; back without transverse rows of enlarged spinous tubercles ....
.... *Uromastyx aegyptius* Forsskål, 1775 (p. 290)
1b. Whorls of spinous scales on upper surface of tail separated by small scales; back with more or less regular transverse rows of enlarged spinous tubercles 2
2a. 9–15 femoral and preanal pores on each side; 7–10 tubercles across base of tail; enlarged denticulate scales along anterior border of ear; 20–25 transverse rows of scales on middle of belly, within a space equal to length of head (tip of snout to angle of jaw) .... *Uromastyx asmussi* (Strauch, 1863) (p. 291)
2b. 15 or more femoral and preanal pores on each side; 12 tubercles across base of tail; no enlarged denticulate scales along anterior border of ear; 30–40 transverse rows of scales on middle of belly, within a space equal to length of head ....
.... *Uromastyx loricatus* (Blanford, 1875) (p. 293)

---

[2] It has been suggested (Frost and Etheridge, 1989:34; Ananjeva and Dujsebayeva, 1997:46–49) that Fitzinger's 1843 family group name "Fam. Leiolepides" is a senior family-group synonym for the family group referred to here as Uromastycidae Theobald (1863). These authors neglected to note that under Section IV, Criteria of Availability, Article 11e(iii) of ICZN (1964:11), a "family-group name published before 1900 . . . but not fully latinized, is available with its original date and authorship, *provided that it has been latinized by later authors and that it has been generally accepted by zoologists interested in the group concerned as dating from its first publication in vernacular form* (italics mine)." To the best of my knowledge, neither Leiolepidae nor Leiolepinae, as either a family- or subfamily-group name has been used before, while the family-group name Uromastycidae has appeared in print, although not frequently. Nonetheless, under the rules, Uromastycidae remains the available and valid group name for those who choose to recognize the family group as a valid phylogenetic unit or clade. See de Queiroz and Gauthier (1994:29) for a nomenclatural proposal whereby, in phylogenetic nomenclature, Chamaeleonidae Rafinesque (1815) would not have priority over Agamidae Spix (1825) and Chamaeleonidae, Agamidae, and Leiolepidae (or Uromastycidae) would be nested within the Agamidae.

Uromastyx aegyptius (Forsskål, 1775) (Plate 19A)
Egyptian spiny–tailed lizard, thub

Lacerta aegyptia Forsskål, 1775:13 (Type locality: Egypt; Holotype: not located).
Uromastix aegyptius: J. Anderson, 1896:79.
Uromastyx aegyptius: Mertens, 1962:425. — Joger, 1986:187–192. — Moody, 1987:285–288.
Uromastix microlepis Blanford, 1875a:658–660 (Type locality: vicinity of Basrah, Iraq; Syntypes: BMNH 74.8.11.1/1946.8.14.55, 74.8.11.1/1946.8.11.67); 1876:334–337. — Boulenger 1885a: 407. — F. Werner, 1917:200; 1936:200. — S. Anderson, 1963:475. — Schleich 1977:127, 129.
Uromastyx microlepis: Wermuth, 1967:103–104. — S. Anderson, 1968:332; 1974:36, 42. — Welch, 1983:20.
Uromastyx aegyptius microlepis: Arnold, 1986:416. — Leviton, *et al.*, 1992:23–24, col. pl. 3E–F.

Diagnosis: Whorls of spinous scales on upper surface of tail not separated by small scales; scales on body minute, more than 300 around midbody, no enlarged scales on back or flanks; two or more transverse rows of scales on lower surface of tail correspond to one on upper surface; fourth toe with lateral fringe; preanal and femoral pores present.

Color pattern: (Fig. 135) Dorsum olive gray (in alcohol), with small, rather indistinct darker spots on back; venter dirty white.

Size: Snout–vent length to 320 mm, tail 220. The smallest juvenile examined was collected in July or August and measures 78 mm snout-vent, tail 54 mm.

FIGURE 135. *Uromastyx aegyptius*. (From J. Anderson, 1898, pl. 14.)

Habitat: This species has not been studied in Iran. John Anderson (1898:130) described the habitat in Egypt; there it occurs along the lines of drainage in the desert, where it finds the sparse vegetation on which it feeds. These drainages include the deep wadis characteristic of that region. It digs burrows to a depth of over a meter in hard-packed sand. Anderson's description is in accord with the observations of Gallagher and Arnold (1988:409), who mentioned "typical, colonial burrows" on firm ground in Oman.

Distribution: North Africa (from Algeria through the Sahara to northern Egypt); southwest Asia (Israel, Sinai Peninsula, northern Saudi Arabia and along the Persian Gulf

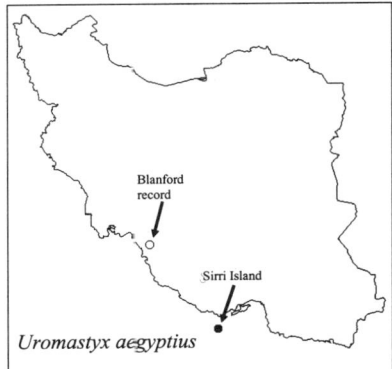

coast, Iraq, Kuwait, southwestern Iran). An early Blanford record for this species in Iran has recently been supplemented by its discovery on the Iranian island of Sirri in the Persian Gulf (Plate 19A).

Remarks: This species probably occurs along the gulf coast of Khuzestan and Fars at least as far as Bushire, inhabiting sandy stretches, while *Uromastyx loricatus* occupies the silty plains and foothills.

Mertens (1956:93) felt that *U. microlepis* probably should be considered a subspecies of *U. aegyptius*, and Schmidt (1939;59) also commented that distinguishing between the two forms by means of presence or absence of enlarged ventrolateral tubercles is difficult when the tubercles are reduced or few. Moody (1987:287) considers them synonymous, and states that the only character used to diagnose *aegyptius* — spiny scales on flanks — is variable throughout the ranges of both nominal forms. He was able to discover no other diagnostic characters to separate the two. Joger (1986:190) was unable to distinguish between them on the basis of immunological data.

In none of the specimens I have examined from gulf coastal Arabia are there any tubercles on the flanks. There are 41–61 scales along the midline of the belly in a space equal to the length of the head. Two syntypes of *Uromastyx microlepis* (BMNH 74.8.11.1/1946.8.11.67 and 74.8.11.1/1946.8.14.55) from Basrah, Iraq have 39 and 42 scales along midline of belly in the length of the head. A specimen from "Arabia " labelled "*U. spinipes*" (AMNH 44592) has 39 and also has scattered tubercles on the flanks. The Basrah syntypes have 13–19 femoral pores on each side; the Arabian specimens have 17–20 femoral and preanal pores on each side.

Moody (1987:287–288) states that *Uromastyx aegyptius* is the sister species to the *acanthinurus* group (which includes *U. geyri* and *dispar*) of North Africa. Joger (1986:191) places it as a sister species to a clade which includes *U. ornatus* and *U. ocellatus*.

Uromastyx asmussi (Strauch, 1863) (Plate 19B–C)
Iranian spiny-tailed lizard

Centrotrachelus asmussi Strauch, 1863:479 (Type locality: Sar-i-Tschah [= Sar Chah], Iran; Holotype: originally in Dorpater Zoologischen Museum). — Blanford, 1876:337–340, pl. 21. — Bedriaga, 1879:38–39.
Uromastix asmussi: Boulenger, 1885a:409. — Nikolsky, 1897:325. — Zarudny, 1903:15. — F. Werner, 1936:197. — Forcart, 1950:147. — S. Anderson, 1963:475. — Schleich, 1977:127, 129.
Uromastyx asmussi: Mertens, 1956:93. — Wermuth, 1967:102. — S. Anderson, 1968:332; 1974:36, 42. — Nilson and Andrén, 1981:136–137. — Welch, 1983:20. — Moody, 1987:285–288. — Welch, et al., 1990:60.

Diagnosis: Whorls of spinous scales on upper surface of tail separated by small scales; 20 transverse rows of scales on middle of belly, on space corresponding to length of head; back with transverse series of large pointed tubercles; fourth toe with distinct lateral fringe; enlarged denticulate scales along anterior border of ear; 2–3 preanal and 6–8 femoral pores on each side; 7–8 enlarged spinous scales on widest whorl of tail; series of tuberculate scales on nape and occiput surrounded by tiny subgranular scales. (Moody, 1987:285).

Color pattern: (Fig. 136) Back (in alcohol) buff or pale brown with greenish tinge;

FIGURE 136. *Uromastyx asmussi*. (From Blanford, 1876, pl. 21.)

upper surface of head, limbs, and tail much darker; venter yellowish brown, heavily mottled with dark olive, except on belly (Smith, 1935:247–248). Males have dorsal tubercles scarlet in life (perhaps seasonally; most species of *Uromastyx* capable of considerable color change). Female uniform dirty brownish (Blanford, 1876:339–340).

Size: Snout-vent length 260 mm, tail 210.

Natural history: Known food items include leaves, seeds, and stems of herbaceous plants (Blanford, 1876:340).

Habitat: Inhabits stony ground and gravelly plains where there are scattered patches of low brush, chiefly barilla and tamarisk. It lives in large burrows, which it excavates itself. (Blanford, 1876:340).

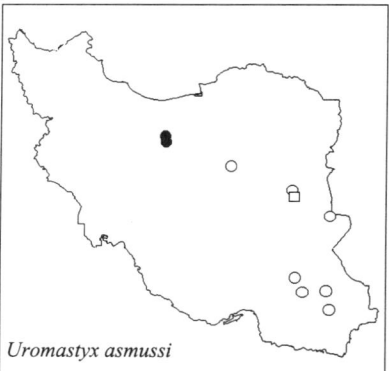

Uromastyx asmussi

Distribution: Eastern portion of Central Plateau of Iran and adjacent areas of Pakistan and southern Afghanistan; 550–1200 m elevation on Iranian Plateau.

Remarks: The affinities of this species are with *Uromastyx loricatus*, a morphologically similar species of the Mesopotamian Plain and western Zagros foothills. Mertens (1956:93) has suggested that the two forms may be only subspecifically distinct. Moody (1987:286) considered them distinct species, based on an analysis of morphological characters. Their present ranges are separated by the interruption of the coastal lowlands by the mountains in the vicinity of Bandar-e Lingeh. Immunological evidence discussed by Joger (1986) suggests that the two taxa form a natural group with *U. hardwickii*, the easternmost form of the genus, but Moody (1987:286) takes issue with this, based on his morphological study, in which he finds *U. hardwickii* to be the most primitive species in the genus, a sister clade to all other species groups (see Moody's [1987:288] cladogram).

The specimens I have examined have about 24 scales along the midline of the belly in a space equal to the length of the head (tip of snout to angle of jaw), 10–11 femoral pores and seven tubercles across base of tail. Contrary to Blanford's statement, I find the keels of the plantar scales lie along the axis of the scale.

Uromastyx loricatus (Blanford, 1875) (Plate 19D)
Mesopotamian spiny-tailed lizard, small-scaled spiny-tailed lizard

Centrotrachelus loricatus Blanford, 1875a:660–661 (Type locality: Bushire, Iran; Holotype: BMNH 1946.8.11.59 [?]); 1876:340–342. — Bedriaga, 1879:39. — Blanford, 1881:677.
Centrotrachelus asmussi: Murray, 1884b:101 (in error).
Uromastix loricatus: Boulenger, 1885a:409–410, pl. 32. — F. Werner, 1917:200–201; 1936:200. — S. Anderson, 1963:453–455.
Uromastyx loricatus: Clark, *et al.*, 1966:6. — Wermuth, 1967:103. — S. Anderson, 1968:332. — Haas and Y. Werner, 1969:341. — Tuck, 1971a:55. — Welch, 1983:20. — Joger, 1986:187–191. — Moody, 1987:285–288.
Uromastyx loricata: Frynta, *et al.*, 1997:8.

Diagnosis: Whorls of spinous scales on upper surface of tail separated by small scales; 30–40 transverse rows of scales on middle of belly in space equal to length of head; back with transverse series of large pointed tubercles; no enlarged denticulate scales along ear margin; 3–5 preanal and 11–15 femoral pores on each side; series of enlarged scales on nape not surrounded by subgranular scales.

Color pattern: (Fig. 137) Back (in alcohol) yellowish gray or cream, with small brown spots; sometimes series of round, yellow, brown-margined ocelli; venter yellowish white. In life, dorsal coloration changes from dark slate gray in cool morning hours to nearly white at midday, with large areas of bright orange across dorsum.

Size: Snout-vent length to 252 mm, tail 190 in males; largest female examined measures 233 mm snout-vent, tail 189.

Natural history: These animals were seen from mid-April through mid-August in Khuzestan, usually at midday when no other reptiles were active during the summer heat. They were most often seen basking on mounds of excavated soil at the mouths of their burrows. Easily alarmed, they quickly retreated into burrows when approached, but they were sometimes seen foraging some 25 m from the burrow. They defend themselves by

FIGURE 137. *Uromastyx loricatus*. (From Boulenger, 1885, pl. 32.)

whipping the heavy spiny tail from side to side with rapid, violent strokes. Their burrows often exceed one and a half meter in length, more than 30 cm below the ground surface at their terminus, and usually having two or more sharp bends. Observations on temperature relationships are to be found in an earlier paper (S. Anderson, 1963:454–455).

Females collected in January, June, and July have small eggs (up to 4 mm diameter in January) in the ovaries. The smallest juvenile examined, collected in January, measures 57 mm snout-vent, tail 40 mm. Fecal pellets and digestive tracts examined contained only plant material.

Habitat: (Plates 21E, 22A–C) In the foothills of Khuzestan these large lizards live in the plains and valleys, where they excavate their burrows in hard-packed silty alluvium. I found one individual basking near its burrow on the platform of the ziggurat at Choga Zanbil. East of Agha Jari I saw them sitting on small rocks and rock piles from which they could quickly reach their burrows. According to Major O. St. John (Blanford, 1876:661), near Bushire they dig their burrows in areas of sandy clay, where there are small bushes such as wormwood or barilla.

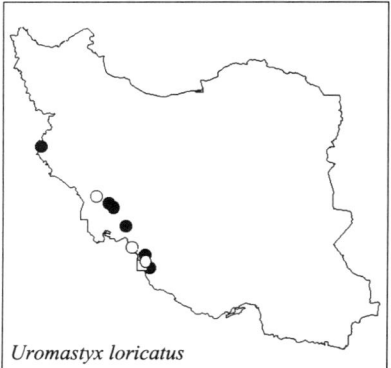

Uromastyx loricatus

Distribution: Apparently confined to the Mesopotamian Plain and Zagros foothills of Iraq and southwestern Iran, and the gulf coastal plain of southwestern Iran. Sea level to about 300 m.

Remarks: In all specimens examined, there are more than 15 femoral and preanal pores (usually 16 or 17) on each side; *Uromastyx asmussi* has 9–15. There are 12 tubercles across base of tail (7–10 in *U. asmussi*). Adults usually have about 30 scales along midline of belly in a space equal to length of head, while small juveniles with proportionally larger heads have about 40. There is no trace of a lens-like structure in the scale over the pineal foramen as there is in *U. aegyptius*.

The holotype, probably a female, had apparently rotted before preservation. It is eviscerated, the eyes are gone and there is other damage to the head. Scale characters of the body are intact. There is no indication of femoral pores; there are 10 tubercles across base of tail, 11 in second row, 12 in third row; 33 transverse scale rows in middle of belly in one head length. Snout-vent length 235 mm, tail 177 mm.

Family VARANIDAE

Genus *Varanus* Merrem
Monitors

Varanus Merrem, 1820:58 (Type species: *Lacerta varia* White, 1790, by subsequent designation of Gray 1827:55 [see note in Zhao and Adler, 1993:200]).

Definition: Postorbital arch incomplete; temporal arch complete; supratemporal fossa not roofed over; premaxillary single, narrow; nasals narrow, paired or united; infraorbital vacuity bounded by pterygoid, palatine, and ectopterygoid bones, maxillary excluded; epipterygoid present. Teeth dilated at base, pleurodont; palatine and pterygoids toothless. Clavicle slender, not dilated, interclavicle anchor-shaped. Tongue smooth, very long, bifid, retractile into sheath at base. Pupil round; eyelids well developed. Head covered with small, polygonal juxtaposed scales; back with roundish or oval scales, surrounded by rings of granules, except in very young; ventral scales quadrangular, in transverse series. Tail long, not fragile. Preanal pores usually present, a single pair just in front of vent. (Smith 1935; Mertens 1942).

Distribution: Africa, southern Asia, East Indies, Australia, and islands of the western Pacific. About 30 living species, two in Iran.

Remarks: The systematics, morphology, and ecology of the genus are discussed in detail by Mertens (1942) in his large monograph of the family. More recently, the intrageneric relationships have again become the subject of study, using molecular techniques as well as reexamination of morphology (see, for example, Ast, 1997; King, *et al.*, 1997; Kluge, 1997).

Key to the Species of *Varanus* in Iran

1a. Tail compressed throughout its length, with low double-toothed crest above; abdominal scales in 88–110 transverse series from collar fold to groin .... *Varanus bengalensis bengalensis* (Daudin, 1802) (p. 295)
1b. Tail round in cross-section, or slightly compressed posteriorly, without double-toothed crest above; abdominal scales in 110–125 transverse series from collar fold to groin....2
2a. Tail round in cross-section throughout its length; back with five to eight (usually six) gray bars in addition to one or two nuchal crossbars, pattern becoming indistinct in older animals, pattern of dots predominating; tail patterned nearly to tip with 19–28 dark crossbars .... *Varanus griseus griseus* (Daudin, 1803) (p. 297)
2b. Posterior half of tail narrow in cross-section, compressed, distinct keel above; back with five to eight (usually six) sepia bars in addition to nuchal crossbar; tail with 13–19 dark crossbars, end of tail light in color, without pattern .... *Varanus griseus caspius* (Eichwald, 1841) (p. 297)

Varanus bengalensis (Daudin, 1802)
Bengal monitor, Indian monitor

Varanus bengalensis bengalensis (Daudin, 1802) (Plate 19E)
Bengal monitor, Indian monitor

Tupinambis bengalensis Daudin, 1802:3:67 (Type locality: Bengal; Holotype: MNHN 2179).
Varanus bengalensis: Duméril and Bibron, 1836:480. — Nikolsky, 1899:394. — Zarudny, 1903:15–

17. — F. Werner, 1936:200. — Wettstein, 1951:438. — S. Anderson, 1968:333. — Schleich, 1977:127, 129.
Varanus dracaena: Blanford, 1876:360–361. — Bedriaga, 1897:41.
Varanus monitor: Smith, 1935:402–403, figs. 1–2, p. 399 (not *Lacerta monitor* Linnaeus [based on Seba's plates, which are of *V. salvator* and *V. niloticus, fide* Mertens 1942]). — S. Anderson, 1963:475.
Varanus (Indovaranus) bengalensis bengalensis: Mertens, 1942:13; 1959:225–226; 1963:5. — S. Anderson, 1974:34. — Welch, 1983:59. — Welch, *et al.*, 1990:123.

Diagnosis: Supraoculars not enlarged; abdominal scales in 88–110 transverse series; tail compressed throughout its length, with low, double–toothed crest above.

Color pattern: Young dark olive or dull orange to light brown in life, with numerous light spots or ocelli more or less transversely arranged to form 10–13 narrow crossbars, and often alternating with dark spots or bars, rarely with dark bars only; top of head with light spots; more or less distinct dark temporal streak; venter whitish, with narrow dark transverse bars, which may be broken up into spots. Adult brownish or olive above, usually with blackish dots; venter yellowish, uniform or mottled or spotted with black, or with scattered light and dark scales; spots most numerous on throat; tip of tail whitish. (Smith 1935:402–403; Minton 1966:113). Auffenberg (1994:39–47) describes color variation throughout the distributional range.

Size: Snout–vent to 750 mm, tail 1,000 mm (Smith 1935:403); most specimens much smaller, however, Minton (1966:113) recording a maximum snout-vent length of 380 mm for an adult female from Pakistan.

Natural history: Nothing has been published on the natural history of this species in Iran. Minton (1966:113) summarizes information on diet and reproduction in Pakistan. See also Auffenberg (1979, 1981, 1983, 1994), Auffenberg, *et al.* (1991), Jacob and Ramaswami (1976) for additional extralimital studies, and Auffenberg (1994) for a thorough coverage of its biology.

Habitat: Most frequently found in areas where some water is available, but does occur in all but the most arid habitats of the desert to the southeast of the Iranian Plateau. Minton (1966:113) discusses its habitat in Pakistan; Clark, *et al.* (1969:311) remark on the habitat in eastern Afghanistan. A live specimen from Afghanistan maintained in the herpetology laboratory of the California Academy of Sciences appeared to suffer from water loss if kept a few days without water applied to the skin. Any water poured over the skin quickly disappeared, but whether the water was absorbed by the skin, as it appeared to be, or was carried by capillarity to the mouth, was not determined. This animal also drank water in captivity. Blanford (1876:360–361) stated that in his experience this species was a thoroughly terrestrial lizard, living in dry places far from water. His single Iranian specimen was from a region where the only water occurs in small streams, which are dry except in a few pools. I saw only one monitor in Iran which I identified as *Varanus bengalensis*. This adult was on a road where it crossed a small stream near the village of Dahaneh Nais, 37 km by road in a northerly direction from Rudan [23°26'N, 57°12'E] on the road to Jiroft in southern Kerman (Plate 20F). This is an area of date cultivation where water is permanently available. *Bufo surdus* and *Euphlyctis cyanophlyctis* were collected at this local-

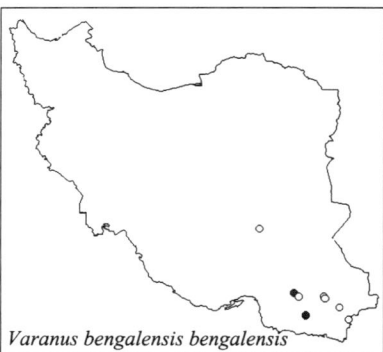
Varanus bengalensis bengalensis

ity. The monitor escaped and was not positively identified. See Auffenberg (1994:74–75) for a discussion of hypothetical habitats in Iran and (pp. 71–137) for habitats throughout the range.

Distribution: From about 35°N (Afghanistan) to 7°S (Java), about 57°E (Iran) to 113°E (Java), including all countries within these extremes (Auffenberg, 1994:71). In Iran it occurs only in southern Kerman and Baluchistan provinces; however, there is a record for Zizi, at Vik River (ZIL 12354) (? a tributary of Rud-e Zahreh, about 30 km W Gach, Saran) (see Auffenberg, 1994:72–75, for an extended discussion of distribution in Iran).

Remarks: Auffenberg's (1994) monograph of the Bengal monitor is one of the most exhaustive studies of all aspects of the biology of a single lizard species yet published.

Varanus griseus (Daudin, 1803)
Desert monitor

Tupinambis griseus Daudin, 1803:8:352 (Type locality: Egypt; Type[s]: whereabouts unknown).
Monitor griseus: W. Peters, 1870:109.
Varanus (Psammosaurus) arenarius: Bedriaga, 1879:40.
Psammosaurus scincus: Blanford, 1881:677.
Varanus griseus: Boulenger, 1885b:306–307. — Nikolsky, 1897:325– 326. — Zarudny, 1897:355; 1903:17. — Nikolsky, 1905:89. — Annandale, 1906:197. — Nikolsky, 1915:259–262. — Morich, 1929:30–31. — F. Werner, 1936:197. — Mertens, 1942:338–347. — Terentjev and Chernov, 1949:161–162. — S. Anderson, 1963:455–456; 1968:333. — Tuck, 1971a:60. — Schleich, 1977:127–129.
Varanus griseus caspius: Mertens, 1956:94; 1957:231; 1963:11. — S. Anderson, 1974:34. — Nilson and Andrén, 1981:137, fig. 6. — Welch, 1983:59. — Welch, et al., 1990:124.

Diagnosis: Nostril oblique slit, much nearer orbit than end of snout; tail circular in cross–section, or slightly compressed posteriorly; abdominal scales in 110–125 transverse rows.

Varanus griseus griseus (Daudin, 1803) (Plate 19F)
Desert monitor

Tupinambis griseus Daudin, 1803:8:352 (Type locality: Egypt).
Varanus (Psammosaurus) griseus griseus: Mertens, 1954:354, pl. 33, fig. 1. — Leviton, et al., 1992:76, col. pl. 12A–B.
Varanus griseus griseus: S. Anderson, 1974:34.

Diagnosis: Tail circular in cross–section throughout its length.
Color pattern: Back with five to eight (usually six) narrow gray bars in addition to one or two nuchal crossbars; pattern becomes indistinct in older animals, pattern of dots predominating. Tail patterned nearly to tip with 19–28 dark crossbars. (Mertens, 1954:354).
Size: To 1.5 m total length (Mertens, 1942:344).

Varanus griseus caspius Eichwald, 1831 (Plate 19G–H)
Transcaspian desert monitor

Psammosaurus caspius Eichwald, 1831:190 (Type locality: Dardscha Peninsula, east coast of Caspian Sea [restricted by Mertens 1959:231]; Type[s]: whereabouts unknown).
Psammosaurus caspius: Eichwald, 1841:48, pls. 7–9. — Blanford, 1876:359–360.
Varanus arenarius: De Filippi, 1865:352.
Varanus (Psammosaurus) griseus caspius: Mertens, 1954:355, pl. 33, fig. 3.

Diagnosis: Anterior half of tail more or less circular in cross-section, posterior half narrow in cross-section, compressed, a distinct keel above.

Color pattern: Back with five to eight (usually six) narrow sepia bars in addition to nuchal crossbars; tail with 13–19 dark crossbars, end of tail (usually posterior third) without pattern in adults (young specimens may have tail barred throughout length).

Size: Snout–vent length to 600 mm.

Natural history: In Khuzestan, the characteristic track of the desert monitor (continuous mark of tail between large footprints) is to be seen in the morning on sand dunes. During the hottest months of the year they forage in the early daylight hours, hunting systematically over several hundred meters, entering one burrow after another in search of rodents, reptiles, and large arthropods. Nothing has been published on the biology of these lizards in Iran; Terentjev and Chernov (1949:161–162) summarize information about reproduction and diet in the Central Asian republics. Tsellarius and coworkers (1991, 1994, 1995, 1996, 1997) continue field studies of this species in Central Asia.

Habitat: (Plate 21F) In Khuzestan, I saw these monitors in sandy areas and on rocky slopes and hillsides; they use rodent burrows, small caves, and crevices as refuge and may forage over a fairly large area (at least more than half a kilometer from their burrows). They are often seen many kilometers from any water. Because they are relatively large foraging predators, their population densities are low relative to those of other lizards; however, their tolerance for a wide range of desert conditions accounts for their wide distribution through the warm deserts of the Palearctic. Opportunistic feeders independent of the need for drinking water, they occupy habitats which probably are not available to mammalian and avian predators. Whereas low temperatures may limit their distribution in the mountainous areas of Southwest Asia, their apparent absence from Mazanderan, Gilan, and Azarbaijan may be related to competition with other predators, rather than to temperature and humidity.

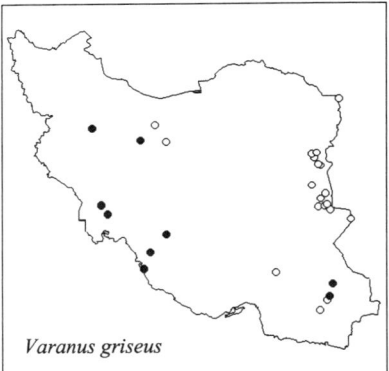

Varanus griseus

Distribution: The subspecies *Varanus griseus caspius* is found throughout the Plateau of Iran, from the Zagros Mountains in the west, eastward through Afghanistan and Baluchistan and north through the Central Asian republics. The systematic status of populations in Khuzestan, Lorestan, and Fars in Iran and perhaps in southeastern Anatolia is not yet clear (see below); *V. g. griseus* is the lowland subspecies ranging across the Arabian Desert, Iraq, Syria, Jordan, and North Africa (records for Lebanon are doubtful; see Martens and Kock, 1992:10). The subspecies to the east of the Plateau of Iran in Pakistan is *V. g. koniecznyi* Mertens.

Remarks: Two juvenile specimens from Khuzestan (CAS 86630–86631) agree with Mertens' (1954:354) diagnosis of the subspecies *griseus* in the color pattern of the tail, the crossbars continue throughout the length of the tail (22–23 dark caudal bars), but the tail possesses a distinct keel and is very slightly compressed. Presence of crossbars throughout length of tail is not diagnostic in any case, since a juvenile from central Iran also has very distinct crossbars the length of tail. These two Khuzestan specimens are predominantly dark above (in preservative) with transverse rows of light spots. There are five distinct crossbars on the snout. USNM 160302 is a large adult from east of Gach Saran; it lacks bars on the posterior half of tail and the tail is compressed (however, the specimen is emaciated). A hatchling from Khuzestan (UMMZ 133282) measuring 122 mm snout-vent, 167 mm tail, agrees with the diagnosis of *Varanus g. griseus*, and has 5 narrow distinct dark bars across the dorsum between the shoulder and pelvic girdles, 24 dark bars on the tail; the tail is only very slightly compressed toward the tip and lacks any keel. Of Arabian specimens examined, a very small juvenile from Shimal (CAS 84422) most closely agrees with the diagnosis and photograph of *V. griseus* (Mertens 1954: pl. 33, fig. 1), except that the crossbars are very dark and distinct. A subadult from Doha Dalum is intermediate between the Shimal and the Khuzestan lizards, with the seven dark crossbars being very distinct and contrasting sharply with the light spaces. The tail, while not compressed, has a slight dorsal keel. Preservation of color pattern in material from Abqaiq (CAS 84298, 84300–84302, 84437) is very poor, but the specimens appear to agree generally with Mertens' diagnosis of *V. griseus*. Blanford (1881:677) stated that the specimens from Ghainak and Konar Takhteh are indistinguishable from Egyptian specimens.

On biogeographical grounds, one would expect the monitors of Khuzestan, Lorestan, and Fars to resemble more closely those of the Arabian Peninsula (*Varanus g. griseus, fide* Mertens) than those of the Iranian Plateau (*V. g. caspius*). The available material does not clearly support this expectation, however. Too few adult specimens from southwestern Iran are available for the question of their status to be explored satisfactorily at present.

PLATES

Plates		Text pages
1	SATELLITE PHOTOGRAPH OF IRAN	42
2–5	FAMILY AGAMIDAE *Calotes*, plate 2; *Laudakia*, plate 2; *Phrynocephalus*, plates 3–4; *Trapelus*, plate 5	67–110
6	FAMILY ANGUIDAE *Anguis*, plate 6; *Ophisaurus*, plate 6	111–115
6	FAMILY EUBLEPHARIDAE *Eublepharis*, plate 6	116–124
6–10	FAMILY GEKKONIDAE *Agamura*, plate 6; *Assacus*, plate 6; *Bunopus*, plate 7; *Carinatogecko*, plate 7; *Crossobamon*, plate 7; *Cyrtopodion*, plates 7–8; *Hemidactylus*, plate 8; *Pristurus*, plates 8–9; *Stenodactylus*, plate 9; *Teratoscincus*, plates 9–10; *Tropiocolotes*, plate 10	125–198
11–16	FAMILY LACERTIDAE *Acanthodactylus*, plate 11; *Eremias*, plates 11–12; *Lacerta*, plates 12–15; *Mesalina*, plates 15–16; *Ophisops*, plate 16	199–261
16–18	FAMILY SCINCIDAE *Ablepharus*, plate 16; *Chalcides*, plate 17; *Eumeces*, plate 17; *Mabuya*, plates 17–18; *Ophiomorus*, plate 18; *Scincus*, plate 18	262–288
19	FAMILY UROMASTYCIDAE *Uromastyx*, plate 19	289–294
19	FAMILY VARANIDAE *Varanus*, plate 19	295–299
20–25	HABITATS Baluchistan-Sistan Province, plates 20–21; East Azarbaijan Province, plates 24–25; Fars Province, plates 21–22; Kerman Province, plate 20; Khorasan Province, plate 24; Khuzestan-Lorestan Province, plates 21–22; Mazanderan Province, plates 22–25; Persian Gulf, plate 20; Tehran Province, plate 23; Turkmenistan, plate 25; West Azarbaijan Province, plates 24–25	44–51; 58–59

PLATE 1

SATELLITE PHOTOGRAPH OF IRAN

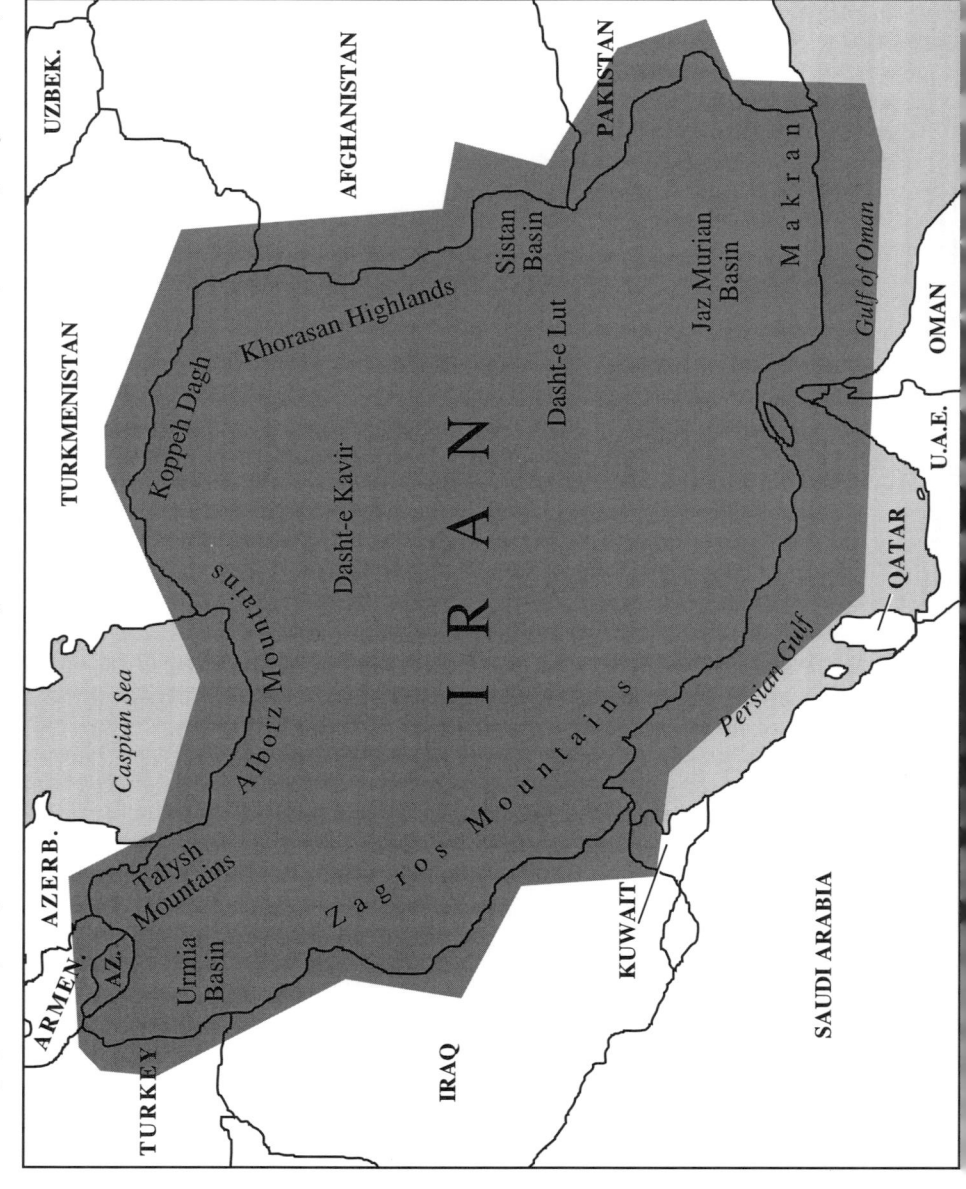

This satellite mosaic image of Iran was assembled in the mid-1970s from Landsat MSS data acquired in the early 1970s. The MSS sensors on board Landsats 4 and 5 detect reflected radiation from the Earth's surface in the visible and IR wavelengths. The satellites orbit at an altitude of 705 km and are designed and operated to collect data over a 185 km swath. The geographic limits and primary physical features in this photograph can be interpreted from the outline map. The dark shaded block corresponds to the satellite image.

(Photograph courtesy of Byron Loubert, Earth Satellite Corporation, 6011 Executive Boulevard, Rockville, Maryland 20852.)

PLATE 1

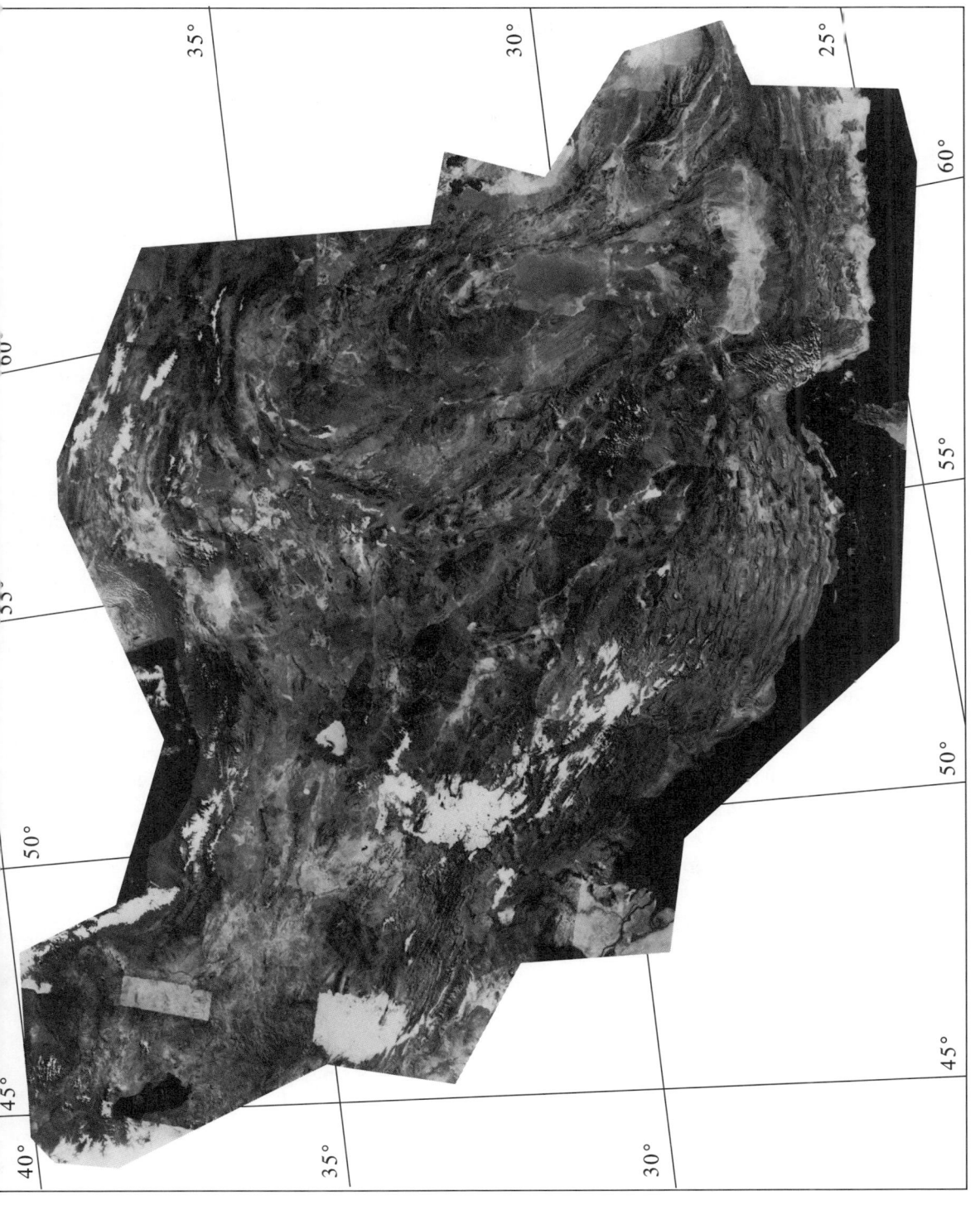

PLATE 2

AGAMIDAE I

Family Agamidae

Calotes versicolor (Text, pages 68–69).

A: Adult male, Karachi, Karachi District, Pakistan. Photograph by Sherman A. Minton.

Laudakia caucasia caucasia (Text, pages 70–73. For habitat, see plates 23C–D, 24B, D–E, 25D).

B–C: Adult male (B) and adult female (C), about 37°19'N, 56°16'E, 3 km W of Qara Bil, (former) Mohammad Reza Shah National Park (now Golestan National Park), Mazanderan Province, Iran, 1000–1300 m, 31 May 1975. Photographs by Steven C. Anderson.

Laudakia erythrogastra (Text, pages 73–75).

D: Adult, 37°12'N, 59°33'N, within 2 km of Khiveabad, along Turkmenistan-Iran control border and Laensuv River, foothills of the Kopet-Dag, Ashkhabad Region, Turkmenistan, 690 m, 25 April 1992. CAS specimen. Photograph by J. Robert Macey.

Laudakia microlepis (Text, pages 76–78).

E: Adult male, about 39°29'N, 59°12'E, 8 km S of Robat-e Sang on road to Heydariyeh, Khorasan Province, Iran, 1760 m, 6 June 1975. Photograph by Steven C. Anderson.

F: Adult female, about 34°05'N, 58°50'E, 36 km S of Bidokht on road to Birjand, Khorasan Province, Iran, 1540 m, 6 June 1975. MMTT 1169. Photograph by Steven C. Anderson.

Laudakia nupta nupta (Text, page 80. For habitat, see plate 22A–C, E).

G: Adult male, about 29°24'N, 51°02'E, 41 km SE of Bandar-e Rig on road to Borazjan, Fars Province, Iran, 40 m, 18 May 1975. MMTT 1035. Photograph by Steven C. Anderson.

Laudakia nupta fusca (Text, page 80).

H: Adult male, Khadeji Falls, Dadu District, Pakistan. Photograph by Sherman A. Minton.

PLATE 2

PLATE 3

AGAMIDAE II

Family Agamidae

Phrynocephalus arabicus (Text, pages 84–85).

A: Adult male, Abqaiq, Eastern Province, Saudi Arabia. Photograph by William Ross.

B: Juvenile, Dhahran area, Eastern Province, Saudi Arabia. Photograph by William Ross.

Phrynocephalus clarkorum (Text, pages 85–86).

C: Adult, near Nushki, Chagai District, Pakistan. Photograph by Sherman A. Minton.

Phrynocephalus helioscopus helioscopus (Text, pages 87–88).

D–E: Adult, 39°12'N, 55°00'E, 36 km W of the 2-km-long road S to Kazandjik on the Ashkhabad-to-Krasnovodsk road, Krasnovodsk Region, Turkmenistan, 40 m, 7 May 1992. CAS specimen. Photographs by J. Robert Macey.

Phrynocephalus interscapularis (Text, pages 88–89).

F: Adult, N of Mary, Karakum Desert, Turkmenistan, May 1993. GNM 93–18.881. Photograph by Göran Nilson.

Phrynocephalus luteoguttatus (Text, pages 89–90).

G: Adult, near Nushki, Chagai District, Pakistan. Photograph by Sherman A. Minton.

Phrynocephalus maculatus maculatus (Text, pages 90–92. For habitat, see plate 20A).

H: Adult, about 30°15'N, 57°14'E, 13 km E of Kerman, Kerman Province, Iran, 2000 m, 5 April 1975. MMTT 732. Photograph by Steven C. Anderson.

I: Adult, about 30°15'N, 57°12'E, 10 km E of Kerman, Kerman Province, Iran, 1700–2000 m, 5 April 1975. Photograph by Steven C. Anderson.

PLATE 3

PLATE 4

AGAMIDAE III

Family Agamidae

Phrynocephalus maculatus maculatus (Text, pages 90–92. For habitat, see plate 20A).

A: Adult, W of Zabol, Sistan Province, Iran, 500 m, 27 April 1975. Photograph by Steven C. Anderson.

Phrynocephalus mystaceus (Text, pages 92–93. For habitat, see plate 24A).

B: Adult, about 34°39'N, 58°47'E, 35 km N of Gonabad on road to Torbat-e Heydariyeh, Khorasan Province, Iran, 850 m, 9 June 1975. Photograph by Steven C. Anderson.

C: Adult, 38°33.522'N, 63°09.175'E, Repetek Biosphere Reserve, Turkmenistan, 300 m, 31 August 1995. Specimen not collected. Threat display. Photograph by Steven C. Anderson.

Phrynocephalus ornatus (Text, pages 93–94. For habitat, see plate 24A).

D: Adult, about 34°39'N, 58°47'E, 35 km N of Gonabad, Khorasan Province, Iran, 850 m, 9 June 1975. Photograph by Steven C. Anderson.

Phrynocephalus persicus (Text, pages 94–96).

E: Adult, Iğdir, N of Mt. Ararat, Kars Province, Turkey, 1988. ZIG specimen. Photograph by Göran Nilson.

Phrynocephalus raddei raddei (Text, pages 96–97. For habitat, see plate 25E).

F: Adult, Karakum, 55 km N of Ashkhabad on Ashkhabad-to-Bakhardok road, then 21 km WNW on dirt road, Ashkhabad Region, Turkmenistan, 24 May 1989. CAS specimen. Photograph by J. Robert Macey.

Phrynocephalus scutellatus (Text, pages 97–99. For habitat, see plate 20B–G).

G: Adult male, about 36°21'N, 56°43'E, salt flats 5 km W of Kahak on road to Shahrud, Turan Protected Region, Khorasan Province, Iran, 800 m, 10 June 1975. Photograph by Steven C. Anderson.

H: Adult female, about 29°06'N, 56°00'E, 50 km S of Sirjan, Kerman Province, Iran, 1680 m, 6 April 1975. CAS 141032. Photograph by Steven C. Anderson.

I: Juvenile. Same data as for plate 4G.

PLATE 4

PLATE 5

AGAMIDAE IV

Family Agamidae

Trapelus agilis (Text, pages 100–104. For habitat, see plates 20A–B, 21B, 22A–C, E, 23E).

A: Adult male, about 29°28'N, 51°21'E, 5 km northerly of Dalaki, Fars Province, Iran, 110 m, 20 May 1975. CAS 141149. Photograph by Steven C. Anderson.

B: Adult male, about 30°15'N, 57°14'E, 13 km E of Kerman, Kerman Province, Iran, 2000 m, 5 April 1975. Photograph by Steven C. Anderson.

C: Gravid female, S of Shah Reza, Esfahan Province, Iran, 24 June 1968. Photograph by Josef Eiselt.

D: Adult, about 34°44'N, 52°10'E, near Shah 'Abbas Caravanserai near foot of Siah Kuh, Kavir Protected Region, Tehran (Central) Province, Iran, 1000 m, 13 June 1975. Typical posture in habitat. Photograph by Steven C. Anderson.

Trapelus persicus persicus (Text, pages 104–107. For habitat, see plate 21F–G).

E: Adult male, 9 km southerly of Borazjan on road from Bushire, Fars Province, Iran, 40 m, 19 May 1975. CAS 141147. Photograph by Steven C. Anderson.

F: Adult male, about 32°01'N, 48°16'E, 2 km W of Andimeshk-to-Ahvaz road on road to park headquarters, Karkheh National Park, Khuzestan Province, Iran, 14 May 1975. MMTT 1066. Photograph by Steven C. Anderson.

G: Adult female, about 28°53'N, 51°02'E, near Alchangi, 33 km northeasterly of Bushire, Fars Province, Iran, 0–50 m, 19 May 1975. Photograph by Steven C. Anderson.

Trapelus ruderatus ruderatus (Text, pages 107–110).

H: Adult male, 10 km N of Khorramabad, Lorestan Province, Iran, 28 March 1975. MMTT 684. Photograph by Steven C. Anderson.

I: Adult female, Bamu National Park, Fars Province, Iran, 15 June 1975. MMTT 1406. Photograph by Steven C. Anderson.

PLATE 5

PLATE 6

ANGUIDAE, EUBLEPHARIDAE, GEKKONIDAE I

Family Anguidae

Anguis fragilis colchicus (Text, pages 111–113. For habitat, see plate 23A).

A: Adult, about 36°17'N, 52°21'E, 26 km S of Amol on road to Tehran, Mazanderan Province, Iran, 450 m, 28 May 1975. Photograph by Steven C. Anderson.

B: Juvenile, about 100 m from shore of Caspian Sea, about 3 km W of Ramsar, Gilan Province, Iran. Photograph by Eskandar Firouz.

Ophisaurus apodus (Text, pages 113–115. For habitat, see plates 23D, 25A).

C: Adult male, about 37°19'N, 56°16'E, 3 km W of Robat-e Qareh Bil, (former) Mohammad Reza Shah National Park (now Golestan National Park), Mazanderan Province, Iran, 1000–1300 m, 31 May 1975. CAS 141171. Photograph by Steven C. Anderson.

D: Adult, 5 km SE of Maku, West Azarbaijan Province, Iran, 14 June 1968. Photograph by Josef Eiselt.

Family Eublepharidae

Eublepharis angramainyu (Text, pages 118–120. For habitat, see plate 22A–C).

E: Adult female, Shahbazan, Lorestan Province, Iran, 500 m, 15 April 1968. J. F. Schmidtler collection. Photograph by Josef F. Schmidtler.

Eublepharis turcmenicus (Text, pages 122–124. For habitat, see plate 25F).

F: Adult male, 39°06'N, 55°08'E, vicinity of Temen Spring, 2.5 km W of Danata on paved road from Danata to the Ashkhabad-to-Krasnovodsk road, then 5.4 km S on dirt road, Krasnovodsk Region, Turkmenistan, 300 m, 6 May 1992. CAS 184771. Photograph by J. Robert Macey.

Family Gekkonidae

Agamura persica (Text, pages 130–132. For habitat, see plates 21C, 23E).

G: Adult, 34°44'N, 52°10'E, 2–4 km SE of Shah 'Abbas Caravanserai, north foot of Siah Kuh, Kavir Protected Region, Tehran (Central) Province, Iran, 1000+ m, 13 June 1975. MMTT 1316. Photograph by Steven C. Anderson.

Assacus elisae (Text, pages 134–136. For habitat, see plate 21D).

H: Adult female, 33°19'N, 47°53'E, Afrineh, 77 km southerly from Khorramabad, on road to Andimeshk, Lorestan Province, Iran, 1100 m, 12 May 1975. CAS 141111. Photograph by Steven C. Anderson.

PLATE 6

PLATE 7

GEKKONIDAE II

Family Gekkonidae

Bunopus tuberculatus (Text, pages 140–143. For habitat, see plates 20G, 21B, F).

A: Adult, 52–56 km NW of Ganaveh on road to Ganaveh from Agha Jari-to-Behbehan road, Khuzestan Province, Iran, 70 m, 17 May 1975. Photograph by Steven C. Anderson.

Carinatogecko aspratilis (Text, pages 144–145).

B: Adult, Shiraz, Fars Province, Iran, 2 June 1995. ZMGU 190. Photograph by Daryanabard Kami.

Crossobamon eversmanni (Text, pages 147–148. For habitat, see plate 24A).

C: Adult male, about 34°47'N, 60°47'E, 4 km E of communications station at outskirts of Tayyebat, Khorasan Province, Iran, about 900 m, 4 June 1975. CAS 141178. Photograph by Steven C. Anderson.

Cyrtopodion agamuroides (Text, pages 153–154. For habitat, see plate 21A).

D: Adult, 26°09'N, 61°27'E, Hudar, Sarbaz River, Baluchistan Province, Iran, 320–380 m, 23 April 1975. MMTT 875. Photograph by Steven C. Anderson.

Cyrtopodion caspium caspium (Text, pages 155–156).

E: Adult, 39°12'N, 55°00'E, 36 km W of the 2-km-long road S to Kazandjik on the Ashkhabad-to-Krasnovodsk road, Krasnovodsk Region, Turkmenistan, 40 m, 7 May 1992. CAS specimen. Photograph by J. Robert Macey.

Cyrtopodion gastrophole (Text, pages 156–158. For habitat, see plate 22D).

F: Adult, about 29°28'N, 51°21'E, 5 km northerly from Dalaki on road to Shiraz, at base of foothills, Fars Province, Iran, 100 m, 20 May 1975. MMTT 1049. Photograph by Steven C. Anderson.

Cyrtopodion kachhense (Text, pages 159–150).

G: Adult, Karachi, Karachi District, Pakistan. Photograph by Sherman A. Minton.

Cyrtopodion longipes longipes (Text, pages 163–164. For habitat, see plate 21B).

H: Adult female, about 31°03'N, 61°38'E, abandoned village SE of road between Zabol and Dust-e Mohammad Khan, 10 km SW of Rud-e Hirmand, Sistan Province, Iran, 450 m, 27 April 1975. CAS 141095. Photograph by Steven C. Anderson.

PLATE 7

PLATE 8

GEKKONIDAE III

Family Gekkonidae

Cyrtopodion russowii (Text, pages 165–166).

A: Adult, 38°24'N, 63°11'E, Repetek Desert Reserve Station, Repetek, Chardjou Region, Turkmenistan, 11 May 1989. CAS specimen. Photograph by J. Robert Macey.

Cyrtopodion sagittifer (Text, pages 166–167. For habitat, see plate 20G).

B: Adult, about 27°10'N, 60°09'E, Jaz Murian depression, 31 km W of Bampur, Baluchistan Province, Iran, 470 m, 25 April 1975. Photograph by Steven C. Anderson.

Cyrtopodion scabrum (Text, pages 167–169).

C: Adult male, Abadan, Khuzestan Province, Iran, 27 March 1975. MMTT specimen. Photograph by Steven C. Anderson.

Cyrtopodion spinicauda (Text, pages 169–170. For habitat, see plate 25F).

D: Adult, 39°06'N, 55°08'E, vicinity of Temen Spring, 2.5 km W of Danata on paved road from Danata to the Ashkhabad-to-Krasnovodsk road, then 5.4 km S on dirt road, Krasnovodsk Region, Turkmenistan, 300 m, 5 May 1992. CAS specimen. Photograph by J. Robert Macey.

Cyrtopodion turcmenicum (Text, pages 170–171).

E: Adult, Baluchistan Province, Iran (no other data provided). Photograph by Milan Kaftan.

Hemidactylus flaviviridis (Text, pages 172–173).

F: Adult male, 25°18'N, 60°37'E, on walls of Department of Environment headquarters, Chah Bahar, Baluchistan Province, Iran, near sea level, 23 April 1975. CAS 141078. Photograph by Steven C. Anderson.

Hemidactylus persicus (Text, pages 173–174).

G: Adult female, between Masjed Soleyman (31°57'N, 49°16'E) and Haft Gel (31°28'N, 49°30'E), Khuzestan Province, Iran, 12 May 1958. CAS 86377. Photograph by Steven C. Anderson.

Hemidactylus turcicus (Text, pages 174–175).

H: Gravid female, about 26°31'N, 54°47'E, ruined buildings on beach, 5 km W of Bandar-e Lengeh, Kerman Province, Iran, near sea level, 11 April 1975. Photograph by Steven C. Anderson.

Pristurus rupestris (Text, pages 176–177. For habitat, see plate 20C–E).

I: Adult male, about 26°32'N, 54°43'E, 9 km W of Bandar-e Lengeh airport, Kerman Province, Iran, 0–50 m, 12 April 1975. Agonistic tail display. Photograph by Steven C. Anderson.

PLATE 8

PLATE 9

GEKKONIDAE IV

Family Gekkonidae

Pristurus rupestris (continued) (Text, pages 176–177. For habitat, see plate 20C–E).

A: Adult female, 27°09'N, 57°05'E, on date palm in hotel garden, Minab, Kerman Province, Iran, 30 m, 8 April 1975. CAS 141033. Photograph by Steven C. Anderson.

Stenodactylus affinis (Text, pages 182–183).

B: Adult, 52–56 km NW Ganaveh on track to Agha Jari-Behbehan road, Khuzestan Province, Iran, 70 m, 17 May 1975. Photograph by Steven C. Anderson.

Stenodactylus doriae (Text, pages 183–184. For habitat, see plate 21F–G).

C: Adult, about 31°16'N, 49°10'E, 1 km E of Kupal, Khuzestan Province, Iran, 25–50 m, 16 May 1975. Photograph by Steven C. Anderson.

D: Adult, about 27°17'N, 56°29'E, 12 km E of Shaqu on road between Bandar 'Abbas and Minab, Kerman Province, Iran, 0–50 m, 10 April 1975. Photograph by Steven C. Anderson.

Teratoscincus bedriagai (Text, pages 186–187. For habitat, see plate 21B).

E–F: Adult (E) and juvenile (F), about 31°03'N, 61°38'E, abandoned village SE of road between Zabol and Dust-e Mohammad Khan, 10 km SW of Rud-e Hirmand, Sistan Province, Iran, 450 m, 27 April 1975. Photographs by Steven C. Anderson.

Teratoscincus microlepis (Text, page 187. For habitat, see plate 21B).

G: Adult, about 31°03'N, 61°38'E, abandoned village SE of road between Zabol and Dust-e Mohammad Khan, 10 km SW of Rud-e Hirmand, Sistan Province, Iran, 450 m, 28 April 1975. CAS 141096. Photograph by Steven C. Anderson.

Teratoscincus scincus keyserlingii (Text, pages 188–190. For habitat, see plates 20G, 21B, 24A).

H: Adult, about 27°17'N, 56°29'E, 12 km E of Shaqu on road between Bandar 'Abbas and Minab, Kerman Province, Iran, 0–50 m, 10 April 1975. Photograph by Steven C. Anderson.

PLATE 9

PLATE 10

GEKKONIDAE V

Family Gekkonidae

Teratoscincus scincus keyserlingii (continued) (Text, pages 188–190. For habitat, see plates 20G, 21B, 24A).

A: Juvenile, about 31°03'N, 61°38'E, abandoned village SE of road between Zabol and Dust-e Mbohammad Khan, 10 km SW Rud-e Hirmand, Sistan Province, Iran, 450 m, 27 April 1975. Photograph by Steven C. Anderson.

Teratoscincus scincus scincus (Text, pages 189–190).

B–C: Adult (B) and juvenile (C), 38°33.522'N, 63°09.175'E, Repetek Biosphere Reserve, Turkmenistan, 31 August 1995. Specimens not collected. Adult exhibits characteristic defensive tail display. Photographs by Steven C. Anderson.

Tropiocolotes helenae helenae (Text, pages 191–193).

D: Adults, near Kazerun, Khuzestan Province, Iran. Photograph by Jean Vasserot.

E: Adult, Iran (no other data). Photograph by Milan Kaftan.

Tropiocolotes helenae fasciatus (Text, page 193).

F: Adult female, Sorkh-e Dize, 125 km W of Kermanshah on road to Baghdad, Lorestan Province, Iran. ZSM 501/68 (holotype). Photograph by Josef F. Schmidtler.

Tropiocolotes latifi (Text, pages 193–194).

G: Adult, 31°36'49"N, 53°49'52"E, 5 km S of Aliabad, Kerman Province, central Iran. MNHP specimen. Photograph by Jiří Moravec.

Tropiocolotes persicus bakhtiari (Text, pages 195–196. For habitat, see plate 22A–C).

H: Adult female, 31°57'N, 49°21E, Sar-e Gach, Khuzestan Province, Iran, 13 May 1958. CAS 86408 (holotype). Photograph by Steven C. Anderson.

Tropiocolotes* cf. *steudneri (Text, pages 196–198. For habitat, see plate 20D).

I: Gravid adult female, about 26°32'N, 54°43'E, 9 km W of Bandar-e Lengeh airport, Kerman Province, Iran, 0–50 m, 12 April 1975. MMTT 1048. Photograph by Steven C. Anderson.

PLATE 10

PLATE 11

LACERTIDAE I

Family Lacertidae

Acanthodactylus blanfordi (Text, pages 201–203. For habitat, see plate 20G).

A: Adult, about 27°10'N, 60°10'E Jaz Murian depression, 30 km W of Bampur, Baluchistan Province, Iran, 500–540 m, 21 April 1975. Photograph by Steven C. Anderson.

Acanthodactylus grandis (Text, pages 203–205. For habitat, see plate 21G).

B: Adult, about 28°53'N, 51°02'E, near Alchangi, 33 km northeasterly of Bushire, Fars Province, Iran, 0–50 m, 19 May 1975. Photograph by Steven C. Anderson.

Acanthodactylus micropholis (Text, pages 205–206).

C: Adult, about 23°01'N, 55°43'E, 109 km W of Bandar 'Abbas on road to Bandar-e Lengeh, Kerman Province, Iran, 0–50 m, 11 April 1975. MMTT 813. Photograph by Steven C. Anderson.

Acanthodactylus schmidti (Text, pages 206–207. For habitat, see plate 21F).

D: Adult, about 31°16'N, 49°10'E, 1 km E of Kupal, Khuzestan Province, Iran, 25–50 m, 16 May 1975. Photograph by Steven C. Anderson.

Eremias acutirostris (Text, page 211. For habitat, see plate 21B).

E: Adult, about 31°03'N, 61°38'E, abandoned village SE of road between Zabol and Dust-e Mohammad Khan, 10 km W of Rud-e Hirmand, Sistan Province, Iran, 450 m, 28 April 1975. MMTT specimen. Photograph by Steven C. Anderson.

Eremias arguta (Text, pages 212–213).

F: Adult, 45°49'N, 44°38'E, 15 km SSW (airline) of Iki Burul on the Iki Burul-to-Manych road, Iki Burul District, Kalmyk Autonomous Republic, Russia. Photograph by J. Robert Macey.

Eremias fasciata (Text, pages 213–215. For habitat, see plate 21B).

G: Adult, about 36°09'N, 56°08'E, 34 km S of 'Abbasabad on dirt track into Turan Protected Region, Khorasan Province, Iran, 870 m, 11 June 1975. Photograph by Steven C. Anderson.

Eremias grammica (Text, pages 215–216. For habitat, see plate 24A).

H: Adult male, about 34°41'N, 58°48'E 35 km N of Gonabad on road to Torbat-e Heydariyeh, Khorasan Province, Iran, 850 m, 9 June 1975. CAS 141202. Photograph by Steven C. Anderson.

Eremias intermedia (Text, pages 216–217).

I: Adult, 38°10'N, 54°44'E, 14.2 km SW of Madau (65.8 km SW of Bugdaili) on Bugdaili-to-Kizyl-Atrek road, Krasnovodsk Region, Turkmenistan, 40 m. Photograph by J. Robert Macey.

PLATE 11

PLATE 12

LACERTIDAE II

Family Lacertidae

Eremias lineolata (Text, pages 218–219. For habitat, see plate 24A).

A: Adult, about 34°41'N, 58°48'E, 35 km N of Gonabad on road to Torbat-e Heydariyeh, Khorasan Province, Iran, 850 m, 9 June 1975. Photograph by Steven C. Anderson.

Eremias persica (Text, pages 221–223. For habitat, see plates 20B, 21C, 24C).

B: Adult male, about 36°24'N, 56°17'E, 16 km S of Abbasabad on dirt track into Turan Protected Region, Khorasan Province, Iran, 800–870 m, 11 June 1975. Photograph by Steven C. Anderson.

Eremias scripta (Text, pages 224–225).

C: Adult, near Nushki, Chagai District, Pakistan. Photograph by Sherman A. Minton.

D: Adult, Karakum, N of Ashkhabad, Turkmenistan, May 1993. GNM 93-18.881. Photograph by Göran Nilson.

Eremias strauchi strauchi (Text, pages 225–226. For habitat, see plate 25D).

E: Adult, about 38°52'N, 45°10'E, 21 km E of Maku on road to Marand, West Azarbaijan Province, Iran, 1090 m, 29 June 1975. Photograph by Steven C. Anderson.

Eremias strauchi kopetdaghica (Text, pages 226–227).

F: Adult, about 35°29'N, 59°12'E, 8 km S of Robat-Sang on road to Torbat-e Heydariyeh, Khorasan Province, Iran, 1760 m, 6 June 1975. Photograph by Steven C. Anderson.

Eremias velox velox (Text, pages 227–228. For habitat, see plate 23D).

G: Adult, about 37°19'N, 56°16'E, 3 km W of Robat-e Qarabil, (former) Mohammad Reza Shah National Park (now Golestan National Park), Mazanderan Province, Iran, 1000–1300 m, 31 May 1975. Photograph by Steven C. Anderson.

Lacerta brandtii (Text, pages 231–232. For habitat, see plate 24C).

H: Adult male, 45 km SE of Tabriz, East Azarbaijan Province, Iran, 2000 m. Photograph by Josef Eiselt.

I: Adult male, about 37°22'N, 46°30'E, Maragheh paleontological site, 4 km SW Chahlilvan, 29 km E of Maragheh, East Azarbaijan Province, Iran, 1800 m, 3 July 1975. Photograph by Steven C. Anderson.

PLATE 12

PLATE 13

LACERTIDAE III

Family Lacertidae

Lacerta brandtii (continued) (Text, pages 231–232. For habitat, see plate 24C).

A: Adult male, about 37°22'N, 46°30'E, Maragheh paleontological site, 4 km SW Chahlilvan, 29 km E of Maragheh, East Azarbaijan Province, Iran, 1800 m, 3 July 1975. Photograph by Steven C. Anderson.

Lacerta cappadocica urmiana (Text, pages 232–234. For habitat, see plate 25G).

B: Adult, about 37°25'N, 44°56'E, 5 km W of Band (15 km W of Reza'iyeh) on dirt track to waterfalls, West Azarbaijan Province, Iran, 1500 m, 1 July, 1975. Photograph by Steven C. Anderson.

C: Adult male, 5 km E of Sero, West Azarbaijan Province, Iran, 7 July 1968. Photograph by Josef Eiselt.

Lacerta chlorogaster (Text, pages 234–235. For habitat, see plate 23B).

D–E: Adult male (D) and in (E) adult male on left, others females, about 37°21'N, 56°01'E, near stream S of road, 10 km W of park headquarters, 7 km E of park entrance, (former) Mohammad Reza Shah National Park (now Golestan National Park), Mazanderan Province, Iran, 450 m, 30 May 1975. CAS and MMTT specimens. Photographs by Steven C. Anderson.

Lacerta defilippii (Text, pages 236–237. For habitat, see plate 24B).

F: Adult male, about 36°12'N, 51°20'E, 64 km S of Chalus on road to Karaj, Mazanderan Province, Iran, 1670 m, 20 June 1975. Photograph by Steven C. Anderson.

G: Adult female (above) and adult male (below), about 36°12'N, 51°20'E, 64 km S of Chalus on road to Karaj, Mazanderan Province, Iran, 1670 m, 20 June 1975. CAS 141224 (above), CAS 141223 (below). Photograph by Steven C. Anderson.

Lacerta media media (Text, pages 237–238).

H–I: Adult male (H) and adult female (I), E of Qasr-e Shirin, Kermanshahan Province, Iran, 1968. ZSMH 15/68 (male) and J. F. Schmidtler personal collection (female). Photographs by Josef F. Schmidtler.

PLATE 13

PLATE 14

LACERTIDAE IV

Family Lacertidae

Lacerta praticola (Text, pages 239–240).

A: Adult, no data. Photograph by Ilya S. Darevsky.

B: Adult, Sochi, Russia, 1990. GNM 93–18.883. Photograph by Göran Nilson.

Lacerta princeps princeps (Text, pages 240–241. For habitat, see plate 22F).

C–D: Adult male (C) and juvenile (D), 8 km E of Dasht Arjan, 57 km W of Shiraz, Fars Province, Iran, about 2100 m, 24 June 1968. NMW 19296:1 (male). Photographs by Josef Eiselt.

Lacerta princeps kurdistanica (Text, pages 241–243).

E: Adult male, high pass about 8 km W of Semdinli, 50 km SSE of Yüksekova, Hakkari Province, Turkey, about 2000 m, 25 May 1966. NMW 18546. Photograph by Josef Eiselt.

Lacerta raddei raddei (Text, pages 243–244. For habitat, see plate 24E).

F–G: Adult male, about 38°17'N, 46°57'E, 31 km S of Ahar on road to Tabriz, East Azarbaijan Province, Iran, 2150 m, 27 June 1975. CAS 141261. Photographs by Steven C. Anderson.

Lacerta raddei vanensis (Text, pages 244–245. For habitat, see plate 24D).

H: Adult male, about 39°14'N, 47°29'E, 1 km S of Qaraqachid on road to Meshkin Shahr, East Azarbaijan Province, Iran, 900 m, 26 June 1975. CAS 141254. Photograph by Steven C. Anderson.

PLATE 14

PLATE 15

LACERTIDAE V

Family Lacertidae

Lacerta raddei vanensis (continued) (Text, pages 244–245. For habitat, see plate 24D).

A: Adult male, about 39°14'N, 47°29'E, 1 km S of Qaraqachid on road to Meshkin Shahr, East Azarbaijan Province, Iran, 900 m, 26 June 1975. CAS 141254. Photograph by Steven C. Anderson.

Lacerta steineri (Text, pages 245–246).

B: Adult, Mazanderan Province, Iran (no other data provided). Photograph by Petr Kodym.

Lacerta strigata (Text, pages 246–248. For habitat, see plate 22G).

C: Adult male, about 36°41'N, 52°29'E, public beach on Caspian Sea, 14 km W of Babolsar, Mazanderan Province, Iran, below sea level, 29 May 1975. MMTT 1094. Photograph by Steven C. Anderson.

Lacerta valentini valentini (Text, pages 248–249).

D: Adult male (above) and adult female (below), Armenia (Rasdan?). Photograph by Herman J. in den Bosch.

E: Adult male, Mt. Legli, Armenia, June 1992. Photograph by Göran Nilson.

***Lacerta valentini* ssp.** (Text, pages 248–249).

F: Adult, no data. Photograph by Ilya S. Darevsky.

Mesalina brevirostris (Text, pages 249–250).

G: Adult, N of Al Khobar, Eastern Province, central Saudi Arabia. Photograph by William Ross.

H: Adult, Miani Hor, Las Bela District, Pakistan. Photograph by Sherman A. Minton.

PLATE 15

PLATE 16

LACERTIDAE VI AND SCINCIDAE I

Family Lacertidae

Mesalina guttulata (Text, page 253).

A: Adult, eastern Saudi Arabia, 19 November 1981. Photograph by William Ross.

Mesalina watsonana (Text, pages 251–254. For habitat, see plates 20A–B, 21C, 22A–C, E).

B: Adult male, about 30°15'N, 57°14'E, 13 km E of Kerman, Kerman Province, Iran, 2000 m, 5 April 1975. CAS 141021. Photograph by Steven C. Anderson.

Ophisops elegans (Text, pages 254–256. For habitat, see plate 25G).

C: Adult, about 29°06'N, 57°56'E, 75 km N of Jiroft on road to Bam, Kerman Province, Iran, 1850 m, 15 April 1975. Photograph by Steven C. Anderson.

D: Adult, Gulutshul, E of Tabriz, East Azarbaijan Province, Iran. Photograph by Herman A. J. in den Bosch.

Family Scincidae

Ablepharus bivittatus (Text, pages 263–264. For habitat, see plate 25B).

E: Adult, 38°00'N, 48°34'E, Neur Lake, 35 km by air ESE of Ardabil, 16 km easterly by dirt road from Bandalalu, which is 34 km S of Ardabil on road to Khalkhal, East Azarbaijan Province, Iran, 2400 m, 23 June 1975. MMTT 1277. Photograph by Steven C. Anderson.

F: Adult, 45 km SE of Tabriz, East Azarbaijan Province, Iran, 2000 m, 15 June 1968. Photograph by Josef Eiselt.

Ablepharus pannonicus (Text, pages 264–267).

G: Adults, Ziarat, Sibi District, Baluchistan, Pakistan. AMNH 86873–86874. Photograph by Sherman A. Minton.

H: Adult, 38°30'N, 56°47'E, 2 km by air SE of Saivan, Ashkhabad Region, Turkmenistan, 1200–1400 m, 21 May 1992. CAS specimen. Photograph by J. Robert Macey.

PLATE 16

PLATE 17

SCINCIDAE II

Family Scincidae

Chalcides ocellatus (Text, page 268).

A: Adult, Perge, W of Antalya, Antalya Province, Turkey, 1981. ZIG specimen. Photograph by Göran Nilson.

Eumeces schneideri princeps (Text, pages 269–271. For habitat, see plate 25C).

B: Adult, shore of Lake Reza'iyeh, E of Reza'iyeh, West Azarbaijan Province, Iran, 8 July 1968. Photograph by Josef Eiselt.

C: Adult, S of Kaaka, Kopet Dag, Turkmenistan, at Iran border, May 1993. GNM 93–18.881. Photograph by Göran Nilson.

D: Juvenile, Bishapur archeological excavations W of Shiraz, Fars Province, Iran. Photograph by Herman A. J. in den Bosch.

Eumeces taeniolatus parthianicus (Text, pages 272–274).

E: Adult, Murgab River, Turkmenistan, May 1993. GNM 93–18.881. Photograph by Göran Nilson.

F–G: Adult, near Duchak, southern Turkmenistan. Photographs by Herman A. J. in den Bosch.

Mabuya aurata septemtaeniata (Text, page 275. For habitat, see plates 21G, 22A–C).

H: Adult, east bank of Rud-e Dez, Khuzestan Province, Iran, 23 March 1975. MMTT 692. Photograph by Steven C. Anderson.

PLATE 17

PLATE 18

SCINCIDAE III

Family Scincidae

Mabuya aurata transcaucasica (Text, pages 275–277. For habitat, see plates 23D, 25D).

A: Adult male, about 38°52'N, 45°10'E, 21 km E of Maku on road to Marand, West Azarbaijan Province, Iran, 1090 m, 29 June 1975. CAS 141270. Photograph by Steven C. Anderson.

Mabuya vittata (Text, pages 277–278).

B: Adult, Sar-e Pol, 20 km W of Sorkh-e Dize, Lorestan Province, Iran, 500 m, 28 May 1970. J. F. Schmidtler specimen. Photograph by Josef F. Schmidtler.

Ophiomorus brevipes (Text, pages 280–282).

C: Adult female, near Minab, Kerman Province, Iran, about 100 m, 21 October 1958. CAS 86593. Photograph by Steven C. Anderson.

Ophiomorus nuchalis (Text, pages 282–283).

D: Adult and subadult females, 34°44'N, 52°11'E, near Cheshmeh Shah, N foot Siah Kuh (mountain), Kavir Protected Region, Tehran (Central) Province, Iran, May 1976. Preserved specimens. GNM Re.ex. 4418 (holotype, with broken tail) and GNM Re.ex. 4419 (paratype). Photograph by Göran Nilson.

Ophiomorus persicus (Text, page 283).

E: Adult, Sivand, Fars Province, Iran. Photograph by Milan Kaftan.

Ophiomorus tridactylus (Text, pages 285–287. For habitat, see plate 21B).

F: Adult, about 31°03'N, 61°38'E, abandoned village SE of road between Zabol and Dust-e Mohammad Khan, 10 km SW Rud-e Hirmand, Sistan Province, Iran, 450 m, 28 April 1975. Photograph by Steven C. Anderson.

G: Characteristic tracks of this species, near Rud-e Hirmand, Sistan Province, Iran, 100 m, 27 April 1975. Photograph by Steven C. Anderson.

Scincus scincus conirostris (Text, pages 287–288. For habitat, see plate 21F).

H: Adult male, 30°43'N, 49°49'E, Agha Jari, Khuzestan Province, Iran, about 300 m, 16 August 1958. CAS 86479. Photograph by Steven C. Anderson.

PLATE 18

PLATE 19

UROMASTYCIDAE AND VARANIDAE

Family Uromastycidae

Uromastyx aegyptius (Text, pages 290–291).

A: Adult, 25°55'N, 54°32'E, Jazireh-ye Sirri (island), Persian Gulf, Iran. Photograph by Houshang Ziaii, courtesy Eskander Firouz.

Uromastyx asmussi (Text, pages 291–293).

B: Adult male, Kavir Protected region, Tehran (Central) Province, Iran, June 1975. MMTT specimen. Dark pattern in response to low temperature. Photograph by Steven C. Anderson.

C: Adult male, 40 km W of Caravanserai, Kavir Protected region, Tehran (Central) Province, Iran, 29 May 1975. MMTT specimen. Light pattern in response to high temperature. Photograph by Steven C. Anderson.

Uromastyx loricatus (Text, pages 293–294. For habitat, see plates 21E, 22A–C).

D: Adult female (left), road between Masjed Soleyman (31°57'N, 49°16'E) and Haft Gel (31°28'N, 49°30'E), Khuzestan Province, Iran, 6 July 1958. CAS 86470; and adult male (right), 31°45'N, 49°08'E, near Dar-e Khazineh, Khuzestan Province, Iran, 6 July 1958. CAS 86469. Photograph by Steven C. Anderson.

Family Varanidae

Varanus bengalensis bengalensis (Text, pages 295–297. For habitat, see plate 20F).

E: Subadult, Hab Chowki, Karachi District, Pakistan. Photograph by Sherman A. Minton.

Varanus griseus griseus (Text, page 297. For habitat, see plate 21F).

F: Young adult, Salisil, Eastern Province, Saudi Arabia. Photograph by Jeffrey L. Briggs.

Varanus griseus caspius (Text, pages 297–299).

G: Adult, eastern Fars Province, Iran. Photograph by Bijan Darrehshuri.

H: Juvenile, Kavir Protected Region, Tehran (Central) Province, Iran, 24 April 1975. MMTT 771. Photograph by Steven C. Anderson.

PLATE 19

PLATE 20

HABITATS I

A: 30°15'N, 57°14'E, terrain ENE of Kerman, Kerman Province, Iran, 2000 m. Habitat of *Phrynocephalus m. maculatus, Trapelus agilis*, and *Mesalina watsonana*. 5 April 1975. Photograph by Steven C. Anderson.

B: 29°06'N, 56°00'E, 50 km S of Sirjan on Kerman-Bandar 'Abbas road, Kerman Province, Iran, 1680 m. Habitat of *Phrynocephalus scutellatus, Trapelus agilis, Eremias persica*, and *Mesalina watsonana*. 6 April 1975. Photograph by Steven C. Anderson.

C: 27°23' 56°16'E, foothills of Kuh-e Genu, 20 km N of Shaqu to turnoff to Kuh-e Genu, westerly 8 km on track to base of foothills, Kerman Province, Iran, 300 m. Rocky bed of seasonal stream. Habitat of *Pristurus rupestris*. 9 April 1975. Photograph by Steven C. Anderson.

D: 26°32'N, 54°43'E, 9 km W of Bandar-e Lengeh airport, Kerman Province, Iran, 0–50 m. Habitat of *Pristurus rupestris, Tropiocolotes steudneri, Echis carinatus,* and *Bufo surdus*. 12 April 1975. Photograph by Steven C. Anderson.

E: 26°57'N, 56°14'E, Qeshm Island, above Qeshm (town), Persian Gulf, Iran, about 25 m. Rocky outcrops. Habitat of *Pristurus rupestris*. 13 April 1975. Photograph by Steven C. Anderson.

F: Dahaneh Nais, 37 km northerly from Rudan on road to Jiroft, Kerman Province, Iran, 450 m. Habitat of *Varanus bengalensis, Bufo surdus,* and *Euphlyctis cyanophlyctis*. 14 April 1975. Photograph by Steven C. Anderson.

G: 27°10'N, 60°09'E, Jaz Murian depression, 31 km W of Bampur, Baluchistan-Sistan Province, Iran, 470 m. *Acacia* sp. on sandy waste near active dunes. Habitat of *Phrynocephalus scutellatus, Bunopus tuberculatus, Cyrtopodion sagittifer* (on and under bark of *Acacia* trees), *Teratoscincus scincus keyserlingi,* and *Acanthodactylus blanfordi*. 24 April 1975. Photograph by Steven C. Anderson.

PLATE 20

PLATE 21

HABITATS II

A: 26°09'N, 61°27'E, along Rud-e Sarbaz across from Hudar, Baluchistan, Baluchistan-Sistan Province, Iran, 380 m. Large cobbles on flat terrain above river. Habitat of *Cyrtopodion agamuroides* and *Coluber rhodorhachis*. 23 April 1975. Photograph by Steven C. Anderson.

B: 31°03'N, 61°38'E, abandoned village SE of road between Zabol and Dust-e Mohammad Khan, 10 km SW of Rud-e Hirmand, Sistan Basin, Baluchistan-Sistan Province, Iran, 450 m. Aolian deposits of loess inside and outside of mud-brick habitations. Habitat of *Trapelus agilis, Bunopus tuberculatus, Cyrtopodion l. longipes* (on walls of buildings), *Teratoscincus bedriagai, T. microlepis, T. scincus keyserlingii, Eremias acutirostris, E. fasciata, Ophiomorus tridactylus*, and *Eryx t. tataricus*. 27 April 1975. Photograph by Steven C. Anderson.

C: 29°28'N, 60°41'E, 32 km W of Zahedan on road to Cheshmeh Ziarat, Baluchistan, Baluchistan-Sistan Province, Iran, 1900 m. *Artemisia* vegetation on dry, disintegrating shale on hillside. Habitat of *Agamura persica, Eremias persica, Mesalina watsonana*, and *Testudo graeca zarudnyi*. (See text, page 59). 18 April 1975. Photograph by Steven C. Anderson.

D: 33°19'N, 47°53'E, Afrineh, 77 km southerly from Khorramabad on road to Andimeshk, Lorestan, Khuzestan-Lorestan Province, Iran, 1100 m. Habitat of *Asaccus elisae* (under exfoliating sandstone above stream). 12 May 1975. Photograph by Steven C. Anderson.

E: Ziggurat at Choga Zanbil, Khuzestan, Khuzestan-Lorestan Province, Iran. Habitat of *Uromastyx loricatus*; burrows in compacted, fine-grain alluvial sediments. 13 May 1975. Photograph by Steven C. Anderson.

F: 31°16'N, 49°10'E, 1 km E of Kupal, along Ahvaz-to-Behbehan road, Khuzestan, Khuzestan-Lorestan Province, Iran, 25–50 m. Active sand dunes. Habitat of *Trapelus persicus, Bunopus tuberculatus, Stenodactylus doriae, Acanthodactylus schmidti, Scincus scincus conirostris, Varanus griseus, Diplometopon zarudnyi*, and *Eryx jayakari*. (See text, page 58). 16 May 1975. Photograph by Steven C. Anderson.

G: 28°53'N, 51°02'E, near Alchangi, 33 km SE of Bushire, Fars Province, Iran, 0–50 m. Stabilized sand on coastal plain. Habitat of *Trapelus persicus, Stenodactylus doriae, Acanthodactylus grandis*, and *Mabuya aurata septemtaeniata*. 19 May 1975. Photograph by Steven C. Anderson.

PLATE 21

PLATE 22

HABITATS III

A: Western foothills and valleys of the Zagros Mountains along Masjed Soleyman-to-Haft Gel road, Khuzestan, Khuzestan-Lorestan Province, Iran, 300 m. Wheat under cultivation in valley. Habitat of *Laudakia n. nupta, Trapelus agilis, Eublepharis angramainyu, Tropiocolotes persicus bakhtiari, Mesalina watsonana, Mabuya aurata septemtaeniata, Uromastyx loricatus, Coluber rhodorachis, Psammophis schokari, Pseudocyclopodion persica, Telescopus tessellatus, Walterinnesia aegyptius, Bufo viridis,* and *Rana ridibunda.* (See text, pages 49 and 58). 7 March 1958. Photograph by Steven C. Anderson.

B: Western Zagros foothills, Zagros Mountains in the distance, along Masjed Soleyman-to-Lali road, Khuzestan, Khuzestan-Lorestan Province, Iran, 300 m. Habitat for same species as plate 22A. 21 March 1975. Photograph by Steven C. Anderson.

C: Western foothills of Zagros Mountains, along Masjed Soleyman-to-Godar Landar road, Khuzestan, Khuzestan-Lorestan Province, Iran, 300 m. Vertical sedimentary ridges, hills, and valleys. Habitat for same species as plate 22A. 10 April 1975. Photograph by Steven C. Anderson.

D: 29°28'N, 51°21'E, 5 km northerly from Dalaki on road between Bushire and Shiraz, at point where coastal plain meets foothills of Zagros Mountains, Fars Province, Iran, 100–110 m. Habitat of *Cyrtopodion gastropholis*, which was found on inner walls of this building over a covered well or spring on the plain. 20 May 1975. Photograph by Steven C. Anderson.

E: 29°28'N, 51°21'E, 5 km northerly from Dalaki on road between Bushire and Shiraz, at point where coastal plain meets foothills of Zagros Mountains, Fars Province, Iran, 100–110 m. Habitat of *Laudakia n. nupta, Trapelus agilis, Mesalina watsonana,* and *Rana ridibunda.* 20 May 1975. Photograph by Steven C. Anderson.

F: 8 km E of Dasht-e Arzhan, W of Shiraz, Fars Province, Iran, 2100 m. Mountain pass through Zagros Mountains, dense *Pistacia-Amygdalus* scrub forest on hillsides. Habitat of *Lacerta p. princeps*. 24 June 1968. Photograph by Josef Eiselt.

G: 36°41'N, 52°29'E, public beach on Caspian Sea, 14 km W of Babolsar, Mazanderan Province, Iran, below sea level. Sandy marsh and beach dunes. Habitat of *Lacerta strigata, Pelobates syriacus, Bufo viridis,* and *Rana ridibunda.* 29 May 1975. Photograph by Steven C. Anderson.

PLATE 22

PLATE 23

HABITATS IV

A: 36°17'N, 52°21'E, 26 km S of Amol, Mazanderan Province, Iran. Hyrcanian forest, 450 m. Habitat of *Anguis fragilis*. (See text, page 59). 28 May 1975. Photograph by Steven C. Anderson.

B: 37°21'N, 56°01'E, (former) Mohammad Reza Shah National Park (now Golestan National Park), 7 km E of entrance, 10 km W of park headquarters, near stream S of road, Mazanderan Province, Iran, 450 m. Hyrcanian forest understory in valley of Rudkhaneh-ye Chashmeh, western Kopet Dagh. Habitat of *Lacerta chlorogaster* and *Rana macrocnemis pseudodalmatina*. (See text, page 59). 30 May 1975. Photograph by Steven C. Anderson.

C: 37°21'N, 56°12'E, (former) Mohammad Reza Shah National Park (now Golestan National Park), 9 km E of park headquarters, Mazanderan Province, Iran, 800 m. Eastern margin of Hyrcanian Forest in valley of Rudkhaneh-ye Chashmeh, western Kopet Dagh. Habitat of *Laudakia c. caucasia, Rana macrocnemis pseudodalmatina,* and *Rana ridibunda*. (See text, page 59). 30 May 1975. Photograph by Steven C. Anderson.

D: 37°19'N, 56°16'E, (former) Mohammad Reza Shah National Park (now Golestan National Park), 3 km S on road 22 km W of Robat-e Qareh Bil in valley of Rudkhaneh-ye Chashmeh, western Kopet Dagh, Mazanderan Province, Iran, 1000–3000 m. Habitat of *Laudakia c. caucasia, Ophisaurus apodus, Eremias v. velox, Mabuya aurata transcaucasica, Natrix tessellata,* and *Rana ridibunda*. 31 May 1975. Photograph by Steven C. Anderson.

E: 34°49'N, 52°10'E, 2–4 km SE of Shah 'Abbas Caravanserai at foot of Siah Kuh, Kavir Protected Region, Tehran Province, Iran, 1000+ m. Central Plateau. Habitat of *Trapelus agilis, Agamura persica,* and *Pseudocerastes persicus*. (See text, pages 58–59). 13 June 1975. Photograph by Steven C. Anderson.

PLATE 23

PLATE 24

HABITATS V

A: 34°39'N, 58°47'E, 35 km N of Gonabad on road to Torbat-e Heydariyeh, Khorasan Province, Iran, 850 m. Small dune area on flatlands of Central Plateau. Habitat of *Phrynocephalus mystaceus, P. ornatus, Crossobamon eversmanni, Teratoscincus scincus keyserlingii, Eremias grammica*, and *E. lineolata*. 9 June 1975. Photograph by Steven C. Anderson.

B: 36°12'N, 51°20'E, central Alborz Protected Region, 64 km S of Chalus on road between Chalus and Karaj, Mazanderan Province, Iran, 1670 m. Upper margins of Hyrcanian Forest region. Habitat of *Laudakia c. caucasia* and *L. defilippii*. 20 June 1975. Photograph by Steven C. Anderson.

C: 37°22'N, 46°30'E, Maragheh paleontological site, 4 km SW of Chalilvan, 29 km E of Maragheh, East Azarbaijan, Iran, 1800 m. Hills and valley south of Kuh-e Savalan. Habitat of *Eremias persica, Lacerta brandtii,* and *Testudo graeca ibera*. 3 July 1975. Photograph by Steven C. Anderson.

D: 39°14'N, 47°29'E, 1 km S of Qaraqachid on road to Meshkinshahr, East Azarbaijan Province, Iran, 900 m. Sedimentary rock ridges of Moghan Steppe. Habitat of *Laudakia c. caucasia, Lacerta raddei vanensis,* and *Bufo viridis*. 26 June 1975. Photograph by Steven C. Anderson.

E: 39°20'N, 44°17'E, 11 km NE of 'Arab-e Dizehsi on road from Maku, West Azarbaijan Province, Iran, 1900 m. Mountain pass near Turkish border, eastern Anatolian Highlands. Habitat of *Laudakia c. caucasia, Lacerta raddei,* and *Vipera raddei*. 29 June 1975. Photograph by Steven C. Anderson.

PLATE 24

PLATE 25

HABITATS VI

A: 36°19'N, 51°16'E, 39 km S of Chalus on road between Chalus and Karaj, northern slope of central Zagros Mountains, Mazanderan Province, Iran, 600 m. Riparian vegetation along Rudkhaneh-ye Chalus (stream). Habitat of *Ophisaurus apodus*. 20 June 1975. Photograph by Steven C. Anderson.

B: 38°00'N, 48°34'E, Neur Lake, 35 km by air ESE of Ardabil, East Azarbaijan Province, Iran, 2400 m. Mountain lake surrounded by rocky slopes on western side of Talesh Mountains. Habitat of *Ablepharus b. bivittatus, Natrix tessellata,* and *Rana camerani*. 24 June 1975. Photograph by Steven C. Anderson.

C: Shore of Lake Reza'iyeh, E of Reza'iyeh, West Azarbaijan Province, Iran. Habitat of *Eumeces schneideri princeps, Testudo graeca ibera,* and *Mauremys c. caspica*. 8 July 1968. Photograph by Josef Eiselt.

D: 38°52'N, 45°10'E, 21 km E of Maku on road to Marand, West Azarbaijan Province, Iran, 1050 m. Stony uplands of northern Reza'iyeh Basin. Habitat of *Laudakia c. caucasia, Eremias s. strauchi,* and *Mabuya aurata transcaucasica*. 29 June 1975. Photograph by Steven C. Anderson.

E: Southern Karakum, 55 km N of Ashkhabad on Ashkhabad-to-Bakhardok road, then 21 km WNW on dirt road, Askhabad Region, Turkmenistan. Habitat for *Phrynocephalus r. raddei*. 24 May 1989. Photograph by J. Robert Macey.

F: 39°06'N, 55°08'E, vicinity of Temen Spring, 2.5 km W of Danata on paved road from Danata to the Ashkhabad-to-Krasnovodsk road, then 5.4 km S on dirt road, Krasnovodsk Region, Turkmenistan, 300 m. Habitat for *Eublepharis turcmenicus* and *Cyrtopodion spinicauda*. 6 May 1992. Photograph by J. Robert Macey.

G: 37°25'N, 44°56'E, 5 km W of Band, 15 km W of Reza'iyeh, on dirt track to waterfalls, West Azarbaijan Province, Iran, 1500 m. Mountains forming western margin of Reza'iyeh Basin. Habitat of *Lacerta cappadocica urmiana, Ophisops elegans,* and *Rana ridibunda*. 1 July 1975. Photograph by Steven C. Anderson.

PLATE 25

BIBLIOGRAPHY

The following bibliography is intended to serve the needs of anyone researching any aspect of Iranian herpetology; consequently, it includes titles not specifically cited in the text. I have cited authors' names as fully as known to me for the benefit of future historians and librarians. For these names I have drawn heavily on Adler (1989) and Applegarth (1989) as sources.

Abel, Erich. 1952. Zur Biologie von *Agama agilis* Ol. und *Agama ruderata* Ol. Zoologischer Anzeiger, 149:125–133.
Adams, C. G., A. W. Gentry, and P. J. Whybrow. 1983. Dating the terminal Tethyan event. *In* J.E. Meulenkamp (ed.). Reconstruction of Marine Paleoenvironments. Utrecht Micropaleontological Bulletin, 30:273–298.
Adle, Ahmed Hossein. 1960a. Climats de l'Iran. Publications de l'Université de Téhéran, no. 444. University of Tehran, Tehran, 239 pp., 51 tables, 30 maps (In Farsi, 38 pp. French résumé).
Adle, Ahmed Hossein. 1960b. Regions Climatiques et Végétation en Iran. Publications de l'Université de Téhéran, no. 626. University of Tehran, Tehran, 73 pp., 4 tables, 10 maps (In Farsi; 36 pp. French résumé).
Adler, Kraig Kerr. 1989. Herpetologists of the past, pp. 5–141. *In* Kraig Adler (ed.). Contributions to the History of Herpetology. Contributions to Herpetology, no. 5. Society for the Study of Amphibians and Reptiles, Oxford, Ohio.
Aellen, Paul. 1950. Ergebnisse einer botanisch-zoologischen Sammelreise durch den Iran 1948/49. Verhandlungen der Naturforschenden Gesellschaft in Basel, 61:128–140.
Aitchison, J. E. T. 1887. The zoology of the Afghan Delimitation Commission. Transactions of the Linnean Society, London, 5:53–142.
Akhmedov, S. B. 1983. K sistematike solotnistoi mabuyi zakavkaziya i srednei Azii [On the taxonomy of *Mabuya aurata* in Transcaucasia and Middle Asia]. Vestnik Zoologii, 1983(2):84–85 (In Russian).
Akhmedov, S. B., and Nikolai Nikolaevich Szczerbak. 1978. *Gymnodactylus caspius insularis* ssp. n. (Reptilia, Sauria), Novii podvid kaspiiskogo gekkona s ostrova Bulf v Kaspiiskom more. Vestnik Zoologii, 1978(2):80–82.
Akhmedov, S. B., and Nikolai Nikolaevich Szczerbak. 1987. Geographic variation and intraspecific systematics of *Mabuya aurata* (Sauria, Scincidae). Vestnik Zoologii, 1987(5):20–24 (In Russian).
Alcock, Alfred William, and Frank Finn. 1896. An account of the amphibians and reptiles collected by Dr. F. P. Maynard, Captain A. H. McMahon, C.I.E. and the members of the Afgan-Baluch Boundary Commission of 1896. Journal of the Asiatic Society of Bengal, 65, pt. 2(4):550–566, pls.11–15.
Alekperov, Abdulla Mustafaevitch. 1978. Zemnovodnnie i Presmikayushchiesya Azerbaidzhana SSR [Amphibians and Reptiles of Azerbaidzhan SSR]. Akademija Nauk Azerbaidzan SSR, Elm, Baku, 264 pp. (In Russian).
Allouse, Bashir E. 1955. Bibliography of the Vertebrate Fauna of Iraq and Neighboring Countries, vol. 3. Reptiles and Amphibians. Natural History Museum, Baghdad, 21 pp.
Ananjeva, Natalia Borisovna. 1981a. *Phrynocephalus helioscopus* (Pallas 1771) — Sonnengucker, pp. 191–202. *In* Wolfgang Böhme (ed.). Handbuch der Reptilien und Amphibien Europas, vol. 1. Echsen (Sauria) I. Akademische Verlagsgesellschaft, Wiesbaden.
Ananjeva, Natalia Borisovna. 1981b. *Phrynocephalus mystaceus* (Pallas 1776) — Bärtiger Krötenkopf, pp. 203–216. *In* Wolfgang Böhme (ed.). Handbuch der Reptilien und Amphibien Europas, vol. 1. Echsen (Sauria) I. Akademische Verlagsgesellschaft, Wiesbaden.
Ananjeva, Natalia Borisovna. 1981c. Structural characteristics of skull, dentition and hyoid of lizards of the genus *Agama* from the fauna of the USSR. Academy of Sciences of the USSR, Proceedings of the Zoological Institute, 101:3–20.
Ananjeva, Natalia Borisovna. 1998. Department of Herpetology, Zoological Institute, Russian Acad-

emy of Sciences, St. Petersburg, Russia: history and current research. Herpetological Review, 29(3):136–140, figs. 1–5.

Ananjeva, Natalia Borisovna, and Chary Ataevich Atayev. 1984. *Stellio caucasius triannulatus* ssp. nov. – A new subspecies of the Caucasian *Agama* from south-western Turkmenia. Trudy, Zoological Institute, Akademii Nauk USSR, Leningrad, 124:4–11, pls. 1–3.

Ananjeva, Natalia Borisovna, and Tatyana N. Dujsebayeva. 1997. SEM study of skin sense organs in two *Uromastyx* species (Sauria: Agamidae) and *Sphenodon punctatus* (Rhynchocephalia: Sphenodon). Russian Journal of Herpetology, 4:46-49, 1 fig.

Ananjeva, Natalia Borisovna, Leo Jakovlevich Borkin, Ilya Sergeevich Darevsky, and Nikolai Lutseanovich Orlov. 1988. Dictionary of Animals in Five Languages. Amphibians and Reptiles. Russky Yazyk Publ., Moscow, 554 + (2) pp. (In Latin, Russian, English, German, and French).

Ananjeva, Natalia Borisovna, Leo Jakovlevich Borkin, Ilya Sergeev Darevsky, and Nikolai Lutseanovich Orlov. Amphibians and Reptiles of Russia and Adjacent Territories. Encyclopedia of the Nature of Russia. ABF, Moscow, 415 pp. (In Russian).

Ananjeva, Natalia Borisovna, and Valentina Fyodorovna Orlova. 1979. Distribution and geographic variability of *Agama caucasia* (Eichwald, 1831). Proceedings of the Zoological Institute, USSR Academy of Sciences, 89:4–17 (In Russian; English abstract).

Ananjeva, Natalia Borisovna, and Tatyana Mikhailovna Sokolova. 1990. [The position of the genus *Phrynocephalus* Kaup, 1825 in agamid systems]. Trudy, Zoological Institute, Akademii Nauk USSR, Leningrad, 207:12–21 (In Russian; English abstract).

Ananjeva, Natalia Borisovna, and Boris S. Tuniyev. 1992. Historical biogeography of the *Phrynocephalus* species of the USSR. Asiatic Herpetological Research, 4:76–98.

Ananjeva, Natalia Borisovna, and Boris S. Tuniyev. 1994. Some aspects of historical biogeography of Asian rock agamids. Russian Journal of Herpetology, 1(1):42–52.

Anderson, John. 1872. On some Persian, Himalayan, and other reptiles. Proceedings of the Zoological Society of London, 1872(2):371–404.

Anderson, John. 1894. On two new species of agamoid lizards from the Hadramut, south-eastern Arabia. Annals and Magazine of Natural History, ser. 6, 14:376–378.

Anderson, John. 1896. A Contribution to the Herpetology of Arabia with a Preliminary List of the Reptiles and Batrachians of Egypt. R. H. Porter, London, 122 pp. (Reprint edition, Soc. Study Amphib. Reptiles, 1984).

Anderson, John. 1898. Reptilia and Batrachia. Zoology of Egypt, vol. 1. Bernard Quaritch, London, lvx + 371 pp., 60 col. pls., map. (Reprint 1965 Wheldon & Wesley & J. Cramer).

Anderson, John. 1984. Herpetology of Arabia. Facsimile Reprints in Herpetology. Society for the Study of Amphibians and Reptiles, Oxford, Ohio, xxxv + 122 pp. (Reprint of 1896 ed. Introduction and checklist of the Arabian herpetofauna by Alan E. Leviton and Michele L. Aldrich).

Anderson, Steven Clement. 1961. A note on the synonymy of *Microgecko* Nikolsky with *Tropiocolotes* Peters. Wassman Journal of Biology, 19(2):287–289.

Anderson, Steven Clement. 1963. Amphibians and reptiles from Iran. Proceedings of the California Academy of Sciences, ser. 4, 31(16):417–98, 15 figs.

Anderson, Steven Clement. 1966a. The turtles, lizards, and amphisbaenians of Iran. Ph.D., Stanford University, 660 p. (University Microfilms International, Order no. 66–14, 629. Ann Arbor).

Anderson, Steven Clement. 1966b. A substitute name for *Agama persica* Blanford. Herpetologica, 22(3):230.

Anderson, Steven Clement. 1966c. The lectotype of *Agama isolepis* Boulenger. Herpetologica, 22(3):230–231.

Anderson, Steven Clement. 1968. Zoogeographic analysis of the lizard fauna, chapter 10, pp. 305–371. *In* W. B. Fisher (ed.). The Land of Iran. The Cambridge History of Iran, vol. 1. Cambridge University Press, Cambridge, UK.

Anderson, Steven Clement. 1970. Checklist and key to the lizards of Iran. Year book of the American Philosophical Society, vol. for 1969. American Philosophical Society, Philadelphia, pp. 284–285.

Anderson, Steven Clement. 1972. Adaptations in geckos. Pacific Discovery, 25(1):1–11.

Anderson, Steven Clement. 1973. A new species of *Bunopus* (Reptilia: Gekkonidae) from Iran and a key to the genus *Bunopus*. Herpetologica, 29(4):355–358.

Anderson, Steven Clement. 1974. Preliminary key to the turtles, lizards and amphisbaenians of Iran. Fieldiana: Zoology, 65(4):27–44.

Anderson, Steven Clement. 1977. [Key to the Lizards and Amphisbaenians of Iran]. Iran National Museum of Natural History, Tehran, 25 pp. (In Farsi [Persian], translated and adapted by Yahdolah Seirani).

Anderson, Steven Clement. 1978a. Geographic distribution: *Pelobates syriacus syriacus*. Herpetological Review, 9(1):21.

Anderson, Steven Clement. 1978b. Geographic distribution: *Rana macrocnemis pseudodalmatina*. Herpetological Review, 9(1):22.

Anderson, Steven Clement. 1978c. Geographic distribution: *Eremias acutirostris*. Herpetological Review, 9(1):22.

Anderson, Steven Clement. 1978d. Geographic distribution: *Eryx jayakari*. Herpetological Review, 9(1):22.

Anderson, Steven Clement. 1979. Synopsis of the turtles, crocodiles, and amphisbaenians of Iran. Proceedings of the California Academy of Sciences, ser. 4, 41(22):501–528.

Anderson, Steven Clement. 1984. (Review of) Herpetology of Europe and Southwest Asia: A checklist and bibliography of the orders Amphisbaenia, Sauria and Serpentes by K. R. G. Welch. Association of Systematics Collections Newsletter, 12(2):19–21.

Anderson, Steven Clement. 1985. Amphibians [of Iran], pp. 987–990. *In* Ehsan Yarshater (ed.). Encyclopaedia Iranica, vol. 1, fasc. 9. Routledge and Kegan Paul, London.

Anderson, Steven Clement. 1993. A note on the syntopic occurrence of three species of *Teratoscincus* in eastern Iran. Dactylus, 1(4):8–10, figs. 1–3.

Anderson, Steven Clement. 1995. Preliminary observation and comments on lizard ecology in Iran, pp.12–13. *In* Natalia Ananjeva (ed.). Abstracts of the Second Asian Herpetological Meeting, 6–10 September, 1995, Asgabat, Turkmenistan. Folium Publishing Co., Moscow.

Anderson, Steven Clement. 1997. (Review of) Gecko fauna of the USSR and contiguous regions by Nikolai N. Szczerbak and Michael L. Golubev. 1996. Translated from the Russian by Michael L. Golubev and Sasha A. Malinsky. Editorial supervision by Alan E. Leviton and George R. Zug. Society for the Study of Amphibians and Reptiles. xii + 233 pp. Herpetological Review, 28(4):221–222.

Anderson, Steven Clement, and Alan Edward Leviton. 1966a. A new species of *Eublepharis* from southwestern Iran (Reptilia: Gekkonidae). Occasional Papers of the California Academy of Sciences, (53):1–5.

Anderson, Steven Clement, and Alan Edward Leviton. 1966b. A review of the genus *Ophiomorus* (Sauria: Scincidae) with descriptions of three new forms. Proceedings of the California Academy of Sciences, ser. 4, 33(16):499–534.

Anderson, Steven Clement, and Alan Edward Leviton. 1967a. A new species of *Eremias* (Reptilia: Lacertidae) from Afghanistan. Occasional Papers of the California Academy of Sciences, (64):1–4.

Anderson, Steven Clement, and Alan Edward Leviton. 1967b. A new species of *Phrynocephalus* (Sauria: Agamidae) from Afghanistan, with remarks on *Phyrnocephalus ornatus* Boulenger. Proceedings of the California Academy of Sciences, ser. 4, 35(11):227–234.

Anderson, Steven Clement, and Alan Edward Leviton. 1969. Amphibians and reptiles collected by the Street Expedition to Afghanistan, 1965. Proceedings of the California Academy of Sciences, ser. 4, 37(2):25–56, 8 figs.

Anderson, Steven Clement, and Alan Edward Leviton. 1989. Late Tertiary compressional deformation of Southwestern Asia and its impact on the present-day distribution of amphibians and reptiles: A case study in plate tectonics and biogeography. First World Congress of Herpetology, 11–19 September 1989, University of Kent at Canterbury, Abstracts (no pagination).

Andersson, Lars Gabriel. 1900. Catalogue of Linnean type-specimens of Linnaeus's Reptilia in the

Royal Museum in Stockholm. Kunglica Svenska Vetenskapsakademiens Handlingar, 26, pt. 4(1):1–29.

Andrén, Claes, and Göran Nilson. 1979a. A new species of toad (Amphibia, Anura, Bufonidae) from the Kavir Desert, Iran. Journal of Herpetology, 13(1):93–100, figs.1–3.

Andrén, Claes, and Göran Nilson. 1979b. *Vipera latifi* (Reptilia, Serpentes, Viperidae), an endangered viper from Lar Valley, Iran, and remarks on the sympatric herpetofauna. Journal of Herpetology, 13(3):335–341.

Angel, Fernand. 1936. Reptiles et batrachiens de Syrie et de Mésopotamie récoltés par M. P. Pallary. Bulletin de l'Institut d'Égypté, 18:107–116.

Annandale, Thomas Nelson. 1906a. *Testudo baluchiorum*, a new species. Journal of the Asiatic Society of Bengal, new series, 2(3):75–76.

Annandale, Thomas Nelson. 1906b. Notes on the fauna of a desert tract in southern India. Part I. Batrachians and reptiles, with remarks on the reptiles of the desert region of the Northwest Frontier. Memoirs of the Asiatic Society of Bengal, 1:183–202, pls. 9–10.

Annandale, Thomas Nelson. 1913. The Indian geckos of the genus *Gymnodactylus*. Records of the Indian Museum, 9:309-326, 2 pls.

Anonymous. 1905a. Eminent living geologists: W. T. Blanford. Geological Magazine, Decade 5, 2(1):1–15, pl. 1.

Anonymous. 1905b. William Thomas Blanford, 1832–1905. Ibis, 1905:643–647.

Anonymous (probably Malcolm A. Smith). 1945. Reptiles and Amphibians. Persia. Geographical Handbook Series. B.R. 525 (restricted). Naval Intelligence Division, Oxford University Press, Oxford, pp. 18–21.

Applegarth, John Stirling.1989. Index of authors in taxonomic herpetology, pp. 143–178. *In* Kraig Kerr Adler (ed.). Contributions to the History of Herpetology. Contributions to Herpetology, no. 5. Society for the Study of Amphibians and Reptiles, Oxford, Ohio.

Arnold, Edwin Nicholas. 1972. Lizards with northern affinities from the mountains of Oman. Zoologische Mededelingen Leiden, 47:111–128, 2 pls.

Arnold, Edwin Nicholas. 1973. Relationships of the Palearctic lizards assigned to the genera *Lacerta*, *Algyroides* and *Psammodromus* (Reptilia: Lacertidae). Bulletin of the British Museum (Natural History), Zoology, 25(8):291–366.

Arnold, Edwin Nicholas. 1977. Little-known geckoes (Reptilia: Gekkonidae) from Arabia with descriptions of two new species from the Sultanate of Oman. Journal of Oman Studies, (1) Special Report 1:81–110, map.

Arnold, Edwin Nicholas. 1980a. Reptiles of Saudi Arabia: A review of the lizard genus *Stenodactylus* (Reptilia: Gekkonidae). Fauna of Saudi Arabia, 2:368–404, 14 figs.

Arnold, Edwin Nicholas. 1980b. The reptiles and amphibians of Dhofar, Southern Arabia. Journal of Oman Studies, (Special Report 2. Scientific Results of the Flora and Fauna Survey 1977 [Dhofar]):273–332, 24 col. pls., 1 foldout map.

Arnold, Edwin Nicholas. 1983. Osteology, genitalia and the relationships of *Acanthodactylus* (Reptilia: Lacertidae). Bulletin of the British Museum (Natural History), Zoology, 44(5):291–339, 9 text-figs., 9 tables.

Arnold, Edwin Nicholas. 1984a. Evolutionary aspects of tail shedding in lizards and their relatives. Journal of Natural History, London, 18(1):127–169.

Arnold, Edwin Nicholas. 1984b. Ecology of lowland lizards in the eastern United Arab Emirates. Journal of Zoology, London, 204(3):329–354.

Arnold, Edwin Nicholas. 1984c. Variation in the cloacal and hemipenial muscles of lizards and its bearing on their relationships. Symposia of the Zoological Society, London, (52):47–85.

Arnold, Edwin Nicholas. 1986a. Mite pockets of lizards, a possible means of reducing damage by ectoparasites. Biological Journal of the Linnean Society, 29:1–21.

Arnold, Edwin Nicholas. 1986b. New species of semaphore gecko (*Pristurus*: Gekkonidae) from Saudi Arabia and Socotra. Fauna of Saudi Arabia, 8:352–377.

Arnold, Edwin Nicholas. 1986c. A new spiny-footed lizard (*Acanthodactylus*: Lacertidae) from Saudi Arabia. Fauna of Saudi Arabia, 8:378–384, pl. 1, figs. 1–5.

Arnold, Edwin Nicholas. 1986d. A key and annotated check list to the lizards and amphisbaenians of Arabia. Fauna of Saudi Arabia, 8:385–435.

Arnold, Edwin Nicholas. 1986e. The hemipenis of lacertid lizards (Reptilia: Lacertidae): structure, variation and systematic implications. Journal of Natural History, 20:1221–1257.

Arnold, Edwin Nicholas. 1989. Towards a phylogeny and biogeography of the Lacertidae: Relationships within an Old-World family of lizards derived from morphology. Bulletin of the British Museum (Natural History), Zoology, 55(2):209–257.

Arnold, Edwin Nicholas. 1992. The Rajasthan toad-headed lizard, *Phrynocephalus laungwalaensis* (Reptilia: Agamidae), represents a new genus. Journal of Herpetology, 26(4):467–472.

Arnold, Edwin Nicholas. 1993a. Comment — Function of the mite pockets of lizards: an assessment of a recent attempted test. Canadian Journal of Zoology, 71(4):862–864.

Arnold, Edwin Nicholas. 1993b. Historical changes in the ecology and behaviour of semaphore geckos (*Pristurus*, Gekkonidae) and their relatives. Journal of Zoology, London, 229(3):353–384.

Arnold, Edwin Nicholas, and John A. Burton. 1978. A Field Guide to the Reptiles and Amphibians of Britian and Europe. Collins, London, 272 pp., 40 col. pls., 126 maps.

Arnold, Edwin Nicholas, and Michael D. Gallagher. 1977. Reptiles and amphibians from the mountains of northern Oman with special reference to the Jebel Akhdar region. Journal of Oman Studies, (Special Report 1. Scientific Results of the Oman Flora and Fauna Survey):59–80, map.

Arnold, Edwin Nicholas, and Andrew S. Gardner. 1994. A review of the Middle Eastern leaf-toed geckoes (Gekkonidae: *Asaccus*) with descriptions of two new species from Oman. Fauna of Saudi Arabia, 14:424–441.

Arnold, Edwin Nicholas, and Alan Edward Leviton. 1977. A revision of the lizard genus *Scincus* (Reptilia: Scincidae). Bulletin of the British Museum (Natural History), Zoology, 31:187–248.

Ast, Jennifer C. 1997. A phylogenetic analysis of *Varanus* using mitochondrial DNA sequences. *In* Zbyněk Roček and Scott Hart (eds.). Herpetology '97. Abstracts of the Third World Congress of Herpetology 2–10 August 1997 Prague, Czech Republic. Third World Congress of Herpetology, Prague (no pagination).

Atayev, Chary Ataevich. 1985. Presmikajushchiesja gor Turkmenistana. [Mountain Reptiles of Turkmenistan]. Ylym Publ., Ashkhabad, 344 pp. (In Russian).

Atayev, Chary Ataevich. "1993" (1994). On finding *Cyrtopodion longipes* Nikolsky, 1896 on the north slope of east Kopetdag. Izvestiya Akademii Nauk Turkmenistana, 4:62–63.

Atayev, Chary Ataevich, Oleg Pavlovich Bogdanov, and Sakhat Muradovich Shammakov. 1968. On the finding of *Alsophylax spinicauda* (Squamata: Gekkonidae) in the USSR. Zoologicheski'i Zhurnal, Akademiya Nauk SSSR, 47(9):1420–1421, 1 fig. (In Russian with English abstract).

Atayev, Chary Ataevich, Boris S. Tuniyev, and Sakhat Muradovich Shammakov. "1993" (1994). Data on status of some rare and little-known species of reptiles in Kopetdag. Izvestiya Akademii Nauk Turkmenistana, 4:15–21.

Atayev, Chary Ataevich, Anver Kejusevich Rustamov, and Sakhat Muradovich Shammakov. 1994. Reptiles of Kopetdagh, chapter 20, pp. 329–350. *In* Victor Fet and Khabibulla Atamuradov (eds.). Biogeography and ecology of Turkmenistan. Monographiae Biologicae, vol. 72. Kluwer Academic Publishers, Dordrecht, Boston, and London.

Auffenberg, Walter. 1979. Intersexual differences in behavior of captive *Varanus bengalensis* (Reptilia, Lacertilia, Varanidae). Journal of Herpetology, 13(3):313–315.

Auffenberg, Walter. 1981. Combat behavior in *Varanus bengalensis*. Journal of the Bombay Natural History Society, 78(1):54–72.

Auffenberg, Walter. 1983a. Courtship behavior in *Varanus bengalensis* (Sauria: Varanidae), pp. 535–551. *In* Anders G. J. Rhodin and Kenneth Miyata (eds.). Advances in herpetology and evolutionary biology. Essays in honor of Ernest E. Williams. Museum of Comparative Zoology, Harvard University, Cambridge, Massachusetts.

Auffenberg, Walter. 1983b. The burrows of *Varanus bengalensis*: characteristics and use. Records of the Zoological Survey of India, 80(3–4):375–385.

Auffenberg, Walter. 1989. Exploitation of monitor lizards in Pakistan. World Wildlife Fund Traffic Bulletin, 11(1):8–12

Auffenberg, Walter. 1994. The Bengal Monitor. University of Florida, Gainesville, xxvii + 560 pp.
Auffenberg, Walter, and Naeem Ahmed Khan. 1991. Studies of Pakistan reptiles. Notes on *Kachuga smithi*. Hamadryad, 16(1–2):25–29.
Auffenberg, Walter, and Hafeezur Rehman. 1991. Studies on Pakistan reptiles, pt. 1. The genus *Echis* (Viperidae). Bulletin of the Florida Museum of Natural History. Biological Sciences, 35(5):263–314.
Auffenberg, Walter, and Hafeezur Rehman. 1993. Studies on Pakistan reptiles, pt. 3. *Calotes versicolor*. Asiatic Herpetological Research, 5:14–30.
Auffenberg, Walter, Hafeezur Rehman, Fehmida Iffat, and Zahida Perveen. 1989. A study of *Varanus flavescens* (Hardwick & Gray) (Sauria: Varanidae). Journal of the Bombay Natural History Society, 86(3):286–307, figs. 1–10.
Auffenberg, Walter, Hafeezur Rehman, Fehmida Iffat, and Zahida Perveen. 1990. Notes on the biology of *Varanus griseus koniecznyi* Mertens (Sauria:Varanidae). Journal of the Bombay Natural History Society, 87(1):26–36, figs. 1–4.
Auffenberg, Walter, Q. N. Arain, and N. Khurshid. 1991. Preferred habitat, home range and movement patterns of *Varanus bengalensis* in southern Pakistan. Mertensiella, (2):7–28.
Autumn, Kellar, and Dale F. Denardo. 1995. Behavioral thermoregulation increases growth rate in a nocturnal lizard. Journal of Herpetology, 29(2):157–162.
Avery, Peter, Gavin Hambly, and Charles Melville. 1991. From Nader Shah to the Islamic Republic. The Cambridge History of Iran, vol. 7. Cambridge University Press, London, xxiii + 1072 pp.
Avery, Roger Anthony. 1982. Field studies of body temperatures and thermoregulation, chapter 4, pp. 93–166. *In* Carl Gans and F. Harvey Pough (eds.). Biology of the Reptilia, Physiology C. Physiological Ecology. Biology of the Reptilia, vol. 12. Academic Press, New York.
Baig, Khalid Jared, and Wolfgang Böhme. 1991. Callous scalation in female agamid lizards (*Stellio* group of *Agama*) and its functional implications. Bonner Zoologische Beitrage, 42(3–4):275–281.
Balland, Daniel. 1989. Boundaries iii. Boundaries of Afghanistan, pp. 404–415. *In* Ehsan Yarshater (ed.). Encyclopaedia Iranica, vol. 4, fasc. 4. Routledge and Kegan Paul, New York and London.
Balletto, Emilio M., Maria Adelaide Cherchi, and John Gasperetti. 1985. Amphibians of the Arabian Peninsula. Fauna of Saudi Arabia, 7:318–392.
Baloutch, Mohammad. 1972. *Lytorhynchus maynardi* (Alcock and Finn 1896) trouve en Iran. Bulletin of the Faculty of Science, Tehran University, 4:79–82.
Baloutch, Mohammad. 1976. Une nouvelle espèce de *Lacerta* (Lacertilia, Lacertidae) du sud-est de l'Iran. Bulletin du Muséum National d'Histoire Naturelle, Paris, 294:1379–1384.
Baloutch, Mohammad, and Haji Gholi Kami. 1995. Amphibians of Iran. Tehran University Publications, Tehran, 177 pp. (In Farsi [Persian]).
Baloutch, Mohammad, and Michel Thireau. 1986. Une espèce nouvelle de gecko *Eublepharis ensafi* (Sauria, Gekkonidae, Eublepharinae) du Khouzistan (Sud ouest de l'Iran). Bulletin Mensuel de la Société Linnéene de Lyon, 55(8):281–288.
Bannikov, Andrei Grigoryevich, Ilya Sergeevich Darevsky, Vladimir G. Ischchenko, Anver Kejusevich Rustamov, and Nikolai Nikolaevich Szczerbak. 1977. Opredelitel Zemnovodnikh i Presmikayushchikhsya Fauni SSSR [Guide to the Reptiles and Amphibians of the USSR]. Enlightenment Publ., Moscow, 414 pp. (In Russian).
Banta, Benjamin Harrison 1961. Herbivorous feeding of *Phrynosoma platyrhinos* in southern Nevada. Herpetologica, 17(2):136–137.
Baran, İbrahim. 1986. Bibliographie der Amphibien und Reptilien der Türkei, pp. 79–118. *In* Max Kasparek (ed.). Zoologische Bibliographie der Türkei. Pisces, Amphibia, Reptilia. Max Kasparek Verlag, Heidelberg (Parallel texts in German and English).
Baran, İbrahim, and Mehmet Kutsay Atatür. 1998. Turkish Herpetofauna (Amphibians and Reptiles). Republic of Turkey, Ministry of Environment, Ankara. (13) + 214 pp., 122 col. pls.
Baran, İbrahim, and Ulrich F. Gruber. 1981. Taxonomische Untersuchungen an türkischen Inselformen von *Cyrtodactylus kotschyi* (Steindachner 1870) (Reptilia: Gekkonidae). Teil I. Die Popu-

lationen der nördlichen Ägäis, des Marmarameeres und des Schwarzen Meeres. Spixiana, 4(3):255–270.
Baran, İbrahim, and Ulrich F. Gruber. 1982. Taxonomische Untersuchungen an türkischen Gekkoniden. Spixiana, Munich, 5(2):109–138, 5 figs.
Barbour, Thomas. 1908. Some new reptiles and amphibians. Bulletin of the Museum of Comparative Zoology, Harvard College, 51(12):315–325.
Başoğlu, Muhtar, and İbrahim Baran. 1977. Kisim I. Lamplumbaga ve Kertenkeleler [Part I. The Turtles and Lizards.]. Turkiye Surungenleri [The Reptiles of Turkey]. Ilker Matbaasi, Bornova-Izmir, 272 pp. (In Turkish with English summary, pp. 191–233, 235–236).
Başoğlu, Muhtar, and İbrahim Baran. 1980. Kisim II. Yilanlar [Part II. The Snakes.]. Turkiye Surungenleri [The Reptiles of Turkey]. Ilker Matbaasi, Bornova-Izmir, ix + 218 pp. (In Turkish with English summary).
Başoğlu, Muhtar, and Neclâ Özeti. 1973. Turkiye Amfibleri. [The Amphibians of Turkey]. Ege Üniversitesi Fen Fakültesi Kitaplar, ser. 50. Ege Üniversitesi Matbaasi, Bornova-İzmir, (x) + 155 pp. (In Turkish).
Bastani, B. 1979. A clinical review of snakebite with emphasis on species in Fars Province. Iranian Journal of Medical Science, 10(1–4):163–200.
Bauer, Aaron Matthew. 1987. (Review of) The gekkonid fauna of the USSR and adjacent countries by N. N. Szczerbak and M. Golubev. Copeia, 1987(2):525–527.
Bauer, Aaron Matthew, and Rainer Günther. 1991. An annotated type catalogue of the geckos (Reptilia: Gekkonidae) in Zoological Museum, Berlin. Mitteilungen aus dem Zoologischen Museum in Berlin, 67(2):279–310.
Bauer, Aaron Matthew, and Rainer Günther. 1995. An annotated type catalogue of the lacertids (Reptilia: Lacertidae) in the Zoological Museum, Berlin. Mitteilungen aus dem Zoologischen Museum in Berlin, 71(1):37–62.
Bauer, Aaron Matthew, and Anthony Patrick Russell. 1991. Pedal specialisations in dune-dwelling geckos. Journal of Arid Environments, 20:43–62.
Bauer, Aaron Matthew, and Anthony Patrick Russell. 1994. Is autotomy frequency reduced in geckos with "actively functional" tails? Herpetological Natural History, 2(2):1–15.
Bauer, Aaron Matthew, Anthony Patrick Russell, and Robert E. Shadwick. 1989. Mechanical properties and morphological correlates of fragile skin in gekkonid lizards. Journal of Experimental Biology, 145:79–102.
Bauer, Aaron Matthew, Anthony Patrick Russell, and Norman R. Dollahon. 1990. Skin folds in the gekkonid genus *Rhacodactylus*: a natural test of the damage limitation hypothesis of mite pocket function. Canadian Journal of Zoology, 68:1196–1201.
Bauer, Aaron Matthew, Anthony Patrick Russell, and Norman R. Dollahon. 1993. Function of the mite pockets of lizards: a reply to E. N. Arnold. Canadian Journal of Zoology, 71(4):865–868.
Bazin, M., Eckart Ehlers, and Bernard Hourcade. 1985. Alborz iii. Geography of the Alborz, pp. 813–821. In Ehsan Yarshater (ed.). Encyclopaedia Iranica, vol. 1, fasc. 8. Routledge and Kegan Paul, New York and London.
Bedriaga, Jacques Vladimir von. 1879a. Herpetologische Studien (Fortsetzung). Archiv für Naturgeschichte, 45(1):243–339, pls. 17–18.
Bedriaga, Jacques Vladimir von. 1879b. Verzeichniss der Amphibien und Reptilien Vorder-Asiens. Bulletin de la Société Imperiale des Naturalistes de Moscou, 54(3):22–52.
Bedriaga, Jacques Vladimir von. 1886. Beiträge zur Kenntnis der Lacertiden-Familie (*Lacerta, Algiroides, Tropidosaura, Zerzumia* und *Bettaia*). Abhandlungen der Senckenbergischen Naturforschenden Gesellschaft, 14(1):17–144.
Bedriaga, Jacques Vladimir von. "1905" (1906). Neue Saurier aus Russisch-Asien. Annuaire du Musée Zoologique de l'Académie Impériale des Sciences de St.-Pétersbourg, 10(3–4):210–243.
Bernor, L. R. 1983. Geochronology and zoogeographic relationships of Miocene hominoids, pp. 21–64. In Russell L. Ciochon and R. Corruccini (eds.). New Interpretations of Ape and Human Ancestry. Plenum Press, New York.
Beutler, Axel. 1981. *Cyrtodactylus kotschyi* (Steindachner, 1870): Ägäischer Bogenfingergecko, pp.

53–74. *In* Wolfgang Böhme (ed.). Handbuch der Reptilien und Amphibien Europas, vol. 1. Echsen (Sauria) I. Akademische Verlagsgesellschaft, Wiesbaden.

Bibron, Gabriel, and Jean Baptiste Geneviève Marcellin Bory de Saint Vincent. 1833. Vertébrés à sang froid. Reptiles et Poissons, pp. 57–80. *In* Isidore Geoffroy Saint-Hilaire and Étienne-François Geoffroy Saint-Hilaire (eds.). Expédition Scientifique de Mond, vol. 3, pt. 1. Zoologie. Paris, (Reptiles, pp. 57–74, pls. 6–14; Amphibians, pp. 74–76, pl. 15).

Bischoff, Wolfgang, and Wolfgang Böhme. 1980. Der Systematische Status der türkischen Wüstenrenner des Subgenus *Eremias* (Sauria: Lacertidae). Bonner Zoologische Beiträge, 26(2):297–306, 3 figs.

Black, Jesse. 1997. Keeping and breeding leopard geckos (*Eublepharis macularius*). Reptiles, 5(3):10–18.

Blanford, William Thomas. 1874a. Description of new lizards from Persia and Baluchistan. Annals and Magazine of Natural History, ser. 4, 13(78):453–455.

Blanford, William Thomas. 1874b. Descriptions of new reptiles and amphibians from Persia and Baluchistan. Annals and Magazine of Natural History, ser. 4, 14:31–35.

Blanford, William Thomas. 1874c. Note on *Ablepharus pusillus*. Annals and Magazine of Natural History, ser. 4, 14:461.

Blanford, William Thomas. "1874" (1875a). Description of two uromasticine lizards from Mesopotamia and southern Persia. Proceedings of the Zoological Society of London, 1874:656–661.

Blanford, William Thomas. 1875b. List of Reptilia and Amphibia collected by the late Dr. Stoliczka in Kashmir, Ladak, Eastern Turkestan, and Wakkan, with descriptions of new species. Journal of the Asiatic Society of Bengal, 44(2):191–196, 201, 2 pls.

Blanford, William Thomas. 1875c. On some lizards from Sind. Journal of the Asiatic Society of Bengal, 44(2):232–233.

Blanford, William Thomas. 1876a. Eastern Persia. An Account of the Journeys of the Persian Boundary Commission, 1870–1872, vol. 2. The Zoology and Geology. Macmillan and Co., London, viii + 516 pp.

Blanford, William Thomas. 1876b. On some lizards from Sind, with descriptions of new species of *Ptyodactylus*, *Stenodactylus* and *Trapelus*. Journal of the Asiatic Society of Bengal, 45(2):18–26, 2 pls.

Blanford, William Thomas. 1879. Notes on Reptilia. Journal of the Asiatic Society of Bengal, 48, 2(3):127–132.

Blanford, William Thomas. 1881. On a collection of Persian reptiles recently added to the British Museum. Proceedings of the Zoological Society of London, 1881:671–682, pl. 59, textfigs. 1–4.

Bloom, R. A., K. W. Selcer, and W. K. King. 1986. Status of the introduced gekkonid lizard, *Cyrtodactylus scaber*, in Galveston, Texas. Southwestern Naturalist, 31:129–131.

Blyth, Edward. 1853. Notices and descriptions of various reptiles, new or little known. Journal of the Asiatic Society of Bengal, 22(7):639–655.

Blyth, Edward. 1854. Report of the Curator, Zoological Department. Journal of the Asiatic Society of Bengal, 23(7):737–740.

Bobek, Hans. 1952a. Die Klimaökologische Gliederung von Iran, pp. 244–248, 1 fig. Proceedings of the VIIIth General Assembly — XVIIth Congress International Geographical Union, Washington, DC 1952. Washington, DC.

Bobek, Hans. 1952b. Beiträge zur Klima-ökologischen Gliederung Irans. Erdkunde, Archiv für Wissenchaftliche Geographie, 6:65–84, 6 figs., map suppl.

Bobek, Hans. 1963. Nature and implications of Quarternary climatic changes in Iran. Arid Zone Research, 20:403–413.

Bobek, Hans, 1968, Vegetation, chapter 8, pp. 280–293. *In* W. B. Fisher (ed.). The Land of Iran. The Cambridge History of Iran, vol. 1. Cambridge University Press, Cambridge, UK.

Böhme, Wolfgang. 1985. Zur Nomenklatur der Paläarctischen Bogenfingergeckos, Gattung *Tenuidactylus* Ščerbak & Golubev, 1984 (Reptilia: Gekkonidae). Bonner Zoologische Beiträge, 36(1–2):95–98.

Böhme, Wolfgang. 1993. Hemipenial microornamentation in *Lacerta brandtii* De Filippi, 1863:

Falsification of a systematic hypothesis? (Squamata: Sauria: Lacertidae). Herpetozoa, 6(3–4):141–143.

Boettger, Oskar. 1876. Über *Hemidactylus turcicus* von Türkei. Bericht über die Thätligkeit des Offenbacher Vereins für Naturkunde, 1876:57.

Boettger, Oskar. 1880. Die Reptilien und Amphibien von Syrien, Palestina, und Cypern. Berichte über die Senckenbergische Naturforschende Gesellschaft, Frankfurt am Main, 1879–1880:132–219.

Boettger, Oskar. 1886. Reptilia et Amphibia, pp. 30–82. *In* G. Radde (ed.). Die Fauna und Flora des Südwestlichen Caspi-Gebietes. Leipzig.

Boettger, Oskar. 1888a. Die Reptilien und Batrachien Transkaspiens. Zoologischen Jahrbüchern. Abtheilung für Systematik, Geographie und Biologie der Thiere, 3:870–972, pl. 34, figs. 1–4.

Boettger, Oskar. 1888b. Über die Reptilien und Batrachier Transcaspiens. Zoologischer Anzeiger, 11(279):259–263.

Boettger, Oskar. 1892a. Katalog der Batrachier-Sammlung im Museum der Senckenbergischen Naturforschenden Gesellschaft in Frankfurt am Main. Senckenbergischen Naturforschenden Gesellschaft, Frankfurt am Main, x + 73 pp.

Boettger, Oskar. 1892b. Wissenschaftliche Ergebnisse der Reise Dr. Jean Valentins im Sommer 1890. Berichte über die Senckenbergische Naturforschende Gesellschaft Frankfurt am Main, 1892:131–150.

Boettger, Oskar. 1893. Katalog der Reptilien-Sammlung im Museum der Senckenbergischen Naturforschenden Gesellschaft in Frankfurt am Main. I. Teil (Rhynchocephalen, Schildkröten, Krokodile, Eidechsen, Chamäleons). Senckenbergischen Naturforschenden Gesellschaft, Frankfurt am Main, x + 140 pp.

Boettger, Oskar. 1898. Katalog der Reptilien-Sammlung im Museum der Senckenbergischen Naturforschenden Gesellschaft in Frankfurt am Main. II. Teil. Schlangen. Senckenbergischen Naturforschenden Gesellschaft, Frankfurt am Main, ix + 160 pp.

Bogdanov, Anatoly P. 1891. Materials for the history of scientific activity in Russia on zoology. Newsbulletin of the Imperial Moscow University, no.70 (Transactions of the Department of Zoology, no. 6), Moscow. (In Russian).

Bogdanov, Oleg Pavlovich 1962. Presmikayuschiesya Turkmenii [Reptiles of Turkmenia]. Izdatel 'stvo Akademii Nauk Turkmenskoi SSR, Ashkhabad, 235 pp. (In Russian).

Börner, Achim-Rüdiger. 1976. Second contribution to the systematics of the southwest Asian lizards of the geckonid genus *Eublepharis* Gray 1827: Materials from the Indian subcontinent. Saurologica, (2):1–15, 11 figs., 10 tabs., 3 pls.

Börner, Achim-Rüdiger. 1981a. The genera of Asian eublepharine geckos and a hypothesis of their phylogeny. Miscellaneous Articles in Saurology [privately printed], Cologne, (9):1–14.

Börner, Achim-Rüdiger. 1981b. Third contribution to the systematics of the southwest Asian lizards of the geckonid genus *Eublepharis* Gray 1827: Further materials from the Indian subcontinent. Saurologica, (3):1–7.

Bornmüller, J. 1893. Reisebriefe aus Persien. Mitteilungen des Thuringischen botanischen Vereins, 1893(3–4):31–49.

Bosch, Marc, Steven Bullock, and Wayne Kinunen. 1970. Crocodiles; second survey. Job Completion Report. Division of Research and Development, submitted to Iran Game and Fish Department, F-6–49, 26 pp. (Unpublished).

Bosuk-Białynicka, Magdelena, and Scott Michael Moody. 1984. Priscagaminae, a new subfamily of the Agamidae (Sauria) from the late Cretaceous of the Gobi Desert. Acta Paleontologica Polonica, 29:51–81.

Boulenger, Edward George. 1920. On some lizards of the genus *Chalcides*. Proceedings of the Zoological Society of London, 1920:77–83, 4 figs.

Boulenger, George Albert. 1880. On the Palearctic and Aethiopian species of *Bufo*. Proceedings of the Zoological Society of London, 37:545–547.

Boulenger, George Albert. 1882a. Catalogue of the Batrachia Salientia s. Ecaudata in the Collection of the British Museum. Trustees of the British Museum, London, xvi + 503 pp. (published 25

March, 1882, *fide* Sherborn, 1934; reprint edition, Wheldon and Wesley, Codicote, and Verlag J. Cramer, Lehre, 1966).

Boulenger, George Albert. 1882b. Catalogue of the Batrachia Gradientia s. Caudata and Batrachia Apoda in the Collection of the British Museum. Trustees of the British Museum British Museum, London, viii + (1) + 127 pp., 9 pls. (published 9 December 1882, *fide* Sherborn, 1934; reprint edition, Wheldon and Wesley, Codicote, and Verlag J. Cramer, Weinheim, 1965).

Boulenger, George Albert. 1885a. Catalogue of the Lizards in the British Museum (Natural History). Geckonidae, Eublepharidae, Uroplatidae, Pygopodidae, Agamidae, vol. 1 [2nd ed.]. Trustees of the British Museum, London, xii + 497 pp. Reprint edition, Wheldon and Wesley, Codicote, and Verlag J. Cramer, Weinheim, 1965).

Boulenger, George Albert. 1885b. Catalogue of the Lizards in the British Museum (Natural History). Iguanidae, Xenosauridae, Zonuridae, Anguidae, Anniellidae, Helodermatidae, Varanidae, Xantusiidae, Teiidae, Amphisbaenidae, vol. 2 [2nd ed.]. Trustees of the British Museum, London, xiii + 497 pp. (Reprint edition, Wheldon and Wesley, Codicote, and Verlag J. Cramer, Weinheim, 1965).

Boulenger, George Albert. 1887a. Catalogue of the Lizards in the British Museum (Natural History) Lacertidae, Gerrhosauridae, Scincidae, Anelytropidae, Dibamidae, Chamaeleonidae, vol. 3 [2nd ed.]. Trustees of the British Museum, London, xii + 575 pp., 40 pls. (Reprint edition, Wheldon and Wesley, Codicote, and Verlag J. Cramer, Weinheim, 1965).

Boulenger, George Albert. 1887b. Les espèces du genre *Opiomore*. Bulletin de la Société Zoologique de France, 12:519–534.

Boulenger, George Albert. 1889a. Catalogue of the Chelonians, Rhynchocephalians, and Crocodiles in the British Museum (Natural History) [new ed.]. Trustees of the British Museum, London, x + 311, pls. 1–6 pp. (Reprint edition, Wheldon and Wesley, Codicote, and Verlag J. Cramer, Lehre, 1966).

Boulenger, George Albert. 1889b. Reptiles and batrachians. *In* J. E. T. Aitchison, The zoology of the Afghan Delimitation Commission. Transactions of the Linnean Society, London, ser. 2, 5(3):94–106, pls. 8–11.

Boulenger, George Albert. 1890a. Reptilia and Batrachia. The Fauna of British India, including Ceylon and Burma. Taylor and Francis, London, xviii + 541, 142 pp., figs.

Boulenger, George Albert. 1890b. On the occurrence of *Eublepharis macularius* in Transcaspia. Annals and Magazine of Natural History, ser. 6, 5:352.

Boulenger, George Albert. 1893. Catalogue of the Snakes in the British Museum. Containing the Families Typhlopidae, Glauconiidae, Boidae, Ilysiidae, Uropeltidae, Xenopeltidae, and Colubridae Aglyphae, vol. 1. British Museum (Natural History), London, xiii + 448 pp., 28 pls. (Reprint edition, J. Cramer, Weinheim, Wheldon and Wesley, Codicote, and Hafner, New York, 1961).

Boulenger, George Albert. 1894. Catalogue of the Snakes in the British Museum (Natural History). Containing the conclusion of the Colubridae Aglyphae, vol. 2. British Museum (Natural History), London, xi + 382 pp., 20 pls. (Reprint edition, J. Cramer, Weinheim, Wheldon and Wesley, Coldicote, and Hafner, New York, 1961).

Boulenger, George Albert. 1896a. Catalogue of the Snakes in the British Museum (Natural History). Containing the Colubridae (Opisthoglyphae and Proteroglyphae), Amblycephalidae, and Viperidae, vol. 3. British Museum (Natural History), London, xiv + 727 pp., 25 pls.

Boulenger, George Albert. 1896b. On the lizards of the genus *Eremias*, section *Boulengeria*. Proceedings of the Zoological Society of London, 1896:920–930.

Boulenger, George Albert. 1899. Reptilia and Amphibia of Lake Urmi and its neighborhood. Journal of the Linnean Society of London, 27:378–381.

Boulenger, George Albert. 1904. On the *Lacerta depressus* of Camerano. Proceedings of the Zoological Society of London, 1904(part 2):232–239.

Boulenger, George Albert. 1909a. Description of a new lizard of the genus *Acanthodactylus* from Syria. Annals and Magazine of Natural History, ser. 8, 4:188–189.

Boulenger, George Albert. 1909b. Description of a new species of *Lacerta* from Persia. Proceedings of the Zoological Society of London, 1908:934–936, 1 pl.

Boulenger, George Albert. 1913. Second contribution to our knowledge of the varieties of the wall-lizard (*Lacerta muralis*). Transactions of the Zoological Society of London, 20:135–230, pls. 16–23.

Boulenger, George Albert. 1917. Descriptions of new lizards of the family Lacertidae. Annals and Magazine of Natural History, ser. 8, 19:277–279, 2 figs.

Boulenger, George Albert. 1918a. On the varieties of the lizard *Ophiops elegans* Men. Annals and Magazine of Natural History, ser. 9, 2:158–162.

Boulenger, George Albert. 1918b. Sur les lézards de genre *Acanthodactylus* Wiegm. Bulletin de la Société Zoologique de France, 43:143–155.

Boulenger, George Albert. 1918c. Description of a new lizard of the genus *Acanthodactylus* from Mesopotamia. Journal of the Bombay Natural History Society, 25:373–374.

Boulenger, George Albert. 1918d. A synopsis of the lizards of the genus *Eremias*. Journal of Zoological Research, 3:1–12.

Boulenger, George Albert. 1920a. Monograph of the Lacertidae, vol. 1. British Museum (Natural History), London, x + 352 pp. (Reprint edition, Johnson, New York, 1966.)

Boulenger, George Albert. 1920b. Descriptions of four new snakes in the collection of the British Museum. Annals and Magazine of Natural History, ser. 9, 6:108–111.

Boulenger, George Albert. 1920c. Description of a snake of the genus *Zamenis* from Persia. Journal of the Bombay Natural History Society, 27:251.

Boulenger, George Albert. 1920d. Description of a new land tortoise from northern Persia. Journal of the Bombay Natural History Society, 27:251–252.

Boulenger, George Albert. 1920e. A list of snakes from Mesopotamia, collected by members of the Mesopotamian Expeditionary Force 1915–1919. Journal of the Bombay Natural History Society, 27(2):347–350.

Boulenger, George Albert. 1920f. A list of lizards from Mesopotamia collected by members of the Mesopotamian Expeditionary Force, 1915–1919. Journal of the Bombay Natural History Society, 27(2):351–353.

Boulenger, George Albert. 1920g. A monograph of the South Asian, Papuan, Melanesian, and Australian frogs of the genus *Rana*. Records of the Indian Museum, 20:1–226.

Boulenger, George Albert. 1921a. Monograph of the Lacertidae. vol. 2. British Museum (Natural History), London, viii + 451 pp. (Reprint edition, Johnson, New York, 1966.)

Boulenger, George Albert. 1921b. Liste des publications ichthyologiques et herpetologiques (1877–1920) de G. A. Boulenger. Annales de la Société Royale Malacologique Belgique, 52:11–56.

Boutan, Louis Marie August. 1893. Mémoire sur les reptiles rapportés de Syrie par Theodore Barrois: I, genre *Ptyodactyle*. Revue Biologique du Nord de la France, 5:329–45, 368–84, 444–8.

Brandt, Johann Friedrich. 1838. Note sur quatre nouvelles espèces de serpents de la côte occidentale de la mer Caspienne et la Perse septentrionale, découvertes par M. Kareline. Bulletin Scientifique de la Académie Impériale de St. Pétersbourg, 3:241–244.

Brongniart, Alexandre. "8" (1800a). Essai d'une classification naturelle des reptiles. Iere. Partie. Etablissement des orders. Bulletin des Science, Société Philomathique de Paris, 11(2[35]):81–82.

Brongniart, Alexandre. "8" (1800b). Essai d'une classification naturelle des reptiles. IIe. Partie. Formation et disposition des genres. Bulletin des Science, Société Philomathique de Paris, 11(2[36]):89–91, pl.

Bullock, Steve, and Wayne Kinunen. 1971. Lavan island aquatic survey. Job Completion Report Division of Research and Development submitted to Iran Game and Fish Department, F-4— 49, F-7-49, 9 pp. (Unpublished).

Busack, Stephen Dana, B. G. Jericho, Linda R. Maxson, and Thomas Marshall Uzzell. 1988. Evolutionary relationships of salamanders in the genus *Triturus*: the view from immunology. Herpetologica, 44(3):307–316.

Butzer, Karl W. 1956. Late glacial and postglacial climatic variation in the Near East. Erdkunde, Archiv für Wissenchaftliche Geographie, 11:21–34.

Butzer, Karl W. 1957. Mediterranean pluvials and the general circulation of the Pleistocene. Geografiska Annaler, Stockholm, 39:48–53.

Butzer, Karl W. 1958a. Quarternary Stratigraphy and Climate in the Near East. Bonner Geographische Abhandlungen, 157 pp.
Butzer, Karl W. 1958b. The Near East during the last glaciation: a palaeogeographical sketch. Geographical Journal, 124:367–369.
Butzer, Karl W. 1961. Climatic change in arid regions since the Pliocene, pp. 31–56. *In* L. Dudley Stamp (ed.). A History of Land Use in Arid Regions. UNESCO, Paris.
Butzer, Karl W. 1963. Climatic-geomorphologic interpretation of Pleistocene sediments in the Euroafrican subtropics, pp. 1–27. *In* F. Clark Howell and Francois Bourliere (eds.). African Ecology and Human Evolution. Aldine, Chicago.
Camerano, Lorenzo. 1877. Considerazioni sul genere *Lacerta* Linn. e descrizione di due nuove specie. Atti della Accademia delle Scienze di Torino, 13:79–98.
Capocaccia, Lilia. 1961. Catalogo dei tipi di rettili del Museo Civico di Storia Naturale di Genova. Annali del Museo Civico di Storia Naturale di Genova, 72:86–111.
Casimir, Michael J. 1970. Zur Herpetofauna des Iran und Afghanistans. Aquarien und Terrarien Zeitschrift (DATZ), 23(5):150–154.
Casimir, Michael J. 1971. Zur Herpetofauna der Provinz Badghis (NW-Afghanistan). Aquarien und Terrarien Zeitschrift (DATZ), 24:244–246.
Chapman, Randolph W. 1971. Climatic changes and the evolution of landforms in the Eastern Province of Saudi Arabia. Bulletin of the Geological Society of America, 82:2713–2728.
Chapman, Randolph W. 1978. General information on the Arabian Peninsula, 1.1. geology, 1.2 geomorphology, pp. 4–30. *In* Saad S. Al-Sayari and Josef G. Zötl (eds.). Quaternary Period in Saudi Arabia, vol. 1: Sedimentological, Hydrogeological, Hydrochemical, Geomorphological, and Climatological Investigations in Central and Eastern Saudi Arabia. Springer-Verlag, Vienna and New York.
Cherlin, Vladimir A. 1981. The new saw-scaled viper, *Echis multisquamatus* sp. nov. from Southwestern and Middle Asia. Trudy, Zoological Institute, Akademii Nauk USSR, Leningrad, 101:92–95 (In Russian with English summary).
Cherlin, Vladimir A. 1983a. Novie dannie o sistematke zmei roda *Echis* [New facts on the taxonomy of snakes of the genus *Echis*]. Vestnik Zoologii, 1983(2):42–46 (In Russian).
Cherlin, Vladimir A. 1983b. Dependence of scale pattern in snakes of the genus *Echis* from climatic factors. Zoologicheski'i Zhurnal, Akademiya Nauk SSSR, 62(2):252–258 (In Russian with English summary).
Cherlin, Vladimir A. 1990. Taxonomic revision of the snake genus *Echis* (Viperidae). II. An analysis of taxonomy and description of new forms. Trudy, Zoological Institute, Akademii Nauk USSR, Leningrad, 207:193–223 (In Russian, English abstract).
Cherlin, Vladimir A., and Leo Jakovlevich Borkin. 1990. [Taxonomic revision of the snake genus *Echis* (Viperidae). I. An analysis of the history of study and synonymy]. Trudy, Zoological Institute, Akademii Nauk USSR, Leningrad, 207:175–192 (In Russian, English abstract).
Cherlin, Vladimir A., Aleksey Yuri Tsellarius, and A. V. Gromov. 1983. On the thermobiology of the lizard *Teratoscincus scincus* in the Karakumi. Ekologiya, Sverdlovsk, 14(3):84–87 (In Russian).
Chernov, Sergius Aleksandrovich. 1926. Sur la connaissance de la faune herpetologique d'Armenie et de la contree du Nakhieczevan. Bulletin Scientifique de l'Institut de l'Exploration Regional du Caucase du Nord, 1:63–72. (In Russian).
Chernov, Sergius Aleksandrovich. 1937. [A Field Guide to the Snakes, Lizards, and Turtles of Armenia.]. Akademia Nauk SSSR, Biological Institute, Armenian Branch, Erevan, 54 pp.
Chernov, Sergius Aleksandrovich. 1939. Herpetological fauna of Armenian SSR and Nakhichevan ASSR. *In* Zoological Collected Papers of Armenian Branch, Academy of Sciences of USSR, 1:79–194. (In Russian).
Chernov, Sergius Aleksandrovich. 1959. [Reptilia. The fauna of Tadjik SSR]. Proceedings of the Tadjik SSR Academy of Science, 48:1–203 (In Russian).
Clark, Richard J. 1991. Contribution to the reptile fauna of northern Iran. Bulletin of the British Herpetological Society, (35):35–46.
Clark, Richard J. 1992. Notes on the distribution and ecology of *Phrynocephalus clarkorum* Anderson

& Leviton 1967 and *Phrynocephalus ornatus* Boulenger 1887 in Afghanistan. Herpetological Journal, 2:140–142, fig. 1 (map).

Clark, Richard J., and Erica D. Clark. 1973. Report on a collection of amphibians and reptiles from Turkey. Occasional Papers of the California Academy of Sciences, (104):1–62.

Clark, Richard J., Erica D. Clark, and Steven Clement Anderson. 1966. Report on two small collections of reptiles from Iran. Occasional Papers of the California Academy of Sciences, (55):1–9.

Clark, Richard J., Erica D. Clark, Steven Clement Anderson, and Alan Edward Leviton. 1969. Report on a collection of amphibians and reptiles from Afghanistan. Proceedings of the California Academy of Sciences, ser. 4, 36(10):279–316, 5 figs.

Clergue-Gazeau, Monique Andréa, and J. P. Farcey. 1979. Un *Batrachuperus* adulte dans une grotte d'Iran. Espèce nouvelle? International Journal of Speleology, 10:185–193.

Clergue-Gazeau, Monique Andréa, and Robert Thorn. 1979. Une nouvelle espèce de salamandre du genre *Batrachuperus*, en provenance de l'Iran septentrionale (Amphibia, Caudata, Hynobiidae). Bulletin de la Société d'Histoire Naturelle, Toulouse, 114:455–460.

Cope, Edward Drinker. 1862. On *Neurergus crocatus* from Iran and Iraq. Proceedings of the Academy of Natural Sciences of Philadelphia, 1862:343.

Corkill, Norman L., and J. A. Cochrane. 1966. The snakes of the Arabian Peninsula and Socotra. Journal of the Bombay Natural History Society, 62:475–506, map.

Curzon, George N. 1892a. Persia and the Persian Question, vol. 1. Frank Cass and Co., Ltd., London, xxiv + 639 pp.

Curzon, George N. 1892b. Persia and the Persian Question, vol. 2. Frank Cass and Co., Ltd., London, xii + 653 pp.

Cuvier, Georges Jean-Léopold-Nicolas-Frédéric Dagobert. 1816. Le Règne Animal Distribué d'après son Organisation. Tome II, Contenant les Reptiles, les Poissons, les Mollusques et les Annélides. vol. 2. Déterville, Paris, xviii + 532 pp. (Reprinted 1969 Culture et Civilisation, Bruxelles).

Cuvier, Georges Jean-Léopold-Nicolas-Frédéric Dagobert. 1829. Le Règne Animal, vol. 2 [nouvelle edition, revue et augmentée ed.]. Paris, xv + 406 pp.

Cyrén, Carl August Otto. 1924. Klima und Eidechsen-verbreitung. Eine Studie der geographischen Variation und Entwicklung einer Lacerten, insbesondere unter Berücksichtigung der klimatischen Faktoren. Meddelanden från Göteborgs Musei Zoologiska Avdelning, 29:1–82.

Czarevsky, Sergius Fedorovich. 1915. Aperçu des représentants du genre *Eryx*, principalment de l'Empire Russe et des pays limitrophes. Annuaire du Musée Zoologique de l'Académie Impériale des Sciences de St.-Pétersbourg, 20:340–388.

Czarevsky, Sergius Fedorovich. 1923. A note on the identity of *Phyllodactylus elisae* Werner and *P. eugeniae* Nikolsky. Annuaire Musée Zoologique, Petrograd, 24:101–103.

Dalj, S. K. 1954. Animal Kingdom of the Armenian SSR, vol. 1. Vertebrates. Armenian Academy of Science, Yerevan (= Erivan), 415 pp. (Herpetology on pp. 281–302 [In Russian]).

Darevsky, Ilya Sergeevich. 1960. [Poisonous Snakes of Armenia]. Aipetrat Ed., Erivan (= Yerevan), 47 pp. (In Armenian).

Darevsky, Ilya Sergeevich. 1967. [Rock Lizards of the Caucasus (Systematics, Ecology, and Phylogenesis of Polymorphic Groups of Caucasian Lizards of the Subgenus *Archaeolacerta*)]. Nauka Press, Leningrad, 214 pp. (In Russian).

Darevsky, Ilya Sergeevich. 1970. Systematic status of *Rhynchocalamus melanocephalus satunini* Nik. (Serpentes, Colubridae) previously included in the genus *Oligodon*. Zoologicheski'i Zhurnal, Akademiya Nauk SSSR, 49(11):1685–1689 (In Russian, English summary).

Darevsky, Ilya Sergeevich. 1978a. Rock Lizards of the Caucasus. Indian National Scientific Documentation Centre, Published for the Smithsonian Institution and the National Science Foundation, Washington, D.C., New Delhi, v + 276 pp. (Translation of 1967 Russian edition).

Darevsky, Ilya Sergeevich. 1978b. Kakoi vid eublaphar (Sauria, Gekkonidae) vstrechaetsya b srednei Azii? [Which species of *Eublepharis* (Sauria, Gekkonidae) inhabits Middle Asia?]. Trudy Zoologicheskogo Instituta, Akademiya Nauk SSSR, 61(Novie vidi zhivotnih [New species of Animals]):204–209 (In Russian).

Darevsky, Ilya Sergeevich. 1981a. *Eumeces schneiderii* (Daudin 1802) — Tüpfelskink, pp. 355–365.

In Wolfgang Böhme (ed.). Handbuch der Reptilien und Amphibien Europas, Echsen (Sauria) I. Handbuch der Reptilien und Amphibien Europas, vol. 1. Akademische Verlagsgesellschaft, Wiesbaden.

Darevsky, Ilya Sergeevich. 1981b. A first record of the ocellated skink, *Chalcides ocellatus* (Forskål) in the fauna of the USSR. Trudy Zoologicheskogo Instituta, Akademiya Nauk SSSR, 101:49–51,fig.7.

Darevsky, Ilya Sergeevich. 1984a. *Lacerta saxicola* Eversmann 1834 — Felseidechse, pp. 82–99. *In* Wolfgang Böhme (ed.). Handbuch der Reptilia und Amphibia Europas, vol. 2/I. Echsen II. AULA-Verlag, Wiesbaden.

Darevsky, Ilya Sergeevich 1984b. *Lacerta strigata* Eichwald 1831 — Kaspische Smaragdeidechse, pp. 345–361. *In* Wolfgang Böhme (ed.). Handbuch der Reptilien und Amphibien Europas, vol. 2/I, Eschen II. AULA-Verlag, Wiesbaden.

Darevsky, Ilya Sergeevich, and Axel Beutler. 1981. *Ophisops elegans* Ménétriés 1832 – Schlangenauge, pp. 461–477. *In* Wolfgang Böhme (ed.). Echsen (Sauria) I. Handbuch der Reptilien und Amphibien Europas, vol. I. AULA-Verlag, Wiesbaden.

Darevsky, Ilya Sergeevich, and Ivan Sergeevich Chumakov. 1962. A new Pleistocene species of the lizards *Ablepharus* from Rudny Altai. Paleontologicheskii Zhurnal, 1:127–130 (In Russian).

Darevsky, Ilya Sergeevich, and Nikolai Nikolaevich Szczerbak. 1978. *Eremias andersoni*, a new lizard (Reptilia, Lacertilia, Lacertidae) from Iran. Journal of Herpetology, 12(1):13–15.

Darevsky, Ilya Sergeevich, Josef Eiselt, and Galina Pantelejmonovna Lukina. 1984. Rock lizards of the *Lacerta saxicola* Eversmann group of northern Iran. Trudy, Zoological Institute, Akademii Nauk USSR, Leningrad, 124:102–108 (In Russian with English summary).

Das, Indraneil, Basudeb Dattagupta, and Nimai Charan Gayen. 1998. History and catalogue of reptile types in the collection of the Zoological Survey of India. Journal of South Asian Natural History, 4(1):1–66.

Daudin, François-Marie. 1801–1803. Histoire Naturelle, Générale et Particulière des Reptiles; Ouvrage faisant Suite, a l'Histoire Naturelle, Générale et Particulière composée par Leclerc de Buffon, et rédigée par C. S. Sonnini, vols. 1–8. Paris (volumes dated as follows: Vols. 1 and 2, 1801; Vols. 3, 4, and 5, 1802; Vols. 6, 7, and 8, 1803 [See Harper {1940}, Vanzolini {1977:14}, and Zhao and Adler {1993:363} for details]).

DeBlase, Anthony. 1980. The bats of Iran: systematics, distribution, ecology. Fieldiana: Zoology, (4):xvii + 424 pp.

De Filippi, Filippo. 1843. Intorno ad alcune specie di rettili. Giornale Instituto Lombardo Scienza Litteratura Arti Bibliotheca Italiana, 6:407–415.

De Filippi, Filippo. 1863. Nuove o poco note specie di animali vertebrati raccolte in um viaggio in Persia nall'estate dell' anno 1862. Archivo Zoologico Anatomia Fisiologia Modena, 2:15.

De Filippi, Filippo. 1864. Riassunto del catalogo degli animali vertebrati delle Provincie Caucasiche e della Persia occidentale. Attivo della Societa Italiana Scienza Naturali, 1864(September):184–186.

De Filippi, Filippo. 1865. Note di un Viaggio in Persia nel 1862. G. Daelli & C. Editori, Milano, xi + 396 + (2) pp. (Reptiles [general discussion section] 18, 80–81, 87–88, 98, 105–106, 108, 114, 117, 143, 161, 196, 213, 255, 276, 301–302 [Checklist, see Jan *in* DeFilippi, 1865] 352–357).

De Filippi, Filippo. 1868. Sulla struttura della cute dello *Stellio caucasicus*. Memorie dell' Accademia dello Scienze di Torino, 23:363–373, 1 pl.

Dely, Olivér György. 1981. *Anguis fragilis* Linnaeus 1758 — Blindschleiche, pp. 241–258. *In* Wolfgang Böhme (ed.). Handbuch der Reptilien und Amphibien Europas, Echsen (Sauria) I. Handbuch der Reptilien und Amphibien Europas, vol. 1. Akademische Verlagsgesellschaft, Wiesbaden.

Demidoff, Anatoly Nikolaevich de. 1840–1842. Voyage dans la Russie Méridionale et la Crimée, par la Hongrie, la Valachie et la Moldavie, exécuté en 1837, sous la Direction de M. Anatole de Demidoff par MM. de Sainson, le Play, Huot, Léveille, Raffet, Rousseau, de Nordmann et du Ponceau. Ernest Bourdin et Ce, Éditeurs, Paris.

Department of Defense. 1992. Venomous Snakes of the Middle East (Identification Guide). Defense Intelligence Agency, Washington, D.C., viii + 160 pp.

Deperno, Christopher S., and William E. Cooper. 1996. Labial-licking for chemical sampling by the leopard gecko (*Eublepharis macularius*). Journal of Herpetology, 30(4):540–543.

Disi, Ahmad M. 1996. A contribution to the knowledge of the herpetofauna of Jordan. VI. The Jordanian herpetofauna as a zoogeographic indicator. Herpetozoa, 9(1–2):71–81, fig. 1.

Disi, Ahmad M., and Wolfgang Böhme. 1996. Zoogeography of the amphibians and reptiles of Syria, with additional new records. Herpetozoa, 9(1–2):63–70, fig. 1.

Dixon, James Ray, and Steven Clement Anderson. 1973. A new genus and species of gecko (Sauria: Gekkonidae) from Iran and Iraq. Bulletin of the Southern California Academy of Sciences, 72(3):155–160.

Dotsenko, I. B. 1985. [Revision of the genus *Eirenis* (Reptilia, Colubridae). Communication 1. Resurrection of the genus *Pseudocyclophis* Boettger, 1888]. Vestnik Zoologii, 1985(4):41–44 (In Russian).

Dotsenko, I. B. 1989. Revizija roda *Eirenis* (Rept., Colubr.) Soovshchenie 2. Struktura roda *Eirenis*. Vestnik Zoologii Kiev, 1989:23–29.

Dresch, J. 1975. Bassins arides iraniens. Bulletin de l'Association des Géographes Français, (430):337–351.

Duméril, André-Marie-Constant, and Gabriel Bibron. 1836. Erpétologie Générale ou Histoire Naturelle Complète des Reptiles, vol. 3. Librairie Encyclopédique de Roret, Paris, iv + 517 pp.

Duméril, André-Marie-Constant, and Gabriel Bibron. 1837. Erpétologie Générale ou Histoire Naturelle Complète des Reptiles, vol. 4. Librairie Encyclopédique de Roret, Paris, ii + 571 pp.

Duméril, André-Marie-Constant, and Gabriel Bibron. 1839. Erpétologie Générale ou Histoire Naturelle Complète des Reptiles, vol. 5. Librairie Encyclopédique de Roret, Paris, viii + 854 pp.

Duméril, André-Marie-Constant, and Gabriel Bibron. 1844. Erpétologie Générale ou Histoire Naturelle Complète des Reptiles, vol. 6. Librarie Encyclopédique de Roret, Paris, xii + 609 pp.

Duméril, André-Marie-Constant, and Auguste-Henri-André Duméril. 1851. Catalogue Méthodique de la Collection des Reptiles du Muséum d'Histoire Naturelle de Paris. Gide et Baudry, Paris, iv + 224 pp.

Duméril, Auguste-Henri-André. 1856. Description des reptiles nouveaux ou imparfaitment connus de la collection du Muséum d'Histoire Naturelle, et remarques sur la classification et les caractères des reptiles. Archiv du Muséum d'Histoire Naturelle, Paris, 8:437–586, pls.17–22.

Duncan, F. Martin. 1937. On the dates of publication of the society's 'Proceedings,' 1859–1926. Proceedings of the Zoological Society of London, 1937(1):71–84.

Ehlers, Eckart. 1992. Climate, pp. 707–713. *In* Ehsan Yarshater (ed.). Encyclopaedia Iranica, vol. 5, fasc. 7. Mazda, Costa Mesa, California.

Eichwald, Carl Eduard Ivanovich von. 1831. Decima Classis. Amphibia. V. *In* Zoologia Specialis quam Expositis Animalibus tum Vivis, tum Fossilibus Potissimum Rossiae in Universum, et Poloniae in Specie, in USUM Lectionum Publicarum in Universitate Caesarea Vilnensi Habendarum. Pars Posterior [= volume 3], Specialem Expositionem Spondylozoorum Continens (pp. 116–197), Explicatio Tabularum (pp. 395–396). Josephi Zawadzki, Vilnae, (3), 404 pp., 2 folding pls.

Eichwald, Carl Eduard Ivanovich von. 1834–1838. Reise auf dem Caspischen Meere und in den Caucasus. Bd. 1, Abt. 2. Reise in den Kaukasus. x, 894 p., 5 pls., 1837, Stüttgart and Tübingen.

Eichwald, Carl Eduard Ivanovich von. 1838. Faunae Caspii Maris primitiae. Bulletin de la Société des Naturalistes de Moskou, 11(2):125–174.

Eichwald, Carl Eduard Ivanovich von. 1839. De duabus novis amphibiorum speciebus. Bulletin de la Société des Naturalistes de Moskou, 2:303–307.

Eichwald, Carl Eduard Ivanovich von. 1841. Fauna Caspio-Caucasia nonnullis observationibus novis. Petropoli, iv + 233 pp, 40 col. pls.

Eiselt, Josef. 1940. Der Rassenkreis *Eumeces schneideri* Daudin. Zoologischer Anzeiger, 131(9–10):209–228.

Eiselt, Josef. 1967. Nachrufe a. o. Universitätsprofessor Dr. phil. Otto Wettstein-Westersheimb. Annalen des Naturhistorischen Museums in Wien, 70:1–18.

Eiselt, Josef. 1968. Ergebnisse zoologischer Sammelreisen in der Türkei: Ein Beitrag zur Taxonomie der Zagros-Eidechse, *Lacerta princeps* Blanf. Annalen des Naturhistorischen Museums in Wien, 72:409–434, pl. 1 (col.), pls. 2–5.

Eiselt, Josef. 1969. Zweiter Beitrag zur Taxonomie der Zagros-Eidechse, *Lacerta princeps* Blanford. Annalen des Naturhistorischen Museums in Wien, 73:209–220.

Eiselt, Josef. 1970. Die Suche nach der Zagroseidechse. AquaTerra (Solothurn), 7(11):109–114, col. pl.

Eiselt, Josef. 1971a. *Eirenis rechingeri* n. sp. (Colubridae, Serpentes) aus dem Iran. Annalen des Naturhistorischen Museums in Wien, 75:375–381, pls. 1–2.

Eiselt, Josef. 1971b. Forschungsarbeit des Naturhistorischen Museums Wien im und für den Iran. Österreiche Zeitschrift für Kultur Politik und Wirtschaft des Islamischen Länder, 1970–71:29–33.

Eiselt, Josef. 1973. Ein neuer Blattfinger-Gecko (*Phyllodactylus*, Sauria Rept.) aus dem Iran und Bemerkungen zu *Phyllodactylus elisae* Werner 1895. Annalen des Naturhistorischen Museums in Wien, 77:173–179.

Eiselt, Josef. 1979. Ergebnisse zoologischer Sammelreisen in der Türkei: *Lacerta cappadocica* Werner, 1902 (Lacertidae, Reptilia). Annalen des Naturhistorischen Museums in Wien, 82:387–422, pls. 1–6.

Eiselt, Josef. 1995. Ein Beitrag zur Kenntnis der Archaeolacerten (sensu Méhelÿ, 1909) des Iran (Squamata: Sauria: Lacertidae). Herpetozoa, 8(1–2):59–72.

Eiselt, Josef, and Josef Friedrich Schmidtler. 1971. Vorläufige Mitteilung über zwei neue Subspezies von Amphibia Salientia aus dem Iran. Annalen des Naturhistorischen Museums in Wien, 75:383–385.

Eiselt, Josef, and Josef Friedrich Schmidtler. 1973. Froschlurche aus dem Iran unter Berücksichtigung ausseriranischer Populationsgruppen. Annalen des Naturhistorischen Museums in Wien, 77:181–243, pls. 1–4.

Eiselt, Josef, Josef Friedrich Schmidtler, and Ilya Sergeevich Darevsky. 1993. Untersuchungen an Felseidechsen (*Lacerta saxicola*-Komplex) in der östlichen Türkei. 2. Eine neue Unterart der *Lacerta raddei* Boettger, 1892 (Squamata: Sauria: Lacertidae). Herpetozoa, 6(1–2):65–70, 7 figs.

Eiselt, Josef, and Hans M. Steiner. 1970. Erstfund eines hynobiiden Molches in Iran. Annalen des Naturhistorischen Museums in Wien, 74:77–90, pls. 1–2.

Elpatjevsky, Vladimir Sergeevich. 1901. Genus *Ablepharus* Fitz. b kollektsiyakh Zoologicheskogo Muzeya Moskovskogo Universiteta [The genus *Ablepharus* Fitz., in the collections of the Zoological Museum of Moscow University]. Dnevnik Zoologicheskogo Otdeleniya ob-va Lyubitelei Estestvoznania, Antropologii i Etnografii, 3(2):37–39.

El-Toubi, M. R. 1938. The osteology of the lizard *Scincus scincus*. Bulletin of the Faculty of Science, Egyptian University, Cairo, (14):1–38.

Eremchenko, Valery Konstantinovich, and Nikolai Nikolaevich Szczerbak. 1980. Concerning ancestral species of ablepharine lizards (Reptilia, Sauria, Scincidae) of the USSR. Vestnik Zoologii Kiev, 1980(4):10–15 (In Russian with English abstract).

Eremchenko, Valery Konstantinovich, and Nikolai Nikolaevich Szczerbak. 1986. Ablefaridnie Yashcheritzi Fauni SSSR i Sopredelnikh Stran [Ablepharine Lizards in the Fauna of the USSR and Neighboring Countries]. Akademiya Nauk Kirgizkoi SSR, Ylym, Frunze, 171 pp. (In Russian).

Ernst, Carl Henry, and Roger William Barbour. 1989. Turtles of the World. Smithsonian Institution, Washington, DC, xii + 313 pp., 16 pls.

Estes, Richard Dean, Kevin De Queiroz, and Jacques Armand Gautier. 1988. Phylogenetic relationships within Squamata, pp. 119–281. *In* Richard Dean Estes and Gregory Kent Pregill (eds.). Phylogenetic Relationships of the Lizard Families. Stanford University Press, Stanford.

Eversmann, Eduard Friedrich. 1823. Reise von Orenburg nach Buchara ... nebst einem naturhistorischen Anhänge und einer Vorrede von H. Lichtenstein. Christiani, Berlin, viii + 150 pp.

Fejérváry von Komlós-Keresztes, Géza Gyula Imre de. 1936. Notes on a very little known lizard, *Lacerta princeps* Blanford, with description of the male specimen preserved in the Vienna Natural

History Museum. Annales de Historico-Naturales Musei Nationalis Hungarici, Budapest, 30:1–21, pl. 1

Felber, Hans, Heinz Hötzl, Viktor Maurin, Heribert Moser, Werner Rauert, and Josef G. Zötl. 1978. Gulf coastal region and its hinterland, 2.1.2, Sea level fluctuations during the Quaternary Period, pp. 50–57. *In* Saad S. Al-Sayari and Josef G. Zötl (eds.). Quaternary Period in Saudi Arabia, vol. 1: Sedimentological, Hydrogeological, Hydrochemical, Geomorphological, and Climatological Investigations in Central and Eastern Saudi Arabia. Springer-Verlag, Vienna and New York.

Fet, Victor. 1994a. Introduction: one hundred years of natural history in Turkmenistan, chapter 1, pp. 1–4. *In* Victor Fet and Khabibulla Atamuradov (eds.). Biogeography and Ecology of Turkmenistan. Monographiae Biologicae, 72. Kluwer Academic publishers, Dordrecht, Boston, and London.

Fet, Victor. 1994b. Biogeographic position of Khorassan-Kopetdagh, chapter 12, pp. 197–204. *In* Victor Fet and Khabibulla Atamuradov (eds.). Biogeography and Ecology of Turkmenistan. Monographiae Biologicae, 72. Kluwer Academic Publishers, Dordrecht, Boston, and London.

Fet, Victor, and Khabibulla Atamuradov. 1994. Biogeography and Ecology of Turkmenistan. Monographiae Biologicae, 72. Kluwer Academic Publishers Group, Dordrecht, Boston, and London, 616 pp.

Field, Henry. 1953. Bibliography on Southwestern Asia. University of Miami Press, Coral Gables, xvi + 106 pp.

Field, Henry. 1955. Bibliography on Southwestern Asia: II. A Second Compilation. University of Miami Press, Coral Gables, xviii + 126 pp.

Field, Henry. 1956a. Bibliography on Southwestern Asia: III. A Third Compilation. University of Miami Press, Coral Gables, Florida, xxviii + 230 pp.

Field, Henry. 1956b. An anthropological reconnaissance in the Near East, 1950. Papers of the Peabody Museum (Harvard University), 48:1–119.

Field, Henry. 1957. Bibliography on Southwestern Asia: IV. A Fourth Compilation. University of Miami Press, Coral Gables, Florida, xlvi + 464 pp.

Field, Henry. 1958. Bibliography on Southwestern Asia: V. A Fifth Compilation. University of Miami Press, Coral Gables, Florida, xxxii + 275 pp.

Field, Henry. 1959a. Bibliography on Southwestern Asia: VI. A Sixth Compilation. University of Miami Press, Coral Gables, Florida, xxxvi + 328 pp.

Field, Henry. 1959b. An anthropological reconnaissance in West Pakistan, 1955, with appendixes on the archaeology and natural history of Baluchistan and Bahawalpur. Papers of the Peabody Museum (Harvard University), 52:1–332.

Field, Henry (ed.). 1960. Subject Index to Bibliographies on Southwestern Asia: I–V. (Pt. II. Zoology, index compiled by Bernard J. Clifton.) University of Miami Press, Coral Gables, Florida, vii + 83 pp.

Field, Henry. 1962. Bibliography on Southwestern Asia: VII. A Seventh Compilation. University of Miami Press, Coral Gables, Florida, xxxix + 305 pp.

Field, Henry. 1963. Subject Index to Bibliographies on Southwestern Asia: VI-VII. Pt. II. Zoology, By Bernard J. Clifton. University of Miami, Coral Gables, Florida, 48 pp.

Field, Henry. 1966. Bibliography: 1926–1966. Nos. 1–631. Privately printed by author, vii + 112 pp.

Field, Henry, and Bernard J. Clifton. 1970. Bibliography on Southwestern Asia. Supplement VI. Botany and Zoology. Field Research Projects, Coconut Grove, Florida, ii + 50 pp.

Field, Henry, and Edith M. Laird. 1968. Bibliography on Southwest Asia. Supplement I. Anthropogeography, Botany and Zoology Field Research Projects, Coconut Grove, Florida, iii + 92 pp.

Field, Henry, and Edith M. Laird. 1968. Bibliography on Southwest Asia. Supplements II–III Anthropogeography, Botany and Zoology (in 1 vol.). Field Research Projects, Coconut Grove, Florida, (II):iv + 50; (III):ii + 8.

Field, Henry, and Edith M. Laird. 1969. Bibliography on Southwestern Asia: Supplement IV. Anthropogeography, Maps, Botany and Zoology. Field Research Projects, Coconut Grove, Florida, 78 pp.

Field, Henry, and Edith M. Laird. 1970. Bibliography on Southwestern Asia: Supplement V. Anthropogeography. Field Research Projects, Coconut Grove, Florida, 78 pp.

Field, Henry, Edith M. Laird, and Bernard J. Clifton. 1972. Bibliography on Southwestern Asia: Supplement VII. Anthropogeography, Botany and Zoology. Field Research Projects, Coconut Grove, Florida, iv + 97 pp.

Firouz, Eskandar. 1974. Environment Iran. National Society for the Conservation of Natural Resources and Human Environment, Tehran, 51 pp.

Firouz, Eskandar. 1976. Environmental and nature conservation in Iran. Environmental Conservation, 3(1):32–42, 9 figs.

Firouz, Eskandar. 1998. Environmental protection, i. in Persia, pp. 465–472, figs. 1–8. In Ehsan Yarshater (ed.). Encyclopaedia Iranica, vol. 8, fasc. 5. Mazda Publishers, Costa Mesa, California, pls. 1–8.

Firouz, Eskandar, and Fred A. Harrington, Jr. 1976. Iran: concepts of biotic community conservation. International Union for the Conservation of Nature Occasional Papers, (15):1–31, 2 figs., 1 tab.

Firouz, Eskandar, and W. H. Wambold. 1976. Environmental Protection in Iran. Department of the Environment, Tehran (not seen).

Firouz, Eskandar, Jerry D. Hassinger, and D. A. Ferguson. 1970. The wildlife parks and protected regions of Iran. Biological Conservation, 3(1):37–45, 10 figs.

Fisher, W. B. 1968a. The Land of Iran. The Cambridge History of Iran, vol. 1. Cambridge University Press, Cambridge, England, xix + 784 pp.

Fisher, W. B. 1968b. Physical geography, chapter 1, pp. 3–110, 28 figs. In W. B. Fisher (ed.). The Land of Iran. The Cambridge History of Iran, vol. 1. Cambridge University Press, Cambridge (Bibliography, p. 743).

Fitzinger, Leopold Josef Franz Johann. 1826. Neue Classification der Reptilien nach ihren natürlichen Verwandtschaften. Nebst einer Verwandtschafts-Tafel und einen Verzeichnisse der Reptilien-Sammlung des K. K. Zoologischen Museums zu Wien. J. G. Heubner, Wien, viii + 66 pp., folding table. (See note on publication date in Zhao and Adler, 1993:368) (Reprint 1997 Society for the Study of Amphibians and Reptiles, Ithaca, New York).

Fitzinger, Leopold Josef Franz Johann. 1843. Systema Reptilium. Fasciculus Primus: *Amblyglossae* (Conspectus Geographicus). Braumüller und Seidel, Vindobonae (=Vienna), 106 + iv pp. [no further parts published] (Reprint 1973 Society for the Study of Amphibians and Reptiles, Oxford, Ohio).

Flower, Stanley Smyth. 1933. Notes on the Recent reptiles and amphibians of Egypt, with a list of the species recorded from that kingdom. Proceedings of the Zoological Society of London, 1933:735–851, map.

Forcart, Lothar Hendrich Emil Wilhelm. 1950. Amphibien und Reptilien von Iran. Verhandlungen der naturforschenden Gesellschaft in Basel, 61:141–156.

Forsskål, Petrus. 1775. Descriptiones Animalium, Avium, Amphibiorum, Piscium, Insectorum, Vermium, quae in Itinere Orientale Observavit Petrus Forskål. Post Mortem Auctoris Editit C. Niebuhr, adjuncta est Materia Medica Kahirina atque Tabula Maris Rubri Geographica. Heineck and Faber, Havniae (= Copenhagen), (19) + xxxiv + 164 pp., map, pls. 1–43. (Amphibia, pp. 12–21, map; also pls. 21–43.)

Frank, Norman, and Erica Ramus. 1995. A Complete Guide to Scientific and Common Names of Reptiles and Amphibians of the World. NG Publishing Inc, Pottsville, Pennsylvania, 377 pp.

Fritz, Uwe. 1994. Zur innerartlichen Variabilität von *Emys orbicularis* (Linnaeus, 1758). 4. Variabilitat und Zoogeographie im pontokaspischen Gebiet mit Beschreibung von drei neuen Unterarten. (Reptilia: Testudines: Emydidae). Zoologische Abhandlungen (Dresden), 48(1):53–93.

Frost, Darrel Richmond. 1985. Amphibian Species of the World. A Taxonomic and Geographical Reference. Allen Press, Inc. and the Association of Systematic Collections, Lawrence, Kansas, v + 732 pp.

Frost, Darrel Richmond, and Richard Emmett Etheridge. 1989. A phylogenetic analysis and taxonomy of iguanian lizards (Reptilia: Squamata). University of Kansas Museum of Natural History Miscellaneous Publications, (81):ii + 65, figs. 1–24.

Frynta, Daniel, Jiří Moravec, Jovana Čiháková, Jiří Sadlo, Zdena Hodková, Milan Kaftan, Petr Kodym, David Král, Václav Pitule, and Ladislav Sejna. 1997. Results of the Czech Biological Expedition to Iran. Part 1. Notes on the distribution of amphibians and reptiles. Acta Societatis Zoologicae Bohemicae, 61:3–17.

Fuhn, Ion Eduard. 1969a. Revision and redefinition of the genus *Ablepharus* Lichtenstein, 1823 (Reptilia, Scincidae). Revue Roumaine de Biologie, ser. zool., 14:23–41.

Fuhn, Ion Eduard. 1969b. The "polyphyletic" origin of the genus *Ablepharus* (Reptilia: Scincidae): A case of parallel evolution. Zeitschrift für zoologische Systematik und Evolutionsforschung, 7:67–76.

G-A, H. H. 1905. Obituary. William Thomas Blanford, CIE, LLD, FRS. Ibis, 1905:643–647.

Gallagher, Michael D. 1971. The Amphibians and Reptiles of Bahrain. Privately printed by author, Bahrain, ii + 40 pp.

Gallagher, Michael D. 1990. Snakes of the Arabian Gulf and Oman, 2nd ed. Privately printed by author, Muscat, Oman, 17 pp. (with numerous color photos).

Gallagher, Michael D., and Edwin Nicholas Arnold. 1988. Reptiles and amphibians from the Wahiba Sands, Oman. Journal of Oman Studies, (Special Report 3):405–413.

Ganji, M. H. 1968. Climate, chapter 5, pp. 212–245. *In* W. B. Fisher (ed.). The Land of Iran. The Cambridge History of Iran, vol. 1. Cambridge University Press, Cambridge, UK.

Gans, Carl. 1960. Studies on amphisbaenids (Amphisbaenia, Reptilia). 1. A taxonomic revision of the Trogonophinae, and a functional interpretation of the amphisbaenid adaptive pattern. Bulletin of the American Museum of Natural History, 119:129–204.

Gans, Carl, and Paul F. A. Maderson. 1973. Sound producing mechanisms in Recent reptiles: review and comment. American Zoologist, 13:1195–1203.

Gans, Carl, and F. Harvey Pough (eds.). 1982a. Biology of the Reptilia. vol. 12, Physiology C. Physiological Ecology. Academic Press, New York and London.

Gans, Carl, and F. Harvey Pough. 1982b. Physiological ecology: its debt to reptilian studies, its value to students of reptiles, chapter 1, pp. 1–13. *In* Carl Gans and F. Harvey Pough (eds.). Biology of the Reptilia, vol. 12. Physiology C. Physiological Ecology. Academic Press, New York and London.

Gardner, Andrew S. 1994. A new species of *Asaccus* (Gekkonidae) from the mountains of northern Oman. Journal of Herpetology, 28(2):141–145.

Gasperetti, John. 1988. The snakes of Arabia. Fauna of Saudi Arabia, 9:169–450, 29 pls., 135 figs.

Gasperetti, John, Andrew Francis Stimson, J. D. Miller, James Perran Ross, and Patricia R. Gasperetti. 1993. Turtles of Arabia. Fauna of Saudi Arabia, 13:170–367.

Geoffroy Saint-Hilaire, Étienne-François, and Isidore Geoffroy Saint-Hilaire. 1827. Description des reptiles qui se trouvent en Égypte. *In* Étienne-François Geoffroy Saint-Hilaire (ed.). Description de l'Égypte, ou Recueil des Observations et des Recherches qui ont été Faites en Egypte pendant l'Expédition de l'Armée Française. Histoire Naturelle, vol. 1, part 1. Imprimerie Impériale, Paris, 8 pls. (see note in Zhao and Adler, 1993:369).

Girdler, R. W. 1984. The evolution of the Gulf of Aden and Red Sea in space and time. Deep-Sea Research, 31:747–762.

Gmelin, Johann Friedrich. 1789. Caroli a Linné Systema Naturae. Editio Decima Tertia, Aucta Reformata. vol. 1(3). G. E. Beer, Lipsiae (= Leipzig), (1), 1033–1516 pp. (herpetology is on pp. 1033–1125 [see note in Zhao and Adler, 1993:369–370]).

Gmelin, Samuel Gottlieb. 1774. Reise durch Russland zur Untersuchung der drey Natur-Reiche, vol. 3. Reise durch das Nördliche Persien in den Jahren 1770, 1771 bis April 1772. Academy of Sciences, St. Petersburg, 508 pp., 51 pls.

Gmelin, Samuel Gottlieb. 1784. Reise durch Russland. Reise von Astrachan; Ungleichen Zweite Persiche Reise: in den Jahren 1772 und 1773, bis im Frühling 1774. Nebst dem Leben des Verfassers (edited by P. S. Pallas), vol. 4. Academy of Sciences, St. Petersburg, 218 pp.

Golay, Philippe. 1985. Checklist and Keys to the Terrestrial Proteroglyphs of the World. Elapsoïdea Fondation Culturelle, Genève, (4):ix + 91.

Golay, Philippe, Hobart Muir Smith, Donald G. Broadley, James Ray Dixon, Colin McCarthy,

Jean-Claude Rage, Beat Schätti, and Michihisa Toriba. 1993. Endoglyphs and Other Major Venomous Snakes of the World: A Checklist. Azemiops S.A. Herpetological Data Center, Aire-Geneva, Switzerland, xv + 478 pp.

Goldfuss, Georg August. 1820. Handbuch der Zoologie, vol. 2. Johann Leonard Schrag, Nürnberg, xxiv + 506 pp., 4 pls.

Goldsmid, Frederic John. 1876. Eastern Persia. An Account of the Journeys of the Persian Boundary Commission 1870–71–72, vol. 1. The Geography with Narratives. Macmillan and Co., London, lviii + 443 pp.

Golubev, Mikhail Leonidovich. 1985. Structura roda *Tropiocolotes* (Reptilia, Gekkonidae) [Structure of the genus *Tropiocolotes* (Reptilia, Gekkonidae)]. Vestnik Zoologii, 1984(6):12 (In Russian).

Golubev, Michael L. (=Mikhail Leonidovich Golubev), Muhammad Sharif Khan, and Steven Clement Anderson. 1995. On the systematics of some Palearctic geckos. Abstracts of the Second Asian Herpetological Meeting, 6–10 September, 1995. Ashgabat, Turkmenistan (no pagination).

Golubev, Mikhail Leonidovich, and T. S. Sattorov. 1992. O vnutrividovoi strukture i mezhvidovikh otnosheniyakh ushastoi kruglogolovki *Phrynocephalus mystaceus* (Reptilia, Agamidae) [On intraspecific structure and intraspecific relations of the ear-folded toad agama *Phrynocephalus mystaceus* (Reptilia, Agamidae)]. Vestnik Zoologii, 1992(3):26–32, (In Russian with English abstract).

Golubev, Mikhail Leonidovich, and Nikolai Nikolaevich Szczerbak. 1981a. Novij vid roda *Gymnodactylus* (Reptilia, Gekkonidae) iz Pakistana [A new species of the genus *Gumnodactylus* Spix 1823 (Reptilia, Sauria, Gekkonidae) from Pakistan]. Vestnik Zoologii, 1981(3):40–45 (In Russian with English summary).

Golubev, Mikhail Leonidovich, and Nikolai Nikolaevich Szczerbak. 1981b. *Carinatogecko* Gen. N. (Reptilia, Gekkonidae) — Novi'i rod gekkonovikh yascherich iz yugo-zapadnoi Azii [*Carinatogecko* Gen. N. (Reptilia, Gekkonidae) — new genus of gekkonid lizard from Southwest Asia.]. Vestnik Zoologii, 1981(5):34–41, figs. 1–3 (In Russian with English summary).

Golubev, Mikhail Leonidovich, and Nikolai Nikolaevich Szczerbak. 1985. O vzaimootnosheniyah dvuh rodov palearcticheskih gekkonov: *Tenuidactylus* i *Agamura* (Reptilia, Gekkonidae) [On the relationships of two Palearctic gecko genera, *Tenuidactylus* and *Agamura* (Reptilia, Gekkonidae)]. Voprosi Gerpetologii: Avtoref. Dokl. VI-j Vsesoyuz. Gerpetol. Konf. [Problems of Herpetology. Sixth All-Union Herpetological Conference Abstracts], Tashkent. Nauka Press, Leningrad (In Russian).

Gorelov, Yu. K., and V. S. Lukarevsky. 1990. [On occurence of *Stellio erythrogaster* in the Soviet part of Eastern Kopet-Dagh]. Izvestia Akademii Nauk Turkmenskoi SSR, (6):63 (In Russian).

Gorelov, Yu. K., Ilya Sergeevich Darevsky, and Nikolai Nikolaevich Szczerbak. 1974. Dva novikh dlya fauni SSSR vida yashcherits semeistva gekkonov [Two new for the USSR fauna lizard species from the family Gekkonidae]. Vestnik Zoologii, 1974(4):33–38 (In Russian).

Gray, John Edward. 1825. A synopsis of the genera of reptiles and Amphibia, with a description of some new species. Annals of Philosophy, n.s., 10:193–217. (Reprint 1966 by Society for the Study of Amphibians and Reptiles, Ithaca, New York.)

Gray, John Edward. 1827. A synopsis of the genera of the saurian reptiles in which some new genera are indicated, and the others reviewed by actual examination. Philosophical Magazine, London, ser. 2, 2(7):54–58.

Gray, John Edward. "1830–1834" [1830–1835]. Illustrations of Indian Zoology; chiefly selected from the collection of Major-General Hardwicke, vols. 1 (1830–1832) and 2 ("1832–1835" [1832–1835]). Treuttel, Wurtz, Treuttel Junior, and Richter; Parbury, Allen and Co., London, 102 pls., no text (see note in Zhao and Adler, 1993:370).

Gray, John Edward. 1831. Synopsis Reptilium; or short descriptions of the species of reptiles. Part 1. Cataphracta. Tortoises, Crocodiles, and Enaliosaurians. Treuttel, Wurtz; G. B. Sowerby; and W. Wood, London, viii + 85 + (2) pp., 10 pls. (and extra plate [11] added in some copies).

Gray, John Edward. 1838a. Catalogue of the slender-tongued saurians with descriptions of many new genera and species. Annals and Magazine of Natural History, ser. 1, 1:274–283, 388–394.

Gray, John Edward. 1838b. Catalogue of the slender-tongued saurians with descriptions of many new genera and species [part 3]. Annals and Magazine of Natural History, ser. 1, 2(10):287–293.

Gray, John Edward. 1839. Catalogue of the slender-tongued saurians, with descriptions of many new genera and species [part 4]. Annals and Magazine of Natural History, ser. 1, 2(11):331–337.

Gray, John Edward. 1842a. Synopsis of the species of prehensile-tailed snakes, or family Boidae. The Zoological Miscellany, (Part 2):41–46. (Reprint 1971 by Society for the Study of Amphibians and Reptiles, Ithaca, New York.)

Gray, John Edward. 1842b. Monographic synopsis of the water snakes, or the family Hydridae. The Zoological Miscellany, (Part 2):59–68. (Reprint 1971 by Society for the Study of Amphibians and Reptiles, Ithaca, New York.)

Gray, John Edward. 1842c. Monographic synopsis of the vipers, or the family Viperidae. The Zoological Miscellany, (Part 2):68–71. (Reprint 1971 by Society for the Study of Amphibians and Reptiles, Ithaca, New York.)

Gray, John Edward. 1844. Catalogue of the Tortoises, Crocodiles, and Amphisbaenians in the Collection of the British Museum. British Museum, London, viii + 80 pp.

Gray, John Edward. 1845. Catalogue of the Specimens of Lizards in the Collection of the British Museum. British Museum, London, xxvii + 289 pp.

Gray, John Edward. 1849. Catalogue of the Specimens of Snakes in the Collection of the British Museum. British Museum, London, xv + 125 pp.

Gray, John Edward. 1855. Catalogue of the Shield Reptiles in the British Museum. Part I. Testudinata. British Museum, London, v + 79 pp., pls. 1–42.

Gray, John Edward. 1864. Revision of the species of Trionychidae found in Asia and Africa, with the descriptions of some new species. Proceedings of the Zoological Society of London, 1864:76–98.

Gray, John Edward. 1869. Notes on the families and genera of tortoises (Testudinata), and on the characters afforded by the study of their skulls. Proceedings of the Zoological Society of London, 1869(12):165–225, 1 pl.

Gray, John Edward. 1870. Supplement to the Catalogue of Shield Reptiles in the Collection of the British Museum. Part I. Testudinata. Trustees of the British Museum, London, x + 120 pp., figs. 1–40.

Gray, John Edward. 1873. Hand-List of the Specimens of Shield Reptiles in the British Museum. Trustees of the British Museum, London, iv + 124 pp.

Greer, Allen E. 1970. A subfamilial classification of scincid lizards. Bulletin of the Museum of Comparative Zoology, Harvard University, 139:151–183.

Grismer, Larry Lee. 1988. The phylogeny, taxonomy, classification, and biogeography of eublepharid geckos (Reptilia: Squamata). *In* Richard Dean Estes and Gregory Kent Pregill (eds.). Phylogenetic Relationships of the Lizard Families. Stanford University Press, Stanford.

Grismer, Larry Lee. 1989. *Eublepharis ensafi* Baloutch and Thireau, 1986: A junior synonym of *E. angramainyu* Anderson and Leviton, 1966. Journal of Herpetology, 23(1):94–95.

Grismer, Larry Lee. 1991. Cladistic relationships of the lizard *Eublepharis turcmenicus* (Squamata:Eublepharidae). Journal of Herpetology, 25(2):251–253.

Gronovius, Laurentius Theodorus. 1763. Zoophylacium Gronoviani Fasciculus exhibens animalia quadrupeda, amphibia atque pisces quae in museo suo adservat, rite examinavit, systematice disposuit, descripsit, atque iconibus illustravit, pp. 10–26. *In* Zoophylacium Gronoviani Fasciculus exhibens animalia quadrupeda, amphibia, pisces, insecta, vermes, mollusca, testacea, et zoophyta, quae in museo suo adservavit, examini subjecit, systematice disposuit atque descripsit. T. Haas and Co. and S. and J. Luchtmans, Lugduni Batavorum [= Leiden].

Guibé, Jean. 1954. Catalogue des Types de Lézards du Muséum National d'Histoire Naturelle. Muséum National d'Histoire Naturelle, Paris, 119 pp.

Guibé, Jean. 1957. Reptiles d'Iran récoltes par M. Francis Petter. Description d'un vipèride nouveau: *Pseudocerastes latirostris*. Bulletin du Muséum National d'Histoire Naturelle, Paris, ser. 2, 29(2):136–142.

Guibé, Jean. 1966a. Reptiles et amphibiens récoltes par la Mission Franco-Iranienne. Bulletin du Muséum National d'Histoire Naturelle, Paris, ser. 2, 38(2):97–98.

Guibé, Jean. 1966b. Contribution a l'étude des genres *Microgecko* Nikolsky et *Tropiocolotes* Peters (Lacertilia, Gekonidae). Bulletin du Museum National d'Histoire Naturelle, Paris, ser. 2, 38(2):337–346.

Günther, Albert Carl Ludwig Gotthilf. 1858a. Catalogue of Colubrine Snakes. Trustees of the British Museum, London, xvi + 281 + (3) pp. (reprinted 1971 British Museum [Natural History]).

Günther, Albert Carl Ludwig Gotthilf. "1858" (1859). Catalogue of the Batrachia Salientia in the Collection of the British Museum. Trustees of the British Museum, London, xvi + 160 + (8) pp., 12 pls. (Title page bears the year 1858. According to C. D. Sherborn, 1934, Ann. Mag. Nat. Hist., ser. 10, 13:309, it was published 12 February 1859.).

Günther, Albert Carl Ludwig Gotthilf. 1864a. The Reptiles of British India. Ray Society, London, xxvii + 444 pp., 26 pls. (Reprint, about 1982, Oxford & IBH Publ., New Dehli.)

Günther, Albert Carl Ludwig Gotthilf. 1864b. Description of a new species of *Eublepharis*. Annals and Magazine of Natural History, ser. 3, 41:429–430.

Günther, Robert T. 1889. Contribution to the natural history of Lake Urmi, N. W. Persia, and its neighborhood. Journal of the Linnean Society of London, 27:345–373, map pl. 21.

Gvirtzman, G., and B. Buchbinder. 1978. The late Tertiary of the coastal plain and continental shelf of Israel and its bearing on the history of the eastern Mediterranean. Initial Reports of the Deep Sea Drilling Project Leg 2A, 42(2):1195–1222.

Gvirtzman, G., and B. Buchbinder. 1977. The desiccation events in the eastern Mediterranean during Messinian time as compared with other Miocene desiccation events in basins around the Mediterranian, pp. 411–420. *In* B. Biju-Duval and L. Montader (eds.). Structural History of the Mediterranean basins. Editions Technip, Paris.

Haas, Georg. 1951. On the present state of our knowledge of the herpetofauna of Palestine. Bulletin of the Research Council of Israel, 1:67–95.

Haas, Georg. 1957. Some amphibians and reptiles from Arabia. Proceedings of the California Academy of Sciences, ser. 4, 29(3):47–86, 12 text-figs.

Haas, Georg, and Yehudah Leopold Werner. 1969. Lizards and snakes from southwestern Asia collected by Henry Field. Bulletin of the Museum of Comparative Zoology, Harvard University, 138(6):327–405, 21 pls.

Hahn, Donald Edgar. 1978a. Liste der rezenten Amphibien und Reptilien. Scolecophidia, Anomalepididae, Leptotyphlopidae, und Typhlopidae. Das Tierreich, 101:1–65.

Hahn, Donald Edgar. 1978b. A brief review of the genus *Leptotyphlops* (Reptilia, Serpentes, Leptotyphlopidae) of Asia, with description of a new species. Journal of Herpetology, 12(4):477–489.

Harding, Keith A., and Kenneth Reginald George Welch. 1980. Venomous Snakes of the World. A Checklist. Pergamon Press, Oxford, x + 188 pp.

Hardwicke, Thomas, and John Edward Gray. 1827. A synopsis of the species of saurian reptiles, collected in India by Major-General Hardwicke. Zoological Journal, 3(10):213–229.

Harper, Francis. 1940. Some works of Bartram, Daudin, Latreille, and Sonnini, and their bearing upon North American herpetological nomenclature. American Midland Naturalist, Notre Dame, 23(3):692–723.

Hassinger, Jerry D. 1968. Introduction to the mammal survey of the Street Expedition to Afghanistan. Fieldiana: Zoology, 55(1):1–81.

Heimes, Peter. 1987. Beitrag zur Systematik der Fächenfinger (Sauria: Gekkonidae: *Ptyodactylus*). Salamandra, 23(4):212–235.

Hellmich, Walter. 1959. Bemerkungen zu einer kleinen Sammlung von Amphibien und Reptilien aus Süd-Persien. Opuscula Zoologica, herausgegeben von der Zoologischen Staatssammlung in München, 1959(35):1–9.

Hemming, Francis, and Diana Noakes. 1958. Official Index of Rejected and Invalid Generic Names in Zoology: I, Names 1–1169. International Trust for Zoological Nomenclature.

Henle, Klaus. 1995. A brief review of the origin and use of '*stellio*' in herpetology and a comment on the nomenclature and taxonomy of agamids of the genus *Agama* (*sensu lato*). Herpetozoa, 8(1–2):3–9.

Herrmann, Hans-Werner, Ulrich Joger, and Göran Nilson. 1992a. Molecular phylogeny and systematics of viperine snakes. I. General phylogeny of European vipers (*Vipera* sensu stricto). Proceedings of the 6th Ordinary General Meeting of Societas Europaea Herpetologica, Budapest.

Herrmann, Hans-Werner, Ulrich Joger, and Göran Nilson. 1992b. Phylogeny and systematics of viperine snakes. III: resurrection of the genus *Macrovipera* (Reuss, 1927) as suggested by biochemical evidence. Amphibia-Reptilia, 13:375–392, figs. 1–5.

Heyden, Carl Heinrich Georg von. 1827. Reptilien, pp. 1–24, 6 pls. *In* Wilhelm Peter Eduard Simon Rüppell (ed.). Atlas zu der Reise im nördlichen Afrika, vol. I. Zoologie. Heinrich Ludwig Brönner, Frankfurt am Main.

Hoberlandt, Ludvík. 1981. Results of the Czechoslovak-Iranian entomological expeditions to Iran. Introduction to the Second expedition 1973. Acta entomologica Musaéi. Nationalis Pragae, 40:5–32.

Höggren, Mats, Gören Nilson, Claes Andrén, Nikolai Lutseanovich Orlov, and Boris S. Tuniyev. 1993. Vipers of the Caucasus: natural history and systematic review. Herpetological Natural History, 1(2):11–19.

Hoofien, Jaccb Haim. 1958. An addition to the fauna of Sinai, *Eremias brevirostris* Blanf. (Reptilia, Lacertidae). Annals and Magazine of Natural History, 10:719–720.

Horowitz, A. 1979. The Quaternary of Israel. Academic Press, New York, xv + 394 pp.

Huey, Raymond Brunson. 1982. Temperature, physiology, and the ecology of reptiles, chapter 3, pp. 25–91. *In* Carl Gans and F. Harvey Pough (eds.). Biology of the Reptilia. Physiology C. Physiological Ecology. Biology of the Reptilia, vol. 12. Academic Press, New York.

Inger, Robert Frederich. 1972. *Bufo* of Eurasia, pp. 102–118, 357–360. *In* William Frank Blair (ed.). Evolution of the genus *Bufo*. University of Texas Press, Austin, Texas.

International Commission on Zoological Nomenclature. 1950. "*Mabuya*" Fitzinger, 1826 (class Reptilia): correction in the "Official List of Generic Names in Zoology" of entry relating to: correction of error in Opinion 92. Bulletin of Zoological Nomenclature, London, 4(13–15):356.

International Commission on Zoological Nomenclature. 1966. Opinion 794. *Spalerosophis* Jan, 1865 (Reptilia): validated under the plenary powers. Bulletin of Zoological Nomenclature, 23(5):229–231.

International Commission on Zoological Nomenclature. 1985. International Code of Zoological Nomenclature, adopted by the XX General Assembly of the International Union of Biological Sciences. International Trust for Zoological Nomenclature, London, 338 pp.

International Union for the Conservation of Nature. 1987. Directory of Wetlands of International Importance. IUCN, Gland, Switzerland and Cambridge, UK, 460 pp.

Issar, A. 1969. The groundwater provinces of Iran. Bulletin of the International Association of Scientific Hydrology, 14(1):87–99.

Iverson, John Burton. 1992. A Revised Checklist with Distribution Maps of the Turtles of the World. Privately printed by the author, Richmond, Indiana, xiii + 363 pp.

Jacob, D., and L. S. Ramaswami. 1976. The female reproductive cycle of the Indian monitor lizard *Varanus monitor*. Copeia, 1976(2):256–260.

Jahanbani, Amanollah, and Manouchehr Saram. 1960. Crocodile hunting in Baluchistan. Game and Nature, 15:18–22 (In Farsi [Persian]).

Joger, Ulrich. 1984. The Venomous Snakes of the Near and Middle East. Tübinger Atlas des vorderen Orients, ser. A, (12):1–115, (A wall chart of distribution maps, authored by Joger and issued by Richert Verlag in Wiesbaden, was published in 1983. These colored maps cover all Middle East venomous snakes).

Joger, Ulrich. 1991. A molecular phylogeny of agamid lizards. Copeia, 1991(3):616–622.

Joger, Ulrich, Hans-Werner Herrmann, and Göran Nilson. 1992. Molecular phylogeny and systematics of viperine snakes. II. A revision of the *Vipera ursenii* complex. Proceedings of the 6th Ordinary General Meeting of Sociétas Europaea Herpetologica, Budapest.

Kami, Haji Gholi. 1997a. Rediscovery of the southern crested newt, *Triturus (cristatus) karelini* (Salamandridae), from its easternmost locality in Iran. Zoology of the Middle East, Heidelberg, 15:37–40, figs. 1–2.

Kami, Haji Gholi. 1997b. First record of the olive ridley turtle, *Lepidochelys olivacea*, in Iranian coastal waters (Testudines, Cheloniidae). Zoology of the Middle East, Heidelberg, 15:67–70, fig. 1.

Kami, Haji Gholi, and Ebrahim Vakilpoure. 1996a. Geographic distribution: *Bufo bufo*. Herpetological Review, 27(3):148.

Kami, Haji Gholi, and Ebrahim Vakilpoure. 1996b. Geographic distribution: *Pelobates syriacus*. Herpetological Review, 27(3):149.

Kami, Haji Gholi, and Ebrahim Vakilpoure. 1996c. Geographic distribution: *Rana camerani*. Herpetological Review, 27(3):150.

Kami, Haji Gholi, and Ebrahim Vakilpoure. 1996d. Geographic distribution: *Rana macrocnemis pseudodalmatina*. Herpetological Review, 27(3):150.

Kami, Haji Gholi, and Ebrahim Vakilpoure. 1996e. Geographic distribution: *Tropiocolotes latifi*. Herpetological Review, 27(3):153.

Kassler, P. 1973. The structural and geomorphic evolution of the Persian Gulf, pp. 11–33. *In* B. H. Purser (ed.). The Persian Gulf. Berlin.

Kaup, Johann Jakob. 1825. Einige Bemerkungen zu Merrems Handbuch. Isis von Oken, Jena, 1825(5):cols.589–593, pl. 3.

Kaverkin, Yu. I., and Nikolai Lutseanovich Orlov. 1996. Experience of captive breeding of *Eublepharis turcmenicus* Darevsky, 1978 (Eublepharidae, Sauria). Russian Journal of Herpetology, 3(1):99.

Kechichian, Joseph A. 1989. Boundaries iv. With Iraq, pp. 415–417. *In* Ehsan Yarshater (ed.). Encyclopaedia Iranica, vol. 4, fasc. 4. Routledge and Kegan Paul, New York and London.

Kessler, Karl Fodorovich. 1878. Puteschestvie po Zakavkazkomu kraju v 1875 godu s zoologicheskoi teslya [Travels in Transcaucasia in 1875 for zoological purposes]. Trudy St. Petersburgskogo Obshchestva Estestvolspitatelei, 8(Prilozhenie trudam):1–200 (in Russian).

Khalaf, Kamel T. 1959. Reptiles of Iraq, with Some Notes on the Amphibians. Ar-Rabitta Press, Baghdad, v + 96 pp.

Khalaf, Kamel T. 1960. Notes on a collection of lizards and snakes from Iraq. Iraq Natural History Museum Publications, (18):12–18.

Khan, Muhammad Sharif. 1968. Amphibian fauna of Dist. Jhang with notes on habits. Pakistan Journal of Science, 20(5–6):227–233.

Khan, Muhammad Sharif. 1972. Checklist and key to the lizards of the Jhang District, West Pakistan. Herpetologica, 28(2):94–98.

Khan, Muhammad Sharif. 1976. An annotated checklist and key to the amphibians of Pakistan. Biologia Lahore, 22(2):201–210.

Khan, Muhammad Sharif. 1977. A checklist and key to the snakes of District Jhang, Punjab, Pakistan. Biologia Lahore, 23(2):145–157.

Khan, Muhammad Sharif. 1980. Affinities and zoogeography of herpetiles of Pakistan. Biologia Lahore, 26(1–2):113–171.

Khan, Muhammad Sharif. 1982a. An annotated checklist and key to the reptiles of Pakistan. Part III: Serpentes. Biologia Lahore, 28(2):215–254.

Khan, Muhammad Sharif. 1982b. Key for the identification of amphibian tadpoles from the plains of Pakistan. Pakistan Journal of Zoology, 14(2):133–145.

Khan, Muhammad Sharif. 1982c. Collection, preservation and identification of amphibian eggs from the plains of Pakistan. Pakistan Journal of Zoology, 14(2):241–243.

Khan, Muhammad Sharif. 1983. Venomous terrestrial snakes of Pakistan. The Snake, 15(2):101–105.

Khan, Muhammad Sharif. 1986. Checklist and distribution of the lizard fauna of Pakistan. Hamadryad, 11(1–2):18–21.

Khan, Muhammad Sharif. 1987. Checklist and distribution of amphibians in Pakistan. Hamadryad, 12(1):10.

Khan, Muhammad Sharif. 1990. Venomous terrestrial snakes of Pakistan and snake bite problem, pp. 419–445. *In* Ponnampalam Gopalakrishnakone and Loke Ming Chou (eds.). Snakes of Medical Importance (Asia-Pacific Region). National University of Singapore & International Society of Toxicology, Singapore.

Khan, Muhammad Sharif. 1993. A checklist and key to the gekkonid lizards of Pakistan. Hamadryad, 18:35–41.

Khan, Muhammad Sharif. 1994. A revised checklist and key to the amphibians of Pakistan. Hamadryad, 19:11–14.

Khan, Muhammad Sharif. 1997. Taxonomic notes on Pakistani snakes of the *Coluber karelini-rhodorachis-ventromaculatus* species complex: a new approach to the problem. Asiatic Herpetological Research, 7:51–60.

Khan, Muhammad Sharif, and Khalid Javaid Baig. 1988. Checklist of the amphibians and reptiles of District Jhelum, Punjab, Pakistan. The Snake, 20:156–161.

Khan, Muhammad Sharif, and M. Muhammad R. Z. Khan. 1997. A new skink from the Thal Desert of Pakistan. Asiatic Herpetological Research, 7:61–67.

Khan, Muhammad Sharif, and Muhammad Ramzan Mirza. 1976. An annotated checklist and key to the reptiles of Pakistan, Part I: Chelonia and Crocodilia. Biologia, Lahore, 22(2):211–219.

Khan, Muhammad Sharif, and Muhammad Ramzan Mirza. 1977. An annotated checklist and key to the reptiles of Pakistan, Part II: Sauria (Lacertilia). Biologia, Lahore, 23(1):41–64.

Khan, Muhammad Sharif, and Rashida Tasnim. 1987. A Field Guide to the Identification of Herps of Pakistan. Part I: Amphibia. Biological Society of Pakistan Monographs, no.14, 28 pp.

Khaza'ie, Reza, and Robert G. Tuck, Jr. 1974. Geographic distribution: *Lacerta trilineata media*. Herpetological Review, 5(4):107–108.

King, Dennis, Peter Baverstock, and Susan Fuller. 1997. Biogeographic origins of varanid lizards: a molecular perspective. *In* Zbyněk Roček and Scott Hart (eds.). Herpetology '97. Abstracts of the Third World Congress of Herpetology 2–10 August 1997 Prague, Czech Republic. Third World Congress of Herpetology, Prague (no pagination).

King, Frederic Wayne, and Russell Louis Burke. 1989. Crocodilian, Tuatara, and Turtle Species of the World. A Taxonomic and Geographic Reference. Association of Systematic Collections, Washington, DC., xxii + 216 pp.

Kinunen, Wayne, and Steve Bullock. 1970. Baluchistan crocodile investigations. Job Report, Division of Research and Development, Tehran, 20 pp. (Unpublished).

Kinunen, Wayne, and Peter Walczak. 1971. Persian Gulf sea turtle nesting surveys. Job Completion Report, Division of Research and Development (submitted to Iran Game and Fish Department, F-7-50), Tehran, 16 pp. (Unpublished).

Klawinski, P. W., R. K. Vaughan, D. Saenz, and W. Godwin. 1994. Comparison of dietary overlap between allopatric and sympatric geckos. Journal of Herpetology, 28(1):25–30.

Klembara, Josef. 1981. Beitrag zur Kenntnis der subfamilie Anguinae (Reptilia, Anguidae). Acta Universitatis Carolinae (Geologica), 1981(2):121–168.

Kluge, Arnold Girard. 1967. Higher taxonomic categories of gekkonid lizards and their evolution. Bulletin of the American Museum of Natural History, 135(1):1–59, pls. 1–5.

Kluge, Arnold Girard. 1983. Cladistic relationships among gekkonid lizards. Copeia, 1983(2):465–475.

Kluge, Arnold Girard. 1985. Notes on gekko nomenclature (Sauria: Gekkonidae). Zoologische Mededelingen Leiden, 59(10):95–100.

Kluge, Arnold Girard. 1987. Cladistic relationships in the Gekkonoidea (Squamata, Sauria). Miscellaneous Publications, Museum of Zoology, University of Michigan, (173):iv + 1–54.

Kluge, Arnold Girard. 1991. Checklist of gekkonid lizards. Smithsonian Herpetological Information Service, (85):1–35.

Kluge, Arnold Girard. 1993a. Gekkonoid Lizard Taxonomy. International Gecko Society, San Diego, 245 pp.

Kluge, Arnold Girard. 1993b. *Calabaria* and the phylogeny of erycine snakes. Zoological Journal of the Linnean Society, 107:293–351.

Kluge, Arnold Girard. 1994. Principles of phylogenetic systematics and the informativeness of the karyotype in documenting gekkotan lizard relationships. Herpetologica, 50(2):210–221.

Kluge, Arnold Girard. 1997. A review of varanid lizard phylogeny: old hypotheses versus new data. *In* Zbyněk Roček and Scott Hart (eds.). Herpetology '97. Abstracts of the Third World Congress of Herpetology 2–10 August 1997 Prague, Czech Republic. Third World Congress of Herpetology, Prague (no pagination).

Kotschy, Theodor. 1861. Der westliche Elbrus bei Teheran in Nord-Persien. M. Auer, Vienna, 46 pp. (with a map of the mountains).

Král, Bohumil. 1969. Notes on the herpetofauna of certain provinces of Afghanistan. Zoologicke Listy, 18(1):55–66.

Kryshtofovich, A. N. 1929. Evolution of the Tertiary flora in Asia. New Phytologist, 28:303–312.

Kulagin, N. 1888. Spiski i opisaniya kollektsi zemnovodnykh i presmykayushchkhsya zoologicheskogo muzeya Moskovkogo universitets [Lists and descriptions of collections of amphibians and reptiles of the Zoological Museum of the Moscow University]. Izvestiya Obshchestva Lyubitelei Estestvoznaniya, 56(2):1–39 (In Russian).

Lacepède, Bernarde Germain Étienne. 1788. Histoire Naturelle des Quadrupèdes Ovipares et des Serpens, vol. 1. Hôtel de Thou, Paris, 17, (1), 651 pp.

Lacepède, Bernarde Germain Étienne. 1789. Histoire Naturelle des Quadrupèdes Ovipares et des Serpens, vol. 2. Hôtel de Thou, Paris, (1), (1), 8, 5–9, (1), 144 (="Discours"), 527 pp., 22 pls.

Lantz, Louis Amédée. 1918. Reptiles from the River Tajan (Transcaspia). Proceedings of the Zoological Society of London, 1918(7):11–7, pl. 1.

Lantz, Louis Amédée. 1928a. Les *Eremias* de l'Asie occidentale. Bulletin du Muséum de Géorgie, 4(3):1–72, pls. 1–3.

Lantz, Louis Amédée. 1928b. Les *Eremias* de l'Asie occidentale. Bulletin du Muséum de Géorgie, 5:1–64.

Lantz, Louis Amédée. 1928c. Les *Eremias* de l'Asie occidentale. Muséum de Géorgie, Tiflis, 1–136, pls. 1–3 ("Extrait du *Bulletin du Muséum de Géorgie*, t.t. IV et V").

Lantz, Louis Amédée. 1931. Note sur la forme typique d'*Ophisops elegans elegans*. Bulletin du Muséum de Géorgie, 6:34–42.

Lantz, Louis Amédée, and Carl August Otto Cyrén. 1920. Note sur les *Lacerta viridis* du Caucase. Bulletin de la Société Zoologique de France, 45:33–37.

Lantz, Louis Amédée, and Carl August Otto Cyrén. 1936. Contribution à la connaissance de *Lacerta saxicola* Eversmann. Bulletin de la Société Zoologique de France, 61:159–181.

Lantz, Louis Amédée, and Carl August Otto Cyrén. 1939. Contribution à la connaissance de *Lacerta brandtii* De Filippi et de *Lacerta parva* Boulenger. Bulletin de la Société Zoologique de France, 64:228–243, pl.

Lantz, Louis Amédée, and G. F. Suchow. 1934. *Apathya cappadocica urmiana* subsp. nov., eine neue Eidechsenform aus dem persischen Kurdistan. Zoologische Anzeiger, 106:294–299, fig.

Lanza, Benedetto. 1978. On some new or interesting East African amphibians and reptiles. Monitore Zoologico Italiano, (14):229–297, figs. 1–41.

Latifi, Mahmoud. 1975. Commercial production of anti-snakebite serum (antivenin), pp. 561–588. *In* Carl Gans and K. A. Gans (eds.). Biology of the Reptilia, vol. 8: Physiology B. Academic Press, New York and London.

Latifi, Mahmoud. 1984a. [The Snakes of Iran]. Iran Department of the Environment, Tehran, 221 pp. (In Farsi [Persian]).

Latifi, Mahmoud. 1984b. Variation in yield and lethality of venoms from Iranian snakes. Toxicon, 22(3):373–380.

Latifi, Mahmoud. 1991. The Snakes of Iran. Society for the Study of Amphibians and Reptiles, Contributions to Herpetology no. 7, viii + 159 pp., 25 pls. (Translated from the Iranian edition by S. Sajadian; volume editors, A. Leviton and G. Zug).

Latifi, Mahmoud. 1992. The Snakes of Iran. [2nd ed.]. Iran Department of the Environment, Tehran, 231 pp. (In Farsi [Persian]).

Latifi, Mahmoud, and R. Farzanpay. 1973. Yield of venom and distribution of Iranian venomous snakes. Pahlavi Medical Journal, 4:556–564.

Latifi, Mahmoud, and H. Manhouri. 1966. Antivenin production. Memorias do Instituto de Butantan, 33(3):893–897.

Latifi, Mahmoud, Alphonse Richard Hoge, and M. Eliazan. 1968. The poisonous snakes of Iran. Memorias do Instituto de Butantan, 33 [1966](3):735–744.

Laurenti, Josephus Nicolaus. 1768. Specimen Medicum, Exhibens Synopsin Reptilium Emendatam cum Experimentis circa Venena et Antidota Reptilium Austriacorum, quod Authoritate et Consensu. J. Thomae Trattnern, Vienna, (8) + 214 + (3) pp., 5 pls. (Reprint 1966, A. Asher, Amsterdam.)

Lay, Douglas M. 1967. A study of the mammals of Iran resulting from the Street Expedition of 1962–63. Fieldiana: Zoology, 54:1–282.

Leptien, Rolf. 1993a. An uncommon day gecko from the United Arab Emirates, *Pristurus rupestris* (Blanford 1874). Dactylus, 1(4):34–37, 3 figs.

Leptien, Rolf. 1993b. Observations on the Arabian Desert gecko *Bunopus tuberculatus* Blanford, 1874 from the United Arab Emirates. Dactylus, 2(2):56–58.

Leviton, Alan Edward, and Michele L. Aldrich. 1984. Introduction: John Anderson (1833–1900): a zoologist in the Victorian Period; checklist of the amphibians and reptiles of the Arabian Peninsula; bibliography of John Anderson; recent references, pp. v-xxxv, figs. 1–4. *In* John Anderson. Herpetology of Arabia. Facsimile Reprints in Herpetology. Society for the Study of Amphibians and Reptiles, Oxford, Ohio.

Leviton, Alan Edward, and Steven Clement Anderson. 1967. Survey of the reptiles of the Sheikhdom of Abu Dhabi, Arabian Peninsula. Part II. Systematic account of the collection of reptiles made in the Sheikhdom of Abu Dhabi by John Gasperetti. Proceedings of the California Academy of Sciences, ser. 4, 35(9):157–192, 12 figs., 8 tabs.

Leviton, Alan Edward, and Steven Clement Anderson. 1970a. Review of the snakes of the genus *Lytorhynchus*. Proceedings of the California Academy of Sciences, ser. 4, 37(7):249–274, 25 figs.

Leviton, Alan Edward, and Steven Clement Anderson. 1970b. The amphibians and reptiles of Afghanistan, a check list and key to the herpetofauna. Proceedings of the California Academy of Sciences, ser. 4, 38(10):163–206.

Leviton, Alan Edward, and Steven Clement Anderson. 1972. Description of a new species of *Tropiocolotes* (Reptilia: Gekkonidae) with a revised key to the genus. Occasional Papers of the California Academy of Sciences, (96):1–7.

Leviton, Alan Edward, and Steven Clement Anderson. 1984. Description of a new species of *Cyrtodactylus* from Afghanistan with remarks on the status of *Gymnodactylus longipes* and *Cyrtodactylus fedtschenkoi*. Journal of Herpetology, 18(3):270–276.

Leviton, Alan Edward, Steven Clement Anderson, Kraig Kerr Adler, and Sherman Anthony Minton. 1992. Handbook to Middle East Amphibians and Reptiles. Contributions in Herpetology, No. 8. Society for the Study of Amphibians and Reptiles, Oxford, Ohio, vii + 252 pp.

Leviton, Alan Edward, and Robert H. Gibbs, Jr. 1988. Standards in herpetology and ichthyology: standard symbolic codes for institution resource collections in herpetology and ichthyology. Supplement no. 1: Additions and corrections. Copeia, 1988(1):280–282.

Leviton, Alan Edward, Robert H. Gibbs, Jr., Elizabeth Heal, and C. E. Dawson. 1985. Standards in herpetology and ichthyology: Part I. Standard symbolic codes for institutional resource collections in herpetology and ichthyology. Copeia, 1985(3):802–832.

Lichnova, O. P. 1992. [Biochemical polymorphism, systematics, and phylogeny of lizards of genus *Phrynocephalus* (Agamidae, Reptilia)]. Doctoral dissertation, Moscow State University, Moscow, 24 pp. (in Russian).

Lichtenstein, Martin Hinrich Carl. 1823a. Verzeichniss der Doubletten des zoologischen Museums der Königlichen Universität zu Berlin nebst Beschreibung vieler bisher unbekannter Arten von Säugethieren, Vögeln, Amphibien und Fischen. T. Trautwein, Berlin, x + 118 pp., pl. 1.

Lichtenstein, Martin Hinrich Carl. 1823b. Naturhistorischer Anhang, pp. 112–147. *In* Edward Eversmann (ed.). Reise von Orenburg nach Buchara. E. H. G. Christiani, Berlin.

Lichtenstein, Martin Hinrich Carl. 1856. Nomenclator Reptilium et Amphibiorum Musei Zoologici Berolinensis (Namenverzeichniss der in Zoologischen Sammlung der Königlichen Universität zu Berlin aufgestellten Arten von Reptilien und Amphibien nach ihren Ordnungen, Familien und Gattungen. Koenigliche Akademie der Wissenschaften, Berlin, iv + 48 pp.

Linnaeus, Carolus. 1758. Systema Naturae per Regna Tria Naturae, Secundum Classes, Ordines, Genera, Species, cum Characteribus, Differentiis, Synonymis, Locis, vol. 1, Part 1 (10th ed.). Laurenti Salvii, Stockholm, (4) + 823 + (1) pp. (Reprint 1956, British Museum [Natural History], London.)

Linnaeus, Carolus. 1766. Systema Naturae per Regna Tria Naturae, Secundum Classes, Ordines, Genera, Species, cum Characteribus, Differentiis, Synonymis, Locis, vol. 1, Part 1 (12th ed.). Laurenti Salvii, Stockholm, 532 pp. (Herpetology sections reprinted 1963, Society for the Study of Amphibians and Reptiles, Ithaca, New York.)

Loveridge, Arthur. 1925. A mite pocket in the gecko, *Gymnodactylus lawderanus* Stoliczka. Proceedings of the Zoological Society of London, 1925, pt. 4:1431.

Loveridge, Arthur. 1936. New geckos of the genus *Hemidactylus* from Zanzibar and Manda Islands. Proceedings of the Biological Society of Washington, 49:59–62.

Loveridge, Arthur. 1947. Revision of the African lizards of the family Gekkonidae. Bulletin of the Museum of Comparative Zoology, Harvard University, 98(1):1–469.

Macey, J. Robert, Allan Larson, Natalia Borisovna Ananjeva, and Theodore Johnstone Papenfuss. 1995. A mitochondrial DNA-based phylogenetic hypothesis of the Asian Agamidae, p. 38. *In* Natalia Borisovna Ananjeva (ed.). Abstracts of the Second Asian Herpetological Meeting, 6–10 September 1995, Asgabat, Turkmenistan. Folium Publishing Co., Moscow.

Macey, J. Robert, James A. Schulte II, Allan Larson, Natalia Borisovna Ananjeva, Nasrullah Rastegar-Pouyani, Yao-Zhao Wang, and Theodore Johnstone Papenfuss. 1997. A phylogenetic hypothesis for arid West Asian and African agamine acrodont lizards based on mitochondrial DNA sequences. *In* Zbyněk Roček and Scott Hart (eds.). Herpetology '97. Abstracts of the Third World Congress of Herpetology 2–10 August 1997 Prague, Czech Republic. Third World Congress of Herpetology, Prague (no pagination).

Macey, J. Robert, James A. Schulte II, Natalia Borisovna Ananjeva, Allan Larson, Nasrullah Rastegar-Pouyani, Sakhat M. Shammakov, and Theodore Johnstone Papenfuss. 1998. Phylogenetic relationships among agamid lizards of the *Laudakia caucasia* species group: testing hypotheses of biogeographic fragmentation and an area cladogram for the Iranian Plateau. Molecular Phylogenetics and Evolution, 10(1):118–131.

Madden, C. T., and J. A. Van Couvering. 1976. The proboscidean datum event: Early Miocene migration from Africa. Geological Society of America Abstracts Program:992–993.

Mahendra, Beni Charan. 1936. Contribution to the bionomics, anatomy, reproduction and development of the Indian house-gecko, *Hemidactylus flaviviridis* Rüppell. Part 1. Proceedings of the Indian Academy of Science, 4:250–281, pls. 14–18.

Malnate, Edmond Virgil. 1971. A catalogue of primary types in the herpetological collections of the Academy of Natural Sciences (ANSP). Proceedings of the Academy of Natural Sciences, Philadelphia, 123(9):345–375, errata.

Martens, Harald, and Dieter Kock. 1991. Erstnachweise für drei Gecko-Gattungen in Syrien (Reptilia: Sauria: Gekkonidae). Senckenbergiana Biologie, 71(1–3):15–21, map.

Martens, Harald, and Dieter Kock. 1992. The desert monitor, *Varanus griseus* (Daudin 1803), in Syria. Senckenbergiana Biologie, 72(1–3):7–11, map.

Marx, Hymen. 1953. The elapid genus of snakes *Walterinnesia*. Fieldiana: Zoology, 34:189–196.

Marx, Hymen. 1959. Review of the colubrid snake genus *Spalerosophis*. Fieldiana: Zoology, 39(30):347–361, figs. 58–59, 1 map, 1 tab.

Marx, Hymen, and George Bernard Rabb. 1965. Relationships and zoogeography of the viperine snakes (Family Viperidae). Fieldiana: Zoology, 44(21):161–206, figs. 32–46.

Marx, Hymen, and George Bernard Rabb. 1972. Phyletic analysis of fifty characters of advanced snakes. Fieldiana: Zoology, 63(1153):1–321.

Mautz, William J. 1982. Patterns of evaporative water loss, chapter 10, pp. 443–481. *In* Carl Gans and F. Harvey Pough (eds.). Biology of the Reptilia, Physiology C. Physiological Ecology. Biology of the Reptilia, vol. 12. Academic Press, New York.

Maxson, Linda R. 1981. Albumin evolution and its phylogenetic implications in toads of the genus *Bufo*. 2. Relationships among Eurasian *Bufo*. Copeia, 1981:579–583.

Mayer, Werner, and G. Benyr. 1994. Albumin-Evolution und Phylogenese in der Familie Lacertidae (Reptilia: Sauria). Annalen des Naturhistorischen Museums in Wien, 96B:621–648.

Mayer, Werner, and Wolfgang Bischoff. 1996. Beiträge zur taxonomischen Revision der Gattung *Lacerta* (Reptilia:Lacertidae). Teil1: *Zootoca, Omanosaura, Timon*, and *Teira* als eigenständige Gattungen. Salamandra, 32(3):163–170.

Mayer, Werner, and Franz Tiedemann. 1982. Chemotaxonomical investigations in the collective genus *Lacerta* (Lacertidae; Sauria) by means of protein electrophoresis. Amphibia-Reptilia, 2:349–355.

Mayhew, Wilbur W. 1968. Biology of desert amphibians and reptiles, chapter 6, pp. 195–356. *In* G. W. Brown, Jr. (ed.). Desert Biology, vol. 1. Academic Press, New York and London.

Mazurmovich, B. N. 1983. Alexandr Mikhailovich Nikolsky. Nauka, Moscow, 77 pp. (In Russian).

McLachlan, Keith. 1989. Boundaries, i. With the Ottoman Empire, pp. 401–403. *In* Ehsan Yarshater (ed.). Encyclopaedia Iranica, vol. 4, fasc. 4. Routledge and Kegan Paul, New York and London.

McMahon, A. Henry. 1897. The southern borderlands of Afghanistan. Geographical Journal, 9:393–415.

McMahon, A. Henry. 1906. Recent survey and exploration in Seistan. Geographical Journal, 28(3):209–228.

Mearns, Barbara, and Richard Mearns. 1998. The Bird Collectors. Academic Press, San Diego, London, Boston, New York, Sydney, Tokyo, and Toronto, xviii + 472 pp.

Méhelÿ, Ludwig von. 1907a. DeVries fajkeletkezesi elmeletenek kritikaja. Termeszettudomanye Kozlony, 39 (Potfuzetek 85–86):28 pp.

Méhelÿ, Ludwig von. 1907b. Archaeo-és Neolacertók (Volasz Boulenger G. A. es Dr. Werner, F. urak birólatói). Allattani Közlemenyek, 6:97–120 (In Hungarian).

Méhelÿ, Ludwig von. 1907c. Zur Lösung der "Muralis-" Frage. Vorläufige Mitteilung. Annales Musei Nationalis Hungarici, 5:84–88, pl. 3.

Méhelÿ, Ludwig von. 1907d. Archaeo- und Neolacerten. (Erwiderung an die herren G. A. Boulenger, FRS und Dr. F. Werner.). Annales Musei Nationalis Hungarici, 5:469–493.

Méhelÿ, Ludwig von. 1909. Materialien zu einer Systematik und Phylogenie der muralisähnlichen Lacerten. Annales Musei Nationalis Hungarici, 7:409–621, 25 pls.

Méhelÿ, Ludwig von. 1910. Ueber vermeintliche Mauereidechsen aus Persien. Zoologische Anzeiger, 35:592–596.

Ménétriés, Edouard. 1832. Catalogue Raisonné des Objets de Zoologie Recueillis dans un Voyage au Caucase et jusqu'aux Frontières Actuelles de la Perse, Entrepris par Ordre de S. M. l'Empereur. Impériale Académie des Sciences, St. Petersbourg, (4) + 271 + [xxxiii] + iv + (1) pp.

Merrem, Blasius. 1820. Tentamen Systematis Amphibiorum [Versuch eines Systems der Amphibien], vol. 1. J. C. Krieger, Marburg, xv + 191 [German] + 191 [Latin] pp., 1 pl.

Mertens, Robert Friedrich Wilhelm. 1921a. Zur Kenntnis der geographischen Formen von *Chalcides ocellatus* Forskål (Rept., Lac.). Senckenbergiana Biologie, 3(3–4):116–120.

Mertens, Robert Friedrich Wilhelm. 1920b. Über die geographischen Formen von *Eumeces schneideri* Daudin. Senckenbergiana Biologie, 2:176–179.

Mertens, Robert Friedrich Wilhelm. 1922a. *Lacerta strigata wolterstorffi* subsp. n. from Syria. Archiv für Naturgeschichte, 88a(3):193–195.

Mertens, Robert Friedrich Wilhelm. 1922b. Verzeichniss der Typen in der herpetologischen Sammlung des Senckenbergischen Museums. Senckenbergiana Biologie, 4:162–183.

Mertens, Robert Friedrich Wilhelm. 1924a. Amphibien und Reptilien aus dem nördlichen Mesopotamien. Abhandlungen und Berichte Museum für Naturkunde und Vorgeschichte Naturwissenschafte Verein, Magdeburg, 3:349–390, pl. 12.

Mertens, Robert Friedrich Wilhelm. 1924b. Ein neuer Gecko aus Mesopotamia. Senckenbergiana Biologie, 6:84.

Mertens, Robert Friedrich Wilhelm. 1924c. Herpetologische Mitteilungen: V. Zweiter Beitrag zur Kenntnis der geographischen Formen von *Eumeces schneideri* Daudin. Senckenbergiana Biologie, 6:177–185.

Mertens, Robert Friedrich Wilhelm. 1925. Amphibien und Reptilien aus dem nördlichen und östlichen Spanien, gesammelt von Dr. F. Haas. Abhandlungen der Senckenbergischen Naturforschenden Gesellschaft, 39:27–129, pls. 2–4.

Mertens, Robert Friedrich Wilhelm. 1940. Bermerkungen über einige Schlangen aus Iran. Senckenbergiana Biologie, 22(3–4):244–259.

Mertens, Robert Friedrich Wilhelm. 1942. Die Familie der Warane (Varanidae). Abhandlungen der Senckenbergischen Naturforschenden Gesellschaft, 162, 165–166:1–391, 34 pls., 12 text-figs.

Mertens, Robert Friedrich Wilhelm. 1946. Dritte Mitteilung über die Rassen der Glattechse *Eumeces schneiderii*. Senckenbergiana Biologie, 27(1–3):53–62.

Mertens, Robert Friedrich Wilhelm. 1952a. Amphibien und Reptilien aus der Türkei. Revue de la Faculté des Sciences de l'Université d'Istanbul, ser. B, 17(1):41–75.

Mertens, Robert Friedrich Wilhelm. 1952b. Nachtrag zu 'Amphibien und Reptilien aus der Türkei'. Revue de la Faculté des Sciences de l'Université d'Istanbul, ser. B, 17(4):353–355.

Mertens, Robert Friedrich Wilhelm. 1952c. Über den Glattechsen-Namen *Ablepharus pannonicus*. Zoologische Anzeiger, 149:48–50.

Mertens, Robert Friedrich Wilhelm. 1954. Über die Rassen des Wüstenwarans (*Varanus griseus*). Senckenbergiana Biologie, 35(5–6):353–357, pl. 33.

Mertens, Robert Friedrich Wilhelm. 1956. Amphibien und Reptilien aus S.O.-Iran, 1954. Jahreshefte des Vereins für vaterländische Naturkunde in Würtemberg, 111(1):90–97.

Mertens, Robert Friedrich Wilhelm. 1957. Weitere Unterlagen zur Herpetofauna von Iran 1956. Jahreshefte des Vereins für vaterländische Naturkunde in Würtemberg, 112(1):118–128, 1 text-fig.

Mertens, Robert Friedrich Wilhelm. 1959a. The World of Amphibians and Reptiles. McGraw-Hill, New York, 207 pp.

Mertens, Robert Friedrich Wilhelm. 1959b. Liste der Warane Asiens und der Indoaustralischen Inselwelt mit systematischen Bemerkungen. Senckenbergiana Biologie, 40:221–240, pls. 25–29.

Mertens, Robert Friedrich Wilhelm. 1962. Bemerkungen über *Uromastyx acanthinurus* als Rassenkries (Rept. Saur.). Senckenbergiana Biologie, 43(6):425–432, fig. 1.

Mertens, Robert Friedrich Wilhelm. 1963. Liste der rezenten Amphibien und Reptilien. Helodermatidae, Varanidae, Lanthanotidae. Das Tierreich, 79:x + 26.

Mertens, Robert Friedrich Wilhelm. 1965. Bermerkungen über einige Eidechsen aus Afghanistan. Senckenbergiana Biologie, 46(1):1–4.

Mertens, Robert Friedrich Wilhelm. 1967. Die herpetologische Sektion des Natur-Museums und Forschungs-Instituts Senckenberg in Frankfurt am Main nebst einem Verzeichnis ihrer Typen. I. Senckenbergiana Biologie, 48:1–106.

Mertens, Robert Friedrich Wilhelm. 1969. Die Amphibien und Reptilien West-Pakistans. Stüttgarter Beiträge zur Naturkunde, 197:1–96.

Mertens, Robert Friedrich Wilhelm, Ilya Sergeevich Darevsky, and Konrad Klemmer. 1967. *Vipera latifi*, eine neue Giftschlange aus dem Iran. Senckenbergiana Biologie, 48(3):161–168.

Mertens, Robert Friedrich Wilhelm, and Lorenz Müller. 1928. Liste der Amphibien und Reptilien Europas. Abhandlungen der Senckenbergischen Naturforschenden Gesellschaft, 41(1):1–62.

Mertens, Robert Friedrich Wilhelm, and Heinz Wermuth. 1960. Die Amphibien und Reptilien Europas. Verlag Waldemar Kramer, Frankfurt am Main, xi + 264 pp.

Meszoely, Charles Aladar Maria. 1970. North American fossil anguid lizards. Bulletin of the Museum of Comparative Zoology, Harvard University, 139(2):87–150, 17 figs., 2 pls.

Mezhzherin, S. V., and Mikhail Leonidovich Golubev. 1989. [The genetic divergence of *Phrynocephalus* Kaup (Reptilia, Agamidae) of the USSR fauna]. Reports of Ukraine SSR Academy of Sciences. Series B Geology, Chemistry and Biological Sciences, 12:72–74 (In Russian).

Minton, Sherman Anthony, Jr. 1966. A contribution to the herpetology of West Pakistan. Bulletin of the American Museum of Natural History, 134(2):29–184, 28 pls.

Minton, Sherman Anthony, Jr, Steven Clement Anderson, and Jeromie A. Anderson. 1970. Remarks on some geckos from Southwest Asia, with descriptions of three forms and a key to the genus *Tropiocolotes*. Proceedings of the California Academy of Sciences, ser. 4, 37(9):333–362.

Moody, Scott Michael. 1980. Phylogenetic and Historical Biogeographical Relationships of the Genera in the Family Agamidae (Reptilia: Lacertilia). Doctoral dissertation, University of Michigan, Ann Arbor, 373 pp.

Moody, Scott Michael. 1987. A preliminary cladistic study of the lizard genus *Uromastyx* (Agamidae, *sensu lato*), with a checklist and diagnostic key to the species; Proceedings of the 4th Ordinary General Meeting of the Societas Europaea Herpetologica, Nijmegen, pp. 285–288.

Moravec, Jiří. 1994. A new lizard from Iran, *Eremias* (*Eremias*) *lalezharica* sp. n. (Reptilia: Lacertilia: Lacertidae). Bonner Zoologische Beiträge, 45(1):61–66, 2 figs.

Moravec, Jiří, and Michal Černy. 1994. Second finding of the Iranian gecko *Tropiocolotes latifi*. Casopis Narodniho Muzea, Rada Prirodovedna, 163(1–4):88.

Morich, L. D. 1922. Spisok presmikaiushchikhsya, sobrannikh v 1921 g. v Zakakaspiiskoi oblasti, Turkmenistane i severnoi Persii [A list of reptiles collected in 1921 in the Transcaspian region and North Persia]. Trudy Stavropolskogo selskogo khoziaistva Instituta, 1(11):35–47 (In Russian).

Morich, L. D. 1929. Presmykaiushchiesya Turmenistan i sopredelnoi Persii [Reptiles of Turkmenistan and contiguous Persia]. Turkmenovedenie, 1929(4):30–35 (In Russian).

Moritch (see Morich).

Moritz (see Morich).

Mosauer, Walter. 1934. The reptiles and amphibians of Tunisia. Publications of the University of California at Los Angeles in Biological Sciences, 1(3):49–64, 1 text-fig., 1 map.

Müller, Lorenz. 1939. Über die von den Herren Dr. v. Jordans und Dr. Wolf im Jahre 1938 in Bulgarien gesammelten Amphibien und Reptilien. Mitteilungen Königlichen naturwissenschaften Institut Sofia, 13:1–17.

Mulder, John. 1998. Die Zagros-Eidechse *Timon princeps* (Blanford, 1874) im natürlichen Lebensraum. Die Eidechse (Beiträge zur Kenntnis der Lacertiden), 8(3):75–79, figs. 1–3.

Murray, James A. 1884a. The Vertebrate Zoology of Sind: A Systematic Account, with Descriptions of all the Known Species of Mammals, Birds, and Reptiles Inhabiting the Province, etc. Education Society's Press, Byculla; Richardson & Co., 13 Pall Mall, Bombay and London, xvi + 424 pp.

Murray, James A. 1884b. Additions to the present knowledge of the vertebrate zoology of Persia. Annals and Magazine of Natural History, ser. 4, 14:101–106.

Murray, James A. 1884c. Additions to the reptilian fauna of Sind. Annals and Magazine of Natural History, ser. 4, 14:106–111.

Murray, James A. 1892. The Zoology of Baloochistan and Southern Afghanistan. Education Society, Bombay, 83 pp.

Nader, Iyad A., and Suad Z. Jawdat. 1976. Taxonomic study of the geckos of Iraq (Reptilia: Gekkonidae). Bulletin of the Biological Research Center, University of Baghdad, (5):1–41.

Nagy, Kenneth A. 1982. Field studies of water relations, chapter 11, pp. 483–501. *In* Carl Gans and F. Harvey Pough (eds.). Biology of the Reptilia, Physiology C. Physiological Ecology. Biology of the Reptilia, vol. 12. Academic Press, New York.

Neronov, V. M., and T. Y. Semenova. 1980. [Material on the biogeography of Iran]. Itogi Nauki i Tekhniki, ser. Seriaiia Biogeografieiia, 2:85–127 (In Russian).

Nesterov, P. V. 1916. [Trois formes nouvelles d'amphibiens (Urodela) du Kurdistan]. Annuaire du Musée Zoologique de l'Académie Impériale des Sciences de St.-Pétersbourg, 21(1916):1–30, figs.1–6, pls.1–3 (In Russian).

Nikolsky, Aleksandr Mikailovich. 1886. Contribution to the knowledge of the fauna of vertebrate animals of northeastern Persia and Transcaucasia. Travaux St. Pétersbourg Société Naturalistes, 17:403–406 (In Russian).

Nikolsky, Aleksandr Mikailovich. 1896. Diagnosis reptilium et amphibiorum in Persia orientali a N. Zarudny collectorum. Annuaire du Musée Zoologique de l'Académie Impériale des Sciences de St.-Pétersbourg, 4(1):369–372.

Nikolsky, Aleksandr Mikailovich. 1897. Les reptiles, amphibiens, et poissons recueillis par Mr. N. Zarudny dans la Perse orientale. Annuaire du Musée Zoologique de l'Académie Impériale des Sciences de St.-Pétersbourg, 2(3):306–348, pl. (In Russian).

Nikolsky, Aleksandr Mikailovich. 1899. Deux nouvelles espèces de *Teratoscincus* de la Perse orientale. Annuaire du Musée Zoologique de l'Académie Impériale des Sciences de St.-Pétersbourg, 4(2):145–147 (In Russian).

Nikolsky, Aleksandr Mikailovich. "1899" (1900). Reptiles, amphibiens et poissons, recueillis pendant le voyage de Mr. N. A. Zarudny en 1898 dans la Perse. Annuaire du Musée Zoologique de l'Académie Impériale des Sciences de St.-Pétersbourg, 4:375–417 (In Russian).

Nikolsky, Aleksandr Mikailovich. 1903. [On three new species of reptiles collected by Mr. N. Zarudny in Eastern Persia in 1901]. Annuaire du Musée Zoologique de l'Académie Impériale des Sciences de St.-Pétersbourg, 8:95–98 (In Russian).

Nikolsky, Aleksandr Mikailovich. 1907a. Presmykaiushchiiasia i zemnovoddnyia, sobrannyia N. A. Zarudnym v Persii v 1903–1904 gg. [Reptiles et amphibiens recueillis par M. N. A. Zarudny en Perse en 1903–1904]. Annuaire du Musée Zoologique de l'Académie Impériale des Sciences de St.-Pétersbourg, 10 [1905](3–4):260–301, pl. 1, 10 figs. (In Russian; cover bears 1906 as year of issue).

Nikolsky, Aleksandr Mikailovich. 1907b. *Alsophylax laevis* sp. nov. (Geckonidarum). Annuaire du Musée Zoologique de l'Académie Impériale des Sciences de St.-Pétersbourg, 10(3–4):333–335.

Nikolsky, Aleksandr Mikailovich. 1915. Faune de la Russie. Reptiles, vol. 1. Chelonia et Sauria. Petrograd, 532 pp. (In Russian). (Translated into English as: Fauna of Russia and Adjacent Countries, 1963, by L. and E. Kochva, Israel Program for Scientific Translation, for the National Science Foundation and Smithsonian Institution; 352 pp., 9 pls. [original pagination indicated in margins]).

Nikolsky, Aleksandr Mikailovich. 1916. Faune de la Russie. Reptiles, vol. 2. Ophidia. Petrograd, 349 pp. (In Russian). (Translated into English as: Fauna of Russia and Adjacent Countries, 1963, by A. Mercado, Israel Program for Scientific Translation, for the National Science Foundation and Smithsonian Institution; 247 pp., 8 pls. [original pagination indicated in margins]).

Nikolsky, Aleksandr Mikailovich. 1918. Faune de la Russie. Amphibiens. Petrograd, 309 pp. (In Russian. (Translated into English as: Fauna of Russia and Adjacent Countries, 1962, by F. Por, Israel Program for Scientific Translation, for the National Science Foundation and Smithsonian Institution; 225 pp., 4 pls. [original pagination indicated in margins]).

Nilson, Göran, and Claes Andrén. 1978. A new species of *Ophiomorus* (Sauria: Scincidae) from Kavir Desert, Iran. Copeia, 1978(4):559–564.

Nilson, Göran, and Claes Andrén. 1981. Die Herpetofauna des Kavir-Schutzgebietes, Kavir-Wüste, Iran. Salamandra, 17(3–4):130–146.

Nilson, Göran, and Claes Andrén. 1984a. A taxonomic account of the Iranian ratsnakes of the *Elaphe longissima* species-group. Amphibia-Reptilia, 5(2):157–171.

Nilson, Göran, and Claes Andrén. 1984b. Systematics of the *Vipera xanthina* complex (Reptilia: Viperidae). An overlooked viper within the *xanthina* species-group in Iran. Bonner Zoologische Beiträge, 35(1–3):175–184.

Nilson, Göran, and Claes Andrén. 1985a. Systematics of the *Vipera xanthina* complex (Reptilia: Viperidae). I. A new Iranian viper in the *raddei* species-group. Amphibia-Reptilia, 6(2):207–214.

Nilson, Göran, and Claes Andrén. 1985b. Systematics of the *Vipera xanthina* complex (Reptilia: Viperidae) III. Taxonomic status of the Bulgar Dagh viper in south Turkey. Journal of Herpetology, 19(2):276–283.

Nilson, Göran, and Claes Andrén. 1986a. A review of the *Vipera xanthina* complex (Reptilia: Viperidae), pp. 223–225. *In* Zbyněk Roček (ed.). Studies in Herpetology. Proceedings of the European Herpetological Meeting (3rd Ordinary General Meeting of the Societas Europaea Herpetologica), Prague 1985. Charles University for the Societas Europaea Herpetologica, Prague.

Nilson, Göran, and Claes Andrén. 1986b. The mountain vipers of the Middle East: The *Vipera xanthina* complex (Reptilia, Viperidae). Bonner Zoologische Monographien: Zoologisches Forschungsinstitut und Museum Alexander Koenig, Bonn, 20:1–90.

Nilson, Göran, and Claes Andrén. 1992. The species concept in the *Vipera xanthina* complex: reflecting evolutionary history or hiding biological diversity? Amphibia-Reptilia. 13(4):421–424.

Nilson, Göran, and Per Sundberg. 1981. The taxonomic status of the *Vipera xanthina* complex. Journal of Herpetology, 15(3):379–381.

Obst, Fritz Jürgen. 1978. Zur geographischen Variabilität des Scheltopusik, *Ophisaurus apodus* (Pallas) (Reptilia, Squamata, Anguidae). Zoologische Abhandlungen Staatliches Museum für Tierkunde in Dresden, 35(8):129–140.

Obst, Fritz Jürgen. 1981. *Ophisaurus apodus* (Pallas 1775) — Scheltopusik, Panzerschleiche, pp. 259–274. *In* Wolfgang Böhme (ed.). Handbuch der Reptilien und Amphibien Europas, vol. 1. Echsen (Sauria). Akademische Verlagsgesellschaft, Wiesbaden.

Obst, Fritz Jürgen. 1984. A record of the gecko *Hemidactylus turcicus* (Linnaeus, 1758) in Turkmenia. Proceedings of the Zoological Institute, Leningrad,124:142–143.

Oken, Ludwig Lorenz. 1817. Cuviers und Okens Zoologien neben einander gestellt. Isis von Oken, Jena, 1(8):cols.1145–page 1185. (See Zhao and Adler, 1993:389 for comments on pagination.)

Olivier, Guilaume-Antoine. 1804. Voyage dans l'Empire Othoman, l'Égypte et la Perse, vol. 4. H. Agasse, Paris. (Not seen, pagination unknown.)

Olivier, Guilaume-Antoine. 1807. Voyage dans l'Empire Othoman, l'Égypte et la Perse, vol. 10. H. Agasse, Paris. (Not seen, pagination unknown.)

Orlov, Nikolai Lutseanovich. 1981. About the eastern range limit of the gecko *Gymnodactylus longipes* Nikolsky, 1896. Trudy, Zoological Institute, Akademii Nauk USSR, Leningrad, 101:89–91, pl. 13 (In Russian with English summary.)

Orlova, Valentina Fyodorovna. 1978. Geograficheskoie rasprostanenie ivnutrivdovaya izmenchivost lugovoi yascherici na Kavkaze, pp. 204–215. *In* A. M. Sudilovskaya and V. E. Flint (eds.). Ptici i Presmikayushchiesya (Issliedovaniya po Faune Sovietkogo Soyuza). Moscow (In Russian).

Orlova, Valentina Fyodorovna. 1981a. *Agama caucasia* (Eichwald, 1831) — Kaukasus-Agame, pp. 136–148. *In* Wolfgang Böhme (ed.). Handbuch der Reptilien und Amphibien Europas, vol. 1. Echsen (Sauria) I. Akademische Verlagsgesellschaft, Wiesbaden.

Orlova, Valentina Fyodorovna. 1981b. *Agama sanguinolenta* (Pallas 1814) — Steppenagame, pp. 149–160. *In* Wolfgang Böhme (ed.). Handbuch der Reptilien und Amphibien Europas, vol. 1. Echsen (Sauria) I. Akademische Verlagsgesellschaft, Wiesbaden.

Pallas, Peter Simon. 1771. Reise durch verschiedene Provinzen des Russischen Reichs, vol. 1. Kaiserlichen Akademie der Wissenschaften, St. Petersburg, (12) + 504 pp., 25 pls.

Pallas, Peter Simon. 1773. Reise durch verschiedene Provinzen des Russischen Reichs, vol. 2. Kaiserliche Akademie der Wissenschaften, St. Petersburg, (Book 1): (4), 3–368 + (6) pp., 11 pls.; (Book 2): 369–744 pp., 30 pls., 1 map.

Pallas, Peter Simon. 1775. *Lacerta apoda* descripta. Novi Commentarii Academiae Scientarum Imperialis Petropolitanae, Petropoli, 19:435–454, pls. 9–10.

Pallas, Peter Simon. 1776. Reise durch verschiedene Provinzen des Russischen Reichs, vol. 3. Kaiserlichen Akademie der Wissenschaften, St. Petersburg (see pp. 538, 702).

Pallas, Peter Simon. 1778. Reise durch verschiedene Provinzen des Russischen Reichs in einem ausführlichen Auszuge. vol. 3. J. G. Fleischer, Frankfurt and Leipzig, (10), 488, 80, (24) pp., 51 pls.

Pallas, Peter Simon. 1814. Animalia Monocardia seu Animalia Frigidi Sanguinis Imperii Rosso-Asiatici, vol. 3. 428 pp. (published as a section of Pallas, 1811–1831, *q.v.*).

Pallas, Peter Simon. "1811–1831". Zoographia Rosso-Asiatica, sistens Omnium in Extenso Imperio Rossico, et Adjacentibus Maribus Observatorum Recensionem, Domicilia, Mores et Descriptiones, Anatomen atque Icones Plurimorum. Volumen Tertium [= Vol. 3]. Caesarea Academie Scientiarum, Petropoli [= St. Petersburg], vii + 428 + cxxv pp. (see Zhao and Adler, 1993:392 for comments on date of publication).

Panov, E. N., L. Yu. Zykova, M. E. Gauzer, and V. I. Vasil'ev. 1987. [Zone of intergradation of different forms of *Stellio caucasius* complex from Southwest Turkmenia.] Zoologicheski'i Zhurnal, Akademiya Nauk SSSR, 66(3):402–411 (In Russian).

Parker, Hampton Wildman. 1930. Three new reptiles from southern Arabia. Annals and Magazine of Natural History, ser. 10, 6:594–598.

Perveen, Zahida, Fehmida Iffatr, and Walter Auffenberg. 1988. New reptile records for Pakistan. Herpetological Review, 19(3):61.

Peters, Günther. 1962. Studien zur Taxonomie, Verbreitung und Ökologie der Smaragdeidechsen. I. *Lacerta trilineata*, *viridis*, und *strigata* als selbstandige Arten. Mitteilungen aus dem Zoologischen Museum in Berlin, 38(1):127–52, figs. 1–4, pl. 1.

Peters, Günther. 1964a. Studien zur Taxonomie, Verbreitung und Ökologie der Smaragdeidechsen. III. Die orientalischen Populationen von *Lacerta trilineata*. Mitteilungen aus dem Zoologischen Museum in Berlin, 40(2):185–250, 27 figs.

Peters, Günther. 1964b. Sekundäre Geschlechtsmerkmale, Wachstum und Fortpflanzung bei einigen transkaukasischen *Eremias*-Formen. Senckenbergiana Biologie, 45(3–5):445–467, 15 figs.

Peters, Wilhelm Karl Hartwig. 1869. Mitteilungen über neue Saurier und Batrachier. Monatsberichte der Königlichen Preussischen Akademie der Wissenschaften zu Berlin, 1869:786–790.

Peters, Wilhelm Karl Hartwig. 1870. Über die afrikanischen Warneidechsen, Monitors, und ihre geographische Verbreitung. Monatsberichte der Königlichen Preussischen Akademie der Wissenschaften zu Berlin, 1870:106–110.

Peters, Wilhelm Karl Hartwig. 1880. Über die von Hrn. Gerhard Rohlfs und Dr. A. Stecker auf der Reise nach der Oase Kufra gesammelten Amphibien. Monatsberichte der Königlichen Preussischen Akademie der Wissenschaften zu Berlin, 1880:305–309, pl.

Petzold, Hans-Günter. 1971. Blindschleiche und Scheltopusik. Die Familie Anguidae. A. Ziemsen Verlag, Wittenberg, Lutherstadt, 102 pp.

Pianka, Eric Rodger. 1986. Ecology and Natural History of Desert Lizards. Princeton University Press, Princeton, New Jersey, xi + 208 pp.

Planhol, Xavier de. 1987. Azerbaijan i. Geography, pp. 205–215. *In* Ehsan Yarshater (ed.). Encyclopaedia Iranica, vol. 3, fasc. 2. Routledge and Kegan Paul, New York and London.

Planhol, Xavier de. 1989. Boundaries ii. With Russia, pp. 403–406. *In* Ehsan Yarshater (ed.). Encyclopaedia Iranica, vol. 4, fasc. 4. Routledge and Kegan Paul, New York and London.

Por, Francis Dov. 1978. Lessepsian Migration. Ecological Studies, vol. 23. Springer-Verlag, Berlin, London, New York. x + 228 pp.

Pough, F. Harvey, and Carl Gans. 1982. The vocabulary of reptilian thermoregulation, chapter 2, pp. 17–23. *In* Carl Gans and F. Harvey Pough (eds.). Biology of the Reptilia. Physiology C. Physiological Ecology. Biology of the Reptilia, vol. 12. Academic Press, New York.

Prashad, Baini. 1916. Some observations on a common house-lizard (*Hemidactylus flaviviridis*) of India. Journal of the Bombay Natural History Society, 24:834–838.

Presch, William. 1988. Cladistic relationships within the Scincomorpha, pp. 471–492. *In* Richard Dean Estes and Gregory Kent Pregill (eds.). Phylogenetic relationships of the lizard families. Stanford University Press, Stanford.

Pritchard, Peter Charles Howard. 1966. Notes on Persian turtles. British Journal of Herpetology, 3(11):271–275.

Pritchard, Peter Charles Howard. 1979. Encyclopedia of Turtles. T.F.H. Publ., Neptune, New Jersey, 895 pp.

Procter, Joan Beauchamp. 1921. Further lizards and snakes from Persia and Mesopotamia. Journal of the Bombay Natural History Society, 28(1):251–253.

Queiroz, Kevin de, and Jacques Armand Gauthier. 1994. Toward a phylogenetic system of biological nomenclature. Trends in Ecology and Evolution, 9(1):27–31.

Radde, Gustav Ferdinand Richard. 1886. Die Fauna und Flora des südwestlichen Caspi-Gebietes. Wissenschaftliches Beiträge zu den Reisen an der Perisisch-Russischen Grenze. F. A. Brockhaus, Leipzig, x + 425 pp.

Rage, Jean-Claude. 1988. Gondwana, Tethys, and terrestrial vertebrates during the Mesozoic and Cainozoic, pp. 255–273. *In* M. G. Audley-Charles and Anthony Hallam (eds.). Gondwana and Tethys. Geological Society Special Publication, no. 37. Oxford University Press, Oxford.

Rai', Mehdi M. 1965. Recherches sur les colubrides d'Iran. Thesis, Docteur des Sciences Naturelles, Faculté des Sciences de Montpellier, 85 pp.
Rai', Mehdi M. 1969. Sur deux serpents récoltes en Iran, *Hydrophis spiralis* Gray et *Hydrophis ornatus* Günther. Actes de la Société Linnéenne de Bordeaux, 106: 4 pp. (unpaginated in my copy).
Rai', Mehdi M. 1978. Une nouvelle récolte de *Telescopus rhinopoma* Blanford 1874 (Serpentes: Colubridae), une espèce trés rare. Canadian Journal of Zoology, 56(1):146–149.
Rastegar-Pouyani, Nasrullah. 1995. Lizard fauna of Kermanshahan Province (western Iran), p. 48. *In* Natalia Ananjeva (ed.). Abstracts of the Second Asian Herpetological Meeting, 6–10 September, 1995, Asgabat, Turkmenistan. Folium Publishing Co., Moscow.
Rastegar-Pouyani, Nasrullah. 1996. A new species of *Asaccus* (Sauria: Gekkonidae) from the Zagros Mountains, Kermanshahan Province, western Iran. Russian Journal of Herpetology, 3(1):11–17.
Rastegar-Pouyani, Nasrullah. 1997. Systematics and distribution of the *Trapelus agilis* complex. *In* Zbyněk Roček and Scott Hart (eds.). Herpetology '97. Abstracts of the Third World Congress of Herpetology, 2–10 August 1997, Prague, Czech Republic. Third World Congress of Herpetology, Prague (no pagination).
Rastegar-Pouyani, Nasrullah. 1998. A new species of *Acanthodactylus* (Sauria: Lacertidae) from Qasr-e-Shirin, Kermanshah Province, western Iran. Proceedings of the California Academy of Sciences, ser. 4, 50(9):257–265, 9 figs., 1 table.
Rastegar-Pouyani, Nasrullah. 1999 (In press). First record of the lacertid *Acanthodactylus boskianus* (Sauria: Lacertidae) for Iran. Asiatic Herpetological Research.
Rastegar-Pouyani, Nasrullah, and Göran Nilson. "1997" (1998). A new species of *Eremias* (Sauria: Lacertidae) from Fars Province, south-central Iran. Russian Journal of Herpetology, 4(2):94–101, figs. 1–8.
Rastegar-Pouyani, Nasrullah, and Göran Nilson. 1998. A new species of *Lacerta* (Sauria: Lacertidae) from the Zagros Mountains, Esfahan Province, west-central Iran. Proceedings of the California Academy of Sciences, ser. 4, 50(10):267–277, 7 figs.
Reed, Charles A. 1956. Temporary bipedal locomotion in the lizard *Agama caucasica* in Iraq. Herpetologica, 12(2):128.
Reed, Charles A., and Hymen Marx. 1959. A herpetological collection from northeastern Iraq. Transactions of the Kansas Academy of Sciences, 62(1):91–122, figs. 1–6.
Reuss, Adolph. 1833 (?1834). Zoologisches Miscellen. Reptilien. Saurier. Batrachier. Museum Senckenberg, Frankfurt am Main, 1:47, pl. 3, fig. 1.
Rieppel, Olivier. 1980. The Phylogeny of Anguinomorph Lizards. Birkhauser Verlag, Basel, 86 pp.
Robertson, Alastair. 1987. The transition from a passive margin to an Upper Cretaceous foreland basin related to ophiolite emplacement in the Oman Mountains. Geological Society of America Bulletin. 99(5):633–653, 21 figs.
Rögl, F., and F. F. Steininger. 1983. Vom Zerfall der Tethys zu Mediterran und Paratethys. Die Neogene Palaeogeographie und Palinspastik des cirkum-mediterranen Raumes. Annalen des Naturhistorischen Museums in Wien, 85(A):133–153.
Rösler, Herbert, and Nikolai Nikolaevich Szczerbak. 1993. Die Jugendentwicklung von *Eublepharis turcmenicus* Darevskij, 1978 im Terrarium. Salamandra, 28(3–4):275–278.
Rostombekov, V. 1928. Ophidia de l'expedition de l'Ourmie en 1916. Bulletin du Museum de Géorgie, 4:121–133.
Rostombekov, V. 1938. Chelonia, Sauria, and Amphibia of the Urmia Expedition in 1916. Bulletin du Museum de Géorgie, 9A:11–21.
Rüppell, Wilhelm Peter Eduard Simon. 1826–1828. Atlas zu Reise im nördlichen Afrika. I. Zoologie. Heinr. Ludw. Brönner, Frankfurt am Main. (Section on reptiles, pp. 1-24, by C. H. G. von Heyden [*q.v.*] published in 1827.)
Rüppell, Wilhelm Peter Eduard Simon. 1835. Neue Wirbeltiere zu der Fauna von Abyssinien gehörig. III. Amphibien. Siegmund Schmerber, Frankfurt am Main, 18 pp., pls. 1–6. (Published as a section of Rüppell, 1835–1840, *q.v.*)
Rüppell, Wilhelm Peter Eduard Simon. 1835–1840. Neue Wirbeltiere zu der Fauna von Abyssinia gehorig. Siegmund Schmerber, Frankfurt am Main, vi + 140 pp., 95 pls.

Rustamov, Anver Kejusevich, and Nikolai Nikolaevich Szczerbak. 1985. [The herpetogeographical separation of the region of Middle Asia]. The Problems of Herpetology. Sixth USSR Herpetological Conference. Science Press, Leningrad (abstract [In Russian]).

Rustamov, Anver Kejusevich, Chary Ataevich Atayev, O. S. Sopyev, and A. N. Makarov. 1985. On the ecology of *Eublapharis turkmenicus* Darevsky, 1978. Izvestiya Akademiya Nauk Turkmenistan SSR (Biol.), 1:3–7 (In Russian).

Rykena, Silke, and Wolfgang Bischoff. 1997. *Timon princeps kurdistanicus* (Suchov, 1936) – Kurdische Zagros-Eidechse. Die Eidechse, Beiträge zur Kenntnis der Lacertiden, 8(2):41–43.

Rykena, Silke, Hans Konrad Nettmann, and W. Bings. 1977. Zur Biologie der Zagros-Eidechse, *Lacerta princeps* Blanford 1874. I. Beobachtungen im Freiland und im Terrarium an *L. p. kurdistanica* Suchov, 1936 (Reptilia, Sauria, Lacertidae). Salamandra, 13(3–4):174–184.

Salvador, Alfredo. 1981. *Hemidactylus turcicus* (Linnaeus 1758) – Europaischer Halbfingergecko, pp. 84–107. *In* Wolfgang Böhme (ed.). Handbuch der Reptilien und Amphibien Europas, vol. 1. Echsen (Sauria) I. Akademische Verlagsgesellschaft, Wiesbaden.

Salvador, Alfredo. 1982. A revision of the lizards of the genus *Acanthodactylus* (Sauria: Lacertidae). Bonner Zoologische Monographien: Zoologisches Forschungsinstitut und Museum Alexander Koenig, Bonn, 16:1–167.

Sborshchikov, I. M., L. A. Savostin, and L. P. Zonenshan. 1981. Present plate tectonics between Turkey and Tibet. Tectonophysics, 79:45–73.

Scarlatto, O. A. (ed.). 1981. Zoological Institute of the Academy of Sciences of the USSR, 150 Years. Nauka, Leningrad, 242 pp. (In Russian).

Ščerbak, N. N. (see Szczerbak, N. N.).

Schlegel, Hermann. 1858. Handleiding tot de Beoefening der Dierkunde. Natuurkundige Leercursus ten Gebruike der Koninklijke Militaire Akademie, vol. 2. H.-G. Nys for Koninklijke Militaire Akademie, Breda, xx + 628 + (2) pp., pls. 1–27 (herpetology, pls. 1–4).

Schleich, Hans-Hermann. 1976. Über *Phrynocephalus helioscopus* aus Persien (Reptilia, Sauria, Agamidae). Salamandra, 12(4):189–193.

Schleich, Hans-Hermann. 1977. Distributional maps of reptiles of Iran. Herpetological Review, 8(4):126–129.

Schleich, Hans-Hermann. 1979a. Geographic distribution: *Agama agilis*. Herpetological Review, 10(2):59.

Schleich, Hans-Hermann. 1979b. Geographic distribution. Sauria. *Agama caucasica* (Caucasian agama). Herpetological Review, 10(2):60.

Schleich, Hans-Hermann. 1979c. Geographic distribution: *Typhlops vermicularis*. Herpetological Review, 10(2):61.

Schleich, Hans-Hermann. 1979d. Feldherpetologische beobachtungen in Persien nebst morphologischen Daten den Agamen *Agama agilis*, *A. caucasica* und *A. erythrogaster*. Salamandra, 15(4):237–253.

Schmidt, Karl Patterson. 1939. Reptiles and amphibians from southwestern Asia. Field Museum of Natural History, Zoological Series, 24(7):49–92.

Schmidt, Karl Patterson. 1941. Reptiles and amphibians from central Arabia. Field Museum of Natural History, Zoological Series, 24(16):161–165, text-figs. 17–18.

Schmidt, Karl Patterson. 1952. Diagnoses of new amphibians and reptiles from Iran. Natural History Miscellany, (93):1–2.

Schmidt, Karl Patterson. 1953a. A Check List of North American Amphibians and Reptiles, 6th ed. Univ. Chicago Press, Chicago, viii + 280 pp.

Schmidt, Karl Patterson. 1953b. Amphibians and reptiles of Yemen. Fieldiana: Zoology, 34(24):253–261, textfig. 43.

Schmidt, Karl Patterson. 1955. Amphibians and reptiles from Iran. Videnskabelige Meddelesen fra Dansk Naturhistorist Forening, 117(3):193–207.

Schmidtler, Josef Friedrich. 1986a. Orientalische Smaragdeidechsen: 1. Zur Systematik und Verbreitung von *Lacerta viridis* in der Türkei (Sauria: Lacertidae). Salamandra, 22(1):29–46.

Schmidtler, Josef Friedrich. 1986b. Orientalische Smaragdeidechsen: 2. Über Systematik und Synök-

ologie von *Lacerta trilineata*, *L. media* und *L. pamphylica* (Sauria: Lacertidae). Salamandra, 22(2–3):126–146.
Schmidtler, Josef Friedrich. 1986c. Orientalische Smaragdeidechsen: 3. Klinaparallele Pholidoservariaticn. Salamandra, 22(4):242–258.
Schmidtler, Josef Johann, and Josef Friedrich Schmidtler. 1969. Über *Bufo surdus*; mit einem Schlüssel und Anmerkungen zu der übrigen Kröten Irans und West-Pakistans. Salamandra, 5(3–4):113–123.
Schmidtler, Josef Johann, and Josef Friedrich Schmidtler. 1970. Ein Nachtgeist aus Luristan. Aquarien Magazin, 1970(6):239–241.
Schmidtler, Josef Johann, and Josef Friedrich Schmidtler. 1971. Eine Salamander-Novität aus Persien. *Batrachuperus persicus*. Aquarien Magazin, 1971(11):443–445.
Schmidtler, Josef Johann, and Josef Friedrich Schmidtler. 1972. Zwerggeckos aus dem Zagros-Gebirge (Iran). Salamandra, 8(2):59–66.
Schmidtler, Josef Johann, and Josef Friedrich Schmidtler. 1975. Untersuchungen an westpersischen Bergbachmolchen der Gattung *Neurergus* (Caudata, Salamandridae). Salamandra, 11(2):84–98.
Schmidtler, Josef Johann, and Josef Friedrich Schmidtler. 1978. Eine neue Zwergnatter aus der Türkei; mit einer Übersicht über die Gattung *Eirenis* (Colubridae, Reptilia). Annalen des Naturhistorischen Museums in Wien, 81:383–400.
Schofield, Richard N. 1989. Boundaries with Turkey, pp. 417–418. *In* Ehsan Yarshater (ed.). Encyclopaedia Iranica, vol. 4, fasc. 4. Routledge and Kegan Paul, New York and London.
Schultschik, Günter, and Sebastian Steinfartz. 1996a. Die Salamandriden des Iran sowie Anmerkungen zur Herpetofauna (Reisebericht Iran 1995). Elaphe, 4(2):76–77.
Schultschik, Günter, and Sebastian Steinfartz. 1996b. Ergebnisse einer herpetologischen Excursion in der Iran. Herpetozoa, 9:91–95.
Selcer, K. W., and R. A. Bloom. 1984. *Cyrtodactylus scaber* (Gekkonidae): a new gecko to the fauna of the United States. Southwestern Naturalist, 29(4):499–500.
Sengör, A. M. C. 1984. The Cimmeride Orogenic System and the Tectonics of Eurasia. Geological Society of America Special Paper, no. 195, 82 pp.
Shammakov, Sakhat Muradovich. 1981. Presmikayushchiesya Ravninnogo Turkmenistana [Reptiles of the Turkmenian Plains]. Ylim Publ., Ashkhabad, 312 pp. (In Russian).
Sharma, Ramesh Chandra. 1978. A new species of *Phrynocephalus* Kaup (Reptilia: Agamidae) from the Rajastan Desert, India with notes on its ecology. Bulletin of the Zoological Survey of India, 1(3):291–294.
Shaw, George. 1802. General Zoology, or Systematic Natural History, vol. 3. Amphibia. Thomas Davison, London, pt. 1, i-viii + 312, 86 pls.; pt. 2, 313–615 pp., 54 pls.
Shcherbak (see Szczerbak)
Sherborn, C. Davies. 1934. Dates of publication of catalogues of natural history (post 1850) issued by the British Museum. Annals and Magazine of Natural History, ser. 10, 13(74):308–312.
Shockley, Clarence H. 1949. Herpetological notes for Ras Jiunri, Baluchistan. Herpetologica, 5:121.
Smirina, E. M., and Aleksey Yuri Tsellarius. 1996. Aging, longevity, and growth of the desert monitor lizard (*Varanus griseus* Daud.). Russian Journal of Herpetology, 3(2):130–142.
Smith, Malcolm Arthur. 1926. A Monograph of the Sea Snakes (Hydrophiidae). British Museum (Natural History), London, xvii + 130 pp., 2 pls.
Smith, Malcolm Arthur. 1931. The Fauna of British India, Including Ceylon and Burma. Reptilia and Amphibia. The Fauna of British India, vol. 1, Loricata, Testudines. Taylor and Francis, London, xxviii + 185 pp., 2 pls., map.
Smith, Malcolm Arthur. 1933. Remarks on some Old World geckos. Records of the Indian Museum, 35:9–19.
Smith, Malcolm Arthur. 1935. The Fauna of British India, Including Ceylon and Burma. Reptilia and Amphibia. The Fauna of British India, vol. 2, Sauria. Taylor and Francis, London, xiii + 440 pp., 1 pl., 2 maps.
Smith, Malcolm Arthur. 1940. Contributions to the herpetology of Afghanistan. Annals and Magazine of Natural History, ser. 11, 5:382–384.

Smith, Malcolm Arthur. 1943. The Fauna of British India, Ceylon and Burma, Including the Whole of the Indo-Chinese Subregion. Reptilia and Amphibia. The Fauna of British India, vol. 3, Serpentes. Taylor and Francis, London, xii + 583 pp., folding map.

Smith, Malcolm Arthur. 1945. (See Anonymous. 1943 [p. 304]).

Sobolevskii, N. I. 1929. The herpetofauna of the Talysh and of the Lenkoran lowland. Mémoires de Société des Amis des Sciences Naturelles d'Anthropologie et d'Ethnographie, Section Zoologie, 5:1–141 (In Russian with English summary).

Steindachner, Franz. 1867a. Amphibien, pp. 1–70. *In* Franz Steindachner (ed.). Reise der Österreichischen Fregatte Novara um die Erde in den Jahren 1857, 1858, 1859 unter den Befehlen des Commodore B. von Wüllerstorf-Urbair. Zoology, vol. 1, part 4. Kaiserlich-Königlichen Hof-Staatsdruckerei, Vienna, 5 pls. with explanations; (reprint edition, same publisher, 1869).

Steindachner, Franz, 1867b, Reptilien, pp. 1–98. *In* Franz Steindachner (ed.). Reise der Österreichischen Fregatte Novara um die Erde in dem Jahren 1857, 1858, 1859 unter den Befehlen des Commodore B. von Wüllerstorf-Urbair. Zoology, vol. 1, part 3. Kaiserlich-Königlichen Hof-Staatsdruckerei, Vienna, 3 pls. with explanations; (reprint edition, same publisher, 1869).

Steindachner, Franz. 1867c. Über mehrere neue Reptilien aus Chile, Brasilien und Persien. Anzeiger Akademie der Wissenschaften, Wien, 4:40-41.

Steindachner, Franz. 1867d. Herpetologische Notizen. Sitzungsberichte der Österreichische Akademie der Wissenschaften, Wien, Mathematisch-naturwissenschaftliche Klasse, 55(1):265–74, pls.1–4.

Steindachner, Franz. 1870. Herpetologische Notizen II. 1. Reptilien gesammelt während einer Reise in Senegambien (October bis December 1868). Sitzungsberichte der Österreichische Akademie der Wissenschaften, Wien, Mathematisch-naturwissenschaftliche Klasse, 62(1):326–335, fig. 1.

Steiner, Hans M. 1973. Beiträge zur Kenntnis von Verbreitung, Ökologie und Bionomie von *Batrachuperus persicus* (Caudata, Hynobiidae). Salamandra, 9(1):1–6.

Štěpánek, Otakar. 1937a. *Gymnodactylus kotschyi* Steindachner und sein Rassenkreis. Archiv für Naturgeschichte, ser. n.f., 6(2):258–280.

Štěpánek, Otakar. 1937b. *Anguis fragilis pelopenessiacus*, n. ssp. Zoologische Anzeiger, 118:107–110, 1 fig.

Steward, J. W. 1969. The Tailed Amphibians of Europe. David & Charles, Newton Abbot, Devon, UK, 180 pp., 42 pls., 16 maps.

Steward, J. W. 1971. The Snakes of Europe. Newton Abbott, Devon, UK, 238 pp.

Stimson, Andrew Francis. 1969. Liste der rezenten Amphibien und Reptilien. Boidae (Boinae + Bolyeriinae + Loxoceminae + Pythoninae). Das Tierreich, 89:xi + 1–49.

Stoliczka, Ferdinand. 1872a. Notes on the reptilian and amphibian fauna of Kachh. Proceedings of the Asiatic Society of Bengal, 1872:71–85.

Stoliczka, Ferdinand. 1872b. Notes on reptiles, collected by Surgeon F. Day in Sind. Proceedings of the Asiatic Society of Bengal, 1872:85–92.

Strauch, Alexander Aleksandrovich. 1863. Characteristik zweier neuen Eidechsen aus Persien. Bulletin de l'Académie Impériale des Sciences de St. Pétersbourg, 6:477–480.

Strauch, Alexander Aleksandrovich. 1867. Bemerkungen über die Eidechsengattung *Scapteira* Fitz. Mélange Biologie Académie de St. Pétersbourg, 6:403–426.

Strauch, Alexander Aleksandrovich. 1868a. Bemerkungen uber die Eidechsengattung *Scapteira* Fitz. Bulletin de l'Académie Impériale des Sciences de St. Pétersbourg, 12:313–328.

Strauch, Alexander Aleksandrovich. 1868b. Über die Arten Eidechsengattung *Ablepharus* Fitz. Bulletin de l'Académie Impériale des Sciences de St. Pétersbourg, 12:359–372.

Strauch, Alexander Aleksandrovich. 1868c. Ueber die Arten der Eidechsen-Gattung *Ablepharus* Fitz. Mélange Biologie Académie de St. Pétersbourg, 6:553–570.

Strauch, Alexander Aleksandrovich. 1869. Synopsis der Viperiden, nebst Bemerkungen über die geographische Verbreitung dieser Giftschlangen-Familie. Mémoires de l'Académie Impériale des Sciences de St. Pétersbourg, ser. 7, 14(6):1–143.

Strauch, Alexander Aleksandrovich. 1870. Revision der Salamandriden-Gattungen nebst Beschreibung einiger neuer oder weniger bekannter Arten dieser Familie. Mémoires de l'Académie Impériale des Sciences de St. Pétersbourg, ser. 7, 16:1–110.

Strauch, Alexander Aleksandrovich. 1873. Die Schlangen des Russischen Reichs, in systematischer und zoogeographischer Beziehung geschildert. Mémoires de l'Académie Impériale des Sciences de St. Pétersbourg, ser. 7, 20(4):1–287.

Strauch, Alexander Aleksandrovich. 1876. Part III. Reptilia and Amphibia, pp. ii, iv + 55 pp., pls. I-VIII. *In* N. M. Przheval'ski'i. Mongolia and the Tangut Lands, a Three-Year Journey in Eastern High Asia, vol. 2. Russian Geographical Society, St. Petersburg (In Russian). (See Zhao and Adler, 1993:405 for additional comments.)

Strauch, Alexander Aleksandrovich. 1887. Bemerkungen über die Geckoniden-Sammlung in zoologischen Museum der kaiserlichen Akademie der Wissenschaften zu St. Petersburg. Mémoires de l'Académie Impériale des Sciences de St. Pétersbourg, ser. 7, 35(2):72, 1 pl.

Strauch, Alexander Aleksandrovich. 1889. Das Zoologische Museum der Kaiserlichen Akademie der Wissenschaften zu St. Petersburg in seinem funfzigjahrigen Bestehen. St. Petersburg, 372 pp. (In Russian).

Strauch, Alexander Aleksandrovich. 1890. Bemerkungen über die Schildkrötensammlung im zoologischen Museum der kaiserlichen Akademie der Wissenschaften zu St. Petersburg. Mémoires de l'Académie Impériale des Sciences de St. Pétersbourg, ser. 7, 38(2):1–127.

Street, William S., and Janice K. Street, with Richard Sawyer. 1986. Iranian Adventure, the First Street Expedition. Field Museum of Natural History, Chicago, 305 pp.

Stugren, Bogdan. 1984. *Lacerta praticola* Eversmann 1834 — Wieseneidechsen, pp. 318–331. *In* Wolfgang Böhme (ed.). Handbuch der Reptilien und Amphibien Europas, vol. 2/I. Echsen II. AULA-Verlag, Wiesbaden.

Stull, Olive Griffith. 1935. A check list of the family Boidae. Proceedings of the Boston Society of Natural History, 40(8):387–408.

Suchow, G. F. 1936. Eine neue Unterart der Eidechse aus dem persischen Kurdistan. Travaux de l'Institut Zoologique de l'Académie des Sciences URSS, 3:303–308, 2 figs.

Szczerbak, Nikolai Nikolaevich. 1972. [New subspecies of *Eremias strauchi* — *Eremias strauchi kopetdaghica* ssp. nova (Sauria, Reptilia) from Turkmenia]. Vestnik Zoologii, 6(2):83–86 (In Russian).

Szczerbak, Nikolai Nikolaevich. 1974. Yaschurki Palearcktiki [The Palearctic Desert Lizards]. Akadeimya Nauk Ukrainskoi SSR Institut Zoologii. Naukova Dumka, Kiev, 296 pp., 92 photos (In Russian).

Szczerbak, Nikolai Nikolaevich. 1978. *Gymnodactylus turcmenicus* sp. n. (Reptilia, Sauria) — Novi vid gekkona iz yushnoi Turkmenii [A new species of gecko from southern Turkmenia.]. Vestnik Zoologii, 1978(3):39–44, figs.1–2 (In Russian, English summary).

Szczerbak, Nikolai Nikolaevich. 1981a. Grundzüge der herpetogeographischen Gliederung des Paläarktischen Gebietes, p. 42. *In* Olivér György Dely (ed.). Auszuge der Vortage der ersten Herpetologischen Konferenz Sozialischer Lander, Budapest.

Szczerbak, Nikolai Nikolaevich. 1981b. Osnovi gerpetogeographicheskogo rajnonipovaniya territorii SSSR [The fundamentals of herpetogeographic division of the territory of the USSR], pp. 157–158. Voprosi Gerpetologii: Avtoref Doklady v Vsesoyuzn. Gerpetol. Konf. Nauka Press, Leningrad (In Russian).

Szczerbak, Nikolai Nikolaevich. 1981c. *Cyrtodactylus russowii* (Str., 1887) — Transkaspischer Bogenfingergecko, pp. 78–83. In Wolfgang Böhme (ed.). Handbuch der Reptilien und Amphibien Europas, vol. 1. Echsen (Sauria). Akademische Verlagsgesellschaft, Wiesbaden.

Szczerbak, Nikolai Nikolaevich. 1986a. [The Nature of Central Kopetdag]. Ilym, Ashkhabad, 190 pp. (In Russian).

Szczerbak, Nikolai Nikolaevich. 1986b. Review of the Gekkonidae of the fauna of the USSR and neighboring countries, pp. 705–709. *In* Zbyněk Roček (ed.). Studies in Herpetology, Proceedings of the European Herpetological Meeting (3rd Ordinary General Meeting of the Societas Europaea Herpetologica), Prague 1985. Charles University, for the Societas Europaea Herpetologica, Prague.

Szczerbak, Nikolai Nikolaevich. 1988. On the nomenclature of Palearctic *Tenuidactylus* species (Gekkonidae, Reptilia). Vestnik Zoologii, 1988(4):84–85 (In Russian).

Szczerbak, Nikolai Nikolaevich. 1989. Further study of systematics of Palearctic rock geckoes; First World Congress of Herpetology, 11–19 September 1989, University of Kent at Canterbury, United Kingdom. Abstracts (no pagination).

Szczerbak, Nikolai Nikolaevich. 1990. Sistematika i geograficheskaya izmenchivost shchitkogo stsinka — *Eumeces taeniolatus* (Sauria, Scincidae). [Systmatics and geographic variability of *Eumeces taeniolatus* (Sauria, Scincidae)]. Vestnik Zoologii, 1990(3):33–40 (In Russian, English abstract).

Szczerbak, Nikolai Nikolaevich. 1994. Zoogeographic analysis of the reptiles of Turkmenistan, chapter 19, pp. 307–328. *In* Victor Fet and Khabibulla Atamuradov (eds.). Biogeography and Ecology of Turkmenistan. Monograhiae Biologicae, vol. 72. Kluwer Academic Publishers, Dordrecht, Boston, and London.

Szczerbak, Nikolai Nikolaevich, and S. B. Akhmedov. 1990. Sistematika i geograficheskaya izmenchivost dlinnonogogo stsinka (*Eumeces schneideri* Daud., Sauria, Scincidae, Reptilia). [Systematics and geographic variability of *Eumeces schneideri* Daud. (Sauria, Scincidae, Reptilia)]. Vestnik Zoologii, 1990(1):23–28.

Szczerbak, Nikolai Nikolaevich, and Mikhail Leonidovich Golubev. 1977a. Vzaimootnosheniya rodov *Gymnodactylus* i *Alsophylax* i ikh vnutrirodovaya struktura [Relationships between the genera *Gymnodactylus* and *Alsophylax* and their intrageneric structure], pp. 237–238. Voprosi Gerpetologii: Avtoref. Doklady IV Vsesoyuzn. Gerpetol. Konf. Nauka Press, Leningrad (Abstract. In Russian).

Szczerbak, Nikolai Nikolaevich, and Mikhail Leonidovich Golubev. 1977b. [Systematics of the Palearctic geckos (genera *Gymnodactylus*, *Bunopus*, *Alsophylax*)]. Gerpetologicheskii Sbornik. Trudy Zoologicheskogo Instituta, Akademiya Nauk SSSR, 74:120–133 (In Russian with English summary).

Szczerbak, Nikolai Nikolaevich, and Mikhail Leonidovich Golubev. 1981. Novie nakhodki zemnovodnikh i presmikayushchikhsya v Srednei Asii i Kazakhstane [New findings of amphibians and reptiles in Middle Asia and Kazakhstan]. Vestnik Zoologii, 1981(1):70–72 (In Russian).

Szczerbak, Nikolai Nikolaevich, and Mikhail Leonidovich Golubev. 1984. [On generic assignment and generic structure of the Palearctic *Cyrtodactylus* lizard species (Reptilia, Gekkonidae, *Tenuidactylus* gen. n.)]. Vestnik Zoologii, 1984(2):50–56 (In Russian).

Szczerbak, Nikolai Nikolaevich, and Mikhail Leonidovich Golubev. 1986. [The Gecko Fauna of the USSR and Adjacent Regions]. Nauka Dymka, Kiev, 232 pp., 8 pls. (In Russian).

Szczerbak, Nikolai Nikolaevich, and Mikhail Leonidovich Golubev. 1996. The Gecko Fauna of the USSR and Adjacent Regions [English ed., translated from the Russian by Michael L. Golubev and Sasha A. Malinsky; Alan E. Leviton and George R. Zug, eds.]. Society for the Study of Amphibians and Reptiles, Ithaca, New York, 232 pp., 8 pls.

Szczerbak, Nikolai Nikolaevich, V. V. Zhukova, and Evgeny Maksimovich Pisanetz. 1981. Kariotipi gekkonov podroda *Cyrtodactylus* (*Gymnodactylus*, Gekkonidae, Sauria, Reptilia) fuani SSSR [Karyotypes of geckos of the USSR]. Doklady Akademii Nauk UkrSSR, ser. B, 1981(8):85–87 (In Russian).

Szczerbak, Nikolai Nikolaevich, Yu. D. Khomustenko, and Mikhail Leonidovich Golubev. 1986. [Amphibians and reptiles of the Kopetdag Reserve and adjacent territories], pp. 76–110. *In* Nikolai Nikolaevich Szczerbak (ed.). The Nature of Central Kopetdag. Ilym, Ashkabad (In Russian).

Szczerbak, Nikolai Nikolaevich, Anatoly A. Tokar, and I. V. Kyrilenko. 1997. Catalogue of the Zoological Museum, National Natural History Museum, Ukrainian Academy of Sciences: Geckonid Lizards (Reptilia: Sauria: Geckonidae [*sic*]). Zoological Museum, National Natural History Museum, Ukrainian Academy of Sciences, Kiev, 45 pp. (In Russian).

Taylor, Edward Harrison. "1935" (1936). A taxonomic study of the cosmopolitan scincoid genus *Eumeces*. University of Kansas Science Bulletin, (23):1–643, pls.1–43.

Tchernov, Eitan. 1988. The biogeographical history of the southern Levant, pp.159–250. *In* Yoram Yom-Tov and Eitan Tchernov (eds.). The Zoogeography of Israel. The Distribution and Abundance at a Zoogeographical Crossroads. Monographiae Biologicae, vol. 62. Dr. W. Junk Publishers, Dordrecht, Boston, and Lancaster.

Terentjev, Paul Victorovich, and Sergius Aleksandrovich Chernov. 1949. Opredeltel Presmykaiushchikhsa I Zemnovodnykh [Guide to Reptiles and Amphibians]. [3rd ed.]. Soviet Science Press, Moscow and Leningrad, 340 pp. (In Russian).

Terentjev, Paul Victorovich, and Sergius Aleksandrovich Chernov. 1965. Key to Amphibians and Reptiles.(English translation of 1949 edition, Predelitel presmykaiushchikhsia i zemnovodnykh, translated by L. Kochva, Israel Program for Scientific Translations). Smithsonian Institution, Washington, D.C., 315 pp.

Thomas, H. 1985. The Early and Middle Miocene land connection of the Afro-Arabian plateau and Asia: a major event for hominoid dispersal?, pp. 42–50. *In* Eric Delson (ed.). Ancestors: the Hard Evidence. Allan R. Liss, New York, xii + 366.

Thorn, Robert. "1968" (1969). Les Salamandres d'Europe, d'Asie et d'Afrique du Nord. Description et Moeurs de Toute les Espèces et Sous-Espèces d'Urodeles de la Région Paléartique d'après l'état de 1967. Lechevalier, Paris, iv + 376 pp., 16 pls., 11 maps.

Tiedemann, Franz, and Michel Häupl. 1980. Typenkatalog der Herpetologischen Sammlung. Teil II: Reptilia. Katalog der wissenschaftlichen Sammlung des Naturhistorischen Museums in Wien, 4(2 [Vertebrata]):1–79.

Tokar, Anatoly A. 1986. Studien zur taxonomischen Lage der zentralasiatischen Schlangen der Gattung *Eryx* (Daud., 1803), pp. 174–178, 242. *In* Emilia Ivanovna Vorobeva (ed.). Gerpetologicheskie Issledovaniya v Mongolskoi Narodnoi Respublike. [Herpetological studies in Mongolia.]. Academy of Science of the USSR, Moscow (In Russian).

Tokar, Anatoly A. 1989a. Study of systematics of genus *Eryx* (Boidae). First World Congress of Herpetology, 11–19 September 1989, University of Kent at Canterbury, United Kingdom [Abstracts], (no pagination).

Tokar, Anatoly A. 1989b. [Revision of the genus *Eryx* (Serpentes, Boidae), based on osteological data.]. Vestnik Zoologii, 1989(4):46–55 (In Russian).

Tokar, Anatoly A. 1989c. On the discovery of *Eryx elegans* (Gray) outside its normal area of distribution. Vestnik Zoologii, 1989(6):25 (In Russian).

Tokar, Anatoly A. 1991. A revision of the subspecies structure of javelin sand boa, *Eryx jaculus* (Linnaeus, 1758) (Reptilia, Boidae). Herpetological Researches, Leningrad, 1991(1):18–41 (In Russian; English abstract).

Tokar, Anatoly A., and Fritz Jürgen Obst. 1993. *Eryx jaculus* (Linnaeus, 1758) — Westliche Sandboa, pp. 35–53. *In* Wolfgang Böhme (ed.). Handbuch der Reptilien und Amphibien Europas, vol. 3/I. Schlangen (Serpentes) I. AULA-Verlag, Wiesbaden.

Tremper, Ronald L. 1997. Designer leopard geckos. Reptiles, 5(3):16–18.

Tsaruk, Oleg J. 1985. [Abstract] [On systematics of *Trapelus agilis* s. str. complex (Agamidae)]. Pages 225–226 *in* Proceedings of the VI All-Union Herpetological Conference (Tashkent, September 18–20, 1985). 'Nauka' Publishers, Leningrad. (In Russian).

Tsellarius, Aleksey Yuri. 1977. A contribution to the ecology of *Eremias grammica* (Lacertidae, Sauria) in East Karakumy. Zoologicheski'i Zhurnal, Akademiya Nauk SSSR, 56(2):224–231 (In Russian with English summary).

Tsellarius, Aleksey Yuri. 1994. Behaviour and mode of life of the monitor in sand desert. Priroda, Moscow, (5):26–35.

Tsellarius, Aleksey Yuri, and Vladimir A. Cherlin. 1981. [On the ecology of the mottled lizards (Lacertidae, Sauria) in Badkhyz.]. Izvestiya Akademii Nauk Turkmenistan SSR (Biol.), 1981(6):46–52 (In Russian).

Tsellarius, Aleksey Yuri, and Vladimir A. Cherlin. 1991. Individual identification and new method of marking of *Varanus griseus* (Reptilia, Varanidae) in field conditions. Herpetological Researches, Leningrad, 1991(1):104–118.

Tsellarius, Aleksey Yuri, and Vladimir A. Cherlin. 1994. Duration of egg incubation in *Varanus griseus* and emergence of hatchlings to the surface in the sand deserts of Middle Asia. Selevinia, 2(4):43–46 (In Russian).

Tsellarius, Aleksey Yuri, Vladimir A. Cherlin, and Yuri H. Menshikov. 1991. Predvaritelinoe soobshchenie o Rabotakh po izucheniyu biologii *Varanus griseus* (Reptilia, Varanidae) v srednei Asii. [Preliminary report on the study of biology of *Varanus griseus* (Reptilia, Varanidae) in Middle Asia]. Herpetological Researches, Leningrad, 1991(1):61–103 (In Russian; English abstract).

Tsellarius, Aleksey Yuri, and Yuri G. Men'shikov. 1994. Indirect communications and its role in the formation of social structure in *Varanus griseus* (Sauria). Russian Journal of Herpetology, 1(2):121–132.

Tsellarius, Aleksey Yuri, and Yuri G. Men'shikov. 1995. Construction of nest burrows and protection of clutch by females of *Varanus griseus*. Zoologicheski'i Zhurnal, Akademiya Nauk SSSR, 74(1):119–129.

Tsellarius, Aleksey Yuri, Yuri G. Men'shikov, and Elena Yu. Tsellarius. 1995. Spacing pattern and reproduction in *Varanus griseus* of western Kyzylkum. Russian Journal of Herpetology, 2(2):153–165.

Tsellarius, Aleksey Yuri, and Elena Yu. Tsellarius. 1996. Courtship and mating in *Varanus griseus* of western Kyzylkum. Russian Journal of Herpetology, 3(2):122–129.

Tsellarius, Aleksey Yuri, and Elena Yu. Tsellarius. 1997. Behavior of *Varanus griseus* during encounters with conspecifics. Asiatic Herpetological Research, 7:108–130.

Tsellarius, Elena Yu. and Aleksey Yuri Tsellarius. 1997. Thermal conditions of activity of *Varanus griseus*. Zoologicheski'i Zhurnal, Akademiya Nauk SSSR, 76(2):206–211 (In Russian, English summary).

Tuck, Robert G., Jr. 1971a. Amphibians and reptiles from Iran in the United States National Museum Collection. Bulletin of the Maryland Herpetological Society, 7(3):48–86.

Tuck, Robert G., Jr. 1971b. Rediscovery and redescription of the Khuzistan dwarf gecko, *Microgecko helenae* Nikolosky (Sauria: Gekkonidae). Proceedings of the Biological Society of Washington, 83(42):477–482.

Tuck, Robert G., Jr. 1973. Additional notes on Iranian reptiles in the United States National Collection. Bulletin of the Maryland Herpetological Society, 9(1):13–14.

Tuck, Robert G., Jr. 1974a. Some amphibians and reptiles from Iran. Bulletin of the Maryland Herpetological Society, 10:59–65.

Tuck, Robert G., Jr. 1974b. Geographic distribution: *Bufo olivaceus*. Herpetological Review, 5(4):107.

Tuck, Robert G., Jr. 1974c. Geographic distribution: *Rana macrocnemis pseudodalmatina*. Herpetological Review, 5(4):107.

Tuck, Robert G., Jr. 1974d. Geographic distribution: *Cyrtodactylus scaber*. Herpetological Review, 5(4):107.

Tuck, Robert G., Jr. 1974e. Geographic distribution: *Coluber rhodorhachis*. Herpetological Review, 5(4):108.

Tuck, Robert G., Jr. 1974f. Geographic distribution: *Echis carinatus*. Herpetological Review, 5(4):108.

Tuck, Robert G., Jr. 1974g. Geographic distribution: *Lytorhynchus ridgewayi*. Herpetological Review, 5(4):108.

Tuck, Robert G., Jr. 1974h. Marha-ye sammi va khatarnak-e Iran va kelid-e shenasa'i anha. [Poisonous and dangerous snakes of Iran and a key to their identification]. Shekar va Tabi'at [Hunting and Nature], 1353(August-September):44–52 (In Farsi [Persian]).

Tuck, Robert G., Jr. 1975. Geographic distribution: *Bufo bufo*. Herpetological Review, 6(4):115.

Tuck, Robert G., Jr. 1977. The turtles of Iran. Shekar va Tabi'at [Hunting and Nature], 214:20–25, 60–65 (In Farsi [Persian]).

Tuck, Robert G., Jr. 1979. Notes on the Turan Biosphere Reserve herpetofauna, northwestern Iran. Bulletin of the Maryland Herpetological Society, 15(4):95–123.

Tuniyev, Boris S., Chary Ataevich Atayev, and Sakhat Muradovich Shammakov. 1991. *Stellio erythrogaster nurgelddievi* ssp. (Agamidae, Sauria) — new subspecies from eastern Kopet-Dagh. Izvestiya Akademii Nauk Turkmenistan SSR (Biol.), 6:50–60.

Underwood, Garth Leon. 1954. On the classification and evolution of geckos. Proceedings of the Zoological Society of London, 124(3):469–492.

Vaughan, R. Kathryn, James Ray Dixon, and Jerry L. Cooke. 1996. Behavioral interference for perch sites in two species of introduced house geckos. Journal of Herpetology, 30(1):46–51.

Vogel, Zderĕk. 1954. Aus dem Leben der Reptilien. Artia, Prague, 85 pp.

Vølse, Helge. 1939. The sea snakes of the Iranian Gulf and the Gulf of Oman. Danish Scientific Investigations in Iran, part 1:1–45.

Voris, Harold Knight. 1977. A phylogeny of the sea snakes (Hydrophiidae). Fieldiana: Zoology, 70(4):79–169.

Walczak, Peter S. 1971. Green sea turtle nests and turtle sightings at Hormoz. Iran Fishery Co., Fishery Research Institute Progress Report, submitted to Iran Game and Fish Department, F-7-49, 4 pages (unpublished).

Walczak, Peter S., and Wayne Kinunen. 1971. Gulf of Oman turtle nesting survey; Job Progress Report, Division of Research and Development, submitted to Iran Game and Fish Department, F-7-50, 6 pages (unpublished).

Wall, Frank 1908. Notes on a collection of snakes from Persia. Journal of the Bombay Natural History Society, 18(4):795–805, 1 text-fig.

Weber, Neal A. 1960. Some Iraq amphibians and reptiles with notes on their food habits. Copeia, 1960(2):153–154.

Welch, Kenneth Reginald George. 1983. Herpetology of Europe and Southwest Asia: A Checklist and Bibliography of the Orders Gymnophiona and Urodela. Robert E. Krieger Publishing Co., Malabar, Florida, viii + 135 pp.

Welch, Kenneth Reginald George. 1984a. Herpetology of Africa and Southwest Asia: a preliminary checklist of the orders Gymnophiona and Urodela. Herptile, 9(2):50–52.

Welch, Kenneth Reginald George. 1984b. Herpetology of Africa and Southwest Asia: a preliminary checklist of the crocodilians and chelonians. Herptile, 9(2):71–75.

Welch, Kenneth Reginald George, Peter Stuart Cooke, and Adam Stephen Wright. 1990. Lizards of the Orient: a Checklist. Robert E. Krieger Publishing Company, Malabar, Florida, v + 162 pp.

Wermuth, Heinz. 1965. Liste der rezenten Amphibien und Reptilien. Gekkonidae, Pygopodidae, Xantusiidae. Das Tierreich, 80:xxii + 246.

Wermuth, Heinz. 1967. Liste der rezenten Amphibien und Reptilien: Agamidae. Das Tierreich, 86:xiv + 127.

Wermuth, Heinz, and Robert Mertens. 1977. Liste der rezenten Amphibien und Reptilien. Testudines, Crocodylia, Rhynchocephalia. Das Tierreich, 100:xxvii + 174.

Werner, Franz Josef Maria. 1895. Ueber eine Sammlung von Reptilien aus Persien, Mesopotamien und Arabien. Verhandlungen der zoologisch-botanischen Gesellschaft, Wien, 45:14–21 (reprint pp. 1–8), pl. 3.

Werner, Franz Josef Maria. 1903. Über Reptilien und Batrachien aus Westasien (Anatolia und Persien). Zoologische Jahrbüchern, Jena: Abteilung für Systematik, 19:329–346, pls. 23–24.

Werner, Franz Josef Maria. 1913. Neue oder seltene Reptilien und Frosche des Naturhistorischen Museums in Hamburg. Jahrbuch Hamburg Wissenschaft Anstalt, 30:1–51.

Werner, Franz Josef Maria. 1917a. Versuch einer Synopsis der Schlangenfamilie der Glauconiiden. Mitteilungen Zoologisches Museum Hamburg, 34:191–208.

Werner, Franz Josef Maria. 1917b. Reptilien aus Persien (Provinz Fars). Verhandlungen der zoologisch-botanischen Gesellschaft, Wien, 67:191–220.

Werner, Franz Josef Maria. 1929a. Beiträge zur Kenntnis der Fauna von Syrien und Persien. Zoologische Anzeiger, 81:238–242.

Werner, Franz Josef Maria. 1929b. Übersicht der Gattungen und Arten der Schlangen aus der Familie Colubridae. III Teil (Colubrinae). Zoologische Jahrbüchern, Jena: Abteilung für Systematik, 57:1–196, 48 figs.

Werner, Franz Josef Maria. "1936" (1937). Reptilien und Gliedertiere aus Persien, Festschrift 60 Gebrutstage Prof. Embrik Strand, vol. 2. Lizdevnieciba "Latvia," Riga, pp. 193–202.

Werner, Franz Josef Maria. 1938. Reptilien aus Iran and Belutschistan. Zoologische Anzeiger, 121(9–10):265–271.

Werner, Yehudah Leopold. 1967. Regeneration of specialized scales in tails of *Teratoscincus* (Reptilia: Gekkonidae). Senckenbergiana Biologie, 48:117–124.

Wettstein, Otto. 1928a. Amphibien und Reptilien aus Palästina und Syrien (Zoologische Studienreise von R. Ebner, 1928, mit Unterstützung der Akademie der Wissenschaften in Wien). Anzeiger der Akademie der Wissenschaften, Wien, 65:281.

Wettstein, Otto. 1928b. Amphibien und Reptilien aus Palästina und Syrien. Sitzungsberichte der Österreichische Akademie der Wissenschaften, Wien, Mathematisch-naturwissenschaftliche Klasse, 137:773–785, pl. 1.

Wettstein, Otto. 1951. Ergebnisse der Österreichischen Iran-Expedition 1949/1950, Amphibien und Reptilien. Versuch einer tiergeographischen Gliederung Irans auf Grund des Reptilien-verbreitung. Sitzungsberichte der Österreichische Akademie der Wissenschaften, Wien, Mathematisch-naturwissenschaftliche Klasse, ser. 1, 160(5):427–448.

Wettstein, Otto. 1953. *Vipera renardi* aus Persien. Zoologische Anzeiger, 150:266–268.

White, John. 1790. Journal of a Voyage to New South Wales with . . . Plates of Non-Descript Animals, Birds, Lizards, Serpents, Curious Cones of Trees and Other Natural Productions. Debrett, London, xv + 299 + 35 pp., 65 col. pls.

Whiteman, Ronald S. 1978. Evolutionary history of the lizard genus *Phrynocephalus* (Lacertilia, Agamidae). Master Thesis, California State University, Fullerton, viii + 113 pp.

Wiegmann, Arend Friedrich August Heinrich. 1834a. Herpetologia Mexicana, seu Descriptio Amphibiorum Novae Hispaniae. Saurorum Species [pars prima ed.]. Sumptibus C. G. Luderitz, Berlin, 54 pp., 10 pls. (Reprint, Soc. Study Amphib. Reptiles, 1969.)

Wiegmann, Arend Friedrich August Heinrich. 1834b. Amphibien. *In* Beiträge zur Zoologie, gesammelt auf einer Reise um die Erde, von Dr. F. J. F. Meyen, M.d.A.d.N. Nova Acta Academiae Caesareae Leopoldino-Carolinae Germanicae Naturae Curiosorum, 17(1):183–268, 268a–d, 10 pls.

Williams, Kenneth Lee, and Van Wallach. 1989. Snakes of the World., vol. 1. Synopsis of Snake Generic Names. Krieger Publishing Co., Malabar, Florida, viii + 234 pp.

Wischuf, Tilman, and Uwe Fritz. 1996. Eine neue Unterart der Bach Schildkröte (*Mauremys caspica ventrimaculata* subsp. nov.) aus dem Iranischen Hochland. Salamandra, 32(2):113–122.

Witte, Gaston François de. 1973. Description d'un Gekkonidae nouveau de l'Iran (Reptilia: Sauria). Bulletin de l'Institute Royale Sciences Naturelles de Belgique, Biologie, 49(1):1–6.

Witte, Gaston François de. 1980. Note relative à *Rhinogekko misonnei* de Witte et *Agamura femoralis* M. Smith. Bulletin de l'Institute Royale Science Naturelles Belgique, Biologie, 52(8):1–3.

Wolfart, Reinhardt. 1987. Late Cretaceous through Quaternary paleogeographic evolution of the Middle East. *In* Friedhelm Krupp, et al. (eds.). Proceedings of the Symposium on the Fauna and Zoogeography of the Middle East, Mainz 1985. Tübinger Atlas des vorderen Orients, 28:9–22.

Wolterstorff, Willi. 1898. Die Urodelen Südasiens. Blätter für Aquarien und Terrarienkunde, 9(8):90–93, (9):101–104, (10):113–115.

Wolterstorff, Willi. 1925. Katalog der Amphibien-Sammlung im Museum für Natur- und Heimatkunde zu Magdeburg. Ersten Teil: Apoda, Caudata. Abhandlungen der Museum, Magdeburg, 4:231–310.

Wolterstorff, Willi. 1926. Über *Triton crocatus* Cope. Zoologischer Anzeiger, 67:1–6.

Wolterstorff, Willi, and Karl Wolf Herre. 1935. Die Gattungen der Wassermolche der Familie Salamandridae. Archiv für Naturgeschichte, 4(2):217–229.

Zarudny, Nikolai Aleksyevich. 1896. Marschroute der Reise in Ost-Persien im Jahre 1896. Annuaire du Musée Zoologique de l'Académie Impériale des Sciences de St.-Pétersbourg, 1:18–21.

Zarudny, Nikolai Aleksyevich. 1897. Note sur les reptiles et amphibiens de la Perse orientale. Annuaire du Musée Zoologique de l'Académie Impériale des Sciences de St.-Pétersbourg, 2(3):349–361 (In Russian).

Zarudny, Nikolai Aleksyevich. 1898. Marschroute der Reise in Ost-Persien im Jahre 1898. Annuaire du Musée Zoologique de l'Académie Impériale des Sciences de St.-Pétersbourg, 3(3–4):v–xii.

Zarudny, Nikolai Aleksyevich. 1900. Exkursion im nordwestlichen Persien und die Vögel dieser

Gegand. Mémoires de l'Académie Impériale des Sciences de St.-Pétersbourg. (Not seen; pagination unknown.)

Zarudny, Nikolai Aleksyevich. 1901. Ekskursija po vostochnoj Persii [Voyage en Perse orientale.]. Mémoires de la Société Impériale Russe de Geographie, Geographie Génerale, 36(2):1–361, folding map (In Russian).

Zarudny, Nikolai Aleksyevich. 1902a. Marschroute der Expedition der kaiserl. Russ. Geogr. Gesell. in Ost-Persien während der Jahre 1900 und 1901. Annuaire du Musée Zoologique de l'Académie Impériale des Sciences de St.-Pétersbourg, 7:1–9.

Zarudny, Nikolai Aleksyevich. 1902b. Vorläufiger kürzer Bericht über die Reise in Persien in den Jahren 1900 und 1901. Mémoires de la Société Impériale Russe de Geographie, Geographie Génerale. (Not seen; pagination unknown.)

Zarudny, Nikolai Aleksyevich. 1903a. Pticy vostochnoj Persii. Ornitologicheskie rezul'taty ekskursii po vostochnoj Persii v 1898 g. [Les oiseaux de la Perse orientale. Materiaux ornithologiques du voyage fait en 1898]. Mémoires de la Société Impériale Russe de Geographie, Geographie Génerale, 36(2):1–467, pls. 1–8 (In Russian).

Zarudny, Nikolai Aleksyevich. 1903b. O gadakh' i Raybakh' vostochnoj Persii. Gerpetologicheskie i ikhtiologicheskie rezul'taty ekskursii po vostochnoj Persii v' 1898 g. [Les reptiles, amphibiens et poissons de la Perse orientale. Materiaux herpetologiques et ichthyologiques du voyage fait en 1898.]. Mémoires de la Société Impériale Russe de Geographie, Geographie Génerale, 36(3):1–42 (In Russian).

Zarudny, Nikolai Aleksyevich. 1904. Marschroute der Reise in West-Persien in dem Jahren 1903–1904. Annuaire du Musée Zoologique de l'Académie Impériale des Sciences de St.-Pétersbourg, 9:44–51 (In Russian).

Zarudny, Nikolai Aleksyevich. 1911. Verzeichnis der Vögel persiens. Journal für Ornithologie, 59:185–214.

Zhao, Er-mi, Yao-ming Jiang, Qing-yun Huang, Shu-qin Hu, Liang Fei, and Chang-yuan Ye. 1993. Latin-Chinese-English Names of Amphibians and Reptiles. Science Press, Beijing, 329 pp.

Zhao, Er-mi, and Kraig Kerr Adler. 1993. Herpetology of China. Society for the Study of Amphibians and Reptiles, Oxford, Ohio, 522 pp., 42 figs., 48 pls.

Zohary, M. 1963. On the geobotanical structure of Iran. Bulletin of the Research Council of Israel, 11D supplement:1–113, folded map in pocket.

Zykova, L. Yu., and E. N. Panov. 1990. On possible hybridization between the agamas *Stellio caucasius* and *S. erythrogaster*. Zoologicheski'i Zhurnal, Akademiya Nauk SSSR, 69(7):103–106 (In Russian; English summary).

APPENDIX I
ABBREVIATIONS USED IN TEXT
MUSEUMS

AMNH	American Museum of Natural History, New York City, New York
BMNH	British Museum [Natural History], London (now Natural History Museum, London)
BNHM	Bombay Natural History Museum, Bombay
BYU	Brigham Young University, Provo, Utah
CAS	California Academy of Sciences, San Francisco, California
CUP	Charles University, Prague
—	Dorpater Zoologischen Museum, Dorpat
FMNH	Field Museum of Natural History, Chicago
IRSNB	Institut Royal des Sciences Naturelles de Belgique, Bruxelles
MCZ	Museum of Comparative Zoology, Harvard University, Cambridge, Massachusetts
MLSU	Leningrad State University Zoological Museum, Leningrad (now St. Petersburg)
MMSU	Moscow State University Zoological Museum, Moscow
MMTT	Muze-ye Melli-ye Tarikh-e Tabii, Tehran
MNHNP	Museum Nationale d'Histoire Naturelle, Paris
MNHP	National Museum, Prague
MSNG	Museo Civico di Storia Naturale di Genova "Giacomo Doria," Genoa
MSNM	Museo Civico di Storia Naturale, Milano
MSNTO	Museo Civico di Storia Naturale, Torino
MZUT	Università di Torino Museo Zoologico, Torino
NHMG	Naturhistoriska Riksmuseet, Göteborg
NHRM	Naturhistoriska Riksmuseet, Stockholm
NMW	Naturhistorisches Museum, Wien
RMNH	Rijksmuseum van Natuurlijke Historie, Leiden (now Nationaal Natuurhistorisch Museum)
RSM	Royal Museum of Scotland, Edinburgh (formerly Royal Scottish Museum)
SMF	Natur-Museum und Forschungs-Institut Senckenberg, Frankfurt-am-Main
UMMZ	Ruthven (University of Michigan) Museum of Zoology, Ann Arbor, Michigan
USNM	United States National Museum, Washington, DC (now National Museum of Natural History)
ZFMK	Zoologisches Forschungsinstitut und Museum Alexander Koenig, Bonn
ZIK	Ukrainian Academy of Sciences Zoological Institute, Kiev
ZIL	Zoological Institute, Academy of Sciences, Leningrad (now St. Petersburg)
ZMB	Universität Humboldt Museum für Naturkunde, Berlin
ZMGU	Zoological Museum, Gorgan University, Gorgan, Iran
ZMUC	Københavns Universitet Zoologisk Museum, København
ZMUG	Universität Göttingen Zoologische Museum, Göttingen
ZSI	Zoological Survey of India, Calcutta
ZSM	Zoologischen Staatssammlung, München

AUTHORS

AA&FF 96	Alfred W. Alcock and Frank Finn
AN 97; 99; 05; 07; 15; 16	Aleksandr Nikolsky
AS 63; 68; 87	Alexander Strauch
AS 82	Alfredo Salvador
BS 84	Bogdan Stugren
CA&GN 79	Claes Andrén and Göran Nilson
CE 41	Carl E. von Eichwald
DF et al. 97	Daniel Frynta, et al.
EA 80	Edwin N. Arnold
EA&AL 77	Edwin N. Arnold and Alan E. Leviton
FdF 43; 63; 65	Filippo De Filippi
FS 67	Franz Steindachner
FW 95; 03; 17; 29; 36; 38	Franz Werner
GB 85; 87; 89; 99; 09; 18; 20; 21	George A. Boulenger
GdW 73; 80	Gaston F. de Witte
GH&YW 69	Georg Haas and Yehudah L. Werner
GN&CA 78; 81	Göran Nilson and Claes Andrén
GO 07	Guillaume A. Olivier
GP 62; 64	Günther Peters
HB 98	Herman A. J. in den Bosch
HG&EV 96	Haji Gholi Kami and Ebrahim Vakilpoure
HS 76; 79	H. Hermann Schleich
HW 67	Heinz Wermuth
ID 67; 70; 78; 84	Ilya S. Darevsky
ID et al. 84	Ilya S. Darevsky, Josef Eiselt, and Galina P. Lukina
IF 69	Ion Fuhn
JA 72; 84; 96	John Anderson
JB 79	Jacques von Bedriaga
JD&SA 73	James R. Dixon and Steven C. Anderson
JE 68; 69; 70; 73; 79	Josef Eiselt
JE 93 et al.	Josef Eiselt, Josef F. Schmidtler, and Ilya S. Darevsky
JG 57; 66	Jean Guibé
JFS 86	Josef F. Schmidtler
JJS&JFS 70; 72	Josef J. Schmidtler and Josef F. Schmidtler
JM 84	James A. Murray
JM&MC 94	Jiří Moravec and Michal Černy
KK 78	Karl F. Kessler
KS 39; 52; 53; 55	Karl P. Schmidt
LC 77	Lorenzo Camerano
LC 61	Lilia Capocaccia
LF 50	Lothar Forcart
LL 28; 31	Louis A. Lantz
LL&GS 34	Louis A. Lantz and G. F. Suchow
LL&OC 39	Louis A. Lantz and O. Cyrén
LM 29	L. D. Morich
LvM 09; 10	Lajos von Méhelÿ
MB 76	Mohammad Baloutch
MB&MT 86	Mohammad Baloutch and Michel Thireau
MS 35	Malcolm A. Smith
NA 06; 13	Nelson Annandale
NA&VO 79	Natalia B. Ananjeva and Valentina F. Orlova

NR-P 96; 98	Nasrullah Rastegar-Pouyani
NR-P&GN 98	Nasrullah Rastegar-Pouyani and Göran Nilson
NS 74	Nikolai N. Szczerbak
NS&MG 86; 96	Nikolai N. Szczerbak and Mikhail Golubev
NZ 97; 03	Nikolai A. Zarudny
OB 80; 88	Oskar Boettger
OW 51	Otto von Wettstein
RC 91; 92	Richard J. Clark
RC et al. 66	Richard J. Clark, Erica D. Clark, and Steven C. Anderson
RM 56; 57	Robert Mertens
RT 71; 74; 75; 79	Robert G. Tuck, Jr.
SA 61; 63; 78	Steven C. Anderson
SA&AL 66	Steven C. Anderson and Alan E. Leviton
SM et al. 70	Sherman A. Minton, Steven C. Anderson, and Jeromie A. Anderson
VE&NS 86	Valery K. Eremchenko and Nikolai N. Szczerbak
VO 78	Valentina F. Orlova
VR 38	V. Rostombekov
WA	Walter Auffenberg
WB 74; 75; 76; 81	William T. Blanford
WH 59	Walter Hellmich

APPENDIX II

IRANIAN LOCALITIES AND MATERIAL EXAMINED
(See remarks in the Introduction, p. 10)

Calotes versicolor (Daudin, 1802)
Ostan 10 (Baluchistan-Sistan): Eskan, 3000 ft, N Kuh-e Bam Posht (WB 76); Farra (NZ 03); Kalagan, 3500 ft (WB 76; GB 85); Sarbaz (NZ 03).

Laudakia caucasia (Eichwald, 1831)
Elburz [=Reshteh-ye Kuhha-ye Alborz] (FdF 65); Nerion, Reshteh-ye Kuhha-ye Alborz, 3000 m (FW 03); "Ostpersien" (NMW 19458:1); "Persia" (BMNH 69.3.4.1–2). **Ostan 1 (Tehran)**: Ab Ali, 2600 m (RC 91) Abhar (CAS 142228); Alborz, near Damavand, 6000 ft (BMNH 1908.8.7.2–5); 27 km N Asaran on road to Chalus at junction with road to Gajareh, Central Alborz Protected Region, 2135 m (CAS 141222; MMTT 1245); Avaj (DF *et al.* 97); between Bahramabad and Mo'allem Khani, N Qazvin (MMTT 483); Damavand – "Tarsee," about 2000 m (OW 51); 29 km W Firuzkuh to Damavand, 12 km S junction, 6 km to Khomedeh (MMTT 1490); 38 km W Firuzkuh (USNM 149442–3; RT 71); Golezard Valley, 4 km W Polurd, NE Tehran, 2500 m (HS 79); Jijirud (MMTT 189); Hasanabad, NE Qazvin (MMTT 482); Karaj (CAS 143270–75); Lar (CAS 111934, 143270–73); Lar Valley, 2180–2900 m (CA&GN 79); Mt. Damavand, Alborz Mountains, 7000 ft (BMNH 1957.1.4.6); Qareh Dagh (NA&VO 79); Qolleh-ye Damavand, 3658+ m (MMTT 112); Reshteh-ye Kuhha-ye Alborz N Qazvin, 4000–5000 ft (WB 74); Reshteh-ye Kuhha-ye Alborz N Tehran, 5000–8000 ft (WB 76); Rey (FMNH 20972–6; 20978–80; KS 39); village of Rubberek, N Alborz, 9000 ft (BMNH 1957.1.4.1–5); Sarband (MMTT 12); Siah Sang, 10 km N Rudehen (HS 79); "Teheran" (JA 72; type locality for *Stellio persicus*); 10 km E Tehran (UMMZ 129821–2); 20 mi [32 km] N Tehran (BYU 20966; RC *et al.* 66); "40 mi from Tehran" (BMNH 1934.12.16.5). **Ostan 2 (Gilan)**: Gilan (AN 07; 15); Bijar, Kurdistan, 6–7000 ft (BMNH 1969.2600); 2 mi [3.2 km] N Bijar (FMNH 171149, 171162); about 10 mi [16 km] S Chalus, N slope Chalus Pass (AMNH 97587); about 36°12'N, 51°20'E, 64 km S Chalus on road to Karaj, Central Alborz Protected Region, 1670 m (my sight record); Chodschib (AN 07; 15; NA&VO 79); 20 mi [32 km] N Divandereh (FMNH 171163–5); Keroo (NA&VO 79); Mianeh (CAS 143278–81); Pachenar (AN 07; 15). **Ostan 3 (East Azarbaijan)**: about 38°26'N, 47°15'E, 22 km E Ahar on road from Meshkin Shahr, 1800 m (my sight record); about 38°17'N, 46°57'E, 31 km S Ahar on road to Tabriz, 2150 m (my sight record); Dasht-e Moghan (CAS 111932–3, 143286–90); about 38°56'N, 46°35'E, 6 km W Jolfa, Marakan Protected Region, 1350 m (CAS 141266); Karimabad, near Lake Reza'iyeh [possibly Ostan 4, many places of this name] (VR 38); Maragheh (VR 38); Neycharan (VR 38); about 39°14'N, 47°29'E, 1 km S Qaraqashid on road to Meshkin Shahr, 900 m (CAS 141253; MMTT 1295–6); about 38°30'N, 48°02'E, 18 km W Razi on road to Meshkin Shahr, 1150 m (my sight record). **Ostan 4 (West Azarbaijan)**: Ajefan (NA&VO 79); about 39°20'N, 44°17'E, 11 km NE 'Arab-e Dizehsi on road from Maku, 1800 m (my sight record); Bazargan (HS 79); Guch Ali (UMMZ 129779–81, 129783, 129788); Maku (FMNH 141409); 2 km W, 1 km N Maku (FMNH 141408); 2.2 mi [3.5 km] W Maku, 3925 ft [1197 m] (FMNH 171142–8, 171150–52, 171154–60, 171167–9); 2.2 mi [3.5 km] E Maku (FMNH 171166); about 38°52'N, 45°10'E, 21 km E Maku on road to Marand, 1090 m (CAS 141269; MMTT 1329–31); Marakand (UMMZ 129818); Qotur (NA&VO 79); 5.8 mi [9.3 km] SW Reza'iyeh, 3925 ft [1197 m] (FMNH 171170); 38°50'N, 45°00'E, just over Turkey-Iran border on road to Tehran, about 3500 ft [1068 m] (CAS 96259; RC *et al.* 66). **Ostan 5 (Kordestan-Kermanshah)**: Tavile (NA&VO 79). **Ostan 7 (Esfahan)**: Deh Baha, W Yazd, 9500 ft (BMNH 1936.10.12.2); Kohrud (BMNH 74.11.23.15–18; WB 76; AN 07).

Ostan 11 (Khorasan): Bardu Mountains (AN 99; 15); Bezd (GB 85; 89); Bojnurd, 820 m (RC 91); 6 km E Bojnurd (USNM 149391–2; RT 71); 120 km E Bojnurd (JG 66); between Fariman and Torbat-e Jam (NZ 03); Garm Ab (NA&VO 79); Godar Mohammad Mirza, about 1750 m (LF 50); 1.5 km E Khajar (USNM 153817; RT 71); about 36°26'N, 59°53'E, 29 km N Gold Mosque in Mashhad on road to Kalat-e Naderi, 1130 m (my sight record); about 36°48'N, 58°30'E, 22 km S junction Quchan-Mashhad road with road to Soltanabad, 1570 m (CAS 141175); 3 km S Shahrabad Kord (CAS 102495, 102497; FMNH 141411, 141414); 18 km N Torbat-e Heydariyeh (USNM 148587; RT 71). **Ostan 12 (Mazanderan)**: Ab Ask, 1620 m (RC 91); Almeh Valley, Mohammad Reza Shah National Park (MMTT 545); cliffs above Almeh Valley, Mohammad Reza Shah National park (MMTT 479–480); between Ashabad [Turkmenistan] and Astarabad (AN 15); Babol (BYU 20972; RC et al. 66); Beh Cheshme, N Shahrud, Kosh Yeilagh Protected Region (MMTT 27); 150 km W Bojnurd (USNM 154491-4; RT 71); Chahar Deh (AN 15); Chashmeh-Manqur, near Mahmudabad (MMTT 304–8); Chashmeh Zardabeh, Kosh Yeilagh Protected Region (MMTT 310); Darreh-ye Palangi, Mohammad Reza Shah National Park (MMTT 542–4); foothills near Dasht-e Kavir (NA&VO 79); 35 km N Gorgan (USNM 149416–17); Kalaleh, Gorgan Region (MMTT 525); near Kheyrabad, NE Shahrud (MMTT 309); Minudasht, 700 m (RC 91); Mohammad Reza Shah National Park (MMTT 117, 541); about 37°21'N, 56°12'E, Mohammad Reza Shah National Park, 9 km E park headquarters, 800 m (my sight record); about 37°19'N, 56°16'E, 3 km W Robat-e Qareh Bil, Mohammad Reza Shah National Park, 1000–3000 m (CAS 141172; MMTT 1158–62); Sama (FMNH 141443); Shahrud (NA&VO 79); Vali Abad (DF et al. 97); 1.5 km NE Varangrud, 7900 ft [2410 m] (FMNH 141438); 2.5 km N Varangrud, 9000 ft (FMNH 141439–40); 3 km NE Varangrud (FMNH 141441); 7 km E Varangrud (FMNH 141442).

Laudakia erythrogastra (Nikolsky, 1896)
Ostan 11 (Khorasan): Abghale (JG 57); Chodschi-Abad (AN 99; 15); Fariman (FMNH 83496; AN 97, 99, 15, 16; syntype; NZ 03); Hoseynabad (AN 99; 15); Langarak (MNHNP 1957/21–23; JG 57; HW 67; type locality for *Agama caucasica mucronata*); Khorasan Province (ZIL 9894); Mashhad, 1380 m (CAS 111930, 143263–9; RC 91); vicinity of Mashhad (AN 97, 15; type locality for *Stellio erythrogastra pallida*; HS 79); Qalandarabad (AN 97, 15, 16; syntype); Sarakhs (JG 57); 35°10'N, 60°20'E, between Torbat-e Jam and Fariman (CAS 98098; RC et al. 66). Torbat-e Jam, 1120 m (RC 91).

Laudakia microlepis (Blanford, 1874)
Fathabad [many places of this name] (AN 97); Laidaru (NMW 17196); Schah-as-Kuh (FW 36). **Ostan 7 (Esfahan)**: Kuh-i-Hezar [?=Kuh-e Hazar Darreh; may be Kuh-e Hazaran in Ostan 9] (FW 29). **Ostan 8 (Fars)**: Kushk Zar, 8000 ft [2440 m] (BMNH 1946.8.28.74–77; syntypes; WB 76; GB 85); Talaghan [=Shul-e Dalkhan?] (CAS 143276–7); Yazd-e Khvast (FMNH 20981–3); half-way between Yazd-e Khvast and Persepolis (FMNH 20984). **Ostan 9 (Kerman)**: near Deh- i-Schuturun, about 2200 m (FW 95); Gardaneh-ye Khaneh Sorkh, 9000 ft [2745 m] (WB 76; syntype); Kuh-e Laleh Zar (OW 51); Kuh-e Shah, N Baft (MMTT 1395–6). **Ostan 10 (Baluchistan-Sistan)**: Khash, N slope Kuh-e Taftan (MMTT 1398); NE slope Kuh-e Taftan (MMTT 1399); between Zahedan and Chah Bahar (LF 50); Zabol (CAS 111931, see note in text, p. 77). **Ostan 11 (Khorasan)**: Avaz (AN 99); about 34°05'N,58°50'E, 36 km S Bidokht on road to Birjand, 1540 m (CAS 141196; MMTT 1169, 1173); about 33°03'N, 59°19'E, 23 km N Birjand on road to Torbat-e Heydariyeh, 1850 m (CAS 141191; MMTT 1170–72, 1235); between Birjand and Torbat-e Heydariyeh, about 1500 m (LF 50); Golandar (AN 97); 40 km N Gonabad (JG 66); Kuh-e Baqeran (FW 36); Kuh-e Ferdows (AN 97); Kuhistan (NMW 19458:2; FW 36); Qal-eh Manar Mountains [may be *L. c. caucasia*] (AN 97); about 33°49'N, 59°07'E, 12 km N Qayen on road from Torbat-e Heydariyeh to Birjand, 1420 m (CAS 141190); about 35°29'N, 59°12'E, 8 km S Robat-e Sang on road to Torbat-e Heydariyeh, 1760 m (CAS 141184; MMTT 1163–8, 1234); 25 km N Roshkvar (USNM 153818–19); Saman Shah Mountains (AN 97).

Laudakia nupta (De Filippi, 1843)
Almab, Ziarat (FW 38); between Bushire [Ostan 8] and Bandar 'Abbas [Ostan 9] (BMNH 94.11.13.10); Mach, Bolan Pass (BMNH 91.9.14.13); "Persia" (BMNH 69.3.4.3). **Ostan 1 (Tehran)**:

Qom (CAS 142222–5; BYU 20951; RC *et al.* 66). **Ostan 5 (Kordestan-Kermanshah)**: 3 mi [4.8 km] N Amirabad [32 km N Mehran], 1000 ft [305 m] (FMNH 171201–2); Ilam (FMNH 171200); 12 mi [19 km] E Kermanshah, 4290 ft [1308 m] (FMNH 171199); 20 mi [32 km] E Kermanshah, 6000 ft [1800 m] (BMNH 1905.10.14.11); 42 km W Kermanshah (USNM 154481; RT 71); about 34°13′N, 46°41′E, creek near turnoff to microwave station 42 km W Kermanshah on road to Shahabad, 1640 m (CAS 141298); 60 mi [96 km] NW Kermanshah, 5500 ft [1650 m] (BMNH 1905.10.14.10); between Nowsud and Marivan, 2200 m (my sight record); 8 km E Qasr-e Shirin (USNM 153598; RT 71); Quraitu [Iran-Iraq border] (BMNH 1921.3.30.5; MS 35); 30 mi [48 km] W Shahabad (FMNH 130777). **Ostan 6 (Khuzestan-Lorestan)**: mountains 2 km behind Ab Bid (UMMZ 130573); mountains 5 km behind Ab Bid (UMMZ 130574); mountains 8 km behind Ab Bid (UMMZ 130572); 1 km E Ab Bid, Sar Dasht (UMMZ 130575); Ab-e Dam Dalleh (AN 07); Agha Jari (CAS 86623; SA 63); rocky hills NE Ahvaz (BMNH 1905.10.14.8–9); Bard-e Nishunde (CAS 86502, 86504, 86509–10; SA 63); Bid Zard (AN 07); Binak (CAS 86511; SA 63); 50 km SW Borujerd (USNM 153732; RT 71); near Bund-i-Kir, E Ahvaz, 250 ft [75 m] (BMNH 1905.10.14.12); Dezful (BMNH 53.1.6.88; 53.1.6.89; GB 85); hills behind Dezful (UMMZ 130571); Gholaman (DF *et al.* 97, sight record); Godar Landar (CAS 86252, 86266; SA 63); Harmalah (WH 59); Istgah-e Bisheh, 1200 m (KS 55); Khorramabad, 4100 ft [2150 m] (FMNH 171189–93; MMTT 1408, not seen by me); Lali (CAS 86559; SA 63); ridge of valley S Mahi-Dasht Valey, about 50 mi [80 km] SE Kermanshah (FMNH 130772–6); Mahor Birinji (GH&YW 69); Masjed Soleyman (CAS 86250, 86321, 96332, 86334–6, 86512; SA 63); Masjed Soleyman airfield (CAS 86531; SA 63); between Masjed Soleyman and Batvand (CAS 86434; SA 63); 18 km S Masjed Soleyman (USNM 153739; RT 71); between Masjed Soleyman and Haft Gel (CAS 86372; SA 63); between Masjed Soleyman and Naftak (CAS 86508; SA 63); Meshrageh, 53 mi [85 km] S Ahvaz, 200 ft [61 m] (FMNH 171181); Mohammad Zaman (MMTT 221); Nasrie [Ahvaz] (AN 07); Sarab, near Dow Rud (MMTT 476); Shahbazan, 540 m (KS 55); 9 km ENE Shahbazan, 550 m (KS 55). **Ostan 7 (Esfahan)**: Esfahan, 5000 ft [1525 m] (WB 76); Muteh, Muteh Protected Region (MMTT 561); Qamishlu (DF *et al.* 97). **Ostan 8 (Fars)**: Abshar (DF *et al.* 97); about 29°24′N, 51°02′E, 41 km SE Bandar-e Rig on road to Borazjan, 40 m (MMTT 1035); 3 mi [4.8 km] NW Bastak, 1300 ft [397 m] (FMNH 171185); 6 mi [9.6 km] NW Bastak, 1300 ft [397 m] (FMNH 171186); Bushire (JM 84); Chahak (DF et al. 97); Chahar Berkeh, 700 ft [214 m] (FMNH 171213); Chah Moslem, 50–100 ft [15–30 m] (FMNH 171187); 7 mi [11.2 km] E Chah Moslem, 100 ft [30 m] (FMNH 171197); Dasht-e Arzhan (DF *et al.* 97, sight record); 35 km E Gach Saran, 93 km ESE Behbehan (USNM 153675–82, 153711–24, RT 71); Jahrom, 3200 ft [976 m] (FMNH 171198); Kazerun (BMNH 74.8.15.7–9, 19–20; 94.11.13.6–8; WB 81); 18 mi [28.8 km] NW Kazerun (FMNH 171182); "Khunas Tankteh" [=Konar Takhteh] (BMNH 1903.3.14.2); Kushk Zar, 8000 ft [2440 m] (WB 76); Persepolis [Takht-e Jamshid] (CAS 86474, 86476–7; FMNH 20997–21000; FdF 43; 65; KS 55; SA 63; type locality for *Stellio n. nuptus*); 5 km SE Pol-e Abgineh (CAS 102498–503; FMNH 141401–4, 141407, 141415, 141417, 141419, 141421–3); Qar Sharon (DF *et al.* 97, sight record); Rudkhaneh-ye Shapur (FW 17); Shiraz (BMNH 72.3.22.1; CAS 142226, 142232; WB 81; GB 85; FW 17); N Shiraz (WB 76); 50 km S Shiraz (BMNH 1976.1693); 110 km SE Shiraz (USNM 158438; RT 71); Yasuj (DF *et al.* 97). Ostan 9 (Kerman): Bam (OW 51); about 26°57′N, 56°29′E, 136 km W Bandar 'Abbas-Kerman road on road to Bandar-e Lengeh, 0–50 m (MMTT 815–16); Garne River (AN 99); Jupar, 30 mi S Kerman, 6500 ft (BMNH 1951.1.2.6–19); Kerman, 5000 ft [1525 m] (CAS 143252; WB 76); SE Kerman (WB 76); about 27°13′N, 56°08′E, 32 km up road to summit of Kuhha-ye Genu, Kuhha-ye Genu Protected Region, 1000m (CAS 141036); SE Rigan, 3000 ft [915 m] (BMNH 74.11.23.12–14; WB 76; GB 85); Rud-e Khasken (AN 99); Tahrud, 5000 ft [1525 m] (WB 76). **Ostan 10 (Baluchistan-Sistan)**: Bandan (AN 97); near Bazman (AN 99; NZ 03; FW 38); near Jalq (BMNH 74.11.23.11; WB 76; GB 85; syntype of *Stellio n. fuscus*); Kalagan, 3500 ft [1068 m] (WB 76; syntype of *S. n. fuscus*); Khamun-i-Djaori (NZ 03); Kuh-i Rikeshol (AN 99; NZ 03); Lab-e Bareng (AN 99); Podagi (AN 99; NZ 03); near Sib, 4000 ft [1220 m] (WB 76); Tagab (NZ 03); Tag-i-Dorokh (NZ 03). **Ostan 11 (Khorasan)**: Chadschi-du Tschagi, Zirkuch (AN 99); Chah-i-Bena (NZ 03); Chah-i-Ziru (NZ 03). **Ostan 12 (Mazanderan)**: Tejur (RT 79, sight record with photo).

Phrynocephalus arabicus J. Anderson, 1894

Ostan 6 (Khuzestan-Lorestan): 21 km N, 1 km W Ahvaz (MMTT 62).

Phrynocephalus helioscopus helioscopus (Pallas, 1771)

Ostan 12 (Mazanderan): Dash-Bouroun, frontier post on Rud-e Atrak, NE Gonbad-e Kavus (JG 57); Gonbad-e Kavus (JG 57); 42 km N Gorgan (USNM 149415; RT 71); Nardin (AN 15); 5 mi [8 km] N Pahlavi Dezh (FMNH 141429); 10 mi [16 km] N Pahlavi Dezh (CAS 102504); about 25 km N Pahlavi Dezh (MMTT 181; RT 74); 25 mi [34 km] N Pahlavi Dezh (FMNH 141428); between Tochmak and Rud-e Atrak (AN 15).

Phrynocephalus maculatus maculatus J. Anderson, 1872

Alam (FW 36); Aminabad [many places of this name] (FW 36); Cesme Gi (FW 36); Dasht-e Kavir [Ostan 7 or 11] (AN 07; 15); Domdar (FW 36); Grosse Kawir [?Dasht-e Kavir] (FW 36); between Kerman [Ostan 9] and Shiraz [Ostan 8] (BMNH 74.11.23.58; WB 76; GB 85); Sar-i-Jum, 5000 ft [1525 m] [Ostan 8 or 9] (BMNH 74.11.23.56; WB 76; GB 85); Shar Lut (FW 36); Simurgh, Lut (FW 36); Tanaghet Kawir (FW 36). **Ostan 1 (Tehran)**: gravel steppe N 'Aliabad, 14 km N Kashan (LF 50); desert near Aran, about 950 m (LF 50); steppe 3 km N Aran (LF 50); steppe between Argavani and Marinjab (LF 50); 45 km E Daryacheh-ye Namak (GN&CA 81); gypsum and salt desert near Istgah-e Kavir, 88 km SE Tehran (LF 50); about 15 km E Marinjab (LF 50); about 25 km E Marinjab (LF 50); Rezaabad (AN 07; 15); Shah 'Abbas Caravanserai, Kavir Protected Region (MMTT 1385); 50 km E Shah 'Abbas Caravanserai, Kavir Protected Region (MMTT 1321); about 15 km W Siah Kuh (LF 50); Tarlab (AN 15; 16); Tchartaghi sand desert, about 30 km N Kashan (LF 50). **Ostan 7 (Esfahan)**: Rig-i-Djinn [?Rijan] (FW 36). **Ostan 8 (Fars)**: Abadeh (JA 72; WB 76; type locality). **Ostan 9 (Kerman)**: near Bam, 3000 ft (BMNH 74.11.23.55; WB 76; GB 85); Dshalek [?Chahlak] (AN 15); in the Hamun-e Jaz Murian, 120 km E Kahnuj (OW 51); Kerman, 5000 ft (BMNH 74.11.23.57; WB 76; GB 85); about 30°15'N, 57°12'E, 10 km E eastern edge of city of Kerman, 2000 m (CAS 141022; MMTT 741–2); about 30°15'N, 57°14'E, 2 km on dirt track branching north from Kerman-Bam road 13 km E eastern edge of city of Kerman, 2000 m (MMTT 732); eastern Kerman (AN 99; 16). **Ostan 10 (Baluchistan-Sistan)**: Chakh-i- Dzhanu (NZ 03); Gumbez-i-Nowar, Sistan (AN 99; 16); Nasirabad, Sistan (AN 97; 16; type locality for *Phrynocephalus spiniventris*); about 30°15'N, 60°57'E, 11 km NE junction with Zahedan-Mashhad road, on road to Zabol, 500 m (CAS 141086); about 30°16'N, 61°01'E, 12–18 km NE junction with Zahedan-Mashhad road, on road to Zabol, 500 m (CAS 141087; MMTT 905–11); about 30°24'N, 61°08'E, 33–42 km NE junction with Zahedan-Mashhad road, on road to Zabol, 500 m (CAS 141088; MMTT 912–18); Zamin-i-Shile (NZ 03). **Ostan 11 (Khorasan)**: dunes behind Halvan (FW 36); about 36°21'N, 56°43'E, salt flats 5 km W Kahak [110 km W Sabzevar] on road to Shahrud, Turan Protected Region (CAS 1412110; Nusi (AN 97, 15, 16); Saqi (AN 97). **Ostan 12 (Mazanderan)**: Damghan (AN 15); Na'imabad (AN 07).

Phrynocephalus mystaceus (Pallas, 1776)

Ostan 11 (Khorasan): Achangerun (AN 99); about 34°39'N, 58°47'E, 35 km N Gonbad on road to Torbat-e Heydariyeh, 850 m (CAS 141201; MMTT 1193–7); about 34°41'N, 58°48'E, 39 km N Gonbad on road to Torbat-e Heydariyeh, 850 m (CAS 141205); Khwar (FW 36); Magomed-abad, between Achangerun and Sarakhs (NZ 03). **Ostan 12 (Mazanderan)**: vicinity of Baba Kuh, Turan Biosphere Reserve, about 1160 m (RT 79); below Kariz road, Turan Biosphere Reserve (RT 79); Tochah, Turan Biosphere Reserve (RT 79).

Phrynocephalus ornatus Boulenger, 1887

Ostan 11 (Khorasan): between Achangerun and Sarakhs (NZ 03); between Bamrud and Moham-madabad (NZ 03); about 34°39'N, 58°47'E, 35 km N Gonbad on road to Torbat-e Heydariyeh, 850 m (CAS 141200, MMTT 1191–2); about 34°41'N, 58°48'E, 39 km N Gonbad on road to Torbat-e

Heydariyeh, 850 m (CAS 141204; MMTT 1198–1200); between Mozhnabad and Bamrud (NZ 03); between Mozhnabad and Fandokht (NZ 03); Tag-i-Dorokh (NZ 03); Zirkuch desert (AN 97).

Phrynocephalus persicus **DeFilippi, 1863**

From Armenia to Tehran (FdF 63; LC 61; syntypes); N Daryacheh-ye Reza'iyeh [Ostan 3 or 4] (BMNH 99.9.30.2–3); between Hamadan [Ostan 5] and Arak [Ostan 1] (FW 95); between Hamadan [Ostan 5] and Esfahan [Ostan 7] (FW 95); Kach warech, in Arak-Adschemi [probably Ostan 1] (AN 07); between Pesch [?Fash] and Rabot [Robat, several places of this name] in Bechars (AN 15); Tschar-magal (AN 15). **Ostan 1 (Tehran)**: Abhar (CAS 140824); Alvand, 1900–2000 m (FW 95); 50 km E Arak (OW 51); near Asian, 2350 m (FW 95); Ghazangheshlagh steppe, SW Karaj (LF 50); near Mohammedi [many places of this name; probably Mohammadiyeh, near Arak], 2200 m (FW 95); Qazvin (JG 57); about 15 km ESE Qazvin (MMTT 206–7); plains W Qazvin to Aqa Baba, and to foot of Charsang Mountains (FW 03); Qom, Kavir (MMTT 141); from Soltaniyeh to Tehran (FdF 65; syntypes); 6 mi [9.6 km] S Tehran (AMNH 97588–9); between Tehran and Qazvin, 4000 ft (BMNH 74.11.23.59; WB 76; GB 85; FW 03). **Ostan 2 (Gilan)**: Bijar, Kordestan, 6000–7000 ft (BMNH 1969.2599). **Ostan 3 (East Azarbaijan)**: 130 km NW Ardebil, northeastern Dahst-Moghan (ID 70); Dasht-e Moghan (CAS spec); Ghelmansaray (VR 38). **Ostan 4 (West Azarbaijan)**: Ev Oghly (DF *et al.* 97); Jolfa (VR 38); Khvoy (BMNH 99.9.30.1; GB 99); 38°40′N, 45°30′E, near Khvoy, 4500 ft [1223 m] (CAS 96260–1, 96264; RC *et al.* 66); 46 mi [74 km] N Khvoy (AMNH 97590–93); 20 mi [32 km] SE Maku (FMNH 141430); 39°00′N, 45°00′E, 120 km S Maku on road to Tabriz (HS 76). **Ostan 6 (Khuzestan-Lorestan)**: near Khomeyn, 1300 m (FW 95). **Ostan 7 (Esfahan)**: Choshkeri, in Arak-Adschemi (AN 07; 15); S. Esfahan (BMNH 1905.10.14.19–20); between Khera and Mohammadabad (AN 07; 15); 45 km SW Natanz (USNM 164821; RT 71). **Ostan 8 (Fars)**: Abadeh (JA 72); Kushk Zar, 8000 ft [2440 m] (BMNH 74.11.23.60–64; WB 76; GB 85). **Ostan 10 (Baluchistan-Sistan)**: Tasuki (JG 57 [see remarks, p. 96]). **Ostan 12 (Mazanderan)**: Na'imabad (AN 07; 15).

Phrynocephalus scutellatus **(Olivier, 1807)**

Dasht-e Kavir [Ostan 7 or 11] (AN 07; 15; syntype of *Phrynocephalus olivieri brevipes*); around Dast-e Lut [Ostan 7, 9, or 11] (FW 38); between Esfahan [Ostan 7] and Arak [Ostan 1] (FW 95); Grosse Kawir [?Dasht-e Kavir] (FW 36); Khash [Ostan 9] to Kerman [Ostan 10], 3000–5000 ft (BMNH 1951.1.6.55–60); Kuhistan [Ostan 7 or 11] (FW 36); Kuh Tawareh (FW 36); Kureh-e Gez (FW 36); Let Kal (FW 36); between Mohamadiyeh and Najafabad, 1800 m [many places of these names, probably Ostan 1] (FW 95); "Persia" (BMNH 69.3.4.16–18); Pudeschk-kupa (AN 07; 15; syntype of *P. o. carinipes*); Tara Sirkuh (USNM 69902; RT 71); Tschok et Bena (AN 15). **Ostan 1 (Tehran)**: steppe between Argavani and Marinjab (LF 50); 2 km S Chah Qar-Qareh, Kavir Protected Region (MMTT 231); S Djatschm on Kuh-i-Niswa, 2200 m (LF 50); Gir [?Giv] (FW 36); Kadish (FW 36); plain near Kadish, 950 m (LF 50); 10 km N Kashan (USNM 153582–5; RT 71); Kavir Protected Region (MMTT 104–6); Kharakan (CAS 140813–20); loam desert and *Artemisia* steppe between Kush Kuh and Siah Kuh (LF 50); about 25 km E Marinjab (LF 50); Qom (CAS 140758–812); Rud-e Shur, 59 km SW Tehran (LF 50); Safid Ab (FW 36); about 35°07′N, 50°08′E, 24 km W Saveh on road to Hamadan, 1160 m (CAS 141289–90; MMTT 1356–60); 30 km W Shah 'Abbas Caravanserai on Veramin road (MMTT 770); between Shah 'Abbas and Cheshmeh-ye Sefid Ab, Kavir Protected Region (GN&CA 81); Siah Kuh (BMNH 1934.12.16.6) *Artemisia-Ephedra-Aellenia* steppe N Siah Kuh (LF 50); between Tehran and Qazvin (GB 85); 50 km S Tehran (OW 51). **Ostan 7 (Esfahan)**: 'Abbasabad, N Khur, near 'Arusan (FW 29); 9 mi S 'Aqda, 32°25′N, 53°38′E (BMNH 1936.10.12.4); 4 km SSW Anarak (USNM 158435; RT 71); Chah-e Gorg (FW 36); Esfahan, 6800 ft [2074 m] (BYU 20952; RC *et al.* 66); arid steppe between Esfahan and Yazd (OW 51); 100 km N Esfahan (OW 51); SW Esfahan (BMNH 1905.10.14.13–18); Jafar Abad (DF *et al.* 97); Jandaq (AN 07; syntypes *P. o. carinipes, P. o. brevipes*); Kuh-e Ashin (FW 36); Kuh-e Chastab (FW 36); Kuh-e Hazar Darreh (FW 29); Kuh-e Sofeh (GO 07; type locality); steppe 30 km W Na'in (LF 50); 45 km SW Natanz (USNM 153426; RT 71); near Neh (FW 36); Qamishlu (DF *et al.* 97). **Ostan 8 (Fars)**: Abadeh (BMNH 1919.5.2.6–11; BYU 20960; RC *et al.* 66); 27 km N Abadeh (USNM 153586–7; RT 71); Bandar-e

Lengeh (BMNH 1951.1.6.63); Lar (JG 57); Shiraz (JA 72; CAS 140821–3); 240 km N Shiraz (JG 66); Yazd-e Kvast (FMNH 20991–6 [187 specimens]; KS 39). **Ostan 9 (Kerman)**: 20 km SE Anar (USNM 153596–7; RT 71); near Bam, 4000 ft [1220 m] (WB 76); Esma'ilabad (OW 51); Jupa, 20 mi S Kerman, 30°20′N, 59°10′E, 6500 ft (BMNH 1951.1.2.23–25); Kerman (AN 15); desert E Kerman, 30°21′N, 57°05′E (BMNH 1936.10.12.5); NW Kerman (OW 51); steppe between Kerman and Sa'idabad (LF 50); steppe 30 km W Kerman (LF 50); 28 km SW Mashiz (USNM 149153; RT 71); Minab (FW 29); Rayen, 4000 ft (BMNH 74.11.23.73–75; WB 76; GB 85); near Rigan, 2500 ft (BMNH 74.11.23.69–71; WB 76; GB 85); Sabzvaran (OW 51); about 29°06′N, 56°00′E, 50 km S Sirjan on Kerman-Bandar 'Abbas road, 1680 m (CAS 141023); about 31°13′N, 54°53′E, 98 km S Yazd on road to Rafsanjan, 1600 m (CAS 141108; MMTT 962–6); Zar Rud, Rayen area (MMTT 1416). **Ostan 10 (Baluchistan-Sistan)**: near Bampur, 1500 ft (BMNH 74.11.23.72; WB 76; GB 85); about 27°10′N, 60°10′E, 30 km W Bampur, Jaz Murian depression, 500–540 m (CAS 141074); vicinity of Bazman (AN 99; 15); Dizak [Davar Panah], 4000 ft (WB 76); Khash, 1200 m (BMNH 1951.1.6.61–62; MMTT 710–11; JG 66); 6 km N Khash (USNM 148669–71; RT 71); 15 km N Khash (USNM 148672; Rt 71); Khoshtegan, 3000 ft] (BMNH 74.11.23.64–68; WB 76; GB 85); Magas [Zaboli], 4500 ft (WB 76); Mazel-Ab, eastern Kerman (AN 99); N slope Rig Espakeh (RM 56); Shahestan, Shahrestan-e Saravan, 1200 m (RM 56); Tasuki (JG 57); Zahedan (USNM 149142; RT 71); steppe about 20 km S Zahedan (LF 50); about 29°23′N, 60°49′E, 13 km S Zahedan on road to Khash, 1760 m (CAS 141068); between Zahedan and Chah Bahar (LF 50); 85 km N Zahedan (USNM 149141; RT 71). **Ostan 11 (Khorasan)**: about 36°24′N, 56°17′E, 16 km S 'Abbasabad on dirt road into Turan Protected Region, 800–870 m (CAS 141303; MMTT 1203–7; RT 79); 6 km W Ayubi (USNM 148593; RT 71); about 32°49′N, 59°26′E, 20 km S Birjand on road to Zahedan and NNE on dirt track toward mountains, 1610 m (CAS 141192–3; MMTT 1180–88); 55 km ENE Birjand (USNM 148609–11; RT 71); Bohnabad (AN 97; 15); 15 km NW Doruneh (USNM 148579–80; RT 71); 1 km S Esfideh (USNM 148616–19; 148622; RT 71); 2 km NE Esfideh (USNM 148621; RT 71); 5 km NE Esfideh (USNM 148620; RT 71); Golandar (AN 97; 15); Halvan (FW 36); 4 km NNE Jangal (USNM 148596–8; 148603–4; RT 71); 15 km NE Jangal (USNM 148599–602; RT 71); about 36°21′N, 56°43′E, salt flats 5 km W Kahak on road to Shahrud, Turan Protected Region, 800 m (CAS 141209–10; MMTT 1214–24; RT 79); Mahneh (JG 57); 5 mi [8 km] N Mahneh (FMNH 141431); Mazhan, Dasht-e Lut (USNM 148632–6; RT 71); Niazabad, Zirkuch (AN 99; 15); Qusheh (AN 97; 15); Robat-e Khakestari, near Mashhad (FW 38); serpentine mountain 30 km E Sabzevar (USNM 158439–44; RT 71); Saman Shah Mountains (AN 97; 15); Shur Rud, about 60 km E Qayen (MMTT 609); Tun [Ferdows] (AN 97); Torbat-e Heydariyeh (USNM 148588; RT 71); Zahirabad (AN 97; 15). **Ostan 12 (Mazanderan)**: 'Abbasabad, near Sabzevar (AMNH 61774–5); 5 km W 'Abbasabad (USNM 159343–57; RT 71); 39.5 km W 'Abbasabad (USNM 158512–26; RT 71); Chahar Deh (AN 15); Chehok Spring, Turan Biosphere Reserve, 1025 m (RT 79); Damghan (AN 15); Delbar, Turan Biosphere Reserve, 1300 m (RT 79); 80 km S Delbar, Turan Biosphere Reserve (RT 79); Dasht-e Gilan, Khosh Yeilagh Protected region (MMTT 537); Dasht-e Tal, 30 km SSE Shahrud (MMTT 264–78); between Khaneh Khvodi and Delbar, Turan Biosphere Reserve; RT 79); between Miandasht and 'Abbasabad (AN 15); Na'imabad (AN 07; 15; syntype of *P. o. brevipes*); Shahrud (AN 15); between Semnan and Damghan (FW 38).

Trapelus agilis (Olivier, 1804)

Ain u Waher (FW 36); Almab (FW 38); Baschm-Sorshe (FW 38); beyond Bolazam (FW 36); Cheshman Ap (FW 36); Dech et Chimmer (AN 15); Domdar (FW 36); Gjarmaz (AN 15); Hauz Batil (FW 36); Jonstan, southern slope of Talysh Mountains (GN&CA 81); between Lar [Ostan 2] and Damavand [Ostan 1] (CAS 142230); Pir (FW 36); Ziarat (FW 38). **Ostan 1 (Tehran)**: Aryamehr Botanic Garden, W Tehran (MMTT 399–400); near Bahram, about 20 km E Tehran (LF 50); Chah-e Ebrahim Zare (FW 36); Cheshmeh-ye Sefid Ab, Kavir Protected Region (GN&CA 81); Damavand (LF 50); Daryacheh-ye Namak (FMNH 20985–6, 20988[7]; KS 39); 35 km S Delijan (USNM 153563–4); 135 km N Esfahan (OW 51); 150 km N Esfahan (USNM 153565; RT 71); 14 km SE 'Ein or Rashid Caravanserai, Kavir Protected Region (MMTT 1591–2); about 35°16′N, 52°12′E, 14 km W Garmsar on road to Tehran, 900 m (MMTT 1313); Javadabad/Istgah-e Abardezh (HS 79); 10 km

NW Kashan (USNM 153579–81; RT 71); about 33°54′N, 51°30′E, 8 km S Kashan and 6 km W road between Yazd and Kashan, 1175 m (CAS 141109); Kavir Protected Region (MMTT 99–103, 558–9); between Kush Kuh and Siah kuh (LF 50); Mahallat (CAS 142193–210); Maljat-Abad (AN 07); Qazvin (CAS 142182–192); between Qazvin and Tehran (FdF 65); Qom (CAS 142126–60); near Rud-e Shur, 49 km SW Tehran (LF 50); Saveh (CAS 142092–125); about 35°07′N, 50°08′E, 24 km W Saveh on road to Hamadan, 1160 m (CAS 141293); Shah 'Abbas Caravanserai and vicinity, near foot of Siah Kuh, 1000 m (MMTT 2–3, 538–9, 553, 1236–43, 1314–15, 1317–20, 1380–3; GN&CA 81); between Shah 'Abbas and Cheshme-ye Sefid Ab, Kavir Protected Region (GN&CA 81); Siah Kuh, salt desert (BMNH 1934.12.16.3–4); Tarlab (AN 15); Tehran (BMNH 85.5.27.14–19, 1920.1.20.3094); near Tehran (GB 87); 20 mi [32 km] E Tehran (RC, *et al.* 66). **Ostan 2 (Gilan)**: (WB 76); 4.2 mi [6.7 km] N Takestan (FMNH 170987); 4.7 mi [7.5 km] N Takestan (FMNH 170988). **Ostan 5 (Kordestan-Kermanshah)**: "Hamadan Jinjan" (BMNH 1920.3.20.1); Kuh Daschteh [?Kuhdasht] (FW 38). **Ostan 6 (Khuzestan-Lorestan)**: Ab-e Golestan (CAS 86338; SA 63); Agha Jari (CAS 86487; SA 63); Bard-e Nishunde (CAS 86457, 86461–2; SA 63); 8 km W Behbehan on road to Agha Jari, 370 m (MMTT 1070); 21 km W Behbehan on road to Agha Jari, 250 m (MMTT 1069); 34 km W Behbehan on road to Agha Jari, 250 m (MMTT 1068); Haft Gel (CAS 86556; SA 63); 17 km E Haft Gel (CAS 86403–6, 86419; SA 63); Mahor Barinji (GH&YW 69); Masjed Soleyman (CAS 86251, 86320, 86328, 86331, 86466–7; SA 63); between Masjed Soleyman and Batvand (CAS 86348–51; SA 63); between Masjed Soleyman and Haft Gel (CAS 86399–402, 86418; SA 63); Meshrageh, 53 mi [85 km] SE Ahvaz (FMNH 170936); Sar-e Gach (CAS 86447–8, 86459; SA 63); near Shushtar (CAS 86373, 86464; SA 63); Tol-e Bazun (CAS 86341–4, 86346–7, 86389–95, 86422–3, 86425–6, 86458, 86493–8; SA 63); between Yamaha and Ab-e Tembi (CAS 86427–9, 86449; SA 63). **Ostan 7 (Esfahan)**: about 33°26′N, 52°20′E, 6 km W Ardestan (CAS 141019; MMTT 723–4); 'Arusan (FW 36); Dasht-e Kavir (FW 36); Esfahan (BYU 20953–4; FMNH 20989; KS 39; RC *et al.* 66); 20 km N Esfahan (OW 51); 60 km E Esfahan on road to Na'in (MMTT 186; RT 74); 70 km N Esfahan (OW 51); Jafar Abad (DF *et al.* 97); Kochrud, Irak-Adschemi (AN 07; type locality for *Agama kirmanensis brevicauda*); 20 km E Na'in (LF 50); 33°31′N, 51°53′E, 25 km by road N Natanz.

Ostan 8 (Fars): Abadeh (BMNH 79.8.15.14–15, 1919.5.2.2–5; WB 81); Ab-e Shirin, 100 km W Khonj (MMTT 1669 [not seen]); Ahram (CAS 102491–2; FMNH 141392–3, 141395–6, 141398); Bandar-e Lengeh (BMNH 1951.1.6.54); Binak (CAS 86322, 86625; SA 63); Bushire (BMNH 1903.3.14.1); Bushire peninsula (FW 17); Chahak (DF *et al.* 97); 5 km N Dalaki on road to Shiraz, 110 m (CAS 141149); Deh Bid (BMNH 79.8.15.10–12; WB 81); Gach Saran (CAS 86503; SA 63); Kazerun (BMNH 79.8.15.13; WB 81); Safi Dah [?Kuh-e Safidar] (FW 36); Shiraz (JA 72); 240 km N Shiraz (JG 57); Tangistan (FW 17); Yazd-e Kvast (FMNH 20987[10], 20990[10]; KS 39). **Ostan 9 (Kerman)**: Bandar 'Abbas (BMNH 1951.1.6.50–52); 27°30′N, 56°18′E, 35 mi [56 km] N Bandar 'Abbas (CAS 86374; SA 63); about 26°57′N, 55°29′E, 136 km W Bandar 'Abbas-Kerman road on road to Bandar-e Lengeh, 0–50 m (MMTT 816); Fahraj (USNM 149149–51; MMTT 1389, 1410; RT 71); 6 km W Fahraj on road between Bam and Zahedan (MMTT 982); Jask (BMNH 94.11.13.4–5); Jupar, 20 mi [32 km] S Kerman, 6500 ft (BMNH 1951.1.2.20–22); Kerman (WB 76); about 30°15′N, 57°14′E, 2 km on dirt track branching left from Kerman-Bam road, 13 km E eastern edge of city of Kerman, 2000 m (CAS 141020; MMTT 726–8, 730–31); Kurin (AN 99; type locality for *Agama kirmanensis*); Minab (FMNH 170986); about 27°00′N, 57°08′E, 17 km SSE Minab on inland road to Jask, 50–100 m (CAS 141028); 148 km E Neyriz on road to Sa'idabad (BMNH 1966.355–357); Rayen area, Zar Rud (MMTT 1397); Rigan (BMNH 74.11.23.107; WB 76); Rig Mati, about 150 km SE Sabzevaran (OW 51); 21 km N Rudan on road to Jiroft, 575 m (CAS 141051); N Sabzevaran (LF 50; OW 51); Seh Konj, near Kerman, 7000 ft (BMNH 1936.10.12.1); about 27°19′N, 56°33′E, 19 km SE Shaqu on road to Minab, 0–50 m (CAS 141027); about 29°06′N, 56°00′E, 50 km S Sirjan on Kerman-Bandar 'Abbas road, 1680 m (MMTT 744).

Ostan 10 (Baluchistan-Sistan): Bahu Kalat (BMNH 74.11.23.111–112; WB 76); Bampur (NZ 03; AN 15); Bazman (MMTT 1591–2 [not seen by me]); Bent, 108 km W Nikshahr (MMTT 716); Chah Bahar (JG 66); Davar Panah (WB 76); Esfandak (BMNH 74.11.23.104–106; WB 76); 35 km SE Iranshahr (RM 56); Ispidian, near Mand (BMNH 74.11.23.114; WB 76); 5 km NNE Kahoorak

(USNM 149145–8; RT 71); Khash (BMNH 1951.1.6.45–49; MMTT 707–9); about 28°10'N, 61°11'E, 10 km S Khash on road to Iranshahr, 1460 m (MMTT 895); Khoshtegan, Bam Posht (WB 76); W slope Kuh-e Taftan (MMTT 1429); Makran (AN 15); SE Nahug (RM 56); Nikshahr (AN 15); between Nosratabad and Zahedan (LF 50); Perso-Baluch [Pakistan] frontier (BMNH 1906.8.10.25); about 31°03'N, 61°38'E, 10 km SW Rud-e Hirmand, abandoned village SE road from Zabol to Dust-e Mohammad Khan, 450 m (CAS 141097; MMTT 933–4); E Saravan (RM 56); Sargad (AN 15); Sistan (BMNH 73.1.7.3–5); Tamin, Sargad region, 20 km SW Ladiz (AN 15); Sib (WB 76); 15 mi [24 km] SW Zabol (CAS 102484–90; FMNH 141377, 141379–80, 141382, 141384, 141386–9, 141391, 141445, 141447–9); Zaboli (WB 76); beween Zaboli and Bampur (BMNH 74.11.23.113 [lectotype]; WB 76; GB 85; type locality for *Agama isolepis*); Zahedan (MMTT 1425); about 29°23'N, 60°49'E, 11 km NE jct with Zahedan-Mashhad road, on road to Zabol, 500 m (MMTT 903); between Zahedan and Chah Bahar (LF 50); Zamran (WB 76). **Ostan 11 (Khorasan)**: about 36°24'N, 56°17'E, 16 km S 'Abbasabad, on dirt road into Turan Protected Region, 800–870 m (CAS 141213; MMTT 1208; RT 79); about 36°09'N, 56°08'E, 34 km S 'Abbasabad on dirt road into Turan Protected Region, 870 m (MMTT 1226; RT 79); Baba Kuh, 1160 m (RT 79); Baghestan (RT 79); vicinity of Baghestan (RT 79); 1 km N Baghestan (RT 79); about 32°49'N, 59°26'E, 20 km S Birjand on road to Zahedan and NNE on dirt track toward mountains, 1610 m (MMTT 1189–90); between Birjand and Torbat-e Heydariyeh (LF 50); Dasht-e Lut (FW 38); Delbar (RT 79); 7.5 km E Delbar, 100 m (RT 79); 15 km E Doruneh (RT 71); Esfideh (USNM 148615; RT 71); Ferdows (AN 97); Godar Mohammad Mirza, near Robat-e Khakastari between Torbat-e Heydariyeh and Mashhad (LF 50); about 36°21'N, 56°43'E, salt flats 5 km W Kahak on road to Shahrud, Turan protected Region, 800 m (MMTT 1225); 5 km N Kashmar (USNM 148578; RT 71); Between Khaneh Khvodi and Delbar, 1185 m (RT 79); Khorasan (AN 15); Kuhistan (FW 36); Mashhad (CAS 142229; RC 91); 20 km E Mashhad (USNM 154485; RT 71); near Posht-e Aseman, 1260 m (RT 79); about 33°49'N, 59°07'E, 12 km N Qayen on road from Torbat-e Heydariyeh to Birjand, 1420 m (MMTT 1154, 1178); near Sabzevar (RC *et al.* 66); Sjulpenai mountains [?Kuh-e Ferdows] (AN 97); about 36°35'N, 58°11'E, 23 km N Soltanabad on road to Quchan, 1250 m (CAS 141207); Tajan River [Harirud, on Iran-Afghan-Turkmenistan border] (LL 18); Tayyabat (CAS 96270; FW 38; RC *et al.* 66); about 34°47'N, 60°47'E, 4 km E communications station at outskirts of Tayyabat, about 900 m (CAS 141181; MMTT 1175); 9 km E Tayyabat (MMTT 1174); Torbat-e Heydariyeh (USNM 148586; RT 71); Torbat-e Jam, 955 m (RC 91); between Torbat-e Jam and Fariman (RC *et al.* 66). **Ostan 12 (Mazanderan)**: S. Almeh Valley, Mohammad Reza Shah National Park (MMTT 481); 13 km E Dasht (USNM 149399, 149403–5; RT 71); Dasht-e Tal, SE Shahrud (MMTT 317); Gonbad-e Kavus (JG 57); Mian Kaleh Protected Region (MMTT 185); Mohammad Reza Shah National Park (GN&CA 81); Na'imabad (AN 07); 1 mi [1.6 km] N Pahlavi Dezh (FMNH 141399); 10 mi [16 km] N Pahlavi Dezh (FMNH 141426); 25 mi [40 km] N Pahlavi Dezh (FMNH 141425); NE Shahrud, near Kheyrabad (MMTT 316); Turkmen Plains, near Kangali [?Tangali], on Iran-Turkmenistan border, 75 km N Gorgan (USNM 149418–22; RT 71). **Persian Gulf Islands**: Jazireh-ye Qeshm (BMNH 87.12.20.1).

Trapelus persicus (Blanford, 1881)

Ostan 6 (Khuzestan-Lorestan): Ahvaz (AN 07); N. Ahvaz (BMNH 1905.10.14.5–7); dunes on Ahvaz Ridge, 31°18'N, 48°45'E (CAS 86534–5; SA 63); 23 km N, 1 km W Ahvaz (MMTT 58); 35 km N Ahvaz (USNM 153601); 45 km N Ahvaz (USNM 153602–3; RT 71); 31°16'N, 49°11'E, between Ahvaz and Haft Gel (CAS 86486, 86505–6, 86521–3, 86538–40; SA 63); about 32°01'N, 48°16'E, 2 km W Andimeshk-Ahvaz road, on road to Karkheh National Park headquarters (MMTT 1066); 30 km S Andimeshk (USNM 153685; RT 71); Chogha Pahn, 2 km N Jerud Dam (MMTT 235–8); Choqa Zanbil (DF *et al.* 97, sight record); Gurschir (AN 07); 10 km S Haft Tappeh Ridge (MMTT 239); Harmalah, 120 km NW Ahvaz (WH 59); 15 km N Karkheh-Dez channel (MMTT 234); about 31°16'N, 49°10'E, 1 km E Kupal, 25–50 m (my sight record); Meshrageh, 53 mi [85 km] SE Ahvaz (FMNH 170963–85); Qal'eh-ye Tol (AN 07). **Ostan 8 (Fars)**: near Alchangi, 33 km N Bushire, 0–5 m (CAS 141144; MMTT 1064); 9 km S Borazjan on road from Bushire, 40 m (CAS 141147);

Deh Bid (BMNH 79.8.15.39–42/1946.8.11.39–42 syntypes; WB 81); Kazerun 79.8.15.43/1946.8.11.30 syntype; WB 81).

Trapelus ruderatus (Olivier, 1804)

Khullar [Ostan 8?] (FW 17); Nurabad plain [many places of this name; Ostan 8?] (FW 17); between Qom [Ostan 1] and Arak [Ostan 5], 1700 m (FW 95). **Ostan 1 (Tehran)**: Abhar (CAS 142215–17, 142220–21); between Damavand and Rudkhaneh-ye Jajrud (FW 03); Tehran (JA 72); Qazvin (CAS 142211–14). **Ostan 2 (Gilan)**: Chashme Sangi, about 35 km N Bijar (UMMZ 129792, 129794, 129816–17); Mianeh (CAS 142219). **Ostan 3 (East Azarbaijan)**: Ajab Shir (VR 38); 10 km S Arax River [16 km S Alireza-abad] on road to Meshkin Shahr, 220 m (MMTT 1284); Bandar-e Vanalu (VR 38); between Kartevjul and Eshma ['Ammeh] SE Kuh-e Sahand, 7–8000 ft [2135–2440 m] (VR 38); Kuh-e Sahand, 7000 ft [2135 m] (VR 38); vicinity of Maragheh (AMNH 363B4; VR 38); Sagban (VR 38); Schamedinan, near Chelane [?Khelejan] (VR 38); SW Shahi Peninsula [?Jazireh-ye Shahi], E side Daryacheh-ye Reza'iyeh (BMNH 1965.1461–65); Tabriz (AN 15); steppe on pass toward Talkeh Rud, N Tabriz, about 1600 m (LF 50). **Ostan 4 (West Azarbaijan)**: Govarchin Qaleh (VR 38); Guch Ali, about 57 km NE Khvoy (UMMZ 129782); vicinity of Khvoy (VR 38); Marakand (UMMZ 129819). **Ostan 5 (Kordestan-Kermanshah)**: Bisotun, (USNM 154479 ; RT 71); 24 km S Borujerd (FMNH 130783); at crest of pass at Chahar Zebar, about 17 mi [27 km] SW Kermanshah (FMNH 130784); Faraman (FMNH 141333–53); N Faraman (FMNH 130781–2); Ilam, 5000 ft [1525 m] (FMNH 170857, 170859–74); 70 mi [112 km] SW Ilam, 5000 ft [1525 m] (FMNH 170854, 170857–8); 5 mi [7.5 km] N Kermanshah (FMNH 170855); NW Kermanshah, 5500 ft (BMNH 1905.10.14.4); 8 km E Qasr-e Shirin (USNM 153600; RT 71); 12.9 mi [20.6 km] N Ravansar, 5000 ft [1525 m] (FMNH 170856). **Ostan 6 (Khuzestan-Lorestan)**: Dehloran, Ilam (MMTT 84–86, 671); Dow Rud (MMTT 560); 5 km NW Dow Rud (USNM 159342, 164185; RT 71); Istgah-e Ezna, 1800 m (KS 55); divide between Kermanshah and Kuhdasht valleys, S Kermanshah (FMNH 130780); 10 km N Khorramabad (MMTT 684); Lorestan Province (MMTT 1419 [not seen by me]); valley S of E end Mahidasht Valley, about 50 km SE Kermanshah (FMNH 130786). **Ostan 7 (Esfahan)**: Esfahan (FdF 65; type locality for *Agama lessonae*); NW Esfahan, 6600–8000 ft (BMNH 1905.10.14.1–3); near Esfahan, 5000 ft (WB 76); 25 km N Esfahan (OW 51); Qamishlu (DF *et al.* 97). **Ostan 8 (Fars)**: Bamu National Park (MMTT 1406); Bushire (JA 84; GB 85); Bushire peninsula (FW 17); Dorudsand, Marv Dasht Plain, 40 mi [64 km] N Shiraz (BMNH 1966.362); near Emamzadeh (FW 17); Isvand (FW 17); 30°39'N, 51°36'E, Kuh-e Dinar Ridge, 10 km by road N Yasuj (DF *et al.* 97); Pasargat (DF *et al.* 97); Persepolis [Takht-e Jamshid] (CAS 86475; SA 63); Shiraz (BMNH 74.11.23.80–81; CAS 142218; JA 84); near Shiraz, 4000 ft (BMNH 1969.1694–1695; WB 76); 13 km SE Shiraz (USNM 164822–3; RT 71); Shiraz Arza (BMNH 1966.358–361); Tangestan (FW 17); Tepe Jari, 16 km S Persepolis on Marv Dasht plain (FMNH 130778). **Ostan 9 (Kerman)**: between Deh-i-Schuturun and Pariz, 2000 m (FW 95); Kerman (BMNH 95.5.29.1). **Ostan 10 (Baluchistan-Sistan)**: Perso-Baluch border [*Trapelus r. megalonyx?*] (AA&FF 96; NA 06).

Anguis fragilis colchicus (Nordmann, 1840)

Ostan 1 (Tehran): Tehran (FdF 65; JB 79; certainly in error). **Ostan 2 (Gilan)**: 37°44'N, 48°57'E, 12 km by road W Asalem, 280 m (DF *et al.* 97); 80 km NW Bandar-e Pahlavi (JG 66); Lahijan (OW 51); about 3 km W Ramsar (photo record, [Pl. 6B]); Rasht (JA 72; WB 76; JB 79; type locality for *Anguis orientalis*). **Ostan 11 (Khorasan)**: Siaret, Atrek Valley (BMNH 93.11.1.20); Shirvan, Rud-e Atrak valley (OB 88; AN 15). **Ostan 12 (Mazanderan)**: Aber (AN 15); about 36°17'N, 52°21'E, 26 km S Amol on road to Tehran, 450 m (CAS 141160–141161; MMTT 1085–1087); Astrabad [=Gorgan] (AN 07; 15); 36°46'N, 50°30'E, 10 km by road ENE Chorti, 480 m (DF *et al.* 97, sight record); 15 km SW Dasht (USNM 149400; RT 71); between 4.4 mi [7 km] and 5 mi [8 km] NE Gorgan (FMNH 141672); Khorramabad (OW 51); Shahrud (AN 15); 6 km SW Tamishan and 24 km W Mahmudabad (USNM 149441; RT 71).

Ophisaurus apodus **(Pallas, 1775)**
Bala Morghab (BMNH 86.9.21.72); Gubran [Iran?] (BMNH 86.9.21.74–75); Nachduin mountains (AN 97); Rud-e Atrak [Ostan 11 or 12] (LM 29). **Ostan 2 (Gilan)**: 37°44′N, 48°57′E, 12 km by road W Asalem, 280 m (DF *et al.* 97); Mianeh (CAS 142234–42); Rasht (FW 03); Rezvandeh (RM 57); valley of Safid Rud near Rustamabad [=Kalaruz] (FW 03; AN 07). **Ostan 4 (West Azarbaijan)**: 2.2 mi [3.5 km] W Maku, 3925 ft [1200m] (FM 17). **Ostan 6 (Khuzestan- Lorestan)**: Istgah-e Bisheh, 1200 m (FMNH 69297; KS 53). **Ostan 11 (Khorasan)**: Bojnurd (LM 29); E Bojnurd (RC *et al.* 66); 6 km E Bojnurd (USNM 149393; RT 71); 120 km W Bojnurd (JG 66); between Fariman and Torbat-e Jam (NZ 03); 90 km W Quchan (USNM 154487; RT 71); 3 km S Shahrabad Kord (FMNH 141674). **Ostan 12 (Mazanderan)**: 20 km S Amol on road to Tehran (MMTT 720); Astrabad [=Gorgan] (AN 15); 150 km W Bojnurd (USNM 154489–90; RT 71); 36°28′N, 51°24′E, 25 km by road S Chalus, 200 m (DF *et al.* 97, sight record); about 36°19′N, 51°16′E, 39 km S Chalus on road to Karaj, 600 m (MMTT 1250); 36°38′N, 51°25′E, 45 km by road S Chalus, 800 m (DF *et al.* 97); 36°46′N, 50 30′E, 10 km by road ENE Chorti, 480 m (DF *et al.* 97, sight record); 5 km ENE Dasht (USNM 149410; RT 71); Dasteng-kela, near Barferusch [=Babol] (LF 50); Gonbad-e Kavus (JG 57); 5 mi [8 km] NE Gorgan (FMNH 141673); 44 km SSE Sari on road to Kiasar (MMTT 34–35); "Mazanderan" (CAS 142233); about 37°19′N, 56°16′E, 3 km W Robat-e Qareh Bil, Mohammad Reza Shah National Park, 1000–1300 m (CAS 141171); Minudasht, 1120 m (RC 91).

Eublepharis angramainyu **S. Anderson and Leviton, 1966**
Muhommarat [IRAN?] (BMNH 1909.3.20.1). **Ostan 6 (Khuzestan-Lorestan)**: Ali Khosh, Dehloran (USNM 157373; RT 71; NS&MG 86); 2 km W Dehloran, 900 ft (FMNH 170933); Fakke, about 150 km N Ahvaz (MB&MT 86; type locality for *Eublepharis ensafi*); between Masjed Soleyman and Batvand (CAS 86337, 86384; SA 63; SA&AL 66; NS&MG 86; type locality); between Masjed Soleyman and Naftak (CAS 86333, 86361–2, 86366, 86383, 86396–8, 86416; SA 63; SA&AL 66); between Masjed Soleyman and Sar-i-Gach (CAS 86382; SA 63; SA&AL 66); near Shahbazan, 500 m (JJS&JFS 70). **Ostan 8 (Fars)**: Gach Saran (I. S. Darevsky, pers. commun.); Kazerun (FW 17).

Eublepharis cf. *macularius* **(Blyth, 1854)**
Ostan 11 (Khorasan): Chah-e Ziran, on the plain of Tag-i-Dorokh (NZ 03).

Eublepharis turcmenicus **Darevsky, 1977**
Ostan 11 (Khorasan): Koppeh Dagh, NW Mashhad (ID 78).

Agamura persica **(A. Duméril, 1856)**
Garm Ab [many places of this name on Central Plateau; probably 35°36′N, 53°42′E on railway] (FW 36); N Great Kavir [?Dasht-e Kavir] (FW 36); between Sebzar and Lascht (AS 87; AN 15). **Ostan 1 (Tehran)**: Chashmeh Safid Ab, Dasht-e Kavir (MMTT 506; GN&CA 81); between Chashme Safid Ab and Shah 'Abbas (GN&CA 81); 'Eyn or Rashid Caravanserai, Kavir Protected Region (MMTT 1434); vicinity of 'Eyn or Rashid, about 950 m, in chalk zone forming N end Siah Kuh (LF 50); about 35°16′N, 52°12′E, 14 km W Garmsar on road to Tehran, 900 m (MMTT 1312); between Garmsar and Semnan, 50 km S Deh Namak (MMTT 484); Kavir Protected Region (MMTT 107–8); Qom (CAS 140551–87); Shah 'Abbas Caravanserai, Kavir Protected Region (MMTT 1387; GN&CA 81); vicinity Shah 'Abbas Caravanseri, near foot of Siah Kuh, 1000 m (CAS 141220; NS&MG 86); 300 m S Shah 'Abbas Caravanserai (MMTT 523–4); 2–4 km SE Shah 'Abbas Caravanserai to Shekar Ab (MMTT 1316); 29 km E Shah 'Abbas Caravanserai to Shekar Ab (MMTT 1436); Shekar Ab (MMTT 1437); 2 km S Siah Kuh, Kavir Protected Region (MMTT 11); steppe and loam hills, NW edge Siah Kuh (LF 50; NS&MG 86). **Ostan 7 (Esfahan)**: Esfahan (DF *et al.* 97); ? near Esfahan (WB 76); Galatappeh (CAS 102505–7; FMNH 141263, 141265–6, 141268, 141270; NS&MG 86); Pir Bakran, near Esfahan (OW 51; NS&MG 86). **Ostan 8 (Fars)**: Abaded [=Abadeh?] (BMNH 1919.5.2.1); Abadeh (BYU 20959; RC *et al.* 66); 7 mi [11.2 km] E Cheh Mosullum, 100 ft [305 m] (FMNH 170934). **Ostan 9**

(Kerman): S Bam, edge Kuh-e Jebal Barez (LF 50; NS&MG 86); Jupar, 20 mi S Kerman (BMNH spec.); basin S Kerman (FMNH 171249); "an der Dachme der Parsen [Leichenstatte der Feueranbeter] E Kerman, 1900 m" (FW 95); Lut-e Zangi Ahmad (FW 38); Rayen, SE Kerman, 8000 ft (BMNH 74.11.23.51). **Ostan 10 (Baluchistan-Sistan)**: Bahu Kalat (BMNH 74.11.23.53–54 [1946.8.25.34–35]; WB 76; GB 85; NS&MG 86; syntypes *Agamura cruralis*); near Bandan (NZ 03); Duz-Ab [=Zahedan?] (NS&MG 86); Eskan, near Kuh-e Bamposht (BMNH 74.11.23.52 [1946.8.25.33; WB 76; GB 85; NS&MG 86; syntype *A. cruralis*); mountains between Gjuische and Zeinelabad (NZ 03); about 28°10'N, 61°11'E, 10 km S Khash on road to Iranshahr, 1450 m (MMTT 857, 859); Perso-Baluch border (NA 06); Shahr Sukhteh, SW Zabol (MMTT 1550, [not seen by me]); Sib (NS&MG 86); Zahedan, 1400 m (JG 66); about 29°28'N, 60°41'E, 32 km W Zahedan on road to Cheshmeh Ziarat, 1900 m (CAS 141064); between Zahedan and Chah Bahar (LF 50; NS&MG 86), Zamran, Rud-e Nahang (WB 76; syntype *A. cruralis*). **Ostan 11 (Khorasan)**: 6 km W Ayubi (USNM 148590–92; RT 71; NS&MG 86); Basiran (NS&MG 86); Torbat-e Heydariyeh (USNM 148585; RT 71; NS&MG 86). **Ostan 12 (Mazanderan)**: Chahar Deh (AS 87; AN 15).

Asaccus elisae (F. Werner, 1895)
Ostan 5 (Kordestan-Kermanshah): Quraitu, at Iran-Iraq border (BMNH 1921.3.30.1–3); Sar-e Pol-e Zahab (NP 96). **Ostan 6 (Khuzestan-Lorestan)**: Khuzestan (RSM 1969-15–23); Abu Karaniyeh [stream flowing into Rud-e Karun] (AN 07; syntypes of *Phyllodactylus eugeniae*); Afrineh, 77 km S Khorramabad on road to Andimeshk, 1100 m (CAS 141111); Dezful (AN 07; syntypes of *P. eugeniae*); between Masjed Soleyman and Batvand (CAS 86339–40, 86352, 86436–43, 86525–9; SA 63); Mazu (GH&YW 69); Sar-i-Gach (CAS 86432; SA 63); Shahbazan, 540 m (KS 55); Shush (RM 57); Susa [=Shush], tea room of Chateau of French Expedition to Susa (MMTT 243–5); Shush, office of Department of Environment (CAS 141124; MMTT 988–90, 996–7). **Ostan 8 (Fars)**: Pol-e Abgineh (CAS 102520–22; FMNH 141306); 5 km SE Pol-e Abgineh (CAS 102520–22; FMNH 141307–9, 141311).

Asaccus griseonotus Dixon and S. Anderson, 1973
Ostan 5 (Kordestan-Kermanshah): 38.5 mi [61.6 km] from Shahabad [direction not recorded] (FMNH 170817–24; JD&SA 73; type locality; NP 96). **Ostan 6 (Khuzestan-Lorestan)**: 110 km SW Khorramabad, close NW junction with road to Mavi (JE 73; type locality for *Phyllodactylus ingae*); Pol-e Tang, 60 km NW Andimeshk (32°51'N, 47°56'E) (J. Moravec, pers. commun.).

Asaccus kermanshahensis Rastegar-Pouyani, 1996
Ostan 5 (Kordestan-Kermanshah): Mianrahan region, 40 km NE Kermanshah, about 1450 m (NR-P 96 [type locality]).

Bunopus crassicauda Nikolsky, 1907
Ostan 1 (Tehran): Maljat-Abad, Irak-Adschemi (AN 07); Qazvin (CAS 140588–603; NS&MG 86); Qom (AN 07; NS&MG 86); 15 mi [24 km] SW Rey (AMNH 99663; Minton 930; SM *et al.* 70); 3–8 km S Tehran (NS&MG 86). **Ostan 7 (Esfahan)**: between Khara and Mohammadabad [many places of this name — probably 32°40'N, 51°55'E] (AN07; NS&MG 86; type locality).

Bunopus tuberculatus Blanford, 1874
Godar-i-Mishun (NZ 03); Leb-e Kal (NMW 17303:1; syntype of *Gymnodactylus gabrielis*; FW 36). **Ostan 1 (Tehran)**: Qom (OW 51). **Ostan 5 (Kordestan-Kermanshah)**: Qasr-e Shirin (NS&MG 86). Ostan 6 (Khuzestan-Lorestan): 21 km N, 1 km W Ahvaz (MMTT 60, 65); 31°16'N, 49°11'E, between Ahvaz and Haft Gel (CAS 86492, 86524, 86536–7; NS&MG 86); Dehloran (UMMZ 130577; MMTT 672; NS&MG 86); Dezful (AN 07); 52–56 km NW Ganaveh on road from Agha Jari-Behbehan road, 70 m (CAS 141136; MMTT 1076–7; NS&MG 86); 4 mi [6.4 km] W Gatvand (GH&YW 69; NS&MG 86); Gol Gir (NS&MG 86); Gurschir, Arabistan (AN 07); 20 km S Haft

Tappeh Ridge (MMTT 247); Mahor Birinji (GH&YW 69; NS&MG 86); Meshrageh, 100 ft [30.5 m], 53 mi [84.8 km] SE Ahvaz (FMNH 170778); Nasrie [=Ahvaz] (AN 07; NS&MG 86); Rud-e Karun (AN 07; NS&MG 86); 12 mi [19.2 km] S Shush (FMNH 141349–50); Shustar (AN 07; NS&MG 86). **Ostan 8 (Fars)**: Ahram and vicinity (CAS 102511–19; FMNH 141278, 141280–2, 141285–6, 141288, 141301–5; NS&MG 86); Bandar-e Lengeh (JG 66; NS&MG 86); about 26°31'N, 54°47'E, 4 km on road 1 km W Bandar-e Lengeh airport, ruined buildings on beach (CAS 141047; MMTT 829–30); 10–13 km S Borazjan on road from Bushire, 40 m (CAS 141146; MMTT 1084; NS&MG 86); Bushire (KS 55); Bushire peninsula (FW 17); Chahak (DF et al. 97); Cheh Mosullum, 100 ft [30.5 m] (FMNH 170779–81); near Ja'inak and Golaki, Tangistan (FW 17). **Ostan 9 (Kerman)**: 5 km E Bandar 'Abbas (USNM 149156–7; RT 71); 13–40 km N Bandar 'Abbas (NS&MG 86); Jask (BMNH 94.8.30.8); Kahnuj, 100 km S Sabzvaran (OW 51; NS&MG 86); Minab (CAS 86590–91; SA 63; NS&MG 86); Rigan, 2500 ft [762 m] (WB 76); Sabzvaran (OW 51); about 27°17'N, 56°29'E, 12 km E Shaqu on road from Bandar 'Abbas to Minab, 0–50 m (MMTT 803–6). **Ostan 10 (Baluchistan-Sistan)**: Bahu Kalat (BMNH 74.11.23.95 [1946.8.22.86] 74.11.23.94 [1946.8.22.85]; ZSI 3458; WB 76; NS&MG 86); "Baluchistan" (MCZ 7128; WB 76); near Bampur, Jaz Murian depression, 600 m (BMNH 74.11.23.85/1946.8.22.84 [lectotype]; MMTT 863; NS&MG 86; type locality); 27°10'N, 60°10'E, 31 km W Bampur (MMTT 889–91; NS&MG 86); Chah Bahar (USNM 153823–32; RT 71; JG 66; NS&MG 86); Chauzdar, Sistan (AN 99; NS&MG 86); Degak (NS&MG 86); Duz-ab [=Zahedan?] (NZ 03; NS&MG 86); Esfandak, 3200 ft [976 m] (WB 76); between Faisabad and Basiran (NS&MG 86); Gumbez-i-Nowar (NS&MG 86); Gurmuck (AN 99; NZ 03); Halvan (NS&MG 86); Iranshahr (NS&MG 86); 11 mi [17.6 km] W Iranshahr (FMNH 141272); Kalagan (NS&MG 86); 6 km N Khash (USNM 148668; RT 71; NS&MG 86); Neizar, Sistan (AN 99; NS&MG 86); Pishin, 500 ft [152 m] (BMNH 1946.8.22.84; ZSI 5277; WB 76; NS&MG 86); near Rud-e Bampur (AN 99); about 31°03'N, 61°38'E, 10 km SW Rud-e Hirmand, abandoned village SE road from Zabol to Dust-e Mohammad Khan, 450 m (CAS 141098; MMTT 935–6); Sabzevar (NS&MG 86); Sar Chah (NS&MG 86); Shurab, Magas (NS&MG 86); Tagab, between Shurab abd Panjsareh (NZ 03); Tscha-i-Bena, Sistan (AN 99; NZ 03); Tscha-i-Gjuische, Sistan (AN 99); about 29°08'N, 61°20'E, 52 km S on Zahedan-Mirjaveh road from point where railroad crosses road, S Zahedan (CAS 141062; MMTT 849); between Zahedan and Chah Bahar (LF 50); Ziarat, Baluchistan [many places of this name in Iran and Pakistan] (NMW 15548; FW 38; NS&MG 86; type locality for *Bunopus biporus*). **Ostan 11 (Khorasan)**: between Chadschi-du-i Tschagi and Gjuische (NZ 03; NS&MG 86); Chakh-i-Bena (NS&MG 86); Chakh-i-Gyuishe (NS&MG 86); 15 km NW Doruneh (USNM 148577; RT 71; NS&MG 86); between Feyzabad and Basiran (AN 97; NS&MG 86); Niazabad (NZ 03; NS&MG 86); between Zirkuch and Tag-i-Dorokh (NZ 03). **Ostan 12 (Mazanderan)**: Na'imabad (AN 07); Damghan (NS&MG 86); Semnan (NS&MG 86). **Persian Gulf**: Jazireh-ye Tanb-e Bozorg (BMNH 1922.7.10.4–5; WB 76).

Carinatogecko aspratilis (S. Anderson, 1973)

Ostan 8 (Fars): 35 km E Gach Saran (USNM 193961–2; type locality); Shiraz (ZMGU 190; see Plate 7B).

Carinatogecko heteropholis (Minton, S. Anderson, and J. Anderson, 1970)

Ostan 5 (Kordestan-Kermanshah): Ilam, 5000 ft [1525 m] (FMNH 170955).

Crossobamon eversmanni (Wiegmann, 1834)

Ostan 10 (Baluchistan-Sistan): Hormak (NS&MG 86); Sistan (AN 15). **Ostan 11 (Khorasan)**: between Ahangeran and Sarakhs (NZ 03); Chouz, Zirkuch (AN 99; 15; NS&MG 86); Feyzabad (NS&MG 86); about 34°39'N, 58°47'E, 35 km N Gonabad or road to Torbat-e Heydariyeh, 850 m (MMTT 1157); Mondechi (AN 97; 15; NS&MG 86); between Mozhnabad and Fandokht (NZ 03; NS&MG 86); Tag-i-Dorokh (NZ 03; NS&MG 86); about 34°47'N, 60°47'E, 4 km E communications station at outskirts of Tayyebat, about 900 m (CAS 141178; MMTT 499; NS&MG 86).

APPENDIX II

Cyrtopodion agamurodes (Nikolsky, 1900)
 Ostan 8 (Fars): 30°05'N, 52°55'E, 10 km by road E Sivand, 1700 m (DF *et al.* 97); **Ostan 9 (Kerman)**: Kerman (OW 51); 27°33'N, 56°24'E, E Kuhha-ye Genu (CAS 86370; SA 63). **Ostan 10 (Baluchistan-Sistan)**: Bazman (NZ 03); Duz-Abad [=Zahedan?] (AN 99; NZ 03; syntype); Kuh-i-Boz (NZ 03); Morghak (NZ 03); Neizar (AN 99; NS&MG 86; syntype); Panjsareh (AN 99; NZ 03; NS&MG 86; lectotype); 26°09'N, 61°27'E, Sarbaz River at Huvar, 320–380 m (CAS 141075; MMTT 875); Shar-i-Zagedun (NZ 03); Zagan (NZ 03).

Cyrtopodion brevipes (Blanford, 1874)
 Ostan 8 (Fars): Bushire (JM 84; identification?); Tangistan (JM 84; identification?). **Ostan 10 (Baluchistan-Sistan)**: Aptan [=Abtar], near Bampur (WB 76; NA 13; type locality); Iranshahr, Iran Department of Environment office (MMTT 1651, not seen by me).

Cyrtopodion caspium (Eichwald, 1831)
 "Caspian province" (FdF 65). **Ostan 1 (Tehran)**: Qom, Kavir (MMTT 143). **Ostan 5 (Kordestan-Kermanshah)**: ?Kermanshah (NS&MG 86). **Ostan 8 (Fars)**: ?Shiraz (NS&MG 86). **Ostan 10 (Baluchistan- Sistan)**: Neizar, Sistan (NS&MG 86); Sistan (AN 15). **Ostan 11 (Khorasan)**: between Ahangeran and Sarakhs (NZ 03); Angun (NZ 03); Bekhars (NZ 03; AN 15; NS&MG 86); Birjand (AN 97; NS&MG 86); Bojnurd, 1060 m (RC 91); Boz-Chouz-Pain (AN 97; 15; NS&MG 86); Chouz, Zirkuch (AN 99; 15; NS&MG 86); Kerat (NZ 03; NS&MG 86); Khashtadan (NZ 03; NS&MG 86); Khouz-i-Musafir [?=Howz-e Musa] (NZ 03); Kariz (NZ 03; NS&MG 86); Lenger (NZ 03); Marandiz (AN97; NS&MG 86); Mashhad (USNM 154484; NZ 03; RT 71; NS&MG 86); 40 km N Mashhad (NS&MG 86); Mozhnabad (NZ 03); Sarakhs (NS&MG 86); Soltanabad (NS&MG 86); Tajan River [Harirud] (LL 18); Torbat-e Jam (NZ 03). **Ostan 12 (Mazanderan)**: 'Abbasabad (AMNH 61776); Ak-kala [Pahlavi Dezh], near Astrabad [Gorgan] (GB 85; 91; AS 87; AN 15; NS&MG 86); Baba Kuh, abandoned village, Turan Biosphere Reserve, 1275 m (RT 79); Delbar, Turan Biosphere Reserve, 1300 m (RT 79); Gorgan (AN 07; 15; MHNHP 1966.14; JG 66; NS&MG 86); Na'imabad (AN 07; 15); 25 m [40 km] N Pahlavi Dezh (CAS 102508; FMNH 141275–6); 3.5 mi [5.6 km] NW Semnan (FMNH 141277); Shahrud (MMTT 507–14; NS&MG 86).

Cyrtopodion gastrophole (F. Werner, 1917)
 Ostan 8 (Fars): Fars Province (ZMUG 74–75; CAS 100472–3; FW 17; type locality); Chahak (DF *et al.* 97). 29°28'N, 51°21'E, 5 km N Dalaki on road to Shiraz at base of foothills, 100 m (MMTT 1049).

Cyrtopodion heterocercum (Blanford, 1874)
 Ostan 5 (Kordestan-Kermanshah): Hamadan (NMW 7286; WB 74; 76; GB 85; OW 51; SM *et al.* 70; type locality). **Ostan 8 (Fars)**: Persepolis (NS&MG 86; 96).

Cyrtopodion kachhense (Stoliczka, 1872)
 Ostan 8 (Fars): Bushire (NA 13).

Cyrtopodion kirmanense (Nikolsky, 1900)
 Ostan 8 (Fars): Persepolis [=Takht-e Jamshid] (FMNH 21007; KS 39, identification?). **Ostan 10 (Baluchistan-Sistan)**: Bazman (NZ 03); Bid (NZ 03); between Deh Pabid and Ehnarik (NZ 03); Garne (NZ 03); "eastern Kerman" (AN 99; syntypes); Kuh-e Taftan (AN 99; NZ 03; syntypes; lectotype); Kukh-i-Boz-ab (NZ 03); Mazel-ab (NZ 03); Morghak (NZ 03); Rud-i-Tamin (NZ 03).

Cyrtopodion longipes longipes (Nikolsky, 1896)
 Ostan 10 (Baluchistan-Sistan): Bandan (NZ 03); Bazman (NZ 03); Mazel-ab, Sargad Region (NZ 03); Morghak, Sargad Region (NZ 03); Neizar in Sistan (AN 99; NS&MG 86); Tscha-i-Gjuische (AN

99; NZ 03; NS&MG 86); Zagan, Sargad Region (NZ 03); 10 km SE Rud-e Hirmand, abandoned village SE road from Zabol to Dust-e Mohammed Khan (AL&SA 84; CAS 141095, MMTT 931–932). **Ostan 11 (Khorasan)**: Aliabad (NS&MG 86); Baaza River, Zirkuch (AN 99; NZ 03; NS&MG 86); Birjand (NZ 03; NS&MG 86); Chah-e Ziran (AN 99; NZ 03; NS&MG 86); Khouz-i-Musafir (NZ 03); Mozhnabad (NZ 03); Neh [Nehbandan] (AN 97, type locality; AL&SA 84; NS&MG 86; ZIL 8809[3], syntypes); Pegu Rabat (AL&SA 84; ZIL 9787[4]); Tag-i-Dorokh, between Duruh and Chah-e Ziran (NZ 03).

Cyrtopodion russowii zarudnyi (Nikolsky, 1900)

Ostan 10 (Baluchistan-Sistan): Neizar, Sistan (AN 99; NS&MG 86; type locality for *Gymnodactylus zarudnyi*); Shar-i-Zagedun (NZ 03). **Ostan 11 (Khorasan)**: Bekhars (NS&MG 86).

Cyrtopodion sagittifer (Nikolsky, 1900)

Ostan 10 (Baluchistan-Sistan): Bampur (AN 99; NZ 03; NS&MG 86; lectotype); about 27°10'N, 60°10'E, 30 km W Bampur, Jaz Murian depression, 500–540 m (CAS 141072; MMTT 870; NS&MG 86); about 27°10'N 60°09'E, 31 km W Bampur, Jaz Murian depression (CAS 141083; MMTT 892–3); Farra, "eastern Kerman" [32 km from Bampur] (AN 99; NS&MG 86; syntype).

Cyrtopodion scabrum (Heyden, 1827)

Ostan 1 (Tehran): Djavadiyeh, Tehran (MMTT 562, 580); Karaj (MMTT 579); Qom, Kavir (MMTT 143–5); Shahrara (MMTT 92); Tehran (CAS 140546–9; MMTT 196–7, 302, 501, 566, 1378, 1405, 1546–8; NS&MG 86). **Ostan 6 (Khuzestan-Lorestan)**: Abadan (MMTT 676–8); Agha Jari (CAS 86481–3; SA 63; NS&MG 86); Chogha Pahn, 2 km N Jerud Dam, 80 m (MMTT 246); 25 km E Haft Tappeh (MMTT 71); Kurait (CAS 86558; SA 63); Karun Depot, Ahvaz (CAS 96150); Mansu Abad, about 8 km NNE Shahbazan, 1200 m (KS 55); Masjed Soleyman (CAS 96158–9, 86431, 86513, 86626; SA 63; NS&MG 86); near Masjed Soleyman airfield (CAS 86532–3; SA 63); Meshrageh, 53 mi [84.8 km] SE Ahvaz, 200 ft [61 m] (FMNH 170801); Ramhormoz, 200 ft [61 m] (FMNH 170802–3); Shadegan (RM 57); Shalgahi (GH&YW 69); Shush, Department of Environment office (CAS 141121; MMTT 991–5, 998–1000; NS&MG 86). **Ostan 7 (Esfahan)**: Jafar Abad (DF *et al.* 97); Yazd (NS&MG 86). **Ostan 8 (Fars)**: Ahram and vicinity (CAS 102509; FMNH 141287, 141292–3; NS&MG 86); Bandar-e Lengeh (MNHNP 1966.11–12; JG 66; NS&MG 86); about 26°31'N, 54°47'E, 4 km on road 1 km W Bandar-e Lengeh airport, ruined buildings on beach (CAS 141043–4; MMTT 817–19; NS&MG 86); Bushire (JM 84; NA 13; KS 55); Cheh Mosullum (FMNH 170804–8); Kazerun (CAS 140545; NS&MG 86); Shiraz (CAS 140544, 140550; BMNH 46.6.15.81; WB 76; GB 85; NS&MG 86); Tangistan (JM 84). **Ostan 9 (Kerman)**: Bandar 'Abbas (CAS 141026; MMTT 973–4; MNHNP 1966.13; FW 95; JG 66; NS&MG 86). **Ostan 10 (Baluchistan-Sistan)**: Chah Bahar (USNM 148647; JM 84; RT 71; NS&MG 86); Iranshahr (MMTT 1655–8, not seen by me); Zabol (MMTT 1551, not seen by me); Zahedan (RM 56). **Ostan 11 (Khorasan)**: Birjand (NS&MG 86); Boshruyeh (NS&MG 86). **Persian Gulf**: Hormoz Village, Jazireh-ye Hormoz (MMTT 228–9; RT 74; NS&MG 86).

Cyrtopodion spinicauda (Strauch, 1887)

Ostan 12 (Mazanderan): Shahrud (AS 87; AN 15; NS&MG 86; type locality).

Cyrtopodion turcmenicum (Szczerbak, 1978)

Ostan 12 (Mazanderan): Gorgan (NMW 19724; not seen by me).

Hemidactylus flaviviridis Rüppell, 1840

Ostan 6 (Khuzestan-Lorestan): Abadan (MMTT 674–5); Bandar-e Shapur, sea level (FMNH 170825); Khorramshahr (KS 55); Meshrageh, 53 mi [84.4 km] SE Ahvaz, 200 ft (FMNH 170825). **Ostan 8 (Fars)**: Charbagh, near Bushire (JM 84); Tangistan (JM 84). **Ostan 9 (Kerman)**: Bandar

'Abbas, sea level (MMTT 975; FW 29; KS 55); Jask (BMNH 94.8.30.11, 94.11.3.3; JA 96); Minab, 100 ft (FMNH 170827–34); 2 mi [3.2 km] SW Minab, 100 ft (FMNH 170826). **Ostan 10 (Baluchistan-Sistan)**: Chah Bahar (CAS 141078; MMTT 976; JM 84); between Zahedan and Chah Bahar (LF 50).

Hemidactylus persicus J. Anderson, 1872

Ostan 6 (Khuzestan-Lorestan): Dezful (MMTT 83); vicinity of Dezful (AN 07); between Masjed Soleyman and Haft Gel (CAS 86377, 86421; SA 63); between Masjed Soleyman and Naftak (CAS 86499–500; SA 63); Sar-i-Gach (CAS 86414–15, 86424, 86454–5; SA 63). **Ostan 8 (Fars)**: Abshar (DF *et al.* 97); Chahbagh (FW 17 sight record); Tang-e tschekun (FW 17). **Ostan 9 (Kerman)**: Minab, 100 ft (FMNH 170809–10). **Ostan 10 (Baluchistan-Sistan)**: Iranshahr, 800 m (RM 56).

Hemidactylus turcicus (Linnaeus, 1758)

Ostan 1 (Tehran): Qazvin (JG 57). **Ostan 6 (Khuzestan-Lorestan)**: Ahvaz (KS 55); 9 km ENE Shahbazan (KS 55). **Ostan 8 (Fars)**: about 26°31′N, 54°47′E, 4 km on road 1 km W Bandar-e Lengeh airport, ruined buildings on beach (CAS 141045–6; MMTT 820–28). **Ostan 9 (Kerman)**: Bandar 'Abbas (CAS 141034); Jask (BMNH 94.8.30.9–10, 94.11.13.1–2); Rig Mati, 80 km E Kahnuj (OW 51). **Ostan 10 (Baluchistan-Sistan)**: Chah Bahar (USNM 148452–5; CAS 141079; JG 66; RT 71).

Pristurus rupestris Blanford, 1874

Ostan 8 (Fars): about 26°32′N, 54°43′E, 9 km W Bandar-e Lengeh airport, 0–50 m (CAS 141050; MMTT 833–6); Bushire (FMNH 69296; JM 84; FW 17; KS 52; 55; type locality for *Pristurus rupestris iranicus*); NW Bushire (JM 84); Cheh Mossullum, 50–100 ft [15–30 m] (FMNH 170889–910); Tangistan (JM 84). **Ostan 9 (Kerman)**: Jask (JM 84); Minab, 100 ft [30.5 m] (CAS 141033; FMNH 170919–21); 2 mi [3.2 km] SW Minab, 100 ft [30.5 m] (FMNH 170911–18); about 27°23′N, 56°16′E, 20 km N Shaqu on Bandar 'Abbas-Kerman road and 8 km W on track to foothills of Kuhha-ye Genu, Kuhha-ye Genu Protected Region, 300 m (CAS 141035). **Ostan 10 (Baluchistan-Sistan)**: Chah Bahar (USNM 148646; MMTT 1630, several specimens, not seen by me; RT 71). **Persian Gulf**: Jazireh-ye Khark (BMNH 1919.8.11.34; WB 74; 76; GB 85); "Muscat and Island of Karrack, near Busheer" (BMNH 74.11.23.89–90; probably part of syntypic series); about 26°57′N, 56°14′E, sandstone outcrops above town of Qeshm, Jazireh-ye Qeshm (SCA field no. 121[5], specimens lost).

Ptyodactylus sp.

Ostan 6 (Khuzestan-Lorestan): 12 km NNE Shahbazan, km 347 on railroad, near Ab-i-Khornos, 600 m (KS 55; specimens lost).

Rhinogecko misonnei de Witte, 1973

Ostan 9 (Kerman): 23 km S Bam on road to Zahedan (MMTT 1588, not seen by me); Dasht-e Lut, 30°13′N, 58°47′E (GdW 73; GdW 80; NS&MG 86; type locality). **Ostan 11 (Khorasan)**: vicinity of Kuh-e Bakhtu, 31°40′N, 58°21′E, Dasht-e Lut (GdW 73; GdW 80; NS&MG 86).

Stenodactylus affinis (Murray, 1884)

Ostan 6 (Khuzestan-Lorestan): 21 km N and 1 km W Ahvaz (MMTT 61; EA 80); 52–56 km NW Ganaveh on road S from Agha Jari-Behbehan road, 70 m (CAS 141135; MMTT 1075; EA 80). **Ostan 8 (Fars)**: Tangistan (BMNH 84.7.23.33/1946.8.23.33, 87.9.22.2/1946.8.23.60; JM 84; GB 85; EA 80; type locality); 2–3 km SSE Bushire (KS 55).

Stenodactylus doriae (Blanford, 1874)

Ostan 6 (Khuzestan-Lorestan): about 31°16′N, 48°10′E, 1 km E Kupal, dunes near Ahvaz-Behbehan road, 25–50 m (CAS 141130–33; MMTT 1020–31; EA 80). **Ostan 8 (Fars)**: about 28°53′N,

51°01′E, Alchangi, 32 km ESE Bushire (CAS 141137–41; MMTT 1036–47; EA 80); near Ja'inak, Tangistan (FW 17); Tangistan (BMNH 87.9.23.3; JM 84; EA 80). **Ostan 9 (Kerman)**: Bandar 'Abbas (FdF 65; WB 74; JG 57); one day's march N Bandar 'Abbas on road to Kerman (WB 76; EA 80; type locality); about 27°17′N, 56°29′E, 12 km E Shaqu on road from Bandar 'Abbas to Minab, 0–50 m (CAS 141038–41; MMTT 798–802; 810–11; EA 80).

Teratoscincus bedriagai Nikolsky, 1900

Ostan 1 (Tehran): along river at entrance to Kavir Protected Region (MMTT 561). **Ostan 10 (Baluchistan-Sistan)**: Chadschi-du-i Tschagi (NZ 03; NS&MG 86; type locality); Chah-i-Gjuische (NZ 03; NS&MG 86); about 31°03′N, 61°38′E, 10 km SW Rud-e Hirmand, abandoned village SE road from Zabol to Dust-e Mohammad Khan, 450 m (CAS 141090–92; MMTT 920–23); Sistan (AN 99; syntypes); 15 mi [24 km] SW Zabol (FMNH 14175); about 30°51′N, 61°39′E, 7 km SW Zahak, between Zahak and Khamak, 500 m (CAS 141104; MMTT 948–9). **Ostan 11 (Khorasan)**: Bamrud, Zirkuch (NZ 03; NS&MG 86); 10 km E Doruneh (USNM 148569–70; RT 71; NS&MG 86); Niazabad, Khaf (NZ 03; NS&MG 86); Zirkuch (AN 99; syntypes). **Ostan 12 (Mazanderan)**: between Khaneh Khvadi and Delbar (RT 79).

Teratocscincus microlepis Nikolsky, 1900

Ostan 9 (Kerman): Dasht-e Lut, near Kerman (NS&MG 86). **Ostan 10 (Baluchistan-Sistan)**: Bampur (NS&MG 86); Duz-Ab [?Zahedan] (AN 99; NZ 03; NS&MG 86; type locality); about 31°03′N, 61°38′E, 10 km SW Rud-e Hirmand, abandoned village SE road from Zabol to Dust-e Mohammad Khan, 450 m (CAS 141096; MMTT 919); Tasuki, 120 km from Zabol on road to Zahedan (JG 57).

Teratoscincus scincus keyserlingii Strauch, 1863

Ostan 1 (Tehran): steppe between Argavani and Marinjab (LF 50; NS&MG 86); 2 km S Chah Qar Qareh, Kavir Protected Region (MMTT 51). **Ostan 9 (Kerman)**: about 28°59′N, 58°57′E, 26 km W Mil-e Naderi, near Fahraj on road between Bam and Zahedan, 650 m (CAS 141105; MMTT 950–59); Minab (BMNH 1951.1.6.43–44; NS&MG 86); 2 mi [3.2 km] SW Minab, 100 ft [30.5 m] (FMNH 170815); 4 mi [6.4 km] SW Minab, 100 ft [30.5 m] (FMNH 170816); about 27°17′N, 56°29′E, 12 km E Shaqu on road from Bandar 'Abbas to Minab, 0–50 m (CAS 141042; MMTT 807–9, 812); about 31°13′N, 54°55′E, 98 km S Yazd on road between Yazd and Rafsanjan, 1600 m (CAS 141107; MMTT 961). **Ostan 10 (Baluchistan-Sistan)**: Bambur [?=Bampur], SE Persia (BMNH 1951.1.6.41–42); about 27°10′N, 60°10′E, 30 km W Bampur, Jaz Murian depression, 500–540 m (CAS 141070; MMTT 864–8); about 27°10′N, 60°09′E, 31 km W Bampur, Jaz Murian depression, 470 m (CAS 141081); 11 mi [17.6 km] W Iranshahr (FMNH 141271–2); Perso-Baluch border (NA 06); Rud-e Bampur, near Makhmudabad, 12 km SW Iranshahr (RM 56; NS&MG 86); about 31°03′N, 61°38′E, 10 km SW Rud-e Hirmand, abandoned village SE road from Zabol to Dust-e Mohammad Khan, 450 m (CAS 141093–4; MMTT 924–30); Tasuki, 120 km from Zabol on road to Zahedan (JG 57); about 30°51′N, 61°39′E, 7 km SW Zahak, between Zahak and Khamak, 500 m (CAS 141103); Zia-i-Bolokh (NS&MG 86) Zia-i-Lagun, 30°10′N, 60°50′E (NS&MG 86). **Ostan 11 (Khorasan)**: Ahangeran, Zirkuch (AN 99; NZ 03; NS&MG 86); about 34°39′N, 58°47′E, 35 km N Gonabad on road to Torbat-e Heydariyeh, 850 m (CAS 141197; MMTT 1155–6); Rum (AN 97; 15; type locality for *Teratoscincus zarudnyi*); Sar Chah (AS 63; 87; WB 76; AN 15; NS&MG 86; type locality for *T. keyserlingii*). **Ostan 12 (Mazanderan)**: vicinity of Baba Kuh (RT 79); Chahar Deh (AS 87; AN 15; NS&MG 86).

Tropiocolotes helenae helenae (Nikolsky, 1907)

Ostan 6 (Khuzestan-Lorestan): Aguljashker (AN 07; NS&MG 86); Alchorschir (AN 07; NS&MG 86; type locality); Bid Zard (AN 07; NS&MG 86); Choqa Zanbil (DF *et al.* 97); 35 km E Gach Saran (USNM 153693–703, 153705–10; CAS 120795–6; SM *et al.* 70; RT 71; NS&MG 86);

Isfagan (AN 07; NS&MG 86); Masjed Soleyman (NS&MG 86); 16 km S Masjed Soleyman (USNM 153731; SM *et al.* 70; RT 71; NS&MG 86). **Ostan 8 (Fars)**: Mehkuh, 1400 m, 80 km S Shiraz (JJS&JFS 72; NS&MG 86); Mian Kotal, 80 km W Shiraz, 1200 m (JJS&JFS 72; NS&MG 86).

Tropiocolotes helenae fasciatus **(J. J. Schmidtler and J. F. Schmidtler, 1972)**
   Ostan 5 (Kordestan-Kermanshah): Sorkeh Dizeh, 1500 m, 125 km W Kermanshah, on road to Baghdad (JJS&JFS 72; NS&MG 86). **Ostan 6 (Khuzestan-Lorestan)**: 25 km W Khorramabad, 100 m, on road to Ahvaz (JJS&JFS 72; NS&MG 86).

Tropiocolotes latifi **(Leviton and S. Anderson, 1972)**
   Ostan 7 (Esfahan): Qamishlu (DF *et al.* 97, sight record, photo). **Ostan 8 (Fars)**: Shiraz Mountains, around salt lake, about 29°36'N, 52°33'E, 1495 m (HK&EV 96); 30°05'N, 52°55'E, 10 km by road E Sivand, 1700 m (DF *et al.* 97). **Ostan 9 (Kerman)**: 5 km S Aliabad, 31°36'4"N 53°49'52"E, 2560 m (JM&MC 94); 6 km N Deh Shir, 31°30'52"N 53°45'43"E, 2040 m (JM&MC 94); Kerman (CAS 134365; AL&SA 72; NS&MG 86; type locality). Mazraeh, 31°32'34"N, 53°46'55"E, 2160 m (JM&MC 94).

Tropiocolotes persicus persicus **(Nikolsky, 1903)**
   Ostan 10 (Baluchistan-Sistan): Dehak (AN 03; NS&MG 86; type locality); 100 km N Iranshahr (MNHNP 1966.17; JG 66; NS&MG 86); 20 km SW Pip [identification?] (RM 56; NS&MG 86).

Tropiocolotes persicus bakhtiari **(Minton, S. Anderson, and J. Anderson, 1970)**
   Ostan 6 (Khuzestan-Lorestan): between Masjed Soleyman and Sar-i-Gach (CAS 86408; SA 51; 63; SM *et al.* 70; NS&MG 86; type locality).

Tropiocolotes cf. *steudneri* **(W. Peters, 1869)**
   Ostan 8 (Fars): Bandar-e Lengeh (MNHNP 1966.18; JG 66; NS&MG 86); about 26°32'N, 54°43'E, 9 km W Bandar-e Lengeh airport, 0–50 m (MMTT 1048).

Acanthodactylus blanfordi **Boulenger, 1918**
   Ostan 9 (Kerman): near Bam, 300 ft [915 m] (WB 76; GB 87; 18; 21; MS 35; syntypes; AS 82); 27°12'N, 56°21'E, dunes on road between Bandar 'Abbas and Kerman (CAS 86588–9; SA 63; AS 82); Jask (WB 76; GB 21; MS 35); Kerman (AS 82); about 28°59'N, 58°57'E, 26 km W Mil-e Naderi, near Fahraj on road between Bam and Zahedan, 650 m (CAS 141106; MMTT 960; AS 82); Minab (CAS 86592; SA 63; AS 82); near Rigan, Narmashir, 2500 ft (WB 76); Rig Mati, E Kahnuj, Makran (OW 51); Shaqu (CAS 86600–9; SA 63; AS 82). **Ostan 10 (Baluchistan-Sistan)**: Bahu Kalat (WB 76); Bampur, 2000 ft (MMTT 1652–4 [not seen by me]; WB 76; AN99; NZ 03); SW Bampur (RM 56); 6 km E Bampur (RM 56); 7 km W Bampur (MMTT 1638–49, not seen by me); about 27°10'N, 60°10'E, 30 km W Bampur, Jaz Murian depression, 500–540 m (CAS 141073; MMTT 871–3; AS 82); about 27°10'N, 60°09'E, 31 km W Bampur, 470 m (CAS 141084; MMTT 894; AS 82); Bazman (MMTT 1589–90; AN 99); 40 km SE Bazman on road to Iranshahr (MMTT 1595–6, not seen by me); Chah Bahar (USNM 148656–8; MNHNP 1966.39–40; MMTT 1622–3, 1625–33; RM 56; JG 66; RT 71; AS 82); between ruins of Hauzdar and Shile canal (NZ 03); Hormak (AN 99; NZ 03); 11 mi [17.6 km] W Iranshahr (CAS 102523–30; FMNH 141454–9, 141461, 141465, 141467, 141470, 141473–6; AS 82); NW Iranshahr (RM 56); SW Iranshahr (RM 56); Kach (FW 38); Khash, 1400 m (MMTT 704, 706; JG 66); 6 km N Khash (USNM 148673–5; RT 71; AS 82); Suran, 60 km W Saravan (MMTT 701–3); Tagab (NZ 03); about 30°16'N, 61°01'E, on road to Zabol, 12–18 km NE junction with Zahedan-Mashhad road, 500 m (MMTT 904); between Zahedan and Chah Bahar (LF 50); Zagan, Sargad (AN 99).

Acanthodactylus boskianus (Daudin, 1802)

Ostan 5 (Kermanshah Province): 34°17'N, 47°24'E, 2 km W Harsin (NR-P 1999).

Acanthodactylus grandis Boulenger, 1909

Ostan 6 (Khuzestan-Lorestan): 12 mi [19.2 km] S Shush (CAS 102535–6; FMNH 141354, 141356–7); S Shush (FMNH 171252; AS 82). **Ostan 8 (Fars)**: Ahram (CAS 102531–4; FMNH 141479–83, 141485, 141487, 141489–90, 141443; AS 82); about 28°53'N, 51°02'E, near Alchangi, 33 km NE Bushire, 0–50 m (CAS 141143; MMTT 1063; AS 82).

Acanthodactylus micropholis Blanford, 1874

Ostan 9 (Kerman): 27°01'N, 55°43'E, 109 km W Bandar 'Abbas-Kerman road on road to Bandar-e Lengeh, 0–50 m (MMTT 813). **Ostan 10 (Baluchistan-Sistan)**: Bampur, 2000 ft [610 m] (WB 76; GB 21; MS 35); 7 km W Bampur (MMTT 1837, not seen by me); Bampur Dam (MMTT 1597, not seen by me); Bazman (AN 99); Bent (MMTT 717); Davar Panah, 4000 ft (WB 76); Duz-Abad in Sargad (AN 99; NZ 03); Garne (NZ 03); Hormak (AN 99; NZ 03); Kalagan, 3500 ft (WB 76); Khash (MMTT 705); Magas [Zaboli], 4500 ft (WB 76; GB 87; 21; MS 35; type locality); Mian Bazar (NZ 03); Murgak (NZ 03); Negur, between Rask and Chah Bahar (MMTT 1613, not seen by me); Podagi (NZ 03); Qal'eh-ye Bid (NZ 03); near Rigan, 2500 ft (WB 76; GB 87; 21; MS 35; AS 82); Rud-e Bampur (AN 99); Siah Kuh (AN 99); Shurab (NZ 03); between Zahedan and Chah Bahar (LF 50); Zamran, 2000 ft (WB 76). **Persian Gulf**: Qeshm Island (CAS 96151–2; AS 82).

Acanthodactylus nilsoni Rastegar-Pouyani, 1998

Ostan 5 (Kermanshah Province): 34°30'N, 45°33'E, 5 km S Qasr-e Shirin, about 7 km from Iran-Iraq border, 285 m (NR-P&GN 98; type locality).

Acanthodactylus schmidti Haas, 1957

Ostan 6 (Khuzestan-Lorestan): Al Baji, near Ahvaz (MMTT 633–6); 21 km N, 1 km W Ahvaz (MMTT 63–4); 31°16'N, 49°11'E, dunes on road between Ahvaz and Haft Gel (CAS 86488–91, 86515–20, 86541–55; SA 63; AS 82); about 31°16'N, 49°10'E, 1 km E Kupal, dunes near Ahvaz-Behbehan road, 25–50 m (CAS 141125–9; MMTT 1006–18; AS 82). **Ostan 8 (Fars)**: between Abu Tavil and 'Isvand on road from Tschahkata to Borazjan (FW 17).

Eremias acutirostris (Boulenger, 1887)

Ostan 10 (Baluchistan-Sistan): about 31°03'N, 61°38'E, 10 km SW Rud-e Hirmand, abandoned village SE road from Zabol to Dust-e Mohammad Khan, 450 m (CAS 141099; MMTT 937, 940–1; SA 78).

Eremias andersoni Darevsky and Szczerbak, 1978

Ostan 1 (Tehran): 34°30'N, 52°40'E, Dasht-e Kavir, 40–45 km E Daryacheh-ye Namak (ID & NS 78; type locality; GN&CA 81).

Eremias arguta (Pallas, 1773)

Salian (JB 79). **Ostan 3 (East Azarbaijan)**: western slope Bogrovdag [Kuhha-ye Talysh] (NS 74); 45 km SE Tabriz (RC 91). **Ostan 11 (Khorasan)**: Kuhistan (FW 36). **Ostan 12 (Mazanderan)**: Shahrud (JB 79).

Eremias fasciata Blanford, 1874

Kohak, Perso-Baluch Frontier [Pakistan?] (BMNH 1917.3.6.27). **Ostan 9 (Kerman)**: Hamun-e Jaz Murian (NZ 97; NS 74); Kerman (AN 15); W Rigan, 2500 ft [763 m] (WB 76); near Sa'idabad, 5500 ft [1678 m] (BMNH 74.11.23.35–37/1946.8.7.57–59; WB 76; type locality; GB 87; 21; MS 35;

NS 74); Sir-i-Tam, SW Kerman (BMNH 1917.3.6.25–26/1946.8.7.34–35; GB 21). **Ostan 10 (Baluchistan- Sistan)**: Nasirabad (AN 99; LL 28; NS 74); Neizar (AN 99; 15; LL 28); about 31°03'N, 61°38'E, 10 km SW Rud-e Hirmand, abandoned village SE road from Zabol to Dust-e Mohammad Khan, 450 m (CAS 141100; MMTT 938–9); Sistan (AN 97; 15); Zaboli, 4500 ft [1373 m] (WB 76); about 29°23'N, 60°49'E, 13 km S Zahedan on road to Khash, 1500 m (CAS 141067; MMTT 854–5). **Ostan 11 (Khorasan)**: between Ahangeran and Charakhs (NZ 03); Baaza, Zirkuch (AN 99; NS 74); Baaza River, Zirkuch, SE Khorasan (LL 28; NS 74); Bala-Khaf (NZ 97; NS 74); Birjand (NZ 97; NS 74); Chah-e Ziran (AN 99; LL 28; NS 74); between Feyzabad and Nusi, eastern Khorasan (LL 28); Katar-bena (NZ 97; NS 74); Khwar (FW 36); Kuh-e Baqeran (NZ 97); between Mozhnabad and Fandokht (NZ 03); Mahmu'i (NZ 97); Mill-Ajaz (AN 97; 15); Parwand, 1125 m (FW 36); Saman Shah (NZ 97); Saqi (AN 97; 15; LL 28); between Sar Chah and Meigun or Meitun, via Basiran and Rum (NZ 97; NS 74); Tag-i-Dorokh, between Dorokh and Chah-e Ziran (NZ 03); Zul-Penai (NZ 97). **Ostan 12 (Mazanderan)**: about 36°09'N, 56°08'E, 34 km S 'Abbasabad, Turan Protected Region, 870 m (CAS 141215; MMTT 1227); about 36°09'N, 56°03'E, 42 km by road SW 'Abbasabad, Turan Protected Region, 920 m (CAS 141218; MMTT 1229–30).

Eremias grammica (Lichtenstein, 1823)

Ostan 11 (Khorasan): between Ahangeran and Charakhs (NZ 03; NS 74); Baaza (NZ 03); Bamrud, Zirkuch (LL 28); about 34°39'N, 58°47'E, 35 km N Gonabad on road to Torbat-e Heydariyeh, 850 m (CAS 141198; MMTT 1120–25); about 34°41'N, 58°48'E, 39 km N Gonabad on road to Torbat-e Heydariyeh, 850 m (CAS 141202); Hadji-i-du Tschahi, Nehbandan (LL 28; NS 74); Tscharachs, 28 km from Ahangeran (AN 99; type locality for *Scapteira persica*; GB 21; LL 28).

Eremias intermedia (Strauch, 1876)

Ostan 11 (Khorasan): River Tajan [=Harirud] at Perso-Afghan-Turkmen border (LL 18); Sarakhs (MNHNP 1957.38–43; JG 57).

Eremias lalezharica Moravec, 1994

Ostan 9 (Kerman): Laleh Zar, 29°31'N, 56°51'E, N foot of Kuh-e Laleh Zar, 2800–3100 m elevation (JM 94; type locality).

Eremias lineolata (Nikolsky, 1896)

Ostan 11 (Khorasan): between Ahangeran and Charakhs (NZ 03); Baaza River, Zirkuch (AN 99; LL 28); mouth of Birjand valley (NZ 97); Chouz [or Houz], SE Khorasan, Zirkuch (AN 99; LL 28); 15 km NW Doruneh (USNM 148581; RT 71); Feysabad (BMNH 99.4.20.5/1946.8.6.6); between Feysabad and Nusi (AN 97; type locality; GB 21; LL 28; NS 74); about 34°39'N, 58°47'E, 35 km N Gonabad on road to Torbat-e Heydariyeh, 850 m (CAS 141199; MMTT 1126–7); about 34°41'N, 58°48'E, 39 km N Gonabad on road to Torbat-e Heydariyeh, 850 m (CAS 141203; MMTT 1128); environs of Mondechi (NZ 97); between Mozhnabad and Fandokht (NZ 03); near Nusi (NZ 97); Sar Chah (NZ 97); Sekhlabad kevir (NZ 97); about 34°47'N, 60°47'E, 4 km E communications station at outskirts of Tayyebat, about 900 m (CAS 141179; MMTT 1101).

Eremias nigrocellata Nikolsky, 1896

Ostan 10 (Baluchistan-Sistan): Khamun [Daryacheh-ye Sistan] (NZ 97; NS 74). **Ostan 11 (Khorasan)**: Abas (NZ 03); Bamrud (NZ 03); Birjand (LM 29); Chah-e Ziran (NZ 03); Charakhs (NZ 03); Chouz [or Houz], Zirkuch (AN 99; LL 28); Esfideh (USNM 148624–30; RT 71); 5 km NE Esfideh (USNM 148625–6; RT 71); between Feyzabad and Mondechi (AN 96; GB 21; NS 74); 15 km NE Jangal (USNM 148605–8; RT 71); Mahneh (MNHP 1957.95–106; JG 57); 10 km S Mahneh (FMNH spec.); Mohammadabad (NZ 97; NS 74); Mondechi (BMNH 99.4.20.4/1946.8.7.44); Mozhnabad (NZ 03); 30 km E Sabzavar (USNM 158447–8; RT 71); Tabas (AN 99; 15; NZ 03; LL 28; NS 74); between

Torbat-e Heydariyeh and Nusi (NZ 97; NS 74). **Ostan 12 (Mazanderan)**: Astrabad [Gorgan region] (GB 21); Damghan (AN 07; 15; NS 74); Na'imabad, near Damghan (LL 28); Shahrud (KK 78; LL 28).

Eremias nigrolateralis Rastegar-Pouyani and Nilson, 1998

Ostan 8 (Fars): 30°52′N, 53°09′E, 150 km NE Shiraz, 1800 m (NR-P&GN 98; type locality).

Eremias persica Blanford, 1875

Baschm-Sorshe, 2500 m (FW 38); Chastadan (AN 15); between Kerman and Shiraz (BMNH 74.11.23.32/1946.8.17.21–23; WB 76; GB 87); near Senghi Caravanseri, about 1200 m (FW 95). **Ostan 1 (Tehran)**: Arak (BYU 209070–1; RC *et al.* 66); 50 km E Arak (OW 51); between Chashmeh Safid Ab and Shah 'Abbas, Kavir Protected Region (GN&CA 81); W Damavand village on road to Tehran (FW 03); 35 km S Delijan (USNM 153562; RT 71); 70 km N Esfahan (OW 51); 138 km N Esfahan (OW 51); 150 km N Esfahan (USNM 153566–78; RT 71); Gavrabad (AN 07; 15; LL 28); Maljat-abad in Irak-Ajemi (AN 07; 15; LL 28); plain E Qazvin, 1400 m (FW 03); Qom, Kavir (MMTT 149); between Qom and Arak (FW 95); near Rahjerd, 1700 m (FW 95); near Rud-e Shur railway station, 59 km SW Tehran (LF 50); Shah 'Abbas Caravanserai, Kavir Protected Region (MMTT 1348, not seen by me); Shamsabad farm, 30 km E Tehran (CAS 100980); Tehran (NS 74); near Tehran (GB 21). **Ostan 2 (Gilan)**: Chashme Sangi, about 35 km N Bijar (UMMZ 129812–15); plains between Manjil and Pachenar, 600–700 m (FW 03); 8.8 mi [14 km] N Takestan (FMNH 170878–80). **Ostan 7 (Esfahan)**: Esfahan, 6800 ft [2074 m] (BMNH 1917.3.6.21; BYU 20956; RC *et al.* 66; NS 74); near Esfahan (WB 76, type locality; GB 87; 21); W Esfahan (GB 21); NW Esfahan (GB 21); between Khara and Mohammadabad (AN 07; 15; LL 28); 33°31′N, 51°54′E, 25 km by road N Natanz, 800 m (DF *et al.* 97). **Ostan 8 (Fars)**: Abadeh (GB 21; NS 74); about 30°51′N, 53°06′E, 28 km N Deh Bid on road between Shiraz and Esfahan, 2160 m (CAS 141154; MMTT 1058–62); Yazd-e Khvast (FMNH 21009[15]; KS 39). **Ostan 9 (Kerman)**: Fahraj (MMTT 1411, not seen by me); Kerman, 5000 ft (BMNH 1917.3.6.22; WB 76; GB 21; NS 74); Kuh-e Hazaran, 10,000 ft (WB 76); Laleh Zar region, S Kerman (OW 51); Rayen, 7000–8000 ft (BMNH 74.11.23.28–29/1946.8.7.42–43; WB 76; GB 87; 21); 40 km SE Sa'idabad (USNM 149154–5; RT 71); about 29°06′N, 56°00′E, 30 km S Sirjan on Kerman-Bandar 'Abbas road, 1680 m (CAS 141024; MMTT 739). **Ostan 10 (Baluchistan-Sistan)**: Bandan (NS 74); Bazman (AN 99; 15; NZ 03; LL 28; NS 74); Dak-i-Do (NZ 03); Duz-Ab, Sargad [Zahedan?] (AN 99; NZ 03; LL 28; NS 74); Farra (NZ 03); Gal-i-Chakh (NZ 03); Garne (NZ 03); Kach, Baluchistan (FW 38); Khash, 1200 m, 40 km N Gonabad (JG 66); environs of Khash, Sargad (NS 74); about 28°02′N, 60°55′E, 39 km S Khash on road to Iranshahr, 1460 m (MMTT 896–902); about 28°01′N, 60°51′E, 49 km S Khash on road to Iranshahr, 1400 m (MMTT 862); Kurin (NZ 03); Mazel-Ab (NZ 03); Nasirabad, Sistan (AN 99; 15; GB 21; LL 28;NS 74); Nech-i-Bendan (AN 15); Sistan, Perso-Baluch border (NA 06); Tscha-i-Gjuische [or Tsha-i-Huishe] (AN 99; LL 28); 15 mi [24 km] SW Zabol (CAS 102548–54; FMNH 141358–60, 141362–7, 141369–72, 141374, 141452); Zaboli, 4500 ft (WB 76; LL 28); Zahedan (USNM 149143–4; RT 71); about 20 km S Zahedan (LF 50); about 29°28′N, 60°41′E, 32 km W Zahedan on road to Chashmeh Ziarat, 1900 m (CAS 141063; MMTT 850–51). **Ostan 11 (Khorasan)**: Baaza, Zirkuch (AN 99; LL 28); Bekhars (AN 15); about 32°49′N, 59°26′E, 20 km S Birjand on road to Zahedan and NNE on dirt track toward mountains, 1610 m (CAS 141195); 55 km ENE Birjand (USNM 148614; RT 71); between Birjand and Torbat-e Heydariyeh (LF 50); Chah-e Ziran, Nehbendan (LL 28); Chouz [or Houz], Zirkuch (AN 99; LL 28); Fariman (NS 74); Kerat, Khashtadan (LL 28); 1.5 km E Khajar (USNM 153816; RT 71); Khorasan (FW 38); Kushkek on road to Bekhars (NZ 03); Mashhad, 1380 m (RC 91); Mil-Ajaz-Khan near Sang Bast (LL 28); Nehbandan (NS 74); Niazabad, Zirkuch (AN 99; 15; LL 28; NS 74); Nusi (AN 97; 15); Pesuk [probably Pazuk] (AN 97; 15; LL 28); Razeh (LL 28); about 34°47′N, 60°47′E, 4 km E communications station at outskirts of Tayyebat, about 900 m (CAS 141180; MMTT 1102–3); Torbat-e Jam, 955 m (RC 91). **Ostan 12 (Mazanderan)**: about 36°24′N, 56°17′E, 16 km S 'Abbas-abad, Turan Protected Region, 800–870 m (CAS 141214; MMTT 1201–2; RT 79); Chahar Deh (AN

15); Damghan (AN 15; NS 74); Delbar, Turan Biosphere Reserve, 1205 m (RT 79); 2.5 km NE Delbar, 1400 m (RT 79); 3.5 km E Delbar (RT 79); 7.5 km E Delbar, 1200 m (RT 79); Koshm Abad near Damghan (LL 28); Na'imabad (AN 07); near Posht-e Aseman, Turan Biosphere Reserve, 1260 m (RT 79); Shahrud (NS 74); Tochah, Turan Biosphere Reserve (RT 79).

Eremias pleskei Bedriaga, 1907

Ostan 3 (East Azarbaijan): Tabriz (NS 74); Talkeh Rud near Tabriz (LL 28). **Ostan 4 (West Azarbaijan)**: 38°50'N, 45°00'E, just E Iran-Turkey border (CAS 96263; RC *et al.* 66); Marand, 1090 m (RC 91).

Eremias strauchi strauchi Kessler, 1878

Ostan 3 (East Azarbaijan): Dasht-e Moghan (NS 74); about 38°56'N, 45°35'E, 6 km W Jolfa, Makran Protected Region, 1350 m (CAS 141264–5; MMTT 1310–11, 1322–25, 1472); Kartevul, near Maragheh, about 1800 m (LL 28; VR 38); Maragheh (AMNH 36383); about 37°22'N, 46°30'E, Maragheh paleo site, 4 km SW Chahlilvan, 29 km E Maragheh, 1800 m (CAS 141285–6; MMTT 1349–51); 500 m N Mordaq (MMTT 551); Tabriz (AN 15; LL 28; VR 38; NS 74); 45 km SE Tabriz, 2060 m (RC 91); Talkheh Rud, near Tabriz (LL 28). **Ostan 4 (West Azarbaijan)**: 46 mi [73.6 km] N Khvoy (AMNH 97594–5); 21 km E Maku on road to Marand, 1090 m (CAS 141271–2; MMTT 1332–4); 25 mi [40 km] E Maku, 3925 ft [1196 m] (FMNH 170877); Marand, 1090 m (RC 91); Peter and Paul Landing, NW corner Daryacheh-ye Reza'iyeh (LL 28; VR 38); NW shore Daryacheh-ye Reza'iyeh (NS 74).

Eremias strauchi kopetdaghica Szczerbak, 1972

Ostan 11 (Khorasan): Qal'eh Manar (AN 97; 15; LL 28; NS 72; 74); about 35°29'N, 59°12'E, 8 km S Robat-e Sang on road to Torbat-e Heydariye, 1760 m (CAS 141185–6; MMTT 1104–6). **Ostan 12 (Mazanderan)**: between Ashkhabad, Turkmen and Gorgan (AN 15; LL 28; NS 72); Dash Baroun, frontier post on Rud-e Atrak, NE Gonbad-e Kavus (JG 57); Gonbad-e Kavus (JG 57); Gorgan (NS 74).

Eremias velox velox (Pallas, 1771)

Ostan 1 (Tehran): Arak (CAS 140661–8); Hesarak (CAS 140654); Kharakan (CAS spec.); Latian Dam area (MMTT 9); Qazvin (CAS 140669–70); Qom (CAS 140655–60). **Ostan 2 (Gilan)**: Gilan, S Rasht (JB 79; WB 76; GB 21; NS 74). **Ostan 9 (Kerman)**: 116 and 148 km E Neyriz on Zaidabad Rd. to Sirjan (BMNH 1966.363–364). **Ostan 11 (Khorasan)**: Bojnurd, 800 m (RC 91); Gouladah (AMNH 61773); 20 km E Mashhad (USNM 154486; RT 71); between Mashhad and Fariman (NZ 97; NS 74); Qalandarabad (NZ 97; NS 74); Tajan River [Harirud], Iran-Afghan-Turkman border (LL 18; 28). **Ostan 12 (Mazanderan)**: about 36°09'N, 56°08'E, 34 km S 'Abbasabad, Turan Protected Region, 870 m (CAS 141216; MMTT 1228; RT 79); about 36°09'N, 56°03'E, 42 km by road S 'Abbasabad, Turan Protected Region (CAS 141219; MMTT 1231; RT 79); head of Almeh Valley, Mohammad Reza Shah National Park (MMTT 477); S Almeh Valley, Mohammad Reza Shah National Park (MMTT 478); 3 km SW Dasht (USNM 149401–2; RT 71); Gorgan (NS 74); 1 mi [1.6 km] S Pahlavi Dezh (FMNH 141541); about 37°19'N, 56°16'E, 3 km W Robat-e Qareh Bil, Mohammad Reza Shah National Park (CAS 141173; MMTT 1098–9); Rud-e Gorgan (AN 15; LL 28); Shahrud (NS 74); 34 km NNE Shahrud (USNM 158527; RT 71).

Lacerta brandtii De Filippi, 1863

Ostan 3 (East Azarbaijan): Ardabil (An 16; GB 20; LL&OC 39); about 4 km from Ardabil at foot of Kuhha-ye Sabalan; 1380 m (LL&OC 39); Salt lake near Ardebil (BMNH 1913.2.4.1–3); Basmenj (FdF 63; 65; type locality; WB 76); Kartevjul, 1820 ft [555 m], near Maragheh (VR 38; LL&OC 39); Kuh-e Sahand, 3500 ft [1086 m]; Kurosch-gol, 40 km SE Tabriz (OW 51); about 37°22'N, 46°30'E, Maragheh paleo site, 4 km SW Chahlilvan (24 km E Maragheh), 1800 m (CAS

141287–8; MMTT 1352–4); Maragheh Shilvand-Rud village (MMTT 1392–4); Rasano, Talysch Mountains (BMNH 1973.3492); 45 km SE Tabriz, 2000 m (RC 91); 30 mi [48 km] SE Tabriz, 6800 ft [2074 m] (CAS 96265–9; RC et al. 66). **Ostan 7 (Esfahan)**: 6 mi [9.6 km] NW Kuh Rang, 8500 ft [2593 m] (FMNH 170956–62).

Lacerta cappadocica urmiana (Lantz and Suchow, 1934)

Ostan 4 (West Azarbaijan): about 37°25′N, 44°56′E, 5 km W Band (15 km W Reza'iyeh) on dirt track to waterfalls, 150 m (CAS 141279; MMTT 1340–41); 17 km W Daryacheh-ye Reza'iyeh (JE 79); "Hapkins Bridge, Ruwandiz Rd., 15 mi [24 km] W of Jindian, S Kurdistan" [possibly in Iraq, close to Iranian border] (BMNH 1934.12.12.22); 61 km NNW Kermanshah, about 1440 m (JE 79); Kherra, gorge of Rud-e Bardeh Sur (LL&GS 34; type locality); Nazlu Rud, 9 km SW Daryacheh-ye Reza'iyeh (FMNH 170932; JE 79); 10 km SW Rez'iyeh (FMNH 141587–8); 6 km E Sero, on Nazlu Chay, about 1650 m (JE 79).

Lacerta chlorogaster Boulenger, 1909

Gouladah (AMNH 61766–72); Rabat-i-Setid (AN 15). **Ostan 1 (Tehran)**: Qom (MMTT 150). **Ostan 2 (Gilan)**: Bandar-e Pahlavi (BMNH 1908.8.7.14–15/1946.9.1.87–88, 1908.8.7.29–34/ 1946.9.2.28–33, syntypes; GB 09; 20; type locality); about 37°42′N, 48°52′E, 8 km W Bandar-e Pahlavi-Astara road on road to Khalkhal 2 km N Asalom, 110 m (CAS 141230); 2 km S Chalus (JE 95); Rasht (BMNH 1915.8.13.1–10; GB 20; JE 95); Shafa Rud (RM 57); Sisangan, ENE Chalus (JE 95). **Ostan 11 (Khorasan)**: Ziaret, near Shirvan (LvM 09; AN 15). **Ostan 12 (Mazanderan)**: 36°46′N, 50°30′E, 10 km by road NNE Chorti, N slope Alborz Mountains (DF et al. 97); Goleh Loveh, near Minudasht (JE 95); Gorgan (AN 07; 15; LvM 10; GB 20); between 5 mi [8 km] NE and 4.4 mi [7 km] SW Gorgan (FMNH spec); 20 km W Mahmudabad (MMTT 1510–12, not seen by me); Mohammad Reza Shah National Park (MMTT 182; RT 74); about 37°21′N, 56°01′E, 10 km W headquarters Mohammad Reza Shah National Park, 7 km E entrance to park, near stream S of road, 450 m (CAS 141168; MMTT 1095–7); Pahlavi Dezh (AN 10; 15); Rudkhanehyeh Talar, between Sorchkela and Pul-i-Dcheverarem, about 300 m (LF 50); 18 km SW Shah Pasand (USNM 149435–7; RT 71; JE 95); 22 km SW Shah Pasand (USNM 149439–40; RT 71).

Lacerta defilippii (Camerano, 1877)

Alborz Mountains N Tehran [Ostans 1, 2, 12] (BMNH 1908.8.7.16–17 WB 76; GB 87; 20); Anau, Vazanderan [?=Mazanderan] (BMNH 74.11.23.43–50); Asadbar, 2500 m, Alborz Mountains (FW 03); "Persia" (BMNH 1928.12.8.870/1946.9.1.75; holotype of *Lacerta muralis fusca* var. *persica*). **Ostan 1 (Tehran)**: Ab Ali, 2545 m (RC 91); Baraghan (CAS 140744–9); 50 km S Chalus, about 800 m (JE 95); Damavand region (OW 51; ID et al. 84); 4 km E Dow Ab, 11000 ft [3355 m] (CAS 102510; FMNH 141584); Hafthos Valley, Tochal Mountains N Tehran (LF 50; ID 67); about 35°59′N, 51°05′E, 37 km N Karaj on road to Chalus, above Karaj Dam Lake, Central Alborz Protected Region, 1800+ m (CAS 141221; ID et al. 84); Koladasht Valley, Alborz Mountains, 9000 ft [2700 m] (BMNH 1957.1.4.8.10); Lar Valley NW Tehran (CAS 140705–21; 111935–41; FdF 65; LC 77; ID et al. 84; JE 95; type locality); Lur Valley [Damavand] (FW 03); Mt. Demavand (BMNH 1957.1.4.7); Pasghaleh, N Tehran (LF 50; ID 67); pass above Polur, 2400 m (JE 95); Polur (MMTT 1513–22, not seen by me; JE 95); 15 km NE Polur, confluence Lar River and Safid Ab (MMTT 52); Qazvin [? mountains N Qazvin] (CAS 140722–43; ID et al. 84); Siah Paulas River, 100 m from Lar River (MMTT 97–8); Taleqan (CAS 140750–2; ID et al. 84); Tehran, 8000 ft [2440 m] [? mountains N Tehran] (BYU 20947–50; RC et al. 66); 40 km E Zanjan (FMNH 141536). **Ostan 2 (Gilan)**: 36 km W Bandar-e Pahlavi along trail following west bank Shafa Rud (MMTT 440); Chalus (JE 95); 12 km W Chalus (FMNH 141594); 1 km S Varangrud village (MMTT 152–4). **Ostan 12 (Mazanderan)**: Amol, 350 m (RC 91); about 36°12′N, 51°20′E, 64 km S Chalus on road to Karaj, 1670 m (CAS 141223–4; MMTT 1246–8; ID 84); Dasht-e Naz National Park (MMTT 420); Gorgan (AN 07; LvM 10; ID 67; ID et al. 84); Mohammad Reza Shah National Park (MMTT 181–2); Tang-e Rah parking area,

Mohammad Reza Shah National Park (MMTT 158–9); Rud-e Atrak, Kopet Dagh (OB 88; ID *et al.* 84); Vali Abad (DF *et al.* 97).

Lacerta media media Lantz and Cyrén, 1920

Nergy, Desht [probably Ostan 3 or 4] (VR 38); Mt. Sitaver [Ostan 3 or 4] (VR 38). **Ostan 1 (Tehran)**: Ahar (CAS 140673–99). **Ostan 2 (Gilan)**: Zanjan (CAS 140700–04). **Ostan 3 (East Azarbaijan)**: Cheljane [?Khelejan], Shamsdinan (VR 38); Neycharan, Kordestan (VR 38); Qerkh Bolagh, foot Kuh-e Sahand, 8000 ft (VR 38). **Ostan 4 (West Azarbaijan)**: 2.2 mi [3.5 km] W Maku (FMNH 170996–7); Reza'iyeh (FMNH 170995); 5.8 mi [9.3 km] SW Reza'iyeh (FMNH 170994, 170998–171000); 10 km SW Reza'iyeh (FMNH 141589); 23 mi [36.8 km] SSE Reza'iyeh (FMNH 141590); 50 km by road SW Reza'iyeh between Zharabad and Agh Bolagh (MMTT 397–8); 59 km SW Reza'iyeh on road to Cherikhabad, 2 km by winding road, up mountain (MMTT 394–6); Rud-e Bardeh Sur, Cherra (VR 38). **Ostan 5 (Kordestan-Kermanshah)**: Chahar Zebar, E Shahabad, 1500 m (JFS 86); 42 km W Kermanshah (USNM 154482–3; RT 71); Sorkh-e Dize, W Kermanshah (JFS 86). **Ostan 6 (Khuzestan- Lorestan)**: Dare Asbar, Ostrankuh Protected Region (MMTT 458–9; RT 75). **Ostan 7 (Esfahan)**: NW Esfahan, 5500–6000 ft (BMNH 1905.10.14.23–24; GB 20; GP 62; 64).

Lacerta mostoufi Baloutch, 1976

Ostan 10 (Baluchistan-Sistan): Deh Salm, 31°12'N, 59°19'E, about 200 km N Kuh-e Malek Mohammad, Dasht-e Lut, 800 m (MB 76; type locality); Kuh-e Malek Mohammad, 30°28'N, 59°10'E, about 100 km NW Nosratabad, 1000 m (MB 76).

Lacerta praticola praticola Eversmann, 1834

Ostan 2 (Gilan): Ardabil (ID pers. commun.); Chejran, between Astara and Ardabil (VO 78; BS 84, ID pers. comm.); Gilan lowlands on the Enzeli [=Anzali] Bay of the Caspian Sea (VO 78; BS 84); Saferud [=Sefid Rud?], Alborz Mountains (ZIL 12630; VO 78; BS 84; ID pers. commun.).

Lacerta princeps princeps Blanford, 1874

Ostan 5 (Kordestan-Kermanshah): 33°43'23"N, 46°57'52"E, 2 km NW Shah Bodagh, 1400 m (HB 98). **Ostan 7 (Esfahan)**: 30°39'N, 51°36, Kuh-e Dinar ridge, 10 km by road E Yasuj, 1800–2300 m (DF *et al.* 97); Sarchun, or Sarkhun, about 125 km SW Esfahan (LM 10; JE 68). **Ostan 8 (Fars)**: 8 km E Dasht-e Arzhan, 57 km W Shiraz, about 2100 m (JE 69; 70); 13 km W Dasht-e Arzhan (JE 69); hills near Neyriz, 7000 ft (WB 74; 76; type locality; GB 87; 20); near Shiraz (BMNH 1903.3.14.3; GB 20; JE 68).

Lacerta princeps kurdistanica Suchow, 1936

Ostan 5 (Kordestan-Kermanshah): Balkcha [or Balkhhah] approximately 35°12'N, 46°08'E, near Halabja, Iraq (JE 68); Beljaki (JE 68); Beydarvaz (JE 68; type locality); 35°15'28"N, 46°29'21"E, 7 km S Negel, 1250 m (HB 98).

Lacerta raddei raddei Boettger, 1892

Ostan 3 (East Azarbaijan): 38°26'N, 47°15'E, 22 km E Ahar on road to Meshkin Shahr, 1800 m (CAS 141258; MMTT 1300); 38°17'N, 46°57'E, 31 km S Ahar on road to Tabriz, 2150 m (CAS 141261–2; MMTT 1303–09; ID *et al.* 84); Altalykh, Kuhha-ye Talysh (ID 67); Ardabil (ID 67; ID *et al.* 84); 67 km SW Ardabil (ID *et al.* 84); 30 km SW Astara (ID *et al.* 84); Bogrovdag ridge (ID *et al.* 84); Karadag ridge (ID *et al.* 84); NW slope Kuhha-ye Talysh (ID 67); 39°14'N, 47°29'E, 1 km S Qaraqachid on road to Meshkin Shahr, 900 m (CAS 141254–6; MMTT 1297–9); Salavat Mt. (ID 67; ID *et al.* 84); Sarka Daria, Qareh Dagh (ID 67; ID *et al.* 84; JE *et al.* 93); 83 km SE Tabriz (ID *et al.* 84).

Lacerta raddei vanensis Eiselt, J. F. Schmidtler, and Darevsky, 1993

Ostan 4 (West Azarbaijan): 39°20'N, 44°17'E, 11 km NE 'Arab-e Dizehsi on road to Maku, 1900 m (CAS 141273; ID *et al.* 84; JE *et al.* 93); 25 km N Khoi, 1700 m (ID *et al.* 84; JE *et al.* 93); 5 km SE Maku, 1250 m (ID *et al.* 84; JE *et al.* 93); 40 km NW Reza'iyeh (ID *et al.* 84); 10 km E Sero, 1900 m (JE *et al.* 93)

Lacerta steineri Eiselt, 1995

Ostan 12 (Mazanderan): Gole-Loweh near Minou-dasht [37°11'N, 55°21'E], SE Gonbad-e Gavous (JE 95; type locality).

Lacerta strigata Eichwald, 1831

Aschar-Ade Island, Caspian Sea (GB 20); Bash Nurashin, Reza'iyeh basin [Ostan 3 or 4] (BMNH 99.9.30.4; GB 99; 20; type locality for *Lacerta viridis woosnami*). **Ostan 2 (Gilan)**: Astara (USNM 154446–78; RT 71); Bandar-e Pahlavi (BMNH 1919.11.24.6, 1919.12.16.1); near Bandar-e Pahlavi (FW 03); south coast of Caspian Sea, NW Iran (GB 17; 20; type locality for *L. v. woosnami*); 6 km W Chalus (CAS 102557; FMNH 141591); 10 km W Chalus (CAS 100981); 12 km W, 1 km S Chalus (CAS 102558); Galugah (RM 57); Kaluraz (AN 07; 15); Khorramabad (OW 51); Kopur Chah (RM 57); Langarud (DF *et al.* 97); Rezvandeh (RM 57); Shafa Rud (RM 57). **Ostan 3 (East Azarbaijan)**: Ne'matabad (VR 38). **Ostan 8 (Fars)**: Shiraz (BMNH 1916.9.7.1; JA 72; WB 76; GB 20; GP 62). **Ostan 11 (Khorasan)**: Talayu (LM 29). **Ostan 12 (Mazanderan)**: 1.5 mi [2.4 km] W Babol Sar (AMNH 97610–11); about 36°41'N, 52°29'E, 14 km W Babol Sar, public beach on Caspian Sea (MMTT 1094); 3 mi [4.8 km] W Behshahr (AMNH 97612); Chahar Deh (AN 15); Chalus (DF *et al.* 97, sight record); 36°46'N, 50°30'E, 10 km by road ENE Chorti, N slope Alborz Mountains, 480 m (DF *et al.* 97, sight record); Gharareh-su River, 28 km W Gorgan (USNM 149424–6; RT 71); Gonbad-e Kavus (JG 57); Gorgan (CE 41; type locality for *L. v. astrabadensis*; JG 66); shore of Caspian near Gorgan (WB 76); Mahmudabad, sea level (RC 91); environs of Mazanderan (JG 57); 1 km NW Sama (FMNH 141585); 10 km NW Sama (FMNH 141586); 22 km SW Shah Pasand (USNM 149438; RT 71); Vali Abad (DF *et al.* 97).

Lacerta valentini valentini Boettger, 1892

(See text, p. 249.)

Lacerta zagrosica Rastegar-Pouyani and Nilson, 1998

Ostan 7 (Esfahan Province): 32°58'N, 50°40'E, 3 km NW Fereydin Shahr, about 140 km NW Esfahan, Zagros Mountains, 2450 m (NR-P&GN 98; type locality).

Mesalina brevirostris Blanford 1874

Ostan 6 (Khuzestan-Lorestan): Ahvaz, E of town at ruins of old city (KS 55); Andimeshk (KS 55); Gatvand (GH&YW 69); Harmalah, 120 km NW Ahvaz (WH 59); Mahor Birinji (GH&YW 69; type locality for *Mesalina brevirostris fieldi*); Meshragheh, 53 mi [84.8 km] SE Ahvaz, 200 ft [61 m] (FMNH 170789–95); Rudkhaneh-ye Jarrahi (GB 21). **Ostan 8 (Fars)**: Ahram (FMNH 141492, 141495–7, 141500, 141502, 141505, 141509, 141511; CAS 102537–40); Bushire (GB 21). **Persian Gulf**: Jazireh-ye Tanb-e Bozorg (WB 74; 76); Qeshm Island (CAS 96153).

Mesalina watsonana (Stoliczka, 1872)

Chaschtadan (AN 15); Cesme Beiman (FW 36); Darpahan (FW 29); Dastgirt [probably Dastgerd, Ostan 11] (AN 97); Gjarmaz (AN 97; 15); Hauz Patil (FW 36); Herat (OW 51); Kuhistan [Ostan 7, 11] (FW 36); Milan-i-Dzhekun-Khamun [Ostan 10 or 11] (NZ 03); Nertschistan (AN 15); Pasengku (AN 07; 15); Paskimer (LM 29); Sendjiri (FW 36); between Shiraz and Tehran (GB 21). **Ostan 1 (Tehran)**: between Argavani and Marinjab (LF 50); Baraghan (CAS 140643); Kadish (FW 36); between Chashmeh Safid Ab and Shah 'Abbas, Kavir Protected Region (GN&CA 81); about 33°59'N,

51°25'E, Kashan, just W of city behind Bagh-e Shah, enclosed public gardens (MMTT 721); about 33°54'N, 51°30'E, 8 km S Kashan and 6 km W road between Yazd and Kashan, 1175 m (MMTT 967–9); Kavir Protected Region (MMTT 109–10); between Kush Kuh and Siah Kuh (LF 50); Qom (CAS 140644–8); between Qom and Arak, 1000–1600 m (FW 95); Rey (FMNH 21011–12; KS 39); 15 mi [24 km] S Rey (AMNH 97598); near Rud-e Shur, 59 km SW Tehran (LF 50); Safid Ab (FW 36); about 35°07'N, 50°08'E, 24 km W Saveh on road to Hamadan, 1160 m (CAS 141291–2; MMTT 1361); near Tehran (WB 76; GB 87; 21); 6 mi [9.6 km] S Tehran (AMNH 97596–7), 50 km S Tehran (OW 51); between Tehran and Karaj, 3800 ft [1159 m] (BYU 20940; RC et al. 66); Varamin (CAS 140649–50). **Ostan 6 (Khuzestan-Lorestan)**: 1 km E Ab Bid, Sar Dasht (UMMZ 130579–80); 2 km E Ab Bid, Sar Dasht (UMMZ 130581–2); Agha Jari (CAS 86484–5; SA 63); Ahvaz (AN 07; 15); Alchorschir (AN 07; 15); 30 km S Andimeshk (USNM 153686; RT 71); Bard-i-Nishunde (CAS 86453; SA 63); 10 km WNW Behbehan (USNM 153683; RT 71); Dezful (AN 07; 15); 7 km NE Gol Gir (MMTT 69–70); Gurschir (AN 07; 15); Isfagan (AN 07; 15); Izeh (MMTT 1670, not seen by me); Mahor Birinji (GH&YW 69); Masjed Soleyman (CAS 86248, 86326, 86329, 86640; SA 63); Masjed Soleyman airfield (CAS 86249; SA 63); 18 km S Masjed Soleyman (USNM 153740; RT 71); between Masjed Soleyman and Ahvaz (CAS 86465; SA 63); between Masjed Soleyman and Haft Gel (CAS 86323, 86386–7, 86445; SA 63); Parchestan (AN 07); Pervanda (AN 07; 15); between Pol-e Jeh Jeh and tar pit site at base of Kuh-e Mordeh (MMTT 224–6); Rud-e Karun (AN 07); Salmiah (AN 07); Sar-i-Gach (CAS 86452; SA 63); Sar-i-Naftak, near Ab-e Golestan (CAS 86444; SA 63); Zeloi (CAS 86325; SA 63). **Ostan 7 (Esfahan)**: Chah-e Gorg (FW 36); Esfahan (BYU 20955; FMNH 21013, 2168[7]; RC et al. 66; KS 39); near Esfahan (WB 76); N Esfahan (WB 76; GB 21); SW Esfahan (GB 21); between Esfahan and Yazd (OW 51); Galatappeh (CAS 102543–7; FMNH 141516, 141518–19, 141521–2, 141526–7, 141532, 141534); Jafar Abad (DF et al. 97); Kuh-rang, 200 km W Esfahan (OW 51); 20 km E Na'in (LF 50); 30 km W Na'in (LF 50). **Ostan 8 (Fars)**: Abadeh (BYU 20961–2; RC et al. 66; GB 21); Ahram (FMNH 141501, 141506, 141510; CAS 102541–2); 3 mi [4.8 km] NW Bastak, 1300 ft [397 m] (FMNH 170783); 6 mi [9.6 km] NW Bastak, 1300 ft [397 m] (FMNH 170784); NE shore Daryacheh-ye Bakhtegan (OW 51); about 29°28'N, 51°21'E, 5 km N Dalaki on road to Shiraz, where foothills begin, 110 m (CAS 141150); Hane Houre (DF et al. 97); Jahrom, 3200 ft [976 m] (FMNH 170782); Naqsh-e Rustam, 7 km N Persepolis (USNM 153593; RT 71); Persepolis [Takht-e Jamshid], 5500 ft [1678 m] (CAS 86478; SA 63); Shiraz (CAS 140651); 13 km SE Shiraz (USNM 164824–7; RT 71); Yazd-e Khvast (FMNH 21010 [33 specimens]; KS 39).

Ostan 9 (Kerman): about 26°32'N, 54°43'E, 9 km W Bandar-e Lengeh airport, 0–50 m (CAS 141048); Hoseynabad, about 20 km SE Kerman (MMTT 623–4); Jask (KS 55); below Jupar (MMTT 1); Kerman, 5000 ft [1525 m] (WB 76; GB 87; 21); about 30°15'N, 57°14'E, 2 km on dirt track branching NE from Kerman-Bam road 13 km E eastern edge city of Kerman, 2000 m (CAS 141021); 60 km from Kerman on road to Bam (MMTT 1586–7, not seen by me); between Kerman and Sa'idabad (LF 50); Khvorgu (CAS 86367; SA 63); about 27°33'N, 56°24'E, E Kuhha-ye Genu (CAS 86368–9; SA 63); Laleh Zar (OW 51); 28 km SW Mashiz (USNM 149152; RT 71); 40 mi [64 km] SE Minab, 100 ft [30.5 m] (FMNH 170785); 63 mi [101 km] SE Minab, 2000 ft [610 m] (FMNH 170786–7); Rayen, 8000 ft [2440 m] (WB 76; GB 87; 21); Sa'idabad, 5500 ft [1678 m] (WB 76; gb 87; 21); N Shaqu (CAS 86587; SA 63); about 27°19'N, 56°33'E, 19 km SE Shaqu on road to Minab, 0–50 m (MMTT 786–7); about 29°06'N, 56°00'E, 50 km S Sirjan [Sa'idabad] on Kerman-Bandar 'Abbas road, 1680 m (MMTT 735–8); about 28°54'N, 55°32'E, 80 km S Sirjan on Kerman-Bandar 'Abbas road, 1860 m (MMTT 785). **Ostan 10 (Baluchistan-Sistan)**: Chah Bahar (JG 66); near Eskelabad, SE edge of Dasht-e Lut basin, 145 km S Zahedan (MMTT 610); Hormak (AN 99); Kalagan (AN 15); Khash (MMTT 712–13); about 28°10'N, 61°11'E, 10 km S Khash on road to Iranshahr, 1450 m (CAS 141069; MMTT 860–1); Kuh-e Bam Posht, 3000 ft [915 m] (WB 76); Kuh-e Taftan (AN 99; 15; NZ 03); Mazel-ab (NZ 03); Nasirabad (GB 21); Neizar (AN 99; 15); 4 km toward Nikhshahr from Chah Bahar (MMTT 1636, not seen by me); 20 km S Pip (RM 56); between Sistan and Bampur (NZ 03); Tagab (NZ 03); Tasuki (JG 57); Zaboli, 4500 ft [1373 m] (WB 76); Zahedan (JG 57); about 29°28'N, 60°49'E, 32 km W Zahedan on road to Chashmeh Ziaret, 1900 m (MMTT 852); about 28°59'N, 60°42'E, 66 km S Zahedan on road to Khash, 1760 m (MMTT 856); 99 km S Zahedan on road to Khash, 1825 m (SCA field 209, lost); between Zahedan and Chah Bahar (LF 50); Zamran, 2000 ft [610 m] (WB 76).

Ostan 11 (Khorasan): between Akhala-Atek and Kuchan-Mashhad (NZ 03); 6 km W Ayubi (USNM 148594; RT 71); Basiran (AN 97; 15); Baz-Chouz-Pain (AN 99); Bekhars (AN 15); about 34°05'N, 58°50'E, 36 km S Bidokht on road to Birjand, 1540 m (MMTT 1111–12); Birjand (LM 29); about 32°49'N, 59°26'E, 20 km S Birjand on road to Zahedan and NNE on dirt track toward mountains, 1610 m (CAS 141194; MMTT 1115–19); 27 km S Birjand on road to Zahedan, 1680 m (MMTT 1114); 55 km ENE Birjand (USNM 148613; RT 71); Bohnabad (AN 97; 15); Chadschi-Abad, Zirkuch (AN 99); Doruneh (USNM 148582–3; RT 71); 1 km S Esfideh (USNM 148623; RT 71); Fariman, 410 m (RC 91); between Feyzabad and Nusi (AN 97; 15); Godar Mohammad Mirza, between Torbat-e Heydariyeh and Mashhad, about 1750 m (LF 50); Golandar (AN 97; 15); Mashhad, 1300 m (RC 91); about 36°26'N, 59°53'E, 29 km N Gold Mosque in Mashhad on road to Kalat-e Naderi, 1130 m (CAS 141177; MMTT 1100); about 36°21'N, 56°43'E, salt flats 5 km W Kahak (110 km W Sabzevar) on road to Shahrud, Turan Protected Region, 800 m (CAS 141208; MMTT 1212); Kerat (AN 15); Langarak (JG 57); between Mashhad and Torbat-e Jam (NZ 03); Mill-Ajaz (AN 97; 9 mi [14.4 km] N Qayen (FMNH 141546); about 33°49'N, 59°07'E, 12 km N Qayen (50 km S Shahabad) on road between Torbat-e Heydariyeh and Birjand, 1420 m (MMTT 1113); 30 km E Sabzevar (USNM 158445–6; RT 71); Saman Shah Mountains (AN 97; 15); about 36°25'N, 58°11'E, 23 km N Soltanabad on road to Quchan, 1250 m (CAS 141206; MMTT 1210–11); Tajan River [Harirud, Perso-Afghan-Turkmenistan border] (LL 18); Tayyebat (CAS 96274; RC et al. 66); 18 km N Torbat-e Heydariyeh (USNM 148589; RT 71); Torbat-e Jam, 955 m (RC 91); Tscha-i-Gjuische (AN 99); between Zirkukh and Tag-i-Dorokh (NZ 03). **Ostan 12 (Mazanderan)**: vicinity Ab-e Ragn spring, Turan Biosphere Reserve, 1530 m (RT 79); Damghan (AN 15); Dasht-e Tal, SE Shahrud (MMTT 252–62); Delbar, Turan Biosphere Reserve, 1205 m (RT 79); 11 km N Delbar, Turan Biosphere Reserve, 1350 m (RT 79); between villages of Kariz, Shakhbaz, and Baghestan, Turan Biosphere Reserve (RT 79); near Kheyrabad, NE Shahrud (MMTT 251); Na'imabad (AN 07). **Persian Gulf**: Jazireh-ye Hengam (WB 74; 76; type locality for *Mesalina pardaloides*; GB 87).

Ophisops elegans Ménétriés, 1832

Near Bag-e Taj [Ostan 8?] (FW 17); Dasht-e Kavir (FW 36); Fiab'y-Deery, near Gyrdykh (VR 38); Kalaker Mountains (FW 36); Khullar [Ostan 8?] (FW 17); Nergy-Desht [Ostan 3 or 4?] (VR 38); plain near Nurabad [Ostan 8?] (FW 17); Pasengku (AN 07; 15); between Rasht [Ostan 2] and Tehran [Ostan 1]; Shemane, Shamsdinan [Ostan 3 or 4] (VR 38). **Ostan 1 (Tehran)**: 50 km E Arak (OW 51); foothills Alborz Mountains N Tehran (CAS 86471; SA 63); near Bahram, about 900 m (LF 50). Baraghan (CAS 140634–6); desert W town of Damavand on road to Tehran (FW 03); 10 km by road E jct Damavand to Firuzkuh, 110 m N "by climbing" (MMTT 1481–6, not seen by me); Darrous, N Tehran (CAS 96139–46, 98116); Daryacheh-ye Namak (FMNH 21015[3]; KS 39); 132 km N Esfahan (OW 51); Ghazangheshlagh steppe SW Karaj (LF 50); Hesarak (CAS 140638–42); Karaj, 3800 ft (BYU 20941–6; RC et al. 66); 5 mi [8 km] N Karaj (AMNH 97606); Latian Dam area (MMTT 7–8); steppe near Mahmudieh (LF 50); Qazvin (JG 57); Qom (CAS 140637; OW 51); Qom, Kavir (MMTT 146–8); 15 mi [24 km] S Rey (AMNH 97599–601); Tehran (FMNH 21014; KS 39); near Tehran, 4000 ft (WB 76); hills E Tehran on Ab 'Ali road (CAS 86472–3; SA 63); 6 mi [9.6 km] S Tehran (AMNH 97607–9); Veramin (CAS specimen). **Ostan 2 (Gilan)**: 2 mi [3.2 km] N Bijar (FMNH 171019); Chashme Sangi, about 35 km N Bijar (UMMZ 129796, 129798–811); Divandarreh, 5800 ft [1769 m] (FMNH 171023–32, 171034); 1 mi [1.6 km] S Divandarreh (FMNH 171011, 171013–18); 3 mi [4.8 km] S Divandarreh (FMNH 171072); 9 mi [14.4 km] SE Divandarreh (FMNH 171071); Ghala Raihand Mountains, W Divandarreh (FMNH 171033); Mianeh, 1545 m (RC 91); Zanjan, 1728 m (RC 91). **Ostan 3 (East Azarbaijan)**: 10 km S Arax River (16 km S 'Alirezaabad) on road to Meshkin Shahr, 220 m (CAS 141251, MMTT 1285–93, 1294 [11 specimens]); Bandar-e Danalu (LL 31; VR 38); shore of Daryacheh-ye Reza'iyeh at Ramanlu (MMTT 13); Ismail-aga, in valley of Nazlu Rud (LL 31, VR 38); 38°25'N, 45°46'E, 25 km by road SE Marand, 900 m (DF et al. 97); SW Shähi Peninsula, Lake Urmia [?=Jazireh-ye Shahi] (BMNH 1965.1466–84); Kartevjul, 6000 ft [1830 m] (LL

31; VR 38); between Kartevjul and 'Ammeh, near Maragheh, 6000–8000 ft [1830–2440 m] (LL 31; VR 38); Kirjawa (GB 99; 21); western part of Kuh-e Mishu Dagh, about 1700 m (LF 50); Kuh-e Sahand, 7000 ft [2135 m] (LL 31; VR 38); Maragheh (AMNH 39375); about 38°30'N, 48°02'E, 18 km W Razi on road to Meshkin Shahr, 1150 m (CAS 141257; MMTT 1280); Sharafkhaneh (LL 31; VR 38); near Sharafkhaneh, 1220 m, on shore of Daryacheh-ye Reza'iyeh (LF 50); Shazalan Island, Daryacheh-ye Reza'iyeh (BMNH 99.9.30.9; GB 99; 21; LF 50); 80 km SE Tabriz, 1758 m (RC 91); Tshalma, Dasht-e Moghan (LL 31). **Ostan 4 (West Azarbaijan)**: about 37°25'N, 44°56'E, 5 km W Band (15 km W Reza'iyeh) on dirt track to waterfalls, 1500 m (CAS 141280; MMTT 1342); Bernarve, Shamsdinan, W Daryacheh-ye Reza'iyeh (LL 31; VR 38); Daradez, between Jolfa and Tabriz (LL 31); Dagh (GB 99; 21); Golmankhaneh, near Reza'iyeh (LF 50); Guch Ali, about 57 km NE Khvoy, 30 km S Jolfa (UMMZ 129784); Gyali-Deeri valley, above Girdyk, Mergever, W Daryacheh-ye Reza'iyeh (LL 31); Jazireh-ye Arzu, Lake Reza'iyeh Protected Region (MMTT 115–16, 198; GB 99; 18; 21; RT 74; type locality for *Ophisops e. persicus*); Jolfa (VR 38); Karym-Abad, W shore Daryacheh-ye Reza'iyeh (LL 31; VR 38); Khosrowabad (LL 31; OB 93); Khvoy (LL 31; VR 38); near Khvoy, 38°40'N, 45°30'E (CAS 96262; RC *et al.* 66); 46 mi [73.6 km] N Khvoy (AMNH 97602–5); Mahabad (BMNH 99.9.30.10; GB 99; 21); Maku (FMNH 141537–8); 4 km W, 3 km S Maku (FMNH 141539–40); Markan (DF *et al.* 97); about 36°56'N, 46°17'E, 19 km ESE Miandow Ab on road to Shahindezh, 1350 m (MMTT 1344–5); Seir (BMNH 99.9.30.12; GB 99; 21); Sopurghan (GB 99; 18; 21; type locality for *O. e. persicus*); between Sopurghan and Reza'iyeh (GB 99; 21).

Ostan 5 (Kordestan-Kermanshah): 41 km SW Bijar, 1940 m (CAS 141302); Faraman (CAS 102562–7; FMNH 141314–16, 141319–22, 141325, 141327–8, 141330–2); 20 km N Hamadan (USNM 154480; RT 71); Ilam, 5000 ft [1525 m] (FMNH 171059); about 34°28'N, 47°46'E, 22 km W Kangavar on old abandoned road parallel to road to Hamadan, 1640 m (CAS 141294); Kermanshah (FMNH 171020–22, 171036); 4 mi [6.4 km] N Kermanshah (FMNH 171037); 5 mi [8 km] N Kermanshah (FMNH 171035); about 34°13'N, 46°41'E, creek near turnoff to microwave station 42 km W Kermanshah on road to Shahabad, 1640 m (CAS 141297; MMTT 1362–4); Quraitu (BMNH 1921.3.30.8–11); 32 km WNW Sanandaj (USNM 158466–511; RT 71); Tappeh Asiab, near Bijaneh, 6.5 km E Kermanshah (FMNH 130768). **Ostan 6 (Khuzestan-Lorestan)**: 1 km E Ab Bid, Sar Dasht (UMMZ 130586); 2 km W Ab Bid, Sar Dasht (UMMZ 130585); near Ab Bid, Sar Dasht (UMMZ 130584); Ahvaz (AN 07; 15); 45 km N Ahvaz (USNM 153604–74; RT 71); Alchorschir (AN 07; 15); 3 mi [4.8 km] SE Aligudarz (FMNH 171067–9, 171083); 25 mi [40 km] SW Aligudarz (FMNH 171084); Andimeshk, 150 m (KS 55); 30 km S Andimeshk (USNM 153687–90; RT 71); 34 km S Borujerd (FMNH 130759–61, 130769); 50 km SW Borujerd (USNM 153733–7; RT 71); Dare Astane, Ostrankuh Protected Region (MMTT 460); Dech-i Diz [?Rud-e Dez] (AN 07; 15); Dehloran (USNM 157374; RT 71); Dezful (AN 07; 15); 60 km S Dezful (UMMZ 130583); Dow Rud (GH&YW 69); 5 km NW Dow Rud (USNM 158450–1; 164814; RT 71); Gholaman (DF *et al.* 97); Harmalah, 120 km NW Ahvaz (WH 59); Istgah-e Bisheh, 1200 m, on Ab-e Sezar (KS 55); about 32°01'N, 48°16'E, headquarters Karkheh National Park, 2 km W Andimeshk-Ahvaz road (CAS 141122; MMTT 1001–4); about 32°01'N, 48°16'E, *Tamarix* thicket near headquarters Karkheh National Park, 6 km W Andimeshk-Ahvaz road (CAS 141123; MMTT 1005); Khorramabad (FMNH 171060–66, 171074); 10 km W Khorramabad (USNM 153738; MMTT 679–83; RT 71); Kuhdasht Valley (FMNH 130770); S side Kuhdasht Valley (FMNH 130771); Mahor Birinji (GH&YW 69); Mansu Abad, 1200 m, about 8 km NNE Shahbazan, upper Tuba Creek (KS 55); Mazu (GH&YW 69); Meshrageh, 200 ft [61 m], 53 mi [84.8 km] SE Ahvaz (FMNH 170835–52, 171106); Qal'eh-ye Tol (AN 07; 15); near ferry on Rud-e Dez, 6 km E road to Choga Zanbil, Dez Protected Region (CAS 141120; MMTT 987); Rud-e Karun (AN 07; 15); Sarab, near Dow Rud (MMTT 475); Sar Dasht, 7000 ft [2135 m] (FMNH 171104); Shalgahi (GH&YW 69); Shustar (AN 07; 15); about halfway between Shushtar and Ahvaz (MMTT 191–2); Tchambachi Valley, 1300 m, Karun, Ab-e Sezar (KS 55); Zagheh (GH&YW 69). **Ostan 7 (Esfahan)**: 21 km E Damaneh (USNM 164816–20); Esfahan, 6800 ft (BYU 20957–8; GB 18; type locality for *Ophisops elegans persicus*; RC *et al.* 66); N Esfahan (BMNH 74.11.23.103); 100 mi [160 km] NW Esfahan, 6000 ft (BMNH 1905.10.14.27–29); 90 mi [144 km] S Esfahan, 7500 ft (BMNH 1905.10.14.25–6); Galatappeh (CAS 102568–70; FMNH 141512, 141514, 141517, 141524–5,

141529, 141535); Isfagan [?Esfahan] (AN 07; 15); between Khara and Mohammadabad (AN 07; 15); Kohrud, 7000 ft (WB 76); near Kuh-rang, 200 km W Esfahan (OW 51); 45 km SW Natanz (USNM 164816–20; RT 71); Qamishlu (DF *et al.* 97); between Yazd and Esfahan (KS 55). **Ostan 8 (Fars)**: Abadeh (BMNH 1919.5.2.12; BYU 20963–4; GB 21; RC *et al.* 66); 93 km ESE Behbehan, 35 km E Gach Saran (USNM 153684, 153725–6; RT 71); steppe near Daryacheh-ye Bakhtegan (OW 51); W end Daryacheh-ye Maharlu, 17 mi E Shiraz (BMNH 1966.365–7); 29°40'N, 51°59'E, 10 km by road E Dasht-e Arzhan, 1800 m (DF *et al.* 97); about 29°42'N, 52°03'E, 12 km E Dasht-e Arzhan on road to Shiraz, 1800 m (CAS 141153; MMTT 1050–3); 10 km S Deh Bid (USNM 153588–92; RT 71); 30°39'N, 51°36'E, Kuh-e Dinar ridge, 10 km by road E Yasuj, 1800–2300 m (DF *et al.* 97); about 29°34'N, 51°53'E, crest of Mian Kotal Pass, about 20 km W Dasht-e Arzhan on road from Bushire to Shiraz (CAS 141151); Naqsh-e Rustam, 7 km N Persepolis (USNM 153594; RT 71); Niriz, E Shiraz (BMNH 74.11.23.101); Pasargat (DF *et al.* 97); 30°14'N, 52°12'E, 8 km by road SSW Qader Abad, 2100 m (DF *et al.* 97); Sarab-e Bahram (FW 17); Shiraz (BMNH 49.3.4.13–15; BMNH 1920.1.20.3090; CAS specimen; GB 18; 21; type locality for *O. e. persicus*); Tang-e Gurguda, about 4500 ft, near Gach Saran (CAS 96149; SA 63); Yasuj (FMNH 171105). **Ostan 9 (Kerman)**: Bag-e Tschinar [?Bagh-e Chenar] (FW 17); Gardaneh-ye Khaneh Sorkh (BMNH 1951.1.6.66); about 29°06'N, 57°56'E, 75 km N Jiroft on road to Bam, 1850 m (CAS 141061; MMTT 839–48); Kerman, 5000 ft (BMNH specimen; WB 76; GB 87; 18; 21; type locality for *O. e. persicus*); Kuh-e Hazaran, 8000–10,000 ft (BMNH 74.11.23.98; WB 76); Rayen area, Zar Rud (MMTT 1414). **Ostan 10 (Baluchistan-Sistan)**: Bazman (AN 99; NZ 03).

Ablepharus bivittatus **(Ménétriés, 1832)**
From Tabriz (Ostan 3) to Qazvin (Ostan 1) (FdF 65). **Ostan 1 (Tehran)**. Lar (CAS 140604–5); Qazvin (IF 69b; VE&NS 86). **Ostan 2 (Gilan)**: 1 m [1.6 km] S. Divandarreh (FMNH 170935). **Ostan 3 (East Azarbaijan)**: 38°00'N, 48°34'E, Daryacheh-ye Neur, 35 km by air ESE Ardabil (16 km E by dirt road from Budalalu which is 34 km S Ardabil on road to Khalkhal), 2400 m (MMTT 1277); between Kartevjul and 'Ammeh, 7000 ft [2135 m] (VR 38; VE&NS 86); Kuh-e Sahand, 11,000 ft [3355 m] (VR 38); Qareh Dagh (VE&NS 86); foot of Kuh-e Sahand, 8000 ft [2440 m] (VR 38; VE&NS 86); Tabriz (AS 68; IF 69b; VE&NS 86). **Ostan 5 (Kordestan-Kermanshah)**: Kermanshah (IF 69b; VE&NS 86). **Ostan 8 (Fars)**: Kusk Zar, 8000 ft (BMNH 74.11 24–25; WB 76;IF 69b; VE&NS 86). **Ostan 12 (Mazanderan)**: Chahar Deh (AN 15); Shah Rud (AN 15; IF 69b; VE&NS 86)).

Ablepharus pannonicus **Fitzinger, 1823**
Nakshi-i-Bahran (BMNH 1903.3.14.4); Rud-e Atrak [Ostan 11 or 12] (LM 29); Sarvestan [probably Ostan 8] (JG 66); Zergend (AN 15). **Ostan 1 (Tehran)**: 6 km N Damavand toward Tizab spring (MMTT 1474–5, not seen by me); Iraq Ajemi region (AN 07); Shahsavaran (FW 36). **Ostan 2 (Gilan)**: Firuzabad, 20 km S Chalus (JG 66). **Ostan 6 (Khuzestan-Lorestan)**: 30 km S Andimeshk (USNM 153691; RT 71); Arabistan [Khuzestan Prov.] (AN 07); Bak-Emishchir (AN 15; VE&NS 86); Dech-i-Diz [?Rud-e Dez] (AN 07; 15; type locality for *Ablepharus brandti brevipes*; VE&NS 86); Harmalah, 120 km NW Ahvaz (WH 59); Rud-e Karun (AN 07; type locality for *A. b. brevipes*; VE&NS 86); 10 km along Rud-e Karun down from Ziarat-Khadir (AN 15; VE&NS 86); Shalgahi (BMNH 1969.1531); Siah Mansur (AN 07; 15; VE&NS 86). **Ostan 7 (Esfahan)**: Qamishlu (DF *et al.* 97, sight record); Sarchun, about 125 km SW Esfahan. **Ostan 8 (Fars)**: Baliz preserve, Shiraz (AN 15; VE&NS 86); Fars Province (FW 17); 29°49'N, 51°59'E, 10 km by road E Dasht-e Arzhan, 1800 m (DF *et al.* 97); 29°40'N, 51°59'E, Kuh-e Dinar ridge, 10 km by road E Yasuj, 1800–2300 m (DF *et al.* 97, sight record); 30°05'N, 52°55'E, 10 km by road E Sivand, 1700m (DF *et al.* 97). **Ostan 9 (Kerman)**: Fahraj. **Ostan 10 (Baluchistan-Sistan)**: Bazman (AN 99; NZ 03; IF 69); Deh Pahbid (NZ 03); Duz Ab [?Zahedan] (NZ 03); Hormak (NZ 03); Iranshahr (MMTT 1650, 1659–68, not seen by me); Kuh-e Taftan (NZ 03); Ladiz (NZ 03); Mazel-Ab (NZ 03); Rud-e Bampur (AN 99; NZ 03; IF 69). **Ostan 11 (Khorasan)**: Gardaneh-ye 'Alam 'Ali, 2000 m, near Kuchan in Kopet Dagh (LF 50); Kerat, Khashtadan (AN 99; NZ 03; IF 69). Kerosh (AN 15; VE&NS 86); Khoshtadan (NZ 03); 3 km S Shahrabad

Kord (FMNH 141477; VE&NS 86); Shahdorani, Sargad (AN 99; IF 69). **Ostan 12 (Mazanderan)**: Nardin (AN 15); Shahrud (AN 07; 15; type locality for *Ablepharus persicus*).

Chalcides ocellatus ocellatus (Forsskål, 1775)

Ostan 8 (Fars): Bushire (BMNH 87.12.20.9–12; JA 72; WB 76; JB 79; FW 17; KS 55); Tangistan (FW 17). **Ostan 9 (Kerman)**: Jask (BMNH 94.11.13.22–23; JA 98). **Ostan 10 (Baluchistan-Sistan)**: Chah Bahar (USNM 148659).

Eumeces schneiderii princeps (Eichwald, 1839)

Bendesun, Kopet Dagh [possibly in Turkmenistan](BMNH 8.5.25.28); Jazireh-ye Qowyun, Daryacheh-ye Reza'iyeh [Ostan 3 or 4] (BMNH 99.9.30.28–29; GB 99); SW Shahi Peninsula, Daryacheh-ye Reza'iyeh (BMNH 1965.1485); Shazelan Island, Daryacheh-ye Reza'iyeh [Ostan 3 or 4] (BMNH 99.9.30.30; GB 99); Vizastra, Azarbaijan [Ostan 3 or 4] (GB 99). **Ostan 1 (Tehran)**: Abhar (CAS 140630); between Chashmeh Safid Ab and Shah 'Abbas, Kavir Protected Region (GN&CA 81); Hesarak (CAS 140631–2); Karaj (CAS 140613–19); Maljat-Abad, Iraq-Ajemi (AN 07); Qazvin (CAS 140620–8); Qom (CAS 140629). **Ostan 2 (Gilan)**: Anguran Protected Region (MMTT 780); Divandarreh, 5800 ft [1769 m] (FMNH 170954); Pachenar (AN 07). **Ostan 3 (East Azarbaijan)**: Jonstan, S slopes Talysh Mountains (GN&CA 81); Maragheh (VR 38). **Ostan 4 (West Azarbaijan)**: Markan (DF *et al.* 97). **Ostan 5 (Kordestan-Kermanshah)**: Karand (BMNH 53.10.2.1); **Ostan 6 (Khuzestan-Lorestan)**: Bid Zard (AN 07); Dech-i-Diz [?Rud-e Dez] (AN 07); Malamir (AN 07); Qal 'eh-ye Tol (AN 07). **Ostan 8 (Fars)**: 93 km ESE Behbehan, 35 km E Gach Saran (USNM 153729–30; RT 71); near Neyriz, E Shiraz, 4000–5000 ft (BMNH 74.11.23.22–23; WB 76; JB 79; GB 87); Persepolis [Takht-e Jamshid] (FMNH 21008; KS 39; type locality for *Eumeces schneideri variegatus*); 38 km N Shiraz, just N Persepolis (USNM 153595; RT 71); Tang-e Tschekun (FW 17). **Ostan 11 (Khorasan)**: between Akhalo-Atek and Quchan (NZ 03); Askalabad, in Quchan (NZ 03); Gerri-Shotur, Khastadan (AN 99); Gululi Dag (AN 97); Kerat, Khastadan (NZ 03); Tajan River (Harirud), Afghan/Iran/Turkmenistan border (LL 18). **Ostan 12 (Mazanderan)**: high in the Persian mountains near the Aul of Aber [Gorgan] (AN 15); Mohammad Reza Shah National Park (GN&CA 81); Turkmen plains 42 km N Gorgan (USNM 149414; RT 71).

Eumeces schneiderii zarudnyi Nikolsky, 1900

Kuh-e Safid (FW 38). **Ostan 8 (Fars)**: Sarjan, SW Kerman, 5500 ft (WB 76; JB 79). **Ostan 10 (Baluchistan-Sistan)**: Baluchistan (CAS 140633); Bandan (NZ 03); Bampur (NZ 03); Bazman (AN 99; type locality); 11 mi [17.6 km] W Iranshahr (FMNH 141566–8, 141571–3, CAS 102559–61); Lab-e Bareng (AN 99); Pishin, 700 ft (WB 76); Shurab (AN 99).

Eumeces taeniolatus parthianicus Szczerbak, 1990

Ostan 11 (Khorasan): Sarakhs (LM 29); Tajan River [Harirud], Iran-Afghanistan-Turkmenistan border (LL 18).

Mabuya aurata septemtaeniata (Reuss, 1834)

Ostan 6 (Khuzestan-Lorestan): 90 mi [144 km] NE Ahvaz, 600 ft (BMNH 1905.10.14.43); Bid Zard (AN 07; 15); Dezful (AN 07; 15); Gholaman (DF *et al.* 97); Gurschir (AN 07; 15); Mahor Birinji (GH&YW 69); between Masjed Soleyman and Haft Gel (CAS 86388; SA 63); Pain-Gyatsch (AN 07); Qal'eh-ye Tol (AN 07; 15); E bank Rud-e Dez, about 3 km S game-guard post, Dez Protected Region (MMTT 692); Shadegan on road to Gorgar (RM 57); 12 mi [19 km] S Shush (FMNH 141348); ruins of Susa [Shush] (BMNH 1904.10.14.41–42); Tappeh Sharafabad (MMTT 515); Tol-e Bazun (CAS 86417; SA 63). **Ostan 8 (Fars)**: Abshar (DF *et al.* 97); 28°53′N, 51°02′E, near Alchangi, 33 km NE Bushire, 0–50 m (CAS 141145; MMTT 1065); Bushire (BMNH 87.12.20.2–8); Bushire peninsula (FW 17); 29°40′N, 51°59′E, 10 km by road E Dasht-e Arzhan, 1800 m (DF *et al.* 97, sight record); 35 km E Gach Saran (USNM 153727–8; RT 71); 30°39′N, 51°36′E, Kuh-e Dinar ridge, 10 km by road

E Yasuj, 1800–2300 m (DF *et al.* 97, sight record); Kusk Zar, 8000 ft [2440 m] (WB 76; JB 79; OB 80); Nurabad plain (FW 17); 30°14'N, 52°12'E, 8 km SSW Qader Abad, 2100 m (DF *et al.* 97, sight record); Qar Sharon (DF *et al.* 97, sight record); 30°05'N, 52°55'E, 10 km by road E Sivand, 1700 m (DF *et al.* 97, sight record); Tang-e Tschekun (FW 17).

Mabuya aurata transcaucasica **Chernov, 1926**
Between Esfahan and Tehran [Ostan 1 or 7] (BMNH no register number); Jazireh-ye Qowyun [Ostan 3 or 4] (BMNH 99.9.30.27; GB 99; VR 38); Lyutfabad (NZ 03); Mergen-ulja River, eastern Iran, near Turkmenistan border [probably Ostan 11 or 12] (AN 97; 15); Rud-e Atrak [Ostan 11 or 12] (LM 29); Sanger (LM 29). **Ostan 1 (Tehran)**: Darrous, N Tehran (CAS 96147–8; SA 63); Qazvin (FdF 65; JB 79; OB 80); near Qom (WB76; OB 80; GB 87); Tehran (BMNH 69.3.4.7–9; GB 87; RM 24; OW 51); Tehran, near Shahyad Square (MMTT 565). **Ostan 2 (Gilan)**: Mianeh (CAS 140608–12). **Ostan 3 (East Azarbaijan)**: foot of Kuh-e Sahand, upper reaches of Mjud- i-Chai, 8000 ft (VR 38); SW Shahi Peninsula [?=Jazireh-ye Shahi], Daryacheh-ye Reza'iyeh (BMNH 1965.1486–88). **Ostan 4 (West Azarbaijan)**: Guch Ali, about 30 km S Jolfa (UMMZ 129786); 38°52'N, 45°10'E, 21 km E Maku on road to Marand, 1090 m (CAS 141270); Markan (DF *et al.* 97). **Ostan 5 (Kordestan-Kermanshah)**: 3 mi [4.8 km] E Harivan (FMNH 170856); Quraitu (BMNH 1921.3.30.12). **Ostan 7 (Esfahan)**: Esfahan (FMNH 21016; KS 39). **Ostan 11 (Khorasan)**: Bojnurd, 1060 m (RC 91); Sarakhs, on the Harirud (NZ 03); River Tajan [Harirud], Iran-Afghanistan-Turkmenistan border (LL 18).

Mabuya vittata **(Olivier, 1804)**
Ostan 5 (Kordestan-Kermanshah): Sar-e Pol, about 500 m (JJS&JFS 72); Mahidasht, 25 km W Kermanshah (city) (CAS 207518).

Ophiomorus blanfordi **Boulenger, 1887**
Southern Iran or Pakistan (BMNH 91.9.14.5; WB 76; GB 87); "Persien" (ZMUG 83, 85; FW 17). **Ostan 10 (Baluchistan-Sistan)**: "Baluchistan?" (BMNH 80.11.10.88, holotype; BMNH 91.9.14.5); Chah Bahar (USNM 148660; SA&AL 66; RT 71; type locality).

Ophiomorus brevipes **(Blanford, 1874)**
Ostan 9 (Kerman): Kerman (AN 15; SA&AL 66); Minab (CAS 86593; SA 63; SA&AL 66); Sa'adatabad, 5500 ft [1678 m] (WB 74; 76; type locality; JB 79; SA&AL 66); Sabzavaran (OW 51; SA&AL 66). **Ostan 10 (Baluchistan-Sistan)**: Bampur (NZ 03); Bandan (NZ 03); Bazman (AN 15; SA&AL 66); Gjuische (NZ 03); 11 mi [17.6 km] W Iranshahr (CAS 101792; FMNH 141548–50, 141552–3; SA&AL 66); Kivask, 4500 ft, Persia-Baluchistan (BMNH 1919.5.2.25); Kuh-e Taftan (AN 15; SA&AL 66); Schur, Sargad (AN 15; SA&AL 66); between Zahedan and Chah Bahar (LF 50). **Ostan 11 (Khorasan)**: Cha-e Ziran (NZ 03); Khvaf (NZ 03); Khvaje Dow Chahi (NZ 03); Niazabad (NZ 03); Tag-e Dorokh (NZ 03); Torbat-e Heydariyeh (BMNH 1908.8.5.1).

Ophiomorus nuchalis **Nilson and Andrén, 1978**
Ostan 1 (Tehran): 34°44'N, 52°11'E, near Cheshmeh Shah, N foot Siah Kuh, Kavir Protected Region (GN&CA 78; 81; type locality).

Ophiomorus persicus **(Steindachner, 1867)**
"Persia" (FS 67); "Kurdistan" (BMNH 46.6.15.83 [2 specimens]; WB 76; GB 87). **Ostan 8 (Fars)**: 5 km E Pol-e Abgineh (CAS 101793–8; FMNH 141555, 141557, 141559–61, 141563, 141565, 141577–80, 141582; SA&AL 66; type locality); 30°05'N, 52°55'E, 10 km by road E Sivand, 1700 m (DF *et al.* 97, see pl. 18E).

APPENDIX II

Ophiomorus streeti **S. Anderson and Leviton, 1966**
Ostan 10 (Baluchistan-Sistan): 11 mi [17.6 km] W Iranshahr (FMNH 141551 [holotype]; CAS 100024; SA&AL 66; type locality).

Ophiomorus tridactylus **(Blyth, 1853)**
Ostan 10 (Baluchistan-Sistan): Nasirabad (BMNH 73.1.7.7–8; WB 76; JB 79; SA&AL 66); Neizar (AN 99; SA&AL 66); Perso-Baluch border, Sistan (NA 06; SA&AL 66); about 31°05′N, 61°42′E, dunes at Rud-e Hirmand on road from Zabol to Dust-e Mohammad Khan, 500 m (CAS 141089); about 31°03′N, 61°38′E, 10 km SW Rud-e Hirmand, abandoned village SE road from Zabol to Dust-e mohammad Khan, 450 m (CAS 141101; MMTT 942–3); Schile, Sistan (AN 99; NZ 03; SA&AL 66); Zabol (CAS 140606–7); 15 mi [24 km] SW Zabol (FMNH 141376; SA&AL 66). **Ostan 11 (Khorasan)**: Chadschi-du-i Tschagi, Zirkukh (AN 99; SA&AL 66).

Scincus scincus conirostris **Blanford, 1881**
Ostan 6 (Khuzestan-Lorestan): Agha Jari (CAS 86479–80; SA 63; EA&AL 77); 21 km N, 1 km W Ahvaz (MMTT 59). **Ostan 8 (Fars)**: Bushire (EA&AL 77); Tangistan, coastal area S Bushire (BMNH 87.9.22.41; EA&AL 77); Tangyak, 7 mi [11 km] S Bushire (BMNH 79.8.15.1–3/1946.8.20.55–57; WB 81; EA&AL 77; type locality).

Uromastyx aegyptius **(Forsskål, 1775)**
Ostan 8 (Fars): Rudkhaneh-ye Shapur (FW 17). **Persian Gulf Islands**: Jazireh-ye Sirri (photographic record, Pl. 19A).

Uromastyx asmussi **(Strauch, 1863)**
Ostan 1 (Tehran): 26 km N headquarters, Kavir Protected Region (MMTT 94); Shah 'Abbas Caravanserai, Kavir Protected Region (MMTT 154). **Ostan 7 (Esfahan)**: Shekar Ab, 980 m (FW 36). **Ostan 9 (Kerman)**: near Rigan, 2500 ft (WB 76; GB 85). **Ostan 10 (Baluchistan-Sistan)**: "Baluchistan" (CAS 154357); Bandan, Sistan (NZ 03); 8 km NW Iranshahr, 800 m (RM 56); Khusrin, about 90 mi NW Bampur, 1800 ft (WB 76); between Murgak and Shurab (NZ 03); between Tscha-i-Gjuische and Zeinabad (NZ 03); between Zahedan and Chah Bahar (LF 50). **Ostan 11 (Khorasan)**: between Fedeshk and Chamur (AN 97); Sar Chah (AS 63; AN 97; type locality).

Uromastyx loricatus **(Blanford, 1875)**
Persian Gulf [coast? Ostan 6 or 8] (BYU 20965; RC *et al.* 66). **Ostan 5 (Kordestan-Kermanshah)**: 8 km E Qasr-e Shirin (USNM 153599; RT 71). **Ostan 6 (Khuzestan-Lorestan)**: 25 km E Agha Jari (my sight necord); Choqa Zanbil (my sight record; DF *et al.* 97, sight record); 31°45′N, 49°08′E, near Darkhazineh (CAS 86469); between Masjed Soleyman and Ahvaz (CAS 86379; SA 63); between Masjed Soleyman and Haft Gel (CAS 86463, 86470; SA 63); Naft-e Safid oil field (CAS 86380; SA 63); 15 mi [24 km] S Shalgahi (GH&YW 69); Yamaha (CAS 86468; SA 63). **Ostan 8 (Fars)**: Ahram (CAS 102479–80; FMNH 135999, 141432, 141434–5, 141437); 25 km NW Borazjan (MMTT 498); Bushire (WB 75, 76; JM 84; GB 85; type locality); Bushire peninsula (FW 17); Chahak (DF *et al.* 97); Gurak (WB 81; GB 85); Milak (FW 17); Tangistan (FW 17).

Varanus bengalensis bengalensis **(Daudin, 1802)**
Ostan 8: Zizi, at Vik River [?] (WA 94). **Ostan 9 (Kerman)**: "Lut Desert" (WA 94); Minab (OW 51); Rudan, Minab River Bridge (WA 94). **Ostan 10 (Baluchistan-Sistan)**: Bampur (NZ 03); Bent, 108 km W Nikshahr (MMTT 714–15); Daftan Village near Zahadan (WA 94); Jaz Murian (BMNH 1951.1.6.64); Pishin, 700 ft (WB 76); Rud-e Bampur (AN 99; NZ 03); Rud-e Kyagur (NZ 03); Sarbaz (NZ 03).

Varanus griseus (Daudin, 1803)

Between Bushire [Ostan 8] and Bandar 'Abbas [Ostan 9] (BMNH 94.11.13.13); Ghainak [probably Ostan 8] (BMNH 79.8.15.16–17; WB 81); Ne [Nehbandan?, Ostan 11] (NZ 97). **Ostan 1 (Tehran)**: Kavir Protected Region (MMTT 771, 1370); Qom (CAS 143291); 34°33'N, 52°12'E, between Shah 'Abbas and Cheshmeh-ye Sefid, Kavir Protected Region (GN&CA 81); Varamin plain (FdF 65; WB 76; JB 79). **Ostan 6 (Khuzestan-Lorestan)**: about 31°16'N, 49°11'E, sand dunes on road between Ahvaz and Haft Gel (CAS 86630–1; SA 63); Rud-e Dez, 2 km W Emamzadeh Ibin Ja'far (UMMZ 133282). **Ostan 8 (Fars)**: 35 km E Gach Saran (USNM 160302; RT 71); Konar Takhteh (BMNH 79.8.15.4; WB 81). **Ostan 9 (Kerman)**: 'Ambarabad, valley on W flank Kuh-e Jebal Barez, Jiroft (RM 57). **Ostan 10 (Baluchistan-Sistan)**: Bandan, Birjand range (NZ 97; 03); Espakeh (RM 56); Garne (NZ 03); Gjuishe (NZ 03); 8 km NW Iranshahr (RM 56); 30 km N Iranshahr (USNM 149140; RT 71); about 28°03'N, 60°53'E, 47 km S Khash, 1480 m (MMTT 1443); Perso-Baluch border, Sistan (NA 06). **Ostan 11 (Khorasan)**: 'Aliabad, Birjand range (NZ 97); Ahangeran (NZ 03); Atkul, 15 km N Ahangeran (NZ 03); Desht-i-Gusseinabad, between Chah-e Ziran and Bandan (NZ 03); Fandokht (NZ 03); between Garm Ab and Bohnabad, Zirkukh (NZ 97; AN 05; 15); Khvajeh Dow Chahi (NZ 03); Mohammadabad [several places of this name — 33°27'N, 60°10'E or 33°32'N, 58°47'E] (AN 97; NZ 03); Nus or Nusi, Kavir-e Namak (NZ 97; AN 97; 05; 15); Sarakhs (LM 29; NZ 03); Tag-i-Dorokh, between Dorokh and Chah-e Ziran (NZ 03); Tauran (FW 36); Zirkukh (NZ 03).

APPENDIX III

GAZETTEER

In the gazetteer, I have attempted to include all of the Iranian reference localities mentioned in the herpetological literature. I have been able to associate most of these with place names given in the United States Board on Geographic Names Gazetteer no. 19 *Iran*, 1956. In the list that follows here, where applicable, the reference name used in the literature is given first. Where that name differs from the U.S. Board on Geographic Names Gazetteer listing, the latter is given in parentheses. The latitude and longitude coordinates as shown in the gazetteer follow. Wherever possible, I have also recorded the approximate elevation as given in the original source or estimated as accurately as possible from the maps of the U.S. Army Map Service or the U.S. Air Force Operational Navigation Charts. Elevational ranges, e.g., 1000–3500 ft, indicate that the topographic relief at the reference site ranges from the base at 1000 ft [or m] to a peak or plateau having an elevation of 3500 ft [or m] Elevations should be considered approximations except where a precise maximum is shown (e.g., 3000–9524 ft indicates a base at approximately 3000 ft, but a measured maximum elevation at 9524 ft).

Variations in the spelling of Iranian place names present great difficulties in interpreting the distribution of animals from the literature and from museum records. Many names have been recorded by collectors on the basis of oral information from local people and then spelled phonetically in English, French, German, or Russian, depending on the native language of the collector. Subsequently, these have often been transliterated into another language used in the published report. Often, local names differ from those appearing in atlases, and there are many villages, mountains, streams, etc. of the same name. I have listed the names as they appear on maps and in reports rather than in the inverted order usual in gazetteers: e.g., Rud-e Atrak rather than Atrak, Rud-e, and Kuh-e Ginau rather than Ginau, Kuh-e (*rud* means river, *kuh* means mountain). Insofar as possible, I have relied on itineraries given in the reports or maps published by the authors, but, in many cases, no itinerary or map has been presented. Frequently, villages have been relocated or renamed over the past 100–200 years, further complicating matters. Even the provinces and other political areas vary greatly from one map to another. Since the Islamic Revolution, many place names have been altered in Iran, and I have made no attempt to follow these changes here. Ostan boundaries and names have changed in the years since I began this project, and the process of political reorganization continues. An attempt to bring the literature records into conformity with present usage proved too arduous and time-consuming. I continue to use here the Ostan names and boundaries based on Fisher (1968:4, fig. 2) (see text, fig. 1, p. 11).

The following statement is taken from *Map of Islamic Republic of Iran* (1984; map no. 169), published by Karun: "In accord with The Law of Administrative Divisions and on the basis of the latest administrative divisions, Iran consists of 24 provinces (Ostan) 195 Townships (Shahrestan) and 498 districts (Bakhsh). Each province is divided into several townships, each township into several districts, each district into several Rural Districts, and each Rural district into several villages. Each division is presided over by a Governor-General (Ostandar), each township by a Governor (Farmandar), each district by a district Governor (Bakhsdar) and each Rural District by a Rural District Administrator (Dehdar) and each village by a headman (Kadkhoda)."

The provinces (Ostan) designated on the above-cited 1984 map of Iran are: Azarbayejan-e-Gharbi, Azarbayejan-e-Sharqi, Bakhtaran, Boyerahmad-va-Kohgiluyeh, Bushehr,

Chaharmahal-va-Bakhtiyari, Esfahan, Fars, Gilan, Hamadan, Hormozgan, Ilam, Kerman, Khorasan, Khuzestan, Kordestan, Lorestan, Markazi, Mazandaran, Semnan, Sistan-va-Baluchestan, Tehran, Yazd, Zanjan (see text, fig. 2, p. 12).

It is important that collectors in the future tie all localities to some standardized gazetteer and current detailed reference map, giving latitude, longitude, and elevation as accurately as possible, preferably utilizing the Global Positioning System. Road distances are useful only when citing a specific map, as roads are frequently rerouted.

The following terms occur in the standard names listed in the gazetteer or on maps of Iran or both.

Farsi	*English*	*Farsi*	*English*
ab	stream, spring, well	kaur	stream
anbar	tank	kavir	salt desert, salt waste
bandar	harbor, anchorage, bay	khalij	bay, gulf
bar andaz	halting place	kharabeh, karabehha	ruins
batlaq	marsh	khirr	stream
berkeh	tank, pool	khowr	inlet, stream, channel, bay
biaban	desert	kotal	pass
borj	tower, fort	kowr	stream
chah	well(s), spring(s)	kuh	mountain(s), range, hill, peak
cham	stream, gorge	kuhha	mountains, range, hills
chashmeh	spring, well	lut	desert
chay	stream	nahr	stream, canal
cheng	promontory, hill	namakzar	salt waste
dagh	mountain(s), hill(s)	ostan	province
dahaneh	gorge, defile, pass, water gap	pal	hill, mountain
dakhmeh	burial caves	paskuh	mountain range
daqq	salt flat, salt depression	pereval	pass
darband	gorge	poshteh	hill, mountain
darreh	stream, valley, gorge, ravine	qabr	tomb
daryacheh	lake, marshy lake	qabrestan	cemetery
dasht	plain, desert, steppe	qal'eh	fort
emamzadeh	tomb, shrine	qolleh	peak, hill, mountain
farmandari-ye koll	special province	ramlat	sandy area
gadik, gaduk	pass	ra's	cape, point
galal	stream	reshteh	mountain range, hills
gardan, gardaneh	pass	rud	stream
godar	pass	rudkhaneh	stream
hamun	salt waste, marshy lake	sar	cape
Hawr, howr	marsh	sarab	spring
howz	tank	saray	caravansary (walled camp)
ishan	hill(s)	selseleh	mountain range, mountains
istgah	railroad station	shahrestan	district
jabal	hill(s), mountain	shatt	stream
jazirat, jazireh	island	shebh-e jazireh	peninsula
jebal	hill(s)	shekasteh	mountain, hill
ju, juy	stream	shur	stream
kal	stream	tall	hill, mountain
kalleh	peak	tang	gorge, stream, pass
kamar	mountain, hill, ridge	tappeh	hill, mountain
kani	well	ziarat	shrine, tomb
karavansara	caravansary (walled camp)		

APPENDIX III

Ostan names used below are as follows: 1. Tehran (Central); 2. Gilan; 3. East Azarbaijan; 4. West Azarbaijan; 5. Kordestan-Kermanshah; 6. Khuzestan-Lorestan; 7. Esfahan; 8. Fars; 9. Kerman; 10. Baluchistan-Sistan; 11. Khorasan; 12. Mazanderan.

Most of the diacritical marks have been omitted to facilitate manuscript preparation.

Locality	Ostan	Latitude	Longitude	Elevation
'Abasabad	7	33 53	55 02	3000 ft
'Abbasabad	12	36 22	56 25	3000 ft
'Ajab Shir	3	37 28	45 54	
'Aliabad	11	32 31	59 49	6000 ft
'Aliabad	1	34 06	51 23	3000 ft
'Ammeh	3	37 17	47 07	7000 ft
'Aqda	7	32 25	53 38	
'Anbarabad	9	28 35	57 55	
'Arusan	7	33 29	54 56	928–1065 m
Ab 'Ali	1	35 47	51 58	
Ab Ask (near Amol)	12			1620 m
Ab Bid	6	32 32	48 54	
Ab-bir (Ab Bid)	6	32 32	48 54	
Ab-e Dam Dalleh	6	31 31	49 49	
Ab-e Golestan	6	32 00	49 06	
Ab-e Ragn spring, Turan Biosphere Reserve	12			1530 m
Ab-e Sezar	6	32 43	48 35	
Ab-e Shirin, 100 km W Khonj	8			
Ab-e Tembi	6	31 50	49 28	1200 ft
Ab-i-Cezar (Ab-e Sezar)	6	32 43	48 35	
Ab-i-Khornos	7	32 48	48 47	2000 ft
Ab-i-Marik, near Namivand, 20 mi NW Kermanshah	5			
Abadan	6	30 20	48 16	
Abadeh	8	31 10	52 37	6000 ft
Abardesh Railway Station (Istqah-e Abardezh)	1	35 15	51 47	
Abas, Zirkukh region, vicinity of Gazik	11			
Abas-Abad ('Abbasabad)	7	33 53	55 02	3000 ft
Abbas-Abad ('Abbasabad)	12	36 22	56 25	3000 ft
Abed (Ab Bid)	6	32 32	48 54	
Aber	12			
Abgababa (Aqa Baba)	1	36 20	49 46	4000–5000 ft
Abghale	11			
Abhar	1	36 09	49 13	
Abshar	8	30 23	51 30	1000 m
Abtar	10	27 14	60 53	2000–3000 ft
Abu Karaniyeh	6	31 42	48 50	
Abu Tavil	8?			
Abu-Garia (Abu Karaniyeh)	6	31 42	48 50	
Achangerun (Ahangeran), Zirkuch	11	33 25	60 13	
Adji-Tschay Valley (Talkheh Rud)	3	37 40	45 46	
Adzhabshir ('Ajab Shir)	3	37 28	45 54	5000 ft
Afineh, 77 km S Khorramabad on road to Andimeshk	6			1100 m
Agh Bolagh	4	38 37	45 13	
Agha Jari	6	30 43	49 49	1000 ft

Locality	Ostan	Latitude	Longitude	Elevation
Aghlagan	3			
Aguljaschker	6			
Ahangeran	11	33 25	60 13	
Ahar	3	38 28	47 04	
Ahmadabad	12	35 46	56 37	ca 1000 m
Ahram	8	28 52	51 16	500 ft
Ahvaz	6	31 19	48 42	50–200 ft
Ahwaz (Ahvaz)	6	31 19	48 42	50–200 ft
Ahwaz Ridge	6	31 18	48 45	100–500 ft
Ain u Wahir	?			1194 m
Ain-ar-Rashid ('Eyn or Rashid)	1	34 43	52 10	4000 ft
Ajefan	4	38 00	44 25	
Aji Chai (Talkheh Rud)	3	37 40	45 46	
Ak-Kala (Pahlavi Dezh)	12	37 01	54 30	sea level
Akhalo-Atek	11			
Al Baji, near Ahvaz	6			
Ala Khvorshid	6	31 22	49 52	
Alami Pass (Gardaneh-ye 'Alam 'Ali)	11	37 22	58 30	2000 m
Alborz Mountains (Resteh-ye Kuhha-ye Alborz)	1,2,12	36 00	53 00	
		(ca 36 40	49 30	
		to 37 20	56 00)	
Alchangi	8			
Alchorschir (Ala Khvorshid)	6	31 22	49 52	
Alem	?			
Ali Abad ('Aliabad)	11 ?	32 31	59 49?	
Ali Kosh, Dehloran	6	32 41	47 16	
Aligudarz	6	33 24	49 41	
Alirezaabad	3			
Alma-Saraj (Ghelmansaray)	3	38 18	45 07	5000 ft
Almab	?			
Almeh Valley, Mohammad Reza Shah National Park	12			
Altalykh, Kuhha-ye Talysh	3			
Althangerun (Ahangeran)	11	33 25	60 13	
Alvand	1	36 19	49 10	1900–2000 m
Ambar-Abad ('Anbarabad)	9	28 35	57 55	4000 ft
Aminabad (many places of this name)	?			
Amirabad	5	33 20	46 16	
Amol	12	36 23	52 20	
Anar	9	30 53	55 18	
Anarak	7	33 20	53 42	
Anbar-Abad ('Anbarabad)	9	28 35	57 55	
Andimeshk	6	32 27	48 21	150 m
Angun, a few km NW Torbat-e Jam	11			
Anguran Protected Region	2	36 50	47 30	
Annabu	12	36 00	56 01	1300 m
Anzali	2	37 28	49 27	
Aptan (Abtar)	10	27 14	60 53	2000–3000 ft
Aqa Baba	1	36 20	49 46	4000–5000 ft
Arabistan (Khuzestan Province)	6			
Arak	1	34 05	49 41	5700 ft

APPENDIX III

Locality	Ostan	Latitude	Longitude	Elevation
Arak Adschemi (Iraq Ajemi)	1	34 30	52 00	
Aran	1	34 04	51 29	3000 ft
Ardabil	3	38 15	48 18	4500 ft
Ardebil (Ardabil)	3	38 15	48 18	4500 ft
Arezou Island (Jazireh-ye Arzu)	3	37 32	45 36	
Argavani	1	34 19	51 23	2000–3000 ft
Aryamehr Botanic Garden, W Tehran	1			
Arzu Island (Jazireh-ye Arzu)	4	37 32	45 36	4000–5000 ft
Asadbar, Alborz Mountains	?			2500 m
Asalem (Navrud)	2	37 42	48 57	
Asaran	1	35 52	53 23	
Aschar-Adé Island, Caspian Sea (Ashuradeh-ye Bozorg)	12	36 50	53 56	
Ashin (Kuh-e Ashin)	7	33 31	53 28	4000–6500 ft
Ashuradeh-ye Bozorg	12	36 50	53 56	
Asian	1	35 44	49 17	6000 ft
Askalabad	11			1800 m
Askan (Eskan)	10	26 48	63 09	3000 ft
Assalem, Talish Mountains (Navrud)	2	37 42	48 57	800 m
Astara	2	38 26	48 52	
Astrabad (Gorgan region)	12	37 00	54 30	
Atkul, 15 km N Ahangeran	11			
Atrek River (Rud-e Atrak)	12	37 28	54 03	
Aul of Aber	12?			
Avaj	1			
Avaz	11	32 56	60 15	4000–5000 ft
Awaz (Avaz)	11	32 56	60 15	4000–5000 ft
Ayubi	11	34 18	60 35	
Äzna (Istgah-e Ezna)	6	33 27	49 28	1800 m
Baaza	11			
Baaza River, Zirkukh	11			
Baba Kuh	12	35 37	56 50	1160 m
Babol	12	36 34	52 42	below sea level
Babol Sar	12	36 43	52 39	
Badschistan Kavir (Kavir-e Namak)	11	34 30	57 40	2000–3000 ft
Baft	9	29 14	56 38	
Bag e Taj	?			
Bag e Tschinar (Bagh-e Chenar)	9	28 11	56 54	4000–9060 ft
Bagh-e Chenar	9	28 11	56 54	4000–9060 ft
Baghestan	12	35 34	56 46	1275 m
Bagiran-Kukh (Kuh-e Baqeran)	11	32 43	59 20	
Bahram	1	35 28	51 32	3500 ft
Bahramabad	1	36 32	50 10	
Bahu Kalat	10	25 43	61 25	0–500 ft
Bakaran Mountains (Kuh-e Baqeran)	11	32 43	59 20	6000–8899 ft
Bak Emishchir	6			
Bakhtiari region (Shahrestan-e Bakhtiari va Chahar Mahall)	6	32 00	50 00	
Bala-Khaf	11	ca 34 33	60 08	

Locality	Ostan	Latitude	Longitude	Elevation
Balkcha, or Balkhah, near Halabja, Iraq	5	35 12	46 08	
Bam	9	29 06	58 21	4000 ft
Bampur	10	27 12	60 27	2000 ft
Bampur depression (Hamun-e Jaz Murian)	10	27 20	58 55	
Bampur River (Rud-e Bampur)	10	27 08	59 06	
Bampusht (Kuh-e Bam Posht)	10	26 45	62 30	3000–5625 ft
Bamrud	11	33 39	60 09	
Bam-Rud (Bamrud)	11	33 39	60 09	
Bamu National Park	8	29 43	52 34	
Bandabad (Bohnabad)	11	34 05	59 53	3000 ft
Bandamir Valley (Rud-e Kor)	8	29 36	53 18	5000–6000 ft
Bandan	10	31 23	60 44	
Bandar 'Abbas	9	27 11	56 17	sea level
Bandar-e Danalu	3	37 28	45 49	4200 ft
Bandar-e Gonaveh	8	29 34	50 31	20 m
Bandar-e Kong	8	26 35	54 56	
Bandar-e Lengeh	8	26 33	54 53	sea level
Bandar-e Maqam	8	26 56	53 29	
Bandar-e Pahlavi	2	37 28	49 27	sea level
Bandar-e Shahpur	6	30 25	49 05	sea level
Bandar-e Vanalu	3			
Bandar Langeh (Bandar-e Lengeh)	8	26 33	54 53	sea level
Bandar Pahlavi (Bandar-e Pahlavi)	2	37 28	49 27	sea level
Band Qir	6	31 39	48 53	
Baniabad (Bohnabad)	11	34 05	59 53	3000 ft
Baraghan	1	35 58	50 57	
Bard-e Nishunde	6	31 57	49 21	1500 ft
Bard-i-nishunde (Bard-e Nishunde)	6	31 57	49 21	1500 ft
Bardu	11	35 27	60 07	5000 ft
Barferusch (Babol)	12	36 34	52 42	below sea level
Baschm-Sorshe	?			2500 m
Bash Nurashin, Reza'iyeh Basin	?			
Basiran	11	31 57	59 06	4000–5000 ft
Basmenj	3	37 59	46 29	6000 ft
Basminsk (Basmenj)	3	37 59	46 29	6000 ft
Bastak	8	27 14	54 22	
Batvand	6	32 00	49 08	1000 ft
Batwand (Batvand)	6	32 00	49 08	1000 ft
Bazargan	4	39 23	44 23	
Baz-Chouz Pain, a few km N Sang Bast	11			
Baz-Khouz-Pain, or Baza-Khouz-Kala, a few km N Sang Bast	11			
Bazman	10	27 49	60 12	3000 ft
Bechars (Bekhars)	11	ca 35 00	60 20	
Behbehan	6	30 35	50 14	
Beh Cheshme, Khosh Yeilagh Protected Region	12			
Behshahr	12	36 43	53 34	
Bekhars Region, SW Torbat-e Jam	11	ca 35 00	60 20	
Beljaki	5			
Benarve	4			
Bender Abbas (Bandar 'Abbas)	9	27 11	56 17	sea level

APPENDIX III

Locality	Ostan	Latitude	Longitude	Elevation
Bendun (Bandan)	10	31 23	60 44	
Bent	10	26 17	59 31	
Berdesur River (Rud-e Bardeh Sur)	4	37 27	44 55	
Berdu Mountains (Bardu?)	11	35 27	60 07 ?	
Beris	10	25 09	61 11	
Bernarve, Shamsdinan, W Daryacheh-Ye Reza'iyeh	4			
Beydarvaz	5	35 16	46 09	
Bezd	11	35 05	60 50	
Biare (Beydarvaz)	5	35 16	46 09	
Bid, Sargad region, 27–30 km SW Zahedan				
(? Kala-i Bid)	10	29 09	61 03 ?	
or (Qal'eh-ye Bid)	10	28 40	60 24 ?	
Bidezar (Bid Zard)	6	31 41	49 36	3000 ft
Bid Zard	6	31 41	49 36	3000 ft
Bijar	2	35 52	47 36	
Bijenah, 7 km E Kermanshah (Birjaneh = Parchineh)	5	36 43	48 56	
Binak	8	29 44	50 19	sea level
Birdjend (Birjand)	11	32 53	59 13	4800 ft
Birdschan (Birjand)	11	32 53	59 13	4800 ft
Birdzhand (Birjand)	11	32 53	59 13	4800 ft
Birjand	11	32 53	59 13	4800 ft
Birjand district (Shahrestan-e Birjand)	11	32 30	59 30	
Bishapur Cave	8	29 44	51 34	1000 m
Bisheh Porem (Istgah-e Bisheh)	6	33 20	48 52	1200 m
Bisitun (Bisotun)	5	34 23	47 26	5000 ft
Bisotun	5	34 23	47 26	5000 ft
Bogrov-Dag (Kuhha-ye Talysh)	3	37 35	48 38	
Bohnabad	11	34 05	59 53	3000 ft
Bojnurd	11	37 28	57 19	
Bolazam	?			
Borazgan (Borazjan)	8	29 16	51 12	
Borazjan	8	29 16	51 12	500 ft
Borazjun (Borazjan)	8	29 16	51 12	500 ft
Borujerd	5	33 54	48 46	
Boshruyeh	11	33 53	57 26	
Bostanabad	3	37 50	46 50	
Boz-Chouz-Pain (N Sang Bast, probably Baz Hauz)	11	36 05	59 45 ?	3500 ft?
Bund-i-Kir, E Ahwaz (Band Qir)	6	31 39	48 53	250 ft
Buni-Abad (Bohnabad)	11	34 05	59 53	3000 ft
Buschähr (Bushire or Bushehr)	8	28 59	50 50	0–350 ft
Bushehr (traditionally Bushire)	8	28 59	50 50	0–350 ft
Bushire (Bushehr)	8	28 59	50 50	0–350 ft
Buszhnurt (Bojnurd)	11	37 28	57 19	
Byelyaki, Kurdistan	?			
Cah Gurg (Chah-e Gorg)	7	34 13	52 32	3000 ft
Castab (Kuh-e Chastab)	7	33 40	55 10	3000–4000 ft
Cesme Beiman	?			

Locality	Ostan	Latitude	Longitude	Elevation
Cesmeh Gi	?			
Chadschi-Abad, Zirkukh	11			
Chadschi-du-i Tschagi	11			
Chahak, 15 km NW Bandar-e Gonaveh by road	8	29 40	50 25	20 m
Chahar Berkeh	8	27 05	54 35	700 ft
Chahar Deh	12	36 26	54 18	7000 ft
Chahar Zebar, about 17 mi SW Kermanshah	?			
Chahbagh, near Bushire	8			
Chah Bahar	10	25 18	60 37	20 ft
Chah Bahar Bay (Khalij-e Chah Bahar)	10	25 20	60 30	
Chahbar (Chah Bahar)	10	25 18	60 37	20 ft
Chah-e Ebrahim Zare'	1	34 07	54 57	
Chah-e Gorg	7	34 13	52 32	3000 ft
Chah-e Jangal	11	32 39	58 21	
Chah-e Ziran	11	31 58	60 28	
Chah-i-Gjuische, 40 km NW Bandan in Nehbandan	10			
Chah-i-Sagak, near Duruh	11			
Chahlak	9	27 46	57 02	2000 ft
Chah Moslem	8	26 44	54 35	
Chah Qar Qareh, Kavir Protected Region	1			
Chah Suguti	?			
Chakh-i-Bena, 37 km S Dastgerd	11			
Chakh-i-Dzhanu, or Tepe Dervish, about 35 km SW Hauzdar	10			
Chakh-i-Gyuishe, 40 km NW Bandan	10			
Chakh-i-Ziru (Chah-e Ziran)	11	31 58	60 28	
Chalus	12	36 38	51 26	−20 m
Chamur (about 23 mi SW Qusheh)	?			
Chara (Khara)	7	32 07	52 44	5000 ft
Charakhs, 28 km from Ahangeran	11			
Charbagh, near Bushire	8			
Charbar (Chah Bahar)	10	25 18	60 37	20 ft
Charsang Mountains	1	36 25	49 40	5000–8000 ft
Chartschäng, on the Siyahrud	12 ?			
Chaschtadan (Choscht-Adan Mountains)	11	34 34	60 33	
Chashmeh Manqur, near Mahmudabad	12			
Chashmeh Safid Ab	1	34 20	52 15	
Chashmeh Zardabeh, Khosh Yeilagh Protected Region	12			
Chashme Rowghani	6	31 33	49 41	2000 ft
Chashme Sangi, about 35 km N Bijar	2			
Chauzdar (Hauzdar)	10	30 34	61 03	
Cheh Mosullum or Chah Moslem	8	26 44	54 35	
Chejran, between Astara and Ardabil	2			
Chelane (Khelejan?)	3	38 01	46 08 ?	
Cheljane (Khelejan?)	3	38 01	46 08 ?	
Cheraghabad	5	35 13	47 20	
Cherikhabad, SW Reza'iyeh	4			
Cherra	4			
Cheshman Ap	?			
Cheshmeh Shah	1	34 44	52 11	

APPENDIX III

Locality	Ostan	Latitude	Longitude	Elevation
Cheshmeh Sefied Ab (Chashmeh Safid Ab)	1	34 20	52 15	
Cheshmeh-ye Sefid Ab (Chashmeh Safid Ab)	1	34 20	52 15	
Chodschi-Abad	?			
Chodschib	2	37 ??	49 ??	
Chogha Pahn, 2 km N Jerud Dam	6			
Chogha Zanbil (Choqa Zanbil)	6	32 00	48 31	100 m
Chohok spring, Turan Biosphere Reserve	12			1025 m
Choi (Khvoy)	4	38 33	44 58	4000 ft
Choqa Zanbil	6	32 00	48 31	100 m
Chorramabad (Khorramabad)	12	36 46	50 53	500 ft
Chorti, N slope Alborz Mountains	12			
Choschkeri	7			
Choscht-Adan Mountains	11	34 34	60 33	
Chouz, Zirkukh	11			
Chur (Khur)	7	33 47	55 03	3000 ft
Daftan, near Zahedan	10			
Dagh	4			
Dak-i-Do, Sargad Region	10	28 59	60 29	
Dalaki	8	29 26	51 17	
Damaneh	7	33 01	50 29	
Damavand	1	35 43	52 04	6000 ft
Damdeli (Ab-e Dam Dalleh)	6	31 31	49 49	
Damgan (Damghan)	12	36 09	54 22	4000 ft
Damghan	12	36 09	54 22	4000 ft
Danalu port (Bandar-e Danalu)	3	37 28	45 49	4200 ft
Daradez, between Jolfa and Tabriz	4			
Dare Asbar	6			
Dare Astane, Ostrankuh Protected Region	6			
Daria-i-Namak (Daryacheh-ye Namak)	1	34 45	51 36	2000 ft
Dar-i-Khazineh (Darkhazineh)	6	31 54	48 59	500 ft
Darkhazineh	6	31 54	48 59	500 ft
Darpahan	?			
Darreh-ye Palangi, Mohammad Reza Shah National Park	12			
Darrous, suburb of Tehran	1	35 43	51 26	4500 ft
Daryacheh-ye Bakhtegan	8	29 20	54 05	5070 ft
Daryacheh-ye Maharlu	8	29 25	52 50	
Daryacheh-ye Namak	1	34 45	51 36	2000 ft
Daryacheh-ye Reza'iyeh	3,4	37 40	45 30	4180 ft
Daryacheh-ye Sistan	10	31 00	61 15	
Darya-i-Namak (Daryacheh-ye Namak)	1	34 45	51 36	2000 ft
Dascht-i-Kawir (Dasht-e Kavir)	7,11	34 40	54 30	
Dash-Bouroun (Dashliborun), frontier post on Atrak River NE Gonbad-e Kavus	12			
Dashliborun	12			
Dasht	12	37 17	56 07	
Dasht-arjan (Dasht-e Arzhan)	8	29 39	51 58	
Dasht-e Arzhan	8	29 39	51 58	1800 m
Dasht-e Gilan (Dasht-e Gorgan?)	12			

Locality	Ostan	Latitude	Longitude	Elevation
Dasht-e Gorgan	12	37 30	55 45	
Dasht-e Kavir	7,11	34 40	54 30	
Dasht-e Lut	7,9,11	33 00	57 00	
Dasht-e Moghan	3	39 40	48 15	
Dasht-e Naz	12	ca 36 30	53 10	
Dasht-e Tal, 30 km SSE Shahrud	12			
Dasht-i-Kavir (Dasht-e Kavir)	7,11	34 40	54 30	
Dasht-i-Lut (Dasht-e Lut)	9	30 13	58 47	
Dashti Naz (Dasht-e Naz)	12	ca 36 30	53 10	
Dasteng-Kela, on Rudkhaneh-ye Talar	12			
Dastgerd	11	32 20	59 41	6000 ft
Dastgirt (Dastjerd?)	7 ?	32 11	52 43 ?	
or (Dastgerd?)	11 ?	32 20	59 41 ?	
Dastjerd	7	32 11	52 43	5000 ft
Dastjerd	12	36 11	56 09	900 m
Davar Panar	10	27 21	62 21	4000 ft
Dech et Chimmer	?			
Dech-i-Diz (Rud-e Dez?)	6	31 39	48 52 ?	
Degak (Dehak)	10	27 11	62 37	4000 ft
Dehak	10	27 11	62 37	4000 ft
Deh Baha, W Yazd	7			9500 ft
Deh Barez	9	27 26	57 12	
Dehbid (Deh Bid)	8	30 38	53 13	8000 ft
Deh Bid	8	30 38	53 13	8000 ft
Deh-e Jami	5	34 23	46 14	
Deh-i-Schuturun	9			2200 m
Dehloran	6	32 41	47 16	
Deh Namak	1	35 15	52 44	
Deh Pabid	10	28 37	60 46	
Deh Salm	10	31 12	59 19	800 m
Dekh-i-Pabid (Deh Pabid)	10	28 37	60 46	
Delbar	12	35 57	56 02	1200–1300 m
Delijan	1	33 59	50 40	
Demawend village (Damavand)	1	35 43	52 04	6000 ft
Descht-i-Kewir (Dasht-e Kavir)	7,11	34 40	54 30	
Desht	3 or 4			
Desht-i-Guseinabad, between Chah-e Ziran and Bandan	11			
Dez (Istgah-e Dezh)	6	31 43	48 38	500 ft
Dezful	6	32 23	48 24	500 ft
Dezh Shapur	5	35 31	46 10	
Dilman (Shapur)	4	38 11	44 47	
Divandarreh	2	35 55	47 02	
Dizak (Davar Panah)	10	27 21	62 21	4000 ft
Dizful (Dezful)	6	32 23	48 24	500 ft
Djatschm	1	35 50	53 04	2200 m
Djemal-Bariz-Massiv (Kuh-e Jebal Barez)	9	28 30	58 20	
Djiroft (Jiroft)	9	28 25	57 45	2000 ft
Doab	12	36 07	51 32	
Do-ab (Doab)	12	36 07	51 32	
Domdar	?			768–799 m

Locality	Ostan	Latitude	Longitude	Elevation
Dorokh (Duruh)	11	32 17	60 30	
Doroud (Dow Rud)	6	33 28	49 04	7000 ft
Dorud (Dow Rud)	6	33 28	49 04	7000 ft
Dorudsand, Marv Dasht Plain, 40 mi N Shiraz	8			
Doruneh	11	35 10	57 18	
Dow Ab	1	36 01	52 59	
Dow Rud	6	33 28	49 04	7000 ft
Dschalek (Chahlak?)	9	27 46	57 02 ?	
Dschandak (Jandaq)	7	34 02	54 26	4000 ft
Dschaz Morian (Hamun-e Jaz Murian)	9	27 20	58 55	1400–2000 ft
Dscherdscherud (Rudkhaneh-ye Jajrud)	1	35 30	51 42	
Dudeh (Dudej)	8	29 33	52 59	5000–6000 ft
Dudej	8	29 33	52 59	5000–6000 ft
Duruh	11	32 17	60 30	3000 ft
Dust-e Mohammad Khan, Sistan at Afghan border	10			
Duz-Ab (Zahedan?)	10	29 30	60 52 ?	
Duz-Abad, Sargad (Zahedan?)	10	29 30	60 52 ?	
Dzhulfa (Jolfa)	4	38 57	45 38	2000 ft
Elwend (Alvand)	1	36 19	49 10	1900–2000 m
Ehnarik, Sargad Region	10			
Elburz Mountains (Resteh-ye Kuhha-ye Alborz)	1,2,12	36 00	53 00	
		ca 36 40	49 30	
		to 37 20	56 00	
Emamzadeh	8	29 11	51 05	0–500 ft
Emamzadeh Iben Ja'afar	6			
Emma ('Ammeh)	3	37 17	47 07	7000 ft
Enzeli (Bandar-e Pahlavi)	2	37 28	49 27	Sea level
Eschen (Asian?)	1	35 44	49 17 ?	
Esfahan	7	32 40	51 38	5200 ft
Esfandak	10	27 06	62 50	3000 ft
Esfideh	11	33 39	59 46	
Eshma ('Ammeh?)	3	37 17	47 07 ?	
Eskan	10	26 48	63 09	3000 ft
Eskelabad	10	28 35	60 48	
Eslamabad	5			
Esma'ilabad	9	30 12	56 46	6000 ft
Espakeh	10	26 51	60 14	2000–3000 ft
'Eyn or Rashid	1	34 43	52 10	4000 ft
Ev Oghly	4	38 58	45 01	1000 m
Fahraj	9	28 58	58 52	
Fais-Abad (Feyzabad, many places of this name)	11			
Fakke, about 150 km N Ahvaz	6			
Fandokht	11	33 47	59 54	3000–4000 ft
Fandukh (Fandokht)	11	33 47	59 54	3000–4000 ft
Faraman	5	34 13	47 18	5000 ft
Fariman	11	35 43	59 53	4000–5000 ft
Farra, 32 km from Bampur (? Fahraj)	10	28 58	58 52 ?	

Locality	Ostan	Latitude	Longitude	Elevation
Fash	5	34 34	47 56	5000–6000 ft
Fatabad (Fathabad, many places of this name)	?			
Fedesch (Fedeshk)	11	32 45	58 50	4000 ft
Fedeshk	11	32 45	58 50	4000 ft
Feizabad (Feyzabad)	11	35 01	58 46	3000 ft
Fendukt (Fandokht)	11	33 47	59 54	3000–4000 ft
Ferdows	11	34 00	58 09	4000 ft
Ferimun (Fariman)	11	35 43	59 53	4000–5000 ft
Feyzabad	11	35 01	58 46	3000 ft
Fiab 'y-Deery	3 or 4			
Firuzabad, 20 km S Chalus	2	36 30	51 10	
Firuzkuh	1	35 45	52 47	
Gabrabad	1	33 47	51 30	4000–5000 ft
Gach Saran	8	30 20	50 48	2000–3000 ft
Gajareh	1			
Gahinak, Tangistan (Ja'inak?)	8	28 47	51 03 ?	
Galatappeh	7	33 13	51 45	
Gal-i-Chakh, Sargad Region	10			
Galugah	2	37 30	49 20	sea level
Ganaveh	6	29 32	50 51	
Ganbul Dagh	4	37 45	45 15	5253 ft
Gardaneh-ye 'Alam 'Ali	11	37 22	58 30	2000 m
Gardaneh-ye Khaneh Sorkh	9	29 49	56 06	9000 ft
Garm Ab	11	33 53	59 42	
Garm Ab	11	ca 37 37	58 20	
Garm Sar	1			
Garne River, Sargad Region	10			
Gatvand	6	32 15	48 50	
Gavater Bay (Khalij-e Gavater)	10	25 04	61 30	
Gavrabad	1			
Gayan (Qayen)	11	33 44	59 11	5000 ft
Gazangheschlag steppe (Ghazangheshlagh)	1	35 43	50 48	4000 ft
Gazik	11	32 59	60 13	
Ge (Nikshahr)	10	26 13	60 12	1000–2000 ft
Gebrabad Caravanseri (Gabrabad)	1	33 47	51 30	4000–5000 ft
Gegartshin-Kala (Govarchin Qal'eh)	4	38 04	45 10	5000 ft
Germab (Garm Ab)	11	ca 37 37	58 20	
Germukh	10			
Gerri-Rud (Harirud)	11	37 24	60 38	
Gerri-Schotur	11			
Gerri-Shotur, between Khvaf and Mozhnabad	11			
Gezik (Gazik)	11	32 59	60 13	
Ghainak (Ja'inak?)	8	28 47	51 03?	
Ghala Raihand Mountains, W Divandarreh	2 or 5			
Gharareh-su River, 28 km W Gorgan	12			
Ghazangheshlagh steppe	1	35 43	50 48	4000 ft
Ghelmansaray	3	38 18	45 07	5000 ft
Ghistigan (Khoshtegan)	10	26 39	63 02	3000 ft
Gholaman, 30 km W Khorram Abad by road	6	33 25	48 12	1000 m

Locality	Ostan	Latitude	Longitude	Elevation
Ghorak (Gurak?)	8	28 56	51 03	0–500 ft
Gilmendom	?			
Gir (Giv?)	1	34 33	50 17?	
or (Kuh-e Gireh?)	9	27 48	57 38?	
Giv	1	34 33	50 17	5000–6000 ft
Gjarmaz	?			
Gjuishe, about 40 km NW Bandan	10			
Gjurgen River (Rud-e Gorgan)	12	36 59	54 05	
Godar-e Mishun	?			
Godar Landar	6	32 01	49 23	1000 ft
Godar Mohammad Mirza	11	35 43	59 24	1750 m
Gohinak (Ja'inak)	8	28 47	51 03	sea level
Golaki	8	28 46	51 13	sea level
Golandar	11	32 24	59 40	6000 ft
Goleloveh, near Minudasht	12			
Golezard Valley, 4 km W Polur	1			2500 m
Gol Gir	6	31 46	49 30	
Golläku (Golaki)	8	28 46	51 13	sea level
Golmankhaneh	4	37 36	45 15	4200 ft
Golmiran	11	33 12	60 10	
Gom (Qom)	1	34 39	50 54	3500 ft
Gombad-i-Qabous (Gonbad-e Kavus)	12	37 17	55 17	0–500 ft
Gonabad	11	34 20	58 42	
[or]	11	36 34	59 17	
Gonbad, 35 km SE Hamadan	5	34 40	48 45	2000 m
Gonbad-e Kavus	12	37 17	55 17	0–500 ft
Gorgan	12	36 50	54 29	500 ft
Gorgan Region	12	37 00	54 30	
Gorgar	6	30 45	48 58	0–500 ft
Gotwand (Gatvand)	6	32 15	48 50	
Gouladagh	11	37 37	55 53	
Gouladah (Gouladagh)	11	37 37	55 53	
Govarchin Qal'eh	4	38 04	45 10	5000 ft
Gozal Darreh Post, Lar Valley	2			2600 m
Grosse Kawir (Dasht-e Kavir?)	7,11	34 40	54 30?	
Guch Ali, about 57 km NE Khvoy or 30 km S Jolfa	4			
Gudhar Mohamed Mirza Pass (Godar Mohammad Mirza)	11	35 43	59 24	1750 m
Gug Tappeh	4	35 05	48 20	
Gulega (Galugah)	2	37 30	49 20	sea level
Guljandar (Golandar)	11	32 24	59 40	6000 ft
Gulmirun (Golmiran)	11	33 12	60 10	
Gululi Dag (Gouladagh)	11	37 37	55 53	
Gumbez-i-Nowar, about 28 km N Hormak	10			
Gungan (Gorgan)	12	36 50	54 29	500 ft
Gurak	8	28 56	51 03	0–500 ft
Gurgan (Gorgan)	12	36 50	54 29	500 ft
Gurmeh Khaneh (Golmankhaneh)	4	37 36	45 15	4200 ft
Gurmuck (Hormak)	10	29 58	60 51	
Gurmukh (Hormak)	10	29 58	60 51	
Gurschir	6			

Locality	Ostan	Latitude	Longitude	Elevation
Gussein-Abad (Hoseynabad)	11	35 52	59 49	4000–5000 ft
Gyali-Deeri Valley, above Gyrdykh, Mergever, W Daryacheh-ye Reza'iyeh	4			
Gyrdykh	4			
Hadji-du-i-Tschahi, Nehbandan	?			
Haft Gel	6	31 28	49 30	900–1000 ft
Häfthos Valley	1	35 50	51 16	5000–7000 ft
Haft Kel (Haft Gel)	6	31 28	49 30	900–1000 ft
Haft Tappeh	6	32 04	48 20	
Halvan	11	33 57	56 15	812 m
Halwan (Halvan)	11	33 57	56 15	812 m
Hamadan	5	34 48	48 30	6000 ft
Hamun-e Jaz Murian	9	27 20	58 55	1400–2000 ft
Hamun-e Jowri	10	28 37	60 50	
Hane Houre	8	30 15	53 09	1600 m
Hasanabad	1	36 28	50 17	
Hard-al Mamlah	6	31 53	48 31	
Harirud, Iran-Afghanistan-Turkmenistan border	11	37 24	60 38	
Harivan	5			
Harmalah, 120 km NW Ahvaz (Hard-al Mamlah)	6	31 53	48 31	
Harp, 4 km E Kuh-e Majred Pass, Turan Biosphere Reserve	12			
Hauz-Batil (Hauz Patil)	10	30 24	61 03	803 m
Hauzdar	10	30 34	61 03	
Hauz Patil	10	30 24	61 03	
Henjam Island (Jazireh-ye Hengam)	?	26 39	55 53	0–350 ft
Herat (Kerat, Khashtadan)	11	34 34	60 33	
Hermau (Garm Ab)	11	33 53	59 42	
Hermoz Island (Jazireh-ye Hormoz)	?	27 04	56 28	
Heroabad (Horowabad)	2	37 37	48 32	
Herowabad	2	37 37	48 32	
Hesarak	1	35 51	50 56	
Heydari	7	32 40	50 35	
Hisomi	?			
Hormak	10	29 58	60 51	
Hormoz (town)	?	27 06	56 28	
Hormoz Island (Jazireh-ye Hormoz)	?	27 04	56 28	
Hoseynabad	11	35 52	59 49	4000–5000 ft
Hoseynabad	9	30 15	57 12	
Houz	11			
Howz-e Musa	11	32 10	59 46	
Ibrahim Zahra (Chah-e Ebrahim Zare'?)	1	34 07	54 57 ?	
Ilam	5	33 38	46 26	
Imamzada (Emamzadeh)	8	29 11	51 05	0–500 ft
Irak-Adschemi (Iraq Ajemi)	1	34 30	52 00	
Iranshahr	10	27 13	60 41	1870 ft
Iranshar (Iranshahr)	10	27 13	60 41	1870 ft
Iraq Ajemi region	1	34 30	52 00	

Locality	Ostan	Latitude	Longitude	Elevation
Isavändi (Isvand)	8	29 08	51 06	0–500 ft
Isfagan (? Esfahan)	6 ?			
Isfahan (Esfahan)	7	32 40	51 38	5200 ft
Isfandak (Esfandak)	10	27 06	62 50	3000 ft
Ismailabad (Esma'ilabad)	9	30 12	56 46	6000 ft
Ismail-aga, valley of Nazlu Rud	3			
Ispahan (Esfahan)	7	32 40	51 38	5200 ft
Ispidan	10	27 10	61 13	1000 ft
Istgah-e Abardezh	1	35 15	51 47	
Istgah-e Bisheh	6	33 20	48 52	1200 m
Istgah-e Dezh	6	31 43	48 38	500 ft
Istgah-e Ezna	6	33 27	49 28	1800 m
Istgah-e Kavir	1	35 14	51 59	2000–3000 ft
Istgah-e Qarun	6	33 25	48 59	5000 ft
'Isvand	8	29 08	51 06	0–500 ft
Izeh	6	31 50	49 50	
Jafar Abad, ESE Kashan	7	33 55	51 53	800 m
Jahrom	8	28 31	53 33	3200 ft
Ja'inak	8	28 47	51 03	sea level
Jajarm	?			
Jalk (Jalq)	10	27 36	62 42	3000 ft
Jalq	10	27 36	62 42	3000 ft
Jandaq	7	34 02	54 26	4000 ft
Jangal	11	34 43	59 15	
Jarrahi River (Rudkhaneh-ye Jarrahi)	6	30 44	48 46	
Jashk (Jask)	9	25 38	57 46	20 ft
Jask	9	25 38	57 46	20 ft
Javadabad	1	35 12	51 41	
Jawad-Abad (Javadabad)	1	35 12	51 41	
Jazireh-ye Arzu	4	37 22	45 36	4000–5000 ft
Jazireh-ye Hengam, Persian Gulf	?	26 39	55 53	0–350 ft
Jazireh-ye Hormoz, Persian Gulf	?	27 04	56 28	
Jazireh-ye Khark, Persian Gulf	?	29 15	50 20	10 ft
Jazireh-ye Larak, Persian Gulf	?	26 52	56 22	
Jazireh-ye Qeshm, Persian Gulf	?	26 45	55 45	
Jazireh-ye Qowyun	4	37 29	45 35	4200–5003 ft
Jazireh-ye Shahi	3	37 50	45 30	
Jazireh-ye Sheykh Sho'eyb, Persian Gulf	?	26 48	53 15	
Jazireh-ye Shotur, Persian Gulf	?	26 47	53 25	
Jazireh-ye Sirri, Persian Gulf	?	25 55	54 32	
Jazireh-ye Tanb-e Bozorg, Persian Gulf	?	26 14	55 19	0–175 ft
Jaz Murian (Hamun-e Jaz Murian)	9	27 20	58 55	1400–2000 ft
Jerud Dam	6			
Jezd (Yazd)	7	31 53	54 25	4080 ft
Jijirud	1	34 32	50 16	
Jiroft	9	28 35	57 45	2000 ft
Jolfa	4	38 57	45 38	2000 ft
Jonstan, S slopes Talysh Mountains	3			
Jupar	9	30 04	57 08	

Locality	Ostan	Latitude	Longitude	Elevation
Kach	10	25 27	61 09	0–500 ft
Kachwarech	?			
Kadich (Kadish)	1	34 13	51 25	2000–3000 ft
Kahnuj	9	27 58	57 45	2000 ft
Kahoorak	10	29 26	59 52	
Kalagan	10	27 29	62 47	3500 ft
Kala-i-Bid (Qal'eh-ye Bid)	10	28 40	60 24	
Kalaker Mountains	?			
Kalaleh, Gorgan region	12			
Kale-Minar Mountains (Qal'eh Manar)	11	35 26	59 56	
Kalender-Abad (Qalandarabad)	11	35 36	59 56	4000–5000 ft
Kale-Tol (Qal'eh-ye Tol)	6	31 37	49 53	3000 ft
Kaluraz	2	36 54	49 28	2000 ft
Kamiran	5			
Kangal (Tangali?), on Iran-Turkmenistan border	12	37 27	54 43 ?	
Karadag (Qareh Dagh)	3	36 16	48 17	
Kara-iltschi Mountains	11			
Karaj	1	35 48	50 59	4000–5000 ft
Karand	5	34 16	46 15	
Karez-badak	10	35 30	60 10	
Karim-Abad, near Lake Reza'iyeh (many places of this name)	?			
Kariz	11	34 49	60 47	
Kariz	12	35 35	56 48	1200 m
Karkheh-Dez channel	6	31 27	48 23	
Karman (Kerman)	9	30 17	57 05	6000 ft
Kartevjul	3			
Kartevül	3			1800 m
Karun (Istgah-e Qarun)	6	33 25	48 59	5000 ft
Karun River (Rud-e Karun)	6	30 25	48 12	
Kashan	1	33 59	51 29	3000–4000 ft
Kashmar	11	35 12	58 27	
Kaskir (Rud-e Khasken)	9	28 52	57 12	6000 ft
Kaskin, 39 km SE Bazman	10	27 30	60 22	
Kasuin (Qazvin)	1	36 16	50 00	4000–5000 ft
Katar-bena	11			
Kavir-e Namak	1	34 30	57 40	2000–3000 ft
Kavir Protected Region	1	34 45	52 10	
Kawir railway station (Istgah-e Kavir)	1	35 14	51 59	2000–3000 ft
Kazerun	8	29 37	51 38	3000 ft
Kazrun (Kazerun)	8	29 37	51 38	3000 ft
Kazvin (Qazvin)	1	36 16	50 00	4000–5000 ft
Kehdish (Kadish)	1	34 13	51 25	2000–3000 ft
Kerat, Khashtadan	11	34 34	60 33	
Keredj (Karaj)	1	35 48	50 59	4000–5000 ft
Kerim-Abad, W shore Daryahcheh-ye Reza'iyeh	4			
Kerman	9	30 17	57 05	6000 ft
Kermanschah (Kermanshah)	5	34 19	47 04	4600 ft
Kermanshah	5	34 19	47 04	4600 ft
Keroo	2	ca 37 00	48 48	

APPENDIX III

Locality	Ostan	Latitude	Longitude	Elevation
Kerosh	11			
Kerym, Abad	4			
Khadzhi-i-du-Chagi (Khvajeh Dow Chahi)	11	31 53	60 13	
Khaf (Khvaf)	11	34 33	60 08	
Khajar	11	35 06	59 42	
Khalafabad	6	30 54	49 24	
Khalij-e Chah Bahar	10	25 20	60 30	
Khalij-e Gavater	10	25 04	61 30	
Khal-Khal (Herowabad), 32 km W Asalem by road	2	37 37	48 32	
Khamun (Daryacheh-ye Sistan)	10	31 00	61 15	
Khamun-i-Djaori, Sargad Region (Hamun-e Jowri)	10	28 37	60 50	
Khaneh	4	36 41	45 08	
Khaneh Khowreh	8	30 52	53 09	2180 m
Khaneh Khvodi	12	36 05	56 04	1100 m
Khanekhodi (Khaneh Khvodi)	12	36 05	56 04	1100 m
Khan-i-Khurreh Pass (Khaneh Khowreh)	8	30 52	53 09	2180 m
Khan-i-Surkh pass (Gardaneh-ye Khaneh Sorkh)	9	29 49	56 06	9000 ft
Khanu (Khanuj)	9	27 58	57 45	2000 ft
Khara	7	32 07	52 44	5000 ft
Kharakan	1	36 31	49 45	
Kharestan	11	32 17	59 56	6000 ft
Kharg Island (Jazireh-ye Khark)	?	29 15	50 20	10 ft
Kharistan (Kharestan)	11	32 17	59 56	6000 ft
Khash	10	28 14	61 14	1200 m
Khashtadan Region	11	34 30	60 35	
Khelejan	3	38 01	46 08	5000 ft
Kherra	3 or 4			
Kheyrabad, NE Shahrud	12			
Khiyav	3	38 24	47 40	
Khoi (Khvoy)	4	38 33	44 58	4000 ft
Khomedeh, E Damavand, W Firuzkuh	1			
Khomeyn	6	33 38	50 04	6000 ft
Khonj	8	27 52	53 27	
Khor Askan (Eskan)	10	26 48	63 09	3000 ft
Khorramabad	6	33 30	48 20	4100 ft
Khorramabad	2	36 46	50 53	500 ft
Khorramshahr	6	30 25	48 11	sea level
Khorram Shahr (Khorramshahr)	6	30 25	48 11	sea level
Khoshtadan Mountains	11	34 34	60 33	
Khoshtegan	10	26 39	63 02	3000 ft
Khoshrowa (Khoshrowabad)	4	38 10	44 43	4200–5000 ft
Khoshrowabad	4	38 10	44 43	4200–5000 ft
Khosh Yeilagh Protected Region	12	36 55	55 32	
Khosrova (Khoshrowabad)	4	38 10	44 43	
Khouzdar (Hauzdar)	10	30 34	61 03	
Khouz-i-Musafir (Howz-e Musa?)	11	32 10	59 46	
Khullar	8 ?			
Khunas Tankteh (Konar Takhteh)	8	29 32	51 24	1000–2000 ft
Khur	7	33 47	55 03	3000 ft
Khurgu (Khvorgu)	9	27 33	56 27	1000–4481 ft
Khushkek, in Bekhars, about 18 km NW Kariz	11			

Locality	Ostan	Latitude	Longitude	Elevation
Khusrin, about 90 mi NW Bampur	10			1800 ft
Khvaf	11	34 33	60 08	
Khvajeh Dow Chahi	11	31 53	60 13	
Khvorgu	9	27 33	56 27	1000–4481 ft
Khvoy	4	38 33	44 58	4000 ft
Khwar	11 ?			
Kiasar, about 68 km SSE Sari	12			
Kirjawa	3	37 18	46 22	5000–6000 ft
Kirman (Kerman)	9	30 17	57 05	6000 ft
Kivask	10			
Kivi	3	37 40	48 22	1000 m
Koc (Kach)	10	25 27	61 09	0–500 ft
Kochrud	7	33 38	51 27	7000–8000 ft
Kohrud	7	33 38	51 27	7000–8000 ft
Kojun Island (Jazireh-ye Qowyun)	4	37 30	45 40	4200–5003 ft
Koladasht Valley, Alborz Mountains	1			
Kom (Qom)	1	34 39	50 54	3500 ft
Komein (Khomeyn)	6	33 38	50 04	6000 ft
Konar Takhteh	8	29 32	51 24	1000–2000 ft
Konar Takhti (Konar Takhteh)	8	29 32	51 24	1000–2000 ft
Kondol	4			
Kopet Dagh (Koppeh Dagh)	11,12	37 40 to 36 15	55 20 60 25	
Koppeh Dagh	11,12	37 40 to 36 15	55 20 60 25	
Kopur Chal	2	37 32	49 14	below sea level
Kopurtschal (Kopur Chal)	2	37 32	49 14	
Koshm-Abad, near Damghan	12			
Koshtegan, Bam Posht	10	26 39	63 02	
Koyun Daghi (Jazireh-ye Qowyun)	4	37 30	45 40	4200–5003 ft
Kucan (Gardaneh-ye 'Alam 'Ali)	11	37 22	58 30	2000 m
Kuchan (Quchan)	11	37 06	58 30	4100 ft
Kuchistan (Khuzestan?)	6 ?			
Kuh Daschteh (Kuhdasht?)	6	33 32	47 36 ?	
Kuhdasht	6	33 32	47 36	4000 ft
Kuh-e Ashin	7	33 31	53 28	4000–6500 ft
Kuh-e Bakhtu	9	31 43	58 39	
Kuh-e Bam Posht	10	26 45	62 30	3000–5625 ft
Kuh-e Bang	8	29 45	50 25	500–991 ft
Kuh-e Baqeran	11	32 43	59 20	6000–8899 ft
Kuh-e Chastab	7	33 40	55 10	3000–4000 ft
Kuh-e Ferdows	11	33 55	58 39	
Kuh-e Gireh	9	27 48	57 38	2000–5133 ft
Kuh-e Hazaran	9	29 30	57 18	7000–14500 ft
Kuh-e Hazar Darreh	7	33 50	54 11	3000–4000 ft
Kuh-e Jebal Barez	9	28 30	58 20	4000–13000 ft
Kuh-e Jupar	9	29 55	57 15	2700 m
Kuh-e Khwaja	11	30 57	61 15	
Kuh-e Laleh Zar	9	28 24	56 46	7000–14350 ft
Kuh-e Majred	12	35 04	56 01	1455 m
Kuh-e Malek Mohammad	10	30 28	59 10	

Locality	Ostan	Latitude	Longitude	Elevation
Kuh-e Mishu Dagh	3	38 25	45 25	4000–8750 ft
Kuh-e Mordeh	6	31 48	49 38	
Kuh-e Safid	?			
Kuh-e Safidar	8	28 55	53 00	5000–10460 ft
Kuh-e Sahand	3	37 44	46 27	6000–12172 ft
Kuh-e Shah, N Baft	9			
Kuh-e Sofeh	7	32 35	51 38	5300–7000 ft
Kuh-e Taftan	10	28 36	61 06	5000–13262 ft
Kuh-e Zul	11	33 35	59 05	
Kuhha-ye Genu	9	27 25	56 09	500–7783 ft
Kuhha-ye Sabalan	3	38 15	47 49	
Kuhha-ye Talysh	3	37 35	48 38	
Kuh-i Akhangerun	11	33 25	60 13	
Kuh-i-Atkul	11			
Kuh-i-Bakhtu (Kuh-e Bakhtu)	9	31 40	58 21	
Kuh-i-Bang (Kuh-e Bang)	8	29 45	50 25	500–991 ft
Kuh-i-Boz, or Kukh-i-Boz ab, Sargad Region	10	29 10	60 50	
Kuh-i-Diamel Bariz (Kuh-e Jebal Barez)	9	28 30	58 20	4000–13000 ft
Kuh-i-Dschupar (Kuh-e Jupar)	9	29 55	57 15	2700 m
Kuh-i-Ginau (Kuhha-ye Genu)	9	27 25	56 09	500–7783 ft
Kuh-i-Hazar (Kuh-e Hazaran)	9	29 30	57 18	10000 ft
Kuh-i-Hezar (Kuh-e Hazar Darreh)	7	33 50	54 11	3000–4000 ft
Kuh-i-Khadzi (? Kuh-e Khwaja)	11	30 57	61 15 ?	
Kuh-i-Lalezar (Kuh-e Laleh Zar)	9	29 24	56 46	7000–14350 ft
Kuh-i-Niswa	1	35 52	53 05	6000–11658 ft
Kuh-i-Rickeschol, Sargad	10			
Kuhistan Region	7,11	33 00	57 00	
Kuh-i-Taftan (Kuh-e Taftan)	10	28 36	61 06	5000–13262 ft
Kuh-i-Tuftan (Kuh-e Taftan)	10	28 36	61 06	5000–13262 ft
Kuh-räng	7	32 18	50 13	
Kuh Tawaren	?			
Ku-i-Tuftan (Kuh-e Taftan)	10	28 36	61 06	5000–13262 ft
Kukh-i-Akhangerun, near Ahangeran	11			
Kukh-i-Atkul, vicinity Mohammadabad	11			
Kukh-i-Boz-ab, or Kuh-i-Boz	10	29 10	60 50	
Kukh-i-Kerat	?			
Kukh-i-Khadzhi, Zirkukh Region	11			
Kukh-i-Magomed-abad, Zirkukh Region	11			
Kukh-i-Murgak, Sargad Region (Morghak)	10	27 58	60 48	
Kukh-i-Rikeshol, Sargad Region	10			
Kukh-i-Tuftan (Kuh-e Taftan)	10	28 36	61 06	5000–13262 ft
Kum (Qom)	1	34 39	50 54	3500 ft
Kurait	6	31 17	48 49	400 ft
Kura River	12			
Kureh-e-Gez	?			
Kuretu (Quraitu)	5	34 36	45 30	1000 ft
Kurin, Sargad Region	10			
Kurusch-göl	3	37 54	46 42	6000 ft
Kushkek (Kushkek), in Bekhars, about 18 km NW Kariz	11			
Kushkizard (Kushk Zar)	8	30 50	52 30	8000 ft

Locality	Ostan	Latitude	Longitude	Elevation
Kush Kowr	10	26 39	61 16	
Kush-Kuh	1	34 21	52 13	3000–4738 ft
Kushk Zar	8	30 50	52 30	8000 ft
Kuss (Qusheh)	11	32 33	59 02	
Kyariz (Kariz)	11	34 49	60 47	
Kyrch-Bulach (Qerkh Bolagh)	3	38 28	46 53	7000 ft
Labe-Ab (Lab-e Bareng)	10	31 09	61 09	1800–2000 ft
Lab-e Bareng	10	31 09	61 09	1800–2000 ft
Ladiz	10	28 56	61 19	
Lahidschan (Lahijan)	2	37 12	50 01	sea level
Lahijan	2	37 12	50 01	sea level
Laidaru	?			
Lake Urmia (Daryacheh-ye Reza'iyeh)	3,4	37 40	45 30	4180 ft
Lake Valasht, near Marzunabad	12			
Lalabad	5			
Laleh Zar	9	29 31	56 51	8000–9000 ft
Lalezar (Laleh Zar)	9	29 31	56 51	8000–9000 ft
Lali	6	32 15	49 05	1000 ft
Langarak	11	36 11	60 02	3000 ft
Langarud	2	37 11	50 09	−20 m
Lar	8	27 41	54 17	3000–4000 ft
Lar	2	36 46	48 54	
Lar Valley, NE Tehran, SW summit of Damavand	1			2180–2900 m
Lar River	1	35 51	51 53	
Larak Island (Jazireh-ye Larak)	?	26 52	56 22	
Lascht	?			
Lash Jowain	?			
Latian Dam area	1	35 47	51 41	
Lavan Island (Jazireh-ye Sheykh Sho'eyb)	?	26 48	53 15	
Leb-e Kal	?			
Leb-Kal	?			
Lengeh (Bandar-e Lengeh)	9	26 33	54 53	
Lenger, a few km NW Torbat-e Jam	11			
Ljabe-Ab (Lab-e Bareng)	10	31 09	61 09	1800–2000 ft
Lur Valley	1			
Lut-e Zangi Ahmad	9	29 22	58 48	2000 ft
Lyadis (Ladiz)	10	28 56	61 19	
Lyutfabad	11			
Mach, Bolan Pass	?			
Machnudabad	10	27 12	60 34	
Magas (Zaboli)	10	27 07	61 40	4500 ft
Magommed-Abad (Mohammadabad, many places of this name)	7	32 40	51 56 ?	
Magomed-abad (Mohammadabad)	11	33 27	60 10	
Mahabad	4	36 45	45 43	4272 ft
Mahallat	1	33 55	50 27	
Mahidasht	5	34 16	46 48	
Mahmudabad	12	36 38	52 15	

Locality	Ostan	Latitude	Longitude	Elevation
Mahmudieh	1	35 44	51 19	4000–5000 ft
Mahmu'i	11	33 19	59 21	
Mahmu'i	11	33 44	59 14	
Mahneh	11	34 59	58 51	3000 ft
Mahomed-Abad (Mohammadabad, many places of this name)	11	33 32	58 47 ?	
Mahommedabad (Mohammadabad, many places of this name)	7 ?	32 40	51 56 ?	
Mahor Birinji	6			
Mainé (Mahneh)	11	34 59	58 51	3000 ft
Makhmudabad, 12 km SW Iranshahr	10			
Makhunik	11	32 27	60 24	
Makran	10	26 00	60 00	
Maku	4	39 17	44 31	4000 ft
Malamir (Qal'eh-ye Tol)	6	31 37	49 53	3000 ft
Malavi	6			
Malek Mohammad mountains (Kuh-e Malek Mohammad)	10	30 28	59 10	
Maljat-Abad, Iraq Ajemi	1			
Mamui (Mahmu'i)	11	33 19 or 33 44	59 21 59 14	
Mand	10	26 07	62 02	700 ft
Manjil	2	36 44	49 24	1000 ft
Mansu Abad	6	32 49	48 44	1200 m
Maraga (Maragheh)	3	37 23	46 13	5000 ft
Maragheh	3	37 23	46 13	5000 ft
Marakand	4	38 51	45 16	
Marand	3	38 26	45 46	
Marandiz	11	34 48	58 22	2000–3000 ft
Mardi Rud	3	37 01	46 05	
Marinjab	1	34 18	51 50	2000–3000 ft
Marivan (Dezh Shapur)	5	35 31	46 10	
Marv Dasht	8	29 50	52 40	
Markan, 8 km N Ev Oghly by road	4	38 52	45 18	1000 m
Marzunabad	12	36 27	51 18	
Mashhad	11	36 18	59 36	3200 ft
Mashiz	9	29 56	56 37	
Mashkan	4	36 48	45 04	
Mashu Dagh (Kuh-e Mishu Dagh)	3	38 25	45 25	4000–8750 ft
Masjed Soleyman	6	31 57	49 16	1000 ft
Masjid-i-Suleiman (Masjed Soleyman)	6	31 57	49 16	1000 ft
Masjid-i-Suleiman airfield	6	31 59	49 15	1220 ft
Mazanderan	12			
Mazel-Ab, Sargad Region	10			
Mazhan	11	32 35	59 01	
Mazu	6	32 46	48 31	
Mazuo (Mazu)	6	32 46	48 31	
Meched (Mashhad)	11	36 18	59 36	3200 ft
Mehevah (Mahneh)	11	34 59	58 51	
Mehkuh, 70 km S Shiraz	8			
Mehran	5	33 07	46 10	

Locality	Ostan	Latitude	Longitude	Elevation
Meigun	11	31 49	59 28	
Meitun (Meigun)	11	31 49	59 28	
Mendschil (Manjil)	2	36 44	49 24	1000 ft
Mergen-ulja River, near eastern Iran-Turkmenistan border	11 ?			
Meshed (Mashhad)	11	36 18	59 36	3200 ft
Meshhed (Mashhad)	11	36 18	59 36	3200 ft
Meshkin Shahr (Khiyav)	3	38 24	47 40	
Meshrageh, 53 mi SE Ahvaz, probably near Rud-e Jarrahi at Khalafabad	6	30 54	49 24	200 ft
Mian-bazar (Mian Bazar)	10	29 40	60 24	
Mian Bazar	10	29 40	60 24	
Miandascht (Miandasht)	12	36 26	56 03	4000–5000 ft
Miandasht	12	36 26	56 03	4000–5000 ft
Mianeh	2	37 26	47 42	
Mianeh	5	34 05	47 04	
Mian Kaleh Protected Region	12	36 53	53 35	
Mian Kotal	8	29 33	51 55	
Mianrahan region, 40 km NE Kermanshah	5			about 1450 m
Milak	8	29 35	51 08	500 ft
Milan-i-Dzhekhun-Khamun (Hamun-e Jaz Murian)	9	27 20	58 55	
Mil-Ayaz-Khan, near Sang Bast	11			
Mill-Ajaz	11			
Minab	9	27 09	57 05	0–500 ft
Minoodasht (Minudasht)	12			
Minudasht	12			
Mirindiz (Marandiz)	11	34 48	58 22	2000–3000 ft
Mir-kala (Mir Qal'eh)	11	37 31	59 17	
Mir Qal'eh	11	37 31	59 17	
Mo'allem Khani	1	36 27	49 50	
Mohammadabad	7	32 40	51 56	
Mohammadabad	11	33 27	60 10	
Mohammadabad	11	33 32	58 47	
Mohammadiyeh	1	34 14	49 39	
Mohammad Reza Shah National Park	12	37 21	56 10	
Mohammad Zaman	6			
Mohammedi (Mohammadiyeh?)	1?	34 14	49 39 ?	2200 m
Mondechi	?			
Mordab	2	37 26	49 25	sea level
Mordaq	3	37 20	46 24	
Morghak	10	27 58	60 48	
Mt. Damavand (Qolle-ye Damavand)	1	35 56	52 08	
Mt. Savalan (Kuhha-ye Sabalan)	3	38 15	47 49	
Mt. Sekhend (Kuh-e Sahand)	3	37 44	46 27	
Mt. Sophia (Kuh-e Sofeh)	7	32 35	51 38	5300–7000 ft
Mozhnabad	11	34 07	60 07	2000–3000 ft
Mudzhnabad (Mozhnabad)	11	34 07	60 07	2000–3000 ft
Murcheh Khvort	7	33 06	51 30	
Murdab (Mordab)	2	37 26	49 25	sea level
Murgak, Sargad Region (Morghak)	10	27 58	60 48	
Muteh	7	33 37	50 47	

APPENDIX III

Locality	Ostan	Latitude	Longitude	Elevation
Nabid	9	29 40	57 38	7000 ft
Nachduin Mountain	?			
Naftak	6	31 58	49 15	1000 ft
Naft-e Safid	6	31 40	49 14	1000 ft
Naft Sefid (Naft-e Safid)	6	31 40	49 14	1000 ft
Nahar	12	35 32	56 46	1350 m
Nahu (Nahug)	10	27 37	62 21	5000 ft
Nahug	10	27 37	62 21	5000 ft
Naim-Abad (Na'imabad)	12	36 14	54 39	4000 ft
Na'imabad	12	36 14	54 39	4000 ft
Nain	7	32 52	53 05	5000 ft
Nain-Abad (Na'imabad)	12	36 14	54 39	4000 ft
Najafabad	1	36 27	49 17	5000 ft
Namet-Abad (Ne'matabad)	3	38 01	46 28	6000 ft
Namivand	5	34 22	46 45	
Naqsh-e Rustam, 7 km N Persepolis	8			
Nardin	12	37 03	55 59	4000–5000 ft
Nardyn (Nardin)	12	37 03	55 59	4000–5000 ft
Narmashir (Rud-e Fahraj)	9	29 30	59 11	
Nashratabad (Nosratabad)	10	29 54	59 59	4000 ft
Nasir-Ab (Nasirabad)	10	31 03	61 25	1500–2000 ft
Nasirabad	10	31 03	61 25	1500–2000 ft
Nasir-Abad (Nasirabad)	10	31 03	61 25	1500–2000 ft
Naslu-tshai (Nazlu Rud)	4	37 42	46 16	
Naslu-Tshay River (Nazlu Rud)	4	37 42	46 16	
Nasrie (Ahvaz)	6	31 19	48 42	50–200 ft
Natanz	7	33 31	51 54	800 m
Navrud	2	37 42	48 57	
Nazlu Rud	3	37 42	45 16	
Ne (Nehbandan?)	11	31 32	60 02?	
Nech-i-Bendan, or Ne-i-Bendan Region	10	31 00 to 31 10	60 30 61 30	
Nedschefabad (Najafabad?)	1 ?	36 27	49 17?	
Negel	5	35 18	46 31	
Negur	10	25 22	61 15	
Neh	7	33 01	52 30	
Neh (Nehbandan)	11	31 32	60 02	
Nehbandan	11	31 32	60 02	4000–5000 ft
Nehbendan (Nehbandan)	11	31 32	60 02	
Nehbid (Nabid)	9	29 40	57 38	7000 ft
Neh-i-Bendan (Nehbandan)	11	31 32	60 02	
Neitshalan (Neycharan)	3	38 21	46 37	
Neizar	10	31 10	61 15	1500 ft
Ne'matabad	3	38 01	46 28	6000 ft
Nergy	4			
Nergy-Desht	4			
Nerion, Reshteh-ye Kuhha-ye Alborz	?			3000 m
Nertschistan	?			
Ney	5	35 29	46 08	5000 ft
Neycharan	3	38 21	46 37	

Locality	Ostan	Latitude	Longitude	Elevation
Neyriz	8	29 12	54 19	4000–6000 ft
Niazabad	11	34 14	60 15	2000–3000 ft
Niazabad	11	35 54	60 27	3000 ft
Niaz-Abad (Niazabad)	11	34 14	60 15	
Niaz-Abad (Niazabad)	11	35 54	60 27	
Nihing River (Rud-e Nahang)	10	26 00	62 44	
Nikshahr	10	26 13	60 12	1000–2000 ft
Niris Lake (Daryacheh-ye Bakhtegan)	8	29 20	54 05	5070 ft
Niriz (Neyriz)	8	29 12	54 19	4000–6000 ft
Nosratabad	10	29 54	59 59	4000 ft
Nou-dekh (Now Deh)	11	33 40	59 57	
Now Deh	11	33 40	59 57	
Now Sud	5			
Nurabad (many places of this name)	8 ?			
Nus, Badschistan Kavir	11			
Nusi	11	35 48	58 26	
Ostrankuh Protected Region	6	33 30	48 50	
Pachenar	2	36 36	49 32	3000 ft
Pahlavi Dezh	12	37 01	54 30	sea level
Pain Gjatsch	6			
Païtschinar (Pachenar)	2	36 36	49 32	3000 ft
Panjsareh	10	27 59	60 11	4000 ft
Parchestan	6	31 53	49 54	3000 ft
Paris (Pariz)	9	29 52	55 46	7000 ft
Pariz	9	29 52	55 46	7000 ft
Parwand	11			1125 m
Pasargat	8	30 12	53 10	800 m
Pasarvandan	10	25 05	61 25	
Pasengku	?			
Pasghaleh Valley	1	35 48	51 20	5000–7000 ft
Paskimer	?			
Patschinar (Pachenar)	2	36 36	49 32	3000 ft
Pa-thschinar (Pachenar)	2	36 36	49 32	3000 ft
Pay Kuh	11	33 42	59 49	
Pazuk	11	34 54	57 05	4000 ft
Pegu-Rabat (between Pay Kuh and Robat?)	11			
Pendsch-Sara (Panjsareh)	10	27 59	60 11	4000 ft
Pendzh-Sara (Panjsareh)	10	27 59	60 11	4000 ft
Persepolis (Takht-e Jamshid)	8	29 57	52 52	5500 ft
Pertschistun (Parchestan)	6	31 53	49 54	3000 ft
Pervandia, or Pervanda	6			
Pesch (Fash?)	5?	34 34	47 56 ?	
Pesuk (Pazuk ?)	11?	34 54	57 05 ?	
Peter and Paul Landing, Daryacheh-ye Reza'iyeh	4			
Petropawlrwsky Port, Daryacheh-ye Reza'iyeh	4			
Pip	10	26 38	60 09	3000 ft
Pir	?			
Pirbakran (Pir Bakran)	7	32 27	51 33	6000 ft

Locality	Ostan	Latitude	Longitude	Elevation
Pir Bakran	7	32 27	51 33	6000 ft
Pishin	10	26 06	61 47	700 ft
Podagi, about 37 km N Bazman	10			
Podatschi, Sargad (Podagi)	10			
Poldasht	4	39 24	44 59	
Pol-e Abgineh	8	29 33	51 46	3000 ft
Pol-e Jeh Jeh	6			
Pol-e Tang	6	31 51	47 56	
Polur	1	35 52	52 03	
Polurd (Polur)	1	35 52	52 03	
Posht-e Aseman	12	35 36	57 01	1260 m
Pudesch-kupa	?			
Pul-i-Dcheverarem	12 ?			
Qader Abad	8	30 14	52 12	2100 m
Qalandarabad	11	35 36	59 56	4000–5000 ft
Qal'eh Manar Pass	11	35 26	59 56	7000 ft
Qal'eh-ye Bid	10	28 40	60 24	
Qal'eh-ye Tol	6	31 37	49 53	3000 ft
Qamishlu, Zagros Mountains	7	32 02	51 29	2000–2200 m
Qareh Aghaj River, E Dasht-e Arzhan by road	8	29 45	52 09	1500 m
Qareh Dagh	3	38 35	46 30	
Qareh Dagh	3	36 16	48 17	
Qar Qareh, Kavir Protected Region	1			
Qar Sharon	8	29 44	51 34	800 m
Qasr-e Shirin	5	34 31	45 35	
Qayen	11	33 44	59 11	5000 ft
Qazvin	1	36 16	50 00	4000–5000 ft
Qerkh Bolagh	3	38 28	46 53	7000 ft
Qeshm Island, Persian Gulf (Jazireh-ye Qeshm)	?	26 45	55 45	0–1331 ft
Qolleh-ye Damavand	1	35 56	52 08	
Qom	1	34 39	50 54	3500 ft
Qotur	4	38 28	44 25	
Quchan	11	37 06	58 30	
Quraitu	5	34 36	45 30	1000 ft
Qusheh	11	32 33	59 02	
Rabat (Robat, several places of this name)	1 ?	34 01	49 32 ?	
Rabat-i-Sefid (Robat-e Khakestari?)	11 ?	35 46	59 23 ?	
Rabat-i-Setid (Robat-e Khakestari?)	11 ?	35 46	59 23 ?	
Raguird (Rahjerd)	1	34 22	50 22	5000–6000 ft
Rahjerd	1	34 22	50 22	5000–6000 ft
Rahmanlu	3	37 40	45 40	
Ramhormoz	6	31 16	49 36	
Ramsar	12	36 54	50 40	−20 m
Rasano	3			
Rasht	2	37 16	49 36	sea level
Rask	10	26 13	61 25	
Ravansar	5	34 43	46 40	
Rayen	9	29 34	57 26	6000–7000 ft

Locality	Ostan	Latitude	Longitude	Elevation
Rayin (Rayen)	9	29 34	57 26	6000–7000 ft
Razeh	11	32 39	60 18	5000 ft
Reikände	12 ?			
Rescht (Rasht)	2	37 16	49 36	
Resht (Rasht)	2	37 16	49 36	
Reshteh-ye Kuhha-ye Alborz	1,2,12	36 00 (ca 36 40 to 37 20	53 00 49 30 56 00)	
Resvandeh (Rezvandeh)	2	37 33	49 09	sea level
Rey	1	35 35	51 25	3000–4000 ft
Rezaabad	1	34 58	50 29	3000–4000 ft
Reza'iyeh	4	37 33	45 04	4600 ft
Rezvandeh	2	37 33	49 09	sea level
Rigan	9	28 37	58 58	2500 ft
Rig-e-Djinn (Rijan?)	7	33 55	51 43?	3000–4000 ft
Rig Ispakeh	10	26 51	60 14	2000–3000 ft
Rigmati (Rig Mati)	9	27 39	58 10	1500 ft
Rig Mati	9	27 39	58 10	1500 ft
Rijan	7	33 55	51 43	3000–4000 ft
Riza (Razeh)	11	32 39	60 18	5000 ft
Riza-Abad (Rezaabad)	1	34 58	50 29	3000–4000 ft
Robat	1	34 01	49 32	8000 ft
Robat	11	33 17	57 31	
Robat-e Khakestari	11	35 46	59 23	6000 ft
Robat Sefid (Robat-e Khakestari)	11	35 46	59 23	6000 ft
Roshkvar	11	34 58	59 37	
Rubberek, N Alborz	1			9000 ft
Rudan (Deh Barez)	9	27 26	57 12	
Rud-e Atrak	12	37 28	54 03	
Rud-e Bampur	10	27 08	59 06	
Rud-e Bardeh Sur	4	37 27	44 55	
Rud-e Dez	6	31 39	48 52	
Rud-e Fahraj	9	29 30	59 11	
Rud-e Gorgan	12	36 59	54 05	
Rudehen	1	35 44	51 52	
Rud-e Hirmand	10	31 12	61 34	
Rud-e Karun	6	30 25	48 12	
Rud-e Khasken	9	28 52	57 12	6000 ft
Rud-e Kor	8	29 36	53 18	5000–6000 ft
Rud-e Nahang	10	26 00	62 40	
Rud-e Shur	1	35 27	50 56	
Rud-i-Bampur	10	27 18	59 06	
Rud-i-Kyagur	10	27 30	60 03	
Rud-i-Mil, between Chah-i-Sagak and Duruh	11			
Rud-i-Tamin	10	28 30	61 04	
Rudkhaneh-ye Jajrud	1	35 30	51 42	
Rudkhaneh-ye Jarrahi	6	30 44	48 46	
Rudkhaneh-ye Shapur	8	29 39	51 03	
Rudkhaneh-ye Talar	12	36 44 to 35 50	52 44 52 56	
Rud-Shour (Rud-e Shur)	1	35 27	50 56	3000–4000 ft

APPENDIX III

Locality	Ostan	Latitude	Longitude	Elevation
Rum	11	33 26	59 11	6000 ft
Rustamabad (Kaluraz)	2	36 54	49 28	2000 ft
Rustemabad (Kaluraz)	2	36 54	49 28	2000 ft
Rust Kartevjul, Maragheh region	3			ca 2000 m
Sa'adatabad	9	29 40	55 51	5500 ft
Sabzawaran (Sabzvaran)	9	28 37	57 45	2000 ft
Sabzevar	11	36 13	57 42	3500 ft
Sabzewar (Sabzevar)	11	36 13	57 42	3500 ft
Sabzvaran	9	28 37	57 45	2000 ft
Safed Rud (Safid Rud)	2	37 23	50 11	0–1000 ft
Safidab (Safid Ab)	1	34 35	49 50	6000–7000 ft
Safid Ab	1	34 35	49 50	6000–7000 ft
Safid Ab	2	36 33	50 20	5000 ft
Safi Dah (Kuh-e Safidar?)	8	28 55	53 00?	
Safid Rud	2	37 23	50 11	0–1000 ft
Sagban	3	38 20	45 53	5000–6000 ft
Saghy (Saqi)	11	33 08	59 18	7000 ft
Sahi (Saqi)	11	33 08	59 18	7000 ft
Sahra e Bahran (Sarab-e Bahram)	8	30 01	51 34	4000 ft
Sah Tang, near Karand	5			
Sa'idabad	9	29 28	55 42	5500 ft
St George's Hill (Ganbul Dagh)	4	37 45	45 15	5215 ft
Salavat Mountain	3			
Saleh Abad (Andimeshk)	6	32 27	48 21	150 m
Salehabad	12	35 34	56 48	ca 1200 m
Salian	?			
Salmi (Salmiah)	6	30 57	49 10	0–500 ft
Salmiah	6	30 57	49 10	0–500 ft
Salmos (Shahpur)	4	38 11	44 47	4000–5000 ft
Sama	12	36 25	51 25	
Saman-schachi (Saman Shah)	11	33 04	59 17	7000 ft
Saman Shah	11	33 04	59 17	7000 ft
Saman-Shakhi (Saman Shah)	11	33 04	59 17	
Samkhol, between Quchan and Ashkhabad, Turkmen	11			
Sanandaj	5	35 19	47 00	
Sang Bast	11	35 59	59 46	4000 ft
Sanger	?			
Sang-e Sar	12	35 43	53 19	
Sang-i-San (Sang-e Sar)	12	35 43	53 19	
Sangun	10	28 35	61 20	1700 m
Saqi	11	33 08	59 18	7000 ft
Sarab	3			
Sarab, near Dow Rud (Sarab [a spring], many in this area)	6			
Sarab-e Bahram	8	30 01	51 34	4000 ft
Sarabunab	?			
Sarakhs	11	36 32	61 11	750–1000 ft
Sarawan (Shahrestan-e Saravan)	10	27 15	62 10	
Sarband	?			

Locality	Ostan	Latitude	Longitude	Elevation
Sarbaz on Rud-e Bampur (error?)	10			
Sarbaz (on Rudkhaneh-ye Sarbaz)	10	26 39	61 15	
Sar Chah	11	32 16	58 52	4000 ft
Sar Chah	11	33 01	59 33	
Sar Chah	11	33 44	59 04	
Sarchun, about 125 km SW Esfahan	7			
Sar Dasht	6	32 32	48 52	
Sar-e Gach	6	31 57	49 25	1500 ft
Sar-e Pol (Sar-e Pol-e Zahab?)	5	34 28	45 52 ?	400 m
Sar-e Pol-e Zahab	5	34 28	45 52	
Sargad Region	10	30 00 to 27 50	60 50 60 00	
Sargun (Sangun)	10	28 35	61 20	1700 m
Sarhaad (Sarhad-e Bala)	11	33 13	59 29	7000 ft
Sarhadd (Sarhad-e Bala)	11	33 13	59 29	7000 ft
Sarhad-e Bala	11	33 13	59 29	7000 ft
Sari	12	36 34	53 04	
Sar-i-Chah (Sar Chah, many places of this name in Khorasan)	11	32 16	58 52 ?	
Sar-i-Gach (Sar-e Gach)	6	31 57	49 25	1500 ft
Sar-i-jum, between Kerman and Shiraz	?			
Sar-i-Naftak	6	32 00	49 06	
Sar-i-Tschah (Sar Chah)	11	32 16	58 52?	
Sarjan (Sa'idabad)	9	29 28	55 42	5500 ft
Sarka Daria, Karadaq (Qareh Dagh)	3			
Sarkhoun	?			
Sarr-Chakh (Sar Chah), via Basiran and Rum	11	33 01 or 33 44	59 33 59 04	
Sarvestan (several places of this name)	8 ?	29 16	53 13 ?	
Saveh	1	35 01	50 20	
Schachrud (Shahrud)	12	36 25	55 01	4000–5000 ft
Schafarud (Shafa Rud)	2	37 35	49 09	below sea level
Schah-as-Kuh	?			
Schamedinan	3 or 4			
Schecharde (Chahar Deh)	12	36 26	54 18	7000 ft
Schile, about 16 km N Hormak	10			
Schirwan (Shirvan)	11	37 24	57 55	
Schur, Sargad	10			
Schur-Ab (Shurab)	10	28 09	60 18	
Schuster (Shustar)	6	32 03	48 51	500–1000 ft
Sebzar	?			
Sefid Kuh (Kuh-e Safid, many places of this name, particularly in western Iran)	?			
Sefid Rud (Safid Rud)	2	37 23	50 11	
Sehend (Kuh-e Sahand)	3	37 44	46 27	6000–12172 ft
Seh Konj	9	30 00	57 26	7000 ft
Seir	4	37 30	45 03	5000 ft
Seivan (Sagban)	3	38 20	45 53	5000–6000 ft
Sekhend Mountains (Kuh-e Sahand)	3	37 44	46 27	
Sekhlabad Kevir	11			
Semnan	12	35 33	53 24	3500–4000 ft

APPENDIX III

Locality	Ostan	Latitude	Longitude	Elevation
Sendjiri	?			
Senghi Caravanseri	?			
Seng-i-Best (Sang Bast)	11	35 59	59 46	4000 ft
Serakhs (Sarakhs)	11	36 32	61 11	
Seri Tschah (probably Sar-i-Chah on Khanikoff's map, about 150 mi NW Lash Jowain, 150 mi NNE Kerman)	?			
Sero, on Nazlu Chay	4			
Shadegan	6	30 40	48 38	sea level
Shafa Rud	2	37 35	49 09	below sea level
Shagu (Shaqu)	9	27 15	56 25	0–500 ft
Shahabad	5	34 06	46 31	
Shah 'Abbas	1	34 46	52 12	
Shah Abdul Azim (Rey)	1	35 35	51 25	3000–4000 ft
Shahbaz (Shakhbaz)	12	35 34	56 47	about 1200 m
Shahbazan	6	32 47	48 39	540 m
Shah Bazan (Shahbazan)	6	32 47	48 39	540 m
Shah Bodagh	5			
Shahdorani	11	32 12	59 25	6000 ft
Shahestan	10	27 22	62 20	1200 m
Shahneshin	3	36 54	49 02	
Shah Pasand	12	37 07	55 16	
Shahpur	4	38 11	44 17	4000–5000 ft
Shahrabad (Shahrabad Kord)	11	37 29	56 46	4000 ft
Shahrabad Kord	11	37 29	56 46	4000 ft
Shahrara	1			
Shahrestan-e Bakhtiari va Chahar Mahall	6	32 00	50 00	
Shahrestan-e Bijar	1,2,5	36 00	47 00	
Shahrestan-e Birjand	9,10,11	32 30	59 30	
Shahrestan-e Saravan	10	27 15	62 10	
Shahr Sukhteh	10	30 30	61 10	
Shahrud	12	36 25	55 01	4000–5000 ft
Shahsavaran	1	34 10	50 01	6000 ft
Shah Sawaran (Shahsavaran)	1	34 10	50 01	
Shakhbaz	12	35 34	56 47	ca 1200 m
Shakhrud (Shahrud)	12	36 25	55 01	4000–5000 ft
Shalgahi	6	32 15	48 31	
Shamsabad	1	35 14	51 44	3000 ft
Shamsdinan	4			
Shapur (Rudkhaneh-ye Shapur)	8	29 39	51 03	
Shaqu	9	27 14	56 22	0–500 ft
Sharafabad, about 10 km S Dezful	6			
Sharafkhaneh	3	38 11	45 29	4200 ft
Sharif Khaneh (Sharafkhaneh)	3	38 11	45 29	4200 ft
Shar-i-Zagedun	10	31 12	61 40	
Shar Lut	?			
Shaspur River (Rudkhaneh-ye Shapur)	8	29 39	51 03	
Shastun (Shahestan)	10	27 22	62 20	1200 m
Shazaban Island, Daryacheh-ye Reza'iyeh	?	37 22	45 36	
Shazalan Island, Daryacheh-ye Reza'iyeh	?	37 34	45 32	4200 ft
Shekar Ab	1			

Locality	Ostan	Latitude	Longitude	Elevation
Shelkar Ab	7	33 40	55 33	980 m
Shemane	3 or 4			
Sheraf-Khane (Sharafkhaneh)	3	38 11	45 29	
Sherif-Chane 'Sharafkhaneh)	3	38 11	45 29	4200 ft
Sheykhabad	5	33 53	48 02	
Shikarab (Shekar Ab)	7	33 40	55 33	980 m
Shile, about 16 km N Hormak	10			
Shiraz	8	29 36	52 32	5500 ft
Shiraz Arza	8			
Shirvan	11	37 24	57 55	4000 ft
Shitvar Island (Jazireh-ye Shotur)	?	26 47	53 25	
Shul-e Dalkhan	8	30 16	52 07	
Shur-ab (Shurab)	10	28 09	60 18	
Shurab	10	28 09	60 18	
Shur Rud, 60 km E Qayen	11	33 45	59 20	
Shush	6	32 12	48 15	500 ft
Shushtar	6	32 03	48 51	500–1000 ft
Shustar (Shushtar)	6	32 03	48 51	
Siah Kuh	10	25 55	61 36	
Siah Kuh	1	34 38	52 16	3000–6611 ft
Siah Mansur	6	32 20	48 30	500 ft
Siah Paulas River	1			
Siah Sang (10 km N Rudehen)	1			
Sia-Mansur (Siah Mansur)	6	32 20	48 30	
Siaret, Atrek Valley	11			
Sib	10	27 15	62 05	3000–4000 ft
Sija-Kugi (Siah Kuh)	10	25 55	61 36	
Sija-Mansur (Siah Mansur)	6	32 20	48 30	
Simurgh, Lut	?			
Sisangan, ENE Chalus	2			
Sir-i-Tam, SW Kerman	9			
Sirjan (Sa'idabad)	9	29 28	55 42	
Sitaver Mountain, W Daryacheh-ye Reza'iyeh	4			
Sivand	8	30 05	52 55	
Siyahrud	12 ?			
Sjulpenai Mountains (Kuh-e Ferdows?)	11	33 55	58 39	
(or Kuh-e Zul?)	11	33 35	59 05	
Soltanabad	11	36 23	58 02	
Soltaniyeh	1	36 26	48 48	
Sopurghan	4	37 45	45 12	5000 ft
Sorchkela	12 ?			
Sorkh-e Dize (Sorkheh Dizeh)	5	34 24	46 03	
Sorkheh Dizeh	5	34 24	46 03	
Soultanabad (Arak)	1	34 05	49 41	5700 ft
Ssaman-Schahi Mountains (Saman Shah)	11	33 04	59 17	7000 ft
Sujbulak (Mahabad)	4	36 45	45 43	4272 ft
Sultanabad (Arak)	1	34 05	49 41	5700 ft
Sultaniah (Soltaniyeh)	1	36 26	48 48	
Superghan (Sopurghan)	4	37 45	45 12	
Suran	10	27 18	62 00	
Susa (Shush)	6	32 12	48 15	500 ft

APPENDIX III

Locality	Ostan	Latitude	Longitude	Elevation
Tabaghet Kawir	?			
Tabas	11	32 48	60 14	5000 ft
Tabris (Tabriz)	3	38 05	46 18	4600 ft
Tabriz	3	38 05	46 18	4600 ft
Tagab, 17 km SE Bazman (? Talab)	10	28 30	61 59 ?	
Tag-i-Dorokh, between Duruh and Chah-e Ziran	11	32 10	60 27	
Tahrud	9	29 26	57 49	5000 ft
Tajan River (Harirud), Iran-Afghanistan-Turkmenistan border	?			
Tajebad (Tayyebat)	11	34 44	60 45	3000 ft
Takestan	2	36 04	49 43	
Takht-e Jamshid	8	29 57	52 52	5500 ft
Talab (Tarlab)	1	34 33	50 37	4000 ft
Talab	10	28 30	61 59	
Talaghan (Talakhan = Shul-e Dulkhan)	8	30 16	52 07	
Talar River (Rudkhaneh-ye Talar)	12	36 44 to 35 50	52 44 52 56	0–9000 ft
Talar Valley (Rudkhaneh-ye Talar)	12	36 15	52 55	300 m
Talayu	11	38 04	56 23	
Taleqan	1	36 10	50 40	
Talkhab	12	35 33	56 56	ca 1200 m
Talkeh Rud	3	37 40	45 46	
Tallau (Talayu)	11	38 04	56 23	
Tamand	11	32 31	59 30	6000 ft
Tamin, Sargad Region, 20 km SW Ladiz	10			
Tamishan	12			
Tangali	12	37 27	54 43	
Tang-e Karam	8	29 06	53 39	5000 ft
Tang-e Rah	12	37 25	55 55	
Tang-e Tschekun	8			
Tang-i-Golestan (Ab-e Golestan)	6	32 00	49 06	
Tang-i-Gurguda, near Gach Saran	8			4500 ft
Tang-i-Kerim (Tang-e Karam)	8	29 06	53 39	5000 ft
Tangistan	8	ca 29 00	51 00	
Tangistan	8	31 27	53 19	
Tangyak	8	28 55	50 52	0–500 ft
Tanjistan (Tangistan)	8	31 27	53 19	
Tappeh Asiab, near Bijaneh, 6.5 km E Kermanshah	5			
Tappeh Sarab, near Bijaneh	5			
Tappeh Sharafabad	6			
Tara Sirkuh	?			
Tarlab	1	34 33	50 37	4000 ft
Tarsee	?			
Tasuki	10	30 25	61 13	1600 ft
Tauran	11			
Tauris (Tabriz)	3	38 05	46 18	
Taut Shami, near Karand	5			
Tavile	5	34 ??	47 ??	
Tavriz (Tabriz)	3	38 05	46 18	4600 ft
Tayyebat	11	34 44	60 45	3000 ft

Locality	Ostan	Latitude	Longitude	Elevation
Tchambachi Valley	6	33 25	48 59	1300 m
Tchartaghi, about 80 km N Kashan	?			
Tebbes (Tabas)	11	32 48	60 14	5000 ft
Teheran (Tehran)	1	35 40	51 26	4000 ft
Tehran	1	35 40	51 26	4000 ft
Tehrud (Tahrud)	9	29 26	57 49	5000 ft
Tejur	12	35 36	56 30	ca 1100 m
Tembi River (Ab-e Tembi)	6	31 50	49 28	1200 ft
Tepe Dervish, or Chakh-i-Dzhanu, about 35 km SW Hauzdar	10			
Tepe Jari, 16 km S Persepolis on Marv Dasht Plain	8			
Tergever	4			
Tizab, N Damavand	1			
Tochah	12	35 38	56 37	ca 1100 m
Tochal Mountains	1	35 54	51 10	5000–11555 ft
Tochmak	?			
Tol-e Bazun	6	31 55	49 25	1200 ft
Torbat-e Heydariyeh	11	35 16	59 13	5000 ft
Torbat-e Jam	11	35 14	60 36	3000 ft
Tschahkata	8			
Tscha-i-Bena, Sistan	10			
Tscha-i-Divan (Shahdorani)	11	32 12	59 25	6000 ft
Tscha-i-Gjuische, Sistan	10			
Tscha-i-Ziru, Zirkukh	11			
Tscharachs, 28 km from Ahangeran	11			
Tschar-magal	?			
Tschebarde (Chahar Deh)	12	36 26	54 18	7000 ft
Tscheharde (Chahar Deh)	12	36 26	54 18	7000 ft
Tschetchme-rogan (Chashme Rowgani)	6	31 33	49 41	2000 ft
Tschock et Bena	?			
Tsch-i-Huishé	11			
Tshalma, Dasht-e Moghan	3			
Tuba Creek	6	32 45	48 40	
Tul-i-Bazun (Tol-e Bazun)	6	31 55	49 25	1200 ft
Tumb Island (Jazireh-ye Tanb-e Bozorg), Persian Gulf	?	26 14	55 19	0–175 ft
Tun (Ferdows)	11	34 00	58 09	4000 ft
Turbat-e Haidari (Torbat-e Heydariyeh)	11	35 16	59 13	5000 ft
Turbet Sheikh-i-Dzham (Torbat-e Jam)	11	35 14	60 36	3000 ft
Tuveh	5			
Uarangrad (Varangrud)	12	36 07	51 22	
Urmi (Reza'iyeh)	4	37 33	45 04	4600 ft
Vali Abad, N slope Alborz Mountains	12	36 14	51 18	1800–2500 m
Varamin	1	35 20	51 39	3000–3600 ft
Varangrud	2	36 07	51 22	
Veramin (Varamin)	1	35 20	51 39	3000–3600 ft
Vizastra	3 or 4			

Locality	Ostan	Latitude	Longitude	Elevation
Weyser, Alborz Mountains	12			1150 m
Yamaha	6	31 47	49 23	1261 ft
Yasuj	8	30 50	51 10	
Yazd	7	31 53	54 25	4080 ft
Yazd-e Khvast	8	31 31	52 07	7000–8000 ft
Yezd (Yazd)	7	31 53	54 25	4080 ft
Yezd-i-Khast (Yazd-e Khvast)	8	31 31	52 07	7000–8000 ft
Zabol	10	31 02	61 30	1600–2000 ft
Zaboli	10	27 07	61 40	4500 ft
Zagan, Sargad, 17 km S Panjsareh	10			
Zagheh	6	33 30	48 42	
Zagheh	6	33 46	49 01	
Zahedan	10	29 30	60 52	4500 ft
Zahirabad	11	35 08	58 58	4000 ft
Zamanabad	12	35 35	56 46	ca 1200 m
Zamin-i-Shile	10	30 02	61 27	
Zamran	10	26 32	62 19	2000 ft
Zangi Achmed (Lut-e Zangi Ahmad)	9	29 22	58 48	2000 ft
Zanjan	2	36 40	48 29	5000–6000 ft
Zanjian (Zanjan)	2	36 40	48 29	5000–6000 ft
Zar Rud, near Rayen	9	29 34	57 26	
Zeinelabad	10			
Zeloi	6	32 13	49 04	
Zergend	?			
Zeunelabad	?			
Zeurabad (Zahirabad)	11	35 08	58 58	4000 ft
Zharabad, SW Reza'iyeh	4			
Zia-i-Bolokh	10			
Zia-i-Lagun	10	30 10	60 50	
Ziarat (many places of this name)	?			
Ziaret, near Shirvan	11			
Ziaret (Ziarat, in region of Rud-e Karun)	6			
Zirkuch	11			
Zir-Kuh	11			
Zirkukh Region	11	34 01 to 31 00	60 00 60 30	
Zizi, at Vik River, 30 km W Gach Saran	8			
Zul-Penai	11			

Index

Page numbers on which portraits occur and plate numbers
of figured lizards are indicated in **boldfaced** type.

Abel, Erich 100, 108–109
Ablepharus 4, 36, 40, 44–45, 47–49, 51, 54, 59–61, 66, 262–267
   alaicus 263–265
   bivittatus 4, 44–45, 49, 51, 60–61, 262–264, **16E-F**
   bivittatus bivittatus 264
   brandti brevipes 265
   brandtii 265–267
   brandtii festae 266–267
   grayanus 265–267
   kitaibelii 266–267
   kitaibelii kitaibelii 267
   lindbergi 264–265
   Menetriesi 264
   pannonicus 4, 44, 47–49, 51, 60–61, 263, 265–267, **16G**
   pannonicus grayanus 265
   pannonicus pannonicus 265
   persicus 255–267
   pusillus 265–267
abudhabi
   See *Bunopus*
Acanthodactylus 3–4, 33, 40, 42, 44, 47–49, 51–52, 58, 60–61, 64–65, 85, 91, 186, 199–208, 211, 254, 256–259
   arabicus 200, 203, 207
   blanfordi 3, 44, 48, 61, 200–203, 205, 208, 257, **11A**
   boskianus 3–4, 47, 200–201, 205, 257, 259
   boskianus euphraticus 257
   cantoris 202, 206–207
   cantoris blanfordi 202
   cantoris cantoris 202
   cantoris schmidti 206
   felicis 259
   fraseri 203–205
   gongrorhynchatus 200, 204
   grandis 3, 47, 58, 60, 186, 200–201, 203–205, 257, **11B**
   haasi 200 204
   masirae 259
   micropholis 3, 44, 48, 51, 61, 200–201, 205–206, 256, **11C**
   nilsoni 3–4, 33, 49, 256, 258
   opheodurus 3–4, 200–201, 256–257, 259
   schmidti 3, 47, 51, 58, 60–61, 85, 91, 186, 200–201, 203–204, 206–208, 257, **11D**
   schreiberi 205, 259
   scutellatus 204
   tilburyi 200
   yemenicus 259
acutirostris
   See *Eremias*
   See *Eremias* (*Scapteira*)
Adams, C. G. 63
Adle, Ahmed Hossein 13, 43
Adler, Kraig Kerr 13, 17, 20, 27–28, 33, 37–38, 87, 92, 113, 166, 172, 189, 213, 227–228, 274, 296
Adolfus 199, 201
aegyptia
   See *Lacerta*
aegyptius
   See *Uromastix*
   See also *Uromastyx*
Aellen, Paul 22
Aeluroscalabotes 117, 124
Aeluroscalabotinae 116
affinis
   See *Ceramodactylus*
   See also *Euprepis*
   See also *Mabuya*
   See also *Stenodactylus*
   See also *Tiliqua*
Afghan Delimitation Commission 19
Afghan ground agama 107
afghanicus
   See *Eublepharis*
africanus
   See *Holodactylus*
Agama
   agilis 100–101, 104
   agilis agilis 101
   agilis isolepis 101
   agilis sanguinolenta 100–101
   baluchiana 107, 110
   blanfordi 101, 104–105
   blanfordi blanfordi 101, 105
   caucasia 71
   caucasica 71, 73, 75–76
   caucasica caucasica 71
   caucasica microlepis 76
   erythrogastra 73
   erythrogastra pallida 73
   isolepis 19, 100, 104
   kirmanensis 101
   kirmanensis brevicauda 101
   lessonae 108
   megalonyx 107, 110
   melanura 75
   melanura lirata 75
   microlepis 71, 76
   microtympanum 20, 107, 109
   mucronata 73, 75
   mutabilis 99
   nupta 74, 78, 80
   nupta fusca 80
   nupta nupta 80
   persica fieldi 104, 107
   ruderata 107–108, 110
   ruderata megalonyx 107
   ruderata ruderata 108
   sanguinolenta 100–101
   sanguinolenta isolepis 101
   scutellata 97

tuberculata 69
versicolor 68
agamas 68, 99
Agamidae 1, 6, 40, 66, 68–69, 289
agamids 9, 32, 55, 57, 68, 102
Agamura 2, 44–46, 48, 52, 55, 59–61, 125–127, 129–132, 139, 145, 147, 155, 158–159, 180–183
   cruralis 130–132
   femoralis 126–127, 130, 180, 182–183
   gastropholis 158
   misonnei 182
   persica 2, 44–46, 48, 59, 61, 126–127, 129–131, 180, **6G**
   persica cruralis 132
agamuroid thin–toed gecko 154
agamuroides
   See *Cyrtodactylus*
   See also *Cyrtopodion*
   See also *Gymnodactylus*
agamuroides group 151
agilis
   See *Agama*
   See also *Lacerta*
   See also *Trapelus*
agilis agilis
   See *Trapelus*
Aitchison, J. E. T. 19, 94
Akhmedov, S. B. 158, 275–277
alaicus
   See *Ablepharus*
albofasciatus
   See *Gymnodactylus*
Alborz lizard 236
Alborz Mountains
   See Iran
Alcock, Alfred William 21, 107, 212
Aldrovandi
   See *Plestiodon*
Alekperov, Abdulla Mustafaevich 38
algericus
   See *Tropiocolotes*
Algyroides 231
Alsophylax 32, 125–126, 140–141, 146, 170–171, 196–197
   crassicauda 140
   laevis 126, 236
   microtus 126
   persicus 32, 146, 196–197
   pipiens 125–126
   spinicauda 126, 170
   tuberculatus 141
amictopholis
   See *Cyrtodactylus*
   See also *Cyrtopodion*
Ananjeva, Natalia Borisovna 9, 14, 68–69, 71–73, 75–78, 81, 87–88, 93, 289
Anatolian rock lizard 233
Anderson's racerunner 212
Anderson, Jeromie A. 2, 99, 146, 180, 197, 287
Anderson, John 1, 3, **18**, 41, 71, 84–85, 90, 100, 104, 108, 111, 126, 173, 175, 247, 255, 268, 272, 290–291
Anderson, Steven Clement 1–3, 5, 9–10, 13, 19, 23–24, 26–29, 32–33, 37, **38**, 39, 41, 43–44, 53–55, 58–59, 68, 71, 73–76, 79–81, 85–87, 89–90, 92–95, 97–98, 100–101, 104–108, 112–114, 118–120, 125–127, 130, 132–133, 136–138, 140–141, 144–146, 148, 154, 156–158, 160–166, 168–170, 173, 175–177, 179–180, 182, 184–190, 193–195, 197, 202–207, 211, 213–214, 216, 219, 221–222, 224–228, 232–234, 236, 238, 241–242, 244, 247, 250, 252–253, 255, 264–265, 268, 270–273, 275–288, 291, 293–295, 297–298
andersoni
   See *Eremias*
Andersson, Lars Gabriel 275–276
Andreas, Prof. 20, 158
Andrén, Claes 5, **28**–29, 41, 59, 71, 90, 98, 101, 130–131, 212, 222, 252, 270, 278–279, 282–283, 293, 298
Angel, Fernand 250–251
Anglo–Persian Oil Company 15
angramainyu
   See *Eublepharis*
Anguidae 1, 6, 66, 111
anguids 111
Anguis 1–2, 47, 50–51, 54, 56, 59–60, 111–113, 278
   colchica 111–112
   fragilis 2, 47, 50–51, 54, 56, 59–60, 111–113
   fragilis colchica 111–112
   fragilis colchicus 47, 111–113, 160, 164, **6A-B**
   fragilis fragilis 113
   oreintalis 111
   punctatissimus 278
   ventralis 113
Annandale, Thomas Nelson 32, 35, 90, 101, 107, 110, 130, 150, 156, 161, 169, 177, 190, 222, 274, 285, 298
Apathya 33, 199, 233
   cappadocica urmiana 33, 233
apoda
   See *Lacerta*
   See also *Pseudopus*
apodus
   See *Ophisaurus*
apodus apodus
   See *Ophisaurus*
apodus thracius
   See *Ophisaurus*
aporosceles
   See *Eremias*
apus
   See *Ophisaurus*
Arabian desert gecko 141
Arabian toad–headed agama 84
arabicus
   See *Acanthodactylus*
   See also *Eumeces*
   See also *Phrynocephalus*
   See also *Stenodactylus*
   See also *Trigonodactylus*
Aralo–Caspian racerunner 216
Archaeolacerta 40, 199, 236, 246
arenarius
   See *Scincus*
   See also *Varanus*

See also *Varanus* (*Psammosaurus*)
arguta
   See *Eremias*
   See also *Eremias* (*Ommateremias*)
   See also *Lacerta*
   See also *Podarces* (*Eremias*)
armored glass lizards 113
Arnold's fringe-toed lizard 259
Arnold, Edwin Nicholas 3–4, 13–14, 37, 39–41, 81, 85, 112, 114, 116–117, 126–127, 132–133, 136–137, 139–140, 142, 144, 162, 174, 176, 178–179, 183–186, 191, 199–208, 211, 213, 229–231, 233–234, 236, 239–240, 243, 246, 248, 250–252, 254, 257, 259, 262, 276–277, 287–288, 291
Asaccus 2, 28, 32–33, 47, 49, 53, 58, 60–61, 64, 125, 127–128, 132–138, 169, 277
   caudivolvulus 133, 137
   elisae 2, 47, 49, 53, 58, 60, 133–136, 169, 277, **6H**
   gallagheri 33
   griseonotus 2, 28, 33, 49, 60, 133–138
   kermanshahensis 2, 33, 49, 133–135, 137–138
   montanus 121, 133
   platyrhynchus 133, 137
Ascalabotes
   sthenodactylus 183
Asia minor thin-toed gecko 159
Asian snake-eyed skink 265
Asiocolotes 126, 193
asmussi
   See *Centrotrachelus*
   See also *Uromastix*
   See also *Uromastyx*
aspratilis
   See *Bunopus*
   See also *Carinatogecko*
Ast, Jennifer C. 296
astrabadensis
   See *Lacerta*
Asymblepharus 263, 265
   borealis 263
Ataev, Chary Ataevich 38, 51, 71, 73, 75–78, 171
Atamuradov, Khabibula 38
Aucher-Eloy 15
Auffenberg, Walter 38, 69, 297–298
aurata
   See *Lacerta*
   See also *Mabouia*
   See also *Mabuya*
aurata aurata
   See *Mabuya*
aurata septemtaeniata
   See *Mabuya*
aurata transcaucasica
   See *Mabuya*
aurita
   See *Lacerta*
Austrian Expedition to Iran
   See Iran
Autumn, Kellar 122
axial pocket 116–117
Azerbaidzhan lizard 243

Baig, Khalid Jared 69

bakhtiari
   See *Microgecko*
   See *Tropiocolotes*
   See *Tropiocolotes* (*Microgecko*)
Bakhtiari dwarf gecko 197
Balland, Daniel 15
Balletto, Emilio M. 39
Baloutch, Mohammad 4, 29, **32**, 113, 120, 229, 238–240, 260
Baluch plate-tailed gecko 189
Baluch rock gecko 141
baluchiana
   See *Agama*
baluchiorum
   See *Testudo*
Bampur thin-toed gecko 167
banded dwarf gecko 193
Bannikov, Andrei Grigoryevich 36, 38, 70–71, 73, 76, 81–83, 87, 89–90, 92, 95, 97, 100, 108, 111–114, 122, 141, 146, 148, 157, 164–167, 170, 190, 200, 208, 210, 216, 219, 221, 224–228, 230, 232, 234, 237–238, 240–241, 244, 247–248, 252, 255, 262, 264–265, 270, 272, 276
Banta, Benjamin Harrison 56
Baran, Ibrahim 36, 38, 109, 164, 170
Barbour, Thomas 37, 116
Başoğlu, Muhtar 38, 109
basoglui
   See *Cyrtodactylus*
Bastani, B. 32
Batrachuperus
   persicus 26–27, 29, 33
Bauer, Aaron Matthew 13, 41, 117, 125, 130, 148, 174, 186, 192, 198, 213, 216, 221, 227
Bazin, M. 50
Bedriaga's plate-tailed gecko 187
Bedriaga's skink gecko 187
Bedriaga, Jacques Vladimir von 4, 35–**37**, 71, 76, 80, 87, 90, 95, 97, 100, 108, 111, 130, 141, 156, 160, 168, 177, 187, 190, 202, 205, 209, 213–214, 222, 224, 227, 232, 236, 238, 241, 247, 252, 255, 264, 268, 270–271, 275, 280, 283, 285, 293–294, 297–298
bedriagai
   See *Teratoscincus*
Bengal monitor 296, 298
bengalensis
   See *Tupinambis*
   See also *Varanus*
bengalensis bengalensis
   See *Varanus*
   See also *Varanus* (*Indovaranus*)
bent-toed geckos 150–151
Benyr, G. 41, 64, 199, 201, 211, 231, 250, 254
Bernor, L. R. 63
Beutler, Axel 164, 256
beutleri
   See *Cyrtopodion*
Bibron, Gabriel 5, 34, 97, 173, 264, 267, 270, 278, 297
bicolor
   See *Contia*
biporus
   See *Bunopus*

Bischoff, Wolfgang 199, 231, 242–243, 250
bivittatus
  See *Ablepharus*
  See also *Scincus*
black rock agama 75
Black, Jesse 122
black-ocellated racerunner 219
black-sided racerunner 221
black-tailed toad agama 90
Blanford's agama 104
Blanford's rough-scaled gecko 159
Blanford's sandfish 288
Blanford's semaphore gecko 177
Blanford's short-nosed desert lizard 250
Blanford's short-toed gecko 156
Blanford's snake skink 279
Blanford, William Thomas **iii**, 1–5, 10, 18–19, 32, 68, 70–71, 75–76, 78–80, 90–91, 95–98, 100, 104–106, 108–109, 111, 113, 129–132, 139–143, 147, 150, 153–154, 156–157, 159, 168, 171–172, 175–179, 183–185, 190, 199, 201–203, 205–206, 209–210, 214–216, 221–222, 226–228, 230, 232, 236, 241–242, 247, 249–250, 252, 255–256, 261, 263–265, 267–268, 270–271, 273–275, 279–280, 283, 285, 287–288, 291–295, 297–298, 300
blanfordi
  See *Acanthodactylus*
  See also *Agama*
  See also *Bunopus*
  See also *Ophiomorus*
  See also *Ophisops*
Blegvad, H. 22
Blepharosteres
  grayanus 265
blind worm 111
Bloom, R. A. 170
blunt-tailed spider gecko 130
Blyth, Edward 1–2, 4–5, 34, 69, 75, 118, 120, 150, 272, 274, 279, 285, 287
blythianus
  See *Eumeces*
Bobek, Hans 13, 44–45, 47, 49–51, 112
Boettger, Oskar 1–2, 4, 35–**37**, 41, 83, 96, 112, 147–148, 163, 175, 226, 230, 236, 243, 245, 248, 255, 260, 275
boettgeri
  See *Lacerta*
  See also *Phrynocephalus*
Bogdanov, Oleg Pavlovich 38, 171
Böhme, Wolfgang 39, 69, 159, 233
borealis
  See *Asymblepharus*
Borkin, Leo Jakolevich 40
Börner, Achim-Rudiger 116, 121–122
Bornmüller, Josef 20
Bory de Saint Vincent, Jean Baptiste Geneviève Marcellin 267, 278
Bosc's fringe-toed lizard 257
boskiana
  See *Lacerta*
boskianus
  See *Acanthodactylus*
Boulenger, George Albert 1, 3–5, **16**, 19, 32–33, 35–36, 39–41, 68, 71, 75–76, 80–81, 83–85, 89–90, 93–95, 97–98, 100, 104, 107–108, 118, 130, 141–142, 144, 147, 151, 156–157, 160, 168, 177, 184–185, 188, 190, 198–199, 201–203, 205–207, 210–211, 213–216, 219–220, 222, 224, 226–227, 229, 232, 234–236, 238, 241, 244, 247, 250–252, 254–255, 257, 260, 264, 268, 270, 275–277, 279–280, 283, 285, 288, 291, 293–294, 298
Boutan, Louis Marie August 179
Brandt's Persian lizard 232
Brandt, Johann Fridrich 32, 232
brandtii
  See *Ablepharus*
  See also *Lacerta*
brandtii festae
  See *Ablepharus*
brevicauda
  See *Agama*
brevicaudus
  See *Trapelus*
brevipes
  See *Cyrtopodion*
  See also *Gymnodactylus*
  See also *Mediodactylus*
  See also *Ophiomorus*
  See also *Phrynocephalus*
  See also *Zygnidopsis*
  See also *Zygnopsis*
brevirostris
  See *Eremias*
  See also *Mesalina*
brevirostris brevirostris
  See *Mesalina*
brevirostris forma *typica*
  See *Eremias*
bridled skink 277
brilliant agama 100
brookii
  See *Hemidactylus*
Buchbinder, B. 63
Bufo
  kavirensis 28
  oblongus 32
  persicus 32
  surdus 26, 198, 297
  viridis 33, 114, 186
  viridis kermanensis 33
Bufoniceps 81
  laungwalensis 81
Bullock, Steven 27
Bunopus 2, 21, 33, 40, 44, 47–48, 51–52, 58–61, 125–126, 129, 138–149, 186, 188, 197
  abudhabi 125, 140, 142, 144
  aspratilis 33, 40, 126, 144–146
  biporus 21, 141–143
  blanfordi 125, 140, 142, 144
  crassicaudus 2, 125, 139–140, 144
  persicus 197
  tuberculatus 2, 47–48, 51, 58–61, 126, 138–144, 147, 149, 186, 188, **7A**
Burke, Russell Louis 37
Burton, John A. 37, 112, 114, 162, 174, 213, 229
Busack, Stephen Dana 41

Butzer, Karl W. 13
buxtoni
   See *Testudo*
Buxtorf, R. 22

C*abrita* 254
Callisaurus 9
Calotes 1, 48, 61, 66, 68–69
   versicolor 1, 48, 61, 66, 68–69, **2A**
   versicolor nigrigularis 69
   See also *Lacerta*
Camerano, Lorenzo 4, 230, 235–236, 244, 260
cantoris
   See *Acanthodactylus*
cantoris group 200, 206
cappadocica
   See *Lacerta*
cappadocica cappadocica
   See *Lacerta*
Carinatogecko 2, 40, 49–50, 60, 125–126, 129, 140, 144–146
   aspratilis 2, 49, 144–146, **7B**
   heterophoris 2, 49, 60, 146
carinatus
   See *Stellio*
Casimir, Michael J. 26
Caspian bent-toed gecko 157
Caspian green lizard 247
Caspian Region 46
Caspian thin-toed gecko 157
caspium
   See *Cyrtopodion*
caspium caspium
   See *Cyrtopodion*
caspium group 152
caspius
   See *Cyrtodactylus*
   See also *Gymnodactylus*
   See also *Psammosaurus*
   See also *Tenuidactylus*
   See also *Varanus*
   See also *Varanus (Psammosaurus)*
caucasia
   See *Agama*
   See also *Laudakia*
caucasia caucasia
   See *Laudakia*
Caucasian agama 70
caucasica caucasica
   See *Laudakia*
caucasica mucronata
   See *Agama*
caucasicus
   See *Stellio*
caucasius
   See *Stellio*
caucasius caucasius
   See *Stellio*
caudivolvula
   See *Lacerta*
caudivolvulus
   See *Asaccus*

Central Asian racerunner 227
Central Plateau
   See Iran
Centrotrachelus
   asmussi 32, 293–294
   loricatus 32, 294
Ceramodactylus 126, 183–186
   affinis 184
   doriae 185
   major 186
Černy, Michal 29, 196
Chalcides 4, 47–48, 51, 60–61, 262 267–268, 277, 279
   ocellatus 4, 47–48, 51, 60–61, 99, 268, 277–278, 293
   ocellatus ocellatus 268, **17A**
   tridactyla 267
   See also *Lacerta*
Chapman, Randolph W. 65
Charif, G. 22
Cherchi, Maria Adelaide 39
Cherlin, Vladimir A. 40, 190, 253
Chernov, Sergius Aleksandrovich 36, 38, 67, 71, 73–75, 81–82, 88–89, 92–93, 95, 100, 108, 112–114, 126, 147, 157, 166–167, 208, 213, 216–221, 223–227, 232, 234, 236, 238, 240, 244, 247, 252–253, 255, 262, 264–266, 270, 272–277, 298–299
chernovi
   See *Ophiomorus*
chitralensis
   See *Cyrtopodion*
chlorogaster
   See *Lacerta*
ciliciense
   See *Cyrtopodion*
Clark, Erica D. 23
Clark, Richard 23, 69, 71–73, 75, 80–81, 85–87, 90, 94–95, 98, 100–101, 108–110, 113–114, 121, 125–126, 130, 157, 170, 213, 222, 224, 226, 228, 232, 234, 236–238, 243, 247–248, 252, 255, 268, 270, 276–278, 294, 297
clarkorum
   See *Lacerta*
   See also *Phrynocephalus*
Clarks' toad-headed agama 85
Clergue-Gazeau, Monique Andréa 33
Clifton, Bernard J. 21
climate map of Iran 43
Cochrane, J. A. 39
coctaei
   See *Hemidactylus*
cocteaui
   See *Hemidactylus*
Colchican slow worm 111
colchicus
   See *Anguis*
   See also *Cyrtopodion*
Coleonyx 117
Coluber
   najadum 114
   rhodorachis 171
comb-fingered geckos 183
common skink 287
common skink gecko 189

condoni
  See *Contia*
conirostris
  See *Scincus*
consobrinoides
  See *Gymnodactylus*
Contia
  bicolor 32
  condoni 33
Cope, Edward Drinker 22, 34, 109
cordylids 262
Corkill, Norman L. 39
Couvering, J. A. Van 63
crassicauda
  See *Alsophylax*
crassicaudus
  See *Bunopus*
Crossobamon 2, 44, 46, 48, 52, 60–61, 125–126, 128, 140–141, 144, 146–149, 183
  eversmanni 2, 44, 46, 48, 52, 61, 146–149, **7C**
  eversmanni eversmanni 149
  eversmanni lumsdeni 148–149
  eversmanni maynardi 149
  lumsdeni 141
  maynardi 126, 149
  orientalis 126
cruralis
  See *Agamura*
Cumming, W. D. 19
Curzon, George N. 17
Cuvier, Georges Jean-Léopold-Nicolas-Frédéric Dagobert 1, 33, 68, 99, 172, 185
cyanophlyctis
  See *Euphlyctis*
cyanura
  See *Lacerta*
cylindrical skinks 267
Cyrén, Carl August Otto 4, 230, 232–233, 236–238, 241, 244, 261
Cyrtodactylus
  agamuroides 55, 154, 158–159
  amictopholis 126, 152, 171
  basoglui 170
  caspius 157
  fedtschenkoi 150, 152, 164–166
  gastropholis 158
  heterocercus 160
  heterocercus heterocercus 160
  kachhensis 161
  kirmanensis 162
  kotschyi 50, 60, 150, 160, 163–164, 171
  longipes 44–46, 61, 150, 152, 164–165
  longipes longipes 164
  longipes microlepis 165
  macularius 120
  pulchellus 127, 150
  saggitifer 168
  sagittifer 48, 150, 152, 161, 168, 171
  scaber 157, 169
  zarudnyi 166
Cyrtopodion 2, 13, 42, 44–51, 53, 55, 58–61, 125–130, 139–140, 145, 147, 150–171, 175, 180, 188
  agamuroides 2, 44–46, 48, 55, 61, 127, 150–151,
153–154, 158–159, **7D**
  amictopholis 126, 150, 152, 171
  brevipes 2, 48, 150–151, 154–156, 161
  caspium 2, 44–46, 48–51, 60–61, 150–152, 156–157, 170
  caspium caspium 152, 157, **7E**
  caspium insularis 158
  chitralensis 126, 151
  elongatus 150, 152
  fedtschenkoi 150, 152, 166
  gastrophole 2, 47, 55, 151, 154–155, 157–158, 163, **7F**
  heterocercum 2, 49, 60, 128, 145, 150, 152–153, 159–161, 164, 168, 171
  heterocercum heterocercum 168
  heterocercum mardinense 160
  kachhense 2, 61, 150–151, 153, 156, 160–161
  kirmanense 2, 44–45, 151–153, 156, 161–162
  kotschyi 2, 50, 60, 150, 152–153, 160, 162–164, 171
  kotschyi beutleri 164
  kotschyi ciliciense 164
  kotschyi colchicus 160, 164
  kotschyi danilewskii 160, 164
  kotschyi ponticus 164
  longipes 2, 44–46, 48, 59, 61, 150, 152–153, 163–166, 188
  longipes longipes 153, 163–165
  longipes microlepis 48, 153, 165
  longipes voraginosis 165
  mintoni 151–152
  montiumsalsorum 150, 152
  russowii 2, 46, 61, 150, 152–153, 166–167, 171, **8A**
  russowii zarudnyi 153, 166–167
  sagittifer 2, 48, 150, 152–153, 161, 167–168, 171, **8B**
  scaber 169
  scabrum 2, 46–49, 51, 53, 58, 60–61, 129, 150–151, 154, 156, 168, 170, 175, **8C**
  spinicauda 2, 51, 61, 150, 152, 169–170, **8D**
  stoliczkai 126, 150–152
  syriacus 163–164
  tibetanus 151–152
  turcmenicum 2, 51, 150, 152, 170–171, **8E**
  watsoni 150–152
Cyrtopodion (Cyrtopodion)
  gastropholis 158
  scaber 169
Cyrtopodion (Mediodactylus)
  heterocercus heterocercus 160
  kotschyi 163

Dalj, S. K. 38
danfordi
  See *Lacerta*
danilewskii
  See *Cyrtopodion*
  See *Cyrtopodion kotschyi danilewskii*
Danish Scientific Investigations in Iran
  See Iran
Darevsky, Ilya Sergeevich 2–3, 13–14, **28**, 33, 36,

38–41, 81, 95, 117–118, 122–123, 164, 173, 190, 209, 212, 214, 236–237, 240, 244–245, 247–249, 256, 263, 270–271
Das, Indraneil 39–40, 71, 75, 80, 90, 111, 130, 141, 175, 205, 241, 250, 272, 280, 285
Daudin, François-Marie 1–5, 33, 68, 113, 172, 199–201, 257, 269, 290, 296–298
DeBlase, Anthony 24, **26**
deccanensis
  See *Gymnodactylus*
De Filippi, Filippo **iii**, 34, 70, 80, 83, 95, 97, 100, 108, 111–112, 157, 159, 185, 221, 225–226, 229, 232, 236, 252, 255, 260, 264, 270, 275–277, 298
defilippii
  See *Lacerta*
  See also *Podarcis*
Dely, Olivér György 112
Demidoff, Anatoly Nikolaevich de 111
Denardo, Dale F. 122
Department of Defense (United States) 15, 37
depressus
  See *Lacerta*
  See also *Tropiocolotes*
derjugini derjugini
  See *Rhithrotriton*
desert lacertas 208
desert monitor 298
deserti
  See *Eremias*
  See also *Ommateremias*
deserti-transcaucasica
  See *Eremias arguta deserti/transcaucasica*
Diplodactylidae 116
Diplodactylinae 116
Diplometopon
  zarudnyi 52, 58, 85, 186
Disi, Ahmad M. 39
dispar
  See *Uromastyx*
dissimilis
  See *Mabuya*
distributional patterns 1, 59
Dixon, James Ray 2, 28, 33, 132–138
dominicensis
  See *Mabuya*
Doria's comb-fingered gecko 185
Doria, Marquis Giacomo 18, 112, 160, 185
doriae
  See *Ceramodactylus*
  See also *Stenodactylus*
Dotsenko, I. E. 41
dracaena
  See *Varanus*
dugesii
  See *Lacerta*
Dujsebayeva, Tatyana N. 289
Duméril, André-Marie-Constant 2, 5, 34, 80, 97, 100, 129–130, 173, 180, 264, 270, 278, 297
Duméril, August-Henri-André 2, 80, 129–130
dwarf geckos 192

*E**chis* 40–41, 57, 198

carinatus sochurecki 198
Ehlers, Eckart 43
ehrenbergi
  See *Ophisops*
Eichwald, Carl Eduard Ivanovich von 1–2, 4, 16–**17**, 70, 87, 104, 127, 150, 152, 157, 159, 230, 247, 261, 269–271, 296, 298
Eirenis
  persicus 41
  rechingeri 33
Eiselt, Josef 4, 13–14, 23, **27**–28, 33, 40–41, 134–138, 172, 230–231, 233–237, 240–247, 249, 260, 270–271
El-Toubi, M. R. 287
Elaphe
  longissima 29
elegans
  See *Ophiops*
  See also *Ophisops*
  See also *Stenodactylus*
elegans elegans
  See *Ophisops*
elegans forma *typica*
  See *Ophiops*
Eliazan, M. 36
elisae
  See *Asaccus*
  See also *Phyllodactylus*
elongatus
  See *Cyrtopodion*
  See also *Gymnodactylus*
Elpatjevsky, Vladimir Sergeevich 263–265
ensafi
  See *Eublepharis*
Eremchenko, Valery Konstantinovich 40, 263–267
Eremias 3–4, 21, 28–29, 32–33, 39, 42, 44–53, 56, 59–61, 65, 114, 123, 188, 199–200, 208–228, 249–250, 252, 254
  acutirostris 3, 46, 52, 59, 61, 188, 208, 210–211, **11E**
  andersoni 3, 33, 45, 59, 209, 212
  aporosceles 21, 208, 212
  arguta 3, 49, 60–61, 208, 210, 213, 217, 219–220, **11F**
  arguta deserti 214
  arguta deserti-transcaucasica 213
  arguta transcaucasica 275
  brevirostris 250
  brevirostris brevirostris 250
  brevirostris fieldi 250
  brevirostris forma *typica* 250
  fasciata 3, 44–46, 52, 61, 209, 213–215, 224, **11G**
  fasciata pleskei 224
  grammica 3, 44–45, 48, 52, 61, 200, 210, 215–216, **11H**
  guttulata 252
  guttulata guttulata 252
  guttulata watsonana 252
  intermedia 3, 48, 52, 61, 210, 214, 216, 219–221
  lineolata 3, 44, 209, 218–219, **12A**
  nigrocellata 4, 32, 44, 48, 50–52, 61, 210, 214, 217, 219–221, 223
  pardalis 225, 252

pardaloides 252
persica 4, 44–46, 48, 50, 53, 61, 210–211, 216, 221–224, 226–227, **12B**
pleskei 4, 45, 52, 60, 209, 224
scripta 4, 208–209, 219, 225, **12C-D**
scripta scripta 225
strauchi 4, 33, 45, 49, 51–52, 60–61, 123, 210, 223, 225–227
strauchi kopetdaghica 33, 51, 210, 227, **12F**
strauchi strauchi 210, 225–226, **12E**
variabilis 221, 226
velox 4, 47, 114, 209, 222, 226–228
velox strauchi 226–227
velox velox 223–224, **12G**
velox velox-persica 222
Eremias (*Eremias*) 211, 218, 222, 226, 228
  lalezharica 218
  persica 222
  strauchi strauchi 226
  velox velox 228
Eremias (*Ommateremias*) 211, 213, 220
  arguta 213
  nigrocellata 220
Eremias (*Rhabderemias*) 211, 214, 219, 224–225
  fasciata 214
  lineolata 219
  pleskei 224
  scripta 225
Eremias (*Scapteira*) 211, 216, 225
  acutirostris 211
  grammica 216
  zarudnyi 216
Erni, A. 22
Ernst, Carl Henry 21, 37
erythrogaster
  See also *Stellio*
erythrogastra
  See *Agama*
  See also *Laudakia*
Eryx
  jayakari 85, 186
  tataricus tataricus 188
Estes, Richard Dean 68
Etheridge, Richard Emmett 289–290
Eublepharidae 6, 66, 116
eublepharids 41, 116
Eublepharinae 116
Eublepharis 2, 26, 33, 41, 44, 49, 51, 53, 58–61, 116–123
  afghanicus 121
  angramainyu 2, 26, 33, 49, 53, 58, 60, 116–120, 123–124, **6E**
  ensafi 118, 120
  fasciolatus 121–122
  fuscus 121
  hardwickii 116
  macularius 2, 44, 116–124
  macularius montanus 121
  smithi 121
  turcmenicus 2, 61, 116–117, 121–124, **6F**
Eumeces 4, 33, 39, 44–51, 53–54, 60–61, 171, 262, 269–273, 279
  blythianus 272

parthianicus 33, 44, 48, 51, 269, 272, 274
pavimentatus 270–271
schneiderii 4, 44–50, 53, 60–61, 171, 262, 269–273
schneiderii princeps 269–272, **17B-D**
schneiderii schneiderii 270
schneiderii variegatus 150, 270–271
schneiderii zarudnyi 269, 271–272
scutatus 272, 274
taeniolatus 4, 33, 44, 51, 269, 272–274
taeniolatus arabicus 273–274
taeniolatus parthianicus 33, 44, 48, 51, 269, 272, 274, **17E-G**
zarudnyi 271
Euphlyctis
  cyanophlyctis 297
euphraticus
  See *Acanthodactylus*
Euprepis
  affinis 275–277
  fellowsii 276
  princeps 270–271
  septemtaeniata 275
  septemtaeniatus 275
euptilopus
  See *Phrynocephalus*
Eurylepis
  taeniolatus 272, 274
Eversmann, Eduard Friedrich 4, 148, 215, 229, 236, 240, 260, 263, 265
eversmanni
  See *Crossobamon*
  See also *Gymnodactylus*
eversmanni eversmanni
  See *Crossobamon*

Fan-footed geckos 178
Farzanpay, R. 29
fasciata
  See *Eremias*
  See also *Eremias* (*Rhabderemias*)
  See also *Podarces* (*Eremias*)
  See also *Rhabderemias*
fasciatus
  See *Microgecko*
  See also *Tropiocolotes*
  See also *Tropiocolotes* (*Microgecko*)
fasciolatus
  See *Eublepharis*
  See also *Gymnodactylus*
fat–tailed gecko 116, 118, 120, 122
fat–tailed geckos 116
feae
  See *Gymnodactylus*
fedtschenkoi
  See *Cyrtodactylus*
  See also *Cyrtopodion*
  See also *Gymnodactylus*
feeding strategies and feeding niches 56
Fei, Liang 219
Fejérváry von Komlós-Keresztes, Géza Gyula Imre 35, 241
Felber, Hans 64

felicis
  See *Acanthodactylus*
fellowsii
  See *Euprepis*
femoralis
  See *Agamura*
  See also *Rhinogecko*
festae
  See *Ablepharus brandtii festae*
Fet, Victor 33
Field's agama 104
Field's short-nosed desert lizard 250
Field, Henry 14–15, 21, 24, 26, 35–37, 73, 104, 250
fieldi
  See *Agama*
  See also *Eremias*
  See also *Mesalina*
  See also *Trapelus*
Finn, Frank 21, 107, 212
Firouz, Eskandar 13–14, 27
Fisher, W. B. 10–11, 13, 42, 46, 50–51, 64
Fitzinger, Leopold Josef Franz Johann 2–5, 33, 68, 81, 111, 127, 150, 179, 183, 200, 208, 228, 263, 265, 267, 274, 287, 289–290
five-streaked lizard 247
flavimaculatus
  See *Trapelus*
flavipunctatus
  See *Pristurus*
flaviviridis
  See *Hemidactylus*
Flower, Stanley Smyth 287
Forcart, Lothar Hendrich Emil Wilhelm 22, 71, 76, 90, 95, 97, 100, 108, 113, 130, 141, 164, 173, 190, 202, 205, 222, 235–236, 238, 252, 255, 265, 271, 280, 293
Forsskål, Petrus 4–5, 268, 290–291
fraasii
  See *Lacerta*
fragilis
  See *Anguis*
Frank, Norman 9, 33, 35–36
fraseri
  See *Acanthodactylus*
frenarus
  See *Gymnodactylus*
fringe-toed geckos 147
Fritz, Uwe 33
Frost, Darrel Richmond 13, 37, 289–290
Frynta, Daniel 29, 71, 80, 130, 134, 141, 154, 158, 169, 175, 194, 196–197, 211, 222, 232, 234, 236, 241, 247, 252, 255, 265, 270, 276, 283, 294
Führ, Ion Eduard 36, 40, 263–267
fusca
  See *Agama nupta fusca*
  See also *Lacerta*
  See also *Laudakia*
fuscus
  See *Eublepharis*
  See also *Stellio*

G**abriel**, Alfons 20–21, 34

gabrielis
  See *Gymnodactylus*
gaddi
  See *Lytorhynchus*
Gallagher, Michael D. 39, 51, 85, 176, 178, 186, 207, 291
gallagheri
  See *Asaccus*
galli
  See *Phrynocephalus*
Gallotia 199, 231
Gallotinae 199
Ganji, M. H. 13, 43
Gans, Carl 55, 228
Gardner, Andrew S. 132–133, 135–137, 139
Gasperetti, John 39
gasperettii
  See *Scincus*
gastrophole
  See *Cyrtopodion*
gastropholis
  See *Agamura*
  See also *Cyrtodactylus*
  See also *Cyrtopodion*
  See also *Gymnodactylus*
Gauthier, Jacques Armand 289–290
Gecko
  lobatus 178
  tuberculosus 172
geckoides
  See *Gymnodactylus*
geckos 125
Gekkonidae 2, 6, 66, 116, 125, 127
Gekkoninae 116, 125, 187
Gekkota 116
Geoffroy Saint-Hilaire, Étienne-François 99, 178, 269–271
geyri
  See *Uromastyx*
giant green lizard 237–238
Gmelin, Samuel Gottlieb 16, 81, 214
Golay, Philppe 37
gold skink 269
golden grass skink 274
Goldfuss, Georg August 3, 178
Goldsmid, Frederic John 11, 18–19
Golubev, Mikhail Leonidovich 2, 13–14, 40–41, 73, 75, 81–82, 88, 93, 95–97, 99, 118, 120–123, 125–127, 130, 132, 138–142, 144–152, 154–172, 176, 181–182, 187–191, 193–199
gongrorynchatus
  See *Acanthodactylus*
Gongylus
  ocellatus 268
Goniurosaurus
  kuroiwai 116
Gonydactylus
  kirmanensis 162
Gonyurosaurus
  lichtenfelderi 116
Gorelov, Yu. K. 75, 141, 164–165
grammica
  See *Eremias*

See also *Eremias* (*Scapteira*)
See also *Lacerta*
See also *Scapteira*
grandis
See *Acanthodactylus*
granular-scaled lizards 290
gray thin-toed gecko 166
gray toad agama 97
Gray, John Edward 34, 36, 68–69, 97, 116, 127, 150, 154, 156, 173–174, 199, 203, 205, 213, 228, 249–250, 264, 275–277, 296
gray-spotted leaf-toed gecko 137
grayanus
See *Ablepharus*
See also *Blepharosteres*
green-bellied lizard 234
Greer, Allen E. 38, 262
griseonotus
See *Asaccus*
griseus
See *Monitor*
See also *Tupinambis*
See also *Varanus*
griseus caspius
See *Varanus*
griseus griseus
See *Varanus*
See also *Varanus* (*Psammosaurus*)
Grismer, Larry E. 41, 116–118, 120–122, 124, 187
Gronovius, Laurentius Theodorus 287
Gruber, Ulrich F. 164, 170
gubernatoris
See *Gymnodactylus*
Guibé, Jean 14, 23, 68, 71, 73–76, 79, 87, 95–98, 101, 112–114, 130, 141, 157, 169–170, 176, 185, 189–190, 192, 197–199, 202, 214, 217, 219–220, 222, 226–227, 247, 252, 255, 265, 282
Günther, Albert Carl Ludwig Gotthilf 19, 34, 36, 107, 110, 121–122, 148, 198, 202, 213, 216, 221, 227, 270
guttata
See *Lacerta*
guttatus
See also *Ptyodactylus*
See also *Stenodactylus*
guttulata
See *Eremias*
See also *Mesalina*
guttulata guttulata
See *Mesalina*
Gvirtzman, G. 63
Gymnodactylus 20, 32, 117, 127, 129–130, 141–143, 147–148, 150, 152, 154, 156–171, 198
agamuroides 130, 154
albofasciatus 150
brevipes 150, 156
caspius 127, 150, 157, 159
consobrinoides 150
deccanensis 150
elongatus 150
eversmanni 147–148
fasciolatus 150
feae 150
fedtschenkoi 164–165
frenatus 150
gabrielis 20, 141–143
gastropholis 20, 127, 130, 158
geckoides 168
gubernatoris 150
heterocercus 159–160
heterocercus heterocercus 160
himalayicus 150
jeyporensis 150
kachhensis 127, 150, 161
khasiensis 150
kirmanensis 150, 162
kotschyi 150, 163
lawderanus 117
longipes longipes 164
microlepis 165
montiumsalsorum 150
nebulosus 150
oldhamii 150
peguensis 150
persicus 129–130
russowii 166–167
sagittifer 167
scaber 168
steudneri 198
stoliczkai 150
triedrus 150
turcmenicus 171
variegatus 150
zarudnyi 166–167
Gymnodactylus (*Cyrtodactylus*)
heterocercus 160
kirmanensis 162
kotschyi syriacus 163
longipes 164–165
microlepis 165
russowii 166
scaber 169
Gymnodactylus (*Mediodactylus*)
sagittifer 168
spinicauda 170
gymnophthalmids 262

H aas, Georg 3, 14, **22**, 36, 80, 85, 90–92, 101, 104, 107, 126, 136, 140–142, 169, 183–184, 201, 204, 206, 237, 248, 250–252, 255–257, 275, 288, 294
haasi
See *Acanthodactylus*
Hahn, Donald Edgar 40
hamulirostris
See *Leptotyphlops*
Harding, Keith A. 37
Hardwicke, Thomas 69
hardwickii
See *Eublepharis*
See also *Uromastyx*
Hart, Scott 34
hasselquistii
See *Lacerta*
See also *Ptyodactylus*
hasselquistii hasselquistii

INDEX

See *Ptyodactylus*
Hassinger, Jerry D. 24
Heimes, Peter 179
Helen's banded dwarf gecko 193
helenae
　See *Microgecko*
　See also *Tropiocolotes*
　See also *Tropiocolotes* (*Microgecko*)
helenae helenae
　See *Microgecko*
　See also *Tropiocolotes*
helioscopa
　See *Lacerta*
helioscopus
　See *Phrynocephalus*
helioscopus helioscopus
　See *Phrynocephalus*
Hellmich, Walter 23, 80, 104, 250, 255, 265–267
Hemidactylus 2–3, 20, 47–49, 51, 53, 58, 60–61, 125, 127–128, 169–176
　brookii 172–173
　coctaei 173
　cocteaui 173
　flaviviridis 2, 47–48, 51, 60–61, 172–173, **8F**
　karachiensis 176
　leschenaulti 174
　mabouia 172
　maculatus 172
　parkeri 76
　persicus 3, 48–49, 51, 58, 60–61, 172–175, **8G**
　robustus 175–176
　turcicus 3, 47–49, 60–61, 169–170, 172–176, **8H**
　turcicus parkeri 176
　turcicus turcicus 176
Hemipodion
　persicum 17, 32, 283
Hemipodium
　persicum 283
Hemitheconyx
　caudicinctus 117
Hemming, Francis 287
Hemsen, Jens 23
Henle, Klaus 69
Hermann, L. H. 24
Herzfeld, Ernst 21
heterocerca
　See *Cyrtopodion*
heterocercus
　See *Cyrtodactylus*
　See also *Gymnodactylus*
　See also *Gymnodactylus* (*Cyrtodactylus*)
　See also *Tenuidactylus* (*Mediodactylus*)
heterocercus heterocercus
　See *Cyrtodactylus*
　See also *Cyrtopodion*
　See also *Gymnodactylus*
　See also *Mediodactylus*
　See also *Tenuidactylus* (*Mediodactylus*)
heteropholis
　See *Carinatogecko*
　See also *Tropiocolotes*
Heyden, Carl Heinrich Georg von 2, 127, 150, 154, 157, 168, 175–176, 179

himalayicus
　See *Gymnodactylus*
Hoberlandt, Ludvik 29
Hoge, Alphonse Richard 36
Holbrookia 91
Holodactylus
　africanus 117
homolepis
　See *Ptyodactylus*
Hoofien, Jacob Haim 126, 150, 251
horny-scaled agama 108
Horowitz, A. 63
horvathi
　See *Phrynocephalus*
hotsoni
　See *Zamenis*
house geckos 172
Huey, Raymond Brunson 55
Hyrcanian Forest 59

ICZN
　See International Commission on Zoological Nomenclature
Indian garden lizard 68
Indian monitor 296
Inger, Robert Frederick 14, 40
insularis
　See *Cyrtopodion*
intermedia
　See *Eremias*
　See also *Podarces* (*Eremias*)
International Commission on Zoological Nomenclature 81, 179, 287, 289
interscapularis
　See *Phrynocephalus*
interscapularis interscapularis
　See *Phrynocephalus*
Iran
　Alborz Mountains 18, 20, 28, 44, 50, 62, 72, 102, 235, 237, 278
　Austrian Expedition 22
　Caspian Region 46–47, 112, 114
　Central Plateau 11, 22–23, 44–45, 50, 58, 72, 91, 96, 98–99, 102, 109, 131, 140, 162, 170, 188, 191, 196, 223, 228, 231, 254, 256, 278–279, 283, 294
　climate map of 43
　Danish Scientific Investigations 22
　establishing present boundaries of 15
　geography of 1, 42
　Khuzestan Plain 47
　Kopet Dagh 44, 47, 50–51, 62, 73, 75, 88, 97, 99, 112, 114, 122–123, 170–171, 223, 228, 230–231, 235, 237, 260, 266, 271, 276
　Kupal Dunes 58
　Persian Telegraph Commission 19
　physiographic and climatic barriers 1, 65
　reports on travel, collections, and field work 15, 27
　Reza'iyeh Basin 22, 25, 45, 234, 245, 255
　Russo-Iranian wars 16
　Safavid Dynasty 16
　Sistan Arbitration Commission 18–19
　Sistan Basin 46, 212

Treaty of Erzurum 15
Treaty of Golestan 17
Treaty of Tehran 17
Zagros Mountains 20, 23, 49, 58, 61, 72, 79, 85, 91, 99, 102–103, 106–107, 119–120, 133, 138, 145, 158, 195, 197–198, 211, 223, 231, 242, 254, 256, 259, 261–262, 264, 285, 299
Iranian Baluchistan 21, 47–48, 132, 149, 182, 197
Iranian fat-tailed gecko 118
Iranian keel-scaled gecko 146
Iranian skink 288
Iranian spiny-tailed lizard 293
iranicus
　See *Pristurus*
Iraqi keel-scaled gecko 146
Ischchenko, Vladimir G. 38
isolepis
　See *Agama*
　See also *Trapelus*
Iverson, John Burton 37

Jacob, D. 297
Jahanbani, Amanollah 23
Jawdat, Suad Z. 160
jayakari
　See *Eryx*
　See also *Lacerta*
　See also *Trapelus*
Jaz Murian bent-toed gecko 167
jerdoni
　See *Ophisops*
jeyporensis
　See *Gymnodactylus*
Joger, Ulrich 37, 41, 68, 81, 291–294

K*achhense*
　See *Cyrtopodion*
kachhensis
　See *Cyrtodactylus*
　See also *Gymnodactylus*
　See also *Tenuidactylus* (*Cyrtopodion*)
　See also *Tenuidactylus* (*Mesodactylus*)
Kachhi thin-toed gecko 161
Kaiser, E. 22
kaiseri
　See *Neurergus*
Kaltenbach, Alfred 23
Kami, Haji Gholi 13, **32**, 146, 196
karachiensis
　See *Hemidactylus*
Kassler, P. 51, 64
Kaup, Johann Jakob 1, 81
Kaverkin, Yu. I. 122–123
kavirensis
　See *Bufo*
Kechichian, Joseph A. 15
keel-scaled geckos 144
keeled rock gecko 168
Kerman bent-toed gecko 161
kermanensis
　See *Bufo*

kermanshahensis
　See *Asaccus*
Kessler, Karl Fodorovich 4, 210, 219, 225–226
Keyserling's plate-tailed gecko 189
Keyserling's skink gecko 189
Keyserling, Count Eugen von 17, 189
keyserlingii
　See *Teratoscincus*
Khalaf, Kamel T. 91, 267, 288
Khan, Muhammad Sharif 19, 38, 151, 203
Khanikoff, N. de 17
Kharan spider gecko 182
khasiensis
　See *Gymnodactylus*
khobarensis
　See *Pseudoceramodactylus*
Khorasan agama 73
Khuzestan dwarf gecko 193
Khuzestan Plain
　See Iran
King, W. K. 37–38, 296
Kinunen, Wayne 27
Kirman thin-toed gecko 161
kirmanense
　See *Cyrtopodion*
kirmanensis
　See *Agama*
　See also *Cyrtodactylus*
　See also *Gonydactylus*
　See also *Gymnodactylus*
　See also *Gymnodactylus* (*Cyrtodactylus*)
　See also *Stellio*
　See also *Tenuidactylus*
　See also *Trapelus*
kitaibelii
　See *Ablepharus*
Klawinski, P. W. 170
Klembara, Josef 113
Klemmer, Konrad 36
Kluge, Arnold Girard 14, 40, 116, 118, 121–122, 125–127, 130, 136–137, 140–141, 146–151, 154, 156–158, 160–166, 168–170, 172–173, 175–177, 182–185, 187, 189–190, 192–198, 296
Kock, Dieter 299
koniecznyi
　See *Varanus*
Kopet Dagh bent-toed gecko 169
Kopet Dagh racerunner 227
Kopet Dagh
　See Iran
Kotschy, Theodor 17, 163
kotschyi
　See *Cyrtodactylus*
　See also *Cyrtopodion*
　See also *Gymnodactylus*
　See also *Mediodactylus*
　See also *Tenuidactylus* (*Mediodactylus*)
kotschyi group 168
Kulagin, N. 190
Kurdistan lizard 242
kurdistanica
　See *Lacerta*
　See also *Timon*

kuroiwa
   See *Gonaturosaurus*

Lacepède, Bernard Germain Étienne 33, 274
Lacerta 4, 32–33, 40–41, 45, 47–52, 54, 59–61, 64, 68, 70, 81, 87, 92, 100, 112–114, 175, 178, 199–200, 208, 213, 215, 217, 219, 221–222, 226–250, 252, 257, 259–262, 267–268, 270, 275, 287, 291, 296–297
   aegyptia 291
   agilis 228, 230–231
   apoda 13
   arguta 213, 217, 219
   aurata 270, 275
   aurita 92
   boettgeri 234
   boskiana 200, 257
   brandtii 4, 79, 229, 231–233, 243, 256, 260, 265–267, **12H-I, 13A**
   calotes 68
   cappadocica 4, 45, 52, 60, 199, 229, 231, 233–234, 243, 259
   cappadocica cappadocica 234
   cappadocica muhtari 234
   cappadocica schmidtlerorum 234
   cappadocica urmiana 229, 233, 259, **13B-C**
   cappadocica wolteri 234
   caudivolvula 81
   chalcides 267
   chlorogaster 4, 32, 47–48, 51, 54, 59–60, 112, 229, 231, 234–236, 239–240, 246–247, 260, **13D-E**
   clarkorum 236
   cyanura 199, 234
   danfordi 231–232, 234
   defilippi 235
   defilippii 4, 47, 50, 52, 60, 230, 235–237, 240, 245, 247, 260, **13F-G**
   depressa 126, 192–193, 199, 236
   dugesii 99
   fraasii 233
   grammica 215
   guttata 81
   hasselquistii 178
   helioscopa 87
   jayakari 199, 231, 234
   laevis 236
   lepida 199
   media 4, 45, 60, 230–231, 237–239, 248, 261
   media media 230, 238–239, 261
   mostoufi 4, 45, 229, 231, 238–239, 260
   muralis 234, 236, 243–244, 246, 248
   muralis defilippii 236
   muralis fusca 236
   muralis fusca persica 236
   muralis valentini 248
   muricata 70
   ocellata 268
   pardalis 250, 252
   parva 231, 233
   pater 199
   praticola 4, 47, 60, 229, 231, 236, 239–241, 246, 260, **14A-B**
   praticola pontica 241
   praticola praticola 240
   princeps 4, 33, 49, 54, 60, 199, 230–231, 240–243, 261
   princeps kurdistanica 33, 230, 241–242, 261, **14E**
   princeps princeps 230, 241, 261, **14C-D**
   raddei 4, 45, 49, 51–52, 60, 230–232, 236–237, 243–245, 247, 260
   raddei defilippii 237
   raddei nairensis 245
   raddei raddei 45, 49, 51, 83, 96–97, 243–245, **14F-G**
   raddei vanensis 245, **14H, 15A**
   sanguinolenta 100
   saxicola defilippii 236–237, 244
   saxicola raddei 244
   scincus 287
   steineri 4, 33, 51, 230–231, 235–237, 246–247, 260, **15B**
   stincus 287
   strigata 4, 45, 47, 49–51, 54, 59–60, 230–231, 243, 247–248, 261, **15C**
   strigata strigata 247
   trilineata 114, 238
   trilineata media 238
   turcica 175
   valentini 4, 49, 52, 60, 230–231, 246, 248, 260
   valentini valentini 49, 230, 248, 260, **15D-E**
   vanensis 245
   variabilis 208, 221, 226
   velox 222, 227
   viridis astrabadensis 247
   viridis major 238
   viridis strigata 247
   viridis vaillanti 247
   viridis woosnami 247
   vivipara 199, 230–231
   zagrosica 4, 33, 49, 52, 231, 260–262
Lacerta (Archaeolacerta)
   steineri 246
lacertas 228
Lacertidae 3–4, 6, 35, 40–41, 66, 199–201, 231, 256
lacertids 39, 54–55, 57, 149, 199, 262
Lacertinae 64, 199
laevis
   See *Alsophylax*
   See also *Lacerta*
Laird, Edith M. 21
Lake Urmia rock lizard 233
Lake Van lacerta 245
Lalehzar racerunner 218
lalezharica
   See *Eremias (Eremias)*
Lantz, Louis Amédée 4, 21, 33, 39, 100, 153, 157, 165, 208, 211, 214–216, 219, 222, 224–230, 232–233, 236–238, 241, 244, 252, 255, 259, 261, 270, 272, 276
Lanza, Benedetto 176
large-scaled rock agama 78, 80
lateral-fold lizards 111
latifi
   See *Microgecko*
   See also *Tropiocolotes*

See also *Tropiocolotes* (*Microgecko*)
Latifi's dwarf gecko 195
Latifi, Mahmoud 9, 11–12, 14, **29**, 36, 77, 195, 278
Laudakia 1, 28, 42, 44–53, 56–61, 67–81, 102, 114, 259, 262, 277
    caucasia 1, 28, 44–45, 47–48, 50–52, 60–61, 67, 69–73, 75–79, 102, 262
    caucasia caucasia 47–48, 50–51, 70, 72, **2–BC**
    caucasia trianulata 73
    caucasica caucasica 71
    caucasica microlepis 76
    erythrogastra 28, 44–45, 48, 52, 61, 73, 75–76, 79, **2D**
    lirata 76
    melanura 1, 52, 70, 75–76, 80
    melanura lirata 70, 75
    microlepis 1, 44–45, 52, 70, 73, 76–79, **2E-F**
    nupta 1, 44–49, 52, 58, 60–61, 70, 72–73, 75–81, 102, 277
    nupta fusca 78, 80–81, **2H**
    nupta nupta 58, 78, 80, 259, **2G**
    stellio 72
    tuberculata 69
laungwalensis
    See *Bufoniceps*
    See also *Phrynocephalus*
Laurenti, Josephus Nicolaus 4–5, 33, 234, 236, 247, 267, 287
lawderanus
    See *Gymnodactylus*
Lay, Douglas M. 11, 13, **24**, 45, 85, 185
leaf-toed geckos 132, 138, 172
Leiolepis 41, 68, 290
Leiolopisma 263
leopard gecko 116, 120
leopard geckos 116
lepida
    See *Lacerta*
    See also *Timon*
Leptien, Rolf 141–142, 178
leptocosymbotes
    See *Stenodactylus*
Leptotyphlopidae 40
Leptotyphlops
    hamulirostris 277
    vermicularis 171
leschenaulti
    See *Hemidactylus*
lessonae
    See *Agama*
Leviton, Alan Edward 1–3, 5, 10, 13–16, 19, 22–24, 29, 33, **38**–40, 53, 68–69, 71, 74–76, 80–81, 84–86, 91, 93–94, 97, 100–101, 103–105, 108, 115, 117–118, 125–127, 136–137, 139–142, 147, 158, 160, 163–166, 169, 173, 175–177, 179–180, 184–186, 188, 190, 192–195, 203–204, 206–207, 222–223, 233, 238, 242, 250, 252, 255, 265, 268, 270, 272–275, 277–288, 291, 298
levitoni
    See *Tropiocolotes*
lichtenfelderi
    See *Gonyurosaurus*
Lichtenstein, Martin Hinrich Carl 1, 3–4, 83, 89, 183,

200, 210, 215, 249–250, 252, 263, 265
lidless skinks 263
lindbergi
    See *Ablepharus*
lineolata
    See *Eremias*
    See also *Rhabderemias*
    See also *Scapteira*
Linnaeus, Carolus 1–5, 33, 68, 111, 113, 173, 175, 228, 267, 270, 274–275, 287, 297
Liocephalus 91
lirata
    See *Laudakia*
liratus
    See *Stellio*
lobatus
    See *Gecko*
Löffler, Heinz 23, 137
Loftus, W. K. 11, 16
long-legged skinks 269
long-legged thin-toed gecko 164
longicaudatus
    See *Phrynocephalus*
longipes
    See *Cyrtodactylus*
    See also *Cyrtopodion*
    See also *Gymnodactylus* (*Cyrtodactylus*)
    See also *Tenuidactylus* (*Tenuidactylus*)
longipes longipes
    See *Cyrtodactylus*
    See also *Cyrtopodion*
    See also *Gymnodactylus*
    See also *Tenuidactylus*
    See also *Tenuidactylus* (*Tenuidactylus*)
longissima
    See *Elaphe*
loricata
    See *Uromastyx*
loricatus
    See *Centrotrachelus*
    See also *Uromastix*
    See also *Uromastyx*
Loveridge, Arthur 117, 169, 176, 178
Lukina, Galina Pantelejmonova 41
lumsdeni
    See *Crossobamon*
    See also *Stenodactylus*
luteoguttatus
    See *Phrynocephalus*
Lygodactylus 132–133
Lygosominae 262
Lytorhynchus 19, 21, 40, 57, 85, 186
    gaddi 85

Mabouia
    See *Hemidactylus*
Mabouia 270, 276
    aurata 270
    septemtaeniata 276
mabouyas 274
Mabuia 275–277
    septemtaeniata 275–276

INDEX

vittata 277
Mabuya 5, 44–45, 47–51, 53–54, 58, 60–61, 114, 186, 259, 262, 270, 274–278
   aurata 5 44–45, 47–51, 53, 58, 60–61, 114, 186, 259, 262, 270, 274–278
   aurata affinis 276
   aurata aurata 276
   aurata septemtaeniata 47, 49, 51, 58, 274–276, **17H**
   aurata transcaucasica 45, 50, 274–276, **18A**
   dissimilis 277
   dominicensus 274
   vittata 5, 60, 274, 277–278
Macey, J. Robert 13, 68–69, 73, 78, 82, 88, 99, 123, 171, 191, 274
macularius
   See *Cyrtodactylus*
   See also *Eublepharis*
maculatus
   See *Hemidactylus*
   See also *Phrynocephalus*
maculatus maculatus
   See *Phrynocephalus*
Madden, C. T. 63
Mahendra, Beni Charan 174
major
   See *Ceramodactylus*
   See also *Lacerta*
   See also *Stenodactylus*
Makran Coast 47, 76, 98, 272
Manhouri, H 29
mardinense
   See *Cyrtopodion*
Martens, Harald 299
Martin, Richard A. 21
Marx, Hymen 14, 39, 73, 109, 138, 146–147, 234, 270–271
masirae
   See *Acanthodactylus*
Masjed Soleyman 23, 57–58, 118, 197, 254
Mauremys
   caspica ventrimaculata 33
Mautz, William J. 56
Maxson, Linda R. 40
Mayer, Werner 41, 64, 199, 201, 211, 231, 241–243, 250, 254
maynardi
   See *Crossobamon*
McCarthy, Colin 14
McLachlan, Keith 15
McLean, C. S. 19
McMahon, A. Henry 19, 21
meadow lizard 240
media
   See *Lacerta*
media media
   See *Lacerta*
Mediodactylus 127, 140, 145, 150, 156, 160, 163, 166, 168, 170–171
   brevipes 156
   heterocercus heterocercus 160
   kotschyi 163
   kotschyi syriacus 163

russowii 166
russowii zarudny 166
sagittifer 168
spinicauda 170
spinicaudus 170
Mediterranean gecko 175
megalonyx
   See *Agama*
   See also *Trapelus*
Méhelÿ, Lajos 35, 96, 234, 236, 241, 244, 246
meizolepis
   See *Ophisops*
melanura
   See *Agama*
   See also *Laudakia*
Ménétriés, Edouard 4, 16–**17**, 254–255, 263
Menetriesii
   See *Ablepharus*
Meriakri, N. V. 21
Merrem, Blasius 5, 33, 99, 290, 296
Mertens, Robert Friedrich Wilhelm 35–37, 39, 75–76, 87, 92, 97, 101, 111–113, 125, 132, 136, 160, 163, 166, 168–169, 173, 175–176 190–191, 197, 202, 234–236, 238, 242–244, 247–248, 252, 268, 270–274, 276, 291–294, 296–300
Mesalina 4, 39, 44–49, 51–53, 58–61, 65, 198–201, 211, 249–254, 256
   brevirostris 4, 47, 51, 60–61, 249–251, 253
   brevirostris brevirostris 250–251, **15G-H**
   brevirostris fieldi 250–251
   brevirostris microlepis 251
   guttulata 60, 252, 254, **16A**
   guttulata guttulata 254
   guttulata watsonana 252
   pardalis 250, 252
   pardaloides 252
   watsonana 4, 44–49, 51–53, 58–61, 198, 200, 249–254, 256, **16B**
   watsonnana (*sic*) 252
mesalinas 249
Mesodactylus 127, 154, 161, 169
Mesopotamian Expeditionary Force 16, 35
Mesopotamian spiny-tailed lizard 294
Meszoely, Charles Aladar Maria 111
Mezhzherin, S. V. 82, 88, 95–96
Microgecko
   helenae 193–195, 197
   helenae fasciatus 195
   helenae helenae 194
   latifi 196
   persicus 197
   persicus bakhtiari 197
   persicus persicus 197
microlepis
   See *Agama*
   See also *Cyrtodactylus*
   See also *Cyrtopodion*
   See also *Gymnodactylus*
   See also *Gymnodactylus* (*Cyrtodactylus*)
   See also *Laudakia*
   See also *Mesalina*
   See also *Stellio*
   See also *Tenuidactylus*

See also *Tenuidactylus* (*Tenuidactylus*)
See also *Teratoscincus*
See also *Uromastix*
See also *Uromastyx*
micropholis
  See *Acanthodactylus*
microspilotes
  See *Neurergus*
microspilotus
  See *Rhithrotriton*
microtus
  See *Alsophylax*
microtympanum
  See *Agama*
miliaris
  See *Ophiomorus*
Minton, Sherman Anthony, Jr. 2, 13, 33, 37, 39, 68–69, 76, 80, 90–91, 98, 110, 120–121, 127, 130–131, 139–140, 142, 146, 155–156, 160–161, 169, 182–183, 189–190, 192–194, 197, 202–203, 205–206, 212, 225, 251, 253, 266, 268, 272–274, 286, 297
mintoni
  See *Cyrtopodion*
Misonne's spider gecko 182
Misonne's swollen-nose gecko 182
misonnei
  See *Agamura*
  See also *Rhinogecko*
mite pocket 117
mitranus
  See *Scincus*
Moghan Steppe 48
moisture (and its effects) 53–54
moltschanovi
  See *Phrynocephalus*
Monitor
  griseus 298
  See also *Varanus*
  See also *Varanus bengalensis*
monitors 296
montanus
  See *Asaccus*
  See also *Eublepharus*
montiumsalsorum
  See *Cyrtopodion*
  See also *Gymnodactylus*
Moody, Scott Michael 40–41, 68, 81, 99, 290–294
Moravec, Jiří 3, 13, **29**, 33, 161, 196, 209, 217–219
Morich, L. D. 21, 113, 215, 219, 236, 247, 252, 265, 272, 276, 298
Mosauer, Walter 277
mostoufi
  See *Lacerta*
muhtari
  See *Lacerta*
Mulder, John 29, 242–243
Müller, Lorenz 87, 92, 111, 113, 163, 238
muralis
  See *Lacerta*
muricata
  See *Lacerta*
Murray's comb-fingered gecko 184
Murray, James A. 3, 19, 80, 108, 120, 150, 156, 168, 173, 176–178, 183–185, 265, 273, 275, 287, 294
mutabilis
  See *Agama*
  See also *Trapelus*
mystaceus
  See *Phrynocephalus*

Nader Shah 16
Nader, Iyad 16, 160
Nagy, Kenneth A. 56
nairensis
  See *Lacerta*
najadum
  See *Coluber*
naked-fingered geckos 150
Natrix
  tessellata 112, 114, 277
nattereri
  See *Tropiocolotes*
Neal, J. W. 24, 73
nebulosus
  See *Gymnodactylus*
nejdensis
  See *Phrynocephalus*
Nesterov, P. V. 32
Neurergus 26, 29, 34
  kaiseri 29
  microspilotes 29
nigrigularis
  See *Calotes*
nigrocellata
  See *Eremias*
  See also *Eremias* (*Ommateremias*)
  See also *Ommateremias*
Nikolsky's long-toed gecko 164
Nikolsky's spider gecko 154
Nikolsky, Alexandr Mikhailovich 1–4, **20**, 32, 35, 70–71, 73, 76, 80–81, 87–90, 92–95, 97, 99–101, 104, 107–108, 111, 113, 123, 130, 136, 140–141, 146, 148, 150–151, 153–154, 157, 160–162, 164, 166–167, 170, 175, 187, 189–190, 193–197, 202, 205, 209–210, 214–216, 219–221, 224–227, 232, 234–236, 244, 247–248, 252, 255, 264–265, 267, 269–271, 275–276, 280, 283, 285, 293, 297–298
Nilson's spiny–toed lizard 258
Nilson, Göran 4–5, 13, **28**–29, 32–33, 41, 59, 69, 71, 90, 98, 101, 130–131, 209, 212, 221–222, 231, 252, 258, 260–262, 270, 278–279, 282–283, 293, 298
nilsoni
  See *Acanthodactylus*
Noakes, Diana 287
northern rock agama 70
nuchalis
  See *Ophiomorus*
Nucras 254
nupta
  See *Agama*
  See also *Laudakia*
nupta nupta
  See *Laudakia*
nuptus
  See *Stellio*

nurgeldievi
   See *Stellio*

O*blongus*
   See *Bufo*
Obst, Fritz Jürgen 113–115, 176
occidentalis
   See *Tropiocolotes*
ocellata
   See *Lacerta*
ocellated skink 268
ocellatus
   See *Chalcides*
   See also *Uromastyx*
ocellatus ocellatus
   See *Chalcides*
officinalis
   See *Scincus*
Oken, Ludwig Lorenz 2, 172
oldhamii
   See *Gymnodactylus*
Olivier's agama 107–108
Olivier, Guillaume-Antoine 1, 5, 16, 83, 97, 100, 104, 107–108, 274, 277
olivieri
   See *Phrynocephalus*
Omanosaura 199, 211, 231, 254
Ommateremias
   arguta deserti 213
   arguta transcaucasica 213
   nigrocellata 220
opaque-lidded skinks 269
opheodurus
   See *Acanthodactylus*
ophiomores 278
Ophiomorus 5, 28, 39, 42, 44–46, 48–49, 52–55, 59, 61, 188, 262, 278–287
   blanfordi 5, 48, 61, 278–281
   brevipes 5, 44, 48, 61, 278–283, 285, **18C**
   chernovi 283
   miliaris 278
   nuchalis 5, 28, 45, 59, 278–279, 282, **18D**
   persicus 5, 49, 54, 278–279, 283–284, **18E**
   raithmai 278, 286
   streeti 5, 39, 48, 278–279, 283–285
   tridactylus 5, 46, 59, 61, 188, 279, 285–286, **18F-G**
Ophiops
   elegans 255
   elegans forma *typica* 255
   elegans persica 255
   See also *Ophisops*
Ophisaurus
   apodus 2, 47–51, 54, 57, 59–61, 111–114, **6C-D**
   apodus apodus 114
   apodus thracius 115
   apus 113
Ophisops 4, 44–45, 47, 49–50, 58, 60–61, 66, 199, 201, 211, 253–256, 259
   blanfordi 255
   elegans 4, 44–45, 47, 49–50, 58, 60–61, 253–256, 259, **16C-D**
   elegans blanfordi 255–256

   elegans ehrenbergi 255–256
   elegans elegans 255–256
   elegans persicus 256
   jerdoni 254
   meizolepis 254
orientalis
   See *Anguis*
   See also *Crossobamon*
Orlov, Nikolai Lutseanovich 122–123, 165
Orlova, Valentina Fyodorovna 71–73, 75, 100, 240–241
ornatus
   See *Phrynocephalus*
   See also *Uromastyx*

Paleogeographical influences 1, 61
Pallas, Peter Simon 1–4, 16–**17**, 70, 81–83, 87, 92, 100, 111, 113, 208–210, 213, 217, 219, 221–222, 226–227
pallida
   See *Agama*
   See also *Stellio*
Paludan, Knud 22
pannonicus grayanus
   See *Ablepharus grayanus*
   See also *Ablepharus pannonicus grayanus*
Panov, E. N. 73, 75
panther gecko 120
Papenfuss, Theodore Johnstone 82, 123, 171, 191, 274
pardalis
   See *Eremias*
   See also *Lacerta*
   See also *Mesalina*
pardaloides
   See *Eremias*
   See also *Mesalina*
   See also *Podarces* (*Eremias*)
Pareremias 211
Parker, Hampton Wildman 186
parkeri
   See *Hemidactylus*
Parthian skink 272
parthianicus
   See *Eumeces*
   See *Eumeces taeniolatus parthianicus*
parva
   See *Lacerta*
pater
   See *Lacerta*
pavimentatus
   See *Eumeces*
   See also *Scincus*
peguensis
   See *Gymnodactylus*
Persian agama 104
Persian dwarf gecko 196–197
Persian gecko 175
Persian Gulf Coast 47
Persian Gulf Islands 51, 206
Persian long-tailed desert lizard 252
Persian racerunner 221
Persian snake skink 283

Persian spider gecko 130
Persian Telegraph Commission
   See Iran
Persian toad agama 95
persica
   See *Agamura*
   See also *Eremias*
   See also *Eremias* (*Eremias*)
   See also *Lacerta*
   See also *Ophiops*
   See also *Podarces* (*Eremias*)
   See also *Scapteira*
persicum
   See *Hemipodion*
   See also *Hemipodium*
persicus
   See *Ablepharus*
   See also *Alsophylax*
   See also *Batrachuperus*
   See also *Bufo*
   See also *Bunopus*
   See also *Eirenis*
   See also *Gymnodactylus*
   See also *Hemidactylus*
   See also *Microgecko*
   See also *Ophiomorus*
   See also *Ophisops*
   See also *Phrynocephalus*
   See also *Stellio*
   See also *Trapelus*
   See also *Tropiocolotes*
persicus persicus
   See *Microgecko*
   See also *Trapelus*
   See also *Tropiocolotes*
   See also *Tropiocolotes* (*Microgecko*)
Peter the Great 16
Peters, Günther 223, 233, 238, 243, 247–248
Peters, Wilhelm Karl Hartwig 3, 192, 198, 298
Petter, J. Francis 23
Petzold, Hans-Günter 111–113, 115
Phelsuma 174
Phrynocephalus 1, 19, 21, 27, 40, 42, 44–48, 50–52, 56, 58–61, 64–65, 68, 81–99
   arabicus 1, 47, 58, 60, 64, 82, 84–85, **3A-B**
   clarkorum 1, 19, 81–82, 84–87, 89, 94–95, **3C**
   euptilopus 21, 82, 90
   guttatus 82, 88
   helioscopus 1, 27, 48, 51–52, 61, 82–83, 87–88, 95–96
   helioscopus helioscopus 87, **3D-E**
   helioscopus persicus 95
   helioscopus saidalievi 88
   horvathi 95–96
   interscapularis 48, 52, 61, 81–83, 89, **3F**
   interscapularis interscapularis 89
   laungwalensis 81
   luteoguttatus 1, 19, 21, 56, 81–83, 89, **3G**
   maculatus 1, 44–46, 48, 52, 59–61, 64, 82, 84–85, 90–91, **3H-I, 4A**
   maculatus longicaudatus 91–92
   maculatus maculatus 59, 84, 90
   maculatus spiniventris 32, 90, 92
   moltschanovi 88
   mystaceus 1, 44–45, 48, 52, 61, 82, 92–94, **4B-C**
   mystaceus galli 92–93
   nejdensis 85
   olivieri 97
   olivieri brevipes 97
   olivieri carinipes 97
   ornatus 1, 19, 44–46, 52, 56, 61, 81–82, 84–87, 89, 93–94, 293, **4D**
   persicus 1, 44–45, 50, 52, 60, 82–83, 90, 95–96, **4E**
   raddei 1, 52, 61, 81–83, 88, 96–97, 99, **4F**
   raddei boettgeri 81
   raddei raddei 83, 96–97
   reticulatus 81, 88
   reticulatus strauchi 81, 97
   rossikowi 82, 88, 99
   scutellatus 1, 44–46, 48, 50, 52, 59, 61, 83, 97–99, **4G-I**
   sogdianus 82
   spiniventris 90
   strauchi 81, 96
Phyllodactylus 20, 28, 33, 132–133, 135–137
   elisae 20, 132, 135
   eugeniae 136
   ingae 28, 33, 136–138
physiographic and climatic barriers 1, 65
Pianka, Eric Rodger 55, 57, 59
pipiens
   See *Alsophylax*
Planhol, Xavier de 15, 49
plate-tailed geckos 186
plateau snake skink 282
platyrhynchus
   See *Asaccus*
Pleske's racerunner 224
pleskei
   See *Eremias*
   See also *Eremias* (*Rhabderemias*)
   See also *Rhabderemias*
Plestiodon
   Aldrovandi 270
   scutatus 274
Plocederma 69, 75
Podarces 213–214, 216, 219, 222, 225, 227, 231, 252
Podarces (*Eremias*)
   arguta 213
   fasciata 214
   intermedia 216
   pardaloides 252
   persica 222
   velox 227
   watsonana 252
Podarces (*Scapteira*)
   scripta 225
Podarcis 231, 235–236, 244
   defilippii 235–236, 244
pointed-snouted racerunner 211
pontica
   See *Lacerta*
ponticus
   See *Cyrtopodion*
Popp, Johann 23

Por, Francis Dov 63
Pough, F. Harvey 55
Prashad, Baini 174
praticola
   See *Lacerta*
praticola praticola
   See *Lacerta*
predation on lizards 57
Presch, William 262
princeps
   See *Eumeces*
   See also *Euprepis*
   See also *Lacerta*
princeps princeps
   See *Lacerta*
   See also *Timon*
Pristiurus (sic)
   rupestris 177
Pristurus 3, 47, 51–52, 55, 60, 125, 127, 169, 176–178, 198
   flavipunctatus 177–178
   rupestris 3, 47, 51, 60, 169, 176–178, 198, **8-I, 9A**
   rupestris iranicus 177–178
Pritchard, Peter Charles Howard 26, 37
Procter, Joan Beauchamp 16, 21, 152, 223
przewalskii
   See *Teratoscincus*
Psammodromus 199, 231
Psammosaurus
   caspius 298
   scincus 298
Pseudoceramodactylus 126, 183
   khobarensis 51, 126
   See also *Stenodactylus*
Pseudocyclophis 41, 277
   persica 277
pseudodalmatina
   See *Rana*
Pseudopus
   apoda 13
Pseudotrapelus
   sinaitus 102
Ptyodactylin 125
Ptyodactylus 3, 49, 53, 60, 125, 127–128, 178–179
   guttatus 179
   hasselquistii 179
   hasselquistii hasselquistii 179
   homolepis 179
   puiseuxi 179
Puellula
   rubida 150
puiseuxi
   See *Ptyodactylus*
pulchellus
   See *Cyrtodactylus*
punctatissimus
   See *Anguis*
pusillus
   See *Ablepharus*
Pygopodinae 116

Rabb, George Bernard 39

racerunner 208, 211–216, 218–219, 221, 224–225, 227
Radde's toad agama 96
Radde, Gustav Ferdinand Richard 19, 96, 226
raddei
   See *Lacerta*
   See also *Phrynocephalus*
raddei raddei
   See *Lacerta*
   See also *Phrynocephalus*
Rage, Jean-Claude 62
Rai', Mehdi M. 29
raithmai
   See *Ophiomorus*
Ramaswami, L. S. 297
Ramus, Erica 9
Rana
   macrocnemis pseudodalmatina 33, 112
   ridibunda 112, 114
Ranck, G. L. 24, 189
rapid fringe-toed lizard 227
Rastegar-Pouyani, Nasrullah 2–4, 13, **32**–33, 69, 135, 138, 200, 209, 221, 223, 231, 256–262
Rechinger, Karl H. 21–22
rechingeri
   See *Eirenis*
red-marked skink 270
redbelly agama 73
Redding, Richard W., Jr. 11, 28
Reed, Charles A. 73, 109, 138, 145–147, 234, 270–271
Rehman, Hafeezur 69
relationships of the Iranian lizard fauna 1, 60
reports on travel, collections, and field work
   See *Iran*
reticulate desert lacerta 211
reticulate racerunner 215
reticulatus
   See *Phrynocephalus*
Reuss, Adolph 274–276
Reza'iyeh (Urumiyeh) Basin
   See *Iran*
Rhabderemias
   fasciata 214
   lineolata 219
   pleskei 224
   scripta 225
Rhinogecko 3, 33, 45, 48, 60–61, 125, 127–128, 130, 180–183
   femoralis 3, 48, 61, 181–183
   misonnei 3, 33, 45, 130, 180–183
Rhithrotriton
   derjugini derjugini 32
   derjugini microspilotus 32
rhodorachis
   See *Coluber*
Rhotropus 174
Rhynchocalamus
   melanocephalus satunini 33
ridibunda
   See *Rana*
Rieppel, Olivier 111
Robertson, Alastair 63
robustus
   See *Hemidactylus*

rock agamas 69
Rögl, F. 63
rose-shouldered toad agama 89
Rösler, Herbert 123
Ross, James Perran 13
rossikowi
　See *Phrynocephalus*
Rostombekov, V. 20, 71, 95, 108, 226, 232, 238, 247, 255, 264, 270, 276
rough thin-toed gecko 168
rough-tail gecko 168
rubida
　See *Puellula*
rubrigularis
　See *Trapelus*
ruderata
　See *Agama*
　See also *Trapelus*
ruderatus
　See *Trapelus*
ruderatus ruderatus
　See *Trapelus*
rupestris
　See *Pristiurus* (sic)
　See *Pristurus*
Rüppell, Wilhelm Peter Eduard Simon 2–3, 150, 168, 173, 175–177
Russo-Iranian wars
　See Iran
russowii
　See *Cyrtopodion*
　See also *Gymnodactylus*
　See also *Gymnodactylus* (*Cyrtodactylus*)
　See also *Mediodactylus*
Rust, Richard 24
Rustamov, Anver Kejusevich 36, 38
rustamovi
　See *Teratoscincus*
Rykena, Silke 242–243

Safavid Dynasty 16
saggitifer (sic)
　See *Cyrtodactylus sagittifer*
sagittifer
　See *Cyrtodactylus* 168
　See also *Cyrtopodion*
　See also *Gymnodactylus*
　See also *Gymnodactylus* (*Mediodactylus*) 168
　See also *Mediodactylus*
Salamandra
　salamandra semenovi 29
　semenovi 32
Salvador, Alfredo 40, 176, 200, 202–207, 257–259
sand lizards 249
sand racerunner 225
sand skinks 278
sandfish 287–288
sanguinolenta
　See *Agama*
　See also *Lacerta*
sanguinolentus
　See also *Trapelus*

Saram, Manouchehr 23
Sattorov, T. S. 81, 88, 93
satunini
　See *Rhynchocalamus*
savignii
　See *Trapelus*
scaber
　See *Cyrtodactylus*
　See also *Cyrtopodion*
　See also *Gymnodactylus*
　See also *Gymnodactylus* (*Cyrtodactylus*)
　See also *Stenodactylus*
　See also *Tenuidactylus* (*Cyrtopodion*)
　See also *Tenuidactylus* (*Mesodactylus*)
scabrum
　See *Cyrtopodion*
Scapteira
　grammica 216
　lineolata 32, 219
　persica 215–216
Sceloporus 117
Schlegel, Hermann 3, 34, 186–187, 189
Schleich, Hans–Hermann 27–28, 68, 71, 73, 76, 80, 87, 90, 92, 95–96, 98, 100–101, 104, 107–108, 112, 114, 118, 130, 136, 140–141, 154, 157, 160, 162, 164, 169, 173, 175–177, 179, 182, 184–185, 187, 189–190, 194–195, 197–198, 202, 205–206, 213, 219, 222, 224, 226–228, 231–234, 236, 238, 241, 247, 250, 252, 255, 265, 268, 270–271, 275–277, 280, 282–283, 285, 288, 291, 293, 297–298
Schmidt, Karl Patterson **21**–22, 35, 71, 73, 80, 97, 100, 102, 108, 113, 136, 141, 162–163, 169, 173, 175, 177–179, 184, 204, 206, 222, 250–252, 255, 267–268, 270–271, 275–276, 287, 292
schmidti
　See *Acanthodactylus*
Schmidtler, Josef Friedrich **26**, 33, 118, 193–195, 238, 245, 248, 277
Schmidtler, Josef Johann **26**, 118, 193–195, 277
schmidtlerorum
　See *Lacerta*
Schmidtlers' dwarf gecko 195
Schneider's skink 269
schneideri
　See *Eumeces*
schneiderii
　See *Eumeces*
　See also *Scincus*
schneiderii schneiderii
　See *Eumeces*
Schofield, Richard N. 15
schreiberi
　See *Acanthodactylus*
Schultschik, Günter 29
Scincella 263
Scincidae 4, 6, 66, 262
Scincinae 262
scincomorphan lizards 262
Scincus 5, 19, 40, 47, 49–52, 55, 58, 60–61, 85, 186, 262–263, 269–271, 277, 279, 287–288
　arenarius 287, 298
　bivittatus 263
　conirostris 19, 288

gasperettii 288
mitranus 287
officinalis 287
pavimentatus 269–271
schneiderii 269
scincus 5, 40, 47, 49–51, 55, 58, 60–61, 85, 186, 287–288
scincus conirostris 40, 47, 49–51, 55, 58, 85, 186, 287–288
stincus 287
vittatus 277
scincus
 See *Lacerta*
 See also *Psammosaurus*
 See also *Scincus*
 See also *Stenodactylus*
 See also *Teratoscincus*
scincus scincus
 See *Teratoscincus*
Scolecophidia 40
scortecci
 See *Tropiocolotes*
scripta
 See *Eremias*
 See also *Eremias* (*Rhabderemias*)
 See also *Podarces* (*Scapteira*)
 See also *Rhabderemias*
scripta scripta
 See *Eremias*
scutatus
 See *Eumeces*
 See also *Plestiodon*
scutellata
 See *Agama*
scutellatus
 See *Acanthodactylus*
 See also *Phrynocephalus*
Selcer, K. W. 170
semaphore geckos 176
semenovi
 See *Salamandra*
Sengör, A. M. C. 61–62
Seps
 viridis 247
septemtaeniata
 See *Euprepis*
 See also *Mabouia*
 See also *Mabuia*
 See also *Mabuya*
septemtaeniatus
 See *Euprepis*
Shah 'Abbas Caravanserai 58
Shammakov, Sakhat Muradovich 38, 75, 171
Sharma, Ramesh Chandra 81
sharp-tailed spider gecko 182
Shaw, George 70, 160
short-legged snake skink 280
Sistan Arbitration Commission
 See Iran
Sistan Basin
 abandoned village in 59
 See also Iran
Sistan racerunner 214

skink geckos 186
skinks 40, 54, 262–264, 267, 269, 278, 280, 287
slevini
 See *Stenodactylus*
slow worm 111
small-scaled long-toed gecko 165
small-scaled rock agama 76
small-scaled skink gecko 189
small-scaled spiny-tailed lizard 294
small-scaled thin-toed gecko 165
Smith, Malcolm Arthur 3, **22**, 35, 37, 39, 66, 68–69, 76, 80, 90–91, 97–98, 100, 107, 110, 112, 118–119, 126–130, 149, 151–152, 160, 175, 179–182, 200, 202, 205, 214, 221–222, 252, 254–255, 263, 265–268, 273, 280, 285, 293, 296–297
smithi
 See *Eublepharis*
snake skinks 278, 280
snake-eyed lacertas 254
snake-eyed lizard 255
snake-eyed skinks 263
snakes of Iran 9, 29, 36
sochurecki
 See *Echis*
sogdianus
 See *Phrynocephalus*
Sokolova, Tatyana Mikhailovna 68, 81
somalicus
 See *Tropiocolotes*
southern grass skink 275
southern tuberculated gecko 141
Southwest Asian leaf-toed geckos 132
Spalerosophis 39
Sphaerodactylinae 116
Sphenocephalus
 tridactylus 285
spider geckos 129
spinicauda
 See *Alsophylax*
 See also *Cyrtopodion*
 See also *Gymnodactylus* (*Mediodactylus*)
 See also *Mediodactylus*
 See also *Tenuidactylus* (*Mediodactylus*)
spinicaudus
 See *Mediodactylus*
spinipes
 See *Stellio*
 See also *Uromastyx*
spiniventris
 See *Phrynocephalus*
spiny-tailed lizards 290
spiny-tailed thin-toed gecko 170
spotted fat-tailed gecko 120
St. John, Oliver B. 18, 248, 295
Starmülner, Ferdinand 23
Steindachner, Franz 2, 5, 17, 32, 34, 107, 150–151, 153, 163–164, 279, 283
Steiner's lacerta 246
Steiner, Hans M. 27, 33, 172, 246
steineri
 See *Lacerta*
 Lacerta (*Archaeolacerta*) 246
Steinfartz, Sebastan 29

Steininger, F. F. 63
Stellio 32, 69–71, 73, 75–76, 80, 101, 104, 290
   carinatus 80
   caucasicus 71
   caucasius 70–71, 104
   caucasius caucasius 71
   erythrogaster 32, 73, 75
   erythrogaster pallida 73
   kirmanensis 101
   liratus 75
   microlepis 76
   nuptus 80
   nuptus fuscus 80
   nurgeldievi 75
   persicus 71, 104
   spinipes 290
   See also *Laudakia*
stellions 69
Stenodactylus 3, 19, 47–48, 51–52, 58, 60, 125–128, 141, 144, 147–150, 157, 168, 183–187, 189
   affinis 3, 47, 58, 60, 183–185, **9B**
   arabicus 51, 125
   doriae 3, 47–48, 58, 60–61, 128, 183–186, **9C**
   elegans 183
   guttatus 185
   leptocosymbotes 185–186
   lumsdeni 19, 126, 141, 144, 147, 149
   major 9, 33–34, 36–37, 40, 46, 50, 56, 65, 68, 125, 185–186, 209, 211, 222, 238, 268
   scaber 127, 150, 157, 168
   scincus 186, 189
   slevini 51, 184–185
steppe agama 100
steppe lacerta 221
steppe-runner 213
Steudner's dwarf gecko 198
steudneri
   See *Gymnodactylus*
   See also *Tropiocolotes*
   See also *Tropiocolotes* (*Tropiocolotes*)
Steward, J. W. 37
sthenodactylus
   See *Ascalabotes*
Stimson, Andrew Francis 14, 39
stincus
   See *Lacerta*
   See also *Scincus*
Stoliczka, Ferdinand 2, 4, 117, 150, 153, 161, 249, 252, 265, 267, 272
stoliczkai
   See *Cyrtopodion*
   See also *Gymnodactylus*
Strauch's racerunner 225
Strauch, Alexander Aleksandrovich 2–5, 17, 32, 34–35, 126, 130, 140, 150, 152–153, 157, 160, 164–166, 170, 186, 188–190, 209–210, 216–217, 219, 225, 264–265, 267, 291, 293
strauchi
   See *Eremias*
   See also *Phrynocephalus*
strauchi strauchi
   See *Eremias*
   See *Eremias* (*Eremias*)

Street's snake skink 285
Street, Janice K. 10, **24**
Street, William S. 10–11, **24**–25, 131, 136, 189–190, 202, 204, 233–234, 272, 285
streeti
   See *Ophiomorus*
strigata
   See *Lacerta*
strigata strigata
   See *Lacerta*
striped racerunner 219
striped toad agama 93
Stugren, Bogdan 240–241
Stull, Olive Griffith 39
Stutz, Howard 23
substrate 52–53, 58–59, 65, 97, 99, 239, 253, 259
   influence on distribution 52
Suchow, G. F. 33, 229–230, 233, 242, 259, 261
sunwatcher 87
surdus
   See *Bufo*
swollen-nose geckos 180
syntopy 57, 94
syriacus
   See *Cyrtopodion*
   See also *Mediodactylus*
Szczerbak, Nikolai Nikolaevich 2–3, 14, 28, 33, 36, 38–41, 52, 54, 56–57, 118, 120–123, 125–127, 130, 132, 138–142, 144–152, 176, 181–182, 187–191, 193–199, 208–217, 224–228, 249, 252, 263–267, 269, 272–277

*T*aeniolatus
   See *Eumeces*
   See also *Eurylepis*
taeniolatus taeniolatus
   See *Eumeces*
Takydromus 231
tataricus
   See *Eryx*
Taylor, Edward Harrison 39, 270–273
Tchernov, Eitan 63
teiid lizards 262
Teira 199, 231
Telescopus
   fallax 114
   tessellatus 119
temperature 43, 52–56, 58, 64, 79, 94, 101, 109, 112, 119–120, 131, 138, 155, 168, 175, 178, 186, 188, 204, 223, 237–238, 271, 278, 288, 295, 299
   influence on distribution 54
Tenuidactylus 127, 150, 154–155, 157, 160–166, 169–170
   caspius 157
   kirmanensis 162
   longipes longipes 164
   longipes microlepis 165
Tenuidactylus (*Cyrtopodion*)
   kachhensis 161
   scaber 169
Tenuidactylus (*Mediodactylus*)
   heterocercus 160

heterocercus heterocercus 160
kotschyi 153
russowii zarudnyi 166
spinicauda 170
Tenuidactylus (Mesodactylus)
  kachhensis 161
  scaber 169
Tenuidactylus (Tenuidactylus)
  longipes 164
  longipes microlepis 165
Teratoscincinae 116, 187
Teratoscincus 3, 32, 44–46, 48, 52, 56, 59–61, 64, 125, 128, 185–192
  bedriagai 3, 32, 44–46, 56, 59, 61, 185–190, 192, **9E-F**
  keyserlingi 32, 186, 190
  microlepis 3, 32, 45–46, 59, 61, 187–189, 192, **9G**
  przewalskii 188
  scincus 3, 44–46, 48, 56, 59, 61, 64, 128, 185–192
  scincus keyserlingii 191, **9H, 10A**
  scincus rustamovi 191
  scincus scincus 56, 190–192, **10B-C**
  zarudnyi 32, 190–191
Terentjev, Paul Victorovich 36, 67, 71, 73–75, 81–82, 88–89, 92–93, 95, 100, 108, 112–114, 126, 147, 157, 166, 208, 213, 216–221, 223–227, 232, 234, 236, 238, 240, 244, 247, 252–253, 255, 262, 264–266, 270, 272–273, 275–276, 298–299
tessellata
  See *Natrix*
tessellatus
  Telescopus 119
Testudo
  baluchiorum 32
  buxtoni 33
thin-toed geckos 150
Thireau, Michel 32, 118, 120
Thomas, H. ii, 18, 35, 63
Thorn, Robert 33, 40
Thorson, G. 22
three-lined lizard 237–238
three-toed sand skink 285
thub 291
  See also *Uromastyx aegyptius*
tibetanus
  See *Cyrtopodion*
Tibeto-Himalayan group 151, 163
Tiedemann, Franz 14, 163
tilburyi
  See *Acanthodactylus*
Tiliqua
  affinis 275–277
Timon 199, 231, 241–243
  lepida 199
  princeps kurdistanica 242
  princeps princeps 241
toad agamas 81, 96
toad-headed agamas 81, 92
Tokar, Anatoly A. 39
Transcaspian bent-toed gecko 166
Transcaspian desert monitor 298
transcaucasia
  See *Eremias*

Transcaucasian grass skink 275
transcaucasica
  See *Mabuya*
  See also *Ommateremias*
Trapelus 1, 28, 42, 44–54, 58–61, 67–68, 75, 81, 85, 99–110, 186, 188, 221, 259, 277
  agilis 1, 28, 44–51, 53, 58–61, 67, 75, 99–104, 106–108, 188, 221, **5A-D**
  agilis agilis 101
  flavimaculatus 102
  isolepis 102
  jayakari 102, 104
  kirmanensis 104
  kirmanensis brevicaudus 104
  megalonyx 107, 110
  mutabilis 109–110
  persicus 1, 47, 49, 58, 60, 85, 100–107, 110, 186
  persicus fieldi 104
  persicus persicus 104–105, **5E-G**
  rubrigularis 100
  ruderata 108
  ruderatus 1, 44–45, 47, 49, 60. 100, 107–109, 277
  ruderatus megalonyx 107–110
  ruderatus ruderatus 49, 108–109, 259, **5H-I**
  sanguinolentus 51, 60
  savignii 104
Treaty of Erzurum 15
Treaty of Golestan 17
Treaty of Tehran 17
Tremper, Ronald L. 121
triannulata
  See *Laudakia*
tridactyla
  See *Chalcides*
tridactylus
  See *Ophiomorus*
  See also *Sphenocephalus*
triedrus
  See *Gymnodactylus*
Trigonodactylus 126, 183
  arabicus 126
  See also *Stenodactylus*
trilineata
  See *Lacerta*
tripolitanus
  See *Tropiocolotes*
Tropiocolotes 3, 23–24, 26, 29, 33, 40, 42, 45, 48–49, 52, 58, 60, 125–126, 128, 146, 192–198
  depressus 126, 192, 199
  fasciatus 49, 193, 195
  helenae 3, 24, 26, 49, 58, 192–195, 197
  helenae fasciatus 192, 195, **10F**
  helenae helenae 49, 58, 192–195, **10D-E**
  heteropholis 40, 126, 146
  latifi 3, 29, 33, 45, 193–195, 199, **10G**
  levitoni 126, 199
  nattereri 199
  persicus 3, 33, 48–49, 58, 60. 103–104, 106–107, 110, 193–194, 196–197
  persicus bakhtiari 33, 193, 196–197, **10H**
  persicus helenae 194, 197
  persicus persicus 48, 193, 196–197
  scortecchi 199

steudneri 3, 192–193, 197–199, **10-I**
tripolitanus 192, 199
tripolitanus algericus 199
tripolitanus occidentalis 199
tripolitanus somalicus 199
Tropiocolotes (Microgecko)
  helenae 194–195
  helenae fasciatus 195
  latifi 196
  persicus bakhtiari 197
  persicus persicus 197
Tropiocolotes (Tropiocolotes)
  steudneri 198
Tsaruk, Oleg J. 104
Tsellarius, Aleksey Yuri 216, 253, 299
tuberculata
  See *Agama*
  See *Laudakia*
tuberculated geckos 139
tuberculatus
  See *Alsophylax*
  See also *Bunopus*
tuberculosus
  See *Gecko*
Tuck, Robert G., Jr. 14, **24**, 27, 71, 76, 79–80, 87, 92, 95, 98, 100–101, 105, 108, 112, 114, 118, 130, 141, 146, 156–157, 169, 176–177, 187, 189–190, 194, 197, 202, 219, 221–222, 228, 234, 238, 247, 252, 255, 265, 268, 270, 276, 280, 294, 298
Tuniyev, Boris S. 69, 72, 75–76
Tupinambis
  bengalensis 297
  griseus 298
turcica
  See *Lacerta*
turcicus
  See *Hemidactylus*
turcicus turcicus
  See *Hemidactylus*
turcmenicum
  See *Cyrtopodion*
turcmenicus
  See *Eublepharis*
  See also *Gymnodactylus*
Turkestan leopard gecko 122
Turkestan plate-tailed gecko 189
Turkish gecko 175
Turkish warty gecko 175
Turkmen Steppe 48
Turkmenian fat-tailed gecko 122
Turkmenian thin-toed gecko 171
two-streaked snake-eyed skink 263
Typhlopidae 40
typical lizards 199

Underwood, Garth Leon 157, 161, 163, 166, 169
urmiana
  See *Apathya*
  See also *Lacerta*
Uromastix
  aegyptius 291
  asmussi 293

  loricatus 294
  microlepis 291
Uromastycidae 6, 41, 66, 68, 289–290
Uromastyx 5, 32, 41, 44, 47–51, 53–54, 56, 58–61, 65, 68, 259, 290–295
  aegyptius 5, 47, 51, 60, 290–292, 295, **19A**
  aegyptius microlepis 291
  asmussi 5, 44, 48, 53, 59, 61, 291–293, 295, **19B-C**
  dispar 107, 293
  geyri 293
  hardwickii 290
  loricata 294
  loricatus 5, 47, 49–50, 53, 58, 60, 259, 291–294, **19D**
  microlepis 32, 291–292
  ocellatus 293
  ornatus 293
  spinipes 292
Uromastyx acanthinurus group 293

*V**aillanti*
  See *Lacerta*
Vakilpoure, Ebrahim 196
Valentin's lizard 248
valentini
  See *Lacerta*
valentini valentini
  See *Lacerta*
vanensis
  See *Lacerta*
Varanidae 6, 39, 66, 296
Varanus 5, 44–49, 51, 53–54, 56–61, 66, 186, 259, 296–300
  arenarius 298
  bengalensis 5, 48, 61, 66, 296–297
  bengalensis bengalensis 296, **19E**
  dracaena 297
  griseus 5, 44–49, 51, 53–54, 57–61, 186, 296, 298–300
  griseus caspius 44–46, 48, 51, 59, 296, 298–300, **19G-H**
  griseus griseus 259, 296, 298–300, **19F**
  griseus konicznyi 299
  monitor 297
Varanus (Indovaranus)
  bengalensis bengalensis 297
Varanus (Psammosaurus)
  arenarius 298
  griseus caspius 298
  griseus griseus 298
variabilis
  See *Eremias*
  See also *Lacerta*
variable agama 68
variable lizards 68
variegatus
  See also *Gymnodactylus*
vegetation 43–48, 51, 53–54, 56, 58, 72, 85, 88, 90, 96–97, 99, 101, 109, 114, 119, 131, 142, 146, 148, 156–157, 170, 175, 182, 188, 190, 194, 198, 202, 206–207, 212–214, 216–221, 225, 227–228, 231, 234–236, 239, 242–244, 246, 248, 253, 259, 262,

270, 276–278, 291
 influence on distribution 13, 53–54, 59, 98, 114, 190, 223, 226, 232, 237, 253, 256, 282–283
velox
 See *Eremias*
 See also *Lacerta*
 See also *Podarces* (*Eremias*)
velox velox
 See *Eremias* (*Eremias*)
velox–persica
 See *Eremias*
ventralis
 See *Anguis*
ventrimaculata
 See *Mauremys*
vermicularis
 See *Leptotyphlops*
versicolor
 See *Agama*
 See also *Calotes*
Villiers, André 23
Vipera
 xanthina 41
viridis
 See *Bufo*
 See also *Seps*
vittata
 See *Mabuia*
 See also *Mabuya*
vittatus
 See *Scincus*
vivipara
 See *Lacerta*
Vogel, Zdeněk 92
voraginosis
 See *Cyrtopodion*
Voris, Harold Knight 39

Walczak, Peter 27
Wall, Frank 35
Walterinnesia 39
warty rock gecko 161
watsonana
 See *Eremias*
 See also *Mesalina*
watsoni
 See *Cyrtopodion*
watsonnana (*sic*)
 See *Mesalina watsonana*
Weber, Neal A. 73, 109, 136–137, 251, 267
Welch, Kenneth Reginald George 37, 68, 71, 73, 75–76, 80, 84, 86, 89–90, 92–93, 95, 98, 100–101, 105, 107–108, 112, 114, 118, 122, 130, 136, 140–141, 146, 148, 154, 156–157, 160, 162–164, 166, 168–170, 173, 175–177, 179, 182, 184–185, 187, 189–190, 194–198, 202–203, 205–206, 212–214, 216, 219–220, 222, 224–228, 232–234, 237–238, 241–242, 244, 247, 250, 252, 255, 264–265, 268, 270–272, 275–276, 280, 282–283, 285, 288, 290–291, 293–294, 297–298
Wermuth, Heinz 68–69, 71, 73, 75–76, 80, 89–93, 95, 99–101, 104, 107–108, 112, 130, 136, 140–141, 148, 154, 156, 162–164, 166, 169–170, 172–173, 175, 177, 184–185, 187, 189–190, 291, 293–294
Werner's bent–toed gecko 158
Werner's leaf-toad gecko 135
Werner, Franz Josef Maria 2, 4, **20**, 71, 73, 76, 80, 89–90, 92–93, 95–97, 100–101, 104, 107–109, 111, 113, 118, 130, 135, 140–141, 148, 154, 157–158, 160, 162–164, 166, 168–170, 173, 175, 177, 184–185, 187, 189–190, 194, 197, 203, 205, 213–214, 216, 219, 222, 232, 234, 236, 241, 247, 250, 252, 255, 264–265, 268, 270–271, 275–276, 280, 283, 285, 291, 293–294, 297–298
Werner, Yehudah Leopold 14, **22**, 35, 57, 80, 101, 104, 107, 169, 237, 248, 250–252, 255–256, 275, 294
western leopard gecko 118
Wettstein, Otto 22–23, 71, 76, 78, 80, 90, 95, 97, 101, 108, 112, 130, 137, 141, 143, 154, 160–161, 168, 176, 202, 222, 231–232, 236, 247, 252, 255, 264, 276–277, 280, 297
White, John 40, 282, 296
Whiteman, Ronald S. 40
Wiegmann, Arend Friedrich August Heinrich 2, 4, 33, 147–148, 200, 208, 227, 269, 288
Williams, Kenneth Lee 14, 37
Wischuf, Tilman 33
Witte, Gaston François de 3, 33, 127, 130, 180–183
Wolfart, Reinhardt 63–64
wolteri
 See *Lacerta*
Wolterstorff, Willi 36, 39
Womochel, Daniel 24, **26**
woosnami
 See *Lacerta*

Xanthina
 See *Vipera*
Xantusiidae 262
Xenagama 81

Yellow-bellied house gecko 173
yellow-headed agama 78
yellow-speckled toad agama 89
yemenicus
 See *Acanthodactylus*

Zagros Mountains
 western foothills of 49
 See also Iran
Zagros Mountains lacerta 261
Zagrosian lizard 241
zagrosica
 See *Lacerta*
Zamenis
 hotsoni 33
Zarudny's bent-toed gecko 166
Zarudny's gray thin-toed gecko 166
Zarudny's skink 271
Zarudny, Nikolai Aleksyevich **iii**, 17, 20, 35, 68–69, 71, 73, 75–77, 80, 90, 92–94, 97, 100–101, 113, 116, 120–121, 130, 141, 148, 154–155, 157, 162, 164, 166, 168, 173, 187–190, 202, 205, 214, 216,

219, 221, 252, 255, 265, 267, 270–271, 276, 280, 285, 293, 297–298
zarudnyi
   See *Cyrtodactylus*
   See also *Cyrtopodion*
   See also *Diplometopon*
   See also *Eremias (Scapteira)*
   See also *Eumeces*
   See also *Gymnodactylus*
   See also *Mediodactylus*
   See also *Tenuidactylus (Mediodactylus)*
   See also *Teratoscincus*
Zhao, Ermi 87, 92, 113, 166, 172, 189, 213, 227–228, 274, 296
Zohary, M. 13, 44, 112, 236
Zootoca 199, 231
Zug, George Robert 14, 29
Zygnidopsis
   brevipes 279
Zygnopsis
   brevipes 279–280
Zykova, L. Yu. 75

PUBLICATIONS OF THE
SOCIETY FOR THE STUDY OF AMPHIBIANS AND REPTILES

SOCIETY PUBLICATIONS may be purchased from:
Dr. Robert D. Aldridge, Publications Secretary,
Department of Biology, Saint Louis University,
3507 Laclede Avenue, Room 127,
Saint Louis, Missouri 63103–2010, USA.

Telephone: area code 314, 977–3916 or 977–1710.
E-mail: ssar@slu.edu *Fax*: area code 314, 977–3658.
Web: http://www.ukans.edu/~ssar/

Prices are effective through December 2000. Make checks payable to "SSAR." Overseas customers must make payment in USA funds using a draft drawn on American banks or by International Money Order. All persons may charge to MasterCard or VISA (please provide account number and expiration date); items marked "out-of-print" are no longer available from the Society.

Shipping and Handling Costs

- *Shipments inside the USA*: Shipping costs are in addition to the price of publications. Add an amount for shipping of the first item ($3.00 for a book costing $10.00 or more or $2.00 if the item costs less than $10.00) plus an amount for any additional items ($2.00 each for books costing over $10.00 and $1.00 for each item costing less than $10.00).
- *Shipments outside the USA*: Determine the cost for shipments inside USA (above) and then add 5% of the total cost of the order.
- *Large prints* (marked *): For shipments inside the USA, add $3.00 for any quantity; outside the USA, instead add $7.00 for any quantity.

CONTRIBUTIONS TO HERPETOLOGY

Book-length monographs, comprising taxonomic revisions, results of symposia, and other major works. Pre-publication discount to Society members. Missing volumes are out-of-print and no longer available from the SSAR.

Vol. 2. *The Turtles of Venezuela*, by Peter C. H. Pritchard and Pedro Trebbau. 1984. Covers half of the turtle species of South America. 414 p., 48 color plates (25 watercolors by Giorgio Voltolina and 165 photographs of turtles and habitats) measuring 8½ × 11 inches, keys, 16 maps. Regular edition, clothbound $45.00; patron's edition, two leatherbound volumes in cloth-covered box, signed and numbered by authors and artist $300.00. (*Also*: set of 25 color prints of turtle portraits, in protective wrapper $30.00.)

Vol. 3. *Introduction to the Herpetofauna of Costa Rica / Introducción a la Herpetofauna de Costa Rica*, by Jay M. Savage and Jaime Villa R. 1986. Bilingual edition in English and Spanish, with distribution checklist, bibliographies, and extensive illustrated keys. 220 p., map. Clothbound $30.00.

Vol. 4. *Studies on Chinese Salamanders*, by Ermi Zhao, Qixiong Hu, Yaoming Jiang, and Yuhua Yang. 1988. Evolutionary review of all Chinese species with keys, diagnostic figures, and distribution maps. 80 p., 7 plates (including 10 color photographs of salamanders and habitats). Clothbound $12.00.

Vol. 5. *Contributions to the History of Herpetology*, by Kraig Adler, John S. Applegarth, and Ronald Altig. 1989. Biographies of 152 prominent herpetologists (with portraits and signatures), index to 2500 authors in taxonomic herpetology, and academic lineages of 1450 herpetologists. International coverage. 202 p., 148 photographs, 1 color plate. Clothbound $20.00.

Vol. 6. *Snakes of the* Agkistrodon *Complex: A Monographic Review*, by Howard K. Gloyd and Roger Conant. 1990. Comprehensive treatment of four genera: *Agkistrodon*, *Calloselasma*, *Deinagkistrodon*, and *Hypnale*. Includes nine supplementary chapters by leading specialists. 620 p., 33 color plates (247 photographs of snakes and habitats), 20 uncolored plates, 60 text figures, keys, 6 charts, 28 maps. Clothbound $75.00. (*Also*: separate set of the 247 color photographs of snakes and habitats, in protective wrapper. $30.00; limited-edition print of the book's frontispiece illustrating snakes of all four genera, from watercolor by David M. Dennis. Signed individually by Dr. Conant and the artist $25.00.)

Vol. 7. *The Snakes of Iran*, by Mahmoud Latifi. 1991. Review of the 60 species of Iranian snakes, covering general biology, venoms, and snake bite. Appendix and supplemental bibliography by Alan E. Leviton and George R. Zug. 167 p., 22 color plates of snakes (66 figures), 2 color relief maps, 44 species range maps. Clothbound $22.00.

Vol. 8. *Handbook to Middle East Amphibians and Reptiles*, by Alan E. Leviton, Steven C. Anderson, Kraig Adler, and Sherman A. Minton. 1992. Annotated checklist, illustrated key, and identification manual covering 148 species and subspecies found in region from Turkish border south through the Arabian Peninsula and the Arabian (Persian) Gulf. Chapters on venomous snakes and snakebite treatment plus extensive bibliography. 264 p., 32 color plates (220 photographs), maps, text figures. Clothbound $30.00.

Vol. 9. *Herpetology: Current Research on the Biology of Amphibians and Reptiles*. 1992. Proceedings of the First World Congress of Herpetology (Canterbury, 1989), edited by Kraig Adler and with a foreword by H.R.H. Prince Philip, Duke of Edinburgh. Includes the plenary lectures, a summary of the congress, and a list of delegates with their current addresses. 225 p., 28 photographs. Clothbound $28.00.

Vol. 10. *Herpetology of China*, by Ermi Zhao and Kraig Adler. 1993. Comprehensive review of Chinese amphibians and reptiles, including Hong Kong and Taiwan. 522 p., 48 color plates (371 photographs illustrating all 164 genera and half of the 661 species), portraits, text figures, maps, indices. Clothbound $60.00.

Vol. 11. *Captive Management and Conservation of Amphibians and Reptiles*, by James B. Murphy, Kraig Adler, and Joseph T. Collins (eds.). 1994. Results of a Society-sponsored symposium, including chapters by 70 leading specialists. Foreword by Gerald Durrell. 408 p., 35 photographs, 1 color plate. Clothbound $58.00.

Vol. 12. *Contributions to West Indian Herpetology*, by Robert Powell and Robert W. Henderson (eds.). 1996. Results of a Society-sponsored symposium, including research chapters by 59 authors and a checklist of species with complete citations.

Foreword by Thomas W. Schoener. 457 p., 28 photographs, 70 color photographs, index. Clothbound $60.00.

Vol. 13. *Gecko Fauna of the USSR and Contiguous Regions*, by Nikolai N. Szczerbak and Michael L. Golubev. 1996. Covers the systematics, natural history, and conservation of the gecko fauna of the former Soviet Union and related species in surrounding regions from Mongolia through Pakistan, the Middle East, and northern Africa. 245 p., 24 colored and 60 black-and-white photographs, spot distribution maps, bibliography, index. Clothbound $48.00.

Vol. 14. *Biology of the Reptilia, vol. 19 (Morphology G)*, by Carl Gans and Abbot S. Gaunt (eds.). 1998. Chapters by 11 authors cover the major organs situated in the coelom: lungs, heart, liver, and spleen. 660 p., 145 figures, indices. Clothbound $58.00. (A complete list of the earlier volumes in this series [vols. 1–18], with names of publishers, is given in the book.)

Vol. 15. *The Lizards of Iran*, by Steven C. Anderson. 1999. Comprehensive summary, including systematics, natural history, and distribution. 450 p., 103 maps, 190 color photographs of lizards and habitats. Clothbound $65.00.

FACSIMILE REPRINTS IN HERPETOLOGY

Exact reprints of classic and important books and papers. Most titles have extensive new introductions by leading authorities. Prepublication discount to Society members. Missing volumes are out-of-print and no longer available from the Society.

ANDERSON, J. 1896. *Contribution to the Herpetology of Arabia*. Introduction and new checklist of Arabian amphibians and reptiles by Alan E. Leviton and Michele L. Aldrich. 160 p., illus. (one plate in color), map. Clothbound $25.00.

BOGERT, C. M., and R. MARTÍN DEL CAMPO. 1956. *The Gila Monster and Its Allies*. The standard work on lizards of the family Helodermatidae. New preface by Charles M. Bogert and retrospective essay by Daniel D. Beck. 262 p., color plate, 62 photographs, 35 text figures, index. Clothbound $38.00.

BOULENGER, G. A. 1877–1920. *Contributions to American Herpetology*. A collection of papers (from various journals) covering North, Central, and South American species, with an introduction by James C. Battersby. Complete in 18 parts totalling 880 p., numerous illustrations, index. Paperbound. Complete set: 18 parts plus index and two tables of contents (for binding in two volumes), in parts as issued $55.00.

COPE, E. D. 1864. *Papers on the Higher Classification of Frogs*. Reprinted from Proceedings of the Academy of Natural Sciences of Philadelphia and Natural History Review. 32 p. Paperbound $3.00.

COPE, E. D. 1871. *Catalogue of Batrachia and Reptilia Obtained by McNiel in Nicaragua; Catalogue of Reptilia and Batrachia Obtained by Maynard in Florida*. 8 p. Paperbound $1.00.

COPE, E. D. 1892. *The Osteology of the Lacertilia*. An important contribution to lizard anatomy, reprinted from Proceedings of the American Philosophical Society. 44 p., 6 plates. Paperbound $4.00.

COWLES, R. B., and C. M. BOGERT. 1944. *A Preliminary Study of the Thermal Requirements of Desert Reptiles*. The foundation of thermoregulation biology, with extensive review of recent studies by F. Harvey Pough. Reprinted from Bulletin of American Museum of Natural History. 52 p., 11 plates. Paperbound $5.00.

ESCHSCHOLTZ, F. 1829–1833. *Zoologischer Atlas* (herpetological sections). Descriptions of new reptiles and amphibians from California and the Pacific. Introduction by Kraig Adler. 32 p., 4 plates (measuring 8½ × 11 inches). Paperbound $3.00.

ESPADA, M. JIMÉNEZ DE LA. 1875. *Vertebrados del Viaje al Pacifico: Batracios*. A major taxonomic work on South American frogs. Introduction by Jay M. Savage. 208 p. plates, maps. Clothbound $20.00.

FAUVEL, A.-A. 1879. *Alligators in China*. Original descript of *Alligator sinensis*, including classical and natural histo 42 p., 3 plates. Paperbound $5.00.

FITZINGER, L. 1826 & 1835. *Neue Classification der Reptil* and *Systematische Anordnung der Schildkröten*. Import nomenclatural landmarks for herpetology, including A phibia as well as reptiles; world-wide in scope. Introduct by Robert Mertens. 110 p., folding chart. Clothbound $30.

GRAY, J. E. 1825. *A Synopsis of the Genera of Reptiles a Amphibia*. Reprinted from Annals of Philosophy. 32 Paperbound $3.00.

GRAY, J. E. 1831–1844. *Zoological Miscellany*. A privat printed journal, devoted mostly to descriptions of amphibia reptiles, and birds from throughout the world. Introducti by Arnold G. Kluge. 86 p., 4 plates. Paperbound $6.00.

GRAY, J. E., and A. GÜNTHER. 1845–1875. *Lizards of Austra and New Zealand*. The reptile section from "Voyage of H.M Erebus and Terror," together with Gray's 1867 related bc on Australian lizards. Introduction by Glenn M. Shea. 82 20 plates (measuring 8½ × 11 inches). Clothbound $20. (*Also*: set of the 20 plates, in protective wrapper $12.00.)

GÜNTHER, A. 1885–1902. *Biologia Centrali-Americar Reptilia and Batrachia*. The standard work on Midc American herpetology with 76 full-page plates measuri 8½ × 11 inches (12 in color). Introductions by Hobart Smith, A. E. Gunther, and Kraig Adler. 575 p., photograpl maps. Clothbound $50.00. (*Also*: separate set of the 12 co plates, in protective wrapper $18.00.)

HOLBROOK, J. E. 1842. *North American Herpetology*. Fi volumes bound in one. The classic work by the father North American herpetology. Exact facsimile of the definiti second edition, including all 147 plates, measuring 8½ × inches (20 reproduced in full color). Introduction a checklists by Richard and Patricia Worthington and by Krä Adler. 1032 p. Clothbound $60.00.

JUNIOR SOCIETY OF NATURAL SCIENCES (CI] CINNATI, OHIO). 1930–1932. Herpetological papers frc the society's Proceedings, with articles by Weller, Walk Dury, and others. 56 p. Paperbound $3.00.

KIRTLAND, J. P. 1838. *Zoology of Ohio* (herpetological pc tion). 8 p. Paperbound $1.00.

LECONTE, J. E. 1824–1828. *Three Papers on Amphibians*, fro the Annals of the Lyceum of Natural History, New York. p. Paperbound $2.00.

McILHENNY, E. A. 1935. *The Alligator's Life History*. T most complete natural history of the American alligatc Introduction by Archie Carr and a review of recent literatu by Jeffrey W. Lang. 125 p., 18 photographs and a portra Clothbound $20.00.

McLAIN, R. B. 1899. *Contributions to North Americ Herpetology* (three parts). 28 p., index. Paperbound $2.0(

ORBIGNY, A. D' [and G. BIBRON]. 1847. *Voyage dans l'Amériqu Méridionale*. This extract comprises the complete section reptiles and amphibians from this voyage to South America. p., 9 plates measuring 8½ × 11 inches. Paperbound $3.00.

PETERS, W. 1838–1883. *The Herpetological Contributio of Wilhelm C. H. Peters (1815–1883)*. A collection of 1 titles, world-wide in scope, and including the herpetologic volume in Peters' series, "Reise nach Mossambique Biography, annotated bibliography, and synopsis of speci by Aaron M. Bauer, Rainer Günther, and Meghan Klipfe 714 pages, 114 plates, 9 photographs, maps, inde Clothbound $75.00.

RAFINESQUE, C. S. 1820. *Annals of Nature* (herpetologic and ichthyological sections), 4 p. Paperbound $1.00.

SOCIETY PUBLICATIONS

RAFINESQUE, C. S. 1822. *On Two New Salamanders of Kentucky.* 2 p. Paperbound $1.00.

RAFINESQUE, C. S. 1832–1833. *Five Herpetological Papers from the Atlantic Journal.* 4 p. Paperbound $1.00.

SCHMIDT, K. P., and G. K. NOBLE. 1919–1923. *Contributions to the Herpetology of the Belgian Congo.* An essential reference for the Congo rain forest and the Sudanese savanna. Introductions by Donald G. Broadley and John C. Poynton. 780 pages, 141 photographs, maps, indices. Clothbound $65.00.

SHAW, G. 1802. *General Zoology, vol. 3: Amphibia.* Complete herpetological section from the first world summary of amphibians and reptiles in English. Introduction by Hobart M. Smith and Patrick David. 975 p., 141 plates. Clothbound $75.00.

SOWERBY, J. DEC., E. LEAR, and J. E. GRAY. 1872. *Tortoises, Terrapins, and Turtles Drawn From Life.* The finest atlas of turtle illustrations ever produced. Introduction by Ernest E. Williams. 26 p., 61 full-page plates (measuring 8½ × 11 inches). Clothbound $25.00.

SPIX, J. B. VON, and J. G. WAGLER. 1824–1825. *Herpetology of Brazil.* The most comprehensive and important early survey of Brazilian herpetology. Introduction by P. E. Vanzolini. 400 p., 98 plates, one in color (each measuring 8½ × 11 inches), map. Clothbound $36.00.

STEJNEGER, L. 1907. *Herpetology of Japan and Adjacent Territory.* Introduction by Masafumi Matsui. Also covers Taiwan, Korea, and adjacent China and Siberia. 684 pages, 35 plates, 409 text figures, keys, index. Clothbound $58.00.

TROSCHEL, F. H. 1850 [1852]. *Cophosaurus texanus, neue Eidechsengattung aus Texas.* 8 p. Paperbound $1.00.

TSCHUDI, J. J. VON. 1838. *Classification der Batrachier.* A major work in systematic herpetology, with introduction by Robert Mertens. 118 p., 6 plates. Paperbound $18.00.

TSCHUDI, J. J. VON. 1845. *Reptilium Conspectus.* New reptiles and amphibians from Peru. 24 p. Paperbound $2.00.

VANDENBURGH, J. 1895–1896. *Herpetology of Lower California.* Herpetology of Baja California, Mexico (collected papers). 101 p., 11 plates, index. Paperbound $8.00.

VANDENBURGH, J. 1914 *The Gigantic Land Tortoises of the Galapagos Archipelago.* The most extensive review of Galapagos tortoises. Foreword by Peter C. H. Pritchard. 290 pages, 205 photographs, maps, index. Clothbound $55.00.

WAITE, E. R. 1929. *The Reptiles and Amphibians of South Australia.* Introduction by Michael J. Tyler and Mark Hutchinson. 282 p., color plate, portrait, 192 text figures including numerous photographs. Clothbound $35.00.

WILCOX, E. V. 1891. *Notes on Ohio Batrachians.* 3 p. Paperbound $1.00.

WRIGHT, A. H., and A. A. WRIGHT. 1962. *Handbook of Snakes of the United States and Canada, Volume 3, Bibliography.* Cross-indexed bibliography to Volumes 1 and 2. 187 p. Clothbound $18.00.

JOURNAL OF HERPETOLOGY

The Society's official scientific journal, international in scope. Issued quarterly as part of Society membership. All numbers are paperbound as issued, measuring 7 × 10 inches.

Volume 1 (1968), numbers 1–4 combined.
Volumes 2–5 (1968–1971), numbers 1–2 and 3–4 combined, $8.00 per double number.
Volume 6 (1972), numbers 1, 2, and double number 3–4.
Volumes 7–33 (1973–1999), four numbers in each volume, $8.00 per single number.

The following volumes and numbers are out-of-print and are *no longer available*: Volume 1, 2, 3, 4, 5(3, 4), 6, 7(1), 8(1), 9(1, 2, 4), 10(1, 4), 11(4), and 12(1, 2).

Cumulative Index for Volumes 1–10 (1968–1976), 72 pages, $8.00.

HERPETOLOGICAL REVIEW AND H.I.S.S. PUBLICATIONS

The Society's official newsletter, international in coverage. In addition to news notes and feature articles, regular departments include regional societies, techniques, husbandry, life history, geographic distribution, and book reviews. Issued quarterly as part of Society membership or separately by subscription. All numbers are paperbound as issued and measure 8½ × 11 inches. In 1973, publications of the Herpetological Information Search Systems (*News-Journal* and *Titles and Reviews*) were substituted for *Herpetological Review*; content and format are the same.

Volume 1 (1967–1969), numbers 1–9, $5.00 per number.
Volumes 2–30 (1970–1999), four numbers in each volume (except volumes 3–4, with 6 numbers each), $5.00 per number.
The following numbers are out-of-print and *no longer available*: Volume 1(number 7), 3(2), 4(1), 5(1, 2), 6(1, 2, 4), 7(3, 4), and 10(2).
Cumulative Index for Volumes 1–7 (1967–1976), 60 pages, $5.00.
Cumulative Index for Volumes 1–17 (1967–1986), 90 pages, $8.00.
H.I.S.S. Publications: News-Journal, volume 1, numbers 1–6, and Titles and Reviews, volume 1, numbers 1–2 (all of 1973–1974), complete set, $10.00.
Index to Geographic Distribution Records for Volumes 1–17 (1967–1986), including H.I.S.S. publications, 44 pages, $6.00.

CATALOGUE OF AMERICAN AMPHIBIANS AND REPTILES

Loose-leaf accounts of taxa (measuring 8½ × 11 inches) prepared by specialists, including synonymy, definition, description, distribution map, and comprehensive list of literature for each taxon. Covers amphibians and reptiles of the entire Western Hemisphere. Issued by subscription. Individual accounts are not sold separately.

CATALOGUE ACCOUNTS:
    Complete set: Numbers 1–700. $400.00.
    Partial sets: Numbers 1–190. $75.00.
                    Numbers 191–410, $85.00.
                    Numbers 411–700, $270.00.
INDEX TO ACCOUNTS 1–400: Cross-referenced, 64 pages, $6.00; accounts 401–600: Cross-referenced, 32 pages, $6.00.
IMPRINTED POST BINDER: $35.00. (*Note*: one binder holds about 200 accounts.)
SYSTEMATIC TABS: Ten printed tabs for binder, such as "Class Amphibia," "Order Caudata," etc., $6.00 per set.

HERPETOLOGICAL CONSERVATION

A new series of book-length monographs, including symposia, devoted to all aspects of the conservation of amphibian and reptiles. Prepublication discount to Society members.

Vol. 1. *Amphibians in Decline: Canadian Studies of a Global Problem*, by David M. Green (ed.). 1997. Chapters by 52 authors dealing with population dispersal and fluctuations, genetic diversity, monitoring of natural populations, as well as effects of temperature, acidity, pesticides, UV light, forestry practices, and disease. 351 p., numerous photographs, figures, and tables, index. Paperbound $40.00.

Vol. 2. *Amphibian and Reptilian Ecotoxicology: Papers from the Third World Congress of Herpetology*, by Michael R. K. Lambert (ed.), in preparation.

HERPETOLOGICAL CIRCULARS

Miscellaneous publications of general interest to the herpetological community. All numbers are paperbound, as issued. Prepublication discount to Society members.

No. 1. *A Guide to Preservation Techniques for Amphibians and Reptiles* by George R. Pisani. 1973. 22 p., illus. $3.00.

No. 2. *Guía de Técnicas de Preservación de Anfibios y Reptiles* por George R. Pisani y Jaime Villa. 1974. 28 p., illus. $3.00.

No. 3. *Collections of Preserved Amphibians and Reptiles in the United States* compiled by David Wake (chair) and the Committee on Resources in Herpetology. 1975. 22 p. Out-of-print.

No. 4. *A Brief Outline of Suggested Treatments for Diseases of Captive Reptiles* by James Murphy. 1975. 13 p. Out-of-print.

No. 5. *Endangered and Threatened Amphibians and Reptiles in the United States* compiled by Ray E. Ashton, Jr. (chair) and 1973–74 SSAR Regional Herpetological Societies Liaison Committee. 1976. 65 p. Out-of-print.

No. 6. *Longevity of Reptiles and Amphibians in North American Collections* by J. Kevin Bowler. 1977. 32 p. Out-of-print. (see also number 21.)

No. 7. *Standard Common and Current Scientific Names for North American Amphibians and Reptiles* (1st ed.) by Joseph T. Collins, James E. Huheey, James L. Knight, and Hobart M. Smith. 1978. 36 p. $3.00. (See also numbers 12, 19, and 25.)

No. 8. *A Brief History of Herpetology in North America Before 1900* by Kraig Adler. 1979. 40 p., 24 photographs, 1 map. $3.00.

No. 9. *A Review of Marking Techniques for Amphibians and Reptiles* by John W. Ferner. 1979. 42 p., illus. $3.00.

No. 10. *Vernacular Names of South American Turtles* by Russell A. Mittermeier, Federico Medem, and Anders G. J. Rhodin. 1980. 44 p. $3.00.

No. 11. *Recent Instances of Albinism in North American Amphibians and Reptiles* by Stanley Dyrkacz. 1981. 36 p. $3.00.

No. 12. *Standard Common and Current Scientific Names for North American Amphibians and Reptiles* (2nd ed.) by Joseph T. Collins, Roger Conant, James E. Huheey, James L. Knight, Eric M. Rundquist, and Hobart M. Smith. 1982. 32 p. $3.00. (See also numbers 7, 19, and 25.)

No. 13. *Silver Anniversary Membership Directory*, including addresses of all SSAR members, addresses and publications of the herpetological societies of the world, and a brief history of the Society. 1983. 56 p., 4 photographs. $3.00.

No. 14. *Checklist of the Turtles of the World with English Common Names* by John Iverson. 1985. 14 p. $3.00.

No. 15. *Cannibalism in Reptiles: A World-Wide Review* by Joseph C. Mitchell. 1986. 37 p. $4.00.

No. 16. *Herpetological Collecting and Collections Management* by John E. Simmons. 1987. 72 p., 6 photographs. $6.00.

No. 17. *An Annotated List and Guide to the Amphibians and Reptiles of Monteverde, Costa Rica* by Marc P. Hayes, J. Alan Pounds, and Walter T. Timmerman. 1989. 70 p., 32 figures. $5.00.

No. 18. *Type Catalogues of Herpetological Collections: An Annotated List of Lists* by Charles R. Crumly. 1990. 50 p. $5.00.

No. 19. *Standard Common and Current Scientific Names for North American Amphibians and Reptiles* (3rd ed.) compiled by Joseph T. Collins (coordinator for SSAR Common and Scientific Names List). 1990. 45 p. Out-of-print. (See also numbers 7, 12, and 25.)

No. 20. *Age Determination in Turtles* by George R. Zug. 1991. 32 p., 6 figures. $5.00.

No. 21. *Longevity of Reptiles and Amphibians in North American Collections* (2nd ed.) by Andrew T. Snider and J. Kevin Bowler. 1992. 44 p. $5.00.

No. 22. *Biology, Status, and Management of the Timber Rattlesnake* (Crotalus horridus): *A Guide for Conservation* by William S. Brown. 1993. 84 p., 16 color photographs. $12.00.

No. 23. *Scientific and Common Names for the Amphibians and Reptiles of Mexico in English and Spanish / Nombres Científicos y Comunes en Ingles y Español de los Anfibios y los Reptiles de México* by Ernest A. Liner. Spanish translation by José L. Camarillo R. 1994. 118 p. $12.00.

No. 24. *Citations for the Original Descriptions of North American Amphibians and Reptiles*, by Ellin Beltz. 1995. 48 p. $7.00.

No. 25. *Standard Common and Scientific Names for North American Amphibians and Reptiles* (4th ed.) compiled by Joseph T. Collins (coordinator for SSAR Common and Scientific Names List). 1997. 44 p. $9.00. (see also numbers 7, 12, and 19.)

No. 26. *Venomous Snakes: a Safety Guide for Reptile Keepers* by William Altimari. 1998. 28 p. $6.00.

PUBLICATIONS OF
THE OHIO HERPETOLOGICAL SOCIETY

OHS was the predecessor to the Society for the Study of Amphibians and Reptiles. All publications international in scope. Paperbound as issued.

Volume 1 numbers 1–4, plus Special Publications 1–2 (all 195 facsimile reprint, out-of-print.

Volume 2 (1959–1960), four numbers, $2.00 per number numbers 3 and 4 out-of-print.

Volume 3 (1961–1962), four numbers, $1.00 per number numbers 1 and 3 out-of-print.

Volume 4 (1963–1964), four numbers; double number 1–2, $4.00 numbers 3 and 4 $2.00 each.

Volume 5 (1965–1966), four numbers, $2.00 per number.

Special Publications 3–4 (1961–1962), $2.00 per number; number 3 out-of-print.

OTHER MATERIALS
AVAILABLE FROM THE SOCIETY

The following color prints and brochures may be purchased from the Society. (*Extra postage required; see "Shipping and Handling Costs" at the beginning of this list.)

*SILVER ANNIVERSARY COMMEMORATIVE PRINT. Full-color print (11½ × 15¼ inches) of a Gila Monster (*Heloderma suspectum*) on natural background, from watercolor by David M. Dennis. Issued as part of Society's 25th Anniversary in 1982. Edition limited to 1000. $6.00 each or $5.00 in quantities of 10 or more.

*WORLD CONGRESS COMMEMORATIVE PRINT. Full color print (11½ × 15 inches) of an Eastern Box Turtle (*Terrapene carolina*) in a natural setting, from a watercolor by David M. Dennis. Issued as part of SSAR's salute to the First World Congress of Herpetology, held at Canterbury, United Kingdom, in 1989. Edition limited to 1500. $6.00 each or $5.00 in quantities of 10 or more.

*WATER SNAKE PRINT. Full-color print (9½ × 12 inches) of endangered water snakes (*Nerodia erythrogaster neglecta* and *N. sipedon insularum*) described by Roger Conant. Edition limited to 450 copies, each individually signed and numbered by Dr. Conant and the artist, David M. Dennis. $25.00.

GUIDELINES FOR THE USE OF LIVE AMPHIBIANS AND REPTILES IN FIELD RESEARCH, by George R. Pisani, Stephen D. Busack, Herbert C. Dessauer, and Victor H. Hutchison, representing a joint committee of ASIH, HL, and SSAR. 1987. Brochure covers animal care, regulations collecting, restraint and handling, marking, housing and maintenance in field, and final disposition of specimens. 14 p. $2.00 ($1.00 each in quantities of five or more copies).

GRANTS AND AWARDS FOR HERPETOLOGISTS, by Joan C. Milam. 1997. A detailed listing, with descriptions and addresses, for about 100 research award programs. 106 p. $8.00

Production Specifications

WORD PROCESSING AND FORMATTING OF TEXT: California Academy of Sciences, San Francisco, California, USA (Alan E. Leviton). Type was set in Adobe Systems Times face using Ventura Publisher 4.2 on an ALR VEISA 486 computer and camera-ready copy printed on a Lexmark Optra R+ laser printer.

PRINTING OF TEXT AND BINDING: Thomson-Shore, Inc., Dexter, Michigan, USA (Ned Thomson, Lana Paton, Rebecca Dawson). The text is printed on 60-pound Domtar Windsor Smooth (Joy) White Offset paper, a recycled stock. The book is covered in Roxite C cloth (vellum finish) with Multicolor Antique endpapers.

PRINTING OF COLOR PLATES: Four Colour Imports, Ltd., Louisville, Kentucky, USA (Kathy Copas, Julie Cundiff), agents for Everbest Printing Co., Ltd., Hong Kong, Peoples Republic of China. The plates were laser-scanned from 35-mm slides and color prints by All Systems Colour, Inc., of Lebanon, Ohio, USA (Craig Dudley) and printed on 128-gsm Korean Chon Ju gloss art paper.

PAPER: All paper in this book is acid- and groundwood-free. It meets the guidelines for permanence and durability of the Committee on Publication Guidelines for Book Longevity of the Council on Library Resources.

DATE OF PUBLICATION: 30 September 1999.

PLACE OF PUBLICATION: Ithaca, New York, USA.

NUMBER OF COPIES: 1000.